Periodic Table of the Elements

METALS

NONMETALS

TRANSITION METALS

PERIODS	IA	IIA	IIIB	IVB	VB	VIB	VIIB	VIII			IB	IIB	IIIA	IVA	VA	VIA	VIIA	O
1	1.0079 H 1																1.0079 H 1	4.00260 He 2
2	6.94 Li 3	9.01218 Be 4											10.81 B 5	12.011 C 6	14.0067 N 7	15.9994 O 8	18.9984 F 9	20.179 Ne 10
3	22.9898 Na 11	24.305 Mg 12											26.9815 Al 13	28.086 Si 14	30.9738 P 15	32.06 S 16	35.453 Cl 17	39.948 Ar 18
4	39.098 K 19	40.08 Ca 20	44.9559 Sc 21	47.90 Ti 22	50.9414 V 23	51.996 Cr 24	54.9380 Mn 25	55.847 Fe 26	58.9332 Co 27	58.71 Ni 28	63.546 Cu 29	65.38 Zn 30	69.72 Ga 31	72.59 Ge 32	74.9216 As 33	78.96 Se 34	79.904 Br 35	83.80 Kr 36
5	85.4678 Rb 37	87.62 Sr 38	88.9059 Y 39	91.22 Zr 40	92.9064 Nb 41	95.94 Mo 42	98.9062 Te 43	101.07 Ru 44	102.9055 Rh 45	106.4 Pd 46	107.868 Ag 47	112.40 Cd 48	114.82 In 49	118.69 Sn 50	121.75 Sb 51	127.60 Te 52	126.9046 I 53	131.30 Xe 54
6	132.9054 Cs 55	137.34 Ba 56	57–71 *	178.49 Hf 72	180.9479 Ta 73	183.85 W 74	186.2 Re 75	190.2 Os 76	192.22 Ir 77	195.09 Pt 78	196.9665 Au 79	200.59 Hg 80	204.37 Tl 81	207.2 Pb 82	208.9804 Bi 83	(210) Po 84	(210) At 85	(222) Rn 86
7	(223) Fr 87	(226.0254) Ra 88	89–103 †	(261) Rf 104	(262) Ha 105	(263) Unh 106	(262) Uns 107	(266) Une 109										

* LANTHANIDE SERIES

138.9055 La 57	140.12 Ce 58	140.9077 Pr 59	144.24 Nd 60	(145) Pm 61	150.4 Sm 62	151.96 Eu 63	157.25 Gd 64	158.9254 Tb 65	162.50 Dy 66	164.9304 Ho 67	167.26 Er 68	168.9342 Tm 69	173.04 Yb 70	174.97 Lu 71

† ACTINIDE SERIES

(227) Ac 89	232.0381 Th 90	231.0359 Pa 91	238.029 U 92	237.0482 Np 93	(242) Pu 94	(243) Am 95	(245) Cm 96	(245) Bk 97	(248) Cf 98	(253) Es 99	(254) Fm 100	(256) Md 101	(253) No 102	(257) Lr 103

OPERATIONAL
ORGANIC CHEMISTRY

A Laboratory Course

SECOND EDITION

John W. Lehman
Lake Superior State University

Allyn and Bacon, Inc.
Boston   London   Sydney   Toronto

To My Parents

Series Editor: James Smith
Cover Administrator: Linda Dickinson
Cover Designer: Susan Hamand
Manufacturing Buyer: William Alberti
Composition Buyer: Linda Cox
Production Editor: Kathy Smith
Production Service: Lifland et al., Bookmakers

Library of Congress Cataloging-in-Publication Data

Lehman, John W.
  Operational organic chemistry.

  Bibliography: p.
  Includes index.
  1. Chemistry, Organic — Experiments. I. Title.
QD261.L39 1988   547'.0028   87-38508
ISBN 0-205-11255-2

Printed in the United States of America.

10 9 8 7 6 5 4 3 2   92 91 90 89 88

EXPERIMENTS

MINILABS

PART III   SYSTEMATIC ORGANIC QUALITATIVE ANALYSIS

PART IV   ADVANCED PROJECTS

PART V   THE OPERATIONS

Preface to the Second Edition

Operational Organic Chemistry has been thoroughly reviewed and extensively rewritten for the second edition. The experimental sections have been reorganized; Part I now contains eleven experiments that cover all the major laboratory operations, and Part II contains thirty-nine experiments that are correlated with topics from a typical organic chemistry lecture course. Such correlations can be made for Part I experiments as well, but it is suggested that most of them (particularly the first five) be done early to teach the basic operations. Nine experiments are completely new: a simple introductory experiment that teaches acid-base chemistry and stoichiometry (Experiment 1); two experiments that explore different kinds of isomerism with the help of molecular models (12 and 16); a synthesis of a urea-inclusion complex (14); an aromatic bromination without Br_2 (29); a synthetic experiment that also introduces linear free-energy relationships (34); a structure determination involving an uncommon reduction reaction (36); a synthetic/mechanistic experiment involving a molecular rearrangement (38); and a literature search that culminates in an organic synthesis (50). Eleven experiments (8, 11, 17, 18, 19, 26, 27, 28, 31, 33, and 47) have been extensively revised to include completely new procedures or yield different products. For example, the Williamson synthesis of phenetole that appeared in the first edition has been changed to a synthesis of phenacetin with a simplified procedure (26), and the aromatic nitration experiment (28) now uses nitronium fluoborate and includes a competitive rate study. The Procedure sections (and usually the Methodology sections as well) of the remaining experiments have been rewritten to reduce the amounts of chemicals, add more safety information, provide additional experience with spectral analysis, or in some cases make the experiments work better. Thirty new Minilabs have been added, and most of them are now described briefly in their associated experiments.

Most students liked the humor in the Situation sections in the first edition, but not all instructors were amused. In order that the Situations would not imply a casual attitude toward the task of teaching or learning organic chemistry in the laboratory, they were rewritten so that they coincide more closely with the given experimental objectives. The Experimental Variations that were included with each experiment in the first edition have been revised to include detailed procedures when appropriate, and these are presented under the new heading Collateral Projects. References to related experiments described elsewhere have been moved to the *Instructor's Manual*. Some of the Topics for Report have been moved to the Report sections that now follow the Procedure sections, and others will be found in the Exercises sections, which have been augmented by many new exercises.

The emphasis on laboratory safety in the first edition has been reinforced by the more frequent use of hazard warnings and handling precautions within

the experimental procedures. Several carcinogenic or particularly hazardous chemicals have been eliminated from experiments in this edition. Carbon tetrachloride has been replaced by Freon TF (1,1,2-trichlorotrifluoroethane), an essentially nontoxic solvent that can be purchased in quantity at a modest price (see the *Instructor's Manual* for suppliers). Chromium(VI) oxide has been replaced by aqueous sodium hypochlorite in the oxidation of isoborneol (Experiment 8), and two other experiments that used chromium(VI) reagents have been deleted. Thiamine hydrochloride has replaced sodium cyanide as the catalyst for the benzoin condensation (49). Chloroform is required for only one major experiment (19), and ethylene bromide, dioxane, and peroxybenzoic acid are not used at all.

In response to suggestions from many users of the first edition, infrared and nuclear magnetic resonance spectroscopy are employed to a much greater extent in this edition. IR or PMR spectra of starting materials are given in many experiments, and information about spectral interpretation has been added to the Methodology sections. Spectra of many products are included in the *Instructor's Manual;* they can be reproduced and distributed if it is impractical for students to record their own spectra. Part III, Qualitative Analysis, has been expanded to provide additional information about the interpretation of IR and PMR spectra. The operations dealing with IR and NMR spectrometry ⟨OP–33 and OP–34⟩ have been expanded to include descriptions of Fourier transform IR, carbon-13 NMR, and chemical-shift reagents. Experiment 48 now includes the interpretation of a mass spectrum, and the corresponding operation ⟨OP–36⟩ has been augmented accordingly. Most of the other descriptions of operations have required only minor revisions, but a section on HPLC has been added to the column chromatography operation ⟨OP–16⟩, and the gas chromatography operation ⟨OP–32⟩ has been expanded to cover open-tubular columns and McReynolds constants.

The Appendixes dealing with laboratory notebooks and checklists have been extensively revised in response to preferences expressed by users of the first edition. Appendix IV on stoichiometric calculations has been rewritten to reflect the widespread use of dimensional analysis techniques in general chemistry. The Bibliography has been updated to include many new sources published since 1980, and the description of bibliographic sources in Appendix VII has been revised accordingly. Experiment 50 contains an extensive section on searching the chemical literature, with detailed information about *Chemical Abstracts* and *Beilstein.*

Throughout the text, certain cross-references are given in angle brackets; these items can be found in the following parts of the book:

⟨OP–00⟩: Part V, Operations, pages 597–775

⟨Bib–X00⟩: Bibliography, pages A-37–A-49

⟨Test C–0⟩: Part III, Classification Tests section, pages 551–565

⟨Procedure D–0⟩: Part III, Derivatives section, pages 565–580

Instructors and laboratory coordinators should be aware that the *Instructor's Manual,* which is available free of charge from Allyn and Bacon, Inc., is an essential adjunct to any laboratory course that uses *Operational Organic Chemistry.*

Acknowledgments

I wish to acknowledge the valuable contributions of Glenda Knigge and Nasser Kamazani, who lab-tested many new procedures; Bonnie Hoyer, who helped develop some procedures using Freon TF; Anne Marie Torcoletti and Linda Clancy, who tested procedures and catalysts for Experiment 33; Professor William Haag, who assisted me with the revision of Experiment 47; and Professor Gerald Weatherby, who provided a number of helpful suggestions. I am indebted to the many users of the first edition of *Operational Organic Chemistry* who completed a questionnaire; their responses and comments have been particularly helpful in the preparation of the second. Thanks are due to the administration of Lake Superior State University for approving a leave that provided the time to complete the manuscript and to the chemistry department of the University of Michigan for granting me visiting faculty appointments. I also wish to thank James M. Smith and Kathy Smith of Allyn and Bacon and Jane Hoover of Lifland et al., Bookmakers, who assisted me during the manuscript preparation and later stages of publication. Most of the spectra in this book and the *Instructor's Manual* are reproduced from the spectral libraries of the Aldrich Chemical Company, whose helpfulness and generosity are gratefully acknowledged.

J.W.L.

Introduction

Purpose and Organization of the Book

Operational Organic Chemistry is about organic chemistry as it is practiced in the laboratory. It is intended to be more than just a compilation of procedures that, if followed step by step, eventually result in the accumulation of data or the preparation of organic substances. It is a *textbook* of experimental chemistry that will help you learn how to perform the fundamental operations of organic chemistry in the laboratory and how to apply them intelligently in new situations. A cook must learn such basic techniques as peeling, slicing, and grating before he or she can prepare an adequate meal; you will have to learn such operations as extraction, distillation, and recrystallization before you can prepare an organic compound such as aspirin or isoamyl acetate. A *good* cook does not follow a recipe mechanically but possesses sufficient understanding of the methodology of cooking to improvise. If you unthinkingly follow a "recipe" for aspirin, you are not going to learn much about aspirin in the process, nor will you understand the reasons for what you are doing or the chemistry involved in the synthesis. You may have gone through the motions of making aspirin, but you will have no idea how to apply what you have done to a different preparation. Some students approach a chemistry experiment as if it were a one-of-a-kind phenomenon — they follow the directions, write up a report, and promptly forget about it. But no experiment in this book is unique; each consists of a set of interrelated operations performed in a logical sequence, and the lessons learned in one experiment will be applied to other experiments.

This book is divided into five parts, preceded by the Laboratory Safety section. Part I contains eleven experiments whose purpose is to help you learn the basic operations by applying them to a preparation or to the solution of a specific problem. For example, you will measure a melting point not only to learn *how* to measure a melting point, but also to establish the identity of a substance you have isolated or synthesized. Many of the operations will be reinforced by repeated application in the experiments in Part I so that you can master them. Once you are familiar with the operations, you will use them in the experiments in Part II, which are correlated with topics from the lecture course. The best way to learn most theoretical concepts is to apply them in the laboratory; in this way, their relationship to the "real world" of organic chemistry can be better understood and appreciated. The thirty-nine experiments in Part II will give you practical experience in dealing with the concepts of organic chemistry and will increase your proficiency in the laboratory. Part III is a comprehensive, self-contained introduction to organic qualitative analysis that will teach you how to identify organic compounds using chemical and spectral methods. Some experiments in Parts I and II

contain references in the forms ⟨Test C–19⟩ and ⟨Procedure D–4⟩. These refer to the chemical tests and preparations of derivatives described in Part III on pages 551–580. Part IV describes some challenging, open-ended research projects that you should undertake only if you are an advanced or highly motivated student. Part V contains the descriptions of all the operations that are referred to by number in the experiments in the form ⟨OP–5⟩. Appendixes I through V contain illustrations of laboratory equipment and information about laboratory notebooks, reports, experimental plans, and calculations. Appendix VI contains tables of properties for qualitative analysis. Appendix VII describes many of the sources mentioned in the Bibliography entries and serves as a guide to the literature of chemistry; in conjunction with Experiment 50, it will help you to carry out a comprehensive literature search. References to entries listed in the Bibliography are made throughout the text in the form ⟨Bib–F7⟩, where the letter refers to a category and the number to a location within that category; for example, F7 is the seventh book listed under category F (Spectroscopy).

Organization of the Experiments

Immediately following the title of each experiment is a list of topics that classify the experiment and correlate its contents with textbook material. For the synthetic experiments, this list generally includes the kind of compound reacting, the kind of compound being synthesized, the reaction type, and significant intermediates.

Each experiment also includes material under some or all of the following headings.

Operations.   Most or all of the operations used in an experiment are listed at the beginning of the experiment. These Operations lists include general separation or analytical methods (such as gas chromatography and NMR spectrometry) as well as basic laboratory techniques (such as vacuum filtration). Each operation is identified by an operation number, such as OP–12 for vacuum filtration, and the description of an operation can be located quickly using the special headings at the top of the odd-numbered pages in Part V. In the Operations lists, the numbers printed in boldfaced type indicate operations being used for the first time; you should read these descriptions thoroughly before you come to the laboratory. Once you have used an operation, you should not have to reread the entire description the next time you use it, but you should at least read the Summary and review the Operational Procedure to refresh your memory. Eventually you should have to refer to the operation descriptions only if you encounter experimental difficulties or are applying an operation in a new situation.

Objectives.   Each experiment is designed to fulfill a number of objectives, which usually include both experimental objectives that will be met by

successful completion of the experiment (such as preparing a specified compound or identifying an unknown) and learning objectives relating to information, methods, and theoretical concepts that will be learned or reinforced as you carry out the experiment.

Situation. Many scientists study more or less familiar objects that can be readily observed, like stars, minerals, or ring-tailed wombats, but chemists deal with fundamental particles that cannot be seen or fully understood. It is easier to empathize with a wombat than with a molecule of *N*,*N*-diethyl-*meta*-toluamide ("Deet" for short; see Experiment 37) — even though the Deet is a more practical companion for a trek through the woods. Thus, nonchemists seldom understand just what motivates chemists to seek out and solve research problems that pertain to the invisible world of atoms and molecules. Each experiment is introduced by a Situation section intended to give you an appreciation of organic chemistry as a problem-solving endeavor rather than a collection of dry facts. Most Situations portray research problems of the kind that are encountered by professional chemists; some are more fanciful and need not be taken too seriously. In all cases, their purpose is to help you approach each experiment as a stimulating challenge to be met rather than an ordeal to be endured.

Background. Chemistry is an *experimental* science — the concepts of chemistry are based on observations of chemical phenomena, and no theory can stand for long without a solid foundation of experimental facts. A major purpose of any laboratory course is to put into practice the theoretical concepts dealt with in lectures. If you do not see the relationship between an experiment and the associated lecture material, the main point of the experiment will be lost. The Background section in each experiment is intended to serve as a link between the experimental work and the related concepts that are presented in the lecture course. This section may also relate historical sidelights or interesting facts that show the relevance of the experiment to the real world.

Methodology. A laboratory course gives you an opportunity to learn by experience, but you will not learn much of value unless you appreciate the significance of the experience at the time. If you follow a procedure mechanically without understanding the purpose of each step, you will be unprepared to deal with unexpected complications when they arise, to modify the procedure when necessary, or to take a shortcut when the situation warrants it. The Methodology section of an experiment describes and explains the experimental approach and may provide additional information that can help you interpret your results or cope with experimental situations as they arise.

Prelab Assignments. You should always read through the entire assigned experiment *before* you come to the laboratory, paying particular attention to

the Methodology and Procedure sections. The Prelab Assignments section of an experiment lists additional tasks to be completed before the lab period, such as reading operations, performing calculations, or writing an experimental plan. Ordinarily you will not be allowed to begin an experiment until your prelab write-up has been approved by the instructor.

Reactions and Properties.   The Reactions and Properties section gives balanced equations for the relevant reactions and tabulates the physical properties of reactants, products, and certain other chemicals. The Properties table provides the data needed for most of the prelab calculations.

Procedure.   Many of the experimental procedures in Part I are spelled out in considerable detail so as to inform or remind you of necessary steps that you might otherwise overlook. In most Part II experiments, the procedures are considerably less detailed. By the time you get to Part II, you will be expected to know how to perform most operations proficiently without the aid of frequent reminders. You should therefore regard the Procedure section more as a guide to the laboratory work than as a set of step-by-step instructions.

Safety Precautions.   The characteristics of some hazardous chemicals and precautions for their use are described under a separate heading that appears in the Procedure sections of experiments in which such chemicals are used. See the Laboratory Safety section (pages 8 – 15) for general information about hazardous chemicals.

Report.   The Report section of an experiment describes specific calculations, analytical data, or other information that should be incorporated into your laboratory report. Your instructor will describe the format you should follow and indicate any additional material that should be included in lab reports. In many cases it will be necessary to read the Background and Methodology sections thoroughly in order to complete your report or the assigned exercises.

Collateral Projects.   The Collateral Projects section describes additional projects that are related to the basic experiment. You are encouraged to complete as many of the projects as you have time for, but you must obtain your instructor's permission before starting any unassigned project.

Minilabs.   A short, self-contained Minilab is included with most of the experiments. Although each Minilab is associated with an experiment that is related in subject matter, you do not have to carry out a Minilab in conjunction with its experiment. Do not start any unassigned Minilab without first receiving permission from your instructor.

Exercises.   Your instructor will assign some of the Exercises to be completed and submitted with your laboratory report.

Library Topics.   The assignments in the Library Topics section are intended to help you become more familiar with the chemical literature, which is described in Appendix VII. Your instructor may make specific assignments or allow you to choose your own. Information on most topics can be found in sources in the Bibliography, particularly under category L.

Laboratory Organization

Because of wide variations in individual working rates, it is usually not possible to schedule experiments so that everyone can be finished in the allotted time; if all laboratories were geared to the slowest student, the objectives of the course could not be accomplished in the limited time available. As a result, some students will invariably get behind during a lab period and find it necessary to put in extra hours outside their scheduled laboratory section in order to complete the course. The students who fall into this group do not necessarily lack ability — some of the brightest students may also be among the slowest — but they are usually not well organized and thus fail to make the best use of their time. The following suggestions should help you work more efficiently in the laboratory:

1.   *Be prepared to start the current experiment the moment you reach your work area.* Don't waste the precious minutes at the start of a laboratory period doing calculations, reading the experiment, washing glassware, or carrying out other activities that should have been performed at the end of the previous period or during the intervening time. The first half hour of any lab period is the most important — if you can use it to collect your reagents, set up the apparatus, and get the initial operation (reflux, distillation, etc.) under way, you should be able to complete the experiment in the designated time period.

2.   *Organize your time effectively.* Set up a regular schedule that allows you to read the experiment and the operation descriptions a day or two before the laboratory period — an hour before the lab is much too late! Plan ahead so that you know approximately what you will be doing at each stage of the experiment. A written experimental plan, prepared as described in Appendix V, is invaluable for this purpose.

3.   *Organize your work area.* Before performing any operation, all of the equipment and supplies you will need to use during the operation should be set out neatly on your bench top in approximately the order in which they will be used. Small objects like spatulas and any item that might be contaminated by contact with the bench top should be placed on a paper towel, laboratory tissue, or mat. After each item is used, it should be removed to an out-of-the-way location (for example, dirty glassware to a washing trough in the sink) where it can be cleaned and then returned to its proper location when time permits. It is important to keep your locker well organized, with each item placed in a specific location when not in use, so that you can

immediately find the equipment you need. This will also enable you to notice when a piece of equipment has been misplaced or stolen, so it can be hunted down or reported to the instructor without delay.

Laboratory Etiquette

Although a little courtesy does help maintain peace and harmony in any work situation, this section is not about "good manners" as such. It is a set of common-sense rules that must be observed if the laboratory is to be a place where students can work together safely and efficiently.

1. *Return all chemicals and supplies to the proper location after use.* You will understand the reason for this rule if you ever experience the aggravation of hunting high and low for a reagent, only to find it at another student's station in a far corner of the lab. Chemicals that are being weighed out may be left at the balance if there are students waiting to use them; otherwise, they should be returned to the reagent shelves. Containers should be taken to the reagents to be filled; reagent bottles should not be taken to your lab station.

2. *Measure out only what you need.* Liquids and solutions should be measured into graduated containers so that you will take no more than you expect to use for a given operation. Solids can usually be weighed directly from their containers or measured from a special solids dispenser.

3. *Prevent contamination of reagents.* Do not use pipets or droppers to remove liquids from reagent bottles, and do not return unused reagent to a stock bottle. Be sure to close all bottles tightly after use — particularly those containing anhydrous chemicals and drying agents.

4. *When you must use a burner in the lab, tell your neighbors, unless they are already using burners.* This will enable them to cover any containers of flammable solvents and to limit or modify their use of such solvents during your operation. In certain circumstances, for example, when ether extractions are being performed, you should use a different heat source, move your operation to a safe location (for instance, under a fume hood) or find something else to do while flammable solvents are in use.

5. *Leave all community property where you found it.* Some items, such as ringstands, steam baths, lab kits, clamps, and condenser tubing, may not be supplied in student lockers. Such items are often part of the standard equipment found at each laboratory station, or they are obtained from the stockroom at the beginning of the laboratory period. Since such items may be needed by students in other lab sections, they should always be returned to the proper storage space at the end of the period.

6. *Clean up for the next person.* There are few experiences more annoying than finding that the lab kit you just checked out is full of dirty glassware, or that your lab station is cluttered with paper towels, broken glass, and

spilled chemicals. The last 15 minutes or so of every laboratory period should be set aside for cleaning up your lab station and the glassware used during the experiment. Put things away so that your station looks uncluttered, clean off the bench top with a towel or wet sponge, remove debris (including condenser tubing and other community supplies) from the sink, and thoroughly wash any dirty glassware that is to be returned to the stockroom, as well as that from your locker. Any spills and broken glassware should be cleaned up immediately, unless the chemical spilled is very toxic if inhaled or is lachrymatory (tear-producing), in which case the instructor should be informed and the area evacuated if necessary.

7. *Heed Gumperson's Second Law,* which states: "In the **laboratory** attention should be directed toward increasing that commodity indicated by the first five letters, and toward diminishing that commodity indicated by the last seven." This does not mean that all conversation must come to a halt, but it suggests that discussion of extraneous subject matter, such as the quality of the food served in the college cafeteria or the mayhem committed at the last football game, should be kept within reasonable proportions. Quiet conversation during a lull in the experimental activity is okay if it doesn't annoy your neighbors; a constant stream of chatter directed at another student during a delicate operation is distracting and may lead to an accident. For the same reason, radios and tape players should not be used in the laboratory.

Laboratory Safety

Preventing Laboratory Accidents

Some of the Procedure sections contain rather long lists of hazard warnings and safety precautions, which may give you the idea that an organic chemistry laboratory is a very dangerous place to work. Actually most organic lab courses are completed without incident, aside from minor cuts or burns, and serious accidents are very rare. Nevertheless, the *potential* for a serious accident is always there, and anyone who fails to observe a few basic safety precautions is taking a serious risk. *You must learn the following safety rules and observe them at all times.* Anyone failing to do so may be expelled from the laboratory.

1. *Wear appropriate glasses or safety goggles in the laboratory at all times.* Prescription glasses should have safety lenses, if possible. Contact lenses should not be worn in the laboratory since corrosive fumes or chemicals may get underneath them. The location and operation of the eyewash fountain should be learned at the first laboratory session.

2. *Never smoke in the laboratory or use open flames in operations involving low-boiling flammable solvents.* Ethyl ether and petroleum ether are particularly hazardous, but other common solvents such as acetone and ethanol can be dangerous as well. The location and operation of the fire extinguishers, fire blanket, and safety shower must be learned at the first laboratory session. Smoking is prohibited in chemistry laboratories because of the fire hazard.

3. *Consider all organic chemicals hazardous to your health and minimize your exposure to them.* Do not inhale the vapors of volatile chemicals or the dust of finely divided solids. Prevent contact between chemicals and your skin, eyes, and clothing; some chemicals cause burns or severe allergic reactions, and others can be absorbed through the skin and cause poisoning. Do not taste any chemicals, and wash your hands thoroughly after handling them. To prevent accidental ingestion of any toxic chemicals, do not bring food or drink into the laboratory or use mouth suction for pipetting. Read the Safety Precautions section for each experiment, and use protective gloves or a fume hood when indicated.

4. *Handle strong acids or bases, bromine, and other corrosive chemicals with great care* and never allow them in contact with skin or eyes. Spills of such chemicals must be cleaned up immediately, using an appropriate neutralizing agent and plenty of water.

5. *Use safe techniques for inserting and removing glass tubing.* Proper procedures for working with glass should be learned by reference to operation OP–4. Great care must be exercised when inserting or removing thermome-

ters and glass tubing from stoppers and thermometer adapters, since severe lacerations can result from accidental breakage.

6. *Wear appropriate clothing.* Serviceable shoes (not sandals) should always be worn in the laboratory to provide adequate protection against spilled chemicals and broken glass. Hairnets for long hair may be advisable if burners are in use (human hair is very flammable!). Clothing should be substantial enough to offer protection against accidental spills of corrosive chemicals, but the laboratory is obviously not the place to wear your best clothes.

7. *Dispose of chemicals in the proper containers and in the manner specified by your instructor.* Most organic chemicals, for reasons of safety and environmental protection, should not be washed down the drain. Whenever possible, they should be disposed of in waste crocks (for solids) or in labeled waste-solvent cans. Some dilute aqueous solutions can be safely washed down the drain, but the instructor should be consulted if there is any question as to the best method for disposing of particular chemicals or solutions.

8. *Never work alone in the laboratory or perform unauthorized experiments.* If you find it necessary to work in the laboratory when no formal lab period is scheduled, you must obtain permission from the instructor and be certain that others will be present while you are working.

Reacting to Accidents: First Aid

Any serious accident involving poisoning or injury should be treated by a competent physician. To minimize the damage from such an accident, all students and instructors should be familiar with some basic first-aid procedures. If you have an accident that requires quick action to prevent permanent injury (flushing chemical burns with water, etc.), take the appropriate action as described below and then, as soon as possible, see that the instructor is informed of the accident. If you *witness* an accident you should summon the instructor immediately and leave the first aid to him or her *unless:* (1) no instructor or assistant is in the laboratory area; (2) the victim requires immediate attention because of stopped breathing, heavy bleeding, etc.; (3) you have had formal training in emergency first-aid procedures and the instructor consents.

If an accident victim has difficulty breathing or goes into shock as a result of any kind of accident, standard procedures for artificial respiration and treating shock should be applied. Detailed first-aid procedures can be found in the *CRC Handbook of Laboratory Safety,* 2nd ed. ⟨Bib – C11⟩, pages 23 – 48, and in various first-aid manuals.

Eye Injuries. If any chemical enters your eyes, they should *immediately* be flooded with water from an eyewash fountain, with the eyelids kept open. If

contact lenses are being worn, they must be removed before irrigation. Irrigation should continue for at least 15 minutes, and the eyes should then be examined by a physician. The use of boric acid or other neutralizing solutions is not recommended for eye injuries, since this sometimes causes more damage than no irrigation at all.

If foreign bodies such as glass particles are propelled into an eye, the injured person should get immediate medical attention. Removal of such particles is a job for a specialist.

Chemical Burns.   The affected area should *immediately* be flushed with water, using a safety shower if the area of injury is extensive or if it is inaccessible to washing from a tap. Speed in washing is the most important factor in reducing the extent of injury. Water should be applied continuously for 10–15 minutes and the burned area should then be covered with sterile gauze or other clean cloth. First-aid procedures for burns caused by specific chemicals are given in the *Sigma-Aldrich Library of Chemical Safety Data* ⟨Bib–C9⟩ and in the ninth edition of the *Merck Index* ⟨Bib–A3⟩. Neutralizing solutions, ointments, or greases should not be used on chemical burns unless specifically called for in a first-aid procedure. Unless the skin is merely reddened over a small area, a chemical burn should be examined by a nurse or physician.

If the burn is extensive or severe, the victim should be made to lie down with the head and chest a little lower than the rest of the body. If the victim is conscious and able to swallow, he or she should be provided with plenty of nonalcoholic liquid to drink until a physician arrives.

Thermal Burns.   For small thermal burns such as those inflicted by handling hot glass, a sterile gauze pad should be soaked in a baking soda solution (2 tablespoons of sodium bicarbonate to a quart of lukewarm water, about $2\frac{1}{2}$% $NaHCO_3$), placed over the burn, then bandaged loosely in place. If the skin is not broken, the burned part can be immersed in clean, cold water, or ice can be applied to reduce the pain. Blisters should not be opened. Unless the skin is merely reddened over a small area, the burn should be examined by a nurse or physician.

In case of an extensive thermal burn, the burned area should be covered with the cleanest available cloth material and the victim should be made to lie down, with the head and chest lower than the rest of the body, until a physician arrives. If the injured person is conscious and able to swallow, he or she should be provided with plenty of nonalcoholic liquid to drink (water, tea, coffee, etc.).

Bleeding, Cuts, and Abrasions.   In case of a minor cut or abrasion, the wound and surrounding skin should be cleansed with soap and lukewarm water that is applied by wiping away from the wound. A sterile pad should be held over the wound until bleeding stops and then be replaced by a clean pad, which should be secured loosely with a triangular or rolled bandage. The pad and bandage should be replaced as necessary with clean dry ones. Contact be-

tween the wound and the mouth, fingers, handkerchiefs, or other unsterile objects should be avoided. Antiseptics should not be used on an open wound.

In case of a major wound that involves heavy bleeding, you should not waste time cleansing the wound; *apply pressure immediately* directly over the wound. Use a pad (clean handkerchief or other cloth), and press firmly with one or both hands to reduce the bleeding as much as possible. The victim should be made to lie down with the bleeding part higher than the rest of the body, and the pad should be held in place with a strong bandage (a necktie may be adequate, or cloth strips torn from a shirt). A physician should be called as soon as possible, and the victim should be kept warm with a blanket or coat in the meantime. If the victim is conscious and able to swallow, he or she should be provided with plenty of nonalcoholic liquid to drink.

Poisoning. Chemical poisoning may have occurred whenever, after chemicals have been handled or accidentally ingested, there is a pain or burning sensation in the throat, discoloration of lips or mouth, stomach cramps, nausea and vomiting, confusion, or loss of consciousness.

If poison is swallowed, a physician or ambulance should be called immediately, and the victim should be given 2–4 glasses of water or milk to drink. If it is safe to do so, vomiting should be induced by placing a finger at the back of the victim's throat or by having the victim drink a glass of warm water in which 2 teaspoons of table salt have been dissolved (10 g in 200 mL), or one ounce (30 mL) of ipecac syrup. Vomiting should *not* be induced if the patient is unconscious, in convulsions, or has severe pain and burning sensations in the mouth or throat, or if the poison is a petroleum product or a strong acid or alkali. When vomiting begins, the victim should be placed face down with head lower than hips, and vomiting should continue to be induced until the vomitus is clear. When there is time, the poison should be identified (if possible) and an appropriate antidote given (see the *Merck Index,* 9th ed., pages MISC–22ff., or call a poison control center). If the poison cannot be identified and a physician is not present, a heaping tablespoon (15 g) of the following *universal antidote* can be given in a half glass of warm water: 2 parts activated charcoal, 1 part magnesium oxide (milk of magnesia), 1 part tannic acid. A sample of the poison should be saved for the physician if possible; if the poison is unknown, the vomitus should be saved for examination.

If a poison has been inhaled, the victim must be taken to fresh air and a physician called immediately. If the victim shows evidence of difficulty in breathing, mouth-to-mouth resuscitation should be administered. The victim should be kept warm and as quiet as possible until professional help is available. If the poison is a highly toxic gas such as hydrogen cyanide, hydrogen sulfide, or phosgene, the persons attempting to rescue the victim should wear gas masks while they are in contact with the vapors.

In case of skin contamination by toxic substances, the same general procedure as for chemical burns should be followed. The *Sigma-Aldrich Library of Chemical Safety Data* ⟨Bib–C9⟩ or other sources should be consulted for specific procedures to be used for certain substances.

Reacting to Accidents: Fires

In case of a serious fire, one's first reaction should be to *get away from it* as quickly as possible. If the fire is small and confined to a container such as a flask or beaker, it can often be extinguished by quickly placing an asbestos pad or watch glass over the mouth of the container. Otherwise, you should get out of the area and let the instructor or an assistant extinguish it. If no instructor or assistant is in the vicinity, you may obtain a fire extinguisher of the appropriate type (dry chemical extinguishers are best for most flammable liquid fires) and attempt to put out the fire by aiming the extinguisher at the base of the fire (maintain a safe distance). Be prepared to call the fire department and arrange to evacuate the area if there is a chance that the fire cannot be controlled.

If your hair or clothing should catch on fire, go directly to the nearest fire blanket or safety shower and attempt to extinguish the fire by wrapping yourself in the blanket or drenching yourself with water. If another person's hair or clothing has caught on fire, try to prevent panic and wrap the person in a coat or fire blanket or lead him or her to a safety shower.

More detailed information on fire and rescue procedures can be found in the *CRC Handbook of Laboratory Safety* ⟨Bib–C11⟩, pages 15–22 and 187–191.

Chemical Hazards

For your own health and safety, it is essential that you *exercise caution while handling chemicals* and *minimize your exposure to them*. Recent information about the health hazards posed by certain chemicals and more restrictive governmental regulations for their disposal have caused most academic chemistry departments to enact stringent policies for dealing with hazardous chemicals. Your instructor will inform you of any departmental or institutional rules relating to the safe handling and disposal of such chemicals. Most of the experimental procedures in this book describe specific chemical hazards and handling precautions under the heading Safety Precautions, and some may also include reminders regarding the disposal of certain chemicals. The hazard descriptions are meant to inform you of the *potential* danger posed by certain chemicals when they are not handled properly. If you observe the third safety rule listed above and follow the recommendations in the Safety Precautions sections, there is little likelihood that you will suffer any ill effects from working with any chemical.

Hazard Signs Used in This Text.   In many experiments specific hazard information on some chemicals is provided in the form of octagonal hazard signs containing index numbers and/or letters. Each hazard index is a number from 0 to 4, where 0 indicates no known hazard and 4 indicates a very great health or safety hazard. The letters are used to warn of other hazards, such as

Contents

reactivity with water and carcinogenic potential. The significance of the numbers and letters is explained in Table 1.

You should learn the location of each category on the hazard signs: going clockwise from the top, the order is Fire → Reactivity → Vapor → Contact. The Fire index number in the upper quadrant rates the fire danger posed by the chemical. The Reactivity index number in the right quadrant assesses the danger of a violent reaction or explosion. The Vapor index number in the lower quadrant rates the potential health effect of inhaling the vapors of the chemical. The Contact index number in the left quadrant rates the adverse health effects that may result from skin contact (keep in mind that the eye contact hazard, which is not rated, may be even greater in some cases). For some chemicals the Vapor and Contact index numbers are replaced by a single Health index number that represents the overall health hazard of the chemical. If any quadrant is left blank, it means that no hazard index number was reported in the sources consulted; it does *not* mean that no hazard exists.

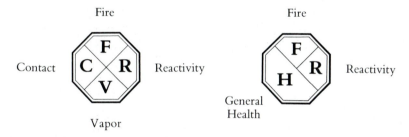

Table 1. Meanings of numbers and letters on hazard symbols

Symbol	Health (Contact or Vapor)	Fire	Reactivity
0	No known hazard	Will not burn	Stable
1	May cause irritation if not treated	Ignites after strong preheating	Unstable only at high temperature and pressure
2	May cause injury; requires treatment	Ignites after moderate heating	Unstable, but won't detonate
3	May cause serious injury despite treatment	Ignites at normal temperatures	Detonates or explodes with difficulty
4	May cause death or major injury despite treatment	Very flammable	Readily detonates or explodes
CA	Carcinogen (cancer-causing agent)		
OX			Strong oxidant; may react violently with combustible material
P			Polymerizes readily
W			Reacts violently with water

The meaning of the hazard signs is illustrated by the examples below.

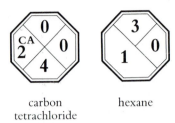

carbon
tetrachloride

hexane

Carbon tetrachloride is nonflammable and thermally stable, so its Fire and Reactivity hazard index numbers are both zero. Inhaling its vapors may be extremely harmful, and there is a moderate risk of injury from skin contact. Carbon tetrachloride also has been identified as a carcinogen in tests with laboratory animals and is suspected of causing cancer in humans. For these reasons, carbon tetrachloride is not used in the experiments in this book. The second hazard sign indicates that hexane is highly flammable but very stable and its health risk is comparatively low. Obviously, you should be alerted whenever you see a high hazard number, and take appropriate precautions to prevent the hazard indicated. For additional information about lab safety and chemical hazards, consult appropriate sources listed under category C in the Bibliography.

About Carcinogens

A few chemicals used in this text are identified as potential carcinogens, that is, agents suspected of causing cancer. Although this label should certainly not be ignored, it is important to point out that such chemicals — *if used only as directed in this book* — should present little (if any) risk of cancer to students handling them. The carcinogenic activity of a compound is generally established by animal tests in which high doses of the chemical are administered by various routes for prolonged periods. For example, oral doses of chloroform and phenacetin have been found to cause cancer in laboratory animals, but ingestion of such chemicals in the lab is easily avoided. The toxicity of chloroform probably represents a greater hazard to chemistry students than any cancer-causing potential it may have. Some halogenated hydrocarbons are suspected of causing cancer if they are inhaled, so you should avoid breathing the vapors of chlorinated solvents such as chloroform and methylene chloride (or any other potentially toxic solvents). Chromium compounds have been shown to cause cancers of the lungs, nasal cavity, and sinuses in humans, but the persons at risk are industrial workers who have been exposed to the dust of chromium compounds continuously in the workplace. In this course, you will use chromic acid *solutions* [prepared from chromium(VI) oxide] a drop or so at a time, so there will be no possibility of inhaling chromium dust. Obviously, there might be some risk if you prepared the solution from powdered chromium(VI) oxide, but such reagents will be made up beforehand.

Compounds that have been designated as carcinogens or suspected carcinogens are identified as such in this book, and the route of administration in positive tests with animals (oral dose, inhalation, etc.) is indicated. Compounds (such as benzene) that might present a significant risk of cancer under conditions likely to be encountered in an undergraduate laboratory are not used, and chloroform is used very sparingly. It is only prudent to exercise particular care when handling potential carcinogens and to wear protective gloves and work under a hood when warranted, but the label "carcinogen" should cause no more apprehension than other hazard warnings.

MASTERING THE OPERATIONS

The experiments in Part I are intended to help you learn the basic operations of organic chemistry and become proficient in their use. The first time an operation is used in this book, its number will be highlighted by appearing in boldfaced type in the Operations list at the beginning of the experiment; it will also be mentioned in the Prelab Assignments. If the experiments of Part I are not assigned sequentially by your instructor, you should examine the Operations lists for operations that you have not encountered before, and study them as part of the Prelab Assignments.

The Conversion of Sodium Benzoate to Benzoic Acid

Laboratory Orientation. Acid-Base Reactions. Reaction Stoichiometry.

Operations

OP–1   Cleaning and Drying Glassware

OP–3   Using Corks and Rubber Stoppers

OP–4   Basic Glass Working

OP–5   Weighing

OP–6   Measuring Volume

OP–12   Vacuum Filtration

OP–21   Drying Solids

Objectives

To prepare benzoic acid from sodium benzoate.

To learn and practice some fundamental laboratory operations.

To become familiar with the laboratory environment and learn where the safety equipment is located.

To learn how to minimize product losses during a laboratory preparation.

To learn how acid-base reactions and solubility differences can be used in organic syntheses.

To learn about some properties and uses of benzoic acid and sodium benzoate.

SITUATION

A resin is actually an organic glass — a noncrystalline solid that usually results from the evaporation of volatile materials such as turpentine from the saps of trees and other plants.

When a resin from the Sumatran tree *Styrax benzoin* is heated to 100°C, white vapors rise and condense to form needlelike crystals of benzoic acid, which got its name from the tree and its resin, gum benzoin. Sodium benzoate, a widely used food preservative, works better in acidic foods like fruits and fruit juices than in more alkaline foods. There is good reason to believe that sodium benzoate is converted to the "natural" plant substance — that is, to benzoic acid — in foods having low pH values. In this experiment, you will prepare benzoic acid by using hydrochloric acid to lower the pH of a solution of sodium benzoate.

sodium benzoate   benzoic acid

The hexagon with a circle in it represents a benzene ring, whose molecular formula in these compounds is C_6H_5.

BACKGROUND

Benzoic acid is so widely distributed in plants that the urine of all plant-eating animals (including humans) contains hippuric acid, a compound synthesized in the kidneys by the combination of benzoic acid with the amino acid glycine. Significant amounts of benzoic acid can be isolated from such diverse natural sources as anise seed, cranberries, prunes, cherry bark, cloves, and the scent glands of the beaver. In cranberries and other plant products, benzoic acid is a natural preservative that inhibits the growth of bacteria, yeasts, and molds and thus retards spoilage.

hippuric acid   glycine

If you read the labels on cans and bottles in a grocery store, you will find sodium benzoate (sometimes called benzoate of soda) listed on many of them. Sodium benzoate is used as a preservative in such food products as jams and jellies, soft drinks, fruit juices, pickles, condiments, margarine, and canned and frozen seafoods, and even in such nonfood items as toothpaste and tobacco. Sodium benzoate is most effective as a preservative in the pH range from 2.5 to 4.5, where much of it is converted to benzoic acid; presumably it is the free acid that has the preservative effect in acidic foods. The property that makes sodium benzoate more popular as a food additive than benzoic acid itself is its water solubility, which makes it easier to blend into water-containing food products.

METHODOLOGY

The Methodology section of this and future experiments outlines and explains the experimental approach and often provides additional information to help you plan and carry out the laboratory work more efficiently. The Exercises at the end of each experiment are often based on material presented in the Background and Methodology sections.

Part of the purpose of this experiment is to help you become familiar with the laboratory environment and with some simple but fundamental laboratory operations. These objectives will be accomplished as you carry out the preparation of an organic compound, benzoic acid, from its salt, sodium benzoate. Both benzoic acid and sodium benzoate are white crystalline solids, but sodium benzoate is nearly 200 times more soluble in water than benzoic acid is (see Table 2). This large solubility difference makes it possible to prepare the acid from its salt in high yield.

When sodium benzoate dissolves in water, it dissociates into benzoate ions, which are weakly basic, and sodium ions.

$$C_6H_5COO^-Na^+ \longrightarrow C_6H_5COO^- + Na^+$$
$$\text{benzoate ion}$$

Hydrochloric acid is a solution of gaseous hydrogen chloride (HCl) in water. The strongest acid present in this solution is the hydronium ion, formed by the transfer of a proton from an HCl molecule to a water molecule.

$$HCl + H_2O \longrightarrow Cl^- + \quad H_3O^+$$
$$\text{hydronium}$$
$$\text{ion}$$

When hydrochloric acid is added to a sodium benzoate solution, protons are transferred from the strong acid H_3O^+ to the basic benzoate ions, forming the weaker acid, benzoic acid.

$$H_3O^+ + C_6H_5COO^- \longrightarrow H_2O + C_6H_5COOH$$
$$\text{benzoic acid}$$

The net reaction is the sum of these three reaction steps:

$$C_6H_5COO^-Na^+ + HCl \longrightarrow C_6H_5COOH + NaCl$$

Note that in these, as in all spontaneous acid-base reactions, protons are transferred from the *stronger* acid to a base, to yield a *weaker* acid and base.

In this experiment, you will start with 25.0 mmol (0.0250 mol, or 3.60 g) of sodium benzoate and precipitate the benzoic acid from an aqueous solution whose total volume is 25 mL. The stoichiometry for the reaction is given in the net equation above, in which sodium benzoate and benzoic acid are present in a 1:1 mole ratio. Thus, the theoretical yield is 25.0 mmol of benzoic acid, which is 3.05 g. (See Appendix IV for a review of stoichiometric calculations.) The solubility of benzoic acid increases rapidly with temperature (see Table 1). Therefore, to improve the recovery of benzoic acid, you must decrease its solubility by cooling down the warm reaction mixture. If it is cooled to 10°C, about 0.05 g of benzoic acid (0.21 g/100 mL times 25 mL) will remain dissolved in the 25 mL of solution. The rest should eventually precipitate from solution and can then be collected by vacuum filtration. At best, you should be able to recover about 3.00 g of benzoic acid.

You will probably recover somewhat less than 3.00 g of benzoic acid. Errors in measuring mass or volume, incomplete precipitation, insufficient cooling, and losses that occur while materials are being transferred from one vessel to another can lower your yield. You should try to minimize such losses by taking great pains to make accurate measurements and to scrape or rinse the last grains of solid from the glassware. From the start, you should get into the habit of doing the kind of careful, meticulous work that will increase your yields and save you time and effort in the long run.

If you come up with an apparent yield that is *greater* than 3.00 g, either you have made a weighing error or your product is still wet. The product of a chemical preparation should always be dried to *constant mass,* meaning that its mass after drying should not change between two successive weighings. For most preparations in this book you can assume that a product is sufficiently dry if—following the initial drying period—its mass does not decrease by more than 0.5% (about 0.01 – 0.02 g for this experiment) after an additional 10 minutes or more of oven drying. If the product is being dried at room temperature in a desiccator, the time between weighings should be greater. Your instructor may require more thorough drying in some instances.

Table 1. Solubility of benzoic acid in water

Temperature, °C	Solubility, g/100 mL of Water
0	0.17
10	0.21
20	0.29
30	0.42
40	0.60
50	0.85
60	1.20
70	1.77
80	2.75
90	4.55

PRELAB ASSIGNMENTS

The prelab assignments for each experiment are to be completed before you come to the laboratory to perform the experiment. This experiment requires more reading than most, so you should begin the reading assignments well in advance of the experiment date.

1. Read the Introduction and Laboratory Safety sections of this book and any Appendixes assigned by your instructor.

2. Read the experiment, and make sure you understand its objectives.

3. Read the sections in Part V describing OP–1, OP–3, OP–4, OP–5, OP–6, OP–12, and OP–21 (most of the descriptions are quite brief).

4. If you are required to keep a formal laboratory notebook (see Appendix II), prepare the notebook as requested by your instructor.

Reactions and Properties

Table 2.   Physical properties

	M.W.	m.p.	Water Solubility
benzoic acid	122.1	122	0.34
sodium benzoate	144.1		61.2

Note: M.W. = molecular weight; melting points are given in °C; solubilities are in grams of solute per 100 mL of water at 25°C.

PRELIMINARIES

Laboratory Safety and Orientation

Prior to the laboratory period, you should have read the Laboratory Safety section of this book. Before you begin to work in the laboratory, the instructor will review the safety rules with you and tell you what safety supplies you must have (safety glasses, protective gloves, etc.). During the first laboratory period, the instructor will show you where safety equipment is located and tell you how to use it. As you locate each item, check it off the following list and make a note of its location. (Your instructor may suggest additions or amendments to the list.) You should also learn the locations of chemicals, consumable supplies (such as filter paper and boiling chips), waste containers, and various items of equipment (such as balances and drying ovens).

Safety Equipment and Supplies

[ ]   Fire extinguishers

[ ]   Fire blanket

[ ]   Safety shower

[ ]   Eyewash fountain

[ ]   Fume hoods

[ ]   First-aid supplies

[ ]   Spill cleanup supplies

Checking In

If you are performing this experiment during the first laboratory session, you should obtain a locker and equipment list and check into the laboratory as directed by your instructor. You will find illustrations of typical locker supplies in Appendix I at the back of this book. Pieces of glassware with chips, cracks, or star fractures should be replaced; they may cause cuts, break on heating, or shatter under stress. If necessary, clean up any dirty glassware in your locker ⟨OP–1⟩ at this time.

Making Useful Laboratory Items

Prepare as many of the following items as your instructor requests. Be sure you have read OP–3 and OP–4 before you begin. Have the instructor inspect and approve the items when you are finished.

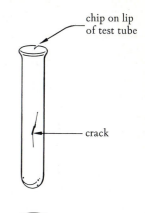

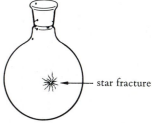

Figure 1.   Glassware defects

Be very careful that you don't burn or cut yourself while working with glass rod and tubing. See OP–4 for safe procedures.

SAFETY PRECAUTION

Flat-Bottomed Stirring Rod.   Make a flat-bottomed stirring rod ⟨OP–4⟩ by cutting a 20–25 cm length of 5-mm or 6-mm soft glass rod, flattening one end so that it flares out to a diameter of 10 mm or more, and rounding the other end. The stirring rod can be used to wash crystals on a Buchner filter or to pulverize solids for melting-point determinations, and for other operations besides stirring.

Vacuum Filtration Apparatus.   Using a 250-mL (or larger) thick-walled bottle, 8-mm soft glass tubing, and a rubber stopper to fit, construct a filter trap like the one shown in Figure 1 on page 627. One length of glass tubing should be about 15 cm long, and the other should be about half that long. Bend both pieces of tubing smoothly ⟨OP–4⟩ so that the horizontal arm on each is 3–4 cm long, and fire-polish the ends. Bore two holes of appropriate diameter in the rubber stopper ⟨OP–3⟩ and insert the glass tubing, using glycerol as a lubricant. **[Be careful! If inserted improperly the glass may break and cut your hands.]** When this assembly is inserted in the bottle, the lower end of the long tube should be 5 cm or so from the bottom; the short tube has only to extend through the stopper. Wrap the bottle with heavy tape to prevent possible injury from implosion (a good grade of transparent plastic tape should be suitable).

Alternatively, you can use a large filter flask as a filter trap.

Select a rubber stopper that will fit your filter flask, and using a cork borer having about the same diameter as the midpoint of your Buchner funnel stem, bore a hole through the middle of the stopper ⟨OP–3⟩. Assemble the vacuum filtration apparatus pictured in Figure 1, page 627. Clamp it to a ringstand and attach it to the water aspirator with thick-walled rubber tubing.

Boiling-Point Tube.   Obtain a piece of 5-mm O.D. (outside diameter) soft glass tubing and carefully seal it at one end ⟨OP–4⟩. Next, cut it to a length of 8–10 cm and fire-polish the open end. Test the tube as described in OP–4 to make sure it is sealed. Save it for use in Experiment 6.

PROCEDURE

See your instructor if you need help to operate the balances used in your laboratory.

SAFETY PRECAUTION

Benzoic acid irritates the skin and eyes, and sodium benzoate is a mild irritant. Minimize your contact with both compounds; do not get them in your eyes or on your skin or clothing.

Reaction.   As accurately as you can, weigh 3.60 g of sodium benzoate ⟨OP–5⟩ and transfer it to a small beaker. Measure 16 mL of distilled or deionized water with a graduated cylinder ⟨OP–6⟩, pour it into the beaker, and stir until the sodium benzoate dissolves. Measure 9 mL of 3 M hydrochloric acid ⟨OP–6⟩ and pour it slowly, with stirring, into the sodium benzoate solution. With the round end of the stirring rod, transfer a drop of the solution to a strip of pH paper; if necessary, add more 3 M HCl to the solution until its pH is 2 or less. Cool the solution to 10°C or below by setting the small beaker in a larger beaker containing cracked ice and a little water and stirring occasionally.

Separation.   Separate the benzoic acid from the reaction mixture by vacuum filtration ⟨OP–12⟩ and wash the benzoic acid on the filter with about 5 mL of ice-cold water. Use the flat end of the stirring rod to mix the crystals well with the wash water, being careful not to displace the filter paper in the process. Let the benzoic acid air-dry on the filter a few minutes while the aspirator is running, then dry it ⟨OP–21⟩ to constant mass. If an oven is used for drying, its temperature should be 90°C or lower, because benzoic acid begins to sublime at around 100°C. Weigh the dry product ⟨OP–5⟩ in a tared (preweighed) vial, and label the vial with the experiment number, the name of the product, its mass, your name, the current date, and any other information required by your instructor. A sample label is shown in the margin.

Cleanup Routine.   After completing this experiment (and all subsequent experiments), you should:

1.   Clean up the glassware you used during the experiment ⟨OP–1⟩.

2.   Clear off your work area, wipe the bench top with a sponge or wet towel, and remove any refuse or equipment from the sink.

3.   Turn in any items you checked out of the stockroom, and return other supplies, such as ringstands and rubber tubing, to their proper locations.

4.   Turn in your labeled product to the instructor, when applicable.

5.   See that all the items on your locker list are safely inside (you may have to pay for missing supplies!), and lock your locker.

When water is specified in experimental procedures, you should generally use distilled or deionized water, not tap water. Tap water should be used only for heating and cooling baths and for washing glassware or other equipment.

Do not use your thermometer as a stirring rod!

Be sure to turn off the aspirator while you wash the crystals and to turn it on again to drain out the wash water.

```
Exp. 1
Benzoic acid
   2.89 g
Cynthia Sizer
  9-20-88
```

Report. Your instructor will let you know what report format to use and what additional items should be included in your report (see Appendix III for suggestions).

Calculate the theoretical yield of benzoic acid to the same number of significant figures as in the starting mass of sodium benzoate, and calculate the percent yield of your preparation. Compare your actual yield of benzoic acid with the attainable yield (discussed in the Methodology section), and try to account for your losses.

Collateral Project

You can carry out the experiment using an equimolar amount of sodium salicylate in place of sodium benzoate. In that case, you will have to calculate the mass of sodium salicylate needed and the theoretical yield of salicylic acid. Physical properties and solubility data for salicylic acid are given in Experiment 4.

Exercises

1. In the Methodology section, the attainable yield of benzoic acid was given as 3.00 g under the conditions described. What would be the attainable yield if: **(a)** you failed to cool the reaction mixture after adding HCl and its temperature was 40°C when it was filtered? **(b)** you cooled the reaction mixture to 0°C instead of 10°C?

2. The water solubilities of oxalic acid and sodium oxalate at 20°C are 10 g/100 mL and 3.7 g/100 mL, respectively. Could you prepare oxalic acid from sodium oxalate by the method used in this experiment? Explain why or why not.

3. Sodium benzoate can be prepared by adding benzoic acid to a hot aqueous solution of sodium hydroxide and evaporating the water. Write a balanced equation for the proton-transfer reaction involved in this preparation. Label the stronger acid, stronger base, weaker acid, and weaker base.

4. Calculate the percentage of sodium benzoate that would fail to precipitate as benzoic acid in this experiment if only enough HCl were added to bring the pH down to 4.00. The ionization constant for benzoic acid is 6.46×10^{-5}.

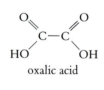

oxalic acid

sodium oxalate

Library Topic

When starting to work on a library topic, check Appendix VII and the Bibliography (particularly category L) for sources related to the topic you are writing on or for general sources such as the *Kirk-Othmer Encyclopedia* ⟨Bib–A13⟩. Such general sources will often tell where you can look for additional information.

Write a short paper about the additives used to retard spoilage or discoloration in foods and to prevent food poisoning. Describe the applications of different food preservatives, and discuss the controversy surrounding such additives as sodium nitrite.

Separation of the Components of "Panacetin"

Separation Methods.

Operations

OP–5 Weighing

OP–7 Heating

OP–11 Gravity Filtration

OP–12 Vacuum Filtration

OP–13 Extraction

OP–14 Evaporation

OP–21 Drying Solids

Objectives

To separate the components of a mixture and determine its percentage composition.

To learn and apply some of the operations used in separating organic compounds.

To learn how flow diagrams are used to outline separation processes.

To learn about the history and uses of some analgesic/antipyretic drugs.

SITUATION

One of the responsibilities of the U.S. Food and Drug Administration is to ensure that over-the-counter drugs are both safe and effective. Testing a drug entails measuring and identifying its components to verify that it has the composition reported by the manufacturer. In this experiment, you will play the role of an FDA investigator and analyze a simulated drug preparation called "Panacetin," which contains aspirin, sucrose, and an unknown constituent. Even though it is chemi-

Admittedly, the FDA would apply somewhat more sophisticated methods than the ones you will use in this experiment.

cally similar to acetaminophen (the active ingredient of Tylenol and other over-the-counter preparations), the unknown compound is suspected to be unsafe for use as a drug. Your assignment, in this experiment and the next, is to determine the composition of the drug preparation and to identify its unknown constituent.

BACKGROUND

The Discovery of Pain-Killing Drugs

Analgesic drugs reduce pain and antipyretic drugs reduce fever; some drugs, such as aspirin, do both. Most of the common over-the-counter analgesic/antipyretic drug preparations contain aspirin, acetaminophen, or combinations of these substances with other ingredients. For example, acetaminophen is the active constituent of Tylenol, and Extra Strength Excedrin currently contains aspirin, acetaminophen, and caffeine.

From their molecular structures, you can see that acetaminophen and the related compound phenacetin are derivatives of acetanilide; their chemical names are *p*-hydroxyacetanilide and *p*-ethoxyacetanilide, respectively. Acetanilide was once a popular drug marketed under the name *antifebrin,* and until very recently phenacetin was used in APC tablets and other analgesic preparations. Phenacetin is no longer approved for medicinal use in the United States because it is rather toxic, causing hemolytic anemia or kidney damage in some patients. It was listed as a carcinogen by the Environmental Protection Agency in 1981. Acetanilide causes a serious form of anemia called methemoglobinemia, in which hemoglobin molecules in the bloodstream are altered in such a way that their ability to transport oxygen is reduced. Even though antifebrin (acetanilide) is now considered too toxic for medicinal use, its discovery did stimulate the development of safer and more effective analgesic/antipyretic drugs.

The discovery of antifebrin came about in 1886 when a pair of clinical assistants, Arnold Cahn and Paul Hepp, were looking for something that would rid their patients of a particularly unpleasant intestinal worm. The trick was to find a drug that would kill the worm but not the patient, and their method — not a very scientific one — was to test the chemicals in their stockroom until they found one that worked. When someone came up with an ancient bottle labeled "naphthalene," they tried it out on a patient who had every malady in the book, including worms. It didn't faze the worms, but it

Some compounds with analgesic/antipyretic properties

aspirin

acetanilide

acetaminophen

phenacetin

naphthalene

Synthesis of phenacetin

NH$_2$

p-aminophenol

OH

NH$_2$

O CH$_2$CH$_3$ ←— ethyl "masking" group

NHCOCH$_3$

acetyl group

OCH$_2$CH$_3$

phenacetin

reduced the patient's fever dramatically. Luckily, before going out on a limb and endorsing naphthalene as a cure-all for fevers, Cahn and Hepp noticed that the chemical didn't have the "mothball" odor characteristic of naphthalene (it was nearly odorless), and they decided to have it tested. Hepp's cousin, a chemist at a nearby dye factory, soon gave them the news — their new drug was not naphthalene at all but acetanilide!

Barely six months after the discovery of antifebrin, a similar drug was developed. It all started with a storage problem. Carl Duisberg, director of research for the Friedrich Bayer Company, had to get rid of fifty tons of *para*-aminophenol — a seemingly useless yellow powder that was a by-product of dye manufacture. Rather than pay a teamster to haul the stuff away, Duisberg decided to change it into something Bayer could sell. After reading about antifebrin, he reasoned that a compound with a similar molecular structure might have the same therapeutic uses. Duisberg knew that a hydroxyl (OH) group attached to a benzene ring is a characteristic of many toxic substances (for example, phenol), so he decided to "mask" the hydroxy group in the *p*-aminophenol with an ethyl (CH$_3$CH$_2$) substituent, as shown in the margin. Subsequent addition of an acetyl (CH$_3$CO) group produced phenacetin, which proved to be a remarkably effective and inexpensive analgesic/antipyretic drug.

Ironically, the substance Duisberg would have obtained had he not first masked the hydroxy group is acetaminophen, which turned out to be a safer drug than phenacetin. Acetaminophen is now as popular as aspirin for treating fevers, headaches, and other aches and pains. In the body, acetanilide and phenacetin are converted to acetaminophen, which is believed to be the active form of all three drugs.

METHODOLOGY

Most natural products and many commercial preparations are mixtures containing a number of different substances. To obtain a pure organic compound from such a mixture, you must separate the desired compound from the other components by taking advantage of differences in physical and chemical properties. Substances having very different solubilities in a given solvent can often be separated by extraction or filtration, and liquids with different boiling points can be separated by distillation. Acidic or basic components are often converted to water-soluble salts, which can then be separated from the water-insoluble components of a mixture.

In this experiment, you will separate the components of a simulated pharmaceutical preparation, "Panacetin," using their solubilities and acid-base properties. Panacetin contains aspirin, sucrose, and an unknown drug that may be either acetanilide or phenacetin. These compounds have the following solubility characteristics:

Some pharmaceutical preparations contain sweeteners to make them more palatable, especially to children.

1. Sucrose is soluble in water but insoluble in the organic solvent methylene chloride (dichloromethane, CH_2Cl_2).

2. Aspirin is soluble in methylene chloride but relatively insoluble in water. Sodium hydroxide converts aspirin to a salt that is insoluble in methylene chloride but soluble in water.

3. Acetanilide and phenacetin, like aspirin, are soluble in methylene chloride and insoluble in water. They are not converted to salts by sodium hydroxide.

Mixing the "Panacetin" with methylene chloride should therefore dissolve the aspirin and the unknown constituent, but leave the sucrose behind as an insoluble solid that can be filtered out. Aspirin (acetylsalicylic acid) can be removed from the methylene chloride solution by extraction with an aqueous solution of sodium hydroxide, which converts the aspirin to its salt, sodium acetylsalicylate, as shown below.

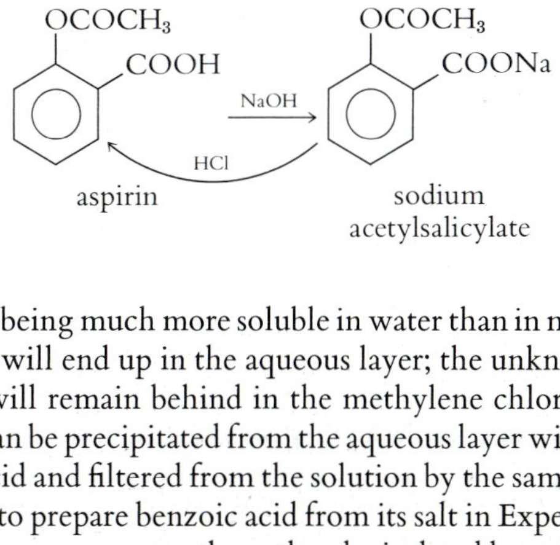

$$aspirin \xrightarrow[\text{HCl}]{\text{NaOH}} sodium\ acetylsalicylate$$

This salt, being much more soluble in water than in methylene chloride, will end up in the aqueous layer; the unknown constituent will remain behind in the methylene chloride layer. Aspirin can be precipitated from the aqueous layer with hydrochloric acid and filtered from the solution by the same method you used to prepare benzoic acid from its salt in Experiment 1. The unknown compound can then be isolated by evaporating the methylene chloride from the remaining solution. To reduce waste and prevent methylene chloride vapors from escaping into the atmosphere, it is advisable to recover the evaporated methylene chloride by means of a cold trap. (Your instructor will inform you of any solvent recovery and recycling procedures to be practiced in your laboratory.)

The separation process described above is summarized by the following flow diagram:

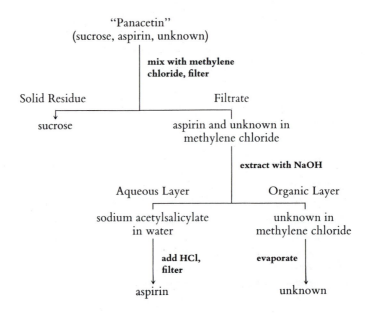

After weighing the dried components, you can estimate the percentage composition of "Panacetin." A comparison of your experimental results with the actual composition (which will be revealed by your instructor) will show how well you carried out the separation. Erroneous results can arise from incomplete mixing with methylene chloride, incomplete extraction or precipitation of aspirin, incomplete drying of the recovered components, and losses in transferring substances from one container to another.

PRELAB ASSIGNMENTS

1. Read the experiment, and make sure you understand its objectives.

The assignment of an operation (such as OP–7) is not meant to include its subsections (such as OP–7a), which are assigned separately.

2. Read the descriptions for OP–7, OP–11, OP–13, and OP–14, and review the other operations listed at the beginning of this experiment.

PROCEDURE

Some components of "Panacetin" may irritate the skin or eyes; phenacetin, when taken orally in large doses, has been found to induce cancer. Although there is little likelihood that momentary exposure to phenacetin will cause cancer, you should minimize your contact with "Panacetin" and its separate components.

Methylene chloride may be harmful if ingested, inhaled, or absorbed through the skin. There is a possibility that methylene chloride may induce cancer if inhaled, but animal tests have been inconclusive. Avoid prolonged or unnecessary contact with the liquid, and do not breathe its vapors.

SAFETY PRECAUTIONS

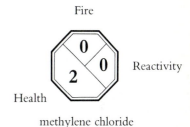

methylene chloride

See Chemical Hazards, pages 12–14, for an explanation of the hazard symbols.

Separation of Sucrose. Accurately weigh ⟨OP–5⟩ about 3 g of "Panacetin" and transfer it to a 125-mL Erlenmeyer flask, then add 50 mL of methylene chloride (dichloromethane) **[vapor hazard! suspected carcinogen!]** to the solid. Stir the mixture *thoroughly* to dissolve as much of the solid as possible, using a flat-bottomed stirring rod to break up any lumps or granules. Using a preweighed fluted filter paper, filter the mixture by gravity ⟨OP–11⟩ to separate the sucrose. Set the filter paper aside (be careful not to spill any of the sucrose) and reweigh it when it is completely dry. Record the mass of the sucrose in your laboratory notebook. If requested, turn in the sucrose to your instructor in a tared and labeled vial.

Separation of Aspirin. Transfer the filtrate to a separatory funnel and extract ⟨OP–13⟩ the aspirin from it with two 25-mL portions of aqueous 1 M sodium hydroxide. Remember that methylene chloride will be on the bottom, so you will have to transfer each layer to a different container and return the methylene chloride layer to the separatory funnel before the second extraction. Save the methylene chloride solution for the next part.

Add 10 mL of 6 M hydrochloric acid slowly, with stirring, to the combined aqueous extracts. Test the pH of the solution as described in Experiment 1 and add more acid, if necessary, to bring it to pH 2 or lower. Cool the mixture in an ice bath, collect the aspirin by vacuum filtration ⟨OP–12⟩, and wash it on the filter with cold distilled water. Let the aspirin dry on the filter a few minutes with the aspirator running, then dry it to constant mass ⟨OP–21⟩. Weigh the aspirin ⟨OP–5⟩ and record its mass in your notebook. If requested, turn in the aspirin to your instructor in a tared and labeled vial.

Isolation of the Unknown Constituent.   Evaporate the solvent ⟨OP–14⟩ from the methylene chloride solution, using a filter flask attached to a trap and aspirator (use a cold trap if the solvent is to be recovered). Heating and swirling the solution over a steam bath or in a hot water bath ⟨OP–7⟩ will increase the evaporation rate. Discontinue evaporation when only a solid residue remains in the flask or when no more solvent evaporates, and put the recovered methylene chloride in a designated solvent recovery container. Scrape out the unknown solid and dry it ⟨OP–21⟩ to constant mass. Weigh this substance in a tared vial and save it for Experiment 3.

If it is quite impure, the unknown may remain liquid after all the solvent is removed. It should solidify after cooling.

Report.   Calculate your percent recovery by dividing the sum of the masses of all components by the mass of "Panacetin" you started with. Calculate the approximate percentage composition of "Panacetin," based on the total mass of components recovered. (These percentages should add up to 100%.) If your instructor provides the true composition of the mixture, try to account for any significant difference between it and your calculated composition.

Exercises

   1.   (a) Why does the sodium acetylsalicylate solution warm up when HCl is added to precipitate the aspirin? **(b)** Why is it important to cool the acidified mixture before filtering the aspirin?

   2.   (a) If the aqueous and organic layers were not thoroughly mixed during the sodium hydroxide extraction, would your calculated percentage of aspirin be too high or too low? Explain. **(b)** Answer the same question with respect to the unknown constituent.

Aspirin is a stronger acid than carbonic acid.

   3.   (a) Acetaminophen is a weaker acid than either aspirin or carbonic acid (H_2CO_3), but a stronger acid than water. What problem would you encounter if the unknown component were acetaminophen rather than acetanilide or phenacetin? **(b)** Suggest a procedure that might be used to separate a mixture of sucrose, aspirin, and acetaminophen.

Library Topic

Write a brief report comparing the medical uses, effectiveness, and side effects of aspirin, acetaminophen, and phenacetin. Describe conditions under which acetaminophen should be used rather than aspirin, and vice versa.

Identification of a Constituent of "Panacetin"

Purification Methods. Melting Points.

Operations

OP-5   Weighing
OP-7   Heating
OP-21 Drying Solids
OP-23 Recrystallization
OP-23c Choosing a Recrystallization Solvent
         ⟨Minilab 1⟩
OP-28 Melting-Point Determination

Objectives

To purify the unknown constituent of "Panacetin" and determine its identity.

To learn how to purify solids by recrystallization and how to measure their melting points.

To learn how melting points can be used to characterize organic compounds.

To learn how to use chemical reference books to obtain information about organic compounds.

BACKGROUND

Physical Properties in the Chemical Literature

Knowing the physical properties of a compound — such as its melting or boiling point, refractive index, and spectral absorption bands — can help you identify the compound and estimate its degree of purity. Whenever a new compound is discovered or synthesized, its physical properties are measured and reported in one or more scientific journals, so that future investigators can recognize that compound when they encounter it.

Selected characteristics of the most frequently encountered compounds are also reported in chemical handbooks and other reference books. For instance, from its entry in the *Merck Index,* you can learn that acetaminophen is known by at least 8 chemical and 45 proprietary drug names, including Alpiny, Bickie-mol, Cetadol, Dial-a-gesic, Enelfa, Finimal, Gelocatil, Homoolan, and so on down the alphabet. The same source lists the compound's important physical properties and uses and tells you where to look for more information about it. This and other general reference books are described under category A in Appendix VII.

Figure 1 reproduces the entry for acetanilide from the 59th edition of the *CRC Handbook of Chemistry and Physics.* The figure reveals that acetanilide has a melting point of 114.3°C (another source reports a slightly higher melting range), that it is only slightly soluble (δ) in room-temperature water but soluble (s) in hot water, and that it crystallizes from water (w) in the form of rhombic crystals (rh) or plates (pl). Additional quantitative solubility information can be found in the *Merck Index,* which reports that a gram of acetanilide dissolves in about 185 mL of cold water (at room temperature or lower) or 20 mL of boiling water and that a gram of phenacetin dissolves in 1310 mL of cold water or 82 mL of boiling water. Such entries often tell you what you need to know in order to purify a compound and identify or characterize the pure substance. These and other properties of acetanilide and phenacetin are summarized in Table 1, where the solubilities are expressed in grams of solute per 100 mL of water.

In more recent editions, the compound is listed under "acetanilide" rather than "acetic acid, N-phenylamide," and the solubility information is less detailed. Always consult the introductory material preceding such tables for explanations of the symbols used.

METHODOLOGY

No separation is perfect; traces of impurities will always remain in a compound that has been separated from a mixture. Therefore, some kind of *purification* process is needed to remove them. Solids can be purified by such operations as recrys-

Table 1.   Physical properties

	M.W.	m.p.	Solubility, Cold Water	Solubility, Boiling Water
acetanilide	135.2	114	0.54	5.0
phenacetin	179.2	135	0.076	1.22

Note: Melting points are given in °C; solubilities are in grams of solute per 100 mL of water.

Figure 1. CRC Handbook entry for acetanilide. (Reprinted with permission from the CRC Handbook of Chemistry and Physics, 59th Edition, pp. C–83 and C–85. Copyright CRC Press, Inc., Boca Raton, FL.)

No.	Name	Synonyms and Formula	Mol. wt.	Color, crystalline form, specific rotation and λ_{max} (log ϵ)	m.p. °C	b.p. °C	Density	n_D	w	al	eth	ace	bz	other solvents	Ref.
	Acetic acid														
Ω a60	—, amide.........	Acetamide. Ethanamide* CH_3CONH_2	59.07	trg mcl (al-eth) λ^{MeOH}	82.3	221.2^{760} 120^{20}	0.9986^{15}_{4} 1.1590^{20}_{4}	1.4278^{78}	s	v	i	......	δ s^h	chls Pys	**B**2,177
Ω a164	—, —, N-phenyl-...	Acetanilide. Antifebrin. $CH_3CONHC_6H_5$	135.17	rh or pl(w) λ^{al} 242(4.16)	114.3 (115–6)	304^{760}	1.2190^{15}	....	δ s^h	v	s	v	s	chl,CCl_4, McOHv to s, v^a	**B**12,137
a165	—, —, N-phenyl-N-propyl-	$CH_3CON(C_6H_5)CH_2CH_2CH_3$	177.25	mcl lf (eth, lig)	49(56)	266^{12}	....	....	i	v	v	......	......	....	**B**12,246

tallization, chromatography, and sublimation; liquids are usually purified by distillation or chromatography.

The solubility information in Table 1 suggests that your unknown compound can be purified by recrystallization from water. Based on the mass of unknown you recovered from Experiment 2, you can easily estimate the volume of boiling water needed to dissolve that mass of either acetanilide or phenacetin. Since these amounts will be quite different, you should obviously begin the recrystallization using the smaller volume of water (don't add it all at once — see OP – 23), and only add more if it is apparent that your compound will not dissolve in that amount of boiling water.

After a compound has been purified, it is usually *analyzed* to establish its identity and degree of purity. Although sophisticated instruments such as NMR spectrometers and mass spectrometers are now used to determine the structures of most new organic compounds, an operation as simple as a melting-point determination can help identify a compound whose properties have already been reported. By itself, the melting point of a compound is not sufficient proof of its identity, since thousands of compounds may share the same melting point. But when an unknown compound is thought to be one of a small number of known compounds, its identity can often be established by mixing the unknown with an authentic sample of each known compound and measuring the melting points of the mixtures. The use of such *mixture melting points* for identification is based on the fact that the melting point of a pure compound is usually lowered, and its melting point range is broadened, when it is combined with another, different compound. For example, if your unknown is phenacetin, it should melt sharply near 135°C, and a mixture of the unknown with an authentic sample of phenacetin should have essentially the same melting point. But a 1:1 mixture of phenacetin with acetanilide should melt at a considerably lower temperature over a broad range. The melting point of a compound can also give a rough indication of its purity. If your compound melts over a narrow range (∼2°C or less) at a temperature close to the literature value, it is probably quite pure. If its melting point range is broad and substantially lower than the literature value, it is probably wet or contaminated by impurities.

For additional experience with recrystallization, you can purify a compound provided by your instructor as described in Minilab 1. Because the recrystallization solvent is not specified, you will have to test several solvents to find a suitable one.

See OP – 23 for a brief discussion of the theory of recrystallization.

PRELAB ASSIGNMENTS

1.   Read the experiment, and be sure you understand its objectives.

2.   Read Appendix VII (particularly about category A) and familiarize yourself with the layout of the Bibliography.

3.   Read OP–23 and OP–28, and review the other operations listed at the beginning of this experiment. If you will be doing Minilab 1, read OP–23*c* as well.

4.   Calculate the approximate volume of boiling water needed to dissolve all of your unknown compound if it is acetanilide and if it is phenacetin.

PROCEDURE

SAFETY PRECAUTION

Acetanilide and phenacetin can irritate the skin and eyes, and phenacetin has been known to induce cancer when taken orally. Minimize contact with your unknown compound.

Use the heat source suggested by your instructor. Remember to use distilled or deionized water.

Purification.   Recrystallize ⟨OP–23⟩ the unknown compound from Experiment 2 by boiling it ⟨OP–7⟩ with just enough water to dissolve it, filtering the hot solution through fluted filter paper, and letting it cool to room temperature. If necessary, induce crystallization by scratching the sides of the flask with a stirring rod, then cool the flask further in ice water to increase the yield of product. Dry the product ⟨OP–21⟩ to constant mass, and weigh it ⟨OP–5⟩ in a tared vial.

Be sure to record the melting-point range as described in OP–23, not a single melting temperature.

Analysis.   Measure the melting points ⟨OP–28⟩ of (1) the purified unknown, (2) a 1:1 mixture of the unknown with acetanilide, and (3) a 1:1 mixture of the unknown with phenacetin. Unless your instructor indicates otherwise, you should carry out at least two measurements with each of these. Turn in the remaining product to your instructor.

Report.   Report the identity of your compound, and explain *clearly* how you arrived at your conclusion, citing all the experimental data that appear to support it. Calculate the percentage of your starting material that you recovered after crystallization and try to account for any significant losses.

Collateral Project

You can calibrate your thermometer by carefully measuring the melting points of the compounds listed in Table 1 of OP–28 (page 723) or other pure compounds suggested by your instructor.

Purification of an Unknown Compound by Recrystallization

MINILAB 1

Inhalation, ingestion, or absorption through the skin of the solvents and the unknown solid may be harmful. Avoid contact with them, and do not breathe their vapors.

SAFETY PRECAUTION

Obtain about 2 g of an unknown solid and try to find a suitable recrystallization solvent for it ⟨OP–23c⟩. Test the solvents water, ethanol, hexane, and 2-butanone, or use other solvents suggested by your instructor. Recrystallize about 1 g of the solid (weigh it accurately!) from the best solvent, and dry it to constant mass. Measure the melting points of the impure and purified solids, and turn the latter in to your instructor. Calculate the percentage recovered, and report your results in tabular form.

Exercises

1. **(a)** How much boiling water is needed to dissolve 1.35 g of acetanilide? **(b)** About how much crystalline acetanilide, at best, can be recovered when the resulting solution is cooled and filtered?

2. A mixture of an unknown compound with benzoic acid melts completely at 89°C, a mixture of it with phenyl succinate melts at 120°C, and a mixture with *m*-aminophenol melts at 102°C. Based on the data in Table 2, give the probable identity of the unknown and explain your reasoning.

Table 2.   Melting points for Exercise 2

Compound	m.p.
o-toluic acid	102°C
benzoic acid	121°C
phenyl succinate	121°C
m-aminophenol	122°C

Examples of journal citations:

J. Org. Chem. **45,** 917 (1980).

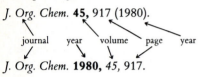

journal   year   volume   page   year

J. Org. Chem. **1980,** *45,* 917.

3. Using one or more of the reference books described in Appendix VII, give the following information about the anti-inflammatory drug ibuprofen: systematic (chemical) names, molecular weight, molecular formula, structural formula, melting point, and a suitable recrystallization solvent.

4. Locate citations for one or more journal articles that give procedures for preparing **(a)** aspirin and **(b)** propoxyphene (Darvon). Give the full name of each journal, the volume number, the page number(s), and the year. Full names of scientific periodicals can be found in the *Chemical Abstracts Service Source Index.* Recent citations often give the year first, followed by the volume number (if any) and page(s); alternatively, the volume number may appear first, followed by the page number(s) and the year in parentheses.

5. Give the names of at least ten compounds (besides acetanilide) that have a reported melting point of 114°C.

The Synthesis of Salicylic Acid from Wintergreen Oil

Preparation and Purification of Solids.

Operations

OP–2 Using Standard-Taper Glassware

OP–5 Weighing

OP–7 Heating

OP–7a Refluxing

OP–12 Vacuum Filtration

OP–21 Drying Solids

OP–23 Recrystallization

OP–28 Melting-Point Determination

Objectives

To prepare salicylic acid from methyl salicylate.

To learn how to carry out a reaction under reflux.

To gain additional experience in applying the separation and purification operations learned in previous experiments.

To learn how to perform calculations for organic syntheses.

To learn how to organize laboratory time efficiently using an experimental plan.

To learn about some properties and uses of methyl salicylate.

SITUATION

Salicylic acid, an important intermediate in the synthesis of aspirin, is usually prepared commercially from benzene, a petroleum product also used in gasoline. Because petroleum is a nonrenewable natural resource, a serious petroleum shortage at some future date seems inevitable. Then we would have to

Proposed route to aspirin

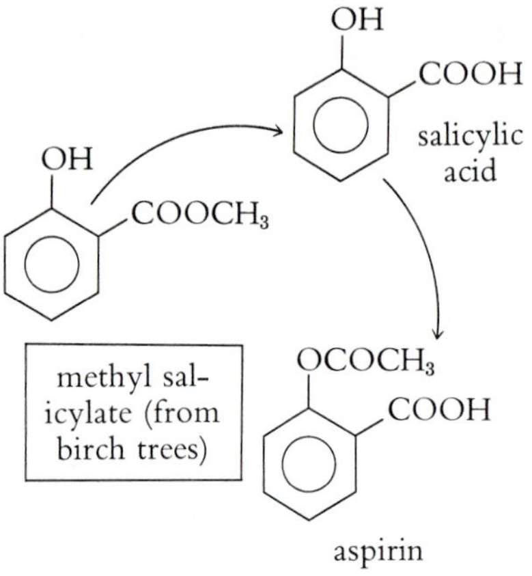

methyl sal-
icylate (from
birch trees)

salicylic
acid

aspirin

Commercial route to aspirin

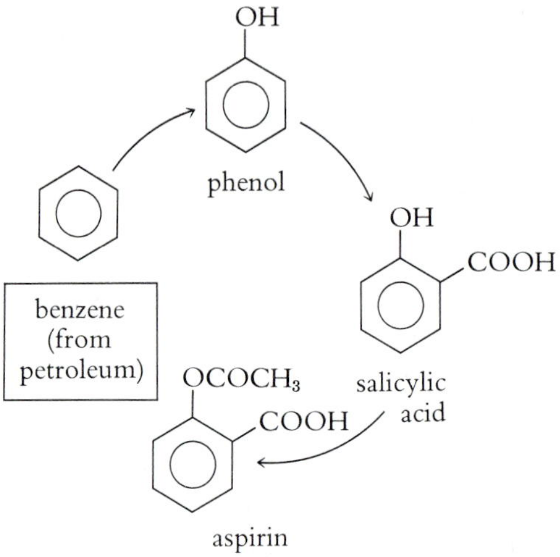

benzene
(from
petroleum)

phenol

salicylic
acid

aspirin

find a new source of salicylic acid or else face the prospect of having no aspirin to cure all the headaches caused by the oil shortage. Until 1874, commercial salicylic acid was made entirely from wintergreen oil, extracted from the leaves of the wintergreen plant or the bark of sweet birch trees. In this experiment, you will reproduce the nineteenth-century synthesis of salicylic acid from the major constituent of wintergreen oil, methyl salicylate. Salicylic acid can be converted to aspirin by the procedure described in Experiment 41.

BACKGROUND

Wintergreen Oil—Natural or Synthetic?

Commercial route to synthetic "wintergreen oil"

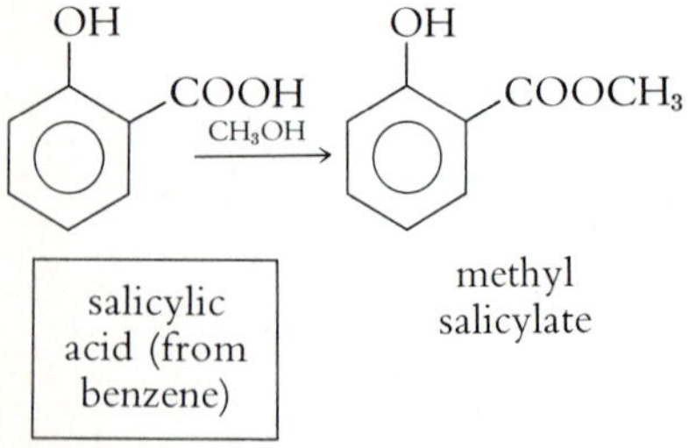

salicylic
acid (from
benzene)

methyl
salicylate

esterification: combination of an organic acid with an alcohol to form an ester

Methyl salicylate was first isolated as a constituent of the oil obtained from the leaves of the wintergreen plant, *Gaultheria procumbens.* This "oil of wintergreen" can be obtained even more readily from other sources such as sweet birch *(Betula lenta)* and yellow birch *(Betula alleghanensis),* and most "natural" methyl salicylate is still distilled from the bark of *Betula lenta.* When cheap raw materials became available from petroleum, a more economical commercial process for synthesizing methyl salicylate was devised that involved the **esterification** of salicylic acid with methyl alcohol. A worldwide petroleum shortage could alter the economics of production so that the natural sources of methyl salicylate would again become the important ones. If this happened, aspirin might someday be manufactured from birch trees!

Regardless of its source, methyl salicylate is methyl salicylate; there is no difference whatever between the **natural** and **synthetic** compounds provided both are of equal purity.

natural: obtained directly (or nearly so) from plant, animal, or mineral sources

synthetic: prepared by chemically altering other compounds, which may themselves be either natural or synthetic

natural methyl salicylate

synthetic methyl salicylate

Both are composed of the same kind of molecules and must therefore have the same properties. Minor differences between oil of wintergreen obtained from *Gaultheria* or *Betula* species and synthetic methyl salicylate are due to the 1% or so of impurities in the natural preparations.

Methyl salicylate, both synthetic and natural, has been used for many years as a flavoring agent because of its very pleasant, penetrating odor and flavor. A good wintergreen tea can be made from either the leaves of the wintergreen plant or the twigs and inner bark of sweet and yellow birch; the distinctive wintergreen flavor is found in everything from root beer to mouthwash. Medicinally, wintergreen oil has some of the pain-killing properties of the other salicylates, including aspirin. When absorbed through the skin, it produces an astringent but soothing sensation, and it is frequently used in preparations for muscular aches and arthritis.

Ben-Gay contains methyl salicylate and menthol.

METHODOLOGY

Organic synthesis can be described as the art of preparing pure organic compounds by chemically modifying other compounds. In the preparation of salicylic acid, for instance, a methyl (CH_3) group must be removed from each molecule of methyl salicylate and replaced by a hydrogen atom. Since we are talking about replacing some 22 sextillion methyl groups by an equal number of hydrogen atoms, the process is going to take some time; it should not be surprising that an extended reaction period is required.

The time required for organic reactions is usually much longer than that for typical inorganic reactions. When silver nitrate and sodium chloride are mixed, their reaction is almost instantaneous, taking place as quickly as silver and chloride

5 g of salicylic acid = 0.036 mol = 22×10^{21} molecules

ions can come together to form silver chloride. In a given organic reaction, the reacting molecules may come together about as frequently as the ions in a precipitation reaction, but they may not collide with enough energy or at the right orientation to react. Ions need only "stick together" to form a product; covalent molecules must undergo a complicated process of bond making and bond breaking to be converted to product molecules.

Completion implies that as many reactant molecules as possible have been converted to product molecules under the conditions of the reaction.

The *reaction,* then, is the crucial step in an organic synthesis — it may be over in a few minutes or require a few weeks, but enough time should be allowed to bring the reaction as nearly as possible to completion before proceeding to the next steps. These steps have already been described in Experiments 2 and 3: the desired product must be *separated* from the reaction mixture, *purified* to remove residual contaminants, and *analyzed* to verify its identity and purity.

Methyl salicylate is the limiting reactant in this synthesis; sodium hydroxide will be used in excess to ensure a reasonably fast, complete reaction. As one of the Prelab Assignments, you will calculate the theoretical yield of salicylic acid from 15 mmol of methyl salicylate. This figure will allow you to tell immediately whether or not the amount of product you recover is close to the maximum yield. Because you probably will not measure out *exactly* 15 mmol of methyl salicylate, you should later recalculate the theoretical yield based on the amount you actually use. It is best to measure limiting reactants that are liquids by mass (the mass of a liquid can be measured more accurately than its volume). However, to avoid waste and minimize spillage at the balance, you should dispense the estimated *volume* of liquid from the reagent bottle before you take it to the balance to weigh it. If the mass of this volume is not close enough to the calculated value, you can add or remove liquid with a dropper. Thus, you should calculate both the mass *and* the volume of a liquid limiting reactant, as directed in the Prelab Assignments.

See OP–5 for additional information about weighing liquids.

The initial reaction will be carried out by boiling a mixture of methyl salicylate and sodium hydroxide in a flask equipped with a reflux condenser. The purpose of the reflux condenser is to return water and methyl salicylate vapors to the reaction flask, so that the reactants can be heated without being boiled away in the process. The heat source must be capable of boiling water, so a steam bath won't work. A heating mantle works well, but a burner is suitable if no mantles are available. When the two reactants are combined, methyl salicylate is immediately converted to a white solid that dissolves when the reaction mixture is heated. Do not add both reactants through the same funnel; if you do, the white solid will form in the neck of the funnel and may plug it up entirely.

In the strongly alkaline reaction mixture, the salicylic acid that forms will be present as the water-soluble salt disodium salicylate. Thus, aqueous sulfuric acid is added to precipitate free salicylic acid. When you recover the product by vacuum filtration, remember that the *solvent* in the reaction mixture is water, not sulfuric acid, so use cold water to wash the product on the filter. The solubility of salicylic acid in water is about 0.14 g/100 mL at 10°C, 0.18 g/100 mL at 20°C, or 6.7 g/100 mL at the boiling point, so it can be purified by recrystallization from water. The melting point of your product should give you a good indication of its purity; the melting range of very pure (99+%) salicylic acid is reported to be 158–160°C.

The written procedure for this experiment is somewhat less detailed than previous ones because by this time you should be familiar with some of the basic operations. You should know, for example, that after recrystallization a solid is collected by vacuum filtration and that a solid collected by vacuum filtration should be washed on the filter with an appropriate amount of the cold solvent. To help you organize your time efficiently, an *experimental plan* for the preparation of salicylic acid, incorporating a list of chemicals and supplies and a lab checklist, is provided below; you should write your own experimental plans for subsequent experiments. Note that the 30-minute reaction period should be used to collect and assemble the supplies needed for other operations.

PRELAB ASSIGNMENTS

For this and all subsequent experiments, "read the experiment, and be sure you understand its objectives" will be an implicit prelab assignment; it will no longer be listed.

1. Read the descriptions for operations OP–2 and OP–7a, and read or review the other operations as necessary.

2. Read Appendix IV on calculations for organic synthesis.

3. Calculate the mass and volume of 15 mmol of methyl salicylate, the theoretical yield of salicylic acid from that much methyl salicylate, and the volume of water needed to recrystallize that much salicylic acid.

Be sure to distinguish millimoles (mmol) from moles (mol); most Prelab Assignments will specify amounts of chemicals in millimoles.

4. After reading the Procedure section carefully, specify all sizes and quantities indicated by asterisks in the Chemicals and Supplies list and the heat sources you will use. Note that boiling flasks come in 25-, 50-, 100-, 250-, and 500-mL sizes; Erlenmeyer flasks in 50-, 125-, and 250-mL sizes; and beakers in 50-, 150-, 250-, and 400-mL sizes.

Reactions and Properties

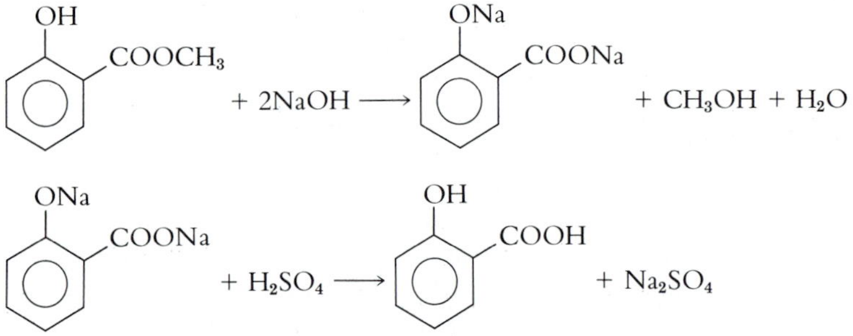

$$+ \; 2NaOH \longrightarrow \qquad + \; CH_3OH + H_2O$$

$$+ \; H_2SO_4 \longrightarrow \qquad + \; Na_2SO_4$$

M.W. = molecular weight
m.p. = melting point (°C)
b.p. = boiling point (°C)
d. = density (g/mL)

Table 1.   Physical properties

	M.W.	m.p.	b.p.	d.
methyl salicylate	152.1	−8	223	1.174
salicylic acid	138.1	159		

Chemicals and Supplies

[An asterisk (*) indicates a number (size or quantity) to be filled in as part of the Prelab Assignments.]

Reflux ⟨OP–2, OP–7, OP–7a⟩
    * g (* mL) of methyl salicylate **[irritant; avoid contact and inhalation]**
    20 mL of 6 M NaOH **[caustic! use gloves and avoid contact]**
    *-mL boiling flask, reflux condenser, boiling chips, ringstand, and heat source (specify)

H₂SO₄ Addition
    22 mL of 3 M H₂SO₄ **[avoid contact]**
    *-mL beaker, pH paper, stirring rod, and dropper

Vacuum Filtration ⟨OP–12⟩
    * mL of cold water for washing
    Buchner funnel, filter flask, filter paper, flat-bottomed
    stirring rod, spatula, and watch glass

Recrystallization ⟨OP–23⟩
    * mL of water for recrystallization
    * mL of cold water for washing
    two *-mL Erlenmeyer flasks, heat source (specify), *-mL
    graduated cylinder, funnel, fluted filter paper, and vac-
    uum filtration apparatus (see above)

Drying and Weighing ⟨OP–21, OP–5⟩
    aluminum dish or other container, oven or desiccator,
    tared vial, and balance

Analysis ⟨OP–28⟩
    capillary m.p. tubes, flat-bottomed stirring rod (or spat-
    ula), watch glass, and melting-point apparatus

Lab Checklist

[ ]   Collect supplies for reflux (clean glassware if necessary)

[ ]   Obtain heat source

[ ]   Assemble reflux apparatus

[ ]   Measure methyl salicylate and 6 M NaOH solution

[ ]   Add reactants to boiling flask

[ ]   Reflux reactants for 30 minutes

[ ]   Collect supplies for H_2SO_4 addition (during reflux)

[ ]   Collect vacuum filtration supplies and assemble apparatus
    (during reflux)

[ ]   Collect supplies for recrystallization (during reflux)

[ ]   Measure H_2SO_4 solution (during reflux)

[ ]   Precipitate product with H_2SO_4

[ ]   Filter and wash product; air dry on filter

[ ]   Measure and boil water for recrystallization

[ ]   Recrystallize product from water (filter hot solution)

[ ]   Filter and wash product; air dry on filter

[ ]   Dry product

[ ]   Weigh product

[ ]   Assemble melting-point apparatus

[ ]   Measure melting point

[ ]   Turn in product

[ ]   Clean up

PROCEDURE

**Sodium hydroxide solutions are very corrosive, capable
of causing severe damage to skin and eyes. Avoid contact
with the sodium hydroxide solution; protective gloves
should be worn, and eye protection is essential. Wash up
any spills *immediately* with large amounts of water, followed by dilute acetic acid or another alkali-spill solution.**

**Methyl salicylate and salicylic acid irritate the eyes and
skin; avoid contact with or inhalation of either.**

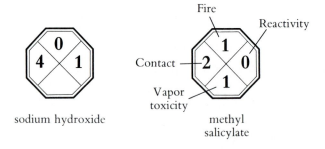

sodium hydroxide   methyl
                                          salicylate

Don't forget the boiling chips!

Reaction.   Obtain an appropriate heat source ⟨OP–7⟩ and
position it correctly. Assemble the apparatus for reflux ⟨OP–
7a⟩, using standard-taper glassware if available ⟨OP–2⟩. Have
the instructor check your apparatus before you begin. Measure
20 mL of aqueous 6 M (20%) sodium hydroxide **[contact
hazard!]** and 15 mmol of methyl salicylate, and combine
them in the boiling flask. Reflux the reactants for about 30
minutes, measuring from the time the reaction mixture starts
to boil.

Let the solution cool to room temperature and transfer it to a beaker. Slowly add 22 mL of aqueous 3 M sulfuric acid with constant stirring, and test the solution with pH paper; if the pH is above 2, add enough sulfuric acid to bring it down to 1–2.

The pH should be 1 or lower to ensure complete precipitation; however, most broad-range pH papers cannot measure pH values that low.

Separation.   Cool the acidified reaction mixture, and collect the salicylic acid by vacuum filtration ⟨OP–12⟩.

Purification and Analysis.   Purify the salicylic acid by recrystallization from water ⟨OP–23⟩, and record the total volume of water used. Dry the product ⟨OP–21⟩, measure its mass ⟨OP–5⟩ and melting point ⟨OP–28⟩, and turn it in.

Report.   Recalculate the theoretical yield of salicylic acid based on the amount of methyl salicylate you used, and calculate your percentage yield. Based on the amount of water you used, estimate the amount of salicylic acid lost during the recrystallization. Try to account for any other significant losses.

Collateral Projects

1.   You can verify the identity of your product by taking a mixture melting point with an authentic sample of salicylic acid.

2.   Dissolve a small amount of your product (about 40 mg) in a mixture containing 1 mL of ethanol and 1 mL of water, add a drop or two of 2.5% ferric chloride solution, and record your observations. Perform the same test with pure samples of aspirin, acetaminophen, phenacetin, and methyl salicylate. What structural feature appears to be necessary for a positive test with ferric chloride?

Exercises

1.   **(a)** Calculate the volume of 6 M NaOH required to react completely with 15 mmol of methyl salicylate. How much of the 6 M NaOH you used was in excess of the theoretical amount? **(b)** What volume of 3 M H_2SO_4 is needed to neutralize all of the disodium salicylate and the excess NaOH present after the initial reaction? How much sulfuric acid was in excess?

2. **(a)** According to the precipitation equation in the Reactions and Properties section, which hydrogen atoms of salicylic acid are acidic? **(b)** Which hydrogen atoms of methyl salicylate (if any) would you expect to be acidic? Draw the structure of the white solid that forms immediately after sodium hydroxide and methyl salicylate are mixed, and write an equation for its formation.

3. **(a)** The pK_a value for the first ionization step of salicylic acid is 3.00. If during the precipitation of salicylic acid, you added only enough sulfuric acid to reduce the pH to 4.00, what percentage of your salicylic acid would have remained in solution as sodium salicylate? **(b)** Answer the question of part (a) for pH values of 3.00, 2.00, and 1.00.

4. From the equations in the Reactions and Properties section, you can see that methanol and sodium sulfate are by-products of the synthesis of salicylic acid. During which step of the synthesis would these compounds have been separated from the final product? Explain your answer, based on any relevant properties of salicylic acid and of the by-products.

Library Topic

Write a report describing the manufacture of salicylic acid and aspirin. Give equations for the important reactions used in the manufacturing processes and, if possible, tell when and how the reactions were discovered.

sodium salicylate

If necessary, consult a reference book for information about the properties of the compounds.

The Preparation of an Artificial
Flavoring Ingredient

Preparation and Purification of Liquids.

Operations

OP–2	Using Standard-Taper Glassware
OP–4	Basic Glass Working
OP–5	Weighing
OP–7	Heating
OP–7a	Refluxing
OP–7b	Semimicro Refluxing ⟨Minilab 2⟩
OP–11	Gravity Filtration
OP–13a	Semimicro Extraction ⟨Minilab 2⟩
OP–19	Washing Liquids
OP–20	Drying Liquids
OP–25	Simple Distillation

Objectives

To prepare isoamyl acetate from acetic acid and isoamyl alcohol.

To learn and apply some operations used for the preparation and analysis of organic liquids.

To learn how a flow diagram is used to outline an organic synthesis.

To learn how to develop an experimental plan.

To learn about the formulation of artificial flavorings.

SITUATION

Flavor chemists combine natural and synthetic compounds to prepare artificial flavorings that can closely reproduce natural flavors. Many fruits, flowers, and spices contain esters that contribute to their characteristic flavors. Such esters make up the basic repertoire of flavor chemists, who have more

The word *flavor* is used to describe the overall sensory effect of a substance taken into the mouth; flavor may involve tactile, temperature, and pain sensations as well as smell and taste.

$$CH_3\overset{\overset{\displaystyle O}{\|}}{C}OCH_2CH_2CH-CH_3$$
$$|$$
$$CH_3$$

isoamyl acetate

$$R-\overset{\overset{\displaystyle O}{\|}}{C}-O-R'$$

acid   alcohol
portion   portion

an ester

At least 168 different compounds, including 36 esters, have been found in a natural vanilla-bean extract.

than two hundred of these odoriferous compounds at their disposal. Although the taste or odor of a single ingredient may seem unrelated to the overall flavor character of a natural substance, the combination of many such ingredients in just the right proportions can yield a close approximation of the natural flavor. In this experiment, you will prepare isoamyl acetate, an important ester used in flavorings that is also known as isopentyl acetate or 3-methylbutyl acetate. Isoamyl acetate has a strong banana odor when it is undiluted, and an odor reminiscent of pears when it is in dilute solution. It is used as an ingredient in artificial coffee, butterscotch, and honey flavorings as well as in pear and banana flavorings.

BACKGROUND

Esters and Artificial Flavorings

Esters are organic compounds that consist of an acid portion and an alcohol (or phenol) portion. In effect esters can be prepared by eliminating water from the molecules of a carboxylic acid and an alcohol or phenol and combining the two fragments. Several of the compounds you worked with in previous experiments were esters: methyl salicylate is the ester of salicylic acid and methanol; aspirin is the ester of acetic acid and salicylic acid.

Most of the volatile esters, including methyl salicylate, have strong, pleasant odors often described as "fruity." Some esters with odors and flavors resembling natural ones are shown in Table 1.

A flavor ingredient is characterized by one or more flavor *notes* that suggest the predominant impact it makes on the senses of taste and smell. Some food products like cola beverages or Juicy-Fruit gum are characterized by "fantasy" flavors that have no counterparts in nature, but most artificial flavorings are meant to resemble natural flavors. Because natural flavors are usually very complex, a cheap artificial flavoring may be only a pale imitation of its natural counterpart. Advances in flavor chemistry have resulted in the production of some superior flavorings that reproduce natural flavors very closely. Artificial flavorings may contain natural products, synthetic organic compounds identical to those found in nature, and synthetic compounds not found in nature but accepted as safe for use in food. Natural products used in flavorings include the *essential oils* distilled from plants, *extracts* obtained by treating plant matter with appropriate solvents, and pure organic compounds isolated from natural materials.

Table 1. Flavor notes of some esters used in artificial flavorings

Name	Structure	Flavor Note
Propyl acetate	$CH_3\overset{\displaystyle O}{\overset{\|}{C}}-OCH_2CH_2CH_3$	pears
Octyl acetate	$CH_3\overset{\displaystyle O}{\overset{\|}{C}}-O(CH_2)_7CH_3$	oranges
Benzyl acetate	$CH_3\overset{\displaystyle O}{\overset{\|}{C}}-OCH_2-\bigcirc$	peaches, strawberries
Isopentenyl acetate	$CH_3\overset{\displaystyle O}{\overset{\|}{C}}-OCH_2CH=\overset{\displaystyle CH_3}{\overset{\|}{C}}-CH_3$	"Juicy-Fruit"
Isobutyl propionate	$CH_3CH_2\overset{\displaystyle O}{\overset{\|}{C}}-OCH_2\overset{\displaystyle CH_3}{\overset{\|}{C}H}-CH_3$	rum
Ethyl butyrate	$CH_3CH_2CH_2\overset{\displaystyle O}{\overset{\|}{C}}-OCH_2CH_3$	pineapples

Superior flavorings may be composed mainly of natural oils or extracts that are fortified with a few synthetic ingredients to enhance the overall effect and replace flavor elements lost during the distillation or extraction process. A high-boiling *fixative* such as benzyl benzoate or glycerine is generally added to retard vaporization of volatile components, and the flavor notes of individual components are blended by dissolving them in a solvent called the *vehicle*. The most frequently used vehicle is ethyl alcohol.

The formulation of artificial flavorings is perhaps as much an art as a science, which means that the components of a given flavoring may vary widely depending on the manufacturer and the specific application. A flavor chemist may formulate a strawberry flavoring by carefully blending several dozen chemicals, whereas the strawberry plant probably manufactures several hundred. Even so, the synthetic and natural flavors may be indistinguishable to all but the most discriminating of strawberry aficionados.

Esters are often used in perfumes as well as in artificial flavorings. Minilab 2 gives the procedure for preparing methyl benzoate, which, as "Oil of Niobe," is used in the formulation of many perfumes.

Fixatives

$$\underset{\text{glycerine}}{\overset{\displaystyle OH\quad OH\quad\ \ OH}{\overset{\displaystyle |\qquad |\qquad\ \ |}{CH_2CH-CH_2}}}$$

$$\underset{\text{benzyl benzoate}}{\bigcirc-\overset{\displaystyle O}{\overset{\|}{C}}OCH_2-\bigcirc}$$

Vehicle

$$CH_3CH_2OH$$
ethyl alcohol
(ethanol)

METHODOLOGY

An ester can often be prepared by heating a carboxylic acid with an alcohol, as shown by this general equation:

$$\underset{\substack{\text{carboxylic} \\ \text{acid}}}{\overset{\displaystyle O}{\underset{\displaystyle \|}{R\overset{}{C}OH}}} + \underset{\text{alcohol}}{HOR'} \longrightarrow \underset{\text{ester}}{\overset{\displaystyle O}{\underset{\displaystyle \|}{R\overset{}{C}OR'}}} + H_2O$$

An acid catalyst is used to increase the rate of the reaction, which would otherwise require a much longer time. You will synthesize isoamyl acetate by combining isoamyl alcohol (3-methyl-1-butanol) with acetic acid and sulfuric acid and refluxing the reaction mixture for an hour. The alcohol is the limiting reactant, so it should be weighed; the acids can be measured by volume. The esterification reaction is reversible, having an equilibrium constant of approximately 4.2. If you were to start with equimolar amounts of acetic acid and isoamyl alcohol, only about two-thirds of each reactant would be converted to isoamyl acetate by the time equilibrium was attained. Your highest attainable yield in that case would be only 67% of the theoretical value. To increase the yield of isoamyl acetate, you will apply Le Châtelier's principle, using an excess of acetic acid — the less expensive reactant — to shift the equilibrium toward the products. Even with this excess, it is impossible to obtain the theoretical yield of product after only one reaction step; in commercial ester syntheses, the unreacted starting materials are recycled until they are completely converted to product.

As discussed in Experiment 2, a pure component can be obtained from a mixture by separating it from all other components of the mixture, using procedures that take advantage of differences in solubility, acid-base properties, boiling points, and other characteristics. Because isoamyl acetate is a liquid, the separation and purification operations will differ from those used previously. At the end of the reflux period, the reaction mixture will contain (in addition to the ester) water, sulfuric acid, unreacted acetic acid, unreacted isoamyl alcohol, and presumably some unwanted by-products. Isoamyl acetate is quite insoluble in water, whereas both sulfuric acid and acetic acid are water-soluble and acidic. This makes it easy to separate the two acids from the product by *washing* the reaction mixture with water and then with aqueous sodium bicarbonate. Water does not remove the acids entirely because they are somewhat soluble in the ester as well, but it removes the bulk

According to Le Châtelier's principle, adding more of one reactant to a reaction at equilibrium will alter the equilibrium so that more reactant molecules are consumed and more product molecules are formed.

Reactions carried out with acid catalysts often yield polymeric, tarlike by-products that are generally high-boiling and insoluble in water.

of them and thus helps prevent a violent reaction with the sodium bicarbonate. Sodium bicarbonate converts the acids to their salts, sodium acetate and sodium sulfate, which are insoluble in the ester but very soluble in water; these salts migrate to the aqueous layer, where they can be removed. Most of the water present in the reaction mixture will be separated from the ester along with the wash liquids; any traces of water that remain are then removed by a drying agent, magnesium sulfate. Note that the rules of thumb given in OP–19 and OP–20 will help you estimate the quantities of wash solvent and drying agent needed.

Some isoamyl alcohol may be removed during the washings; but since it is more soluble in the ester than in water, much of it will remain behind. Because isoamyl acetate and isoamyl alcohol have different boiling points, they can, in principle, be separated by distillation. The isoamyl alcohol should distill first, followed by the ester, and any high-boiling by-products should remain in the boiling flask. This separation is incomplete for the reasons described in OP–25, so you will still have a little of the alcohol in your isoamyl acetate after the purification step. If your instructor permits, you can measure the amount of isoamyl alcohol and other impurities in your product by gas chromatography, as described in Collateral Project 1.

Your instructor may request that you carry out a small-scale preparation (to conserve chemicals); in that case, you should use a semimicro distillation apparatus as described in OP–25a.

The synthesis of isoamyl acetate is summarized in the flow diagram in Figure 1 (on page 58), which illustrates the transformations or separations that occur during each operation. To create such a flow diagram, you should first list all the substances (reactants, solvents, catalysts, etc.) that are present in the reaction mixture before the reaction starts. The equation for the reaction tells you what substances will be formed during the reaction. At the end of the reaction period, the reaction mixture will contain these products as well as the reaction solvent (if any), unreacted starting materials, and usually some by-products formed by various side reactions. The flow diagram should show how all the other substances present in the reaction mixture are separated from the desired product. Each separation or purification operation is represented by a branch in the flow diagram, with the substance(s) being removed on one side and the desired product, along with other components still left behind, on the other side. After the last operation, the product stands alone, all the impurities having been eliminated (on paper, at least!).

You should be aware that isoamyl acetate is a major component of the attack pheromone of the honeybee — the chemi-

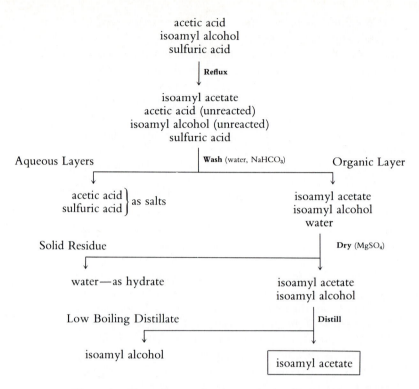

Figure 1.   Flow diagram for the synthesis of isoamyl acetate

cal "messenger" that is released when a worker bee stings someone and that brings other honeybees to the attack. Although isoamyl acetate by itself does not cause the bees to sting, it does put them on guard; so it would be prudent to steer clear of beehives on your way home from lab!

PRELAB ASSIGNMENTS

1. Read the descriptions for OP–19, OP–20, and OP–25, and read or review the other operations as necessary. If you will be doing Minilab 2, read OP–7b and OP–13a.

2. Calculate the mass and volume of 150 mmol of isoamyl alcohol, and the theoretical yield of isoamyl acetate.

3. Read Appendix V and the Procedure section of the experiment carefully. Write an experimental plan specifying such details as the size of the boiling flask (and other glassware), quantities of wash solvents and drying agents, and the expected location of the organic (product) layer during the washings.

Reactions and Properties

$$\underset{\text{acetic acid}}{CH_3\overset{O}{\overset{\|}{C}}-OH} + \underset{\text{isoamyl alcohol}}{HOCH_2CH_2\overset{CH_3}{\overset{|}{C}}HCH_3} \rightleftharpoons \underset{\text{isoamyl acetate}}{CH_3\overset{O}{\overset{\|}{C}}-OCH_2CH_2\overset{CH_3}{\overset{|}{C}}HCH_3} + H_2O$$

Table 2.   Physical properties

	M.W.	b.p.	d.	Solubility
acetic acid	60.1	118	1.049	miscible
isoamyl alcohol	88.1	130	0.815	2.7
isoamyl acetate	130.2	142	0.876	0.25
sulfuric acid	98.1	290	1.84	miscible

PROCEDURE

If your instructor requests that you carry out a small-scale preparation, divide all quantities by three (or more) and carry out the purification by semimicro distillation 〈OP–25a〉.

SAFETY PRECAUTIONS

Acetic acid is corrosive and can cause serious damage to the skin and eyes (especially the eyes); its vapors are highly irritating to the eyes and respiratory tract. Wear protective gloves, and dispense the acid under a hood; avoid contact with the liquid, and do not breathe its vapors. If skin or eye contact should occur, wash the affected area continuously with water for at least 15 minutes, removing any contaminated clothing as you do so. Wash up all spills immediately, and neutralize the spilled liquid with aqueous sodium bicarbonate solution.

Sulfuric acid is very corrosive and can cause very serious damage to the skin and eyes; the concentrated acid reacts violently with water. Wear protective gloves, avoid contact, and do not get water in the acid. Deal with contact or spills as described above for acetic acid.

Isoamyl alcohol and isoamyl acetate are flammable and can irritate the skin, eyes, and respiratory tract. Minimize contact with the liquids, do not breathe their vapors, and keep them away from flames.

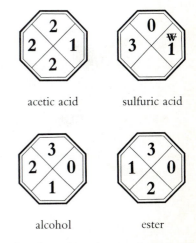

acetic acid   sulfuric acid

alcohol   ester

These hazard numbers are for isobutyl acetate and isobutyl alcohol, but the isoamyl compounds should be similar.

Glacial acetic acid is a pure grade of
acetic acid that freezes at about 17°C.

Reaction. *Under a hood,* measure 17 mL (~300 mmol) of glacial acetic acid **[contact and vapor hazard! use gloves!]** and
combine it with 150 mmol of isoamyl alcohol in a boiling flask
of the appropriate size. Carefully mix in 1 mL of concentrated
sulfuric acid **[contact hazard! use gloves!],** then add some
acid-resistant boiling chips and reflux the mixture gently for 1
hour ⟨OP–7a⟩, using an appropriate heat source ⟨OP–7⟩.

Separation. Wash the reaction mixture ⟨OP–19⟩ with 50 mL
of water, followed by two portions of saturated aqueous sodium bicarbonate, being sure to save the right layer each time.
When washing with aqueous sodium bicarbonate, stir the
layers until gas evolution subsides before you stopper the separatory funnel, and vent it frequently thereafter. Dry the crude
ester ⟨OP–20⟩ with anhydrous magnesium sulfate (or another suitable drying agent) and filter it by gravity ⟨OP–11⟩.

Be sure the thermometer bulb is placed
correctly and that the thermometer reading
has equilibrated before you take a
temperature reading.

Purification and Analysis. Set up an apparatus for simple distillation ⟨OP–2, OP–25⟩ and have your instructor check it
before you start. Distill the crude ester, collect the liquid that
distills between 136°C and 143°C, and weigh it ⟨OP–5⟩.
Submit the purified product to your instructor in a tared and
labeled vial.

Report. Calculate the theoretical yield and your percentage
yield of isoamyl acetate. Calculate the amount of isoamyl acetate at equilibrium, based on the quantities of starting materials
you used and a value of 4.2 for the equilibrium constant. (Use
the quadratic equation; because volumes cancel out, moles can
be used in place of molar concentrations.) Estimate the amount
of isoamyl acetate that was lost (1) as a result of incomplete
reaction, (2) during the washings, and (3) during the distillation (see OP–25). Assume that the ester's solubility in aqueous
$NaHCO_3$ is about the same as in water. Compare the sum of
these estimated losses with your actual product loss and try to
account for any significant differences.

Isoamyl acetate equilibrium

$$i\text{-AmOH} + \text{HOAc} \rightleftharpoons$$
$$i\text{-AmOAc} + H_2O$$

$$K = \frac{[i\text{-AmOAc}][H_2O]}{[i\text{-AmOH}][\text{HOAc}]}$$

Ac = acetyl, CH_3CO-
i-Am = isoamyl, $CH_3CHCH_2CH_2-$
$\qquad\qquad\qquad\quad |$
$\qquad\qquad\qquad CH_3$

Collateral Projects

1. Obtain gas chromatograms ⟨OP–32⟩ of **(a)** your
product and **(b)** a sample of your product spiked with isoamyl
alcohol. Identify the alcohol and ester peaks, and estimate the
purity of your isoamyl acetate as a percentage.

2.   Prepare another ester of a primary alcohol, such as butyl acetate or isobutyl propionate, by the general method described above for isoamyl acetate. Work out a procedure for the synthesis and have it approved by your instructor (see Exercise 3). In some cases, a longer reflux time may be necessary for satisfactory results.

Synthesis of Methyl Benzoate
(Oil of Niobe)

In a 15-cm test tube, place 2.0 g of benzoic acid, 2.0 mL of methanol [**fire and vapor hazard!**], and 3 drops of concentrated sulfuric acid [**contact hazard!**]. Insert a cold-finger condenser and reflux ⟨OP–7b⟩ the reactants very gently (remember the boiling chips!) in a hot-water bath for about 30 minutes. Wash the reaction mixture with water and saturated aqueous sodium bicarbonate according to the procedure for semimicro extraction ⟨OP–13a⟩. Dry the methyl benzoate with a little magnesium sulfate, weigh it, and turn it in to your instructor. Write an equation for the reaction, and calculate the percentage yield. If desired, the methyl benzoate can be purified by semimicro distillation ⟨OP–25a⟩; in that case, collect the product that distills at around 195–201°C.

Be sure to save the right layer!

Exercises

1.   What gas was evolved during the sodium bicarbonate washing? Write balanced equations for the reactions that took place during this operation.

2.   Tell how the procedure for the preparation of isoamyl acetate might be modified to increase the yield and purity of the product.

3.   **(a)** In the Methodology section it was stated that the reaction of an equimolar mixture of isoamyl alcohol and acetic acid will produce, at most, 67% of the theoretical amount of isoamyl acetate. Verify this with an equilibrium constant calculation, using $K = 4.2$. **(b)** Compare this with the corresponding percentage for the conditions used in this experiment (see the Report section). Are your results consistent with Le Châtelier's principle? Explain.

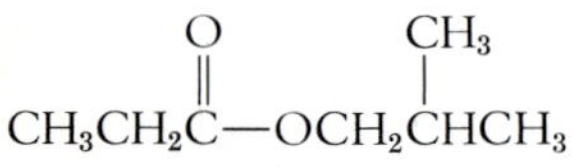

$$CH_3CH_2\overset{\displaystyle O}{\overset{\displaystyle \|}{C}}-OCH_2\overset{\displaystyle CH_3}{\overset{\displaystyle |}{C}}HCH_3$$

isobutyl propionate

4. Write a procedure that would be suitable for the preparation of isobutyl propionate, using the same molar quantities of reactants as in this experiment. Specify the amounts of all materials required and the distillation range for the product. Obtain the necessary physical properties from one of the reference books listed in the Bibliography.

5. A number of esters, such as methoprene, gossyplure, and permethrin, have shown some potential as insect-control agents. Give the structures and chemical names of these esters, and briefly tell how they control insects.

Library Topic

Describe an industrial process for the synthesis of such esters as isoamyl acetate, and compare it to the laboratory synthesis you carried out.

Identification of a Petroleum Hydrocarbon

Physical Properties of Liquids. Alkanes and Cycloalkanes.

Operations

OP – 5   Weighing

OP – 6   Measuring Volume

OP – 25a Semimicro Distillation

OP – 29 Boiling-Point Determination

OP – 29a Micro Boiling-Point Determination

OP – 30 Measuring the Refractive Index

Objectives

To purify and identify an unknown alkane or cycloalkane.

To learn how to carry out a small-scale distillation.

To learn how to measure some physical properties of liquids.

To learn about the production and composition of gasoline.

SITUATION

Petroleum is an extremely complex mixture that contains many different alkanes, cycloalkanes, and other hydrocarbons. Some of these hydrocarbons can be used in motor fuels, either in the form in which they exist in petroleum or after molecular modifications that are brought about by various refining processes. The suitability of a hydrocarbon for use as a component of gasoline depends, in part, on its octane number. In this experiment, you will purify and identify a liquid hydrocarbon found in petroleum and find out whether it would make a good motor fuel.

BACKGROUND

Concocting a Chemical Soup

Gasoline is a kind of "chemical soup" containing an incredibly large number of ingredients, all carefully selected and blended to produce a fuel with the desired properties. A typical high-octane gasoline might contain a mixture of straight and branched chain alkanes, some cycloalkanes (also called naphthenes), a few alkenes, and several different aromatic hydrocarbons as the main fuel components. A dash of tetraethyllead might be added to increase the octane number, along with a lead scavenger such as ethylene bromide, a quick-start additive such as butane to facilitate cold-weather starting, some antioxidants and metal deactivators for stability, antifreeze to prevent carburetor icing, and dyes for identification and eye appeal.

$CH_3CH_2CH_2CH_2CH_2CH_3$
hexane
(straight-chain alkane)

$CH_3CH_2CH_2CH{=}CH{-}CH_3$
2-hexene
(alkene)

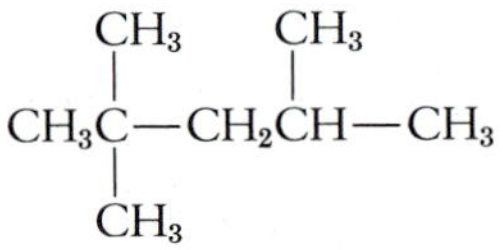

2,2,4-trimethylpentane
(branched alkane)

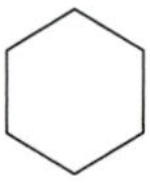

cyclohexane
(cycloalkane)

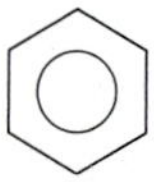

benzene
(aromatic hydrocarbon)

$(CH_3CH_2)_4Pb$
tetraethyllead (antiknock additive)

$BrCH_2CH_2Br$
ethylene bromide (lead scavenger)

$CH_3CH_2CH_2CH_3$
butane (quick-start additive)

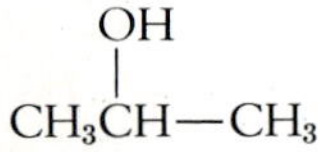

OH
|
$CH_3CH{-}CH_3$
isopropyl alcohol
(de-icer)

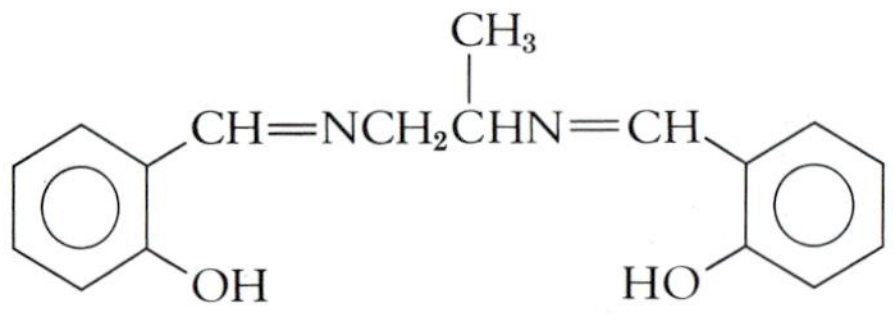

disalicyl-1,2-propanediimine
(metal deactivator)

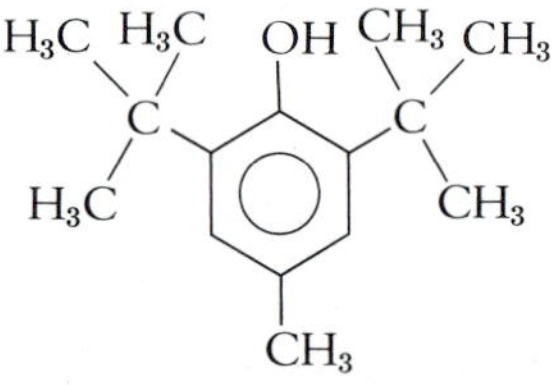

2,6-t-butyl-4-methylphenol (BHT)
(antioxidant)

Nearly all of the major components of gasoline are derived either directly or indirectly from petroleum, which must be refined before a usable fuel is obtained. The word "refine" suggests a simple separation and purification process, but the refining of petroleum is a much more complex operation involving many different chemical reactions.

The first major process to which petroleum is subjected is fractional distillation (fractionation), which separates the components according to their boiling ranges. The fraction boiling between about 50°C and 150°C is often called "straight-run" gasoline. Straight-run gasoline is not a good motor fuel by itself, because it contains a large proportion of unbranched hydrocarbons such as heptane and hexane in addition to branched alkanes and naphthenes. Straight-chain alkanes burn very rapidly, generating a kind of shock wave in the combustion chamber that reduces power and can damage the engine. This "knocking" does not occur with highly branched alkanes, which burn slowly and evenly enough to deliver the optimum amount of power to the cylinder head.

The octane number of a fuel is a measure of its antiknock qualities. The highly branched alkane 2,2,4-trimethylpentane (sometimes called "isooctane") is a very good fuel, and it has arbitrarily been assigned an octane number of 100; *n*-heptane, with no branching, has an octane number of zero. The performance of all grades of gasoline is measured in relation to these two alkanes; for example, a fuel performing as well as a mixture containing 70% 2,2,4-trimethylpentane and 30% heptane is assigned an octane number of 70. Table 1 lists the octane numbers of a selection of petroleum hydrocarbons.

A major objective of petroleum refining is to convert low-octane components of petroleum into high-octane compounds having the proper boiling range. This can be accomplished by chemical reactions such as isomerization, which converts straight-chain alkanes to branched alkanes; cracking,

The chemistry involved in these reactions is described in many textbooks of organic chemistry.

Table 1.   Octane numbers of some petroleum hydrocarbons

Hydrocarbon	Octane Number	Hydrocarbon	Octane Number
nonane	−45	2,4-dimethylpentane	82
octane	−17	methylcyclopentane	82
heptane	0	cyclopentane	83
2-methylheptane	24	2,3-dimethylpentane	89
hexane	26	2-methylbutane	89
3-methylheptane	35	butane	92
2-methylhexane	45	2,3-dimethylbutane	95
pentane	61	2,2-dimethylbutane	96
3-methylhexane	66	2,2,3-trimethylbutane	100
methylcyclohexane	71	2,2,4-trimethylpentane	100
2-methylpentane	73	2,2,3-trimethylpentane	102
3-methylpentane	75	toluene	104
cyclohexane	77	benzene	106

Reactions of tetraethyllead

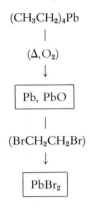

gums: high-molecular-weight substances that form as a result of oxygen-initiated free-radical reactions and that can cause carburetor malfunctions

A small-scale (semimicro) apparatus should be used for distilling small quantities of a liquid.

which breaks down large molecules into smaller fragments; alkylation, which combines short-chain alkane and alkene molecules to form longer, branched molecules; and reformation (also called aromatization), which converts aliphatic hydrocarbons to aromatics. Aromatic hydrocarbons such as toluene have especially high octane numbers and are used to increase the octane rating of unleaded fuels.

Gasoline for use in automobile engines is prepared by combining varying amounts of straight-run gasoline, cracked gasoline, alkylated gasoline, reformate, and other hydrocarbon mixtures in the right proportions to give the desired boiling range and octane number. The properties of the fuel may be further adjusted by adding lead compounds to raise the octane number and low-boiling alkanes to ensure quick starts in cold climates or for winter driving. Leaded gasoline contains either tetramethyl- or tetraethyllead. Both of these are unstable compounds that decompose in the combustion chamber to form lead and lead oxide, which interrupt the fast free-radical chain reactions that lead to knocking. A lead scavenger such as ethylene bromide (1,2-dibromoethane) must be added to convert the excess lead and its oxide to lead bromide, which is volatile enough to pass out of the combustion chamber with the exhaust gases. Antioxidants are used to prevent or retard formation of **gums,** particularly when a high percentage of alkenes is present. Certain metals, such as copper and iron, can catalyze gum-forming reactions; thus, a chelating compound like disalicyl-1,2-propanediimine is usually added to gasoline to combine with and deactivate these metals.

METHODOLOGY

An organic compound is often identified by the preparation of a *derivative,* a solid compound whose melting point can give a clue to its identity. Alkanes and cycloalkanes are comparatively unreactive, making it difficult to convert them to suitable derivatives. They are usually identified by their spectra and physical properties. Among the properties measured are boiling point, refractive index, and density.

A compound's boiling point can be measured either by distilling the compound and recording the observed temperature range during the distillation or by using a special boiling-point apparatus. It is best, in any case, to distill an unknown liquid in order to purify it before measuring its other properties. Then a more precise value for the boiling point can be obtained by a subsequent distillation or, as in this experiment, by a micro boiling-point determination. The refractive index of a liquid can be measured with great accuracy and is therefore valuable for characterizing organic compounds. The density of a liquid can be obtained by accurately weighing a measured volume of the liquid. The volume is measured with a pipet, and the weight is determined to the nearest milligram on an accurate balance.

The physical constants of the hydrocarbons listed in Table 2 are different enough that an accurate determination of all three constants should allow the certain identification of an unknown as one of the nine. Your instructor may add hydrocarbons from Table 1 to the list of possibilities in Table 2. If so, he or she will provide you with the appropriate physical constants or ask you to look them up.

PRELAB ASSIGNMENTS

1. Read the descriptions for OP–25a, OP–29, OP–29a, and OP–30, and read or review the other operations as needed.

2. Review the safety rules and the material under the heading "Reacting to Accidents: Fires" in the Laboratory Safety section.

3. Read the experiment, and prepare a brief experimental plan.

Structures and Properties

Table 2.   List of possible hydrocarbons

Name	Structure	b.p.	n_D^{20}	d^{20}
cyclopentane		49	1.4065	0.746
2,2-dimethylbutane	$CH_3C-CH_2CH_3$ with CH_3 above and CH_3 below	50	1.3688	0.649
2,3-dimethylbutane	$CH_3CH-CH-CH_3$ with CH_3, CH_3 above	58	1.3750	0.662
methylcyclopentane		72	1.4097	0.749

(continued)

Table 2.   List of possible hydrocarbons *(continued)*

Name	Structure	b.p.	n_D^{20}	d^{20}
2,4-dimethylpentane	CH₃CH—CH₂CH—CH₃ (with CH₃ groups on C2 and C4)	80	1.3815	0.673
cyclohexane	(cyclohexane ring)	81	1.4266	0.779
2,3-dimethylpentane	CH₃CH—CH—CH₂CH₃ (with CH₃ groups on C2 and C3)	90	1.3919	0.695
2,2,4-trimethylpentane	CH₃C—CH₂CH—CH₃ (with CH₃ groups, two on C2, one on C4)	99	1.3915	0.692
methylcyclohexane	(methylcyclohexane ring)	101	1.4231	0.769

Note: n_D^{20} = refractive index at 20° using sodium D line; d^{20} = density at 20°

PROCEDURE

The hydrocarbons are all highly flammable, and they may be harmful if inhaled, swallowed, or absorbed through the skin. Keep your unknown hydrocarbon away from hot surfaces (such as hot plates), sparks, and flames. Minimize contact with the hydrocarbon, and do not inhale its vapors.

Be sure the thermometer is placed correctly in the still head.

Boiling-Point Determination. Obtain a sample (about 5 mL) of an unknown hydrocarbon from your instructor and record its number in your laboratory notebook. Select a suitable heat source and purify the hydrocarbon **[fire hazard!]** by semimicro distillation ⟨OP–25a⟩. Measure its boiling-point range during the distillation, and record the "median boiling point" ⟨OP–29⟩ when about half of it has distilled. Carry out a micro boiling-point measurement ⟨OP–29a⟩ on the hydrocarbon. The boiling points should agree within a degree or two; if they don't, repeat the measurement or redistill the hydrocarbon. Record the barometric pressure in the laboratory.

If the hydrocarbon boils over a broad range, it should be redistilled and a pure fraction (collected over a range of 1–2°C) should be used for analysis.

Density and Refractive Index Measurements. See that the temperature of the purified hydrocarbon is close to 20°C, and accurately measure 1 mL into a clean, dry, tared vial using a volumetric pipet ⟨OP–6⟩. **[Do not pipet by mouth!]** Stopper the vial immediately and weigh it to the nearest milligram on an accurate balance ⟨OP–5⟩. Measure the refractive index of the purified hydrocarbon ⟨OP–30⟩, and record the temperature at which the measurement is made.

Report. Make a correction to the refractive index if the temperature of the measurement was not 20°C, and a correction to the boiling point if the atmospheric pressure was below 750 torr. Deduce the identity of your hydrocarbon from the information in Table 2, and explain how you arrived at your conclusion. Look up the octane number of your hydrocarbon in Table 1, and comment on its desirability as a motor-fuel component.

Collateral Project

The great complexity of gasoline can be seen clearly on a gas chromatogram ⟨OP–32⟩. You and your co-workers can obtain and compare gas chromatograms of different grades of gasoline, such as leaded, unleaded, and "gasahol." Refer to the *Journal of Chemical Education,* **49,** 764 (1972) and **53,** 51 (1976) for information and references that will help you interpret the chromatograms and identify some of the components.

Exercises

1. The dry-cleaning solvent 1,1,1-trichloroethane is a colorless, water-insoluble liquid with a boiling point of 74°C, a density of 1.339 g/mL, and a refractive index of 1.4379 at 20°C. Suppose you have two cans of dry-cleaning solvents; one contains a hydrocarbon solvent and the other trichloroethane, but both are missing their labels. How could you tell

CH_3CCl_3
1,1,1-trichloroethane

which is which (other than by odor) using only what is available in any kitchen?

2. The following properties are measured for an unknown hydrocarbon in a laboratory with an ambient temperature of 28°C and a barometric pressure of 28.9 inches of mercury:

boiling point: 78.2°C

refractive index: 1.3780

mass of 5 mL: 3.346 g

Correct the refractive index and boiling point to 20°C and 1 atmosphere, and calculate the density of the unknown. If the unknown is one of the hydrocarbons listed in Table 1, what is its probable identity?

3. Give names and chemical structures for all other isomers of 2,3- and 2,4-dimethylpentane.

4. Show how *n*-butane can be converted to 2,2,4-trimethylpentane using reactions mentioned in the Background section.

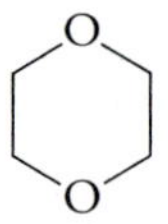

dioxane

5. **(a)** Dioxane has a boiling point of 101°C. Could you separate dioxane from methylcyclohexane by distillation? Explain, based on the liquid and vapor compositions during the distillation. **(b)** How could you separate these liquids? Support your answer with data from the literature.

Library Topic

Write a short report on the petroleum-refining processes used for the production of gasoline. Give some specific examples (with equations) of reactions that occur during each refining step described in the Background section.

Separation of Petroleum Hydrocarbons by Fractional Distillation

Separation Methods. Cycloalkanes.

Operations

OP–5   Weighing

OP–7   Heating

OP–27 Fractional Distillation

OP–32 Gas Chromatography

Objectives

To separate the components of a mixture of hydrocarbons by fractional distillation and analyze the fractions.

To determine the HETP of a fractionating column.

To learn how to carry out a fractional distillation and how to obtain and interpret a gas chromatogram.

To learn how distillation is used in petroleum refining.

SITUATION

Cyclohexane, one of the more abundant hydrocarbons in petroleum, is an important commercial solvent and a starting material for the production of nylon and other commercial products. It is reasonable to expect that pure cyclohexane would be distilled from petroleum directly, but in fact cyclohexane is generally obtained by the hydrogenation of benzene, which is in turn made from cyclohexane-rich petroleum fractions. In this experiment, you will explore some of the uses and limitations of distillation as a separation method by attempting to separate cyclohexane from another petroleum constituent, toluene.

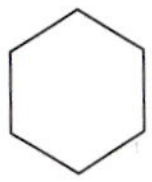

cyclohexane

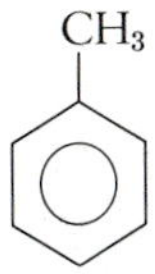

toluene

BACKGROUND

Distillation in Petroleum Refining

Distillation has been used since antiquity to separate the components of mixtures — the ancient Egyptians were making an embalming fluid by distilling wood more than 3500 years ago. In the intervening years, the process of distillation has undergone many improvements and found numerous applications. In one form or another, distillation is used to manufacture perfumes, flavor ingredients, liquors, charcoal, coke, and a host of organic chemicals. One of its most important modern applications is its role in refining petroleum into fuels, lubricants, and petrochemicals. The first step in that process is the separation of petroleum into various hydrocarbon fractions by distilling it through huge fractionating columns, called distillation towers, that are up to 200 feet high. Since components of different molecular weights and carbon structures usually have significantly different boiling points, this process separates the petroleum into portions containing hydrocarbons of similar carbon content and properties. The lowest-boiling hydrocarbons make it all the way to the top of the tower, where they are collected as the top fraction. The top fraction, called straight-run naphtha, is usually upgraded by catalytic reforming and used in gasoline. The less volatile middle fractions are collected part way down the tower; these include kerosene and a gas-oil fraction used to produce diesel fuel, jet fuel, and home heating oil. The bottom fraction is usually subjected to vacuum distillation to produce vacuum gas oils, which can be used as fuel oils or converted to gasoline hydrocarbons by catalytic cracking. Petroleum products such as paraffin wax, lubricating grease, and asphalt also come from this fraction.

With an overall octane number in the range 30–50, straight-run naphtha is not a suitable motor fuel, but its octane number can be increased to close to 100 by catalytic reforming. In this process, the naphtha is heated to about 500°C in the presence of a platinum- or palladium-based catalyst. Catalytic reforming converts alkanes to cycloalkanes and some of these cycloalkanes to aromatic compounds; the result is a mixture rich in high-octane aromatics. For example, *n*-hexane may be cyclized to cyclohexane, which is then dehydrogenated to yield benzene.

$$CH_3CH_2CH_2CH_2CH_2CH_3 \longrightarrow$$

n-hexane   cyclohexane   benzene

Cyclohexane is one of the more abundant components in unrefined petroleum, but it cannot be isolated in pure form by distillation alone. Pure cyclohexane is usually prepared by catalytic hydrogenation of benzene (the reverse of the above dehydrogenation reaction). Unlike cyclohexane, benzene and other aromatics can be obtained from hydrocarbon mixtures in relatively pure form by solvent extraction followed by distillation.

METHODOLOGY

In this experiment, you will separate the components of an equimolar mixture of cyclohexane and toluene by fractional distillation, and assess the completeness of the separation by measuring the composition of each fraction. The experiment should help you appreciate the difficulties associated with the separation of any pure hydrocarbon from petroleum — a mixture containing thousands of hydrocarbons, some with boiling points that differ by only a fraction of a degree.

The degree of separation you attain will depend on such factors as the heat source, the distillation rate, and the efficiency of the column. Electrically heated oil baths provide the most uniform heat, but a heating mantle of the right size should work well enough. Good separation requires a low rate of distillation to maintain a high reflux ratio, so patience will be required if you are to get good results. The efficiency of a given column can be enhanced by packing it carefully or by using a more effective packing material. Another way of improving a separation is to redistill each fraction, as described in Collateral Project 1.

You will determine the composition of each fraction you collect by injecting a very small amount into an instrument called a gas chromatograph. (If you do not have access to a gas chromatograph, you can use a sensitive refractometer instead, as described in Collateral Project 2.) Inside the gas chromatograph, both hydrocarbons will vaporize and their vapors will travel through a packed column at different rates. As the vapors exit the column, their presence will be detected and recorded on a graph called a gas chromatogram, which should display two peaks of different sizes, the cyclohexane peak being the first to appear. The areas of the peaks can be converted to relative masses by multiplying them by the appropriate correction factors from Table 1. (Alternatively, you can measure the correction factors by the method described in OP–32.) From this information, you can compute the masses of cyclohexane and toluene in each of your four main fractions. A graph of

Table 1.   Gas chromatography correction factors for cyclohexane and toluene

	FI Detector	TC Detector
cyclohexane	1.11	0.942
toluene	1.05	1.02

Note: benzene = 1.00

Your instructor will tell you what kind of detector your gas chromatograph has.

boiling point versus composition will indicate the degree of separation; good separation is suggested by a graph that is low in the middle and high on each end.

You will also measure the efficiency of your fractionating column by comparing the composition of the *liquid* in the boiling flask to the composition of the *vapor* that first emerges from the top of the column and condenses into a receiving vial. The number of theoretical plates, *n*, provided by your distillation apparatus can be calculated from the *Fenske equation*, expressed as follows:

See OP–27 for definitions of theoretical plate and HETP and a discussion of the Fenske equation.

$$n = \frac{\log \dfrac{Z_C}{X_C} - \log \dfrac{Z_D}{X_D}}{\log \alpha}$$

X_C and X_D represent the mole fractions of cyclohexane and toluene (respectively) in the liquid; Z_C and Z_D are the corresponding values for the vapor. The volatility factor, α, for the cyclohexane-toluene mixture is 2.33. Keep in mind that the boiling flask furnishes one theoretical plate, so the number of plates provided by the column will be $n - 1$. By measuring the length of your column packing, you can determine the HETP (height equivalent to a theoretical plate) of the column, which is a measure of its efficiency; the lower the HETP, the more efficient the column.

Consult a general chemistry textbook if you need help in converting mass percentages to mole fractions.

You can obtain some additional experience with fractional distillation by distilling turpentine ⟨Collateral Project 3⟩ and with gas chromatography by analyzing a commercial xylene mixture ⟨Minilab 3⟩.

PRELAB ASSIGNMENTS

1. Read OP–27 and OP–32, and read or review the other operations as necessary.

2. Calculate the mass and volume of 0.200 mol of cyclohexane and 0.200 mol of toluene.

3. Prepare a brief experimental plan for the experiment.

Properties

Table 2.　Physical properties

	M.W.	b.p.	d.
cyclohexane	84.2	81	0.774
toluene	92.2	111	0.867

PROCEDURE

SAFETY PRECAUTIONS

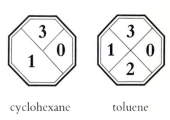

cyclohexane   toluene

Separation.   Accurately weigh out ⟨OP–5⟩ 0.200 mol of cyclohexane **[fire hazard!]** and 0.200 mol of toluene **[fire and vapor hazard!]** and combine the liquids in a 100-mL boiling flask. Pack a distilling column with the column packing provided, and measure the height of the packing to the nearest millimeter. Select an appropriate heat source ⟨OP–7⟩ and assemble an apparatus for fractional distillation ⟨OP–27⟩, using a small beaker as a receiver for the forerun. Clean, dry, weigh, and number four fraction collectors, preferably small bottles or large vials (with a capacity of 20 mL or more) with tight-fitting caps. Have ready a smaller vial to collect the HETP sample.

The column need not be insulated.

Watch the vapors rise in the column as you heat the boiling flask. When they reach the still head, reduce the heating rate enough to keep the ring of condensing vapors between the top of the column packing and the sidearm for several minutes, so that the vapor composition can stabilize before any distillate is collected. Distill the liquid slowly and discard the first few drops of distillate. Collect the next 5–10 drops in the HETP vial and cap it tightly. Quickly replace the vial with the first fraction collector, record the distillation temperature, and distill at a rate of not more than 20 drops per minute. When the distillation temperature reaches 85°C, replace the first fraction collector with the second one and cap the first one tightly. When the distillation temperature reaches 97°C, switch to the third collector. When the temperature reaches 107°C, remove the heat source and let all the liquid that remains in the column drain into the boiling flask. After it has cooled down, transfer the contents of the boiling flask to the fourth collector. At this point, you should have fractions covering the following boiling ranges:

The initial distillate may contain traces of water, which will make it appear cloudy. Be sure the distillate is clear before you collect the first sample.

1. 81–85°C

2. 85–97°C

3. 97–107°C

4. 107–111°C

(At your instructor's request, redistill the fractions as described in Collateral Project 1.)

Analysis. Weigh the four fractions ⟨OP–5⟩, and analyze them and the HETP sample by gas chromatography ⟨OP–32⟩.

Report. Measure the peak areas on each gas chromatogram, and use the appropriate correction factors to convert the areas to relative masses. Calculate the percentage (by mass) and the actual mass of cyclohexane and toluene in each fraction. Plot the component masses for each fraction (on the *y*-axis) as a function of boiling temperature, using the midpoints of the appropriate boiling ranges (on the *x*-axis). Draw a smooth curve connecting the data points for cyclohexane and another for those for toluene. Calculate the number of theoretical plates provided by your fractional distillation apparatus; then calculate the HETP of your column. Comment on the efficiency of your separation and suggest several ways of improving it. Turn in your gas chromatograms (or photocopies) with your report.

Collateral Projects

Wipe out the collector with a clean lab tissue each time you reuse it.

1. To improve the efficiency of the separation, redistill the four fractions as follows. Pour the contents of the first fraction collector into the boiling flask and distill it, collecting the distillate in the same container. When the temperature reaches 85°C, remove the heat source and let any liquid in the column drain into the boiling flask. Next, add the contents of the second fraction collector to the boiling flask. Resume the distillation, collecting the first and second fractions in the corresponding collectors over the boiling ranges specified in the Procedure section. When the temperature reaches 97°C (or when all the liquid has distilled), remove the heat source and add the contents of the third fraction collector to the

cooled boiling flask. Resume the distillation, and collect the three fractions in the corresponding collectors over the specified boiling ranges. When the temperature reaches 107°C, remove the heat source, add the contents of the fourth fraction collector to the cooled boiling flask, and again collect the first three fractions over the appropriate boiling ranges. When the temperature reaches 107°C, stop the distillation and transfer the cooled contents of the boiling flask to the fourth fraction collector.

2.   If a gas chromatograph is not available, estimate the composition of each fraction from its refractive index ⟨OP–30⟩ by reading it from a plot of refractive index versus mole fraction. Construct the calibration "curve" by measuring the refractive indexes of the toluene and cyclohexane you are using, plotting these values at mole fractions of 0 and 1, respectively, and connecting the points by a straight line. The refractive indexes at 20°C of ultrapure cyclohexane and toluene are 1.4260 and 1.4968, respectively, but your measured values may be somewhat different.

3.   For a more challenging separation, carry out the fractional distillation of turpentine to obtain α-pinene and β-pinene, whose boiling points are about 10°C apart. An efficient insulated column providing 5 to 10 theoretical plates is required.

Analysis of Commercial Xylene by Gas Chromatography

MINILAB 3

Obtain a gas chromatogram ⟨OP–32⟩ of a commercial xylene mixture using the column and temperature settings suggested by your instructor. Spike the commercial xylene with at least two of the pure xylenes, and obtain gas chromatograms of the mixtures. Identify the xylene peaks, measure the peak areas on the first chromatogram, and calculate the percentage of each isomer in the commercial xylene. Correction factors for the *ortho, meta,* and *para* compounds are 1.08, 1.04, and 1.04, respectively, if a TC detector is used. No correction factor is necessary with an FI detector.

Table 3.   Temperature-composition data for cyclohexane-toluene

Mol % Cyclohexane		T, °C
Vapor	Liquid	
0	0	110.7
10.2	4.1	108.3
21.2	9.1	105.5
26.4	11.8	103.9
34.8	16.4	101.8
42.2	21.7	99.5
49.2	27.3	97.4
54.7	32.3	95.5
59.9	37.9	93.8
66.2	45.2	91.9
72.4	53.3	89.8
77.4	59.9	88.0
81.1	67.2	86.6
86.4	76.3	84.8
89.5	81.4	83.8
92.6	87.4	82.7
97.3	96.4	81.1
100.0	100.0	80.7

Exercises

1. A certain fractional distillation apparatus consists of a boiling flask surmounted by a 24-cm Vigreux column and a 30-cm glass tube half filled with 4 × 4 mm porcelain saddles. How many theoretical plates does the entire apparatus provide if the HETP of the saddles is 5 cm, the HETP of the Vigreux column is 8 cm, and the HETP of the empty glass tube is 15 cm?

2. Using the data in Table 3, construct a temperature-composition diagram like that in Figure 3 of OP–27 (page 711). **(a)** From your diagram, estimate the initial composition of the distillate obtained by simple distillation of a mixture containing 20 mole percent cyclohexane and 80 mole percent toluene. **(b)** Estimate the initial composition if the same mixture is distilled through a three-plate column.

3. If you were to return your third fraction to the empty boiling flask and redistill it, you might expect it all to distill between 97°C and 107°C as it did the first time. It actually yields some distillate in all four boiling ranges. Explain.

4. Suggest a chemical method that could be used to remove small amounts of toluene from cyclohexane.

5. Show how nylon-66 can be synthesized using cyclohexane as the starting material.

Library Topic

Write a report on the destructive distillation of wood, giving the names, structures, and uses of some compounds obtained from this process.

The Preparation of Camphor

Preparation and Purification of Solids. Oxidation.

Operations

OP–5   Weighing
OP–7c Temperature Monitoring
OP–8 Cooling
OP–9 Mixing
OP–10 Addition
OP–12   Vacuum Filtration
OP–21   Drying Solids
OP–24 Sublimation
OP–28   Melting-Point Determination
OP–31 Measuring Optical Rotation ⟨Minilab 4⟩

Objectives

To synthesize camphor from isoborneol using commercial laundry bleach and assess the purity of the product.

To learn some new operations for carrying out organic reactions.

To learn how to purify a solid by sublimation.

To learn about the history, manufacture, and characteristics of camphor.

SITUATION

The Chinese camphor tree, *Cinnamomum camphora,* is a tall, striking evergreen tree with dark shiny leaves. It is native to Formosa, but it has been introduced into the southern United States and grows well in Florida and California. When steam is blown through the chopped-up wood of a camphor tree, camphor distills with the steam and crystallizes as a gummy white solid. Just as there are left-handed and right-

(+)-camphor   (−)-camphor

isoborneol

Perspective drawings of camphor structure

Construction of molecular models is necessary to demonstrate the almost spherical shape of these molecules.

handed gloves, scissors, and corkscrews, there are left-handed and right-handed camphor molecules. (+)-Camphor from the camphor tree is composed of right-handed molecules, which rotate plane-polarized light to the right (clockwise). Left-handed (−)-camphor has been isolated from feverfew *(Chrysanthemum parthenium),* a daisylike plant completely unrelated to the camphor tree. Camphor can be synthesized from α-pinene, a major constituent of turpentine, by the pathway outlined below. Most synthetic camphor is an equal mixture of left-handed and right-handed molecules.

α-pinene ⟶ pinene hydrochloride ⟶ camphene ⟶
        isobornyl acetate ⟶ isoborneol ⟶ camphor

You will carry out the final step in this synthesis, the oxidation of isoborneol to camphor. Isoborneol is often oxidized by toxic chromium compounds, but you will use a much safer oxidizing agent — a commercial chlorine bleach.

BACKGROUND

Camphor

The history of camphor is longer and more involved than that of perhaps any other natural product. Scientific speculations about camphor have appeared in print since the time of Libavius *(Alchymia,* 1595), and its unusually complicated bicyclic structure stumped scientists for sixty years after the correct molecular formula ($C_{10}H_{16}O$) was determined in 1833. More than thirty different structures were proposed, all of them wrong, before J. Bredt finally came up with the correct one, which was confirmed in 1903 with the total synthesis of camphor from simple open-chain esters.

The penetrating "camphoraceous" odor of camphor is shared by many compounds of similar molecular shape and size but vastly different composition. Compounds as diverse in structure as hexachloroethane, cyclooctane, dimethylpentamethylenesilicon, 2-nitroso-2-methylpropane, and thiophosphoric acid dichloride ethylamide all have the same distinctive odor. This correspondence has provided evidence in support of a stereochemical origin of odor. It has been suggested ⟨Bib–L62, p. 150⟩ that the olfactory receptor transmitting the sensation corresponding to a camphorlike odor has a hemispherical shape and a diameter of about 7 Ångstrom units, to accommodate roughly spherical molecules such as the ones named above. Although a completely satisfactory stereochemical theory of odors has not been developed, there seems little doubt that some relationship exists between molecular shape and odor quality.

Compounds having camphoraceous odors

METHODOLOGY

Most oxidation reactions of organic compounds result in either a gain of oxygen atoms or a loss of hydrogen atoms, or both. Thus, the conversion of the secondary alcohol 2-propanol to acetone is an oxidation reaction because it involves the loss of two hydrogen atoms per molecule.

In a secondary alcohol, two carbon-containing substituents are attached to the carbon that holds the hydroxyl (OH) group.

$$CH_3CHCH_3 \xrightarrow[-2\,H]{\text{oxidizing agent}} CH_3CCH_3$$

2-propanol   acetone

In this experiment, you will oxidize another secondary alcohol, isoborneol, to a ketone, camphor; a balanced equation for the reaction is given in the Reactions and Properties section. Secondary alcohols can be converted to ketones by powerful oxidizing agents such as chromium(VI) oxide and potassium dichromate. Many chromium compounds are highly toxic and corrosive, and some have been known to cause cancer after prolonged exposure. They also present a difficult disposal problem, because they cannot be flushed down the drain and thereby released into the environment. For these reasons, you will use a much safer oxidizing agent, the familiar laundry bleach that is sold under such trade names as Clorox and Javex. Most chlorine bleaches contain about 5.25% sodium hypochlorite in an aqueous solution. Addition of a little acetic acid converts sodium hypochlorite to hypochlorous acid (HOCl), which is probably the active oxidant in the camphor synthesis.

There may also be some elemental chlorine present in the solution containing hypochlorous acid.

In Experiments 4 and 5, you heated the reaction mixture to speed up the rate of reaction; in this experiment, you may have to slow it down instead. The oxidation of isoborneol is exothermic, and the heat evolved can cause the formation of

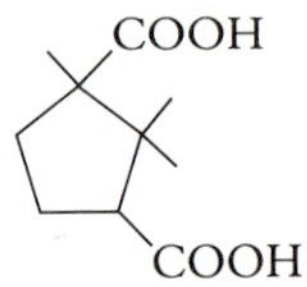

COOH

COOH

camphoric acid

unwanted by-products, such as camphoric acid. Thus, you should keep an ice bath handy to reduce the temperature if necessary. You can also slow down a fast reaction by adding one of the reactants a little at a time, from an addition funnel or an ordinary separatory funnel. To ensure a complete reaction, you will have to add enough sodium hypochlorite solution so that the oxidizing agent is in excess by the end of the reaction. This can be determined by testing the acidic reaction mixture with an indicator paper impregnated with starch and potassium iodide. If HOCl is present, it will oxidize the iodide ions to iodine, which turns the starch a deep blue-violet color. Any excess hypochlorite ion that remains after the reaction can be destroyed by treatment with the reducing agent sodium bisulfite, according to the following reaction:

$$OCl^- + HSO_3^- \longrightarrow Cl^- + HSO_4^-$$

When a reaction takes place under reflux, the boiling action helps mix the reactants; at lower temperatures, the reactants may have to be mixed by other means. Mixing can be effected with a magnetic stirrer or by shaking and swirling the flask after each addition of sodium hypochlorite solution. You will have to monitor the reaction temperature during the reaction by holding a thermometer with the bulb immersed in the reactants as you mix them; the directions for OP–7c tell how to do this without breaking the thermometer.

Because of its compact and symmetrical molecular structure, camphor changes directly from a solid to a vapor when heated, which allows it to be purified by sublimation. Because isoborneol also sublimes at elevated temperatures, the purified camphor may still contain traces of this substance.

Camphor has an unusually large melting-point depression constant of about 40°C kg mol^{-1}. This means you can use the melting points of your crude and purified products to estimate their purity. If you think of impure camphor as a solution, with camphor as the solvent and the impurities as solutes, you can use the equation in the margin to calculate the molal concentration (molality) of the impurities. This is the number of moles of impurities per kilogram of camphor, which can be used to calculate the mole percent (mole fraction × 100) of camphor in each sample.

Most synthetic camphor, including your product, contains equal numbers of left-handed and right-handed molecules and therefore does not rotate polarized light. Wood turpentine ("oil of turpentine") is a natural, optically active liquid distilled from the sap of certain pine trees (of the species *Pinus*

$$\Delta T = K_f \cdot m$$

ΔT = melting-point depression
K_f = molal freezing-point (melting-point) depression constant
m = molal concentration of solute

If necessary, consult a general chemistry text for a review of calculations involving units of concentration.

palustris) that grow in the southeastern United States. Minilab 4 gives a procedure for measuring the optical rotation of some turpentine and estimating the percentages of α- and β-pinene it contains.

PRELAB ASSIGNMENTS

1. Read OP–7c, OP–8, OP–9, OP–10, and OP–24, and read or review the other operations as necessary. Read OP–31 if you will be doing Minilab 4.

2. Calculate the mass of 25 mmol of isoborneol and the theoretical yield of camphor.

3. Prepare an experimental plan for the experiment.

Reactions and Properties

isoborneol + HOCl ⟶ camphor + HCl + H₂O

Table 1.   Physical properties

	M.W.	m.p.	b.p.	d.
isoborneol	154.3	212		
camphor	152.2	179	204	
sodium hypochlorite	74.4	stable only in solution		
acetic acid	60.1	17	118	1.049

PROCEDURE

The reaction should be carried out under a fume hood to keep chlorine fumes out of the laboratory.

Acetic acid is corrosive and can cause serious damage to the skin and eyes (especially the eyes); its vapors are highly irritating to the eyes and respiratory tract. Wear protective gloves, and dispense the acid under a hood; avoid contact, and do not breathe vapors.

SAFETY PRECAUTION

The solution containing sodium hypochlorite and acetic acid may evolve some toxic chlorine gas, which can seriously irritate or damage the eyes and respiratory tract. Use a hood, and do not breathe the vapors of the reaction mixture.

acetic acid   camphor   chlorine

Reaction. Under the hood, combine 25 mmol of isoborneol with 2 mL of glacial acetic acid **[contact and vapor hazard!]** in a 125-mL Erlenmeyer flask. Carefully mix in 5 mL of 5.25% sodium hypochlorite solution (Clorox or another suitable laundry bleach) **[vapor hazard!].** Measure another 40 mL of 5.25% sodium hypochlorite solution into a separatory/addition funnel, stopper the funnel, and support it over the flask. Add this solution ⟨OP–10⟩ to the reaction mixture in small portions, with vigorous swirling or magnetic stirring ⟨OP–9⟩. Monitor the temperature of the solution ⟨OP–7c⟩ and control the rate of addition so that the temperature remains below 50°C. Have an ice bath handy to cool the reaction mixture ⟨OP–8⟩ if it reaches 50°C.

The solution should have a greenish-yellow color while excess hypochlorite is present.

You can test the starch-iodide paper with a solution containing a few drops of sodium hypochlorite solution and a drop of acetic acid in a milliliter of water.

Test the reaction mixture for excess hypochlorite by transferring a drop of the solution to a strip of fresh starch-iodide paper. If the paper does not turn blue within a few seconds, add enough 5.25% sodium hypochlorite to the reaction mixture to give a positive test. Stir the reactants or swirl the flask frequently ⟨OP–9⟩ at room temperature for 30 minutes or more, testing the solution with starch-iodide paper every few minutes and adding more sodium hypochlorite (about 1 mL at a time) if the test is negative. Let the mixture stand until the temperature falls to 30°C or below (testing occasionally), then add saturated sodium bisulfite solution dropwise until the solution is colorless and the starch-iodide test is negative.

Separation. Cool the reaction mixture below 5°C in an ice bath ⟨OP–8⟩. Collect the product by vacuum filtration ⟨OP–12⟩, washing it thoroughly on the filter with several portions of cold water. Dry the crude camphor ⟨OP – 21⟩ at room temperature and save just enough for a melting-point determination.

Purification and Analysis.   Purify the camphor by sublimation ⟨OP – 24⟩, weigh it ⟨OP – 5⟩, and measure the melting points ⟨OP – 28⟩ of the crude and purified camphor. Turn in the final product in a tared and labeled vial.

The melting point may be more accurate if it is obtained in a sealed capillary tube.

Report.   Calculate the percentage yield of camphor. Taking as its melting point the temperature at which your camphor was completely liquefied, calculate the molality of the impurities and the mole percent of camphor in your crude and purified camphor. Assuming that the only significant impurity is isoborneol, estimate the purity of your crude and purified products, expressed as the mass percentage of camphor.

Collateral Project

Oxidize cyclohexanol to cyclohexanone as described in the Reaction section above. Isolate the product by saturating the reaction mixture with solid sodium chloride and extracting it with methylene chloride. Wash the methylene chloride layer with 1 M aqueous sodium hydroxide until the extracts are basic, dry it with magnesium sulfate, evaporate the solvent, and purify the cyclohexanone by distillation. Extracting the combined wash liquids with more methylene chloride will improve the yield.

The Optical Rotation of Turpentine

MINILAB 4

Turpentine irritates the skin and eyes, and its vapors can cause headaches, dizziness, and nausea. Avoid contact, and do not breathe its vapors.

SAFETY PRECAUTION

Using a 25-mL volumetric flask, prepare a solution containing about 2.5 g of turpentine (accurately weighed) in absolute ethanol. Measure the optical rotation of the solution with an accurate polarimeter ⟨OP – 31⟩. Prepare ethanolic solutions of

North American wood turpentine contains (+)-α-pinene and (−)-β-pinene, along with small amounts of other constituents. Your instructor may provide a simulated turpentine made by mixing the two pinenes.

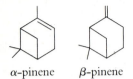

α-pinene   β-pinene

α-pinene and β-pinene in the same way, and measure their optical rotations. Calculate the specific rotation of each solution. Assuming that it contains only α- and β-pinene, calculate the percentage of each constituent in the turpentine.

Exercises

1. In this experiment, you started with a white, strong-smelling solid and ended up with a white, strong-smelling solid. What evidence leads you to conclude that these two substances are in fact different compounds and that you did not just isolate the unreacted starting material? Describe any observations you made during the experiment that support this conclusion.

2. Construct a flow diagram for the synthesis of camphor, using the general format given in Experiment 5.

3. Write a balanced equation for the reaction of the acidified sodium hypochlorite solution with iodide ion (I^-) on the starch-iodide paper, assuming that hypochlorous acid is reduced to chloride ion.

4. Show which carbon-carbon bond of camphor must be broken to form camphoric acid. Use molecular models if necessary.

Library Topic

Write a short report on the history, commercial production, and uses of camphor and camphor oil.

The Isolation and Isomerization of Lycopene from Tomato Paste

Isolation of Natural Products. Ultraviolet-Visible Spectrometry. Geometric Isomers.

Operations

OP–11 Gravity Filtration

OP–13*d* Solid Extraction

OP–14 Evaporation

OP–16 Column Chromatography

OP–18 Paper Chromatography ⟨Minilab 5⟩

OP–19 Washing Liquids

OP–20 Drying Liquids

OP–35 Ultraviolet-Visible Spectrometry

Objectives

To isolate the red pigment lycopene from tomato paste.

To convert *trans*-lycopene to its geometric isomer 13-*cis*-lycopene and verify the reaction by spectrometry.

To learn how to extract organic compounds from solids, how to separate the components of a mixture by column chromatography, and how to record an ultraviolet-visible spectrum.

To learn about carotenoids and the function of Vitamin A in vision.

SITUATION

Ideally, processed foods such as canned tomato products would contain all the nutrients and flavor components present in the fresh fruits or vegetables. In practice, it is not possible to process foods without destroying or altering some of their chemical constituents. However, with careful processing, much of the color, flavor, and nutritional value of a food can be retained. The red pigment that colors ripe tomatoes is an all-*trans* form of lycopene; heat, light, and certain catalysts may change some of this pigment to its 13-*cis* isomer. Although this reaction alone has little or no effect on food value, the presence of 13-*cis*-lycopene in a canned tomato product suggests that the fresh fruit was subjected to excessive heat or light during processing. Your objective in this experiment is to separate the red pigment from commercial tomato paste and find out whether it is pure *trans*-lycopene or a partly isomerized mixture containing 13-*cis*-lycopene.

BACKGROUND

Carotenoids, Vitamin A, and Vision

Lycopene, with its 40 carbon atoms and 13 double bonds, is one of the most unsaturated compounds in nature. Because most of its double bonds are conjugated, lycopene absorbs radiation at long wavelengths in the 400–500-nm region of the visible spectrum. Its resulting deep yellow-red color is responsible for the redness of ripe tomatoes, rose hips, and many other fruits.

An even more important plant pigment is the yellow substance *β*-carotene, which is present not only in carrots but in all green leaves and many flowers as well. Both lycopene and *β*-carotene, along with most of the other natural carotenoids, exist in the stable all-*trans* form.

Figure 1.   Structures of carotenoids. A single straight line branching off from a chain or a ring stands for a methyl group in these and similar formulas. A carbon atom with the requisite number of hydrogens is at each bend of the chain.

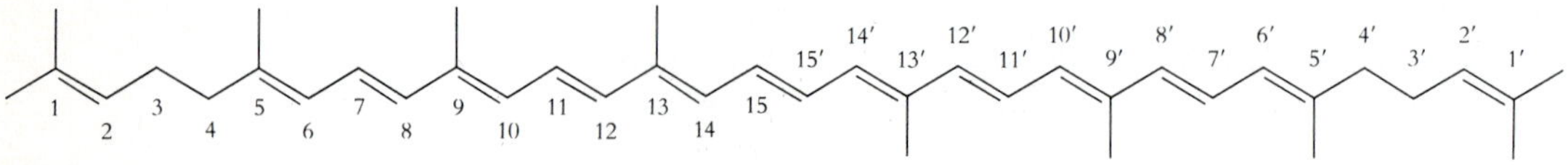

lycopene

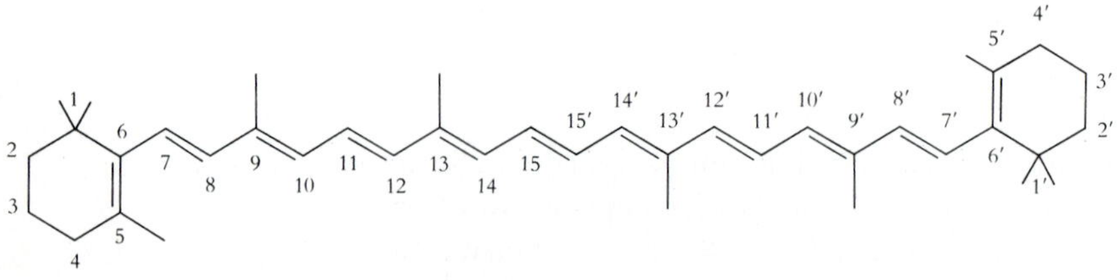

β-carotene

Although the main function of carotenoids in plants remains somewhat of a mystery, the importance of carotenes to animals is clear — β-carotene (and, to a lesser extent, its α- and γ-forms) is converted in the intestinal wall to Vitamin A, which is then esterified and stored in the liver. Generations of children have grown up with the mealtime refrain of "Eat your carrots — they're good for your eyes!" In fact, Vitamin A from carotenes and other sources does play an important role in vision, as well as acting to promote growth.

Cod liver oil is a more direct (if less palatable) source of Vitamin A than carrots and leafy green vegetables.

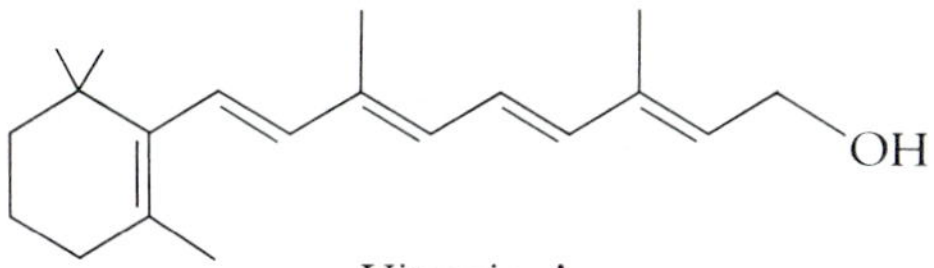

Vitamin A
(all-*trans*)

The process of vision, although extremely complex in its entirety, seems to be based on a simple isomerization of an oxidized form of Vitamin A called retinal. In the rods of the retina, which are responsible for night vision, retinal occurs in combination with the complex protein opsin to form rhodopsin (visual purple). While it is bound to this protein, retinal must assume the shape shown in Figure 2A, with an 11-*cis* double bond and probably a cisoid conformation at the 12–13 single bond as well.

A similar process occurs in the cones of the retina, which are responsible for color vision.

When a photon of light strikes a molecule of this 11-*cis*-retinal, it causes a configurational change that eventually results in the formation of all-*trans*-retinal. Now the *trans* form obviously has a different molecular shape than its *cis* counterpart, being straight and rigid where the *cis* form is bent and twisted. As a result, it no longer fits well into its niche on the opsin molecule. Just as a mother kangaroo might react to a joey that is making a commotion in her pocket by shifting positions to minimize the discomfort and eventually ejecting her unruly

The mechanism by which a single photon can trigger the enormously amplified response of a receptor cell is not yet understood.

A "joey" is a young kangaroo.

Figure 2.   Retinal isomers.

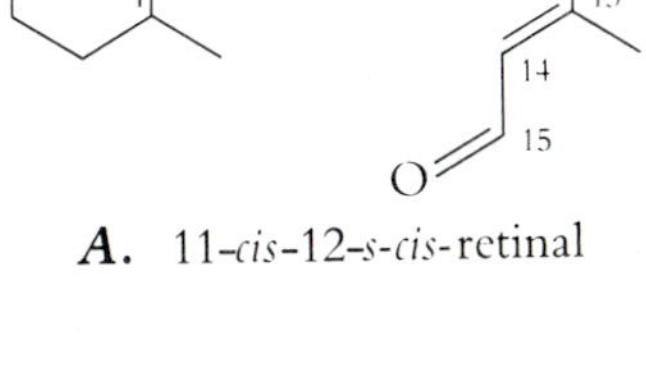

A. 11-*cis*-12-*s*-*cis*-retinal 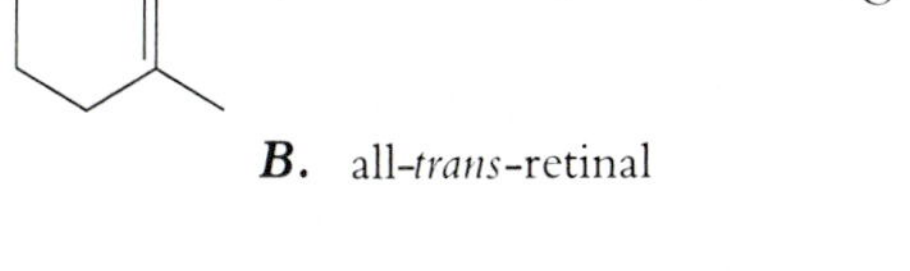   **B.** all-*trans*-retinal

The vision cycle

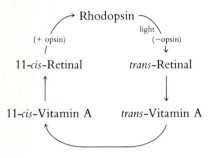

Tangerine tomatoes contain a poly-*cis* isomer of lycopene called prolycopene; adding a drop of iodine to a pale-yellow solution of prolycopene transforms it almost instantly to brilliant orange lycopene.

The eluate is the liquid that drains from a chromatography column.

Petroleum ether is a general term used to describe volatile petroleum distillates having varying compositions and boiling ranges. *Do not confuse it with ethyl ether!*

passenger, so the opsin undergoes a series of configurational and conformational changes that lead it to eject the transformed retinal. This dissociation of the rhodopsin molecule triggers the transmission of a visual message to the brain. Subsequently the *trans*-retinal is enzymatically reduced to all-*trans*-Vitamin A, which isomerizes to 11-*cis*-Vitamin A, which is oxidized back to 11-*cis*-retinal, which promptly combines with another molecule of opsin to regenerate more rhodopsin. At this point, another photon of light can start the cycle all over again.

Although lycopene cannot be converted directly to Vitamin A and thence to retinal, it does undergo a configurational change analogous to those undergone by retinal. Adding a small amount of iodine to a solution of all-*trans*-lycopene can cause its partial conversion to 13-*cis*-lycopene (also called neolycopene A), with substantial changes in its properties. The conversion can easily be followed by spectroscopy, since isomerization causes significant changes in the ultraviolet-visible spectrum of this pigment.

METHODOLOGY

In this experiment, you will extract the carotenoid pigments (carotenes, lycopene, and xanthophylls) from canned tomato paste, separate them by column chromatography, isolate lycopene from the *eluate,* and attempt to observe its isomerization by ultraviolet-visible spectrometry. Although lycopene can be isolated directly from ripe tomatoes, it is easier to extract it from commercial tomato paste, in which the lycopene is about ten times more concentrated—4 grams of tomato paste should yield about a milligram of lycopene. Lycopene will slowly oxidize or isomerize if allowed to stand in solution too long, particularly in the presence of heat, light, or air. For this reason, it is very important to avoid unnecessary delays and to keep the lycopene solutions in a cool, dark place until they are used.

You will extract the carotenoid pigments with a solvent mixture that contains equal volumes of acetone and petroleum ether. Acetone is very soluble in water; thus, washing the extracts with water removes the acetone and leaves the pigments dissolved in the petroleum ether layer. Low-boiling petroleum ether (b.p. $\sim 35-60\,°C$) is preferred for the extraction because it can be evaporated readily at room temperature to produce a concentrated solution of the pigments. This concentrate is then transferred to the top of a chromatography column, which should be packed with neutral alumina (Brockmann grade II–III) by the slurry method described in OP–16. If you use a 25-mL buret as a column, it should be filled to a depth of about 15 cm, which should require about 15–20 g of alumina. You will elute the pigments with high-boiling petroleum ether followed by a more polar mixture containing a little acetone. Lycopene, with its 13 double bonds, adheres to alumina more strongly than do the carotenes, which have 11 or 12 double bonds. Therefore, the yellow carotene band will be eluted ahead of the orange-red lycopene band. Xanthophylls contain polar hydroxyl groups and will thus trail behind lycopene as it passes down the column. The lycopene sample for spectrometric analysis should be collected from the more concentrated center of the orange-red band. If desired, the rest of the lycopene-containing eluate can be evaporated for recovery of the pigment (Collateral Project 1).

Iodine catalyzes the light-induced isomerization of all-*trans*-lycopene to 13-*cis*-lycopene (neolycopene A). The isomerization reaction can be detected by recording the ultraviolet-visible spectrum of the lycopene-containing eluate before and after adding iodine. Because 13-*cis*-lycopene absorbs at slightly lower wavelengths than *trans*-lycopene does, the absorption bands in the visible region of the spectrum should broaden and shift position slightly after isomerization. In the ultraviolet region, a characteristic "*cis* peak" associated with the bent geometry of *cis*-lycopene molecules should appear at a wavelength where all-*trans*-lycopene shows little absorption. After you record the initial spectrum, you can induce isomerization by adding iodine solution to the sample, replacing the cuvette in the spectrometer, and monitoring the sample's absorbance with the wavelength fixed at 475 nm. The 475-nm light should bring the isomerization to completion in a few minutes, and you can record the spectrum when the absorbance reading has stabilized. Alternatively, you can induce isomerization by leaving the iodine-treated solution in a well-lighted place for 15 minutes or so before you record the second spectrum.

If your ultraviolet-visible spectrum does not change as a result of the iodine addition, the lycopene in your tomato paste may have isomerized while the tomatoes were being processed. On the other hand, experimental errors and delays could be responsible for the alteration of the pigment. Before drawing any conclusions about the quality of your tomato paste, you should compare your results with those of other students who used the same brand.

The equilibrium mixture contains nearly equal parts of *trans*-lycopene and 13-*cis*-lycopene.

Green leaves contain a number of colored pigments that can be separated by chromatography. Minilab 5 describes the use of paper chromatography to separate spinach pigments, which may include two kinds of carotene (yellow-orange), four xanthophylls (yellow), chlorophyll a (blue-green), and chlorophyll b (green). With most developing solvents the carotenes move up the paper fastest, followed by the xanthophylls and chlorophylls; individual pigments are not resolved in all cases.

PRELAB ASSIGNMENTS

1.   Read OP – 13*d,* OP – 16, and OP – 35, and review the other operations as necessary. Read OP – 18 and OP – 22 if you will be doing Minilab 5.

2.   Prepare an experimental plan for this experiment.

PROCEDURE

With your instructor's permission, you can choose a brand of tomato paste to use in this experiment. This will allow you to compare your results with those of students using other brands.

SAFETY PRECAUTIONS

Petroleum ether is extremely flammable and may irritate the eyes, skin, and upper respiratory tract. Minimize contact with the liquid, do not breathe its vapors, and keep it away from flames or hot surfaces.

Acetone is very flammable, and its vapor can irritate the eyes and upper respiratory tract. Keep the liquid away from flames, and do not breathe its vapors.

petroleum ether   acetone

The extraction mixture should be prepared using low-boiling petroleum ether (b.p. ~35–60°C). High-boiling petroleum ether (b.p. ~60–80°C) or hexane should be used for the chromatography.

Extraction of Pigments from Tomato Paste.   Weight about 4 g of tomato paste into a small beaker. Extract the solid material ⟨OP – 13*d*⟩ three times with 10-mL portions of a 50% (by volume) mixture of acetone and petroleum ether **[fire hazard!]**, each time filtering the extract through fluted filter paper ⟨OP – 11⟩ into a small Erlenmeyer flask. After each extraction, decant the liquid extract onto the filter, pressing the residue in the beaker with a flat-bladed spatula to squeeze out as much liquid as possible. After the third extraction, transfer the residue to the filter and wash it with 5 mL of the extraction solvent, combining the wash liquid with the extracts. Wash the combined extracts ⟨OP – 19⟩ with 25-mL of saturated sodium chloride solution followed by 25 mL of 10% aqueous potassium carbonate and then 25 mL of water. Dry the petroleum ether layer ⟨OP – 20⟩ with magnesium sulfate, and concentrate the solution to a volume of 1 – 2 mL by evaporating most of the petroleum ether under vacuum *without* heating ⟨OP – 14⟩.

If you evaporate the solution to dryness, dissolve the residue in 1 mL of petroleum ether.

Separation and Isolation of Lycopene.   Prepare an alumina column for chromatography ⟨OP – 16⟩, using high-boiling petroleum ether **[fire hazard!]** as the column-packing solvent. Transfer the pigment concentrate to the column and elute with high-boiling petroleum ether until the yellow carotene band begins to drain out of the column. Change to a 10% solution of acetone in petroleum ether, and elute the orange-red lycopene band, collecting a 5-mL sample of eluate from the center of the lycopene band in a small vial. (Collect the rest in a flask and save it for Collateral Project 1, if desired.) Use this midband sample for the spectral analysis *as soon as possible.*

Keep the column away from strong light.

If you cannot record its spectrum on the same day, store your sample in a tightly closed container in a cool, dark place.

Spectral Analysis and Isomerization of Lycopene.   Record an ultraviolet-visible spectrum ⟨OP – 35⟩ for the midband sample of lycopene over a wavelength range extending from 600 nm in the visible region to 250 nm in the ultraviolet region. If necessary, dilute the lycopene solution with more petroleum ether to keep the strongest peak (at about 475 nm) on scale. Mix a drop of a 0.025% solution of iodine in hexane into the sample, and leave it in strong light for about 15 minutes or in the sample beam at 475 nm until its absorbance remains constant (about 2 minutes). Then record another ultraviolet-visible spectrum over the same wavelength range as before.

Report.   From your spectra, decide whether or not isomerization has taken place. Describe the evidence supporting your conclusion. Try to locate the *cis* peak described in the Methodology section, and give its λ_{max} value. If possible, compare your results with those of other students using different brands of tomato paste; discuss any similarities or differences. Turn in your spectra (or photocopies) with your report.

Collateral Projects

1. Concentrate the lycopene-containing eluate to a small volume, and cool it to obtain crystalline lycopene. Its melting point should be about 175°C.

2. Isolate carotene from strained carrots by the same general procedure, collecting the yellow band (or bands) that elute with petroleum ether. Record ultraviolet-visible spectra for carotene before and after iodine treatment, and see if there is any evidence of isomerization.

MINILAB 5

Separation of Spinach Pigments by Paper Chromatography

SAFETY PRECAUTION

Petroleum ether and acetone are very flammable; toluene is flammable, and its vapors are harmful. Keep the extraction and elution solvents away from flames and hot surfaces; avoid contact, and do not breathe their vapors.

To prevent decomposition of the pigments, the development should be carried out as rapidly as possible in a place away from strong light.

In a mortar, grind about 2 g of dry, finely chopped spinach leaves with 1 mL of acetone until you have a smooth paste. Add 2 mL of acetone and 2 mL of petroleum ether. Extract ⟨OP–13d⟩ the pigments from this mixture, collecting the extract by vacuum filtration (use a small sintered-glass funnel, if available). Wash the extract by *swirling* it with 5 mL of saturated aqueous sodium chloride in a test tube ⟨OP–13a⟩ (avoid vigorous shaking, which may oxidize the pigments). Remove the aqueous (lower) layer with a capillary pipet; then wash the organic layer twice with 5-mL portions of water to remove any remaining acetone. Without delay, use the extract to spot a strip of chromatography paper ⟨OP–18⟩ and develop the chromatogram with 400:1 toluene-methanol. Locate the carotene, chlorophyll, and xanthophyll spots if you can and measure their R_f values.

Exercises

1.   Estimate the concentration of the lycopene solution you analyzed by spectrometry, given that the molar absorptivity of lycopene at 473 nm is 1.86×10^4. Estimate the mass of lycopene in that 5-mL sample.

2.   **(a)** Draw the structure of the 7-*cis*, 11-*cis*, and 13-*cis* isomers of lycopene. **(b)** Linus Pauling predicted that 13-*cis*-lycopene should be considerably more stable than the other two isomers. Explain.

3.   Explain why some hydrocarbons such as lycopene and β-carotene are colored, whereas most other hydrocarbons are not.

4.   Write a reasonable mechanism for the isomerization of lycopene to 13-*cis*-lycopene in the presence of iodine (see Experiment 12).

Library Topic

Write a report describing the formation of natural Vitamin A from carotenes, the production of the synthetic vitamin from β-ionone, and the functions of Vitamin A in the body, giving equations for the pertinent reactions. Explain why it could be hazardous to your health to eat a polar bear's liver, aside from the hazards involved in acquiring one.

Isolation and Identification of a Local Anesthetic from Cloves

Isolation of Natural Products. Infrared Spectrometry.

Operations

OP–5	Weighing
OP–13	Extraction
OP–14	Evaporation
OP–15a	Steam Distillation
OP–20	Drying Liquids
OP–26	Vacuum Distillation ⟨Minilab 6⟩
OP–33	Infrared Spectrometry

Objectives

To isolate and identify the active component of clove oil.

To learn how to separate a volatile oil by steam distillation and how to record and interpret an infrared spectrum.

To learn about some medicinal uses of plant constituents.

SITUATION

A calyx is the outer part of a flower, consisting of green, leaflike sepals.

Cloves are obtained from a small evergreen tree that grows mainly in Zanzibar, an island (now part of Tanzania) off the coast of Africa. The dried calyxes that remain after the flowers have faded are beaten off the clove trees and dried to produce the aromatic spice. Cloves contain about 16–17% by weight of an essential oil that has been used to provide relief from pain, nausea, flatulence, dyspepsia, and sluggish digestion. For centuries, people have chewed on cloves to deaden the pain of a toothache, and dentists still use clove oil as a local anesthetic. One component of clove oil appears to be responsible for most of the pain-killing activity. Your goal in this experiment is to isolate the active component of cloves and determine its chemical structure by infrared analysis.

BACKGROUND

Plants and Healing

Herbal medicine, the use of plants for healing and for preventing disease, has long been associated with witch doctors, medicine men, and old wives' remedies. In this day of modern medical miracles, the arts of the herbalist are often considered mere superstition or quackery — and there may be an element of both in herbal medicine as it is sometimes practiced. Those who seek to cure cancer by eating apricot pits rather than consulting a physician are taking a dangerous gamble. The belief that "if it doesn't cure you, at least it won't hurt you" is not always valid — herbal preparations from *Datura, Aconitum,* and *Lobelia* species are considered dangerously poisonous. Nevertheless, some botanical preparations are undoubtedly effective, and many medicines used by African witch doctors and native American medicine men have found their way onto pharmacists' shelves.

Apricot pits, like cherry pits, apple seeds, and peach pits, can contain dangerous quantities of cyanide.

The widespread notion that all modern drugs and medicines come out of test tubes is wrong on several counts. Nearly half of all prescriptions written contain at least one drug of natural origin, and most synthetic drugs were developed by modifying the molecular structures of natural substances. Many drug companies employ explorers who search for promising botanicals, often basing their explorations on reports of plant usage by native peoples. Once a sufficient quantity of a plant has been collected, an extract is prepared, and the active principles are isolated, analyzed, and tested. Some botanically derived compounds such as reserpine are used in their natural form, but in many cases the molecules are altered to prepare even more effective drugs or drugs with fewer side effects.

By making and testing numerous derivatives of a particular drug, scientists may be able to identify the parts of the molecule that are responsible for the drug's action — there may be specific groups of atoms whose location and configuration are crucial. If so, it may be possible to prepare a purely synthetic compound displaying the same type of activity, or even a useful drug with a completely different therapeutic application, by reshuffling atoms and testing the resulting molecules.

The history of local anesthetics provides a good example of this approach. Certain South American tribes have long been known to chew coca leaves to increase their endurance and allay hunger and thirst. The active principle of these leaves, cocaine, was isolated and found to be an effective local anesthetic. However, it proved much too toxic and habit-forming to use in routine applications. Once it was discovered that the activity of cocaine depends on the presence of both a lipophilic portion and a hydrophilic portion separated by an intermediate structure containing an ester linkage, synthetic analogues were prepared in an effort to develop a drug as effective as cocaine without its disadvantages. This search led through β-eucaine to procaine (Novocaine) and other modern local anesthetics such as tetracaine and lidocaine, whose structures bear little obvious resemblance to their naturally derived prototype.

Structural features of local anesthetics

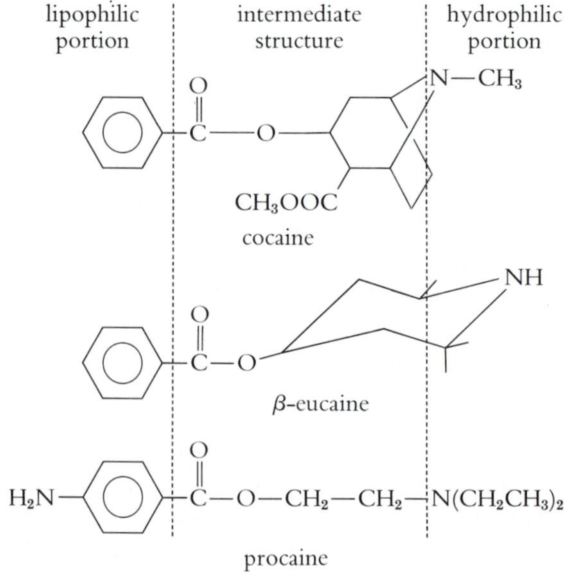

cocaine

β-eucaine

procaine

The snakeroot plant has been used for at least 3000 years as an antidote for snakebite, a cure for insanity and stomach ache, and to soothe grumpy babies.

The total synthesis of reserpine by R. B. Woodward in 1956 is considered one of the outstanding achievements of synthetic organic chemistry.

Another native remedy that was used long before its "discovery" by modern medicine is rauwolfia, a tranquilizer and anti-hypertensive drug derived from the East Indian snakeroot (*Rauwolfia serpentina*). The botanical drug first attracted the attention of medical science when it was found to cure certain kinds of insomnia and to reduce blood pressure. Its use as a tranquilizer for mentally disturbed patients arose from the observation that heart patients treated with rauwolfia acted as if they hadn't a worry in the world. Analysis of the botanical drug led to the discovery that its most active component is the alkaloid now known as reserpine. No commercially feasible process for synthesizing reserpine has yet been developed, so the drug is still derived from snakeroot extracts.

reserpine

Many plants that have been used for years by herbalists have attracted only superficial notice from medical science. The common North American weed called boneset *(Eupatorium perfoliatum)* provides a bitter-tasting tea that was a favorite early American remedy for reducing fevers; it is still used for that purpose by advocates of natural medicines. The closely related Joe Pye weed *(Eupatorium purpureum)* was named after an American Indian who gained fame using it to cure typhus. It is an effective diuretic for the treatment of kidney or bladder ailments. Euell Gibbons, in *Stalking the Healthful Herbs,* called coltsfoot *(Tussilago farfara)* "one of the finest herbal cough medicines." It makes a pleasant-tasting tea with demulcent (soothing) and expectorant properties and has been used as an ingredient in cough drops. Legend has it that Achilles, during the seige of Troy, used yarrow to treat the wounded Greeks; its botanical name *(Achillea millefolium)* takes note that of that tradition. The bruised leaves of yarrow help to stop bleeding, heal cuts, and relieve the pain of a wound. The plant has been used in emergencies by backpackers and other outdoor adventurers.

Botanical medicine is far from a lost art, since there are more different kinds of molecules in the world's natural life forms than have been synthesized in all its chemistry laboratories. Most of them have never been identified, let alone used in medicine. Botanicals should yield many new drugs to researchers who have patience to carefully separate, analyze, and evaluate their constituents.

Like other drugs, botanical medicines may cause undesirable side effects or even death if used improperly. They should be administered *only* by those skilled in their use or with sound professional advice.

In medieval times, yarrow was said to help in preventing baldness, seducing maidens, and conjuring up the devil. Modern science has yet to verify these claims.

METHODOLOGY

Most plants contain a so-called essential oil that consists of relatively volatile and often highly odoriferous components. Although many plant components can be isolated by solvent extraction or other methods, the traditional way of separating

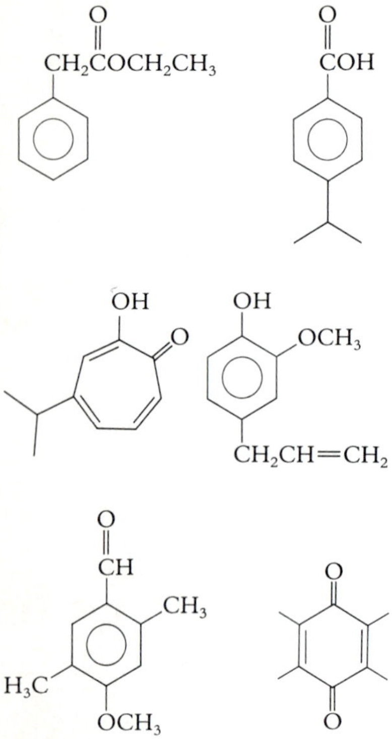

Figure 1.   Compounds with the molecular formula $C_{10}H_{12}O_2$

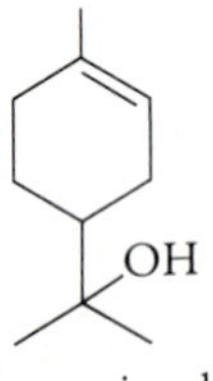

essential oils is by steam distillation. This process is preferable to ordinary distillation because the volatile components distill at temperatures below their normal boiling points, and decomposition due to overheating is reduced or prevented. The volatile oil can then be separated from the water (which codistills with it) by solvent extraction.

Because ground cloves lose their volatile components rapidly if left standing, it is best to grind fresh whole cloves just before using. Cloves can be ground quickly in an electric coffee or spice grinder, or a mortar and pestle will suffice. During the steam distillation, the presence of clove oil in the distillate will be indicated by oily droplets or cloudiness. When all the clove oil has distilled, the distillate should be as clear as pure water. The clove oil is separated from the distillate by extraction with methylene chloride.

Clove oil contains two major components as well as traces of a number of minor constituents, but the active component can be separated from the others by extraction with aqueous sodium hydroxide. This active substance is an oily liquid with the molecular formula $C_{10}H_{12}O_2$. The structures of some natural compounds having this formula are shown in Figure 1. Because these compounds have different sets of functional groups, it is possible to distinguish among them using infrared spectrometry. By observing the presence or absence of IR absorption bands corresponding to specific functional groups, you should be able to arrive at the correct structure for the active component of clove oil. For additional information about the interpretation of IR spectra, read the section titled Infrared Spectra (starting on page 525).

α-Terpineol is a natural alcohol that occurs in a number of plant oils, including pettigrain, cajeput, and long leaf pine oils. It has some therapeutic uses as an antiseptic, and it has also been used as a perfume ingredient. Minilab 6 describes the purification of a technical grade of α-terpineol by vacuum distillation.

PRELAB ASSIGNMENTS

1.   Read the Principles and Applications section of OP–15. Read OP–15a and OP–33, and read or review the other operations as necessary. Read OP–26 if you will be doing Minilab 6.

2.   Write an experimental plan for this experiment.

PROCEDURE

SAFETY PRECAUTION

methylene chloride

Weigh about 5 g of fresh whole cloves, and grind them to a fine powder using a spice grinder or a mortar and pestle. Set up an apparatus for steam distillation ⟨OP–15a⟩ using a large (250–500 mL) boiling flask, and have your instructor check the apparatus. Combine the ground cloves with 50 mL of water in the boiling flask, and steam distill ⟨OP–15a⟩ the mixture to extract the clove oil. Continue the distillation until a drop or two of the distillate, collected on a watch glass, is clear as water, with no oily droplets. Extract the clove oil from the distillate ⟨OP–13⟩ with two 20-mL portions of methylene chloride **[vapor hazard! suspected carcinogen!]** and combine the extracts.

About 150–200 mL of distillate must usually be collected before the distillate becomes completely clear.

Extract the active component of clove oil from the methylene chloride solution with two 15-mL portions of 1 M aqueous sodium hydroxide. Combine the aqueous layers, and place the methylene chloride layer in a designated solvent recovery container. Acidify the aqueous solution to litmus with 10 mL or more of 3 M hydrochloric acid. Extract this aqueous solution with two 15-mL portions of methylene chloride and combine the extracts. Dry the resulting methylene chloride solution ⟨OP–20⟩ with anhydrous magnesium sulfate. Next, evaporate the methylene chloride ⟨OP–14⟩ under vacuum until the boiling stops and the volume of the residue remains constant. Weigh the residue ⟨OP–5⟩, and record its infrared spectrum ⟨OP–33⟩.

To "acidify to litmus" means to add enough acid so that the solution turns blue litmus paper red.

The oily liquid should be reweighed a few minutes after the original weighing. If its mass decreases significantly between weighings, the evaporation should be resumed.

Report.   Using the spectrum and the list of possibilities in Figure 1, deduce the structure of the active component of clove oil. Explain how you arrived at your conclusion. Identify as many bands in the spectrum as you can. Calculate the percentage of the active component based on the mass of cloves you started with. Turn in your IR spectrum (or a photocopy) with your lab report.

Collateral Projects

1. You can isolate clove oil by evaporating the initial methylene chloride extract (omitting the aqueous sodium hydroxide extraction), and analyze it by gas chromatography to determine the approximate percentage of the active component. You can also compare the infrared spectrum of clove oil with that of its active component.

2. You can use steam distillation to obtain the essential oils from anise seed, caraway seed, and cumin seed. The initial methylene chloride extract is evaporated to yield the oil, and the aqueous sodium hydroxide extraction is omitted. Each of these essential oils contains a single major component that can be characterized by infrared spectrometry.

Purification of α-Terpineol by Vacuum Distillation

MINILAB 6

SAFETY PRECAUTION

Before you apply a vacuum, it is essential that you carefully check all glassware components of your apparatus for cracks or star fractures and replace any that appear damaged. Wear safety goggles, and keep well clear of the apparatus when you turn on the aspirator.

A technical-grade chemical is one containing an appreciable amount of impurities. Technical-grade α-terpineol from the Aldrich Chemical Company is 95% pure.

Read OP–26 thoroughly before you start. Measure the temperature of the aspirator water after the aspirator has run for a while, and estimate the minimum pressure the aspirator can attain. Use the nomograph in Figure 1 of OP–29 (page 729) to estimate the boiling temperature of α-terpineol at that pressure, based on its normal boiling point of 220°C. (Keep in mind that this is a *minimum* boiling temperature; the compound will actually distill at a somewhat higher temperature.) Set up a vacuum distillation apparatus ⟨OP–26⟩ and add about 25 mL of technical-grade α-terpineol. If a manometer is available, measure the actual pressure inside the apparatus, and use it to make a more accurate estimate of the boiling point. Purify the α-terpineol by vacuum distillation, collecting the main fraction over a reasonably narrow boiling range. Pour the purified product into a tared vial and cool it in ice (if necessary) to see whether it will solidify; pure α-terpineol is reported to solidify at around 30°C.

Exercises

1. Derive a systematic name for the active component of clove oil, and use this to find its common name in the *Merck Index* or another reference book.

2. **(a)** What property of the active component of clove oil made possible its separation from the other components by the extraction process you used? Is this consistent with the structure you chose for it? Explain. **(b)** Write equations for the chemical reactions involved in the extraction and the subsequent acidification of the extract.

3. Clove oil also contains a compound having the formula $C_{12}H_{14}O_3$, which can be hydrolyzed to yield the active component $(C_{10}H_{12}O_2)$ and acetic acid. Write its structure and a balanced equation for this reaction.

4. Construct a flow diagram for this experiment, showing how the active component is separated from the solid part of the cloves and the other components mentioned in the Methodology section.

Library Topic

Some physiologically active chemicals derived from common plants are podophyllotoxin (from May apple, *Podophyllum peltatum*), sanguinarine (from bloodroot, *Sanguinaria canadensis*), and apocynin (from hemp dogbane, *Apocynum cannibinum*). Write a report on these compounds, giving their structures and describing some of their uses.

Hemp dogbane has also been called American hemp, Canadian hemp, and Indian hemp, but it is not related to true Indian hemp, which is marijuana. An extract of hemp dogbane that was administered during a heart attack to the American president Benjamin Harrison probably saved his life.

Identification of Unknown Ketones by NMR Spectrometry and Thin-Layer Chromatography

Qualitative Analysis. Thin-Layer Chromatography. NMR Spectrometry. Ketones.

Operations

OP–5   Weighing

OP–6   Measuring Volume

OP–12a Semimicro Vacuum Filtration

OP–17 Thin-Layer Chromatography

OP–21   Drying Solids

OP–23a Semimicro Recrystallization

OP–23b Recrystallization from Mixed Solvents

OP–28   Melting-Point Determination

OP–34 Nuclear Magnetic Resonance Spectrometry

Objectives

To identify a ketone by the preparation and TLC analysis of a derivative.

To identify a ketone from its PMR spectrum.

To learn how to carry out a TLC separation and how to record a PMR spectrum.

To learn how to use semimicro laboratory techniques to prepare derivatives and how to perform a mixed-solvent recrystallization.

To learn how chemistry can be used in the solution of crimes.

SITUATION

Crime laboratories are called on to identify a variety of commercial products, such as the ink on a spurious document or the flammable solvent found at the scene of a suspicious fire.

Because aliphatic ketones are excellent solvents for many organic substances, they occur in many consumer products, for example, nail-polish removers and lacquer thinners. Ketones such as 2-butanone and 4-methyl-2-pentanone are used as industrial solvents. Because of their flammability, any of these ketones could be implicated in accidental or deliberately set fires. In this experiment, you will attempt to identify two ketones that might turn up in the laboratory of a forensic chemist.

BACKGROUND

Crime and Chemistry

Forensic chemistry is chemistry applied to the solution of crimes. It deals with the analysis of materials that were used in committing a crime or that were inadvertently left at the scene. Forensic chemistry may also involve the analysis of illicit drugs such as heroin or LSD and their **metabolites** and the measurement of alcohol or other intoxicants in body fluids.

drug metabolites: chemical compounds formed in the body as a result of the chemical or enzymatic breakdown of drugs

An arson accelerant is a flammable substance used to start a fire and cause it to spread rapidly.

A material used in committing a crime might be an arson accelerant used to set a fire, a toxin used in a deliberate poisoning, or an explosive used in a terrorist bombing. Such materials can be identified and sometimes traced to a particular source. Materials found at the scene of a crime might include contact traces, pieces of fiber from clothing, or particles of dust or soil. Contact traces, such as the chips of paint or glass found at the scene of a hit-and-run accident, can be analyzed both chemically and under the microscope to determine the make and model of the car involved. Clothing can be traced by the dyes contained in fibers, and dust and soil particles may link a criminal to a particular occupation or location.

The conviction of a suspect requires that evidence be presented to establish, first, that a crime has actually been committed, and second, that the suspect is connected with it. In an arson case, for example, this ordinarily requires that the incendiary origin of a fire be established before introducing proof showing who did it. One way of establishing that a fire was deliberately set is to prove that an accelerant was used to intensify the fire and allow it to spread rapidly. Because fires burn upward from the point of origin, some accelerant may be recovered later if it has soaked downward into flooring, rags, paper, or other porous materials. If the investigator traces a fire to its point of origin, he or she can often collect samples of material containing the accelerant, which are placed in airtight containers such as large paint cans. A forensic chemist can then separate the accelerant from the debris by, for instance, steam distillation or extraction. Once the accelerant has been isolated, it is usually classified according to chemical type (gasoline, turpentine, etc.) by an instrumental method such as gas chromatography.

In some cases, it may be necessary to match the sample with a liquid in the suspect's possession or with a commercial product. Since the more volatile components of an accelerant are evaporated and burned more rapidly in a fire, a sample of accelerant taken from the scene is not likely to have exactly the same composition as the control material. Therefore, the control is evaporated slowly and tested at various stages to see whether its composition at any point duplicates that of the recovered material. With this procedure it is often possible to determine the brand and grade of gasoline or other accelerant used. Some widely used flammable liquids, such as the industrial solvent methyl ethyl ketone (2-butanone), contain only one major component. These substances can be identified by chemical or spectrometric methods, or a combination of both.

One of the most powerful tools of the forensic chemist is thin-layer chromatography, which provides a rapid, sensitive means of analyzing many different materials. The dyes used to color gasoline can be characterized by the pattern of spots they produce on a TLC plate, making it possible in some cases to trace an accelerant used for arson to its source. The U.S. Treasury Department maintains a library of pen inks catalogued according to their TLC dye patterns, allowing an investigator to match the ink on a document with one on file. In a few cases, TLC analysis has proven that the ink used to fraudulently back-date a document did not even exist on the date in question! Substances suspected of being illicit drugs are frequently screened by TLC; drugs that are mixtures of several substances, such as marijuana, often produce telltale patterns that are easily recognized. A thin-layer chromatogram alone may not be sufficient to establish the identity of a suspect material, but it can narrow down the list of possibilities and thus lead to the positive identification of the material by other means.

METHODOLOGY

In this experiment, you will use thin-layer chromatography to characterize an unknown ketone, whose identity will be confirmed by taking the melting point of a derivative. You will then use an extremely powerful analytical tool, NMR spectrometry, to establish the identity of a second ketone. You can gain additional experience with TLC by characterizing an over-the-counter drug ⟨Minilab 7⟩ and with NMR by identifying an unknown aromatic hydrocarbon ⟨Collateral Project⟩.

Thin-layer chromatography is particularly useful for identifying complex mixtures, but it can also assist in the identification of pure compounds. When a TLC plate spotted with structurally similar organic compounds is developed with an appropriate solvent, the R_f values obtained vary more or less regularly with chain length. This correlation is shown in Table 1 for a series of carboxylic acids. When an unknown compound may be one of a limited number of compounds, a TLC plate is spotted with samples of the unknown and the most likely known compounds. When the plate is developed, it may be possible to match the R_f value of the unknown compound's spot with that of a known compound.

Your first unknown will be a member of the homologous series of methyl ketones represented by the formula $CH_3CO(CH_2)_nCH_3$. Ketones are often converted into 2,4-dinitrophenylhydrazones or other colored derivatives for TLC analysis; the colored spots are easily located after the plate is developed, making a visualizing reagent unnecessary. By measuring the melting point of your derivative and comparing its R_f value with those of known derivatives, you should be able to identify your unknown as one of the ketones in Table 2.

A preparation of a derivative is actually a small-scale organic synthesis, which often produces only a fraction of a gram of product. The usual purpose of such a preparation is to convert an unknown organic compound to a pure crystalline prod-

Each compound in a homologous series differs from the previous member by one structural unit, usually a methylene (CH_2) group. The members of a homologous series are called homologs of one another.

Table 1. R_f values from TLC of homologous carboxylic acids

Carboxylic Acid	Number of Carbon Atoms	R_f
methanoic (formic) acid	1	0.07
ethanoic (acetic) acid	2	0.13
propanoic acid	3	0.30
butanoic acid	4	0.40
pentanoic acid	5	0.50
hexanoic acid	6	0.57
heptanoic acid	7	0.60
octanoic acid	8	0.66

Note: on silica gel, developed with a 19:1 mixture of methyl acetate and 2.5% ammonia

uct having a sharp melting point, so that the melting point can be used to identify the compound. 2,4-Dinitrophenylhydrazones are prepared by simply combining an unknown carbonyl compound with DNPH reagent, which contains 2,4-dinitrophenylhydrazine and sulfuric acid in aqueous ethanol. The crystalline product can be separated from the reaction mixture by vacuum filtration and purified by recrystallization, using small-scale (semimicro) apparatus. Most 2,4-dinitrophenylhydrazones dissolve too readily in ethanol and too sparingly in water for either of these liquids to be a good recrystallization solvent. Thus, you will use a mixture of the two. To obtain the right mixture, you will dissolve the derivative in the better solvent (ethanol) and then add just enough of the poorer solvent (water) to saturate the hot solution.

Modern instruments such as IR and NMR spectrometers can greatly reduce the time and effort required for the positive identification of an unknown. In the second part of this experiment, you will use a proton NMR (PMR) spectrum to identify a saturated ketone that has four to six carbon atoms. If an NMR spectrometer is available, you can record the spectrum yourself, with help from your instructor. Otherwise, your instructor will give you a PMR spectrum of the unknown. In Part III of this book, you will find an introduction to NMR spectral analysis that will help you interpret your spectrum and deduce the structure of the unknown. For additional information, consult the textbook for the lecture course or an appropriate source from category F of the Bibliography.

PRELAB ASSIGNMENTS

1. Read OP–12*a*, OP–17, OP–23*a*, OP–23*b*, and OP–34, and read or review the other operations as necessary. Read the NMR Spectra section of Part III (beginning on page 540).

2. Prepare a brief experimental plan for this experiment.

Reactions and Properties

ketone   2,4-dinitrophenylhydrazine   2,4-dinitrophenylhydrazone

Table 2.   Physical properties and derivative melting points for homologous methyl ketones

Ketone	M.W.	b.p.	d.	Derivative m.p.
2-propanone (acetone)	58.1	56	0.791	126
2-butanone	72.1	80	0.805	117
2-pentanone	86.1	102	0.809	143
2-hexanone	100.2	128	0.811	106
2-heptanone	114.2	151	0.811	89
2-octanone	128.2	173	0.819	58

PROCEDURE

With the instructor's permission, several students can work together on the derivative preparations, each preparing two or more derivatives of known ketones for TLC analysis. Students are responsible for their own unknowns.

The ketones are flammable and may be harmful if ingested, inhaled, or absorbed through the skin. Avoid contact with the liquid ketones, do not breathe their vapors, and keep them away from flames and hot surfaces.

2,4-Dinitrophenylhydrazine is harmful if ingested or absorbed through the skin, and the DNPH reagent contains sulfuric acid, which can burn the skin and eyes. Avoid contact with the DNPH reagent, and wash your hands after using it.

Ethyl acetate and the TLC solvent are very flammable, and inhalation, ingestion, or skin absorption may be harmful. Avoid contact with the liquids, do not breathe their vapors, and keep them away from flames.

Like ordinary chloroform, deuterochloroform is toxic and may be carcinogenic. The amount needed for NMR analysis is small enough that the hazard potential is reduced considerably, but you should take care to avoid contact and inhalation.

SAFETY PRECAUTIONS

ketones

ethyl acetate   TLC solvent

A. Identification of an Unknown Methyl Ketone by TLC

Preparation of 2,4-Dinitrophenylhydrazones. Obtain an unknown methyl ketone from your instructor and record its number in your laboratory notebook. Using a graduated pipet ⟨OP–6⟩, measure 0.20 mL of the unknown into a 15-cm test tube **[do not pipet by mouth!]**, dissolve it in 3 mL of 95% ethanol, and stir in 7.0 mL of the 2,4-dinitrophenylhydrazine reagent **[contact hazard!]**. Set the test tube aside for 15 minutes or until crystallization is complete. Collect the derivative by semimicro vacuum filtration ⟨OP–12a⟩. Using semimicro apparatus, recrystallize ⟨OP–23a⟩ the derivative from an ethanol-water mixed solvent ⟨OP–23b⟩. Dry the derivative ⟨OP–21⟩, weigh it ⟨OP–5⟩, and measure its melting point ⟨OP–28⟩.

Use a cold 3 : 1 mixture of ethanol and water to wash the derivative each time it is collected by vacuum filtration.

Clean and label as many small test tubes as there are known methyl ketones available (see Table 2). Measure 1 mL of the 2,4-dinitrophenylhydrazine reagent into each test tube and add a drop of the appropriate methyl ketone. Set the test tubes aside until crystallization is complete, then collect the crystalline derivatives by semimicro vacuum filtration ⟨OP–12a⟩.

TLC Separation of 2,4-Dinitrophenylhydrazones. Dissolve approximately 10 mg (0.01 g) of the unknown ketone's derivative in 0.5 mL of ethyl acetate, using a spot plate or small labeled test tube. Do the same for each known derivative. Use each solution to spot a silica gel TLC plate ⟨OP–17⟩ and develop the plate using a 3 : 1 mixture of toluene and petroleum ether **[fire and vapor hazard!]** as the developing solvent.

B. Identification of an Unknown Ketone by NMR

Obtain a second unknown ketone (or its PMR spectrum) and record its number in your laboratory notebook. If an NMR spectrometer is available for your use, make up a solution of the unknown in deuterochloroform **[vapor hazard! suspected carcinogen!]** using a TMS standard, and record and integrate its PMR spectrum ⟨OP–34⟩.

Report. Calculate R_f values for all the ketone derivatives from your thin-layer chromatogram. Tabulate the R_f values and all significant spectral parameters from your PMR spectrum. Identify the unknown from part A and tell how you arrived at your conclusion. Deduce the structure of the unknown from part B and justify your conclusion, showing which set of protons is responsible for each signal and explaining the splitting patterns, chemical shift values, etc. Turn in your NMR spectrum and TLC plate (or copies of each) with your report.

Collateral Project

Obtain an unknown aromatic hydrocarbon from your instructor and record its PMR spectrum in deuterochloroform or another suitable solvent. Unless your instructor informs you otherwise, the hydrocarbon will contain 89.49% carbon and 10.51% hydrogen. Deduce its structure, and explain how you arrived at your conclusion.

TLC Analysis of an Analgesic Drug   MINILAB 7

Methylene chloride is suspected of causing cancer in test animals; avoid contact, and do not breathe its vapors. Do not breathe the vapors of the TLC solvent, and minimize your contact with the liquid.

**SAFETY
PRECAUTION**

Obtain an unknown analgesic drug from your instructor and grind one tablet to a fine powder in a mortar. Extract the drug ⟨OP – 13d⟩ with 5 mL of methylene chloride [**suspected carcinogen!**], and use the extract to spot a 10 × 10 cm silica gel TLC plate in two or more places, on a line 1.5 cm from the bottom and 1.5 cm in from each edge ⟨OP – 17⟩. Spot the plate along the same line with previously prepared 1% w/v (weight/volume) solutions of aspirin, acetaminophen, phenacetin, caffeine, and salicylic acid in methylene chloride. Develop the chromatogram with a 2 : 9 : 9 mixture of 1-butanol, 2-butanone, and ethyl acetate. Make the spots visible using either a "black light" (if the TLC plate contains a fluorescent indicator) or iodine vapors, and measure the R_f values of all the spots. Take note of any characteristics of the spots under the ultraviolet light that may assist in identification. Identify as many components of the unknown analgesic as you can.

Better results will be obtained if the TLC plate has a fluorescent indicator.

Exercises

1. **(a)** Write a balanced equation for the reaction of your unknown methyl ketone with 2,4-dinitrophenylhydrazine. **(b)** The DNPH reagent contains 2.9 g of 2,4-dinitrophenylhydrazine in 100 mL of solution. What was the limiting reactant for the preparation of your 2,4-dinitrophenylhydrazone? Calculate the theoretical yield and percentage yield of the reaction.

2. Describe and explain any relationship between chain length and R_f value observed for your TLC separation.

3. Draw structures for all ketones having the molecular formula $C_5H_{10}O$, and sketch the PMR spectrum you would expect to obtain from each one. These spectra should show the relative area, multiplicity, and approximate chemical shift of each signal.

4. Account for the fact that the 2,4-dinitrophenylhydrazone of 2-octanone melts at a lower temperature than does that of acetone, even though the molecular weight of the 2-octanone derivative is much higher.

Library Topic

Write a brief report describing some of the methods used to test for drugs and their metabolites in body fluids. For example, you might tell how the presence of proscribed drugs can be detected in professional athletes or race horses.

CORRELATED LABORATORY EXPERIMENTS

The experiments in Part II are correlated with topics discussed in most introductory textbooks of organic chemistry. The correlations are noted in the list of topics preceding each experiment. In Part II, frequently used elementary operations such as heating and weighing will no longer be listed at the beginning of the experiments or flagged in the procedures. Before each laboratory period, you will be expected to read the experiment, read or review the operations, and prepare an experimental plan. Any additional requirements will be specified in the Prelab Assignments section.

Isomers and Isomerization Reactions

Reactions of Alkenes. Isomerization Reactions. Isomerism.

Operations

OP–11 Gravity Filtration

OP–12 Vacuum Filtration

OP–12*a* Semimicro Vacuum Filtration

OP–21 Drying Solids

OP–28 Melting-Point Determination

Objectives

To convert phenylammonium cyanate to its structural isomer, phenylurea.

To convert dimethyl maleate to its geometric isomer, dimethyl fumarate.

To learn how to construct molecular models of structural and geometric isomers.

To compare the properties of some structural and geometric isomers.

To gain experience in drawing structural formulas and designating configurations of geometric isomers.

SITUATION

In the early years of chemistry, it was thought that no two compounds could have the same elemental composition; there is, after all, only one kind of H_2SO_4 and one kind of NaCl. So when two independent investigators reported the same composition for cyanic acid (HOCN) and fulminic acid (HONC), J. J. Berzelius — the foremost chemist of his time — thought one of them had made a mistake. But after further investigation Berzelius was persuaded that a specific set of atoms can combine in different ways to form compounds having different properties, and he coined the term *isomerism* to describe the

The term *isomerism* is derived from Greek words meaning "equal parts" and refers to the fact that isomers have the same number and kind of atoms.

phenomenon. At about the same time, Friedrich Wohler converted ammonium cyanate to urea in a pivotal experiment that helped to demolish the "vitalistic" theory of organic chemistry. In this experiment, you will explore some of the ways in which different molecules can be built up from the same set of atoms and compare the properties of certain isomers. You will also carry out two isomerization reactions, the conversion of phenylammonium cyanate to its structural isomer, phenylurea, and of dimethyl maleate to its geometric isomer, dimethyl fumarate.

NH_4OCN	H_2NCONH_2
ammonium cyanate	urea

BACKGROUND

Isomers and Molecular Models

Early in the nineteenth century, most scientists believed that plants and animals possessed some "vital force," which allowed them to convert inorganic substances into organic compounds. Since this vital force was presumably absent in inanimate objects, it followed that organic compounds could not possibly be synthesized from inorganic substances in the laboratory. Vitalism received a serious blow in 1828, when Friedrich Wohler tried to make ammonium cyanate by mixing silver cyanate with ammonium chloride and came up with urea instead. This "lucky accident" was the first known synthesis of an organic compound from an inorganic one to take place outside a living organism; as Wohler put it in a letter to Berzelius, "I can make urea without the aid of kidneys, either man or dog!" Wohler's synthesis of urea suggested that there is no essential difference between inorganic and organic compounds, and it paved the way for the flowering of organic chemistry in the last half of the nineteenth century.

Urea is a product of protein metabolism that is excreted in the urine.

Ammonium cyanate and urea share the same molecular formula, CH_4N_2O, and are therefore isomers. Isomers whose molecules differ in the way their atoms are attached to one another are called *structural isomers;* those that differ only in the way their atoms are arranged in space are called *stereoisomers.* *Geometric isomers* are stereoisomers that differ from one another as a result of restricted rotation about the bonds connecting two or more atoms. For example, dimethyl maleate and dimethyl fumarate (see the Reactions and Properties section) are geometric isomers because their carbon-carbon double bonds ordinarily prevent their interconversion. A geometric isomer having comparable groups on the same side of the double bond is called a *cis* isomer, and one with such groups on opposite

cis-1,2-dichloro-
cyclopropane

trans-1,2-dichloro-
cyclopropane

Simple ball-and-stick models may not accurately represent bond lengths, bond angles, or relative atomic sizes; "space-filling" models give a better picture of true molecular shape.

sides is a *trans* isomer. A cyclic compound having two or more substituents other than hydrogen on the ring may also exhibit geometric isomerism, because rotation about the bonds connecting the atoms in the ring cannot occur without breaking the ring apart.

Just as zoology is the study of animals and botany the study of plants, organic chemistry can be regarded as the study of carbon-containing molecules. Whereas certain plants and animals are familiar to everyone, because we see them all around us, molecules are not. We can't see a molecule — the best we can do is to imagine what one looks like, and molecular models can help us do that. All models have their limitations. A plastic model of a Sopwith Camel is a far cry from the real thing, but it can show us what that airplane looks like in three dimensions. A real molecule is obviously not made of colored balls, but an accurate molecular model can depict the three-dimensional structure of a molecule far better than any drawing. In this experiment, you will use molecular models to help you see — in three dimensions — the structural and geometric differences between isomers.

METHODOLOGY

Wohler prepared urea by mixing ammonium chloride and silver cyanate and then evaporating the resulting ammonium cyanate solution to dryness. When heated, ammonium cyanate decomposes to ammonia and cyanic acid, which combine to form urea.

$$NH_4Cl + AgOCN \longrightarrow NH_4OCN + AgCl$$
$$\text{ammonium}$$
$$\text{cyanate}$$

$$NH_4OCN \xrightarrow{\text{heat}} NH_3 + HOC\equiv N \longrightarrow H_2N\overset{\overset{\displaystyle O}{\|}}{C}NH_2$$
$$\text{cyanic} \qquad \text{urea}$$
$$\text{acid}$$

You will attempt to isomerize the phenyl derivative of ammonium cyanate to phenylurea by a similar reaction. Aniline reacts with hydrochloric acid to form the organic salt phenylammonium chloride (see the Reactions and Properties section), which is also called aniline hydrochloride. By combining an aqueous solution of phenylammonium chloride with potassium cyanate, you can prepare a solution containing phenylammonium cyanate, which changes to phenylurea within a few minutes. The precipitated phenylurea is purified by heating, decolorizing, and filtering the reaction mixture. Pure phenylurea forms needlelike crystals whose identity can be verified by a melting-point determination.

Sometimes one of a pair of geometric isomers can be converted to the other by conditions that cause temporary cleavage of a pi bond. When dimethyl maleate is heated with a trace of bromine in the presence of light, some bromine molecules break apart into bromine atoms. A bromine atom then adds to one end of the double bond of a dimethyl maleate molecule, temporarily breaking the pi bond. This allows the molecule to rotate freely about the remaining sigma bond; if the bromine atom is ejected when the $COOCH_3$ groups are in just the right position, the pi bond reforms to yield a molecule of dimethyl fumarate.

$(Me = CH_3)$

You will use molecular models to help you derive and visualize the structures of some isomers. The most popular molecular models are the ball-and-stick type, in which balls of different colors represent different kinds of atoms and rigid or flexible connectors represent the bonds that hold the atoms together in molecules. Each kind of ball is drilled with a number of holes equal to the valence or combining capacity of the atom it represents. Thus, balls representing carbon atoms — colored black — are drilled with four holes arranged tetrahedrally, whereas those representing oxygen atoms have two holes, and those representing hydrogen atoms have one. Building models of all the isomers that have a given molecular formula is simply a matter of putting the appropriate colored balls together in all possible combinations, using just enough connectors to fill all the holes. Often a large number of isomers will share the same molecular formula, and it may take a considerable amount of ingenuity to find them all!

Balls representing nitrogen atoms may have four or five holes to allow the construction of ammonium salts; for most organic compounds only three of the holes are used.

PRELAB ASSIGNMENTS

You are expected to read the experiment, read or review the operations, and prepare an experimental plan before doing this and all subsequent experiments in Part II. Additional assignments will be specified in this section.

1. Calculate the mass of 22 mmol of potassium cyanate and the theoretical yield of phenylurea.

2. If it is available, read the instruction booklet for the molecular model kit you will use in the lab.

Reactions and Properties

A. $PhNH_2 + HCl \longrightarrow PhNH_3Cl$
   aniline   phenylammonium
                                   chloride

   $PhNH_3Cl + KOCN \longrightarrow PhNH_3OCN$
                              phenylammonium
                                   cyanate

$$PhNH_3OCN \longrightarrow PhNH\overset{\overset{\textstyle O}{\|}}{C}NH_2$$
                         phenylurea

B.

dimethyl maleate $\xrightarrow{Br_2}$ dimethyl fumarate

Table 1.   Physical properties

	M.W.	m.p.	b.p.	d.
aniline	93.1	−6	184	1.022
phenylammonium chloride	129.6	198	245	
potassium cyanate	81.1	800 d		
phenylurea	136.2	147	238	
dimethyl maleate	144.1	8	205	1.151
dimethyl fumarate	144.1	102	193	

Note: d = decomposes upon melting

PROCEDURE

Aniline is poisonous and can cause serious injury or death if swallowed, inhaled, or absorbed through the skin; it may also cause allergic skin reactions. Although human tests have so far been negative, aniline is a suspected carcinogen. Wear protective gloves and dispense under a fume hood; avoid contact and do not breathe vapors.

SAFETY PRECAUTION

aniline

A. Isomerization of Phenylammonium Cyanate to Phenylurea

Under the hood, prepare a phenylammonium chloride solution by combining 10 mL of 3 M hydrochloric acid with 10 mL of water in a 150-mL beaker and adding 2.0 mL (~22 mmol) of recently distilled aniline **[poison! suspected carcinogen!]**. Dissolve 22 mmol of potassium cyanate in 20 mL of water in a small beaker, and mix this solution with the phenylammonium chloride solution to form phenylammonium cyanate. Stir for 2–3 minutes, then add 15 mL of water and heat the solution to 80°C to complete the isomerization and dissolve the phenylurea. Add about 0.1 g of decolorizing carbon, and stir for 2 minutes at 80°C. Filter the solution by gravity ⟨OP–11⟩ while it is still hot. Let the solution cool slowly to room temperature while you proceed with part B, then cool it further in ice to complete the crystallization of phenylurea. Collect the crystals by vacuum filtration ⟨OP–12⟩ and dry them thoroughly ⟨OP–21⟩. Weigh the phenylurea and measure its melting point ⟨OP–28⟩.

It may be necessary to filter twice to remove all the carbon, using a more retentive (finer) filter paper the second time. If you filter more than once, keep the solution hot throughout the filtrations.

B. Isomerization of Dimethyl Maleate to Dimethyl Fumarate

Construct molecular models of dimethyl maleate and dimethyl fumarate and show that the two isomers cannot be interconverted without breaking bonds.

Dimethyl maleate and dimethyl fumarate may be harmful if swallowed, inhaled, or absorbed through the skin. Avoid contact with them, and do not inhale their vapors.

SAFETY PRECAUTION

Bromine is highly toxic and corrosive, and its vapors are very harmful. Avoid contact with the solution of bromine in Freon TF, and do not breathe its vapors.

Measure 1 mL of dimethyl maleate into a small test tube and mix in a drop of 1 M bromine in Freon TF [**contact and vapor hazard!**]. Set the test tube in a 150-mL beaker that is half full of boiling water, and place the beaker about 15 cm away from an unfrosted 75- or 100-watt light bulb. Keep the water boiling for 10 minutes or more with the light switched on, then remove the test tube and cool it in ice water until crystallization is complete. Collect the dimethyl fumarate by semimicro vacuum filtration ⟨OP – 12a⟩, wash it on the filter with a little cold 95% ethanol, and dry it. Measure the mass and melting point ⟨OP – 28⟩ of the dimethyl fumarate.

C. Structures and Properties of Isomers

Obtain samples of *para*-toluic acid and methyl benzoate, and describe any differences in properties that you can observe. Note particularly any differences in volatility (as indicated by the intensity of the odor) and physical state. Describe what happens when about 0.1 g of each is stirred for a few minutes with 5 mL of 5% aqueous sodium bicarbonate.

Make a model of benzene using six carbon and six hydrogen atoms. Now replace one hydrogen atom with a methyl (CH₃) group and replace the hydrogen atom on the opposite end of the ring with a carboxyl (COOH) group. (Connect the atoms of the COOH group in such a way, using multiple bonds where necessary, that all holes are occupied.) You now have a model of *para*-toluic acid. Convert this to a model of methyl benzoate by interchanging the methyl group and the hydrogen atom on the carboxyl group. Write structural formulas for both isomers.

Using the same set of atoms as before (eight carbons, eight hydrogens, two oxygens) and keeping the benzene ring intact, construct molecular models for as many additional isomers of *para*-toluic acid as you can. Write their structural formulas. Some of your structures may exhibit geometric isomerism; if so, construct models for the *cis* and *trans* isomers. Structural

benzene ring
(Kekulé structure)

Note that the Kekulé structure shown is only one resonance structure for benzene. Most molecular model sets do not permit construction of an accurate model of the benzene ring.

isomers having the same groups at different relative locations on the benzene ring are sometimes called *positional isomers;* the three possible positional isomers of toluic acid are illustrated. You need not make models for more than one positional variation of each basic structure.

Positional isomers of toluic acid

ortho   meta   para

Report.   Describe and try to explain any differences in properties that you observed for *para*-toluic acid and its isomer, methyl benzoate. Describe and write equations for any reactions you observed with sodium bicarbonate. Your report should include structural formulas for all the models you constructed. Geometric isomers should be designated as *cis* or *trans,* and their structural formulas should show clearly the spatial relationships that distinguish them.

Collateral Projects

1.   Make models for as many isomers having the molecular formula C_3H_6O as you can, and identify any pairs of structural isomers. You should also be able to construct models for a pair of molecules that (like your two hands) are nonsuperimposable mirror images of one another; these are called mirror-image isomers, or enantiomers.

2.   Record the infrared spectra ⟨OP–33⟩ or PMR spectra ⟨OP–34⟩ of *para*-toluic acid and methyl benzoate and of dimethyl maleate and dimethyl fumarate. Interpret the spectra as completely as you can, and try to explain any significant similarities and differences.

Exercises

1. Why is it possible to convert dimethyl maleate to dimethyl fumarate in good yield, but not dimethyl fumarate to dimethyl maleate?

2. It is possible to construct two different molecular models for *ortho*-toluic acid, one with a double bond between the substituted ring carbon atoms and one with a single bond there. In fact, only one form of *ortho*-toluic acid has ever been isolated. Explain.

3. **(a)** Explain why dimethyl maleate has a higher boiling point than dimethyl fumarate. **(b)** Explain why dimethyl fumarate is a solid whereas dimethyl maleate is a liquid at room temperature.

4. Propose a mechanism for the isomerization of phenylammonium cyanate to phenylurea.

Library Topic

Tell how urea is manufactured today and how it is used in the production of such commercial products as urethane plastics, urea-formaldehyde residues, barbiturates, and jet fuel. Give equations for the relevant reactions.

Properties of Representative Carbon Compounds

Functional Group Chemistry. Qualitative Analysis. Infrared Spectrometry.

Operations

OP – 6   Measuring Volume

OP – 25*a* Semimicro Distillation

OP – 29   Boiling-Point Determination

OP – 29*a* Micro Boiling-Point Determination

OP – 33   Infrared Spectrometry

Objectives

To identify the functional group in an unknown organic compound.

To measure the boiling point, density, and water solubility of an organic compound.

To learn how to carry out some classification tests.

To learn how to characterize functional groups from infrared spectra.

To learn how the physical and chemical properties of organic compounds are influenced by functional groups.

SITUATION

Chemists classify organic compounds according to the functional groups they contain. A functional group can be recognized by its chemical and physical behavior — by the way it responds to a chemical reagent or to infrared radiation, for example. In this experiment, you and your co-workers will compare the properties of a number of organic liquids and try to identify the functional group present in each of them.

BACKGROUND

Chemical Taxonomy

A botanist who wishes to identify an unknown flowering plant will generally examine the flowering parts first, to detect features that indicate to which family of plants it belongs. Then by a more detailed examination of the whole plant, he or she can determine the genus and species. For example, a plant having four symmetrical flower petals and six stamens (four long and two short) might be classified with the *Cruciferae,* or mustard, family. White flower petals and the presence of lyre-shaped leaves and a red, globular root classify the plant as a specimen of *Raphanus sativus.* An experienced gardener might immediately identify this plant as a radish that has passed its prime. Similarly, an experienced chemist might recognize some familiar organic compounds by their odors and general appearance; but for most purposes a systematic approach to identification is necessary.

    The identification, or qualitative analysis, of organic compounds is analogous in some ways to **plant taxonomy.** To classify an organic compound into a given family requires detecting a specific **functional group** on the molecules of the compound. Table 1 lists some functional groups found frequently in organic compounds. Because a compound's functional groups influence its chemical and spectral properties, chemists can usually identify those functional groups by observing the compound's chemical behavior with different classification reagents and by studying its infrared spectrum. In addition to affecting a compound's spectra and reactivity, a functional group can greatly affect such physical properties as boiling point, density, and solubility.

Raphanus sativus = garden radish

plant taxonomy: the systematic classification of plants according to their structural features and presumed natural relationships

functional group: the atom or group of atoms that is characteristic of a family of organic compounds and that determines the chemical properties of the family

Figure 1.   Boiling of a liquid

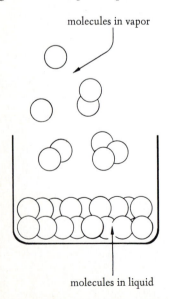

molecules in vapor

molecules in liquid

Molecular Structure and Boiling Points

Boiling occurs in a liquid when the kinetic energy of its component molecules becomes high enough to overcome the forces between them, allowing them to leave the surface of the liquid and enter the gaseous state. Since the kinetic energy of molecules increases with temperature and the energy required to separate molecules depends on the strength of the forces between them, we can expect liquids having strong intermolecular forces also to have high boiling points.

    The three important kinds of intermolecular forces occurring between organic molecules are, in order of increasing strength, (1) London dispersion forces (sometimes referred to

Table 1. Common functional groups and their families

Functional group	Functional group name	Family
—C=C—	carbon-carbon double bond	alkene
—Cl	chlorine atom	alkyl chloride
—OH	hydroxyl group	alcohol (also phenol)
$\overset{O}{\overset{\|}{—C—H}}$ (—CHO)	carbonyl group, with H on carbonyl carbon	aldehyde
$\overset{O}{\overset{\|}{—C—}}$ (—CO—)	carbonyl group, no H on carbonyl carbon	ketone
$\overset{O}{\overset{\|}{—C—OH}}$ (—COOH)	carboxyl group (*carbonyl* + *hydroxyl*)	carboxylic acid
—NH$_2$	amino group	amine (primary)

Note: Condensed representations of some of the functional groups are given in parentheses.

as "van der Waals forces"), (2) dipole-dipole interactions, and (3) hydrogen bonding. *Dispersion forces* are caused by alternating, transient charge separations on the surfaces of molecules. Although dispersion forces occur between all organic molecules, their effect is negligible in many molecules containing oxygen or nitrogen atoms, for which the polar attractive forces predominate. *Dipole-dipole interactions* result when molecules having permanent bond dipole moments line up so that the negative end of one molecule's dipole is opposite the positive end of another's, and vice versa. *Hydrogen bonding* is a special kind of dipole-dipole interaction involving the attraction of a highly polarized hydrogen atom for an electron-donating atom (such as oxygen or nitrogen) on another molecule. In organic chemistry, hydrogen bonding makes important contributions to boiling points only in compounds containing O—H and N—H bonds, and the hydroxyl compounds form the strongest hydrogen bonds.

More detailed information on the various kinds of intermolecular forces can be found in most general chemistry textbooks.

dipole-dipole interaction (formaldehyde)   hydrogen bonding (water)

Molecular Structure and Density

The density of an organic compound depends, to some extent, on the mass/volume ratio of its constituent atoms. Atoms having a high nuclear mass confined within a small atomic vol-

Approximate atomic radii and nuclear masses:

C: 77 pm, 12 u
N: 75 pm, 14 u
O: 74 pm, 16 u
Cl: 100 pm, 35.5 u

u = atomic mass unit; 1 pm = 10^{-12} m.

$CH_3CH_2CH_2CH_2CH_2CH_3$
hexane

benzene

ume, such as those in the top right-hand corner of the periodic table, have a high atomic density, so compounds whose molecules contain such atoms have comparatively high densities. Among the atoms encountered in this experiment (other than hydrogen), oxygen has the highest mass/volume ratio, followed by chlorine, nitrogen, and carbon. The density of an organic liquid will thus depend on the kind of "heavy" atoms (other than carbon and hydrogen) it contains and on the fraction of its molecular mass they contribute. For example, chlorobenzene (C_6H_5Cl) has a higher density than phenol (C_6H_5OH) even though chlorine has a lower atomic density than oxygen, because chlorine contributes 32% of the molecular mass of chlorobenzene whereas oxygen makes up only 17% of the molecular mass of phenol.

Molecular Structure and Solubility

A compound will generally be soluble in a given solvent if the forces between its own molecules are much the same as those between the molecules of the solvent, or if it can form hydrogen bonds with the solvent. Hexane easily dissolves in benzene because both are hydrocarbons whose molecules are held together by dispersion forces. Ethyl alcohol dissolves in water because both compounds form strong intermolecular hydrogen bonds. Some polar compounds that cannot form hydrogen bonds between their own molecules, such as formaldehyde, can hydrogen-bond to hydroxylic solvents such as water and are therefore soluble in those solvents.

Hydrogen bonding interactions

ethanol
and water

formaldehyde
and water

As a general rule, it can be said that polar compounds tend to dissolve in polar solvents and nonpolar compounds in nonpolar solvents; this is stated more succinctly as "like dissolves like." It must be emphasized that "solubility" is a relative term—there are different degrees of solubility. Terms commonly used to indicate the extent to which one compound dissolves in another are, in order of decreasing solubility, *miscible* (∞), *very soluble* (v), *soluble* (s), *sparingly soluble* (δ), and *insoluble* (i).

METHODOLOGY

In this experiment, you will be studying a liquid that belongs to one of the families listed in Table 1; your co-workers will characterize liquids from other families on the list. All the compounds have molecules of about the same size and mass, so differences in their physical, chemical, and spectral properties will depend primarily on their functional groups.

You should first distill your liquid to remove impurities that might affect your results. The boiling point of the liquid can be measured either by distillation or by a micro boiling-point determination. The density is measured by weighing a precisely measured volume of the liquid. The solubility is measured by shaking it with water in a small test tube. For the purposes of this experiment, a liquid will be classified as *miscible* if it dissolves in an equal volume of water and as *soluble* if 0.2 mL of the liquid dissolves in about 6 mL of water. To classify a liquid as miscible or soluble, the solution must be water-clear; separation of a second liquid layer, cloudiness, or the presence of liquid droplets (not air bubbles) in the water indicates that the liquid has not completely dissolved. Since low-boiling liquids may evaporate rapidly, you should keep the test tube stoppered while you make your observations.

You will identify the functional group present in your compound by carrying out a series of classification tests and by analyzing the compound's infrared spectrum. When dissolved in water, compounds containing the carboxyl (COOH) functional group are generally acidic enough to turn blue litmus paper red; those with an amino (NH_2) group turn red litmus blue. Primary and secondary alcohols, which have hydroxyl (OH) functional groups, react with chromic acid solution within 2–3 seconds to form an opaque blue-green suspension. Most compounds containing a carbonyl group (C=O) react with 2,4-dinitrophenylhydrazine (DNPH) to yield yellow or orange precipitates within 15 minutes. Aldehydes, compounds having the carbonyl group attached to a hydrogen atom (HC=O), also react with chromic acid, but more slowly than

Some of the families in Table 1 may not be represented by an unknown; if so, your instructor should provide the necessary data for representative compounds in those families.

Although carboxylic acids contain a C=O bond in the COOH functional group, they are not classified as carbonyl compounds and do not react with the DNPH reagent.

alcohols do. Most compounds containing a carbon-carbon double bond (C=C) react rapidly with dilute aqueous potassium permanganate to form a brown precipitate as the purple color of the permanganate disappears. Aldehydes also give a positive test with aqueous potassium permanganate, and some alcohols react slowly. Several drops of the reagent should be decolorized in a positive test. Heating a copper wire in a flame forms an oxide layer on it that reacts with halogen-containing compounds to yield copper halides, which burn with a green flame.

    Conflicting or ambiguous results may be produced with some tests because of impurities in the unknown or because the tests themselves are open to misinterpretation. In such cases, it may be necessary to repeat the tests or redistill your unknown. For additional information and general equations for the reactions, see the Classification Tests section of Part III.

    Each type of functional group is associated with one or more characteristic infrared bands that can reveal its presence. A carboxyl group (—COOH), for example, contains C=O, C—O, and O—H bonds, all of which give rise to strong infrared absorption bands. Bonds that connect the functional group to another atom or group, such as the C—H bonds adjacent to the double bond of an alkene, may also produce characteristic bands. IR bands that may help you classify your unknown include the two-pronged N—H band near $3400 - 3300$ cm^{-1} characteristic of a primary amine; the broad O—H band centered around 3300 cm^{-1} for an alcohol and 3000 cm^{-1} for a carboxylic acid; the strong C=O band near 1700 cm^{-1}; and the C—Cl band between 600 and 800 cm^{-1}. Alkenes are often characterized by =C—H bending vibrations in the $650 - 1000$ cm^{-1} region and by a C=C band (sometimes weak or absent) near 1650 cm^{-1}. The Infrared Spectra section in Part III gives additional information about the IR spectra of organic compounds.

    To gain more experience in the interpretation of IR spectra, you can identify an unknown terpenoid, as described in Minilab 8.

For example, an aldehyde often contains traces of the corresponding carboxylic acid as an impurity, leading to a positive litmus test.

Approximate positions of alkene =C—H bending bands:

RCH=CH$_2$	910 and 990 cm^{-1} (two bands)
R$_2$C=CH$_2$	890 cm^{-1}
cis-RCH=CHR	700 cm^{-1}
trans-RCH=CHR	970 cm^{-1}
R$_2$C=CHR	815 cm^{-1}

(Frequencies may vary by ± 30 cm^{-1} or more.)

PRELAB ASSIGNMENT

    Review OP–33 and read the Infrared Spectra section of Part III, starting on page 525. Review the other operations as necessary.

PROCEDURE

This experiment can be performed in groups of as many as seven students, each group working with a full set of unknown liquids. Alternatively, students can work individually and report their data to the instructor for posting.

All of the unknown liquids are flammable, and some of them are caustic or have hazardous vapors. Never pipet any liquid by mouth! Avoid contact with the liquids, do not breathe their vapors, and keep them away from flames or hot surfaces. If your liquid has a powerful odor similar to vinegar or household ammonia, it should be distilled under a hood and handled with protective gloves.

2,4-Dinitrophenylhydrazine is harmful if ingested or absorbed through the skin, and the DNPH reagent contains sulfuric acid, which can burn the skin and eyes. Avoid contact with the reagent.

The chromic acid reagent is prepared from chromium(VI) oxide, which is corrosive and very toxic. Inhalation of the dust from solid CrO_3 has been shown to cause cancer in humans; although there is no evidence that brief exposure to chromium(VI) solutions can cause cancer, you should avoid contact with the reagent.

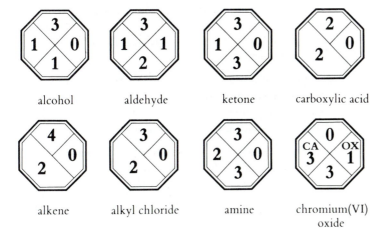

alcohol   aldehyde   ketone   carboxylic acid

alkene   alkyl chloride   amine   chromium(VI) oxide

A. Physical Properties

Obtain a 5-mL sample of an unknown organic compound **[possible contact and vapor hazard!]** and purify it by semi-micro distillation ⟨OP–25a⟩. If it boils over a range of 2°C or less, you can use the median distillation temperature as its boiling point; otherwise, collect a middle fraction over a narrow range and determine its boiling point by distillation ⟨OP–29⟩ or a micro method ⟨OP–29a⟩. Using a volumetric pipet **[do not pipet by mouth!]**, accurately measure ⟨OP–6⟩ 1 mL of the purified unknown into a tared vial, stopper the vial, and weigh it to the nearest milligram on an accurate balance.

Using a measuring pipet **[do not pipet by mouth!]**, measure 0.20 mL of the liquid into a 10-cm test tube. Add 0.20 mL of water, stopper the tube, and shake it vigorously for a few seconds. If two liquid layers separate on standing, or if the mixture is cloudy or contains undissolved droplets, add another 6 mL of water and shake it again. If the unknown dissolves, save the solution for the litmus test.

B. Chemical Tests

Except for the litmus test, all of the following tests should be carried out with the purified unknown liquid. Keep a careful record of your observations in your lab notebook.

Litmus test: If the unknown is miscible or soluble in water, test its aqueous solution with red and blue litmus paper.

2,4-Dinitrophenylhydrazine test: Add 1 drop of the unknown to 1 mL of the DNPH reagent in a test tube. Stopper and shake the test tube, and let the mixture stand for 15 minutes.

Chromic acid test: Dissolve 1 drop of the unknown in 1 mL of acetone in a test tube. Add 1 drop of the chromic acid reagent **[contact hazard! carcinogen!]** and shake the mixture, noting the time of addition.

Potassium permanganate test: Dissolve 1 drop of the unknown in 2 mL of 95% ethanol, and add 10 drops of 0.1 M potassium permanganate with shaking.

Beilstein's test: Make a small loop in the end of a copper wire. Heat the loop to redness in a burner flame. (Do this under a hood or in an area away from the unknowns or other flammable liquids.) Dip the loop into a little of your unknown liquid, and hold it in the lower outside part of the flame.

C. Infrared Spectrum

Record an infrared spectrum ⟨OP–33⟩ of your unknown, using the neat liquid, or obtain the spectrum from your instructor. If the unknown has a boiling point below 75°C, you should use a sealed cell with a thin spacer to prevent evaporation; otherwise, you may use a thin film between two salt plates. If your instructor requests, make copies of your spectrum and distribute them to the other members of your group.

Decide which functional group is present in your unknown. Report the chemical family, boiling point, density, and solubility of your unknown to your co-workers (or instructor), and obtain the same information for the other unknowns.

Report.   Tabulate the results of your classification tests. Describe clearly how you deduced the chemical family of your unknown, telling how you eliminated the other possibilities. Interpret the infrared spectrum of your unknown as completely as you can, and turn in the spectrum (or a copy) to your instructor. Tabulate the physical properties of all the liquids and group together compounds with similar physical properties; you should have one set of groups arranged according to boiling point, a different set for density, and another set for water solubility. Try to explain significant differences between the groups, based on the discussion in the Background section. Where appropriate, specify the type of intermolecular force involved. For example, group together compounds that have boiling points within 5–10°C of one another, and explain why the compounds in one group have higher boiling points than those in another.

Collateral Project

If the reagents are available, you can use additional or alternative classification tests such as Tollens' test ⟨C–23⟩ for aldehydes, sodium iodide in acetone ⟨C–22⟩ for alkyl chlorides, and bromine in Freon TF ⟨C–7⟩ for alkenes.

Identification of a Terpenoid by Infrared Analysis

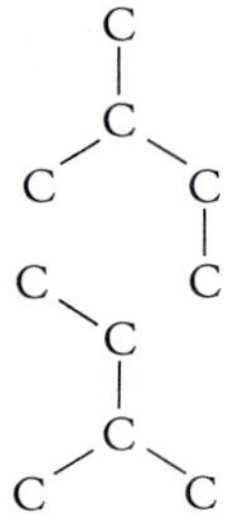

MINILAB 8

isopentane
units

Terpenoids are naturally occurring organic compounds whose molecules contain isopentane "building blocks." The carbon skeletons of all the terpenoids in Figure 2 can be built up by combining two isopentane units arranged as shown.

Obtain an unknown terpenoid from your instructor and record an infrared spectrum of the neat liquid ⟨OP–33⟩. Characterize as many absorption bands as you can and identify any functional groups present. Identify your unknown as one of the six compounds in Figure 2, and tell how you arrived at your conclusion. Turn in the infrared spectrum (or a copy) with your report.

Figure 2.   Structures of terpenoids for Minilab 8

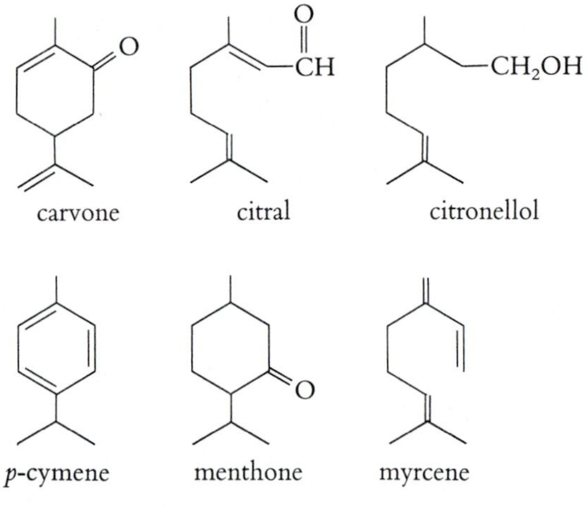

carvone   citral   citronellol

p-cymene   menthone   myrcene

Exercises

1. Compare the infrared spectra of compounds from the families in Table 1, and point out any significant similarities and differences. Pay particular attention to bands associated with specific functional groups.

2. Draw a flow chart that could be used to classify compounds from the families in Table 1, using only the tests given in the Procedure section. There should be at least two branches for each classification test, one leading to compounds that give a positive result and the other leading to compounds that give a negative result.

3. **(a)** Assuming that your unknown compound has a molecular weight between 70 and 79 and only one functional group, draw all possible structures for it and give their systematic names. **(b)** Determine the name and structure of your compound by comparing its infrared spectrum with those in a collection of spectra, such as Bib – F30, Bib – F32, or Bib – F33.

4. Compound X, having the molecular formula $C_4H_6O_2$, is believed to be one of the following:

$$
\begin{array}{ll}
\textbf{1.}\quad \underset{\displaystyle CH_3CH_2\overset{O}{\overset{\|}{C}}-\overset{O}{\overset{\|}{C}}H} &
\textbf{3.}\quad CH_2{=}CH\overset{OH}{\overset{|}{C}}H-\overset{O}{\overset{\|}{C}}H \\[20pt]
\textbf{2.}\quad CH_3\overset{O}{\overset{\|}{C}}-\overset{O}{\overset{\|}{C}}CH_3 &
\textbf{4.}\quad CH_3CH{=}CH\overset{O}{\overset{\|}{C}}-OH
\end{array}
$$

(a) State the chemical family or families to which each compound belongs. **(b)** Draw a flow chart diagramming a procedure that could be used to determine the identity of compound X, using chemical tests described in Part III.

Library Topic

Write a report telling how hydrogen bonding can be detected and studied by spectrometric methods.

Preparation of the Urea-Inclusion Complex of an Alkane

Reactions of Alkanes. Separation of Liquids.

Operations

OP – 9   Mixing

OP – 12   Vacuum Filtration

OP – 13   Extraction

OP – 14   Evaporation

OP – 20   Drying Liquids

OP – 21   Drying Solids

OP – 30   Measuring the Refractive Index

OP – 33   Infrared Spectrometry

Objectives

To prepare the urea-inclusion complex of an alkane and determine the host/guest ratio of the complex.

To separate an alkane from a mixture and identify it.

To learn how to interpret infrared spectra of alkanes.

To learn about the structures and uses of inclusion complexes.

SITUATION

$CH_3(CH_2)_{14}CH_3$
hexadecane

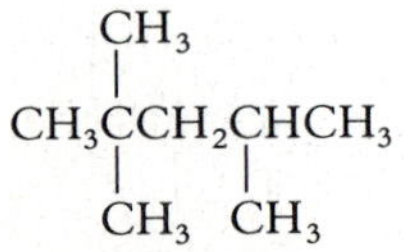

$$CH_3\underset{\underset{CH_3}{|}}{\overset{\overset{CH_3}{|}}{C}}CH_2\underset{\underset{CH_3}{|}}{CH}CH_3$$

2,2,4-trimethyl-
pentane

Long straight-chain alkanes make good diesel fuels but very poor automobile and jet fuels. It should therefore be possible to improve the quality of certain fuels simply by removing their straight-chain components. In 1941, a German chemist discovered that urea combines with straight-chain alkanes having seven or more carbon atoms to form a crystalline complex, but does not combine with most branched alkanes. You will use this property of urea to separate the components of a mixture containing hexadecane and 2,2,4-trimethylpentane.

BACKGROUND

Urea-Inclusion Complexes and Fuel Quality

The quality of a motor fuel is measured by its octane number, which compares its antiknock properties to those of 2,2,4-trimethylpentane. Similarly, the quality of a diesel fuel is measured by its *cetane number,* which compares its ignition properties to those hexadecane (cetane). The cetane number of a fuel is established by matching the fuel's performance to that of a mixture containing hexadecane, whose cetane number is 100, and 2,2,4,4,6,8,8-heptamethylnonane (HMN), whose cetane number is 15. For example, a diesel fuel that has the same ignition properties as a mixture containing 40% hexadecane and 60% HMN is assigned a cetane number of

$$100(0.40) + 15(0.60) = 49$$

Straight-chain alkanes burn more rapidly than branched alkanes; this can be an advantage in a diesel engine, which requires that the fuel ignite when sprayed into hot compressed air. But fuels with very high cetane numbers may ignite before the fuel and air in the combustion chamber are sufficiently mixed, resulting in incomplete combustion and excessive exhaust smoke. Most diesel fuels marketed in the United States therefore have cetane numbers between 40 and 65.

Compounds having long straight carbon chains can combine with urea to form crystalline *inclusion complexes.* In such complexes, the molecules of the *host* compound form a crystal lattice in which there are spaces large enough to hold molecules of the *guest* compound. In urea-inclusion complexes, six or more urea (host) molecules can combine to form a tubular channel around each guest molecule. The channel has a hexagonal cross section, with walls formed by interpenetrating spirals of urea molecules that are hydrogen-bonded to one another. The guest molecule is trapped in this channel and held there by weak van der Waals forces. The longer the guest molecule, the more urea is needed to surround it. A rough estimate of the number of urea molecules per guest molecule (the host/guest ratio) can be made using Equation 1, where n is the number of carbons in the guest molecule.

$$\text{host/guest ratio} \cong 1.5 + 0.65n \qquad (1)$$

Thus, about 6 molecules of urea are sufficient to confine a molecule of heptane, whereas 18 are needed to surround a

2,2,4-Trimethylpentane is sometimes (inaccurately) called isooctane.

$$
\begin{array}{ccc}
CH_3 & CH_3 & CH_3 \\
| & | & | \\
CH_3CCH_2CCH_2CHCH_2CCH_3 \\
| & | & | & | \\
CH_3 & CH_3 & CH_3 & CH_3 \\
& HMN
\end{array}
$$

Originally, the low-quality reference fuel was 1-methylnaphthalene rather than HMN, but HMN can be produced in greater purity at a lower cost.

$$
\begin{array}{c}
O \\
\| \\
H_2NCNH_2 \\
\text{urea}
\end{array}
$$

tetracosane (n-$C_{24}H_{50}$) molecule. Whether a compound qualifies as a guest depends on the size and shape of its molecules, and not on its chemistry. The diameter of the channel formed by urea molecules is about 0.52 nm (5.2 angstrom units), which is large enough to hold straight-chain molecules but not most branched molecules; n-octane fits nicely, but its branched isomers 2-methylheptane and 2,2,4-trimethylpentane do not. Some hydrocarbons having rings or branches attached to long straight chains may form complexes in which the bulky end group stays outside the channel. Urea can also form inclusion complexes with straight-chain primary alcohols, carboxylic acids, esters, and other linear compounds having terminal functional groups and at least seven carbon atoms.

Although straight-chain alkanes make good diesel fuels, they are undesirable components of most automobile and jet fuels. Treating a gasoline-grade petroleum fraction with urea will therefore increase its octane number by removing straight-chain alkanes that can cause knocking. Treating jet fuels with urea helps keep them flowing freely at high altitudes, where the temperature may drop to $-60°C$. Straight-chain alkanes tend to freeze sooner than branched alkanes of the same carbon number; without the urea treatment, wax crystals could form in the fuel at low temperatures.

For example, n-octane freezes 50°C higher than 2,2,4-trimethylpentane.

METHODOLOGY

In this experiment, you will combine urea with a mixture of hexadecane and 2,2,4-trimethylpentane to form an inclusion complex. Then you will determine the host/guest ratio in the complex and identify the guest alkane by infrared spectrometry or refractometry (or both). The complex should form when you combine the alkane mixture with a warm solution of urea in methanol. When this complex is stirred with warm water, the urea dissolves, and the guest alkane forms a separate layer that can be recovered by extraction with methylene chloride.

The identity of the guest alkane can be determined from its refractive index and infrared spectrum. Methyl C—H bonds in alkanes have stretching vibrations around 2960 and 2870 cm^{-1}, whereas the corresponding methylene bands are near 2925 and 2850 cm^{-1}. Often the methyl and methylene bands overlap so that only two or three peaks are observed. Methyl groups also show asymmetrical and symmetrical bending vibrations near 1450 and 1375 cm^{-1}. If there are two or three methyl groups on the same carbon atom, the symmetrical bending band is split into two or more closely spaced peaks near 1385 and 1370 cm^{-1}. Bending vibrations of methylene groups may also give rise to bands near 1465 cm^{-1}, between 1350 and 1150 cm^{-1} (often weak), and around 720 cm^{-1}. The terms used to describe these bond vibrations — scissoring, twisting, wagging, rocking — testify to the fact that molecules are dynamic entities and not the static particles that molecular models might suggest. The intensity of the 720 cm^{-1} methylene rocking band increases in proportion to the number of adjacent methylene groups, so a band in this region is characteristic of unbranched long-chain alkanes.

From the masses of the complex and the recovered alkane, you can determine how many urea molecules are needed to surround each alkane molecule (the host/guest ratio). For this result to be valid, material losses must be kept to a minimum, and the masses of the complex and recovered hexadecane should be measured as accurately as possible. It is also essential that the complex be completely dry and the alkane be entirely free of solvent.

PRELAB ASSIGNMENT

Calculate the mass of 10 mmol of hexadecane and of 200 mmol of urea.

Reactions and Properties

$$n\text{-}C_{16}H_{34} + x\,H_2N\overset{\overset{\displaystyle O}{\|}}{C}NH_2 \longrightarrow n\text{-}C_{16}H_{34} \cdot (H_2N\overset{\overset{\displaystyle O}{\|}}{C}NH_2)_x$$

cetane   urea   cetane-urea complex

Table 1.   Physical properties

	M.W.	b.p.	m.p.	d.	n
hexadecane	226.4	287	18	0.773	1.4345
2,2,4-trimethylpentane	114.2	99	−107	0.692	1.3915
urea	60.06		135	1.323	

Hexadecane $CH_3(CH_2)_{14}CH_3$ 2924.1   720.9
                                                                                     1467.1
                                                                                     1378.0

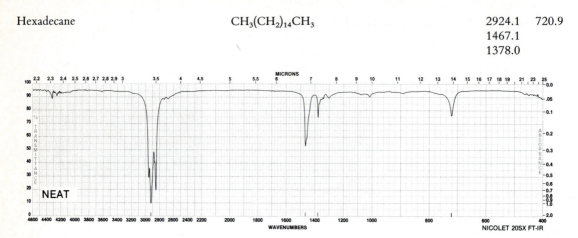

Note: The exact wavenumbers of certain absorption bands (designated by markers at the base of the spectrum) will be listed above most of the IR spectra reproduced in this book.

2,2,4-Trimethylpentane   2956.0   1247.6
                                                                                    1468.6   1168.4
                                                                                    1365.7   979.6

$$
\begin{array}{cc}
CH_3 & CH_3 \\
| & | \\
CH_3CHCH_2\!\!&\!\!CCH_3 \\
& | \\
& CH_3
\end{array}
$$

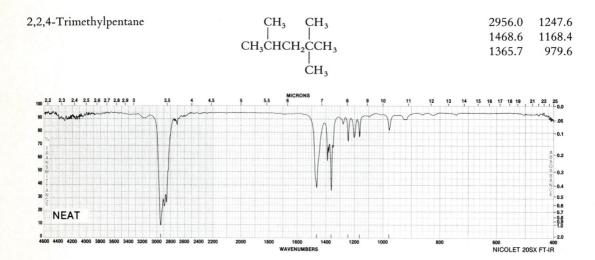

Figure 1.   IR spectra of two alkanes. (These and all other IR spectra in this book are reproduced from *The Aldrich Library of FT-IR Spectra,* by Charles J. Pouchert, with the permission of the Aldrich Chemical Company.)

PROCEDURE

Hexadecane and 2,2,4-trimethylpentane are flammable, and inhalation, ingestion, or skin absorption may be harmful. Avoid unnecessary or prolonged contact and inhalation; keep them away from flames and hot surfaces.

Methanol is very flammable and harmful if ingested, inhaled, or absorbed through the skin. Avoid contact with the liquid, do not breathe its vapors, and keep it away from flames and hot surfaces.

Methylene chloride is harmful if inhaled, ingested, or absorbed through the skin. Animal tests have been inconclusive, but it is suspected to be carcinogenic if it is inhaled. Avoid contact, and do not breathe the vapors.

2,2,4-trimethyl-
pentane

methanol

methylene
chloride

Reaction. Accurately weigh 10 mmol of hexadecane and combine it with 5 mL of 2,2,4-trimethylpentane **[fire hazard!]**. In a 125-mL Erlenmeyer flask, combine 200 mmol of urea with 50 mL of methanol **[vapor and fire hazard!]**. Warm the mixture to 50°C, swirling or stirring ⟨OP–9⟩ until all the urea has dissolved. While the solution is still warm, add the alkane mixture, and stir or swirl until the solid inclusion complex begins to separate. Set the solution aside, and let it cool slowly to 30°C or below; then cool it in an ice bath for 10 minutes or more to complete crystallization.

A magnetic stirrer is convenient, but not essential.

Separation. Collect the inclusion complex by vacuum filtration ⟨OP–12⟩, and wash the crystals on the filter with ice-cold methanol. Let the crystals dry ⟨OP–21⟩ to constant mass at room temperature (heating will destroy the complex).

If the product cannot be left overnight to dry, it should be air-dried thoroughly on the filter, blotted between filter papers, and left to dry in a well-ventilated location.

Weigh the inclusion complex accurately. Then mix it with 25 mL of warm water, and stir the mixture on a steam bath for several minutes until the crystals dissolve and an alkane layer separates. Cool the mixture in an ice bath, and extract ⟨OP–13⟩ the alkane with two portions of methylene chloride **[vapor hazard! suspected carcinogen!].** If the alkane solidifies on cooling, stir in the first portion of methylene chloride and transfer the solution to the separatory funnel after the alkane dissolves. Dry the combined extracts with anhydrous calcium chloride ⟨OP–20⟩. Evaporate the solvent *completely* ⟨OP–14⟩, and weigh the alkane.

Analysis. Identify the alkane from its refractive index ⟨OP–30⟩ or infrared spectrum ⟨OP–33⟩, or both.

Report. Describe clearly how you confirmed the identity of the alkane, and calculate the percentage recovered. Calculate the host/guest ratio in the complex from your experimental results. Estimate the host/guest ratio for hexadecane using Equation 1 and compare the two values. Interpret your infrared spectrum as completely as you can.

Collateral Project

Try to recover the other alkane by adding a large amount of water to the filtrate and separating the alkane layer ⟨OP–13b⟩. After it has been dried with calcium chloride, the alkane can be characterized by obtaining its infrared spectrum and refractive index.

Exercises

1. Compare the infrared spectrum of hexadecane with that of 2,2,4-trimethylpentane, and point out some bands that might help you distinguish straight-chain alkanes from branched ones.

2. Construct a flow diagram for the procedure used in this experiment.

3. What is the cetane number of a fuel that has the same ignition properties as a mixture containing 20% hexadecane and 80% HMN? Would it make a good diesel fuel?

4. Thiourea forms tubular inclusion complexes that are similar in structure to those formed by urea. Combining a mixture of hexadecane and 2,2,4-trimethylpentane with thiourea (H_2NCSNH_2) produces a crystalline solid that decomposes in water to yield 2,2,4-trimethylpentane, but no cetane. Propose an explanation for this result.

5. Outline a possible synthesis of 2,2,4,4,6,8,8-heptamethylnonane (HMN) starting with isobutylene.

Library Topic

Show how urea and thiourea are used in the manufacture of barbiturates, and describe some of the therapeutic uses and undesirable side effects of these drugs.

Preparation of Triphenylmethyl Bromide and the Trityl Free Radical

Reactions of Hydrocarbons. Preparation of Alkyl Halides. Free-Radical Substitution. Free Radicals.

Operations

OP – 6   Measuring Volume

OP – 10a Semimicro Addition

OP – 11   Gravity Filtration

OP – 12a Semimicro Vacuum Filtration

OP – 14   Evaporation

OP – 21   Drying Solids

OP – 22   Drying Gases

OP – 22b Trapping Gases

OP – 23a Semimicro Recrystallization

OP – 28   Melting-Point Determination

Objectives

To prepare triphenylmethyl bromide by the bromination of triphenylmethane.

To prepare the trityl free radical in solution and observe its behavior.

To gain some experience with semimicro synthetic techniques.

To learn about the properties of free radicals and some of their history.

SITUATION

In 1900, Moses Gomberg, a young chemistry instructor at the University of Michigan, published a remarkable paper describing his discovery of the world's first stable free radical, triphenylmethyl. Gomberg's paper also reported the preparation of a hydrocarbon that he believed to be hexaphenyleth-ane, a dimer of triphenylmethyl. In solution, the free radical was in equilibrium with its dimer, which could be isolated by evaporating the solution in an atmosphere of carbon dioxide.

Gomberg equilibrium

triphenylmethyl
(trityl) free
radical

hexaphenylethane (assumed)

tris(4-*t*-butylphenyl)methyl

$$t\text{-Bu} = CH_3\overset{\displaystyle CH_3}{\underset{\displaystyle CH_3}{C}}-$$

In 1968, a group of chemists from the Netherlands reported that they had prepared a similar free radical that would not dimerize. This new radical, tris(4-*t*-butylphenyl)methyl, had *t*-butyl groups at each *para* position. Because the chemists could not explain why such substituents would prevent the formation of a hexaphenylethane-type dimer, they decided to prepare Gomberg's dimer and find out whether it actually had the structure he proposed. In this experiment, you will prepare triphenylmethyl bromide and convert it to the triphenyl-methyl radical to see if you can find evidence for the existence of Gomberg's elusive dimer.

BACKGROUND

The Case of the Disappearing Dimer

In the mid-nineteenth century, many chemists were con-vinced that carbon could exist in a trivalent state in the form of "free" **radicals.** For example, it seemed reasonable to believe that if magnesium chloride could react with sodium metal to yield magnesium, then methyl halides should react with so-dium to form "methyl." However, Charles Wurtz carried out the reaction of methyl iodide with sodium and obtained not methyl but ethane, a synthesis known today as the Wurtz reaction. After many similar attempts ended in failure, chem-ists became increasingly doubtful that truly "free" radicals could exist.

Moses Gomberg never meant to make a free radical. He was trying to synthesize hexaphenylethane to prove a point

radical: a group of atoms such as methyl (CH_3-) that generally exists only in combination with other atoms or groups, such as in methyl bromide, CH_3Br

Reaction of methyl iodide and sodium:

$CH_3I + Na \longrightarrow$ "CH_3" $+$ NaI
(expected)

$2CH_3I + 2Na \longrightarrow CH_3CH_3 + 2NaI$
(actual)

Attempted synthesis of hexaphenyleth-ane:

$$2Ph_3CBr + Zn \longrightarrow$$
$$Ph_3CCPh_3 + ZnBr_2$$

Ph₃COOCPh₃

triphenylmethyl peroxide

The triphenylmethyl (Ph₃C) group is called trityl.

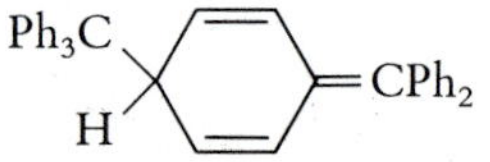

that, had he been successful, would be remembered today by only a handful of specialized chemists. Gomberg's first attempts to prepare this compound using the Wurtz reaction were not productive, so he tried different metals, such as silver and zinc, each time coming up with a snow-white solid that melted at 185°C and gave the wrong analysis for hexaphenyl-ethane. After repeated attempts to prepare this compound, it finally dawned on him that the product was reacting with oxygen in the air and forming triphenylmethyl peroxide. When Gomberg next ran the reaction, he was careful to exclude all air from the reaction mixture, and he obtained a white solid that melted at 147°C and (at last) gave the correct analysis for hexaphenylethane. But this compound behaved very strangely for a hydrocarbon. It reacted with air in solution to form triphenylmethyl peroxide and it rapidly decolorized dilute halogen solutions—something no ordinary hydrocarbon would do.

Gomberg eventually arrived at the conclusion that he had synthesized the world's first stable free radical, triphenyl-methyl. In solution, this colored radical is in equilibrium with its colorless dimer, for which Gomberg proposed the hexa-phenylethane structure. Much later, when researchers learned that the similar tris(4-*t*-butylphenyl)methyl free radical would not form an analogous dimer, they repeated Gomberg's preparation of triphenylmethyl and discovered that its dimer actually has the following structure:

dimer of triphenylmethyl
(Ph = C₆H₅)

So Gomberg's dimer was not hexaphenylethane after all; that elusive hydrocarbon has probably never existed!

METHODOLOGY

Triphenylmethane is more expensive than any of the starting materials you have used so far; consequently you will use less of it. When working with 10 grams or more of starting material, it is possible to be a little sloppy about laboratory technique and still get respectable results. The loss of a few tens of milligrams in each transfer or a few tenths of a gram during washing or recrystallization may not be serious if you have plenty of material to work with. But in an experiment that starts with a gram or so of starting material, losses of that magnitude cut your yield drastically. In this experiment, it will

be very important to avoid unnecessary transfers, to use mini-
mal amounts of wash liquids and recrystallization solvents
(well cooled), and in general to refine your technique so as to
keep losses to a minimum. Another way of cutting down material losses is to reduce the surface area of glass or porcelain in
contact with the chemicals. So you will use **semimicro** apparatus and techniques in operations such as addition, filtration,
and recrystallization.

semimicro: utilizing quantities of
material on the order of a gram or
tenths of a gram

The bromination of triphenylmethane proceeds by a
chain mechanism very similar to that for the chlorination of
methane. Both reactions involve the formation of a free-radical intermediate, which then reacts with halogen to form the
product. Before the reaction can begin, some bromine free
radicals must be formed by irradiation from an appropriate
light source; you can use an ordinary unfrosted light bulb for
this purpose.

Mechanism for the chlorination of
methane:

Initiation

1. $Cl_2 \xrightarrow{h\nu} 2Cl$

Propagation
2. $Cl\cdot + CH_4 \longrightarrow CH_3\cdot + HCl$
3. $CH_3\cdot + Cl_2 \longrightarrow CH_3Cl + Cl\cdot$
(steps 2 and 3 repeated indefinitely)

To cut down material losses, most of the operations will
be carried out in the same vessel, a small side-arm test tube. To
avoid unwanted side reactions, the hydrocarbon will always be
kept in large excess by adding bromine dropwise *to* the triphenylmethane solution. The reaction evolves hydrogen bromide, so it will be necessary to trap this gas by passing it into a
solution of dilute sodium hydroxide. When the reaction is
complete, some excess bromine may remain, imparting color
to the solution. This can be removed by adding a drop or two of
cyclohexene.

Termination
4. $Cl\cdot + Cl\cdot \longrightarrow Cl_2$ (and other
termination steps)

Gas trap reaction:
$HBr + NaOH \longrightarrow H_2O + NaBr$

Removal of excess bromine

cyclohexene   1,2-dibromocyclohexane
                                    (a liquid)

The traditional solvent for bromination reactions is carbon tetrachloride, a very toxic liquid that is now suspected of
causing cancer in humans. You will use a considerably less
toxic substitute, Freon TF (1,1,2-trichlorotrifluoroethane),
whose low boiling point makes it easier to remove from the
reaction mixture. After you have prepared and purified trityl
bromide, you will prepare the triphenylmethyl radical by
Gomberg's method, that is, by treating a triphenylmethyl (trityl) halide with metallic zinc. Exposing the solution to air
should shift the radical-dimer equilibrium and thus provide
evidence for the existence of the dimer.

Freon is the trade name for a class of
alkyl halides containing both chlorine
and fluorine atoms. The *threshold limit
value (TLV)* of a substance is the
maximum concentration in air of that
substance to which workers may be
exposed on a day-to-day basis without
apparent harm. The TLV of Freon 113
is 1000 ppm, indicating that it is about
200 times less toxic than carbon
tetrachloride, whose TLV is 5 ppm.

As a rule, the more stable a free radical is, the faster it will form. If you do Minilab 9, you can compare the stabilities of some free radicals by observing the rates at which four aromatic hydrocarbons react with bromine.

PRELAB ASSIGNMENTS

1. Read OP–10a and OP–22b, and read or review the other operations as necessary.

2. Calculate the mass of 5.0 mmol of triphenylmethane and the theoretical yield of triphenylmethyl bromide.

Reactions and Properties

A. triphenylmethane → triphenylmethyl bromide + HBr

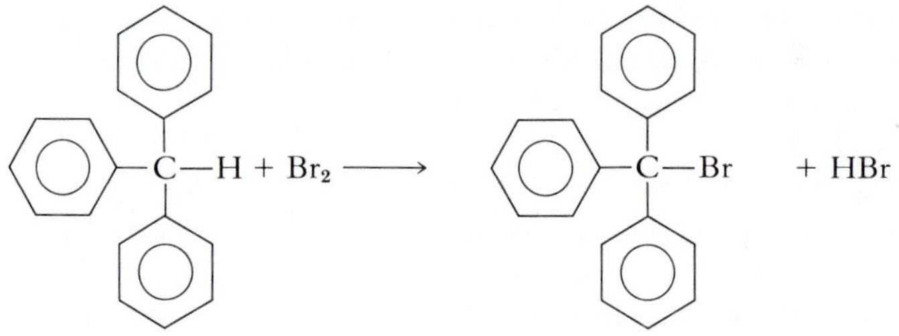

B. triphenylmethyl radical (dimerizes in solution) + ZnBr$_2$

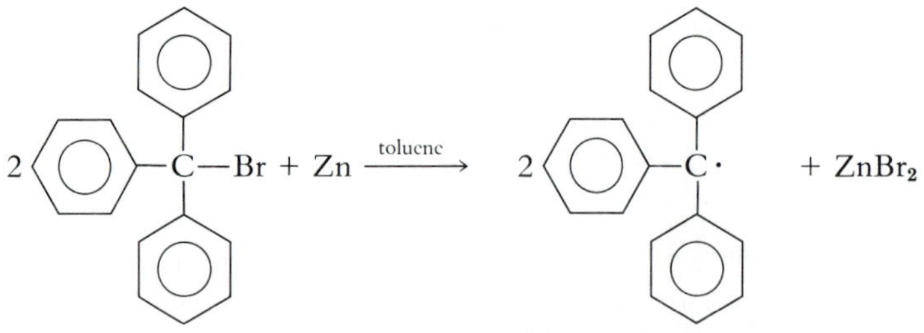

CFCl$_2$CF$_2$Cl
Freon TF
(1,1,2-trichloro-
trifluoroethane)

Table 1. Physical properties

	M.W.	m.p.	b.p.	d.
triphenylmethane	244.3	94	359	
bromine	159.8	−7	59	3.12
Freon TF	187.4	−36	48	1.575
triphenylmethyl bromide	323.2	152–54		

PROCEDURE

Trityl bromide reacts with moisture to form triphenyl-methanol, so it is important to dry any glassware that will be in contact with the product.

Bromine is highly toxic and corrosive, and its vapors can damage the eyes and respiratory tract. Avoid contact with the solution of bromine in Freon TF, and do not inhale the vapors or allow them to escape into the laboratory. Wear protective gloves while handling the solution, and measure it under the hood.

Freon TF (1,1,2-trichlorotrifluoroethane) may be harmful if ingested, inhaled, or absorbed through the skin, so minimize your exposure to it.

Triphenylmethyl bromide may cause serious damage if swallowed, inhaled, or absorbed through the skin. Avoid contact with the product.

Petroleum ether is extremely flammable, so keep it away from flames and hot surfaces.

SAFETY
PRECAUTIONS

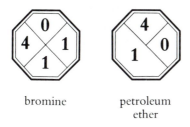

bromine   petroleum ether

A. Preparation of Triphenylmethyl Bromide

Reaction. Assemble the apparatus illustrated in Figure 1, using a 5-mL volumetric pipet as the addition pipet and a 20-cm side-arm test tube as the reaction tube, and clamping the gas-trap tube ⟨OP–22b⟩ to the same ringstand. The tip of the pipet should be about 8 cm (3 inches) above the bottom of the reaction tube when the apparatus is assembled. Fill the gas-trap tube halfway with 1 M aqueous sodium hydroxide, and secure the gas-delivery tubing so that its outlet is about a millimeter *above* the surface of the solution. Clamp an unfrosted 75- or 100-watt light bulb about 10 cm from the reaction tube. Take the reaction tube assembly under the hood and fill the addition

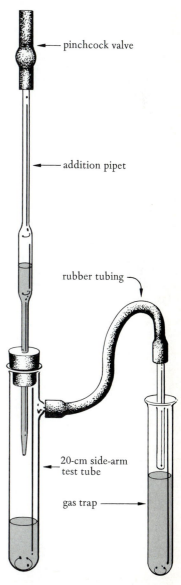

pinchcock valve

addition pipet

rubber tubing

20-cm side-arm test tube

gas trap

Figure 1.   Bromination apparatus

pipet to the mark with 1.0 M bromine in Freon TF **[contact and vapor hazard!]**, using the pipetting assembly pictured in Figure 2 of OP – 6. Set the pipetting assembly into the reaction tube and remove the bulb, leaving the pinchcock valve in place. Check the pipet for leakage; if liquid comes out before you pinch open the valve, use a different pipet. Weigh 5.0 mmol of triphenylmethane on glazed weighing paper, and mix it with 10 mL of Freon TF in the reaction tube.

If some of the Freon TF begins to distill into the gas trap (look for a separate bottom layer), reduce the rate of heating.

Warm the reaction tube in a hot-water bath just enough to dissolve the triphenylmethane, but not enough to boil the solvent. Switch the light on, and add about 1 mL of the bromine solution ⟨OP – 10*a*⟩, shaking gently to mix the reactants. Continue heating with a water bath throughout the addition, but do not heat strongly enough to boil away the solvent. When the red or orange color of the reaction mixture has faded to light yellow, add another 1-mL portion of bromine solution with shaking. Repeat this process until all of the bromine has been added. Let the reaction mixture stand at room temperature for 10 minutes (leave the light on) after the last addition. If the solution remains orange, remove the stopper **[in the hood!]** and add cyclohexene dropwise with swirling until the color disappears or fades to light yellow.

If the aspirator water backs up into the reaction tube, you will have to start over. So don't forget the solvent trap!

Separation.   Cool the reaction mixture, replace the addition pipet assembly by a solid rubber stopper, and attach the side-arm to a water trap and aspirator. Evaporate the solvent ⟨OP – 14⟩ from the reaction tube under vacuum. Protect the product from moisture.

Purification and Analysis.   Recrystallize ⟨OP – 23*a*⟩ the crude trityl bromide **[contact hazard!]** from high-boiling petroleum ether **[fire hazard!].** Collect the product by semimicro vacuum filtration ⟨OP – 12*a*⟩, washing it with a little cold low-boiling petroleum ether, and let it air-dry on the filter. Measure its mass and melting point ⟨OP – 28⟩ when it is completely dry ⟨OP – 21⟩. Save enough trityl bromide for part B and turn in the rest.

B. Preparation and Reactions of the Trityl Free Radical

During this process, make careful notes of your observations and look for evidence suggesting the existence of the trityl free radical and its dimer.

Toluene is flammable and somewhat toxic; avoid contact, do not breathe its vapors, and keep it away from flames.

SAFETY
PRECAUTION

toluene

If an air line is not available, unstopper the tube to admit air, then stopper and shake it.

Dissolve about 0.2 g of recrystallized trityl bromide in 4 mL of toluene **[vapor hazard!]** in a small test tube. Add 0.5 g of 30- or 40-mesh zinc, stopper the tube immediately, and shake vigorously for 10 minutes. Quickly filter the mixture ⟨OP–11⟩ through a thin layer of glass wool into another test tube, and stopper this test tube immediately. Shake the solution, and let it stand (tightly stoppered) for about 5 minutes. Then bubble dry air ⟨OP–22⟩ through it until the color fades, stopper and shake it, and let it stand for a few more minutes. Repeat this process until the solution remains colorless for a minute or so after stoppering.

Report.   Calculate the percentage yield of trityl bromide. Explain what was happening in part B, beginning with the addition of zinc to the trityl bromide solution. Give names and structures of all species for which you saw evidence, and write equations for the reactions you observed.

Collateral Project

Collect the precipitate that formed in part B by semimicro vacuum filtration, and wash it with petroleum ether. Measure the melting point of the dry solid, and give its structure.

Free-Radical Bromination of Hydrocarbons

MINILAB 9

Avoid contact with the hydrocarbons and the bromine solution, and do not breathe their vapors.

SAFETY
PRECAUTION

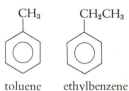

toluene   ethylbenzene

To each of four clean, labeled 10-cm test tubes, add 1 mL of one of four aromatic hydrocarbons (shown in the margin here and on page 150) and 4 mL of Freon TF. Noting the time of addition, quickly add 1 mL of 1 M bromine in Freon TF

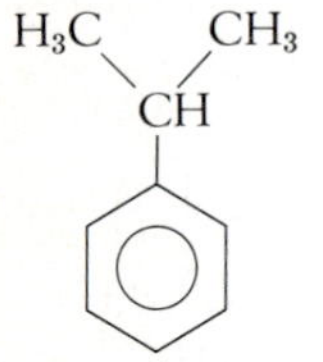

$$H_3C \quad CH_3$$
$$CH$$

isopropylbenzene

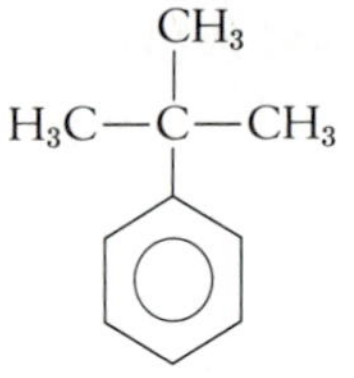

$$CH_3$$
$$H_3C-C-CH_3$$

t-butylbenzene

[contact and vapor hazard! work under the hood!] to each test tube, and stopper and shake it. Set the test tubes in a well-lighted location, and record the approximate time it takes for each solution to become decolorized, observing them closely for the first 5–10 minutes and then at intervals during the lab period (it may be that not all of them will decolorize). Write a balanced equation for each reaction (assuming mono-bromination), and give the structure of the free-radical intermediate. Arrange the free radicals in order of stability, and explain their relative stabilities.

Exercises

1. **(a)** Show how Gomberg's dimer, whose true structure is given in the Background section, can form by the combination of two trityl radicals. **(b)** Explain why tris(4-*t*-butyl-phenyl)methyl does not form an analogous dimer.

2. Write a mechanism for the free-radical bromination of triphenylmethane.

3. Construct a flow diagram for the synthesis of tri-phenylmethyl bromide.

4. Assuming a free-radical mechanism for the bromination of the following compounds, arrange them in order of their expected reactivity, starting with the most reactive.

(a) 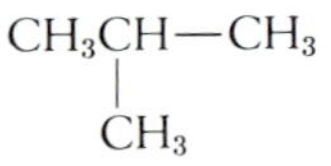 CH_3CH-CH_3
           CH_3

(b) 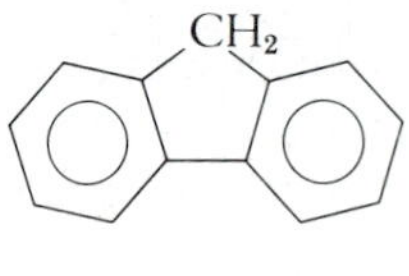 CH_2

(c) 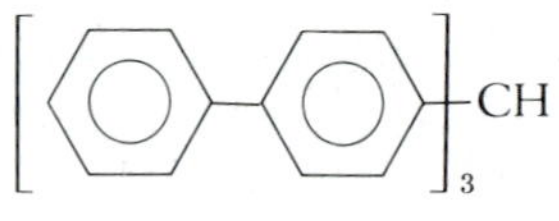 $\left[\bigcirc - \bigcirc \right]_3 CH$

(d) 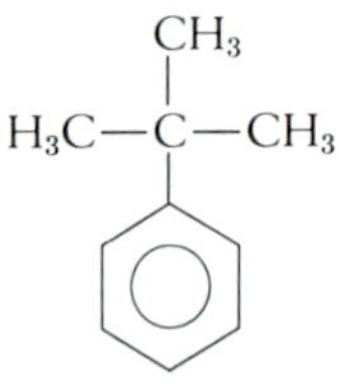 CH_3
$H_3C-C-CH_3$

(e) 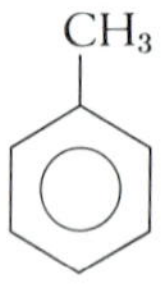 CH_3

(f) 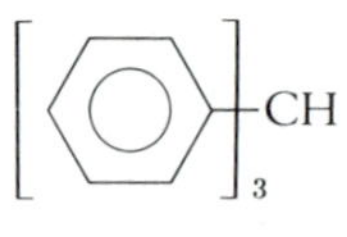 $\left[\bigcirc \right]_3 CH$

Library Topic

Read the selections from Gomberg's work in the reference cited, describe the reasoning that led him to the conclusion that he had prepared the trityl free radical, and tell why he tried to prepare hexaphenylethane in the first place.

H. M. Leicester and H. S. Klickstein, eds., *A Sourcebook in Chemistry, 1400–1900* (Boston: Harvard University Press, 1952), pp. 512–20.

Structures and Properties
of Stereoisomers

Isomerization Reactions. Stereoisomerism.

Operations

OP–7*a*   Refluxing

OP–13   Extraction

OP–14   Evaporation

OP–20   Drying Liquids

OP–29*a* Micro Boiling-Point Determination

OP–30   Measuring the Refractive Index

OP–31   Measuring Optical Rotation

OP–33   Infrared Spectrometry

Objectives

To isomerize (−)-menthone to its diastereomer, (+)-iso-menthone.

To measure and compare the properties of carvone enantiomers.

To determine the structure of a tartaric acid stereoisomer.

To use molecular models to visualize and compare the structures of stereoisomers.

To gain experience in drawing stereochemical formulas and in designating configurations of chiral centers.

To learn about the biochemical significance of molecular dissymmetry.

SITUATION

Imagine yourself on the spaceship *Daedalus* bound for the planetary system of Barnard's star. Arriving on the surface of its Earthlike third planet, called Arret by the inhabitants, your landing party is invited to a feast by some friendly Arretians. You are served their standard banquet fare: roast *krop*, over-

An account of the *Daedalus* project is found in *New Scientist 63*, 522 (1974).

A chiral compound is one exhibiting "handedness." Like your left and right hands, a chiral molecule cannot be superimposed on its mirror image.

done *iloc'corb,* crusty *yawarac*-seed rolls, and an aromatic *t'nim* tea. You had been looking forward to a change from the monotonous space diet, but you soon lose your appetite; most of the food tastes flat or bitter, the seeds on the roll have a minty flavor, and the tea smells like caraway! After politely declining a second helping of *krop,* you return to your shuttlecraft with the rest of the landing party. Soon all of you are suffering from indigestion, causing a serious breach in interplanetary relations.

Earthly organisms are mostly composed of chiral molecules, like the D-monosaccharides (simple sugars) in complex carbohydrates and the L-amino acids that make up proteins and enzymes. It is conceivable that somewhere in the universe there exists a "mirror-image" planet, otherwise similar to Earth, where proteins are composed of D-amino acids and carbohydrates of L-monosaccharides and the configurations of other chiral compounds are reversed as well. On such a planet, spearmint leaves would contain the form of carvone that flavors caraway seeds on Earth, and vice versa. Many foods would taste different from their counterparts here, and we would find them quite indigestible. In this experiment, you and your coworkers will compare the properties of the mirror-image carvones from caraway and spearmint oils. You will also convert (−)-menthone to its diastereomer, (+)-isomenthone, and determine the structure of a tartaric acid stereoisomer.

BACKGROUND

Alice Through the Looking Glass: Dissymmetry and Life

dissymmetric: incapable of being superimposed on its mirror image

One of the great unsolved mysteries of life concerns the origin of optically active compounds. Living systems are composed of **dissymmetric** molecules, and molecular dissymmetry is necessary for life as we know it, but how did such dissymmetry come about? Was there some "molecular Adam," a single dissymmetric molecule that gave rise to all molecular dissymmetry on Earth? Or is molecular dissymmetry a phenomenon that appeared independently at many different locations as a consequence of some kind of fundamental dissymmetry in the universe? Many theories of the origin of **chiral** molecules have been proposed, some of them allowing at least the possibility that life forms on other planets may be made up of molecules that are mirror images of our own. But convincing proof of such theories is hard to come by, and the answer may never be known.

chiral = dissymmetric

See Library Topic.

Life on earth is intimately associated with molecular dis-symmetry. Most of the molecules of life — proteins, carbohydrates, nucleic acids, and enzymes — are chiral and are built up of smaller units that are also chiral. A strand of DNA, for example, consists of two long chains, each having a backbone of linked D-2-deoxyribose molecules twisted into a right-handed double helix. DNA and RNA regulate the synthesis of proteins from L-amino acids, which are combined in specific sequences inside cellular structures called ribosomes. Some of these proteins make up the enzymes that assist in the digestion of carbohydrates, yielding D-glucose to be used by the body for fuel.

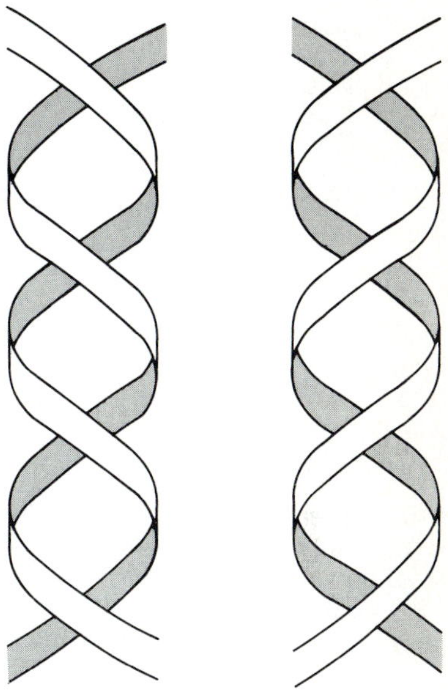

It is conceivable that life could be based on a mirror-image DNA made up of L-2-deoxyribose and twisted into a left-handed double helix; but then protein synthesis could use only D-amino acids and the corresponding enzymes could digest only L-carbohydrates. In other words, if the configuration of one link in the chain of life is reversed, all the rest must be reversed as well. If we could, by some magical contrivance, pass through the looking glass as Alice did, all the people, plants, and other organic matter in the looking-glass world would presumably be constructed of these mirror-image molecules. We could not survive in such a world (though Alice did, in Lewis Carroll's imagination) since digestion, metabolism, reproduction, and other life processes involving chiral molecules would be inhibited or prevented entirely.

Figure 1.   Left-handed and right-handed double helixes

METHODOLOGY

A chiral compound that rotates plane-polarized light in a clockwise direction is designated with a plus sign (+). Its enantiomer, which is designated with a minus sign (−), will rotate plane-polarized light to the same extent but in a counterclockwise direction under the same conditions. Most of the other measurable properties of pure enantiomers are identical. Enantiomers can, however, differ in the way they interact with chiral substrates, and these differences are particularly important in biochemical systems. For example, some kinds of olfactory receptors are apparently chiral, so the (+)- and (−)-enantiomers of a compound that interacts with such receptors may have noticeably different odors.

Carvone is a ketone found in the essential oils of both caraway seeds and the spearmint plant. Both oils contain approximately 55% carvone, along with limonene and some minor components. But the carvone in caraway oil has a very

Experimental proof of the odor differences between enantiomers was lacking until quite recently. The experiments are described in *Science 172,* 1043 (1971).

carvone   limonene

A stereogenic (chiral) atom is an atom having four different substituents, such as the carbon in CHClBrI. See the textbook for your lecture course for an explanation of the D and L designations.

different odor than its spearmint counterpart does, suggesting that the two carvones are enantiomers. You will measure the optical rotation of one form of carvone to find out if it is the (+)- or (−)-enantiomer, and compare its structure, infrared spectrum, and physical properties with those of the other carvone. Unless your instructor suggests otherwise, you can use about 0.5–1.0 g of solute per 10-mL of solution for the optical rotation measurements (some tubes require more solution). If spearmint and caraway oils are available, you can isolate your carvone enantiomer by column chromatography, as described in the Collateral Project.

The three-dimensional structures of stereoisomers are readily visualized and compared by employing molecular models. A molecular model of a compound having one stereogenic atom can be converted to a model of its enantiomer by interchanging any two substituents attached to that atom. The configuration of the atom is reversed by such an interchange; an R configuration changes to an S configuration, and vice versa. If a molecule has two or more stereogenic atoms, the configurations at all of them must be reversed to produce its enantiomer; changing just one yields a diastereomer. The specific rotations of diastereomers may or may not have different signs, but they are ordinarily not equal in magnitude. For example, the specific rotation of (−)-menthone is −30°, whereas that of (+)-isomenthone is about +92°. When (−)-menthone is heated with acid, it is partially isomerized to this diastereomer. You will simulate the isomerization reaction using molecular models, carry it out in the laboratory, and determine its extent by measuring the optical rotation of the product mixture.

Tartaric acid can be produced from potassium hydrogen tartrate (cream of tartar), a by-product of winemaking. This "natural" tartaric acid is the L-(+)-form illustrated. Since tartaric acid has two stereogenic atoms, several other structures are possible. You will construct models for them, and try to deduce the structure of an unknown tartaric acid stereoisomer from its optical rotation.

L-(+)-tartaric acid

PRELAB ASSIGNMENT

Before beginning this experiment, you should know how to draw stereochemical structural formulas, how to designate configurations at stereogenic atoms, how to classify stereoisomers as enantiomers and diastereomers, and how to recognize *meso* compounds. Review the stereochemistry section of the textbook for your lecture course (or see another appropriate source) if you need help in any of these areas.

Reactions and Properties

(−)-menthone   (+)-isomenthone

Table 1.   Physical properties

	M.W.	d.	[α]
(R)-carvone	150.2	0.96	−62°
(S)-carvone	150.2	0.96	+62°
(−)-menthone	154.2	0.895	−30°
(+)-isomenthone	154.2	0.900	+92°

PROCEDURE

A. Isomerization of (−)-Menthone to (+)-Isomenthone

Acetic acid is corrosive and can cause serious damage to skin and eyes; its vapors are highly irritating to the eyes and upper respiratory tract. Wear gloves, and dispense the acid under a hood; avoid contact, and do not breathe its vapors.

SAFETY
PRECAUTION

Ethyl ether is very flammable and may be harmful if inhaled. Do not breathe the vapors, and keep it away from flames and hot surfaces.

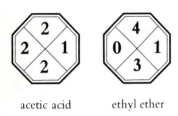

acetic acid   ethyl ether

If your instructor requests, also measure the optical rotation of the (−)-menthone.

Under the hood, mix 1 mL of (−)-menthone with 5 mL of glacial acetic acid **[contact and vapor hazard!]** and 5 mL of 1 M HCl. Reflux the mixture for 30 minutes ⟨OP−7a⟩, and let it cool to room temperature. Add enough 3 M NaOH to raise the pH of the solution to 10 (30 mL or so may be needed), and extract it ⟨OP−13⟩ with two portions of ethyl ether **[fire hazard!]**. Dry the combined ether extracts thoroughly with magnesium sulfate ⟨OP−20⟩, and evaporate the ether ⟨OP−14⟩. Accurately measure the optical rotation ⟨OP−31⟩ of the product in absolute ethanol.

Construct a molecular model of (−)-menthone (in its most stable chair conformation) from the structural formula given in the Reactions and Properties section. Show how it can be converted to a model of (+)-isomenthone by interchanging atoms or groups. Use the models to confirm that the two compounds are diastereomers, not enantiomers, and decide which one is most stable. Determine the configuration (R or S) at each stereogenic atom of each isomer. Construct models for all other stereoisomers having the basic menthone structure, determine the configuration at each stereogenic atom, and designate pairs of enantiomers and diastereomers.

B. Properties of Carvone Enantiomers

Students can work in pairs, each student using a different enantiomer. If desired, the isomers can be isolated from peppermint and spearmint oils as described in the Collateral Project.

Obtain 1−2 mL of a carvone enantiomer from your instructor. Compare its odor with that of the other enantiomer. Measure the refractive index ⟨OP−30⟩ and boiling point ⟨OP−29a⟩ of your carvone, and its optical rotation in 95% ethanol ⟨OP−31⟩. Record its infrared spectrum ⟨OP−33⟩, or obtain the spectrum from your instructor. Compare the properties and infrared spectrum of your enantiomer with those of the other. Construct molecular models for the two enantiomers, and show that they are mirror images of one another. Determine the configuration (R or S) at the stereogenic atom of each model, and decide which model represents your form of carvone.

C. Identification of a Tartaric Acid Stereoisomer

Obtain an unknown tartaric acid stereoisomer, measure its optical rotation in water ⟨OP–31⟩, and calculate its specific rotation. Construct molecular models for all possible tartaric acid stereoisomers and determine the configuration at each stereogenic atom. Using the fact that the specific rotation of L-(+)-tartaric acid is +12°, predict the specific rotations of the other stereoisomers, and deduce the structure of your unknown.

Report. Your report should include drawings representing the stereochemical structures of all the models you constructed, giving the configurations of the stereogenic atoms. Calculate the specific rotations of all compounds whose optical rotations you measured, and the percent composition of the menthone-isomenthone mixture. Explain any similarities or differences in the properties of the carvone enantiomers. Calculate the apparent optical purity of your enantiomer, and interpret its infrared spectrum as completely as you can. Tell how you arrived at the structure for your unknown tartaric acid stereoisomer.

Collateral Project

You can isolate a carvone enantiomer for part B from spearmint or caraway oil by column chromatography. Slurry-pack a chromatography column ⟨OP–16⟩ with silica gel, using high-boiling petroleum ether **[fire hazard!]**. Introduce about 2 g of the essential oil into the top of the column, and elute it with the following solvents: 25 mL of petroleum ether (high-boiling); 50 mL of a 10% solution of methylene chloride in petroleum ether; 25 mL of a 20% solution of methylene chloride in petroleum ether; and 125 mL of a 50% solution of methylene chloride in petroleum ether. Put the first 100 mL of eluate in a designated solvent recovery container; collect the rest in 25-mL fractions. Evaporate the fractions, and measure the refractive indexes of the residues. Use the purest carvone samples ($n \cong 1.499$ at 20°C) in part B.

Exercises

1. **(a)** Draw structures — in the chair form — for the most stable conformations of (−)-menthone and (+)-isomenthone, and indicate which diastereomer should be more stable. **(b)** Is the composition of your isomerization mixture consistent with this conclusion? Explain.

2. Propose a mechanism for the isomerization of (−)-menthone to (+)-isomenthone.

3. A synthetic form of tartaric acid has a specific rotation of 0°, but its melting point is very different from that of *meso*-tartaric acid. Explain.

4. The steroid cholic acid is said to have 2048 possible configurations, of which only one occurs naturally. **(a)** Indicate each stereogenic carbon atom on the cholic acid molecule with an asterisk, and perform a calculation to confirm this isomer number. **(b)** Look up the stereochemical structure of natural cholic acid in the *Merck Index,* and give the *R-S* configuration at each stereogenic atom where possible.

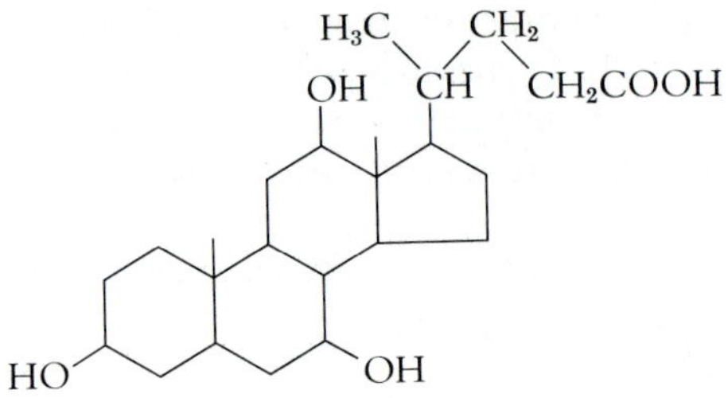

cholic acid

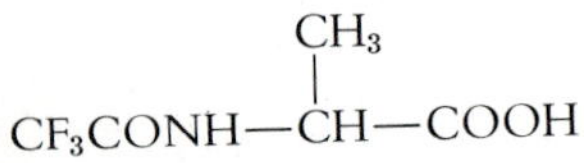

5. The Murchison meteorite, which fell near Murchison, Australia, in 1969, was found to contain traces of the amino acid alanine. When a trifluoroacetyl derivative of this alanine was treated with (S)-2-butanol and the resulting esters were passed through the column of a gas chromatograph, two separate peaks with identical peak areas were observed. **(a)** Draw stereochemical structures of the two compounds formed in the esterification reaction. **(b)** Discuss the probability that the amino acid was biotic (derived from once-living matter).

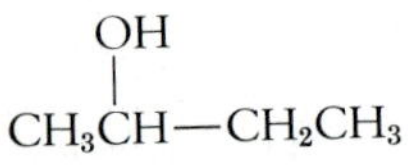

trifluoracetylalanine

2-butanol

Library Topic

Read the articles in *J. Chem. Educ. 49,* 448 and 455 (1972) and write a research paper on the origin and consequences of optical activity. Use the bibliographies included with the articles to locate additional material.

Bridgehead Reactivity in Solvolysis
Reactions of Tertiary Alkyl Bromides

Reactions of Alkyl Halides. Nucleophilic Aliphatic
  Substitution. Reaction Kinetics.

Operations

  OP–6 Measuring Volume
  OP–7 Heating

Objectives

  To measure the solvolysis rate constants for 1-bromoada-
mantane and *t*-butyl bromide in aqueous alcohol solutions.

  To determine the effect of a bridgehead on the rate of an
S_N1 reaction.

  To learn how to conduct a kinetic study and analyze the
data.

  To learn about the mechanisms of solvolysis reactions.

  To learn about adamantane and its derivatives.

SITUATION

  In 1939, Bartlett and Knox discovered that apocamphyl
chloride is surprisingly inert to reagents that usually bring
about nucleophilic substitution reactions. They concluded that
the cagelike structure of this compound inhibits the formation
of a carbenium ion as a reaction intermediate and prevents
nucleophilic attack from the back side at the bridgehead carbon, making both S_N1 and S_N2 reactions difficult or impossible. Highly strained cage compounds such as apocamphyl
chloride would not be expected to form stable carbenium ions,
but adamantane and its derivatives, which also have cagelike
molecules, have normal bond angles. To find out whether such
unstrained cage compounds are also slow to react, you will

See the article by P. D. Bartlett and
L. H. Knox, *J. Amer. Chem. Soc. 61,*
3184 (1939).

apocamphyl
chloride

compare the reactivity at the bridgehead carbon of 1-bromo-adamantane with that at the "uncaged" tertiary carbon of *t*-butyl bromide.

1-bromoadamantane
(1-adamantyl bromide, AdBr)

t-butyl bromide

BACKGROUND

A Gem among Molecules

adamantane

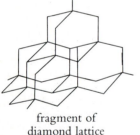

fragment of
diamond lattice

bridgehead: the point at which two or more fused rings are joined in a bicyclic or polycyclic system

Adamantane, whose name is derived from a Greek word meaning "diamond," has molecules of elegant symmetry that can be regarded as fragments of a tetrahedral diamond lattice. Models of this unique molecule reveal that it consists of four interlocking chair-form cyclohexane rings, arranged somewhat like the four planes of a tetrahedron. A space-filling model is nearly spherical, and this molecular shape results in a particularly stable crystal lattice that is responsible for adamantane's melting point of 268°C—unusually high for a hydrocarbon.

The extraordinary structure of adamantane has fascinated chemists for many years, because adamantyl systems have properties that make them almost ideal for the study of certain chemical phenomena. The rigid adamantane skeleton results in a system of known geometry with unstrained, tetrahedral bond angles; the cyclohexane rings making up the adamantane molecule come together at four points, forming a **bridgehead** at each junction; and the cagelike structure prevents certain kinds of interactions and reaction mechanisms, simplifying the analysis of reaction parameters.

When Bartlett and Knox showed that bridgehead compounds such as apocamphyl chloride are quite unreactive with nucleophiles, they suggested that studies of bridgehead reactivity might yield valuable information about reaction mechanisms and the geometry of transition states. That suggestion was taken up by numerous investigators; particularly fruitful results have been obtained in the study of solvolysis reactions. Solvolysis reactions are nucleophilic substitution reactions in which the solvent acts as the nucleophile. The solvolysis of *t*-butyl bromide and similar halides by hydroxylic solvents is believed to proceed by an S_N1 mechanism involving the formation of a carbenium-ion intermediate. Dissociation of the alkyl bromide is the rate-determining step, so the reaction is a first-order one, with the rate equation

$$\frac{-d[\text{RBr}]}{dt} = k[\text{RBr}]$$

Although the solvent does not appear in the rate equation, it can affect the reaction rate by assisting in the formation of the carbenium ion. Some solvent molecules may help push the leaving group off from the rear while others pull it off from the front. Thus, the rate of an S_N1 solvolysis reaction should depend on both the polarity of the solvent and the stability of the carbenium ion formed in the first step; the more polar the solvent and the more stable the cation, the faster the reaction.

Carbenium ions tend to assume a planar geometry; thus, cation formation becomes more difficult as the atom holding the leaving group becomes more restricted in its movement. A *t*-butyl bromide molecule can easily go from a tetrahedral to a planar geometry when it ionizes to form the *t*-butyl carbenium ion, but most bridgehead compounds cannot — norbornyl bromide is about ten *trillion* times less reactive than *t*-butyl bromide!

The carbon framework of the adamantane system is not nearly as rigid as the norbornyl skeleton. In the 1-adamantyl carbenium ion, the bridgehead carbon can flatten out a bit, attaining a bond angle of 113°, which is somewhere between the tetrahedral angle of adamantane itself and the trigonal angle expected for an open-chain carbenium ion. By measuring the relative solvolysis rates for 1-bromoadamantane (1-adamantyl bromide) and *t*-butyl bromide, you should learn whether the nonplanar 1-adamantyl cation is significantly less stable than the planar *t*-butyl carbenium ion.

t-Butyl bromide solvolysis mechanism:

1. $t\text{-BuBr} \longrightarrow t\text{-Bu}^+ + \text{Br}^-$ (slow)

2. $t\text{-Bu}^+ + \text{SOH} \longrightarrow t\text{-Bu}\overset{\text{H}}{\underset{\oplus}{\text{OS}}}$ (fast)

(SOH = hydroxylic solvent)

3. $t\text{-Bu}\overset{\text{H}}{\underset{\oplus}{\text{OS}}} + \text{Base}: \longrightarrow$

$t\text{-BuOS} + \text{Base}:\text{H}^+$ (fast)

Role of solvent in displacement reactions

norbornyl bromide

113°

104°

1-adamantyl carbenium ion

METHODOLOGY

The rate constant for a first-order reaction can be calculated from the following integrated rate equation, where c_0 is the initial concentration of the substrate and c is its concentration at time t:

$$\ln \frac{c}{c_0} = -kt \qquad\qquad \textbf{(1)}$$

The solvolysis reactions will produce hydrogen bromide according to the general equation

$$RBr + SOH \longrightarrow ROS + HBr$$

Therefore, the concentration factor in Equation 1 can be evaluated indirectly by measuring the amount of HBr that forms during the reaction. This will be accomplished by adding portions of a KOH solution and recording the time it takes for the HBr evolved to neutralize the added base, as shown by the color change of an indicator. At any given time during the reaction, the concentration of the alkyl bromide (c) will equal its initial concentration (c_0) minus the concentration of the HBr that has formed by then (c_{HBr}). The concentration of HBr is proportional to the volume of KOH needed to neutralize it (V), and the initial concentration of the alkyl bromide is proportional to the total volume of KOH solution needed to neutralize all the HBr produced (V_∞). Thus, we can write an expression for the concentration factor in terms of the volume of KOH solution added:

$$\frac{c}{c_0} = \frac{c_0 - c_{HBr}}{c_0} = 1 - \frac{c_{HBr}}{c_0} = 1 - \frac{V}{V_\infty}$$

The solvolysis of 1-bromoadamantane will be studied in 40% aqueous ethanol, in which both water and ethanol act as nucleophiles. Because the reaction of t-butyl bromide in 40% ethanol is difficult to measure accurately, the solvolysis will be conducted in 80% ethanol instead. In order to make a meaningful comparison between the two reactions, you will have to estimate the rate constant for t-butyl bromide in 40% ethanol using Equation 2 (the Winstein-Grunwald equation), which relates the rate of a solvolysis reaction to the ionizing power of the solvent:

$$\log \frac{k}{k_0} = mY \tag{2}$$

In this equation, k_0 is the rate constant for a reaction in the reference solvent, and k is the rate constant in another solvent at the same temperature. Y is a measure of the other solvent's ionizing power, and m measures the sensitivity of the substrate to changes in ionizing power. The literature value of m for t-butyl bromide is 0.94, and Y for 40% ethanol is 2.20.

For each kinetic run, you will prepare the reaction solvent by combining 95% ethanol and water in such proportions that, after you have added the alkyl bromide, the solvent will be 40% or 80% aqueous ethanol. It is important to measure the solvents accurately, because an error in solvent composition can markedly affect the solvolysis rate. To the reaction solvent, you will add some bromthymol blue and the alkyl bromide, followed by a measured portion of potassium hydroxide in the appropriate solvent. The indicator should turn blue as each 1-mL portion of KOH solution is added, changing to green when enough HBr is produced by the solvolysis reaction to neutralize the added KOH. As the solution becomes more acidic, its color fades to yellow.

Each portion of KOH will consume the HBr produced by reaction of about 5% of the alkyl halide; to get sufficient data for the rate calculations, you should continue the run until the reaction is at least 50% complete (10 portions or more). After the last portion of KOH has been added, you will heat the reaction flask gently to bring the reaction to completion, then titrate the solution with more KOH to determine V_∞.

If you do Minilab 10, you will compare the relative nucleophilic substitution rates for primary, secondary, and tertiary alkyl bromides with two different reagents and deduce the mechanism of each reaction.

The 1-bromoadamantane is added in 1 mL of ethanol, which affects the solvent composition.

PRELAB ASSIGNMENT

Calculate the mass of 1 mmol of 1-bromoadamantane.

Reactions and Properties

$$RBr + H_2O \longrightarrow ROH + HBr$$

and

$$RBr + CH_3CH_2OH \longrightarrow ROCH_2CH_3 + HBr$$

(R = 1-adamantyl or *t*-butyl)

$$HBr + KOH \longrightarrow H_2O + KBr$$

Table 1.   Physical properties

	M.W.	m.p.	b.p.	d.
1-bromoadamantane	215.1	118		
t-butyl bromide	137.0	−16	73	1.221

PROCEDURE

The alkyl halides must be protected from moisture; be sure your glassware is clean and dry.

A. Solvolysis of 1-Bromoadamantane in 40% Ethanol

Kinetic Run.   Prepare an indicator blank by measuring 25 mL of a pH 6.9 buffer into a 125-mL Erlenmeyer flask and adding 2 drops of bromthymol blue indicator solution. Set this flask aside while you prepare the reaction mixture.

The indicator blank should be green.

Accurately measure ⟨OP–6⟩ 20 mL of 95% ethanol into another 125-mL Erlenmeyer flask, and then add 29 mL of distilled or deionized water. Add four drops of bromthymol blue solution, and swirl to mix. Clamp this reaction flask to a ringstand, and lower it into a water bath containing room-temperature water (20–25°C). Adjust the water level so that the bath is about two-thirds full, and measure the water temperature. Now fill a clean, dry 25-mL buret with a 0.005 M solution of KOH in 40% ethanol (be sure to use the right KOH solution!), and record the initial buret reading ⟨OP–6⟩. Have ready a watch that measures seconds or a timer.

The water temperature should remain nearly constant throughout a kinetic run.

Weigh 1 mmol of 1-bromoadamantane, and dissolve it in 10 mL of *absolute* ethanol. Add 1 mL of this solution to the reaction flask, swirl to mix, and immediately record the time of addition to the nearest second (or start the timer). Add 1 mL of the KOH solution to this reaction mixture, shake to mix, and record the buret reading. (If the solution does not turn blue, add more KOH until it does.) Place the indicator blank beside the reaction mixture, and record the time (to the nearest sec-

The extra bromoadamantane solution can be stoppered and saved for another run or used by co-workers.

ond) when the solution changes to the same shade of green as the blank. Within a minute of the color change, add another 1-mL portion of the KOH solution, and record the time when the solution again turns from blue to green. Repeat the addition of 1-mL portions of KOH, recording the buret reading and the time of the color change after each addition, until 10 portions have been added.

The color change will be easier to see if you set the flasks on a sheet of white paper.

Determination of V_∞. After the last color change, insert a cork loosely into the neck of the flask, and heat the reaction mixture ⟨OP–7⟩ in a water bath at 60°C for 30 minutes. Cool the solution to room temperature, and titrate with the 0.005 M solution of KOH in 40% ethanol to the green endpoint. If the green color fades to yellow after standing, heat the flask in the water bath a few minutes longer, and, after cooling, again titrate the solution to the green endpoint. Subtract your initial buret reading (recorded before you started the kinetic run) from the final reading to get V_∞.

B. Solvolysis of t-Butyl Bromide in 80% Ethanol

Fill a clean, dry 25-mL buret with the 0.05 M solution of KOH in 80% ethanol (be sure to use the right KOH solution!), and record the initial buret reading. Accurately measure 21 mL of 95% ethanol and 4 mL of deionized water into a 125-mL Erlenmeyer flask. Add 2 drops of bromthymol blue solution, swirl to mix, and support the flask in the room-temperature water bath. Pipet in 0.1 mL of *t*-butyl bromide **[suspected carcinogen! do not pipet by mouth!]**, shake to mix, and immediately record the time of addition (or start your timer). Carry out the kinetic run by the same procedure you followed in part A, using at least 10 1-mL portions of the KOH solution. After the last addition, stopper the flask loosely, and heat the reaction mixture in a water bath at 60°C for about 10 minutes. Then titrate it with the 0.05 M solution of KOH in 80% ethanol to determine V_∞.

Use a clean, absolutely dry 1-mL measuring pipet. Do not withdraw the t-butyl bromide directly from the reagent bottle or return any unused liquid to the reagent bottle.

Report. For each run, compute the time (t) of each color change in seconds, measured from the time of addition of the alkyl bromide ($t = 0$). Calculate $\ln(1 - V/V_\infty)$ for each t value, where V is the total volume of KOH that has been added up to that time. Using good graph paper, plot $\ln(1 - V/V_\infty)$ versus t, and determine the value of k (in s^{-1}) for the alkyl bromide. (Alternatively, you may use a calculator or computer that has a linear regression program to determine the least-squares slope from your data.) From its rate constant in 80% ethanol, use Equation 2 to estimate k for t-butyl bromide in 40% ethanol. Calculate the relative solvolysis rates for the two alkyl bromides in 40% ethanol, $k_{t\text{-BuBr}}/k_{\text{AdBr}}$. Indicate which tertiary carbenium ion (1-adamantyl or t-butyl) is more stable, and tell how your experimental results support this conclusion.

Collateral Project

Measure rate constants for the solvolysis of 1-bromoadamantane in different solvents, and calculate their Y values, using $k_0 = 5.10 \times 10^{-7}$ s^{-1} and $m = 1.20$ for 1-bromoadamantane. Suggested solvents are 50% methanol, 50% ethanol, and 40% 2-propanol. You can also measure rate constants at different temperatures, and determine E_{act} from an Arrhenius plot of ln k versus $1/T$.

MINILAB 10

Nucleophilic Substitution Reactions of Alkyl Bromides

SAFETY PRECAUTION

The alkyl bromides are all toxic and have harmful vapors, and some of them are suspected carcinogens. Avoid contact, and do not breathe their vapors. Avoid contact with the silver nitrate reagent.

Measure 1 mL of 15% sodium iodide in acetone into each of three clean, dry 10-cm test tubes. Add 2 drops of n-butyl bromide (1-bromobutane) to the first test tube, 2 drops of sec-butyl bromide (2-bromobutane) to the second, and 2 drops of t-butyl bromide (2-bromo-2-methylpropane) to the third. Stopper and shake the tubes. Observe them closely during the first 15–20 minutes, then at intervals throughout the lab period. Record your observations in your laboratory notebook, indicating the approximate time it takes for each sample to show evidence of a reaction (if it does).

Measure 2 mL of a 0.1 M solution of silver nitrate in ethanol into each of three clean, dry 10-cm test tubes. Add a drop of *n*-butyl bromide to the first test tube, a drop of *sec*-butyl bromide to the second, and a drop of *t*-butyl bromide to the third. Stopper and shake the tubes. Observe them for evidence of a reaction as before.

Write balanced equations for the substitution reactions undergone by each halide (assume that the nucleophile in the second reaction is ethanol). Arrange the alkyl halides in order of reactivity in each reaction, classify each reaction as S_N1 or S_N2, and explain your reasoning. Propose mechanisms for both of the reactions undergone by 2-bromobutane.

Exercises

1. Write a mechanism for the solvolysis reaction of 1-bromoadamantane with ethanol.

2. Which solvent promotes the formation of a carbenium ion more effectively, 40% ethanol or 80% ethanol? Explain.

3. Hydroxide ion can also act as a nucleophile in the substitution reactions of alkyl halides, yet the addition of KOH during the kinetic runs in this experiment has little effect on the rates of the reactions. Explain.

4. In addition to the solvolysis reactions described above, *t*-butyl bromide may undergo an E1 elimination reaction to form 2-methylpropene. Explain why a similar elimination reaction of 1-bromoadamantane is not likely to occur.

5. **(a)** Using your experimental rate constant, calculate the time it should take for 90% of your 1-bromoadamantane to react at room temperature. **(b)** How long should it take for 90% of the *t*-butyl bromide to react in 40% ethanol under the same conditions?

Library Topic

Read the article in *J. Chem. Educ. 50,* 780 (1973) and, using some of the references cited therein, write a short report on the medical uses of adamantane derivatives.

The Dehydration of Methylcyclohexanols

Reactions of Alcohols. Preparation of Alkenes. Elimination Reactions. Carbenium Ions. Regioselectivity.

Operations

OP–19 Washing Liquids

OP–20 Drying Liquids

OP–25 Simple Distillation

OP–27 Fractional Distillation

OP–32 Gas Chromatography

Objectives

To carry out the dehydration of a methylcyclohexanol and determine the composition of the product mixture.

To determine the preferred direction of elimination in the dehydration of 2-methylcyclohexanol.

To look for evidence of carbenium-ion rearrangements in the dehydration of 4-methylcyclohexanol.

To learn about the mechanisms of elimination reactions

To learn how Le Châtelier's principle can be applied to increase the yield of a dehydration reaction.

SITUATION

Cresols are used in Lysol and other disinfectants, solvents, photographic developers, and explosives.

The *cresols* are important commercial chemicals produced by alkaline extraction of coal-tar fractions and various manufacturing processes. Cresols can be converted to the corresponding methylcyclohexanols by catalytic hydrogenation, as in the reduction of *ortho*-cresol to 2-methylcyclohexanol:

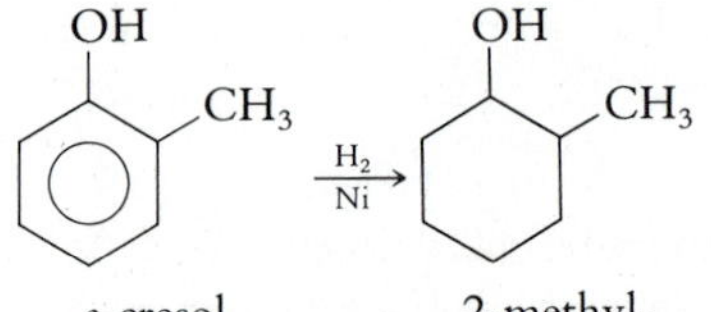

o-cresol   2-methyl-cyclohexanol

Methylcyclohexanols can be dehydrated to methylcyclohexenes, which are produced for sale to research chemists by several chemical companies. Dehydration of 2-methylcyclohexanol might produce 1-methylcyclohexene, 3-methylcyclohexene, or a mixture of both.

1-methyl-   3-methyl-
cyclohexene   cyclohexene

When pure, these alkenes are quite costly; both were priced at around $5 per milliliter in one company's 1987 catalog. In this experiment, you will carry out the dehydration of a methylcyclohexanol and determine the composition of the product by gas chromatography.

BACKGROUND

Regioselective Reactions

More than a century ago at the University of Kazan, Vladimir Vasilevich Markovnikov and Alexander Zaitsev (or Saytzeff) were investigating a chemical reaction both backward and forward: Markovnikov was adding hydrogen iodide to alkenes to prepare alkyl iodides, and Zaitsev was removing the hydrogen iodide from alkyl iodides to prepare alkenes. Markovnikov discovered that hydrogen iodide adds to propene to form mainly isopropyl iodide:

$$CH_3CH{=}CH_2 + HI \longrightarrow CH_3\overset{\overset{\textstyle I}{\textstyle |}}{C}HCH_3$$

From this and other results, Markovnikov formulated his famous rule, which can be expressed as follows for a hydrogen-containing species represented by HZ:

> *Markovnikov's rule:* When HZ adds to the carbon-carbon double bond of an unsymmetrical alkene, hydrogen adds preferentially to the carbon atom that already has more hydrogens.

In the meantime, Zaitsev learned that dehydrohalogenation of *sec*-butyl iodide by alcoholic potassium hydroxide yields mainly 2-butene:

$$CH_3CH_2\overset{\overset{\displaystyle I}{\displaystyle |}}{C}HCH_3 \xrightarrow{\text{KOH}} CH_3CH{=}CHCH_3$$

He proposed an analogous rule for elimination reactions:

Zaitzev's rule: When HZ is removed from a species to form an alkene, hydrogen is lost preferentially from the carbon atom, of those adjacent to the carbon atom bonded to Z, that has fewer hydrogens.

Markovnikov's and Zaitzev's rules together can be paraphrased by the well-known maxim "The rich get richer and the poor get poorer." Molecules often react selectively, favoring some products and not others — Zaitzev's reaction might have yielded as much 1-butene as 2-butene, but it did not. When a reaction could produce two or more different structural isomers but in fact yields mainly one of them, the reaction is said to be *regioselective.* Regioselectivity results from the tendency of chemical reactions to follow the easiest path, which usually leads to the most stable intermediate or product. Zaitzev's rule works reasonably well because, in most cases, it predicts the formation of the most stable alkene. 2-Butene was the major product of Zaitzev's reaction because it is more stable than 1-butene, not because hydrogen-poor carbon atoms have some innate tendency to lose the hydrogens they have.

Although generalizations like Zaitzev's rule can help us predict the products of many organic reactions, organic chemistry remains an empirical science — we cannot be certain that a rule that is valid for one system under a given set of conditions will apply equally well under different circumstances. Each system must be studied experimentally to find out whether or not it behaves in the expected manner. For instance, elimination of hydrogen fluoride from 2-fluorohexene with sodium methoxide yields mainly 1-hexene, in violation of Zaitzev's rule.

Alkenes are stabilized by the presence of alkyl substituents on the carbon atoms connected by the double bond.

$$CH_3CH_2CH_2CH_2\overset{\overset{\displaystyle F}{\displaystyle |}}{C}HCH_3 \xrightarrow{\text{NaOCH}_3} CH_3CH_2CH_2CH_2CH{=}CH_2$$

To understand why Zaitzev's rule applies in some cases but not in others, we must consider some mechanisms of elimination reactions. There are three fundamental ways in which hydrogen (as H^+) and a leaving group (Z^-) can be removed from a molecule: (1) Z^- can leave before H^+; (2) H^+ can leave before Z^-; (3) H^+ and Z^- can leave simultaneously. As illustrated in Figure 1, these different sequences characterize the reaction mechanisms called E1, E1cB, and E2, respectively.

Figure 1.   Generalized mechanisms of elimination reactions

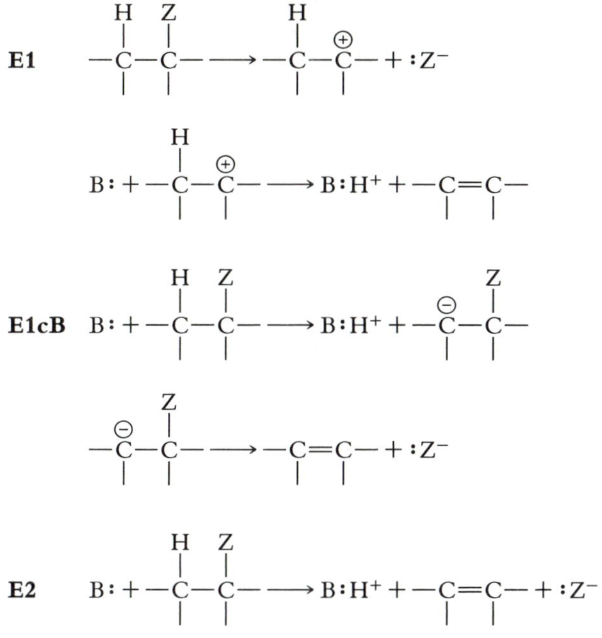

E1 stands for elimination-unimolecular; E2 for elimination-bimolecular; and E1cB for elimination-unimolecular-conjugate base (referring to the carbanion intermediate, a conjugate base of the substrate).

E1

$$-\overset{\overset{\displaystyle H}{|}}{\underset{|}{C}}-\overset{\overset{\displaystyle Z}{|}}{\underset{|}{C}}- \longrightarrow -\overset{\overset{\displaystyle H}{|}}{\underset{|}{C}}-\overset{\oplus}{\underset{|}{C}}- \; + \; :Z^-$$

$$B: + -\overset{\overset{\displaystyle H}{|}}{\underset{|}{C}}-\overset{\oplus}{\underset{|}{C}}- \longrightarrow B:H^+ + -\overset{|}{C}=\overset{|}{C}-$$

E1cB

$$B: + -\overset{\overset{\displaystyle H}{|}}{\underset{|}{C}}-\overset{\overset{\displaystyle Z}{|}}{\underset{|}{C}}- \longrightarrow B:H^+ + -\overset{\ominus}{\underset{|}{C}}-\overset{\overset{\displaystyle Z}{|}}{\underset{|}{C}}-$$

$$-\overset{\ominus}{\underset{|}{C}}-\overset{\overset{\displaystyle Z}{|}}{\underset{|}{C}}- \longrightarrow -\overset{|}{C}=\overset{|}{C}- \; + \; :Z^-$$

E2

$$B: + -\overset{\overset{\displaystyle H}{|}}{\underset{|}{C}}-\overset{\overset{\displaystyle Z}{|}}{\underset{|}{C}}- \longrightarrow B:H^+ + -\overset{|}{C}=\overset{|}{C}- \; + \; :Z^-$$

Note: B may be the reaction solvent or an added base, such as methoxide ion. The leaving group may be protonated during the reaction and leave as a neutral species, HZ.

The E1 mechanism involves the formation of a carbenium-ion intermediate; an E1cB reaction produces a carbanion intermediate; an E2 reaction is a concerted (one-step) process with no intermediate.

The E1 and E1cB mechanisms represent the extremes of a continuous spectrum of mechanistic possibilities, with the E2 mechanism somewhere in the middle. An E2 reaction requires a strong base capable of removing a proton directly from the substrate. Elimination from substrates that have good leaving groups usually follows Zaitzev's rule; with poor leaving groups, it may not. For example, the activated complex for the E2 elimination of hydrogen fluoride from 2-fluorohexane using sodium methoxide apparently has considerable carbanion character:

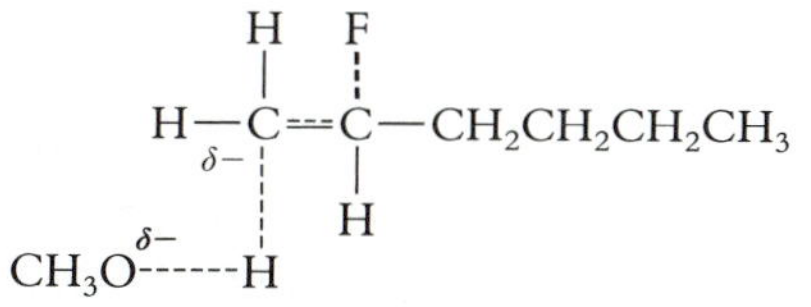

Carbenium ions are positive species containing a trivalent, electron-deficient carbon atom; they are sometimes called carbonium ions or carbocations. Carbanions are trivalent species with negatively charged carbon atoms.

$$H-\underset{\delta-}{\overset{\overset{\displaystyle H}{|}}{C}}\!=\!=\!=\!\overset{\overset{\displaystyle F}{\vdots}}{\underset{\underset{\displaystyle H}{|}}{C}}-CH_2CH_2CH_2CH_3$$

$$\underset{\delta-}{CH_3O}\text{-----}H$$

Because methoxide ion is a strong base and fluoride ion is a poor leaving group, removal of H^+ by the base has proceeded further than ejection of F^- or formation of the double bond in this activated complex. Thus, the reaction can be described as "E1cB-like" even though its mechanism is concerted. In such reactions, the direction of elimination may depend more on the acidity of the proton being removed than on the stability of the incipient double bond. Because alkyl substituents reduce the acidity of adjacent protons, methoxide ion is more likely to remove a proton from carbon-1 than from carbon-3 of 2-fluorohexane, resulting in formation of the less stable alkene.

In an E1 reaction, the substrate must first be ionized to yield a carbenium ion and the leaving group. Then removal of a proton from the carbenium ion by a basic species forms the carbon-carbon double bond. Such a process should be favored by (1) a good leaving group, (2) a substrate that provides a relatively stable carbenium ion, (3) a polar solvent that promotes the ionization of the substrate, and (4) a base (usually the solvent) that is strong enough to remove a proton from the carbenium ion but not strong enough to remove one from the substrate. E1 reactions generally follow Zaitzev's rule, since deprotonation of the intermediate carbenium ion tends to yield the most stable alkene.

E1 eliminations also allow the possibility of rearrangement; concerted reactions do not. In 1944, when Albert Henne and Alfred Matuszak dehydrated 2-heptanol using phosphoric acid, they discovered that the predominant product of the dehydration was the unexpected alkene 3-heptene:

See the article by Albert L. Henne and Alfred H. Matuszak, *Journal of the American Chemical Society, 66,* 1649 (1944).

The carbon-carbon double bond of 3-heptene connects carbon-3 and carbon-4 in the seven-carbon chain; neither of these atoms was bonded to the OH group that was eliminated from 2-heptanol. The formation of 3-heptene can be explained by assuming that the protonated alcohol loses water to form a 2-heptyl carbenium ion, which rearranges to a 3-heptyl carbenium ion, as shown in the following mechanism. Elimination of a proton from carbon-4 of this carbenium ion then forms the observed product.

protonated 2-heptanol   2-heptyl carbenium ion

rearranged (3-heptyl) carbenium ion   3-heptene

METHODOLOGY

In this experiment, you will carry out the dehydration of either 2-methylcyclohexanol or 4-methylcyclohexanol by heating the alcohol in the presence of phosphoric acid. Dehydration of a secondary alcohol proceeds readily with about half a mole of phosphoric acid for every mole of the alcohol. By protonating an alcohol, the acid catalyst converts the poor leaving group OH^- to a much better leaving group, H_2O.

Elimination of H^+ and H_2O from the protonated alcohol yields an alkene, with the unprotonated alcohol serving as the reaction solvent.

According to Le Châtelier's principle, removing a product from a chemical system at equilibrium shifts the equilibrium in the direction favoring the formation of the products.

You will carry out the dehydration reaction in a distillation apparatus so that the products (water and alkene) will continuously distill out of the reaction mixture as they are formed. Their removal will shift the equilibrium to the right and thus increase the yield of alkene.

$$alcohol \rightleftharpoons alkene\uparrow + water\uparrow$$

The upward-pointing arrows in the equation indicate that the products are vaporized under the reaction conditions, not that they are gases at room temperature.

If the reaction mixture is heated to a temperature above the boiling points of the product alkenes but below that of the alcohol, most of the unreacted alcohol will remain in the reaction flask while the alkenes and water distill into the receiving flask. Interposing a vertical column between the boiling flask and the still head will further reduce the amounts of unwanted alcohol, acid, and by-products in the distillate.

The progress of the reaction can be followed by measuring the volume of alkene in the distillate, using a graduated cylinder as a receiver. When the reaction is over, the residue in the reaction flask may begin to foam and emit white vapors. You should remove the heat source at this time, because overheating the residue may form a black tar and generate toxic fumes.

After washing and drying the organic layer, you will purify the product mixture by simple distillation and analyze it by gas chromatography. If you started with 2-methylcyclohexanol, your gas chromatogram may show peaks for both 1- and 3-methylcyclohexene; from the relative areas of these peaks, you can determine the percentage composition of the product mixture and find out whether or not the reaction followed Zaitzev's rule. 4-Methylcyclohexanol can yield only 4-methylcyclohexene by direct elimination, but 1- and 3-methylcyclohexene may be formed by carbenium-ion rearrangements.

4-methyl cyclohexene

3-methyl cyclohexene

1-methyl cyclohexene

no rearrangement required

rearrangement required

The peaks for 3- and 4-methylcyclohexene may not be well separated on your gas chromatogram, but the presence or absence of a 1-methylcyclohexene peak should show whether or not rearrangement has taken place.

If you do Minilab 11, you will prepare a gaseous alkene by the dehydration of *t*-butyl alcohol, and study some of its properties.

PRELAB ASSIGNMENTS

1. Calculate the mass and volume of 150 mmol of both methylcyclohexanols and the theoretical yield of methylcyclohexenes.

2. Estimate the maximum volume of alkene that should be in the distillate when the reaction is complete. Use 0.80 g/mL as the approximate density of the alkene mixture.

Reactions and Properties

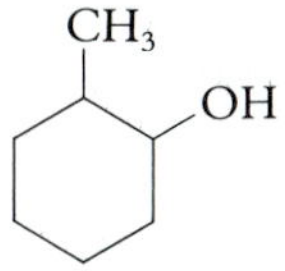

$$\xrightarrow{H_3PO_4} C_7H_{12} + H_2O$$

methyl-
cyclohexenes

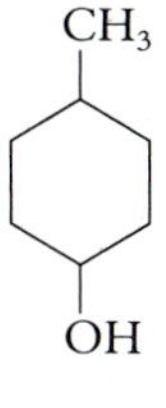

$$\xrightarrow{H_3PO_4} C_7H_{12} + H_2O$$

methyl-
cyclohexenes

Table 1.   Physical properties

	M.W.	b.p.	d.
2-methylcyclohexanol	114.2	166	0.930
4-methylcyclohexanol	114.2	173	0.914
1-methylcyclohexene	96.2	110	0.813
3-methylcyclohexene	96.2	104	0.801
4-methylcyclohexene	96.2	102	0.799
phosphoric acid (85%)	98.0		1.70

Note: The molecular weight given for phosphoric acid is for the pure acid; 85% phosphoric acid is about 14.7 M.

PROCEDURE

Students can work in pairs if desired, each using one of the two cyclohexanols.

The methylcyclohexanols and methylcyclohexenes are flammable, and inhalation, ingestion, or skin absorption may be harmful. Avoid contact, do not breathe their vapors, and keep them away from flames or hot surfaces.

Phosphoric acid can cause serious burns, particularly to the eyes; do not allow it to contact your eyes, skin, or clothing.

Reaction and Separation. Measure 150 mmol of 2-methylcyclohexanol or 4-methylcyclohexanol into a round-bottomed flask, mix in 5 mL of 85% phosphoric acid **[contact hazard!]**, and drop in a few acid-resistant boiling chips. Clamp the flask to a ringstand over a suitable heat source, and assemble the apparatus pictured in Figure 7 of OP–27. Do not put any packing in the column. Use a 25-mL graduated cylinder or another graduated container as the receiver.

If you use a heating mantle or oil bath, support it well above the bench top so that you can lower and remove it when the reaction is over.

Boil the reactants gently so that vapors ascend slowly up the column and begin to condense into the receiver. When the vapors reach the still head and the temperature has stabilized, record the still-head temperature. Observe it at intervals throughout the reaction. Control the rate of heating so that the distillation rate is 1–2 drops per second or less and the temperature stays below 120°C. Estimate the volume of the alkene (top) layer in the distillate, and as it nears the calculated value, monitor the still-head temperature constantly. Lower the heat source and turn it off when you observe a marked temperature change at the still head or foaming and dense white fumes in the reaction flask. Measure the volume of alkene and water in the distillate (see Exercise 2). Carefully discard the residue in the reaction flask as directed by your instructor. Wash ⟨OP–19⟩ the distillate with two 10-mL portions of saturated aqueous sodium bicarbonate. Dry ⟨OP–20⟩ the alkene layer with anhydrous calcium chloride or another suitable drying agent.

The water that codistills with the alkene reduces its boiling temperature, so the still-head temperature may be lower than the expected boiling point of the product.

Purification.   Purify the product by simple distillation ⟨OP –
25⟩, collecting all the liquid that distills above 100°C and
below 115°C. Weigh the product and save it for analysis.

Analysis.   Analyze the product by gas chromatography ⟨OP –
32⟩ as directed by your instructor. Measure the area and retention time of each peak on your gas chromatogram. Identify the
peaks by comparison with a chromatogram provided by your
instructor or by spiking your product mixture with an authentic sample of 1-methylcyclohexene and obtaining a chromatogram of the resulting mixture. Turn in the remaining product
in a labeled vial.

Report.   Assuming that the detector response factors for the
alkenes are equal, calculate the percentage composition of the
product mixture and the mass of each alkene in the product
mixture. Obtain these data from a co-worker who started with
the other alcohol. Tell whether the dehydration of 2-methyl-
cyclohexanol follows Zaitzev's rule and whether the dehydration of 4-methylcyclohexanol involves carbenium-ion rearrangements. Indicate which elimination mechanism (E1,
E1cB, or E2) is most consistent with the experimental conditions and results. Cite the experimental evidence that supports
your conclusion. Propose mechanisms for the dehydration of
both alcohols, showing how each alkene in the product mixture was formed.

*If the peaks for 3- and 4-methylcyclohexene
are not separated on the gas chromatogram,
calculate their combined percentage.*

Collateral Projects

1.   Test the product mixture with bromine ⟨Test C-7,
page 554⟩ or potassium permanganate ⟨Test C-19, page 562⟩
solution, and interpret the results.

2.   You can dehydrate other alcohols, such as cyclohexanol, 3-methylcyclohexanol, and 4-methyl-2-pentanol, by the
same procedure if you adjust the distillation temperature for
the alkene or alkene mixture anticipated. The product mixtures can be analyzed by gas chromatography and the results
interpreted.

Preparation and Properties of a Gaseous Alkene

MINILAB 11

SAFETY PRECAUTIONS   **Avoid contact with sulfuric acid, which can cause serious burns. Avoid contact with *t*-butyl alcohol, and do not breathe its vapors.**

Keep the gas you collect away from flames except when you are testing its flammability.

If the alcohol has solidified, melt it on a steam bath before measuring.

Figure 2.   Apparatus for gas generation

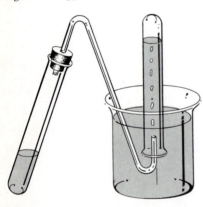

Fill four 15-cm test tubes with water, stopper them tightly, and invert them in a 1-L beaker or pneumatic trough that is about three-quarters filled with water. Remove the stoppers under water so that no water flows out of the test tubes, and leave them inverted in the beaker or trough. Measure 1 mL of *t*-butyl alcohol (2-methyl-2-propanol) into a 15-cm Pyrex test tube, add 5 drops of concentrated sulfuric acid **[contact hazard!]**, mix, and drop in a boiling chip.

Assemble the apparatus illustrated in Figure 2, with the reaction tube clamped to a ringstand and the free end of the gas delivery tube under water. Heat the reaction tube gently on a steam bath until a steady stream of gas bubbles emerges from the delivery tube; then invert one of the water-filled test tubes over the delivery tube until it is filled with gas **[fire hazard!]**. Remove this test tube and repeat the process with the other collecting tubes, stoppering each one as soon as it fills with gas. Turn off the steam, and cool the reaction tube in cold water to stop the generation of gas. Set aside the first collecting tube (which contains air). Add 5 drops of 0.1 M aqueous potassium permanganate to the second tube, and then stopper and shake it. Repeat this procedure with the third tube, using 5 drops of a 1 M solution of bromine in Freon TF. Carefully light one end of a wooden applicator stick or wood splint with a match or burner, and lower the burning end into the fourth tube.

Write the structure of the gas and a balanced equation for its synthesis from *tert*-butyl alcohol. Describe your observations during the three tests, and write balanced equations for the chemical reactions that occurred in each of them.

Exercises

1. **(a)** Under suitable conditions a carbenium ion may (1) combine with a nucleophile or (2) add to the double bond of an alkene to form a new carbenium ion, which may then lose a proton to yield another alkene. Write the structures of some by-products that might form by the combination of 2-methyl-cyclohexyl cations with substances (both reactants and products) present in your reaction mixture. **(b)** Propose mechanisms for the formation of these by-products.

2. Calculate the theoretical volume of water in the distillate and the ratio of the volume of alkene to the volume of water, and compare these with the values you observed.

3. Construct a flow diagram for this experiment, including the by-products you predicted in Exercise 1.

4. In the Situation section, it was stated that the dehydration of 2-methylcyclohexanol may yield 1-methylcyclohexene and 3-methylcyclohexene. Why is 2-methylcyclohexene not a possible product?

5. **(a)** Predict what alkene would be the major product that would result from dehydrating each of the following alcohols, with no carbenium-ion rearrangements.

(1)
$$\underset{\underset{\displaystyle CH_3}{|}}{CH_3}\underset{\underset{\displaystyle CH_3}{|}}{\overset{\overset{\displaystyle CH_3\ OH}{|\ \ \ \ |}}{CHCHCHCH_3}}$$

(2)

(3)
$$CH_3\underset{\underset{\displaystyle CH_3}{|}}{\overset{\overset{\displaystyle CH_3\ OH}{|\ \ \ \ \ |}}{C}}\!-\!CH\!-\!CH_3$$

(b) In each case, predict the most stable dehydration product that could result after a single carbenium-ion rearrangement.

Library Topic

Read the paper by Henne and Matuszak cited in the Background section. Report their conclusions regarding the mechanism of the dehydration reaction. Tell how their preferred mechanism for carbenium-ion rearrangements differs from the currently accepted one.

The Synthesis of 7,7-Dichloronorcarane Using a Phase-Transfer Catalyst

Reactions of Alkenes. Preparation of Alicyclic Compounds. Alkyl Halides. Carbenes. Phase-Transfer Catalysis.

Operations

OP–7c	Temperature Monitoring
OP–9	Mixing
OP–13	Extraction
OP–14	Evaporation
OP–19	Washing Liquids
OP–20	Drying Liquids
OP–25a	Semimicro Distillation
OP–33	Infrared Spectrometry

Objectives

To synthesize 7,7-dichloronorcarane from cyclohexene by a carbene addition reaction.

To learn how phase-transfer catalysts work and how to use one in an organic synthesis.

To learn about the infrared spectra of alkyl chlorides and cyclopropanes.

To learn about some natural and synthetic compounds having cyclopropane rings.

SITUATION

Because hydrocarbons with small, highly strained rings have unusually high heats of combustion, a Monsanto research team has synthesized a number of strained hydrocarbons as possible constituents of high-energy fuels for jet fighters. By adding methylene ($:CH_2$) to the double bonds of β-pinene and bicyclohexylidene, for example, the team was able to prepare the following unusual tricyclic hydrocarbons:

Monsanto synthesis of strained hydrocarbons

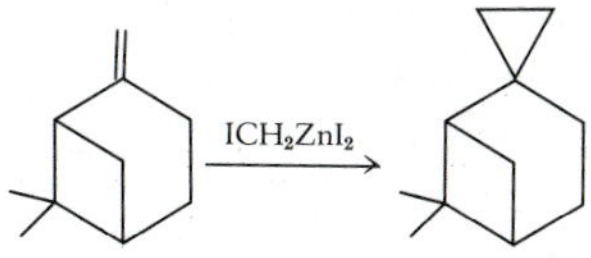

β-pinene

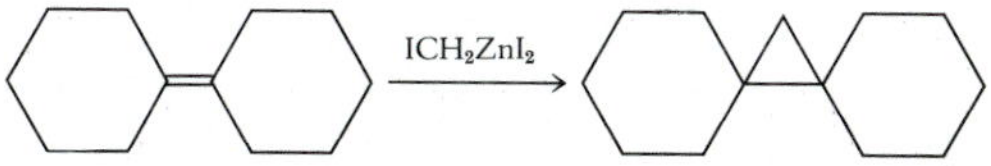

bicyclohexylidene

In 1961, W. R. Moore synthesized tricyclo[4.1.0.0^{2,7}]heptane by treating 7,7-dibromonorcarane with methyllithium in ethyl ether. The reaction apparently involves insertion of the divalent carbon atom of an intermediate carbene into an adjacent C—H bond.

See the article by William R. Moore et al., *J. Am. Chem. Soc. 83,* 2019 (1961).

Moore's preparation of tricyclo[4.1.0.0^{2,7}]heptane

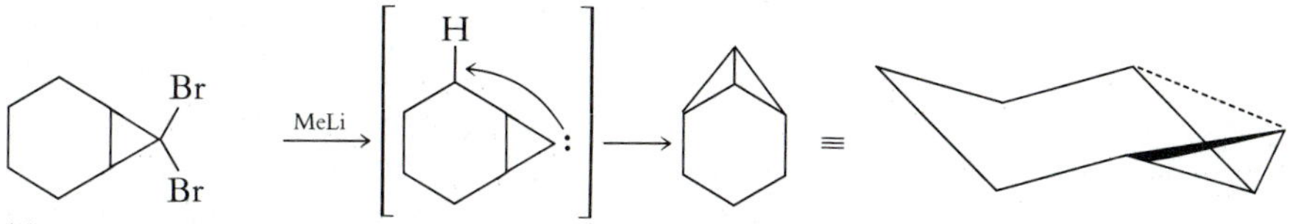

7,7-dibromonorcarane   carbene
                              intermediate

tricyclo[4.1.0.0^{2,7}]heptane

With its two cyclopropane rings fused side by side onto a cyclohexane ring, the exotic product of Moore's synthesis must have very high bond-angle strain, which could make it an excellent—if rather expensive—high-energy jet fuel. It should be possible to reduce the cost of its synthesis by generating the carbene intermediate from a less expensive starting material, 7,7-dichloronorcarane. In this experiment, you will try to synthesize this compound from cyclohexene by a recently developed emulsion technique using a phase-transfer catalyst.

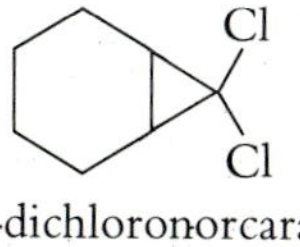

7,7-dichloronorcarane

hexagon

△

triangle

Euell Gibbons, an authority on edible wild plants, tasted arborvitae tea and declared that he "would almost prefer rheumatism."

The sirens of Greek mythology were beautiful female creatures who lured mariners to destruction with their singing.

thujone

3-carene

sirenin

BACKGROUND

The Ubiquitous Triangle

It often appears that nature loves the hexagon, since so many natural compounds contain six-membered rings in their molecules. This might be expected, since both the aromatic benzene ring and the unstrained cyclohexane ring are unusually stable compared to other possible ring structures. So it is surprising to find numerous examples of the triangle — the comparatively unstable cyclopropane ring — in everything from arborvitae to water molds.

Henry David Thoreau reported that northwoods lumbermen of the last century were accustomed to drink "a quart of arborvitae, to make (them) strong and mighty." An extract of the leaves of the arborvitae (white cedar) was thought to impart strength and prevent illness, particularly rheumatism. One of the main constituents of the oil from arborvitae and other *Thuja* species is the bicyclic terpenoid called thujone, which contains a three-membered ring fused onto a cyclopentane ring. Thujone is also found in the essential oils from tansy, sage, and wormwood. A six-and-three ring combination appears in 3-carene, which is obtained from various *Pinus* species and is a major constituent in turpentine oil from Sweden and Finland. Another six-and-three bicyclic is found in the whimsically named sirenin, which is a sperm attractant produced by the female gametes of a water mold, *Allomyces javanicus*. An unusual seven-and-three ring combination occurs in ledol (ledum camphor), a major constituent of the essential oil from a northern shrub, Labrador tea. One of the more startling sights in nature is a pumpkin-colored mushroom that glows in the dark around Halloween. This is the "Jack-O' Lantern" fungus, *Clitocybe illudens,* which has sometimes been mistaken for the edible chanterelle mushroom (with unpleasant consequences). It contains several antibiotic components, illudin-M and illudin-S, which have a cyclopropane ring fused by its apex to a six-membered ring.

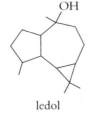

ledol

illudin-S

Pyrethrin is a natural biodegradable insecticide, nontoxic to humans, obtained from the flowers of a daisylike plant, *Chrysanthemum cineariaefolium.* It is made up of a mixture of esters such as cinerin I that contain a cyclopropane ring in the carboxylic acid portion. In recent years, a number of *pyrethroids,* synthetic analogs of the natural compounds, have been developed in an effort to find relatively safe but effective insecticides to replace the environmentally unsound "hard" pesticides. One of the most powerful of these is decamethrin, which is over sixty times more lethal to houseflies than parathion is and over six hundred times more effective against certain mosquitoes than DDT is. Because of their low toxicity and biodegradable properties, it seems quite possible that pyrethroids will become the insecticides of choice in the future.

Renewed interest in "natural" pesticides was stimulated, in part, by Rachel Carson's book *Silent Spring.*

cinerin I

decamethrin

METHODOLOGY

In this experiment, you will generate the highly reactive intermediate dichlorocarbene ($:CCl_2$) by the reaction of chloroform with a strong aqueous solution of sodium hydroxide. The carbene will then combine with cyclohexene present in the reaction mixture to form 7,7-dichloronorcarane. Unfortunately, sodium hydroxide is insoluble in the organic phase, and both chloroform and cyclohexene are virtually insoluble in water, so the reactants could have a hard time getting together. To circumvent this difficulty, you will use a phase-transfer catalyst to "escort" the reactant molecules across the phase boundary.

The principles of phase-transfer catalysis (PTC) can be readily explained with reference to the double-displacement reaction of an alkyl halide with sodium cyanide to form a nitrile. If a high-molecular-weight halide such as 1-chlorooctane is heated with aqueous sodium cyanide, there is no reaction — the cyanide stays in the aqueous layer, the alkyl halide stays in the organic layer, and "never the twain shall meet." If, instead of sodium cyanide, one uses a quaternary ammonium (Q) salt such as tetra-*n*-butylammonium cyanide, the reaction proceeds quite readily and gives a high yield of

Reaction of alkyl halide with sodium cyanide:

$$RCl + Na^+CN^- \longrightarrow RCN + Na^+Cl^-$$

Improved reaction with quaternary ammonium cyanide:

$$RCl + Q^+CN^- \longrightarrow Q^+Cl^- + RCN$$

$$[Q^+ = (CH_3CH_2CH_2CH_2)_4N^+]$$

product. There are several reasons for this enhanced reactivity of the cyanide: First (and most important), the sixteen carbon atoms of the cation make it soluble in the organic phase, and where the cation goes, the anion must follow. Second, cyanide ion is more reactive in the organic phase than it would have been in the aqueous phase, since it is not solvated by water molecules that would shield it from the alkyl halide and decrease its reactivity. Finally, the bulky alkyl groups around the positive nitrogen of the cation decrease the attractive forces between cation and anion and allow the cyanide ion more freedom to attack the alkyl halide. So by substituting a quaternary ammonium ion for the sodium ion in the cyanide salt, we have a good procedure for bringing about the desired transformation. Its main drawback is that the quaternary ammonium cyanide is much more expensive than sodium cyanide.

The PTC technique gets around the high cost of quaternary ammonium salts by recycling them after each reaction step. If the quaternary ammonium cation in the product (tetrabutylammonium chloride) can be made to pick up some more cyanide to react with the alkyl halide, it will function as a true catalyst, accelerating the reaction without being used up. All that is needed is a reservoir of cyanide ion and a small amount of the Q salt to keep the reaction going. The reservoir can be provided by an aqueous layer containing sodium cyanide.

Figure 1 diagrams the process, which occurs as follows: A catalytic amount of Q^+Cl^- combines with cyanide ion in the aqueous phase, and the Q^+CN^- that forms crosses over to the organic layer. There it reacts with the alkyl halide to produce the nitrile (RCN) and form more Q^+Cl^-, which migrates across the interface, picks up some more cyanide, shuttles it back into the organic layer to react with the alkyl halide and form more product and Q^+Cl^-, and so on, until the alkyl halide or the cyanide ion is used up.

You will use the viscous liquid tricaprylmethylammonium chloride (also known as Aliquat 336) as the phase-transfer catalyst for the synthesis of 7,7-dichloronorcarane. In this catalyst, the quaternary cation contains a methyl group and three alkyl chains of eight or ten carbons attached to a

$[CH_3(CH_2)_n]_3NCH_3^+Cl^-$
tricaprylmethylammonium
chloride (Aliquat 336)
$n = 7$ or 9

Figure 1. Phase-transfer process for nucleophilic substitution reaction

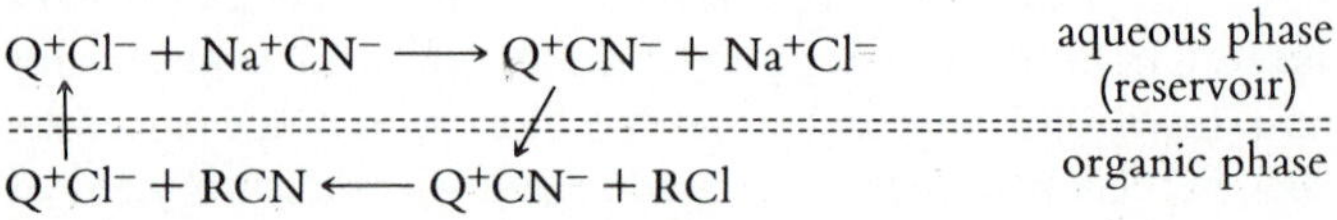

$$Q^+Cl^- + Na^+CN^- \longrightarrow Q^+CN^- + Na^+Cl^-$$

aqueous phase
(reservoir)

$$Q^+Cl^- + RCN \longleftarrow Q^+CN^- + RCl$$

organic phase

nitrogen atom. The initial reaction mixture will consist of an organic phase containing cyclohexene and chloroform and an aqueous phase containing sodium hydroxide, with the catalyst distributed between both phases. The sequence of events in the subsequent reaction is not entirely clear, but the following scenario seems reasonable. A hydroxide ion associated with the quaternary cation can remove a proton from chloroform at the phase boundary to produce CCl_3^-. This ionic species would ordinarily linger in the aqueous phase; but when paired with the quaternary cation, it can invade the organic phase, where it loses a chloride ion to form dichlorocarbene ($:CCl_2$). This highly reactive carbene then attacks a cyclohexene molecule in the organic phase to form the product, and the quaternary cation can return to the aqueous phase to repeat the process.

Some researchers believe that Q^+OH^- passes into the organic phase before reacting with chloroform and that all subsequent reaction steps leading to the product occur in that phase.

In the procedure used for this experiment, the phase-transfer catalyst promotes the formation of an emulsion, in which the organic phase is dispersed throughout the aqueous medium in minute spherical clusters called *micelles*. This greatly increases the area of contact between the phases, which speeds up the reaction by increasing the rate at which reactant molecules cross the phase boundary. The formation of a thick emulsion, with color and texture similar to thick cream, is essential to the success of the reaction. If it does not form, the reaction temperature will not rise much above 40°C and you will recover little, if any, product. A little cyclohexanol will help stabilize the emulsion, and cleaning your glassware thoroughly should remove impurities that might break up the emulsion or prevent its formation. The emulsion can be produced by either manual mixing or magnetic stirring. The manual method requires swirling the reaction flask vigorously enough to "whip" the reactants into a frothy liquid that eventually thickens and becomes completely opaque. With a little practice, you should be able to hold the neck between your thumb and fingers and use the middle finger to rock the flask back and forth with just enough twisting motion to keep the liquid moving around in the flask in a circular path. Your hand and arm should move hardly at all; only the bottom of the flask should be in rapid motion. Shaking the flask with too much enthusiasm may actually break up the emulsion, so try to keep the motion vigorous but not violent. Once the emulsion has stabilized, the swirling need not be as vigorous as before.

Cyclohexene reacts slowly with oxygen to yield a hydroperoxide that is converted by sodium hydroxide to 2-cyclohexenol, which also stabilizes the emulsion. Thus, old, impure cyclohexene may work better than the pure compound!

The reaction is exothermic, and therefore it may be necessary to cool the flask to keep the reactants from boiling away. To monitor the reaction temperature while swirling, you should insert a thermometer into the flask so that its bulb is completely immersed in the emulsion. Keep it in position by carefully holding it against the neck of the flask in the "vee" between your thumb and forefinger. When the reaction is nearly over, the temperature will start to drop spontaneously. The product can be separated from the reaction mixture by extraction and purified by simple distillation. A forerun that distills below 180°C should contain some cyclohexanol and unreacted starting materials.

You will characterize your product by recording its infrared spectrum. The cyclopropane ring of 7,7-dichloronorcarane is highly strained, and bonds to strained ring carbons tend to vibrate at higher frequencies than normal. The carbon-chlorine stretching band for an alkyl halide can occur anywhere between 550 and 850 cm⁻¹, but it tends to be at the higher end of the range when two or more chlorine atoms are bonded to the same carbon; thus, chloroform ($CHCl_3$) has a strong C—Cl band near 750 cm⁻¹ (see Figure 2).

Unlike dichlorocarbene, iodine reacts very slowly (if at all) with most alkenes. However, it does add readily to α- and β-pinene, the major components of turpentine, in a highly exothermic reaction that was used by veterinarians to disinfect the wounds of large animals. The heat of this reaction is sufficient to vaporize the iodine and force the vapors into a wound —with great discomfort for the animal being treated. The pinenes are bicyclic compounds containing a highly strained four-membered ring. Addition of iodine to the double bond involves an intermediate carbenium ion that can undergo skeletal rearrangements, converting the four-membered ring to a five-membered ring. If you do Minilab 12, you will compare the reactivity of one of the pinenes with that of cyclohexene and propose a mechanism explaining your observations.

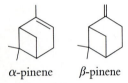

α-pinene β-pinene

See *All Creatures Great and Small* by James Herriot, pp. 27–28 (New York: Bantam Books, 1972) for a graphic description of this procedure.

PRELAB ASSIGNMENT

Calculate the mass and volume of 0.10 mol of cyclohexene and the theoretical yield of 7,7-dichloronorcarane.

Reactions and Properties

cyclohexene $+ CHCl_3 + NaOH \xrightarrow{\text{PTC}}$ 7,7-dichloronorcarane $+ H_2O + NaCl$

Table 1.   Physical properties

	M.W.	b.p.	d.
cyclohexene	82.15	83	0.810
chloroform	119.4	62	1.483
tricaprylmethylammonium chloride	404.2		0.884
7,7-dichloronorcarane	165.1	197–8	

Figure 2.   IR spectrum of chloroform. (Reproduced from *The Aldrich Library of FT-IR Spectra,* by Charles J. Pouchert, with the permission of the Aldrich Chemical Company.)

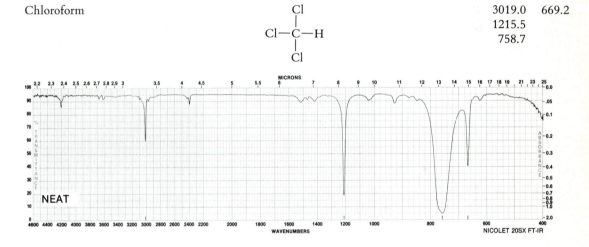

Chloroform

3019.0	669.2
1215.5	
758.7	

PROCEDURE

Chloroform is harmful if inhaled, swallowed, or absorbed through the skin. It is known to induce kidney and liver tumors when ingested by rats or mice and is a suspected carcinogen in humans. Wear gloves, and work under a hood; avoid contact with the liquid, and do not breathe its vapors.

Cyclohexene is very flammable, and inhalation, ingestion, or skin absorption may be harmful. Avoid contact, do not breathe its vapors, and keep it away from hot surfaces and open flames.

SAFETY PRECAUTIONS

chloroform   cyclohexene

SAFETY PRECAUTIONS

sodium hydroxide   ethyl ether

The 50% sodium hydroxide solution can be prepared by measuring 15 g of NaOH pellets [wear gloves!] into an Erlenmeyer flask, adding 15 mL of water [under a hood!], swirling the flask until the solid dissolves, and cooling the hot solution.

Be careful not to swirl any liquid up and out the neck of the flask.

Sodium hydroxide is toxic and very corrosive, causing severe damage to skin, eyes, and mucous membranes. Wear gloves, and avoid contact with the NaOH solution. Wash up any spills immediately.

Ethyl ether is extremely flammable and may be harmful if inhaled. Do not breathe the vapors, and keep it away from flames and hot surfaces.

Reaction. If magnetic stirrers are available, use the alternate procedure below. Carry out this reaction under an efficient fume hood. Combine 0.10 mol of cyclohexene with 9.0 mL (~0.11 mol) of chloroform **[vapor hazard! suspected carcinogen!]** and 20 drops (0.4 – 0.5 g) of tricaprylmethylammonium chloride. *Under a hood,* measure 20 mL of 50% aqueous sodium hydroxide **[contact hazard! wear gloves!]** into a clean 250-mL Erlenmeyer flask, and add the cyclohexene-chloroform mixture. Add 1 mL of cyclohexanol, and swirl the flask vigorously ⟨OP–9⟩ **[wear gloves!]** to whip the liquid into an opaque, creamy emulsion. Insert a thermometer into the flask ⟨OP–7c⟩, and hold it securely so that its bulb is covered by the emulsion while you continue to swirl the flask vigorously enough to maintain the emulsion. If the temperature rises to 60°C, swirl the flask in an ice bath just long enough to bring it down to 55°C. Repeat this operation as necessary to keep the temperature below 60°C. When the temperature drops spontaneously *without* external cooling, continue to swirl until it reaches 45°C; then let the reaction mixture stand until it reaches 35°C or below.

Alternate Reaction Procedure with Magnetic Stirring. Carry out this reaction under an efficient fume hood. Combine the reactants as described above, using a 125-mL Erlenmeyer flask as the reaction flask. Add a stirbar, and adjust the stirring motor ⟨OP–9⟩ to a speed sufficient to whip the mixture into a frothy, cloudy liquid, then add 1 mL of cyclohexanol. When the solution forms an opaque, creamy emulsion, monitor the temperature continually ⟨OP–7c⟩. If the temperature reaches 60°C, remove the flask momentarily (turn off the stirring motor first), and swirl it in an ice bath to bring it down to 55°C, then resume stirring. Repeat this operation as necessary to keep the temperature below 60°C. When the temperature drops spontaneously *without* external cooling, continue stirring until it reaches 50°C. Then remove the reaction mixture from the stirrer, and let it stand until its temperature is 35°C or below.

Separation.   Add 50 mL of saturated aqueous sodium chloride to the reaction mixture, and extract it ⟨OP–13⟩ with two 25-mL portions of ethyl ether. Wash the combined extracts ⟨OP–19⟩ with two portions of saturated aqueous sodium chloride, and dry the ether solution ⟨OP–20⟩ over calcium chloride. Evaporate the solvent ⟨OP–14⟩ under aspirator vacuum using a cold trap, and put the recovered ether in a designated container.

Be sure to save the right layer! Remember that the aqueous phase contains caustic sodium hydroxide, and dispose of it carefully.

Purification and Analysis.   Purify the residue by semimicro distillation ⟨OP–25a⟩, collecting the product around 190–200°C. Weigh the 7,7-dichloronorcarane, and record its infrared spectrum ⟨OP–33⟩ or obtain a spectrum from your instructor.

Discard any forerun that distills below 180°C.

Report.   Identify as many bands as you can in the infrared spectrum of 7,7-dichloronorcarane. In particular, point out the C—Cl and C—H stretching bands involving bonds to the cyclopropane ring. Give the systematic (IUPAC) name for 7,7-dichloronorcarane. Diagram the phase-transfer process for the reaction, using a format like that illustrated in Figure 1.

Collateral Projects

1.   Assess the purity of the product by gas chromatography, or record its PMR spectrum in deuterochloroform. A general-purpose silicone oil/Chromosorb W column can be used at ~110°C for the chromatography.

2.   Students can use different phase-transfer catalysts for the 7,7-dichloronorcarane synthesis, and then compare their results. Some suggested catalysts are tetrabutylammonium bromide, cetyltrimethylammonium bromide, and benzyltrimethylammonium chloride.

The Effect of Ring Strain in an Addition Reaction

MINILAB 12

Be sure to wear your safety glasses! Weigh 0.5 g of iodine into each of two 15-cm test tubes. *Under the hood,* add 1 mL of cyclohexene to the first test tube in one portion, and describe the result. Then add 1 mL of turpentine or α-pinene to the second tube in one portion, step back quickly **[possible violent reaction!],** and describe the result. Dispose of the residue as directed by your instructor. Write an equation for the reaction with α-pinene, showing one or more possible products. Write a mechanism for the formation of a rearranged product from α-pinene, and use it to explain your observations.

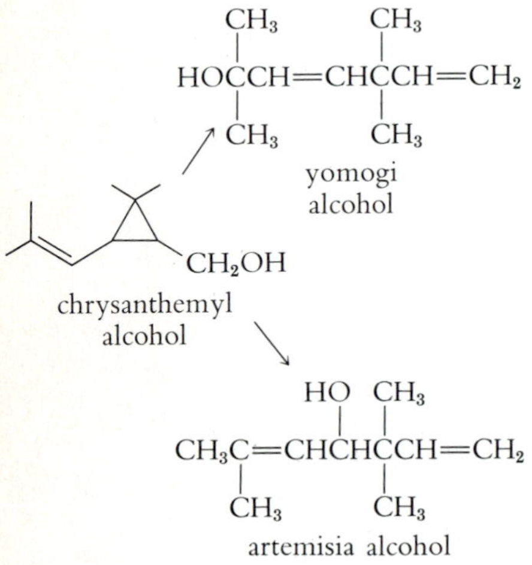

Exercises

1. Write a mechanism for the reaction of cyclohexene with chloroform in the presence of sodium hydroxide and a phase-transfer catalyst, showing the role of the catalyst.

2. Construct a flow diagram for the synthesis of 7,7-dichloronorcarane.

3. Why do you think aqueous sodium chloride rather than water was used in the workup of 7,7-dichloronorcarane?

4. The carbon-carbon sigma bonds of cyclopropane rings exhibit some of the characteristics of pi bonds in other systems. For example, the cyclopropylmethyl cation undergoes reactions that suggest that it is best represented by the resonance structures shown. Explain the isomerization in acidic solution of chrysanthemyl alcohol to yomogi alcohol and artemisia alcohol by writing appropriate mechanisms.

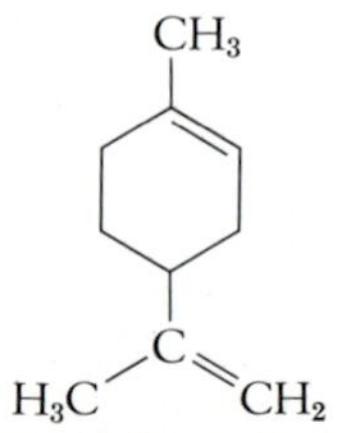

yomogi alcohol

chrysanthemyl alcohol

artemisia alcohol

limonene

Library Topic

Look up the article in *Tetrahedron Letters*, 3013 (1975). Find out which of the double bonds of limonene is attacked first by dichlorocarbene. Give the structure of the major product of this reaction, and summarize the reaction conditions used, indicating how they differ from the conditions used in this experiment.

Stereochemistry of the Addition of Bromine to *trans*-Cinnamic Acid

Reactions of Alkenes. Preparation of Alkyl Halides. Electrophilic Addition. Stereoselectivity.

Operations

OP–7a	Refluxing
OP–9	Mixing
OP–10	Addition
OP–12	Vacuum Filtration
OP–22b	Trapping Gases
OP–28	Melting-Point Determination

Objectives

To carry out the addition of bromine to *trans*-cinnamic acid and identify the product.

To learn about mechanisms of electrophilic addition reactions to carbon-carbon double bonds.

To learn about the stereochemistry of addition reactions.

To learn about the role of cinnamic acid and related compounds in plant biochemistry.

SITUATION

When bromine adds to the double bond of cyclopentene, two bromine atoms end up on opposite sides of the cyclopentane ring.

The mechanisms referred to here are described in the Background section.

This mode of addition is called *anti* addition, and a mechanism involving a bromonium-ion intermediate has been proposed to explain it. Some reactions, such as nucleophilic substitution, follow different mechanisms with different substrates, so it is conceivable that bromine adds differently to other kinds of alkenes. For example, a phenyl-substituted alkene might react via a carbenium ion rather than a bromonium ion, because benzene rings are known to stabilize carbenium ions by resonance. In this experiment, you will carry out the addition of bromine to *trans*-cinnamic acid and determine the stereochemistry of the product, from which you should be able to propose a mechanism for the reaction.

BACKGROUND

The Cinnamic Acid Connection

Cinnamic acid and its close relatives, cinnamaldehyde and cinnamyl alcohol, are naturally occurring compounds that are important as flavoring and perfume ingredients and as sources for pharmaceuticals. Cinnamaldehyde (the major component of cinnamon oil) is used to flavor many foods and beverages and to contribute a spicy, "oriental" note to perfumes. Esters of cinnamic acid are used in perfumery also, but of far greater significance is the role of cinnamic acid in secondary plant metabolism. As an intermediate in the shikimic acid pathway of plant biosynthesis, cinnamic acid is the source of an enormous number of natural substances that (to give only a few examples) contribute structural strength to wood, give flavor to cloves, nutmeg, and sassafras, and produce many of the brilliant colors of nature — the flower pigments that attract insects for pollination, the vivid and delicate shades of a butterfly's wings, and the radiant colors of leaves in autumn.

cinnamic acid

cinnamaldehyde

shikimic
acid

phenylalanine

cinnamic acid

In nature, cinnamic acid is formed by enzymatic **deamination** of the important amino acid phenylalanine. It can then be converted, by a wide variety of biosynthetic pathways, to

deamination: removal of an ammonia (NH_3) molecule

Biosynthesis of flavanone from cinnamic acid

acetates   cinnamic acid   flavanone

coniferyl alcohol (a precursor of lignin) in sapwood, myristicin in nutmeg, safrole in sassafras bark, and to *flavonoids* in a wide variety of plant structures. The flavonoids are natural substances characterized by the 2-aryl-benzopyran structure found in flavanone, which itself is biosynthesized from cinnamic acid by a process involving the linkage of three acetate units to the carboxyl group of the acid. Flavonoids perform no single function in plants. Many are highly colored and attract insects for pollination or animals for seed dispersal; others help to regulate seed germination and plant growth or protect a plant from fungal and bacterial diseases. Certain flavonoids contribute the bitter taste to lemons and the bracing astringency of cocoa, tea, and beer. Although flavonoids and the other derivatives of cinnamic acid have not furnished as many wonder drugs or "useful" chemicals to humanity as have the nitrogen-containing alkaloids, they provide much to delight the eye and stimulate the senses, and the world would be a drearier place without them.

coniferyl alcohol

myristicin

safrole

Stereoselective Reactions

As explained in Experiment 18, a reaction is said to be regioselective if it might produce two or more structural isomers but in fact yields one of them preferentially. Similarly, a reaction is said to be *stereoselective* if it might produce two or more stereoisomers but in fact yields mainly (or entirely) one of them. The electrophilic addition of bromine to cyclopentene yields *trans*-dibromocyclopentane and no *cis*-dibromocyclopentane, indicating that the components of Br_2 must add to opposite sides of the carbon-carbon double bond. This mode of addition is called *anti* addition. (Addition of the components of a reagent to the same side of a double bond is called *syn* addition.) The following scenario has been devised to explain the *anti* addition of bromine to cyclopentene. As a bromine molecule approaches perpendicular to the negatively charged pi cloud of

the carbon-carbon double bond, its bonding electrons are re-
pelled away from the bromine atom closer to the double bond,
leaving it with a partial positive charge. As the positively
charged bromine penetrates the pi cloud, a negative bromide
ion breaks away from it, leaving a cyclic bromonium ion in
which the positive bromine is bonded to two carbon atoms.
Backside attack on the bromonium ion by a bromide ion results
in the observed *trans* product.

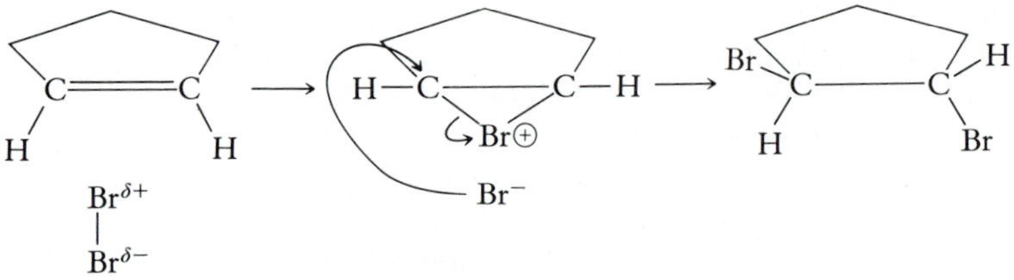

Other electrophilic addition mechanisms would lead to
different products. If addition of an electrophile to a double
bond involves a carbenium-ion intermediate, attack on either
side of the positively charged carbon would produce a mixture
of *syn* and *anti* products, as illustrated in Mechanism 1. This
kind of mechanism should be favored by substituents that stabilize carbenium ions by resonance, such as the phenyl group
of cinnamic acid. A concerted addition reaction might proceed
through a four-centered transition state that yields the *syn*
product, as shown in Mechanism 2.

Mechanism 1:

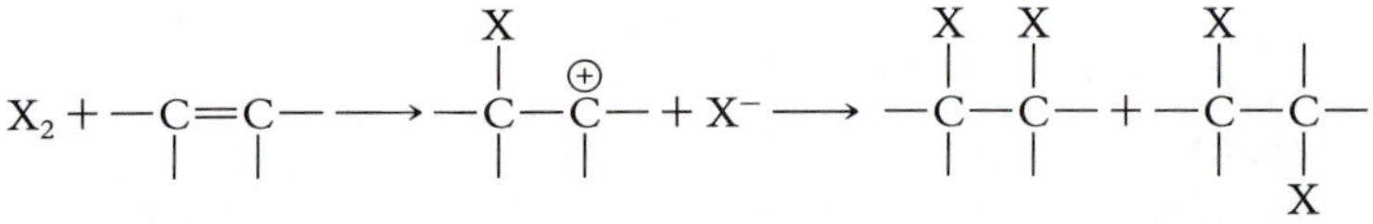

Mechanism 2:

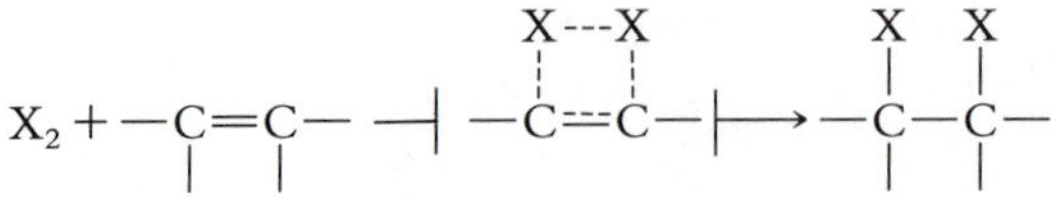

METHODOLOGY

In this experiment, you will carry out the addition of
bromine to *trans*-cinnamic acid and identify the product of the
reaction from its melting point. The product (actually a mix-

ture of enantiomers) could be either *threo*-2,3-dibromo-3-phenylpropanoic acid [whose enantiomers have the (2R,3R) and (2S,3S) configurations], the corresponding *erythro* compound [(2R,3S) and (2S,3R)], or a mixture of the two. The *erythro-threo* nomenclature is used to describe the configurations of compounds having two chiral centers but no plane of symmetry: it is derived from the structures of the two simple sugars, erythrose and threose. The *erythro* compound is analogous to a *meso* compound—when a plane is drawn separating the chiral centers, the maximum number of like substituents can be lined up directly across the plane from each other, and the remaining unlike substituents are also opposite one another. In the *threo* form, when the unlike substituents are opposite one another, the pairs of like substituents are not.

You will perform the reaction by slowly adding a solution of bromine in Freon TF (1,1,2-trichlorotrifluoroethane) to a refluxing solution of *trans*-cinnamic acid in the same solvent. A small excess of bromine is used to ensure that all the cinnamic acid will react; the excess can be consumed by adding a few drops of cyclohexene at the end of the reaction. The dibromide begins to precipitate from solution during the reaction and is separated by filtration after cooling. Because the melting points of the *erythro* and *threo* dibromides are separated by more than 100°C, the product can be easily identified from its melting point. A mixture of both products would melt over a broad range that should not coincide with the melting point of either pure dibromide. From the identity of your product, you should be able to deduce whether the addition of bromine to *trans*-cinnamic acid involves *syn* or *anti* addition or a mixture of the two. Molecular models will help you relate the configurations of the *syn* and *anti* addition products to the Fischer structures shown for the *erythro* and *threo* dibromides.

If you do Minilab 13, you will use the bromine addition reaction to test some commercial products for unsaturation.

Stereoisomers of 2,3-dibromo-3-phenylpropanoic acid

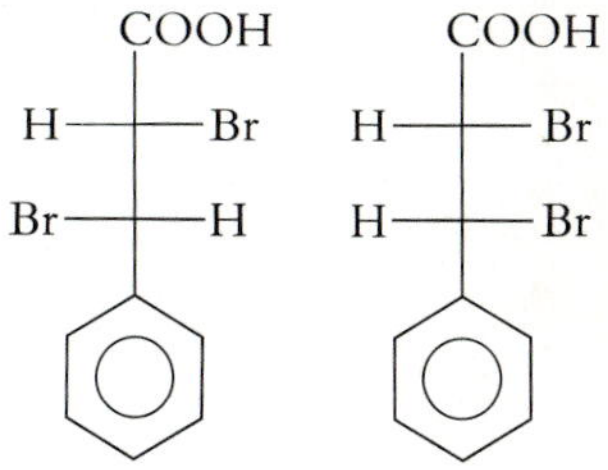

threo
(2S,3S)

erythro
(2S,3R)

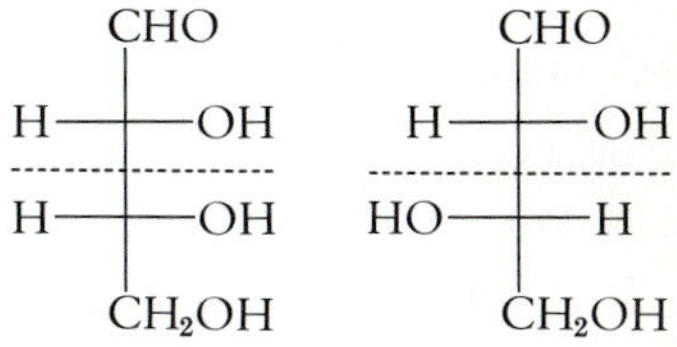

erythrose

threose

PRELAB ASSIGNMENTS

1. Review the parts of OP–10 and OP–22b that apply to full-scale (not semimicro) apparatus.

2. Calculate the mass of 9.5 mmol of *trans*-cinnamic acid and the theoretical yield of 2,3-dibromo-3-phenylpropanoic acid.

Reaction and Properties

$$\underset{\text{trans-cinnamic acid}}{\overset{\text{Ph}}{\underset{\text{H}}{\text{C}}}=\overset{\text{H}}{\underset{\text{COOH}}{\text{C}}}} + Br_2 \longrightarrow \underset{\substack{\text{2,3-dibromo-}\\\text{3-phenylpropanoic acid}}}{Ph-CHBrCHBrCOOH}$$

Table 1.   Physical properties

	M.W.	m.p.	b.p.	d.
trans-cinnamic acid	148.2	135–136		
bromine	159.8	−7	59	3.12
Freon TF	187.4	−36	48	1.575
erythro-2,3-dibromo-3-phenylpropanoic acid	308.0	202–4		
threo-2,3-dibromo-3-phenylpropanoic acid	308.0	93.5–95		

PROCEDURE

If the reaction cannot be carried out under a hood, take care to prevent bromine fumes from escaping into the laboratory. Keep the addition funnel stoppered tightly, and make sure your reflux condenser is working properly.

SAFETY PRECAUTIONS

bromine

Bromine is highly toxic and corrosive, and its vapors can damage the eyes and respiratory tract. Wear gloves when handling the bromine solution, and dispense it under a hood; avoid contact, and do not inhale the vapors.

Freon TF may be harmful if ingested, inhaled, or absorbed through the skin. Minimize your contact with the liquid, and do not breathe its vapors.

Minimize your contact with trans-cinnamic acid and the product.

Reaction.   Assemble an apparatus for addition and reflux ⟨OP–10⟩, clamping it securely to a ringstand. Add a gas trap ⟨OP–22b⟩ containing dilute aqueous sodium thiosulfate if the reaction will not be carried out under a hood. Combine 9.5

mmol of *trans*-cinnamic acid with 20 mL of Freon TF (1,1,2-trichlorotrifluoroethane) in the boiling flask. If a magnetic stirrer ⟨OP–9⟩ is available, put in a stirbar; otherwise add boiling chips. *Under the hood,* measure 10 mL of a 1.0 M solution of bromine in Freon TF **[contact and vapor hazard! wear gloves!]** into the addition funnel. Stopper it immediately, and then replace it on the reaction apparatus.

Bring the reaction mixture to reflux ⟨OP–7a⟩ on a steam bath; then add a small portion (about 1 mL) of the bromine solution. Shake the apparatus gently (or stir magnetically) to mix the reactants thoroughly, and reflux the reaction mixture until the color of the solution fades to pale orange or yellow. Heat gently enough to keep the bromine vapors inside the reflux condenser. Continue the addition of bromine in portions, shaking (or continuing to stir) after each addition and allowing the color to fade before the next addition. Reflux for 15 minutes after all the bromine has been added; then cool the reaction mixture to room temperature. If the orange-red color of bromine persists, add cyclohexene drop by drop through the top of the condenser until the reaction mixture is colorless or light yellow.

Separation and Analysis.   Disassemble the apparatus, transfer the product mixture to a beaker, and cool it in an ice bath for 15 minutes or more to allow complete crystallization of the dibromide. Rinse out the reaction apparatus with sodium thiosulfate solution from the gas trap to destroy any residual bromine. Collect the product by vacuum filtration ⟨OP–12⟩, wash it on the filter with cold Freon TF, and let it air-dry. Measure the mass and melting point ⟨OP–28⟩ of the product when it is completely dry.

Stereochemistry of Bromine Addition.   Construct a molecular model of *trans*-cinnamic acid. Simulate the *syn* addition of bromine by removing one of the C=C connectors and inserting two orange bromine atoms, with connectors, into the vacant holes. Rotate the model around the appropriate carbon-carbon single bond until it corresponds to the Fischer projection for either *threo*- or *erythro*-2,3-dibromo-3-phenyl-propanoic acid. Simulate *anti* addition of bromine by removing the upper connector of the carbon-carbon double bond and moving one end of the lower connector from the hole it occupies to the vacant hole in the same carbon atom. Be careful not to rotate either carbon atom as you do so. Insert two bromine atoms, with connectors, into the vacant holes; then rotate the model as before until it matches one of the Fischer projections.

The cinnamic acid will not dissolve until the reaction mixture is heated.

Because of the high density of Freon TF, it should not be necessary to loosen the stopper on the addition funnel during additions.

Do not add the thiosulfate rinses to the product mixture!

You may want to replace the remaining bond by a rigid connector.

Report. Give the name and structure of your product, and tell whether it was formed by *syn* addition, *anti* addition, or a mixture of both. Cite evidence supporting your conclusion. Tell whether or not the reaction is stereoselective, and write a mechanism that explains its stereochemistry.

Collateral Projects

1. Find a suitable recrystallization solvent as described in OP–23c and use it to purify your product; then dry it and measure its melting point.

2. You can synthesize phenylpropynoic acid (PhC≡CCOOH) from your product by scaling down a procedure given in *J. Am. Chem. Soc. 64,* 2510 (1942). Find a suitable recrystallization solvent to use in place of carbon tetrachloride.

MINILAB 13

Determination of Unsaturation in Commercial Products

Test as many of the following commercial products as possible; your instructor may add more to the list.

olive oil
mineral oil
linseed oil
safflower oil
butter
rubber cement
dry-cleaning solvent
paint thinner (mineral spirits)
turpentine
rubbing alcohol

Using small labeled test tubes, dissolve 3 drops (or 0.1 g of a solid) of each commercial product in 1 mL of Freon TF. *Under the hood,* add a 1 M solution of bromine in Freon TF **[contact and vapor hazard!]** drop by drop to each of the test tubes with swirling. Count the number of drops needed to make the orange-red color persist for at least 30 seconds. Consider a substance unsaturated if it takes more than 2 drops to reach this point. Rank the unsaturated substances in order of the relative amount of bromine consumed.

Using sources such as those cited in the Bibliography, find out what kinds of unsaturated compounds may be present in each of the substances that tested positive. Draw structural formulas for representative components when possible.

Exercises

1. **(a)** A *stereospecific* reaction is one in which reactants that are stereoisomers of one another yield stereochemically different products. What additional reaction would you have to carry out to establish that the addition of bromine to cinnamic acid is stereospecific? **(b)** Write an equation for this reaction, giving a Fischer projection for the product that is expected if the reaction is stereospecific.

For example, if reactants that are diastereomers of one another yield products that are diastereomers of one another, the reaction is stereospecific.

2. **(a)** The product of the bromination of *trans*-cinnamic acid is not a single compound but a racemic mixture of enantiomers. Explain. (Molecular models may be helpful.) **(b)** Draw the structure of each enantiomer and specify the configuration (*R* or *S*) at each chiral carbon.

3. Construct a flow diagram for the synthesis you carried out in this experiment.

4. Draw Fischer projections for the products of bromine addition to maleic acid and fumaric acid (*cis-* and *trans*-HOOCCH=CHCOOH), assuming that bromine adds in the same way to these compounds as it does to cinnamic acid.

5. **(a)** Would you expect the product from this experiment to be optically active? Could it be resolved into optically active constituents? Explain. **(b)** Would the product of the bromination of fumaric acid (see Exercise 4) be optically active? Could it be resolved into optically active constituents? Explain.

Library Topic

Find out how cinnamic acid can be prepared in the laboratory and how it is synthesized industrially, giving equations for the reactions used. Describe at least one industrial process that uses a reaction different from that used in the common laboratory synthesis.

EXPERIMENT 21

Chain-Growth Polymerization of Styrene and Methyl Methacrylate

Reactions of Alkenes. Preparation of Vinyl Polymers. Addition Polymerization. Free Radicals.

Operations

OP–7	Heating	
OP–9	Mixing	
OP–12	Vacuum Filtration	
OP–21	Drying Solids	
OP–25	Simple Distillation	
OP–33	Infrared Spectrometry	

Objectives

To prepare polystyrene by emulsion polymerization and record its infrared spectrum.

To prepare poly(methyl methacrylate) by bulk polymerization.

To learn about some polymerization methods.

To learn about chain-growth polymerization mechanisms.

To learn about the history and uses of synthetic polymers.

SITUATION

Vinyl polymers can be prepared by a variety of experimental methods, including bulk, solution, suspension, and emulsion polymerization. Bulk polymerization can be used to preserve a botanical or zoological specimen (such as a desiccated flower or beetle) by suspending the specimen in a liquid monomer, which is then allowed to polymerize around it. Poly(methyl methacrylate), known by the trade names Lucite and Plexiglas, is often used for that purpose. Polystyrene and other polymers can be used to prepare thin tough films similar to the plastic films used as food wraps. In this experiment, you will prepare poly(methyl methacrylate) by bulk polymerization and polystyrene by emulsion polymerization, and analyze a polystyrene film by infrared spectrometry.

BACKGROUND

Chain-Growth Polymers

Polymers are of enormous importance to humankind—the meat, fruit, and vegetables we eat, the clothing we wear, and the wood we use for housing and furniture all consist partly or entirely of organic polymers. In fact, we are *made* of polymers—proteins in muscles, organs, blood cells, enzymes, and protoplasm; lipids in nerve sheaths, cell walls, and energy-storing fat tissues; nucleic acids in the chromosomes that control our heredity. Compared to these natural polymers, synthetic polymers are newcomers on the scene. Polystyrene was first synthesized in 1839—the same year Charles Goodyear learned how to vulcanize rubber—but its properties were not appreciated. The first commercially useful synthetic polymer did not appear until 1907 when Leo Baekeland synthesized "Bakelite" from phenol and formaldehyde. The synthesis of Nylon 66 and polyethylene in 1939 and the development of synthetic rubbers during World War II (when natural rubber was unavailable) gave added impetus to the search for useful synthetic polymers. The more recent development by Ziegler and Natta of special catalysts for synthesizing stereoregular polymers has, with other advances, made it possible to design "tailor-made" polymers having almost any desired combination of properties.

Synthetic polymers can be divided into the two broad categories of addition and condensation polymers. *Addition polymers* are built up by combining monomer units without eliminating any by-product molecules. The repeating unit of the polymer, therefore, has the same chemical constitution as the monomer. *Condensation polymers* are built up from monomer units containing two or more reactive functional groups that lose a small molecule such as water or HCl as they combine. Thus, the repeating unit of a condensation polymer does not have the same formula as the monomer(s), as illustrated on page 202 for the Nylon 6 synthesized from 6-aminohexanoic acid. But Nylon 6 can also be formed by an addition polymerization reaction from ω-caprolactam, so it is somewhat arbitrary to classify such polymers as either addition or condensation polymers.

In a sense, polystyrene is a "natural" polymer since it is found in styrax from the sweetgum tree. Some modified natural polymers, such as celluloid (produced by nitrating cellulose), were manufactured commercially before the advent of Bakelite, but these are not true synthetics.

Addition polymerization of ethylene

$$n\text{CH}_2\!\!=\!\!\text{CH}_2 \longrightarrow -\!(\text{CH}_2\text{CH}_2)_{\overline{n}}\!-$$
ethylene   polyethylene

repeating unit

(n is a large but indeterminate number.)

Formation of Nylon 6 by
condensation polymerization

$$n\text{H}_2\text{N(CH}_2)_5\overset{\overset{\displaystyle O}{\|}}{\text{C}}\text{OH} \longrightarrow$$

6-aminohexanoic
acid

$$-(\text{NH(CH}_2)_5\overset{\overset{\displaystyle O}{\|}}{\text{C}})_{\overline{n}} + \text{H}_2\text{O}$$

Nylon 6

Formation of Nylon 6 by
addition polymerization

$$n \quad \longrightarrow -(\text{NH(CH}_2)_5\overset{\overset{\displaystyle O}{\|}}{\text{C}})_{\overline{n}}$$

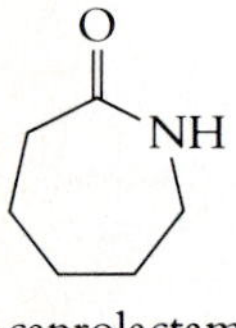

ω-caprolactam   Nylon 6

The terms *chain-growth* and *step-growth polymerization* are used to describe two basic polymerization processes. The former refers to a process in which a few monomer molecules are activated by some initiator, after which a chain of repeating units builds up very rapidly. The latter refers to a process by which the chain length is built up more gradually. The polymerization of styrene with a free-radical initiator is a typical example of chain-growth polymerization. An initiator (usually an organic peroxide) decomposes under the influence of heat or light to form free radicals, which add to styrene molecules according to Markovnikov's rule. Each "activated" styrene molecule then adds in similar fashion to another styrene molecule, leaving the unpaired electron at the end of the chain after each step. This process continues indefinitely until a reaction such as radical coupling or disproportionation occurs between chain ends (or with impurities) and deactivates the chain ends by forming stable products. Throughout a chain-growth polymerization, the bulk of the reaction mixture will consist of finished polymer molecules and unreacted monomers waiting to meet up with a reactive chain end. Because polymerization is so rapid once activation occurs, only one among many millions of molecules is actually involved in the growth process at any given instant.

Mechanism of chain-growth polymerization of polystyrene

1.   Initiator $\xrightarrow[\text{light}]{\text{heat or}}$ R · (free radical)

2.   R· + CH$_2$=CH $\longrightarrow$ RCH$_2$CH·
                    | |
                   Ph   Ph

3. etc. R(CH$_2$CH)$_n$CH$_2$CH· + CH$_2$=CH $\longrightarrow$ R(CH$_2$CH)$_n$CH$_2$CHCH$_2$CH·
                | | | | | |
               Ph   Ph   Ph   Ph   Ph   Ph

Terminating steps:

2R(CH$_2$CH)$_n$CH$_2$CH· $\longrightarrow$ R(CH$_2$CH)$_n$CH=CH + R(CH$_2$CH)$_n$CH$_2$CH$_2$
        | | | | | |
       Ph   Ph   Ph   Ph   Ph   Ph

(disproportionation)

or

R(CH$_2$CH)$_n$CH$_2$CH—CHCH$_2$(CHCH$_2$)$_n$R
        | | | |
       Ph   Ph   Ph   Ph

(radical coupling)

METHODOLOGY

There are a number of different methods used for chain-growth polymerization, each with its advantages and disadvantages. *Bulk polymerization* is the simplest; it is carried out by adding a suitable initiator and using heat or light to promote the reaction. The high heat of reaction makes bulk polymerization of vinyl monomers hard to control, and the method is seldom used commercially except for some polystyrene and poly(methyl methacrylate) products. Dissolving the monomer in an organic solvent alleviates the heat problem, but it is often difficult (if not impossible) to entirely remove the solvent. *Solution polymerization* thus works best for polymers that are commonly used in solution, such as acrylic finishes.

Suspension polymerization is carried out by mechanically dispersing the monomer in water or a similar solvent, so that the polymer is obtained in the form of granular beads that can be easily isolated. *Emulsion polymerization* is similar to suspension polymerization in that the monomer is dispersed in a solvent, usually water. In this method, however, the initiator is dissolved in the aqueous phase, and the monomer is emulsified by a detergent or some other **surfactant.** Polymerization starts in the surfactant micelles rather than in the monomer droplets. At some stage of the process, the polymer particles grow larger than the micelles and absorb all of the surfactant from solution, after which further polymerization occurs within the polymer particle itself. The monomer droplets provide a reservoir of monomer molecules that are continually fed into the growing polymer. The resulting dispersion resembles a rubber latex and can be used in that form, but it is more often coagulated and isolated as a finely divided powder.

surfactant (surface-active agent): a substance whose molecules can concentrate at the interface between an aqueous solution and another phase

In a simple demonstration of the bulk polymerization technique, a sample of methyl methacrylate will be mixed with a little initiator (*t*-butyl peroxybenzoate) in a test tube and placed in the sunlight for a period of time. Since the inhibitor added to a commercial monomer would slow the reaction, the monomer will be prepared by thermal depolymerization of poly(methyl methacrylate) in a distillation apparatus.

Inhibitors are added to most vinyl monomers to stabilize them and prevent premature polymerization.

The emulsion polymerization of styrene will be carried out by dispersing the monomer in an aqueous detergent solution and heating the emulsion in the presence of a water-soluble initiator, potassium peroxydisulfate (potassium persulfate). The polymer forms as a rubbery latex that is broken up by adding an alum solution to precipitate polystyrene. As with most polymerization reactions, precautions must be taken to eliminate atmospheric oxygen.

The polystyrene can be dried in an oven at 110°C.

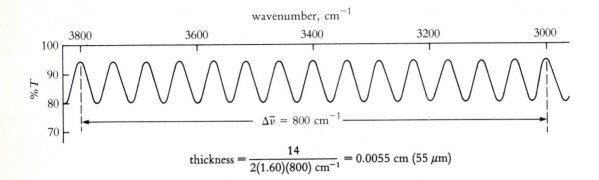

Figure 1.   Determination of film thickness by counting interference fringes

Polystyrene is essentially an aromatic hydrocarbon with a very long alkyl side chain, so its infrared spectrum should resemble that of an ordinary arene. The spectrum of a thin polymer film will often display a number of extraneous small peaks called *interference fringes,* which are caused by interference between beams of infrared radiation reflected from the film's surfaces. As illustrated in Figure 1, the thickness of such a film can be estimated by counting the number of fringes in a given wavenumber interval, $\Delta\bar{\nu}$.

$$\text{Film thickness} = \frac{\text{number of fringes}}{2n(\Delta\nu)}$$

n = refractive index (~ 1.60 for polystyrene)

If you do Minilab 14, you will prepare a step-growth polymer called Nylon 6-10 by the reaction of a diamine with a diacid chloride. (The 6-10 designation indicates the number of carbon atoms in the amine and in the acid chloride, respectively.) The reaction will be carried out by an unusual polymerization technique in which the polymer is formed at the interface between two immiscible liquids, one containing the diamine and the other the diacid chloride. The product can be drawn out of the solution as a "rope" of polymer that is continually lengthened by the formation of new polymer at the interface; this procedure has been aptly called "the nylon rope trick."

Reactions and Properties

A. n (phenyl)—CH=CH$_2$ ⟶ —(CH—CH$_2$)$_n$—

styrene   polystyrene

B. nCH$_2$=CCOOCH$_3$ ⟶ —(CH$_2$—C)$_n$—

CH$_3$ CH$_3$

methyl methacrylate   poly(methyl methacrylate)

Table 1. Physical properties

	M.W.	m.p.	b.p.	d.
styrene	104.15	−31	145−46	0.909
potassium persulfate	270.3	100d		
methyl methacrylate	100.1	−48	100	0.944

Figure 2. IR spectrum of styrene. (Reproduced from *The Aldrich Library of FT-IR Spectra,* by Charles J. Pouchert, with the permission of the Aldrich Chemical Company.)

Styrene

(phenyl)—CH=CH$_2$

3026.8	1494.3	908.6
1630.0	1082.9	775.8
1575.6	991.1	697.1

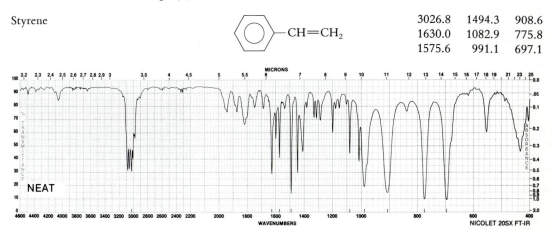

PROCEDURE

A. Emulsion Polymerization of Styrene

If this reaction cannot be carried out under a hood, it should be done under reflux, using a boiling flask in place of the 125-mL Erlenmeyer flask.

styrene   potassium peroxydisulfate

The nitrogen is used to flush out oxygen, which would otherwise inhibit the polymerization reaction.

Look for a bluish opalescence that signals the onset of polymerization. You may work on part B during the reaction period.

Under the hood, combine 5.0 g of styrene with 50 mL of water in a 125-mL Erlenmeyer flask. Bubble nitrogen gently through the solution for about 5 minutes. Add 0.1 g of potassium peroxydisulfate, 0.35 g of sodium lauryl sulfate, and a magnetic stirbar; then insert a cork loosely in the mouth of the flask. Stir the reaction mixture ⟨OP–9⟩ at a rate sufficient to maintain a stable emulsion while heating it in a 40–50°C water bath for 2 hours. Precipitate the polystyrene by adding 10 mL of saturated alum solution and boiling for a few minutes. Recover the polymer by vacuum filtration ⟨OP–12⟩, transfer it to a small beaker, and stir it with small portions of methanol (decant the methanol after each washing) until it is no longer sticky. Filter the polystyrene ⟨OP–12⟩; then dry ⟨OP–21⟩ and weigh it.

Record the infrared spectrum ⟨OP–33⟩ of a polystyrene film provided by your instructor, or prepare a film as described in Collateral Project 1.

B. Bulk Polymerization of Methyl Methacrylate

Methyl methacrylate is flammable and lachrymatory (tear-producing) and can cause serious eye and skin irritation, including allergic skin reactions. Avoid contact, do not breathe the vapors, and keep it away from flames. Use protective gloves and a fume hood.

t-Butyl peroxybenzoate may react violently with strong oxidizing or reducing agents or explode when strongly heated. Avoid contact with it, and keep it away from combustible materials, heat, and flames.

SAFETY PRECAUTIONS

methyl methacrylate

t-butyl peroxybenzoate

Place 10 g of granulated poly(methyl methacrylate) in a simple distillation apparatus ⟨OP–25⟩, and heat the flask with a "soft" burner flame **[be careful — flammable distillate!]**, moving the flame continually over the bottom of the flask. As the polymer softens and starts to depolymerize, heat it just strongly enough to distill the monomer slowly. Stop the distillation when the residue in the flask begins to turn black. Place the distillate in a small test tube, add 5 drops of t-butyl peroxybenzoate **[reactive liquid!]**, and stir to mix. Stopper the tube, and set it in direct sunlight until the next laboratory period. With your instructor's permission, you may break the test tube to recover the clear poly(methyl methacrylate).

Flushing the tube with dry nitrogen will accelerate the polymerization.

Report. Calculate the percentage yield of polystyrene from part A. Interpret the infrared spectrum of polystyrene as completely as you can, and calculate the thickness of the film. Propose a mechanism for the polymerization of methyl methacrylate in the presence of t-butyl peroxybenzoate.

Decomposition of t-butyl peroxybenzoate

Collateral Projects

It may take some trial-and-error experimentation to cast a uniform film.

1.   Prepare a polymer film by dissolving about 0.2 g of polystyrene or poly(methyl methacrylate) in 1–2 mL of a solvent such as toluene or 2-butanone and spreading the viscous mixture uniformly over a *very* clean glass plate. It can be spread with a "doctor knife" constructed by wrapping two lengths of masking tape (several turns each) around a glass rod, as shown in Figure 3. Set the glass plate under a hood until the solvent has completely evaporated, and then carefully scrape off the film with a razor blade.

Figure 3. Apparatus for preparing polymer films

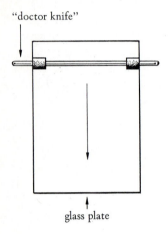

"doctor knife"

glass plate

2.   Try embedding a small, dry object in your bulk polymer (from part B) by suspending it in the methyl methacrylate by a thread. Metal objects should first be coated with clear enamel.

3.   Try to polymerize separate samples of methyl methacrylate as described in part B: **(a)** using about 0.1 g of an inhibitor, hydroquinone, in place of the initiator, and **(b)** using no added initiator or inhibitor. Compare your results with those using the initiator.

MINILAB 14

The Nylon Rope Trick

SAFETY PRECAUTIONS

Wear protective gloves throughout this experiment. Avoid touching the wet polymer with your hands; if you do, wash them immediately with soap and warm water.

Avoid contact with the reactants, and do not breathe their vapors.

Under the hood, dissolve 1.0 mL (4.7 mmol) of pure sebacoyl chloride (decanedioyl chloride) **[contact and vapor hazard! lachrymator!]** in 50 mL of methylene chloride **[suspected carcinogen!]** in a 150-mL tall-form beaker. Combine 10 mL of 1 M aqueous sodium hydroxide with 15 mL of water in a small beaker, then stir in 0.55 g (4.7 mmol) of pure 1,6-hexanediamine **[contact hazard!]** until it dissolves. Tilt the tall-form beaker and slowly pour the hexanediamine solution down the side, taking care not to mix the layers. A polymer film should form at the interface between the layers. Use a small spatula to free the film from the side of the beaker if necessary, and use a long tweezers or a piece of copper wire bent into a hook to grasp the center of the film. Pull up the film slowly and continuously to form a strand of Nylon 6-10, loop it around a cardboard tube (such as the core of a paper-towel or toilet-tissue roll), and rotate the tube to wind it out of the solution. When no more Nylon "rope" can be drawn, stir the solution in the beaker vigorously to form additional polymer, then decant the liquid and place it in a waste solvent container. Unwind the nylon strand into a beaker containing about 100 mL of 50% ethanol, add the recovered polymer, and stir gently to wash the Nylon. Then decant the solvent and lay the polymer on a paper towel to dry (blotting it between two towels will reduce the drying time). When the Nylon 6-10 is thoroughly dry, examine and describe the "rope," then weigh it and calculate the percentage yield. Write the structure of Nylon 6-10 and an equation for its formation.

Exercises

1. Methyl methacrylate can be prepared commercially from acetone, hydrogen cyanide, and methanol. Propose a synthesis of methyl methacrylate from these starting materials, using any necessary inorganic reagents or solvents.

2. Indicate a monomer or pair of monomers that could be used to prepare polymers having the following repeating units:

(a) $-(CHClCHCl)-$ **(b)** $-(CF_2CFCl)-$

(c) $-(CH_2CH_2NH)-$ **(d)** $-(CH_2C=CHCH_2)-$
$$\quad\quad\quad\quad\quad\quad\quad\quad\quad\quad\quad\quad | $$
$$\quad\quad\quad\quad\quad\quad\quad\quad\quad\quad\quad CH_3$$

(e) $-(CH_2CH)-$ **(f)** $-(CH_2CH-CH_2CH)-$
$$\quad\quad\quad | \quad\quad\quad\quad\quad\quad\quad\quad\quad | $$
$$\quad\quad CH_2CH_3 \quad\quad\quad\quad\quad CN$$

3. Compare your polystyrene spectrum with the infrared spectrum of styrene in Figure 2, and try to account for any significant similarities and differences.

4. Show how the two monomers illustrated could combine to form a Diels-Alder addition polymer, and give the structure of the polymeric repeating unit.

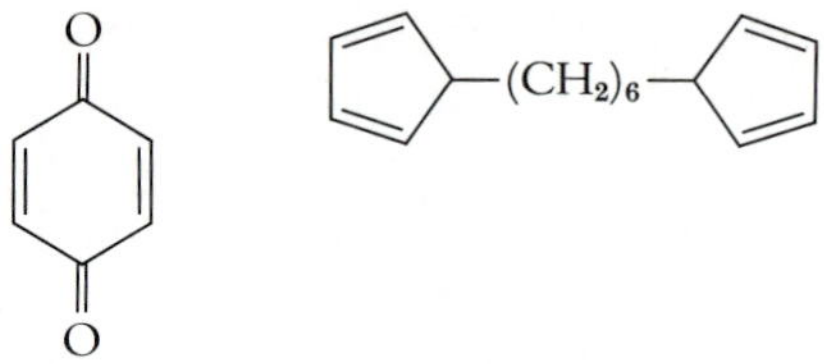

5. Poly(ethylene glycol) can be prepared from either ethylene glycol or ethylene oxide. Write a balanced equation for each reaction, and classify each as an addition or condensation polymerization.

$—(CH_2CH_2O)—$

poly(ethylene glycol)

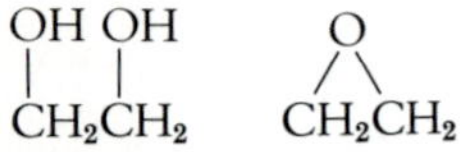

ethylene glycol

ethylene oxide

Library Topic

Write a short report on the stereochemistry of chain-growth polymerization. Illustrate structures for atactic, isotactic, and syndiotactic polymers, tell how they are prepared, and compare their properties.

Hydration of a Difunctional Alkyne

Reactions of Alkynes. Preparation of Carbonyl Compounds. Electrophilic Addition.

Operations

OP–7a   Refluxing

OP–10   Addition

OP–13   Extraction

OP–13c Salting Out

OP–14   Evaporation

OP–15 Codistillation

OP–19   Washing Liquids

OP–20   Drying Liquids

OP–25a   Semimicro Distillation

OP–33   Infrared Spectrometry

Objectives

To synthesize 3-hydroxy-3-methyl-2-butanone by hydration of the difunctional alkyne 2-methyl-3-butyn-2-ol.

To confirm that the hydration reaction follows Markovnikov's rule.

To verify the conversion of an alkyne functional group to a carbonyl group using infrared spectrometry.

To learn about some synthetic sex hormones and other useful alkynols.

SITUATION

One of the most powerful ovulation inhibitors known is the synthetic sex hormone megestrol acetate, a steroid whose structure is shown on page 212. This compound, having both acetyl (CH_3CO) and acetoxy (CH_3COO) groups on the number 17 carbon atom, can be classified as an acetoxyketo steroid.

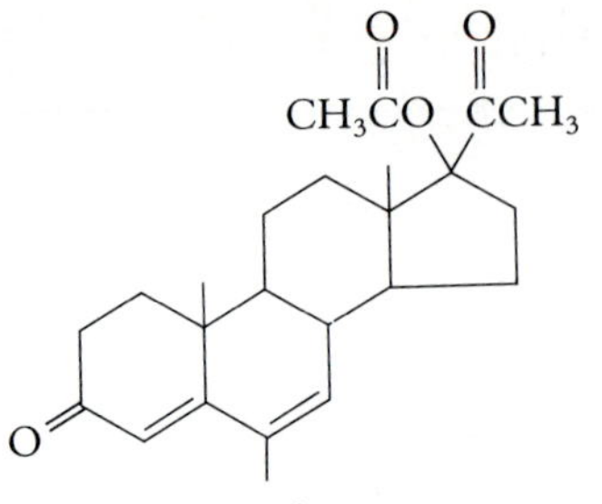

megestrol acetate

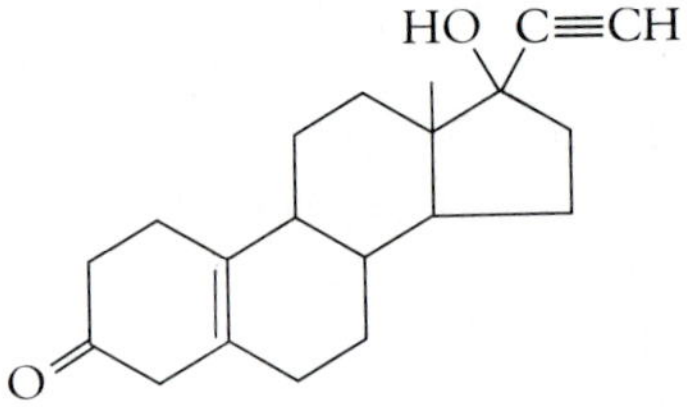

norethynodrel

An alkynol is a compound that contains both alkyne and alcohol functional groups.

A number of synthetic sex hormones such as norethynodrel, mestranol, and ethinyl estradiol are alkynol steroids having an ethynyl ($HC{\equiv}C{-}$) and a hydroxyl function on carbon-17. By hydrating the carbon-carbon triple bond and acetylating the OH group of such a steroid, it should be possible to synthesize the corresponding acetoxyketo steroid:

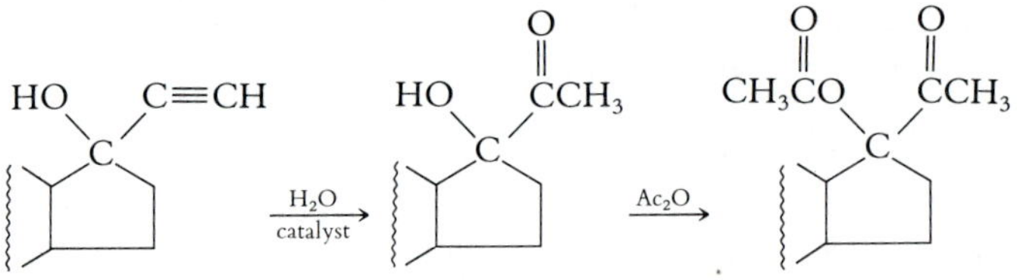

acetoxyketo steroid

If this reaction pathway is feasible, it might be used to produce safer or more effective synthetic hormones. But first you must establish that the hydration reaction actually yields a carbonyl compound and that it follows Markovnikov's rule.

Rather than working with norethynodrel or another expensive synthetic steroid, you will use a much simpler alkynol, 2-methyl-3-butyn-2-ol, as a model compound. By carrying out the hydration of this alkynol and analyzing the product by infrared spectrometry, you will find out if the desired functional-group conversion can be accomplished.

Hydration of the "model compound" 2-methyl-3-butyn-2-ol:

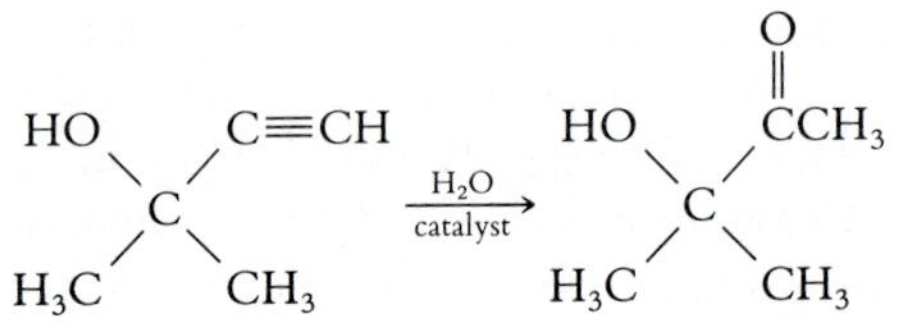

BACKGROUND

From "The Pill" to Oblivon

Natural *estrogens* such as estrone and 17β-estradiol are responsible for promoting the development of secondary sex characteristics in females during puberty; natural *progestins* such as progesterone are necessary to maintain pregnancy in mammals. Both types of hormones belong to an important class of natural products called the steroids, which are characterized by a basic four-ring skeleton, the perhydrocyclopentanophenanthrene nucleus. Cholesterol, the most widely known of the steroids, is essential for the synthesis of the sex hormones and other body regulators.

perhydrocyclopentanophenanthrene

progesterone

17β-estradiol

estrone

In addition to their natural function of promoting sexual maturity in females, the estrogens are very useful therapeutically. They help alleviate the mental and physical discomfort associated with menopause and are believed to prevent coronary atherosclerosis in younger women. Likewise, the progestins have been used to treat menstrual disorders and uterine bleeding and to prevent miscarriages.

Natural hormones cannot be taken orally because they are rapidly deactivated in the liver, so the search for synthetic hormones with similar activity began soon after the natural ones were isolated and characterized. One way to prevent the deactivation of a steroid such as 17β-estradiol is to stabilize the C-17 hydroxy group with an appropriate substituent. The ethynyl (HC≡C—) function seems to fill the bill nicely. By treating estrone with potassium acetylide in liquid ammonia one can synthesize ethinyl estradiol, a potent estrogen that can be taken orally. Some synthetic progestins such as norethynodrel are more effective taken orally than progesterone is by injection.

One of the most intriguing properties of the natural estrogens and progestins is their ability to inhibit ovulation in females. During the 1940s it was a common practice to treat

cholesterol

ethinyl estradiol

The commercial production of synthetic steroids became feasible only after it was discovered that certain Mexican wild yams (of the species *Dioscorea*) contain a natural steroid called diosgenin.

Enovid, the first birth-control pill marketed, contained 9.85 mg of norethynodrel and 0.15 mg of mestranol.

See *J. Chem. Ed. 55,* 591 (1978) for a more thorough discussion of the birth-control pill and its development.

mestranol

methylparafynol
(Oblivon)

menstrual disorders with these hormones, since preventing ovulation stops menstruation. Somewhere along the line it occurred to a few investigators that preventing ovulation also prevents pregnancy. At that time, however, there were no orally active substances suitable for this application, and the idea of having frequent injections did not appeal to many women. By the early 1950s the picture had changed. Several effective progestins (19-norprogesterone, norethindrone, norethynodrel, etc.) had been synthesized by researchers for the Syntex and G. D. Searle companies, and early in 1953 these compounds were evaluated by Gregory Pincus and his colleagues for anti-ovulatory activity. It was soon discovered that the combination of a synthetic progestin with a small amount of synthetic estrogen provided the highest anti-ovulatory activity, and "the pill" was marketed soon afterward. Although the long-term consequences of using oral contraceptives are not entirely known, these drugs are still the most reliable means of preventing pregnancy, and their impact on society has been enormous.

Alkynols having a hydroxy group adjacent to the triple bond (as in norethynodrel and mestranol) are comparatively easy to synthesize, and they appear in a number of pharmaceuticals. The 2-methyl-3-butyn-2-ol used in this experiment is prepared by combining acetone and acetylene with an alkali metal in liquid ammonia. A similar alkynol called methylparafynol has been used in sleeping pills under the trade name Oblivon. Apparently the tertiary alcohol portion of the molecule causes methylparafynol to act as sedative or depressant, while the acetylenic group gives it hypnotic (sleep-producing) properties.

Preparation of 2-methyl-3-butyn-2-ol

METHODOLOGY

Alkyne hydration, the electrophilic addition of water to a carbon-carbon triple bond, is generally accomplished by heating the alkyne with water in the presence of an acid and a mercury(II) salt. Mercuric ion catalyzes the reaction by a process that involves its addition to the triple bond to form a complex (labeled **1** in the reaction mechanism). In one proposed mechanism, the complex is attacked by water to give an intermediate (**2**), which, after losing a proton, is hydrolyzed to an enol (**3**). Isomerization of the enol to the corresponding keto tautomer yields the final product.

Mechanism of alkyne hydration:

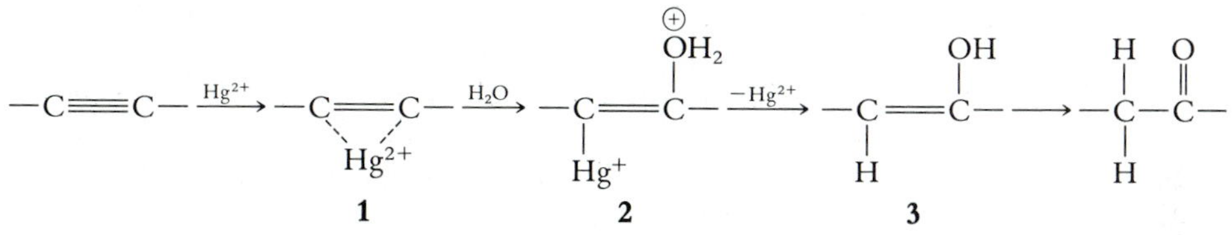

The hydration of alkenes is regiospecific because the intermediate is a carbenium ion that can be stabilized by neighboring alkyl substituents, resulting in Markovnikov addition. The above mechanism for hydration of alkynes does not involve such a carbenium ion, so it is not obvious that such a reaction should follow Markovnikov's rule. If it does, the product should be a methyl ketone; anti-Markovnikov addition to a terminal alkyne yields an aldehyde. An infrared spectrum should reveal which kind of compound you have synthesized.

See Experiment 18 for a discussion of regiospecific reactions and Markovnikov addition.

A terminal alkyne has the structure $RC{\equiv}CH$, in which R can be hydrogen or a carbon-containing substituent.

Markovnikov and anti-Markovnikov hydration of a terminal alkyne:

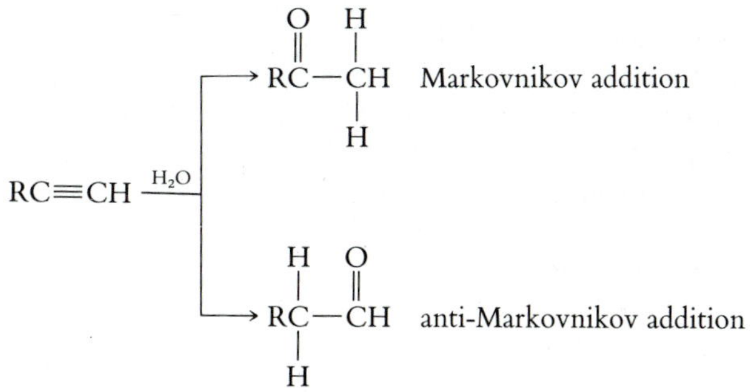

You will carry out the hydration of 2-methyl-3-butyn-2-ol by boiling the reactant with water containing sulfuric acid and mercury(II) sulfate. Because the reaction is strongly exothermic, the alkyne is added slowly to the reaction mixture from an addition funnel. The product can be separated from any nonvolatile impurities in the reaction mixture by codistillation with water. During the reaction, some mercury(II) ion may be reduced to metallic mercury, which will codistill with the product. Because mercury vapor is very toxic, it is important to recover and dispose of this substance properly.

Any acid that distills with the product will be neutralized by adding some potassium carbonate to the distillate. The potassium carbonate, along with some sodium chloride, will be used to "salt out" the product so that it can be extracted more completely. Mercury and other heavy metals can form organometallic salts with terminal alkynes that may explode if heated to dryness. Because such salts cannot form at a low pH, it is unlikely that the product will contain any, but you will wash the methylene chloride extract with dilute acid as a precautionary measure. After evaporating the solvent, you can purify the crude product by semimicro distillation.

If a larger-scale preparation is preferred, all quantities can be doubled.

Infrared spectrometry is particularly useful for detecting functional groups in molecules, so it can readily confirm the conversion of one functional group to another in a synthesis. The IR spectrum of a terminal alkyne is characterized by a strong, narrow $\equiv C-H$ stretching band near 3300 cm^{-1} and a weak $C\equiv C$ stretching band in the 2260–2100 cm^{-1} region. Because the broad O—H band of an alcohol also occurs around 3400–3300 cm^{-1}, it may partially obscure the $\equiv C-H$ band in the spectrum of your reactant. Aldehydes and ketones both give rise to strong carbonyl bands near 1700 cm^{-1}, but the CHO group of an aliphatic aldehyde produces a sharp C—H band of moderate strength near 2720 cm^{-1}. See the Infrared Spectra section of Part III for additional information about interpretation of IR spectra. By comparing the infrared spectrum of your product with that of the reactant, you should be able to ascertain whether the expected functional group conversion has taken place and whether or not the reaction has followed Markovnikov's rule.

PRELAB ASSIGNMENTS

1. Read OP–13c, review OP–15, and read or review the other operations as necessary.

2. Calculate the mass and volume of 0.10 mol of 2-methyl-3-butyn-2-ol and the theoretical yield of the product.

Reactions and Properties

$$\underset{\underset{CH_3}{|}}{\overset{\overset{OH}{|}}{CH_3C}}-C\equiv CH + H_2O \xrightarrow[HgSO_4]{H_2SO_4} \underset{\underset{CH_3}{|}}{\overset{\overset{HO}{|}}{CH_3C}}-\overset{\overset{O}{\|}}{C}-CH_3$$

Table 1.   Physical properties

	M.W.	b.p.	d.
2-methyl-3-butyn-2-ol	84.1	104	0.868
mercuric sulfate	296.7		
3-hydroxy-3-methyl-2-butanone	102.1	140–41	0.953

Figure 1.   PMR spectrum of 3-hydroxy-3-methyl-2-butanone. (This and all other nuclear magnetic resonance spectra in this book are reproduced from *The Aldrich Library of NMR Spectra, Edition II,* by Charles J. Pouchert, with the permission of the Aldrich Chemical Company.)

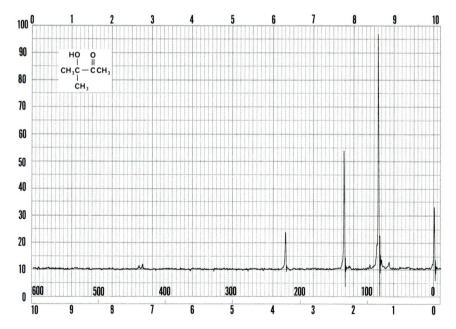

Note: All PMR spectra in this book were run in $CDCl_3$ unless otherwise indicated.

PROCEDURE

SAFETY PRECAUTIONS

Mercury(II) sulfate is *very poisonous* if inhaled or ingested. Do not breathe its dust, or allow it to contact skin or eyes. Wash hands thoroughly after handling the compound.

2-Methyl-3-butyn-2-ol and 3-hydroxy-3-methyl-2-butanone may cause skin or eye irritation, and inhalation, ingestion or skin absorption may be harmful. Avoid contact with the reactant and product, and do not breathe their vapors.

Elemental mercury (formed during the reaction) emits toxic vapors whose concentration can build up to dangerous levels in poorly ventilated areas. Dispose of mercury-containing residues properly, and clean up any spills immediately.

Methylene chloride may be harmful if ingested, inhaled, or absorbed through the skin, and it is suspected of causing cancer if inhaled. Minimize contact with it, and do not breathe its vapors.

Reaction. Assemble an apparatus for addition and reflux ⟨OP–10⟩. Put 0.10 mol of 2-methyl-3-butyn-2-ol in the addition funnel. Carefully weigh out 0.5 g of mercury(II) sulfate [**poison! contact hazard!**]. Mix it with 50 mL of 3 M sulfuric acid, and transfer this solution to the boiling flask. Heat the mixture to boiling; then maintain a gentle reflux ⟨OP–7a⟩ as you slowly add the 2-methyl-3-butyn-2-ol drop by drop, with occasional shaking, over about a 10-minute period. Reflux the mixture for 30 minutes after all the alkyne has been added.

Separation. Assemble the codistillation apparatus illustrated in Figure 1 of OP–15 (page 642), using a graduated cylinder for the receiver. Put 50 mL of water in the addition funnel, and begin distilling the product mixture [*caution:* **foaming may occur**]. Add water during the codistillation to maintain the water level in the boiling flask, until all the water has been added.

If the reaction mixture is not strongly acidic, it might contain an unstable organomercury compound that could explode if heated to dryness. For that reason, monitor both distillations carefully and do not distill to dryness.

Continue distilling until about 75 mL of distillate has been collected **[do not distill to dryness!].** Dispose of the residue in the boiling flask as directed by your instructor. Salt out the product ⟨OP–13c⟩ by shaking the distillate with 15 g of potassium carbonate sesquihydrate, followed by enough solid sodium chloride to saturate the solution. When most of the sodium chloride has dissolved, carefully decant all the liquid into a separatory funnel, leaving behind any droplets of metallic mercury and undissolved sodium chloride. Put this residue in a mercury-wastes container. Extract the liquid mixture in the separatory funnel ⟨OP–13⟩ with two 25-mL portions of methylene chloride and wash the combined extracts ⟨OP–19⟩ with 25 mL of 1 M sulfuric acid, followed by 25 mL of saturated aqueous sodium chloride. Dry the organic layer ⟨OP–20⟩ with about 4 g of *anhydrous* potassium carbonate and evaporate the solvent ⟨OP–14⟩ from the filtered solution.

There should be about 25 mL of liquid in the boiling flask at the end of the codistillation.

Potassium carbonate dihydrate can also be used.

A separate organic layer may form after the salts are added; if so, decant both layers into the separatory funnel.

The methylene chloride can also be removed by simple distillation, using a steam bath.

Purification and Analysis.   Purify your crude product by semi-micro distillation ⟨OP–25a⟩, collecting the 3-hydroxy-3-methyl-2-butanone that distills around 138–141°C **[do not distill to dryness!],** and weigh it. Record infrared spectra ⟨OP–33⟩ of both the reactant and the purified product, or obtain spectra from your instructor.

Report.   Calculate the percentage yield of the synthesis. Interpret the infrared spectra of the reactant and product as completely as you can. Explain clearly how the spectra show that the expected functional group conversion has taken place. Tell whether or not the hydration reaction has followed Markovnikov's rule, and describe the evidence supporting your conclusion.

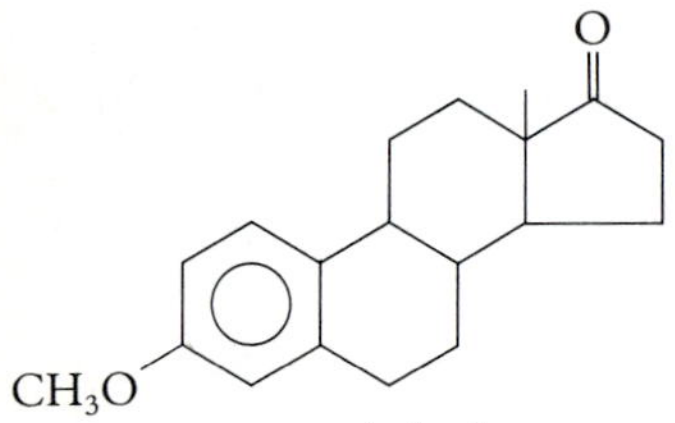

estrone methyl ether

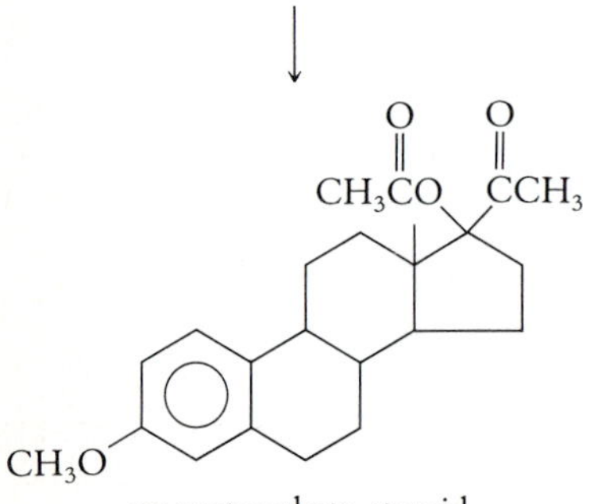

an acetoxyketo steroid

Collateral Projects

1. Test the reactant and product with bromine ⟨Test C–7 in Part III⟩ and potassium permanganate ⟨Test C–19⟩. Explain the results.

2. To further verify the mode of addition, prepare a semicarbazone derivative of the product ⟨Procedure D–4 in Part III⟩ and obtain its melting point, or carry out an iodoform test ⟨Test C–16⟩. The semicarbazone of the Markovnikov product melts at 163°C; that of the anti-Markovnikov product melts at 222–223°C.

Exercises

1. **(a)** Identify each signal in the PMR spectrum of 3-hydroxy-3-methyl-2-butanone (Figure 1) by indicating the proton set responsible for it. **(b)** Sketch the PMR spectrum you would expect to obtain for the product of anti-Markovnikov hydration of 2-methyl-3-butyn-2-ol, showing approximate signal areas, multiplicities, and chemical shifts.

2. Construct a flow diagram for this synthesis.

3. Draw structures for the products that you would expect to obtain from the hydration of acetylene (ethyne), 1-butyne, 2-butyne, and norethynodrel.

4. Write a mechanism for the hydration of 2-methyl-3-butyn-2-ol.

5. Using reactions discussed in this experiment, outline a synthetic pathway for the conversion of estrone methyl ether to the acetoxyketo steroid shown.

Library Topic

"The pill" has been the subject of considerable scientific and moral controversy since its introduction for contraception in the late 1950s. Find out about some of the beneficial and potentially harmful effects of oral contraceptives, describe some new developments in the field of contraception, and discuss the impact of modern birth-control methods on society.

Identification of a Conjugated Diene from Eucalyptus Oil

Reactions of Dienes. Preparation of Bicyclic Compounds. Cycloaddition.

Operations

OP–7a Refluxing

OP–12 Vacuum Filtration

OP–21 Drying Solids

OP–23 Recrystallization

OP–28 Melting-Point Determination

OP–32 Gas Chromatography

OP–33 Infrared Spectrometry

Objectives

To separate a conjugated diene from eucalyptus oil and identify it by preparing its Diels-Alder adduct.

To predict the stereochemistry of the adduct.

To learn how to interpret the infrared spectra of carboxylic anhydrides.

To learn about the mechanism and stereochemistry of the Diels-Alder reaction.

To learn about some natural conjugated dienes.

SITUATION

In 1927, Otto Diels and Kurt Alder treated a constituent of eucalyptus oil with maleic anhydride and obtained a new compound that they described as forming *grosse glasglanzende Krystalle von ungewohnlicher Schonheit.* The reaction Diels and Alder used for that purpose was eventually named for them, and the array of "lustrous crystals" was the Diels-Alder adduct of a natural conjugated diene. In this experiment, you will prepare the same adduct and attempt to identify the diene in eucalyptus oil.

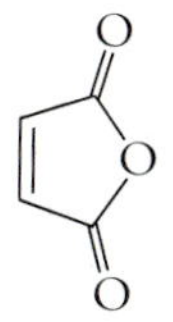

maleic anhydride

The German description translates roughly as "large lustrous crystals of unusual beauty."

limonene

$$CH_3C = CHCH_2CH_2CCH = CH_2$$

$$|||$$

$$CH_3 CH_2$$

β-myrcene

Myrcene is actually a triene, but only two of its double bonds are conjugated, so it is classified here as a conjugated diene.

$$CH_3C = CHCH_2CH = CCH = CH_2$$

$$| |$$

$$CH_3 CH_3$$

β-ocimene

$$CH_3C = CHCH = CHC = CHCH_3$$

$$| |$$

$$CH_3 CH_3$$

allo-ocimene

BACKGROUND

Dienes in Nature

Some natural hydrocarbons are conjugated dienes or trienes, in which two or three double bonds are separated by single bonds. Most conjugated dienes have chemical properties not shared by other dienes, such as the ability to form Diels-Alder adducts with maleic anhydride. These adducts are usually crystalline solids that can be separated from the other components of an essential oil and used to identify the diene.

Dienes and trienes occur in the essential oils of a number of plants and contribute to the flavors and aromas of such plants. For example, limonene has a pleasant lemony odor that enhances the flavor of lemons, oranges, and other citrus fruits. The conjugated diene called β-myrcene is responsible for much of the fragrance and flavor of bay leaves *(Myrcia acris)*. β-Myrcene is also present in hops, verbena, and lemongrass oil. β-Ocimene was first isolated from the Javanese oil of basil *(Ocimium basilicum)* and is usually found in combination with *allo*-ocimene. The latter is also obtained by pyrolysis of α-pinene, the most abundant component of turpentine; myrcene is a major pyrolysis product from β-pinene. Both of the phellandrenes derive their name from the water fennel *Phellandrium aquaticum,* but α-phellandrene apparently doesn't occur in that plant; it was mistaken for its isomer, β-phellandrene, which does. α-Phellandrene *is* found in the oils of bitter fennel, ginger grass, cinnamon, and star anise; β-phellandrene also occurs in lemon oil and Japanese peppermint oil. Another cyclic diene, α-terpinene, is obtained from the essential oils of cardamom, marjoram, and coriander.

α-phellandrene   β-phellandrene   α-terpinene

METHODOLOGY

The Diels-Alder reaction is classified as a 4 + 2 cycloaddition reaction. One component, the *diene,* contributes four atoms to the six-membered ring of the *adduct;* the other, the *dienophile,* contributes two. The four centers of the diene must be connected by two conjugated double bonds, and the dienophile must have a double or triple bond connecting the two

A 4 + 2 cycloaddition reaction

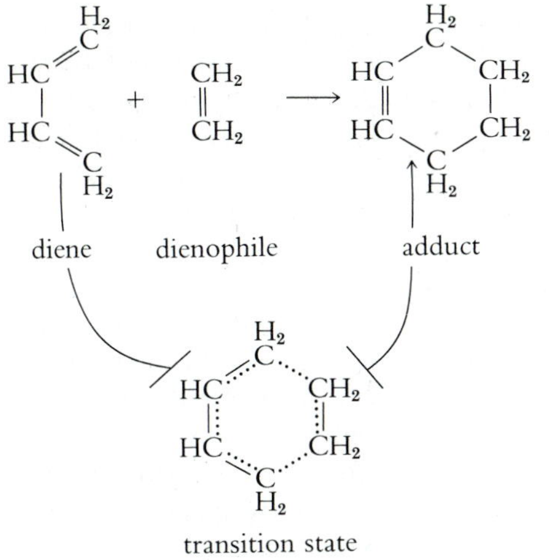

diene   dienophile   adduct

transition state

carbon centers involved in the cycloaddition. The reaction can be viewed as 1,4-addition to a conjugated diene in which the dienophile acts as the electrophile.

The Diels-Alder reaction is stereoselective, usually yielding only one of several possible stereoisomers. For example, maleic acid could react with cyclopentadiene to yield two adducts, designated *exo* and *endo*. In fact, it yields entirely the *endo* adduct, in which the bulkier parts of the dienophile are closer to the carbon-carbon double bond.

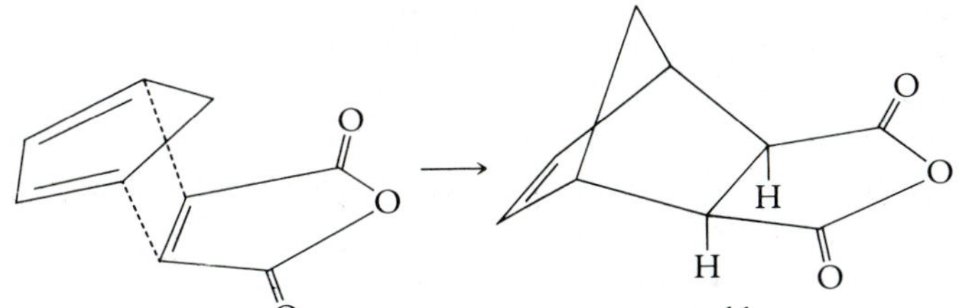

exo-adduct

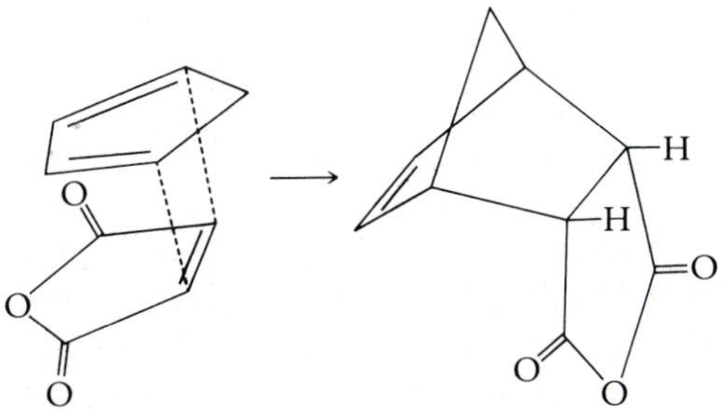

endo-adduct

This orientation results from the fact that overlap between the pi electrons of the diene and those of the dienophile stabilizes the transition state leading to the adduct. Such overlap is possible only when the carbon-carbon double bonds of the diene are in close proximity to the carbonyl groups of the dienophile. Sometimes more than one *endo* adduct is possible; in that event, the dienophile will tend to approach the diene from its less hindered side to give the more stable adduct.

Your objective in this experiment is to separate and identify a conjugated diene from eucalyptus oil by preparing its Diels-Alder adduct with maleic anhydride. The unknown (whose molecular formula is $C_{10}H_{16}$) will be one of four conjugated dienes that were discussed in the Background section. Their names and the melting points of their adducts are listed in Table 2. You will estimate the percentage of diene in the eucalyptus oil using gas chromatography and add enough maleic anhydride to react with that much diene. It is best to avoid using a large excess of maleic anhydride because the unreacted dienophile can be difficult to remove from the product. Both maleic anhydride and the adduct can be hydrolyzed by water, so it is important to use dry glassware and to keep out moisture during the reaction and workup. Because of its moisture sensitivity, maleic anhydride is usually sold in the form of briquettes, which must be pulverized with a mortar and pestle before being used.

A Diels-Alder reaction is sometimes carried out by simply mixing and heating the reactants. The reaction can be very exothermic, however, so a solvent is often used to keep it from proceeding too rapidly. You will prepare the adduct by refluxing maleic anhydride with the eucalyptus oil in ethyl ether, which serves as the reaction solvent. The adduct should precipitate from the reaction mixture as beautiful rectangular crystals; slow cooling may yield crystals several centimeters long. The adduct can then be separated by vacuum filtration and purified by recrystallization from methanol. Because methanol can react with the adduct to form a solvolysis product, you should avoid prolonged boiling during recrystallization. For the same reason, it is not a good idea to leave the crystallized adduct in methanol for more than a few hours.

The melting point of the adduct should reveal the identity of the diene in eucalyptus oil. With the help of molecular models, you should be able to deduce its structure. You can also characterize the adduct by recording its infrared spectrum. The infrared spectra of anhydrides show two carbonyl stretching bands that arise from symmetric and asymmetric stretching

An impure commercial form of the diene may be used to simulate eucalyptus oil.

Hydrolysis of maleic anhydride

maleic acid

modes; maleic anhydride itself has C=O bands near 1780 and 1850 cm⁻¹, as shown in Figure 1. The C—CO—O—CO—C grouping can vibrate as a unit, causing additional bands that occur near 900 cm⁻¹ and 1250 cm⁻¹ for cyclic anhydrides.

Aromatic compounds sometimes behave like conjugated dienes when reacting with certain dienophiles. Furan is an aromatic heterocycle that undergoes Diels-Alder reactions as if it had the diene resonance structure shown. If you do Minilab 15, you will carry out the Diels-Alder reaction of furan and maleic anhydride, which is said to yield the unexpected *exo* adduct.

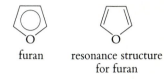

furan   resonance structure
               for furan

The stereochemistry of this adduct is explained in *J. Am. Chem. Soc.* 70, 1161 (1948).

PRELAB ASSIGNMENTS

1. Be prepared to carry out the calculations described in part A of the Procedure.

2. If possible, record the gas chromatogram during the lab period preceding the one for which this experiment is scheduled.

Reaction and Properties

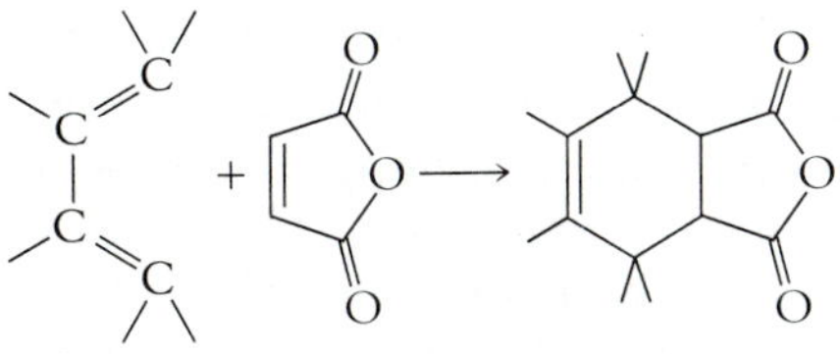

The actual structure of the adduct depends on the structure of the unknown diene.

Table 1.   Physical properties

	M.W.	m.p.	b.p.	d.
diene from eucalyptus oil	136.2		172	0.841
maleic anhydride	98.1	53	202	
ethyl ether	74.1	−116	34.5	0.714

Table 2.   Melting points of maleic anhydride adducts of the dienes

Diene	m.p. of Adduct	Diene	m.p. of Adduct
β-myrcene	33 – 34	*allo*-ocimene	83 – 84
α-terpinene	60 – 61	*α*-phellandrene	126 – 127

Figure 1.   IR spectrum of maleic anhydride. (Reproduced from *The Aldrich Library of FT-IR Spectra,* by Charles J. Pouchert, with the permission of the Aldrich Chemical Company.)

Maleic anhydride

3118.7	1240.1	840.6
1851.1	1058.1	695.9
1778.5	890.8	561.7

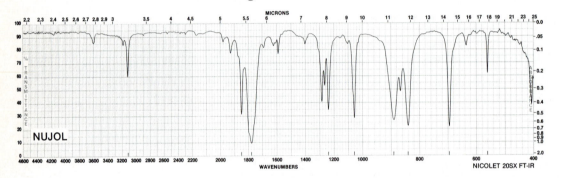

PROCEDURE

SAFETY PRECAUTIONS

Maleic anhydride is corrosive and toxic and can cause severe damage to the eyes, skin, and upper respiratory tract. Wear gloves, avoid contact with skin, eyes, or clothing, and do not breathe the dust. If you must pulverize maleic anhydride briquettes, do it under the hood and wear protective clothing.

Ethyl ether and petroleum ether are extremely flammable and may be harmful if inhaled. Do not breathe the vapors, and keep the solvents away from flames and hot surfaces.

Methanol is harmful if inhaled, swallowed, or absorbed through the skin. Avoid contact with the liquid, do not breathe its vapors, and keep it away from flames and hot surfaces.

maleic anhydride	ethyl ether	petroleum ether	methanol

A. Preliminary Analysis and Calculations

Obtain a gas chromatogram ⟨OP–32⟩ of the eucalyptus oil provided, using the column and conditions recommended by your instructor. Assuming that the unknown diene is responsible for the largest peak on the chromatogram, and that peak areas are proportional to component masses, estimate the mass of the unknown diene in 5.0 g of the oil and the mass of maleic anhydride needed to react with that much diene.

B. Preparation of the Adduct

Dissolve 5.0 g of the eucalyptus oil in 10 mL of anhydrous ethyl ether **[fire hazard!]**, and add the calculated mass of powdered maleic anhydride **[contact and inhalation hazard!]**. Gently reflux the mixture ⟨OP–7a⟩ on a steam bath for 45 minutes or more. While it is still warm, transfer the reaction mixture to a small beaker, cover the beaker with a watch glass, and let it cool slowly to room temperature. Cool it further in an ice bath to allow complete crystallization.

If crystallization does not begin by the time the reaction mixture reaches room temperature, scratch the inside of the flask with a glass stirring rod.

Separation.   Collect the adduct by vacuum filtration ⟨OP–12⟩, washing the crystals on the filter with 10 mL of cold, low-boiling *petroleum* ether.

Purification and Analysis.   Recrystallize the adduct ⟨OP–23⟩ from dry methanol (avoid prolonged boiling). Dry the adduct ⟨OP–21⟩, and measure its mass and melting point ⟨OP–28⟩. Deduce the identity of the adduct from its melting point. Record the infrared spectrum ⟨OP–33⟩ of the adduct in a KBr disc or Nujol mull, or obtain a spectrum from your instructor.

Stereochemistry of the Adduct (Optional).   Construct molecular models for maleic anhydride and the diene in eucalyptus oil. By moving their bonds around, find a way to connect them to make a model representing one form of the adduct. Disconnect and reconnect the diene and dienophile units until you have made models representing all possible structures for the adduct. Based on the discussion in the Methodology section, decide which is the most likely structure.

Report.   Give the name and structure of the diene in eucalyptus oil, and justify your conclusion. Calculate the theoretical yield of the adduct based on the mass of maleic anhydride you used, and calculate your percentage yield. Write all possible structural formulas for the adduct, showing the stereochemistry clearly; designate the structure you believe is the correct one, and justify your choice. Write a balanced equation for the reaction. Interpret the infrared spectrum of the adduct as completely as you can. Compare the adduct's spectrum with that of maleic anhydride; point out and try to explain any significant similarities or differences.

Collateral Project

Record the NMR spectrum of the adduct in deuterochloroform.

Diels-Alder Reaction of Maleic Anhydride and Furan

MINILAB 15

SAFETY PRECAUTIONS

Avoid contact with maleic anhydride, and do not breathe the dust.

Avoid contact with and inhalation of furan.

Keep ethyl ether away from flames and hot surfaces.

Leave the tube in a reasonably cool place so the stopper doesn't pop out. A screw-cap vial can be used in place of the test tube.

Dissolve 1.3 g of finely divided maleic anhydride in 12 mL of anhydrous ethyl ether **[fire hazard!]** in a large test tube by heating gently on a steam bath under a hood (replace any ether that evaporates). Add 1 g (~1.1 mL) of furan, stopper the test tube *tightly* with a rubber stopper, and let the mixture stand for several days or until the next laboratory period. Collect the product by vacuum filtration, weigh it, and measure its melting point. Write an equation for the reaction, showing the correct stereochemistry for the adduct.

Exercises

1.   Which diene named in the Background section will *not* form a Diels-Alder adduct with maleic anhydride? Explain.

2.   Write the structure of the compound that would result if the adduct were heated too long in the recrystallization solvent; also write a balanced equation for its formation.

3.   Construct a flow diagram for the synthesis in this experiment.

4.   Determine which was the limiting reactant in Minilab 15 and calculate the theoretical yield of the adduct. If you prepared the adduct, calculate your percentage yield.

5.   The side reaction most often encountered in Diels-Alder syntheses is dimerization or polymerization in which the diene also acts as a dienophile. For example, butadiene can react with itself to yield 4-vinylcyclohexene as shown. Draw the structures of four possible Diels-Alder dimers of your diene.

Dimerization of 1,3-butadiene

Library Topic

A number of polychlorinated insecticides, such as dieldrin, aldrin, and chlordane, are synthesized using one or more Diels-Alder reactions. Report on the manufacture and use of these insecticides, giving equations for their synthesis from cyclopentadiene.

The Synthesis of Triphenylmethanol and the Trityl Carbenium Ion

Reactions of Carbonyl Compounds. Reactions of Organic Halides. Preparation of Alcohols. Nucleophilic Addition. Organometallic Compounds. Carbenium Ions.

Operations

OP – 1   Cleaning and Drying Glassware

OP – 7*a*   Refluxing

OP – 10   Addition

OP – 12   Vacuum Filtration

OP – 12*a* Semimicro Vacuum Filtration

OP – 13   Extraction

OP – 13*b* Separation of Liquids

OP – 14   Evaporation

OP – 19   Washing Liquids

OP – 20   Drying Liquids

OP – 21   Drying Solids

OP – 22*a* Excluding Moisture

OP – 23   Recrystallization

OP – 28   Melting-Point Determination

Objectives

To synthesize triphenylmethanol by a Grignard reaction.

To prepare a stable carbenium ion, triphenylmethyl.

To learn how to carry out a reaction involving a moisture-sensitive intermediate.

To learn about the preparation and reactions of Grignard reagents.

To learn about the trityl carbenium ion and the related triphenylmethane dyes.

SITUATION

The triphenylmethane dyes come in all colors of the rainbow, from the red of rosaniline through malachite green and Victoria blue to crystal violet. The chromophores responsible for these colors are essentially nitrogen-substituted triphenylmethyl (trityl) cations, but they are not true carbenium ions, because most of the positive charge is distributed to the ring nitrogen atoms, as in the iminium ion form of malachite green.

A chromophore is the part of a molecule responsible for absorption of ultraviolet or visible light.

Alternative structures for malachite green, a triphenylmethane dye

carbenium ion form   iminium ion form

Such charge delocalization stabilizes the dyes so that most of them will keep indefinitely, even in aqueous solution. Unsubstituted trityl salts, which are true carbenium ions, are nearly a *trillion* times less stable than crystal violet, and they hydrolyze rapidly to triphenylmethanol in water. Nevertheless, the trityl cation is unusually stable for a carbenium ion, and its intense orange color reveals its kinship to the triphenylmethane dyes. In this experiment, you will prepare triphenylmethanol by a Grignard reaction and convert it to triphenylmethyl fluoborate, which contains the trityl carbenium ion.

triphenylmethyl
fluoborate

BACKGROUND

The Colorful Career of the Triphenylmethanes

When W. H. Perkin prepared mauve from aniline in 1854, he started a mad race to develop other commercially profitable dyes from aniline and related compounds. Four years later Verguin heated the crude aniline of the day with stannic chloride and produced a beautiful fuchsia-colored substance that was the first of the triphenylmethane dyes. Malachite green,

Repeating this experiment with today's aniline would not produce fuchsin, because its formation was dependent on the presence of *p*-toluidine as an impurity.

Reduced form of malachite green

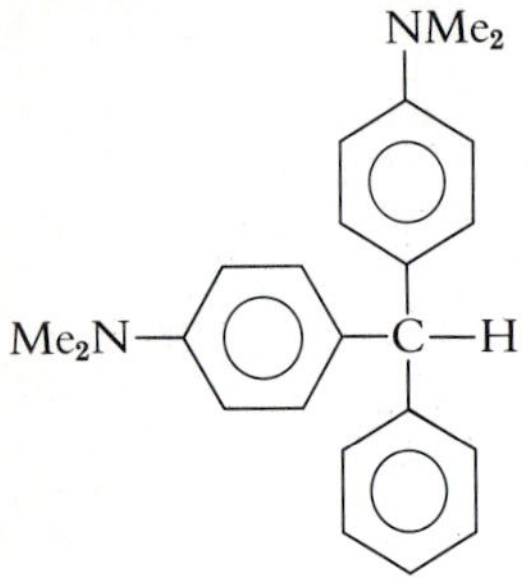

NMe₂

Me₂N—⟨ ⟩—C—H

A resonance structure of the
triphenylmethyl cation

⊕

crystal violet, and similar dyes followed fuchsin onto the scene not long afterward and stimulated research into the molecular basis of color. Why is malachite green, for example, brilliantly colored and its reduced form colorless?

In the early days, the triphenylmethane dyes were represented by quinonoid (iminium ion) structures like the one shown above for malachite green. However, when triphenylmethanol — which, having no nitrogen-containing substituents, cannot form iminium ions — was treated with strong acid, it also produced a colored solution. This led to the suspicion that the colored product, and by extension the triphenylmethane dyes as well, were actually carbenium ions having the positive charge located on the central carbon atom. Some heated arguments ensued between the supporters of the iminium ion structure and those of the carbenium ion structure. Their differences were eventually reconciled by resonance theory, by which the two forms are regarded as contributing structures of a resonance hybrid that has some characteristics of each.

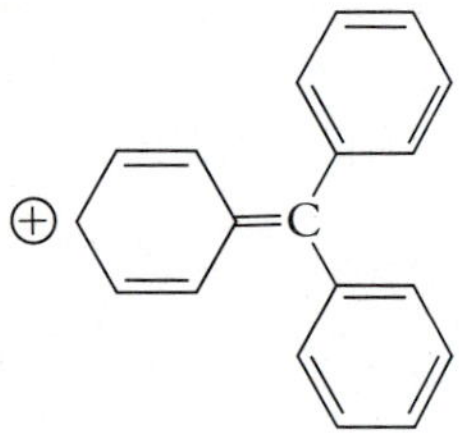

When protected from atmospheric moisture, the trityl cation is stable enough to keep almost indefinitely. It owes this unusual stability to delocalization of the positive charge about the three benzene rings — nine resonance structures of the trityl cation can be drawn with the positive charge on a ring atom. The cation is apparently shaped somewhat like a propeller with the "blades" (benzene rings) pitched at a 32° angle. (Steric interference between the *ortho* hydrogen atoms makes a planar configuration impossible.)

Shape of triphenylmethyl cation

The year 1900 marked two significant milestones in organic chemistry. That was the year Moses Gomberg announced the discovery of the trityl free radical (see Experiment 15) and Victor Grignard reported his discovery and use of Grignard reagents. Just a year later, trityl carbenium ions were being prepared from triphenylmethanol, which, fortuitously, can be synthesized with a Grignard reagent.

METHODOLOGY

Victor Grignard's original procedure for preparing a Grignard reagent was as follows: About a mole of magnesium metal was placed in a dry two-necked round-bottomed flask fitted with a reflux condenser and dropping funnel. A mole of the organic halide was dissolved in ethyl ether and 50 mL of this solution was added to the magnesium. When a white turbidity appeared at the metal surface and effervescence began, more ether was added in portions, with cooling, followed by drop-by-drop addition of the remainder of the halide-ether solution. The reaction was brought to completion by refluxing in a water bath, resulting in a nearly colorless liquid. Essentially the same method is used today for making many Grignard reagents, although extensive studies of the reaction have led to some modifications of the reaction conditions.

It is extremely important that the reagents and apparatus be as dry as possible, since water not only reacts with Grignard reagents but appears to inhibit their formation. In a study using *n*-butyl bromide, it was found that the **induction period** for forming the Grignard reagent was $7\frac{1}{2}$ minutes using sodium-dried ether, 20 minutes using commercial absolute ether, and 2 hours using ether half-saturated with water. It is apparent that careful drying of the reaction apparatus and reagents saves time in the long run; this can increase the yield of Grignard reagent as well.

The type and quantity of reagents and solvents used are also important. For most purposes, an ordinary grade of well-dried magnesium turnings will suffice, if they are rubbed and crushed with a glass rod to remove some of the oxide coating and provide a fresh surface for reaction. Commercial anhydrous ether is suitable for most routine preparations, but the optimum quantity of ether depends on the kind of Grignard reagent. One study showed that the highest yields of phenylmagnesium bromide were obtained with 5 moles of ethyl ether per mole of bromobenzene.

In this experiment, you will prepare a Grignard reagent, phenylmagnesium bromide, by adding a solution of bromobenzene in dry ether to magnesium metal. The onset of the reaction is indicated by the formation of small bubbles at the surface of the magnesium, usually followed by the appearance of a cloudy precipitate; soon afterward the ether should begin to boil spontaneously. If the reaction does not begin within a

See V. Grignard, *Ann. Chim.* [7], *24*, 433 (1901).

See Kharasch and Reinmuth ⟨Bib–L46⟩ for a very thorough discussion of Grignard reactions.

induction period: the time elapsing between the combination of the reactants and the start of a noticeable reaction

Whereas water inhibits formation of Grignard reagents, iodine and other substances may accelerate it. For this reason, it is a common practice to add a small crystal of iodine if the reaction does not start after a few minutes.

Other solvents than ether can be used. Arylmagnesium chlorides form more readily in tetrahydrofuran.

tetrahydrofuran

few minutes, adding a small crystal of iodine and warming the flask with your hand may initiate it. Crushing some of the magnesium against the side of the flask with a flat-bottomed stirring rod may also work, but you must be *very* careful not to poke a hole through the flask! Using a high concentration of bromobenzene helps to get the reaction started, but it can promote the formation of an undesirable by-product, biphenyl, through a coupling reaction at the metal's surface. Thus, the bromobenzene solution should be diluted with ether as soon as the reaction gets underway. Phenylmagnesium bromide reacts rapidly with water to form benzene and more slowly with oxygen to form a magnesium salt of phenol, so the reaction apparatus must be protected from moisture and the Grignard reagent should be used promptly after it is prepared.

If none of these measures is successful, you may have to start over, making certain that all the chemicals and glassware are completely dry.

When benzophenone is added to this Grignard reagent, a magnesium salt of triphenylmethanol precipitates from the reaction mixture. This is converted to triphenylmethanol by pouring the reaction mixture into a dilute, ice-cold solution of sulfuric acid. The acid also dissolves the basic magnesium salts that form along with the triphenylmethanol. After extraction from the reaction mixture with ethyl ether, the crude triphenylmethanol is treated with petroleum ether to remove any biphenyl, and then purified by recrystallization.

Carbenium ions can be prepared by mixing alcohols with a strong acid such as fluoboric acid; the fluoborate anion is a very weak nucleophile that doesn't react with the resulting carbenium ion. You will prepare trityl fluoborate by the reaction of triphenylmethanol with 48% fluoboric acid. The water in the aqueous fluoboric acid solution (as well as that produced during the reaction) could prevent or reverse this reaction, so acetic anhydride is added to consume the water via the following reaction:

Preparation of carbenium ions using fluoboric acid

$$ROH + HBF_4 \rightleftharpoons ROH_2^+ + BF_4^-$$
$$\longrightarrow R^+BF_4^- + H_2O$$

$$\underset{\substack{\text{acetic} \\ \text{anhydride}}}{CH_3\overset{O}{\overset{\|}{C}}O\overset{O}{\overset{\|}{C}}CH_3} + H_2O \longrightarrow \underset{\substack{\text{acetic} \\ \text{acid}}}{2CH_3\overset{O}{\overset{\|}{C}}OH}$$

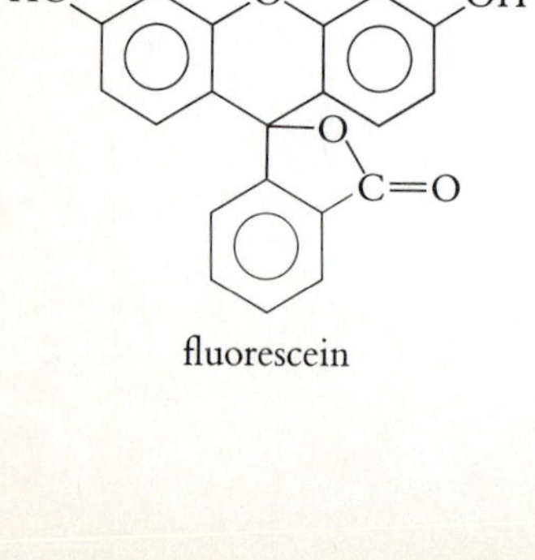

Certain dyes and indicators such as phenolphthalein and fluorescein resemble the triphenylmethane dyes in that their molecules contain three benzene rings attached to a central carbon atom. If you do Minilab 16, you will prepare fluorescein and observe its fluorescence under ultraviolet light.

fluorescein

PRELAB ASSIGNMENTS

1. Read OP–13*b* and OP–22*a* and review the Drying Glassware section in OP–1; read or review the other operations as necessary.

2. Calculate the mass of 22 mmol of magnesium, the mass of 20 mmol of benzophenone, the mass and volume of 22 mmol of bromobenzene, and the theoretical yield of triphenylmethanol. Calculate the theoretical yield of trityl fluoborate from 1.0 g of triphenylmethanol.

Reactions and Properties

A.

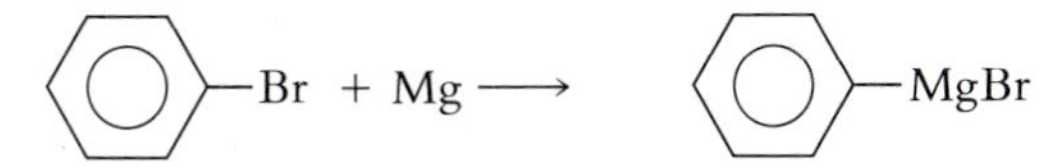

bromobenzene   phenylmagnesium bromide

B.

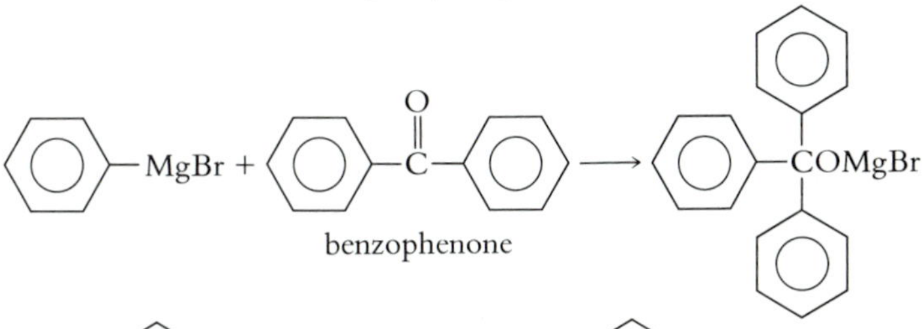

benzophenone

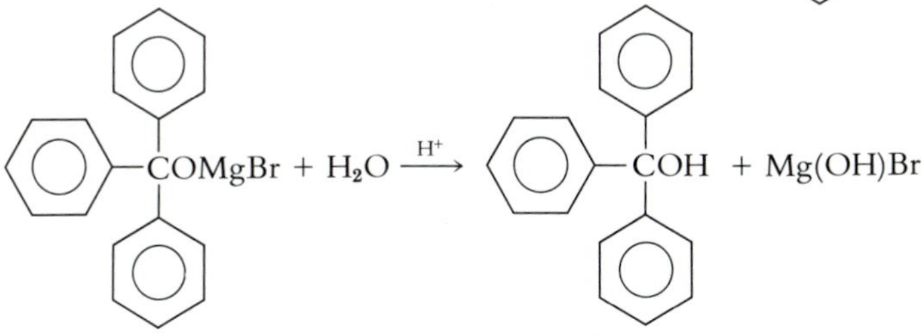

triphenylmethanol

C.

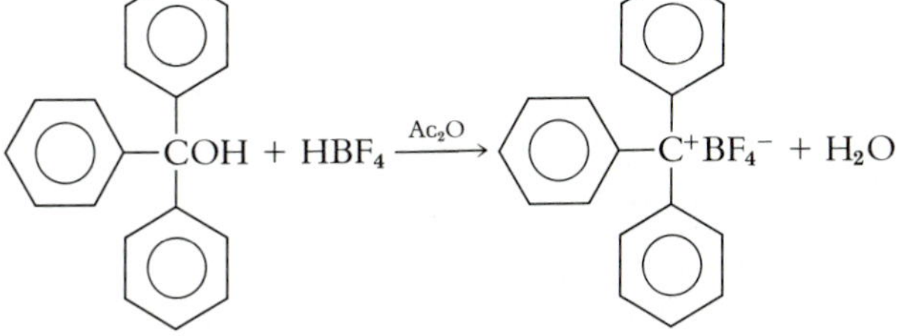

trityl fluoborate

Table 1.   Physical properties

	M.W.	m.p.	b.p.	d.
bromobenzene	157.0	−31	156	1.495
magnesium	24.3			
ethyl ether	74.1	−116	34.5	0.714
benzophenone	182.2	48	306	
triphenylmethanol	260.3	164	380	
fluoboric acid (48%)	87.8			1.41
acetic anhydride	102.1	−73	140	1.082
trityl fluoborate	330.1			

Fluoboric acid is also called fluoroboric acid or tetrafluoroboric acid.

Figure 1.   IR spectra of the starting materials. (Reproduced from *The Aldrich Library of FT-IR Spectra,* by Charles J. Pouchert, with the permission of the Aldrich Chemical Company.)

Bromobenzene

3064.8	1068.7	·733.9
1578.4	999.5	683.8
1474.3	902.6	457.1

Benzophenone

1659.8	1277.4	698.2
1598.4	941.4	638.8
1447.3	763.4	407.3

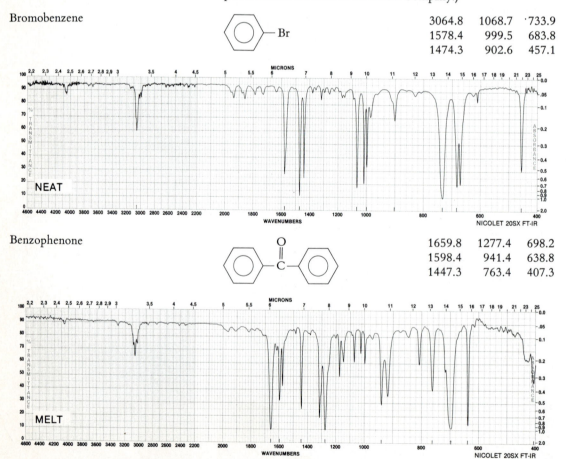

PROCEDURE

If possible, clean and pre-dry the glassware needed for the reaction before the lab period so as to reduce the drying time.

Bromobenzene causes eye and skin irritation, and inhalation, ingestion, or skin absorption may be harmful. Avoid contact with the liquid, and do not breathe its vapors.

Ethyl ether is extremely flammable and may be harmful if inhaled. Do not breathe the vapors, and keep it away from flames and hot surfaces.

Magnesium can cause dangerous fires if ignited; keep it away from flames and hot surfaces.

Petroleum ether is extremely flammable and can be harmful if inhaled, swallowed, or absorbed through the skin. Avoid inhalation or prolonged contact, and keep it away from flames and hot surfaces.

SAFETY PRECAUTIONS

bromobenzene

ethyl ether

magnesium

A. Preparation of Phenylmagnesium Bromide

Reaction.   Clean and *thoroughly* dry ⟨OP–1⟩ all supplies needed for the reaction, including glassware used to measure or transfer chemicals. The reaction apparatus should be oven-dried if possible. As soon as the glassware is cool enough to handle, assemble an apparatus for addition under reflux ⟨OP–10⟩, inserting a drying tube ⟨OP–22a⟩ in the top of the reflux condenser and a stopper or drying tube in the addition funnel. Weigh 22 mmol of clean, dry magnesium turnings into a beaker. Crush some of them with the flat end of a stirring rod to expose fresh metal; then place them in the reaction flask (don't use boiling chips). Dissolve 22 mmol of dry bromobenzene in 5 mL of anhydrous ethyl ether **[fire hazard! no flames in the lab!],** transfer this solution to the addition funnel, and stopper the funnel immediately.

Add the bromobenzene solution all at once to the reaction flask, and replace it in the addition funnel by 5 mL of anhydrous ether **[be sure the stopcock is closed!].** Gently shake the apparatus and hold the flask in the palm of your hand to warm it until the reaction is underway; then remove your hand and let the solution boil unassisted. When it is boiling fairly

If the apparatus is to be flame-dried, ethyl ether and other flammable solvents must not be in use anywhere nearby.

If the apparatus is not assembled promptly, moisture will condense inside it as it cools.

If the reaction does not start within a few minutes, try the procedures suggested in the Methodology section.

vigorously, add the 5 mL of anhydrous ethyl ether. Let the reaction continue until the boiling has virtually stopped. Then *gently* reflux the reaction mixture ⟨OP–7a⟩ on a warm-water · bath (or a steam bath turned very low) for 10 minutes; be sure the condenser is working efficiently. Use the phenylmagnesium bromide solution promptly in part B.

B. Preparation of Triphenylmethanol

Reaction.   Dissolve 20 mmol of benzophenone in 10 mL of anhydrous ethyl ether, and transfer this solution to the addition funnel. Add it to the phenylmagnesium bromide solution drop-by-drop, with occasional shaking, at a rate sufficient to keep the mixture boiling gently with no external heating. When the addition is complete, reflux the mixture gently for 20 minutes; then let it cool to room temperature. In a beaker, mix 10 mL of 3 M sulfuric acid with 25 g of cracked ice; set aside a little of this cold solution to wash out the reaction flask. Pour the reaction mixture slowly, with constant stirring, into the beaker, then wash any additional solid from the reaction flask into the beaker with the reserved acid solution. Continue stirring to dissolve the white solid; any unreacted magnesium should also dissolve eventually.

Add more ether to replace any that evaporates. It may be desirable to add about 1 g of ammonium chloride if the white solid fails to dissolve.

Separation.   Pour the reaction mixture into a separatory funnel and separate the layers ⟨OP–13b⟩, saving them both. Extract the aqueous layer ⟨OP–13⟩ with 10 mL of ethyl ether (not necessarily anhydrous) and combine the ether extract with the original ether layer. Wash the ether solution ⟨OP–19⟩ with one portion each of water and of saturated aqueous sodium bicarbonate, dry it ⟨OP–20⟩ with magnesium sulfate, and evaporate the ether ⟨OP–14⟩. Add 10 mL of high-boiling *petroleum* ether to the solid residue, stir and crush the solid in the solvent for a few minutes, and collect the product by vacuum filtration ⟨OP–12⟩. (At your instructor's request, dry and weigh the crude triphenylmethanol, and measure its melting point.)

You can use a cold trap to recover the ether, which should be returned to a designated solvent-recovery container.

The yield can be improved by evaporating solvent from the hot recrystallization solution until the product just begins to crystallize, allowing the solution to cool to room temperature, and cooling it further in an ice bath.

Purification and Analysis.   Recrystallize the triphenylmethanol ⟨OP–23⟩ from a 2:1 mixture of high-boiling petroleum ether (or hexane) and absolute ethanol. Dry the purified triphenylmethanol ⟨OP–21⟩, weigh it, and measure its melting point ⟨OP–28⟩.

C. *Preparation of Trityl Fluoborate*

SAFETY PRECAUTIONS

acetic anhydride   fluoboric acid

Under the hood, mix 1.0 g of triphenylmethanol with 7 mL of acetic anhydride **[contact and vapor hazard! wear gloves!]** in a small Erlenmeyer flask. Carefully add 1 mL of 48% fluoboric acid **[contact and vapor hazard!]**, and swirl to dissolve the solid. Stopper the flask and let the mixture stand for about 15 minutes; then cool it in ice until crystallization is complete. Collect the product by semimicro vacuum filtration ⟨OP–12a⟩, and wash it on the filter with cold anhydrous ethyl ether. Weigh the trityl fluoborate in a *dry* tared vial and turn it in, along with the rest of your triphenylmethanol.

If the trityl fluoborate is to be stored for any length of time, it should be kept in a desiccator.

Report. Calculate your percentage yield of triphenylmethanol, and try to account for significant losses. Calculate your percentage yield of trityl fluoborate, and draw all possible resonance structures for this compound.

Collateral Projects

1. Record and interpret the infrared spectrum of triphenylmethanol using a KBr disc or Nujol mull. Compare it with the spectra in Figure 1, and describe the evidence indicating that the expected reaction has taken place.

2. Record the ultraviolet–visible spectrum (200–600 nm) of trityl fluoborate in dry acetone, and compare it with the spectrum of triphenylmethane in cyclohexane over the same wavelength interval. Explain any similarities or differences.

3. Dissolve a small amount of trityl fluoborate in dry methanol, and record your observations. Dissolve about 0.1 g of trityl fluoborate in 1 mL of dry acetone; then add a solution of sodium iodide in dry acetone (0.1 g NaI in 1 mL acetone) drop by drop until no more changes are observed. Write balanced equations to explain your observations.

MINILAB 16

Preparation of Fluorescein

SAFETY PRECAUTION

Wear gloves during this experiment, and carry out the reaction under a hood. Avoid contact with zinc chloride, phthalic anhydride, and resorcinol, and do not breathe the dust of any of them.

Do not use your thermometer as a stirring rod!

Under the hood, measure 0.28 g of zinc chloride [**contact hazard!**] into a 10-cm test tube, and heat it gently over a burner flame until no more bubbles of water vapor are evolved from the molten salt. Tip and rotate the tube while stirring the melt, so that the zinc chloride coats the sides of the test tube as it cools. Add 0.40 g of phthalic anhydride [**contact hazard!**] and 0.65 g of resorcinol [**contact hazard!**], and heat the test tube cautiously until the temperature of the melt is 180°C. Keep it near that temperature, stirring constantly, until no more bubbles are evolved. Let the mixture cool for a minute or so; then add 4 mL of 2 M hydrochloric acid and heat the liquid to boiling with constant stirring. Use the stirring rod to break up the solid mass, and continue stirring for several minutes to extract zinc salts and unreacted starting materials. Collect the fluorescein by vacuum filtration, grind it fine in a mortar, and repeat this extraction treatment with another 4-mL portion of 2 M HCl. Dry the fluorescein and weigh it.

Fluorescein stains are very hard to remove, so avoid spilling any and keep it off your skin and clothing.

Dissolve about 10 mg (0.010 g) of fluorescein in 10 mL of 0.1 M aqueous sodium hydroxide. Irradiate the solution with a "black light" in a darkened room, observing both transmitted and reflected light. Add 2 M hydrochloric acid drop by drop until the appearance of the solution under the black light changes markedly. Describe your observations.

Exercises

1. **(a)** Write balanced equations for the reactions of phenylmagnesium bromide and trityl fluoborate with water. **(b)** Predict the products of their reactions with cyclohexanol, and write balanced equations for the reactions.

2. Write a balanced equation for the coupling reaction of bromobenzene at the metal surface to form biphenyl.

3. Construct a flow diagram for the synthesis of triphenylmethanol (parts A and B of the Procedure).

4. Find the structural formula of the basic form of fluorescein called fluorescein sodium, and write a balanced equation for the reaction of fluorescein with aqueous sodium hydroxide.

5. The reaction of phenylmagnesium bromide with benzophenone to form the salt of triphenylmethanol is an example of nucleophilic addition; its reaction with ethyl benzoate yielding the same product involves nucleophilic substitution followed by a nucleophilic addition step. Write reasonable mechanisms for both reactions.

6. Outline a synthetic pathway for each of the following compounds, using the Grignard reaction and starting with benzene or toluene: **(a)** 1,1-diphenylethanol; **(b)** 1,2-diphenylethanol; **(c)** 2,2-diphenylethanol; and **(d)** 2,3-diphenyl-2-butanol.

7. **(a)** Draw all possible resonance structures for malachite green. **(b)** Why *is* malachite green colored and its reduced form colorless (see the Background section)?

ethyl benzoate

Library Topic

Grignard reagents, although usually represented by the general formula RMgX, are actually more complex than this formula indicates. Describe some structures that have been proposed for Grignard reagents, and cite experimental evidence for their existence.

The Reaction of Butanols with Hydrobromic Acid

Reactions of Alcohols. Preparation of Alkyl Halides. Nucleophilic Aliphatic Substitution.

Operations

OP–7a Refluxing

OP–13b Separation of Liquids

OP–15 Codistillation

OP–19 Washing Liquids

OP–20 Drying Liquids

OP–22b Trapping Gases

OP–25a Semimicro Distillation

OP–33 Infrared Spectrometry

Objectives

To prepare an alkyl bromide from 1-butanol or 2-butanol and characterize the product by infrared spectrometry.

To observe the effect of the concentration of a catalyst on a nucleophilic substitution reaction.

To learn more about the mechanisms of nucleophilic substitution reactions.

To learn about some commercially important halogen compounds.

SITUATION

Alkyl halides are used widely in research and industry as chemical intermediates for the synthesis of useful compounds. The companies that manufacture such chemicals must make a profit at it, so their goal is to design processes that afford the highest yield in the least time at the lowest possible cost. In this experiment, you and your co-workers will try to determine the

optimum ratio of catalyst to substrate for the conversion of 1-
and 2-butanol to the corresponding alkyl bromides. Too little
catalyst will limit the amount of alkyl bromide you can recover
in the specified reaction period; too much might reduce the
yield and purity of alkyl bromide by promoting the formation
of by-products.

BACKGROUND

The Problematical Halides

The common expression "you can't get along without them
and you can't get along with them" seems to fit many organic
halogen compounds. Organic halides tend to be very stable.
However, the stability that makes a particular halide a useful
commercial chemical can be a liability if the halide is released
into the environment, where it may present a serious and persistent health hazard.

Halogen-containing organic compounds seldom occur in
nature, unless they are put there by human activity. One of the
few exceptions is Tyrian purple, a bromine-containing dye
derived from spiny carnivorous snails of the murex family.
Because of its rarity — about 9000 snails are needed to produce
a gram of the dye — Tyrian purple was a costly status symbol in
the ancient world. Julius Caesar wore a purple toga as a sign of
office; a Roman senator was only allowed a purple stripe on his
toga.

Organic compounds containing bromine and chlorine are
often used as fire retardants in plastics, textiles, and other polymeric materials. An *additive* fire retardant is combined mechanically with a polymeric material, whereas a *reactive* fire
retardant is actually incorporated into the polymer structure.
Unlike C—H bonds, C—Cl and C—Br bonds are not attacked
by oxygen at ordinary flame temperatures, so incorporating
halogenated components into a material decreases its flammability by reducing its fuel content. Halogenated fire retardants
may also decompose when heated to produce chlorine or bromine atoms, which inhibit burning by combining with the free
radicals involved in combustion chain reactions. Additive fire
retardants include pentabromochlorocyclohexane (PBCCH),
hexabromocyclododecane (HBCD), and decabromodiphenyl
oxide (DBDPO), whose structures are shown on page 244.
Tris-(2,3-dibromopropyl)phosphate (tris-BP), an additive fire
retardant once used in children's sleepwear, was taken off the
market after evidence showed that it is a carcinogen. Vinyl

This so-called royal purple is actually a
rich crimson in color.

Tyrian purple (dibromoindigo)

Fire-retardant chemicals

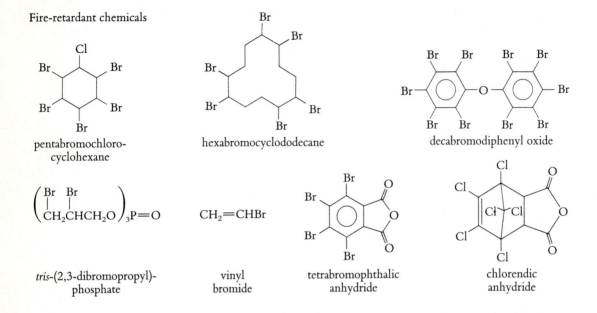

pentabromochloro-
cyclohexane

hexabromocyclododecane

decabromodiphenyl oxide

$$\left(\begin{array}{c} Br \; Br \\ | \quad | \\ CH_2CHCH_2O \end{array}\right)_3 P{=}O$$

$CH_2{=}CHBr$

tris-(2,3-dibromopropyl)-
phosphate

vinyl
bromide

tetrabromophthalic
anhydride

chlorendic
anhydride

bromide, tetrabromophthalic anhydride (TBPA), and chlorendic anhydride are among the reactive fire retardants. TBPA and chlorendic anhydride can be combined with multifunctional alcohols to form fire-resistant polyesters, and vinyl bromide copolymerizes with other unsaturated monomers to produce a variety of fire-resistant addition polymers.

Chlorendic anhydride is manufactured by the Diels-Alder reaction of maleic anhydride with hexachlorocyclopentadiene.

*See Experiment 23 for a discussion of the
Diels-Alder reaction.*

Diels-Alder synthesis of chlorendic anhydride:

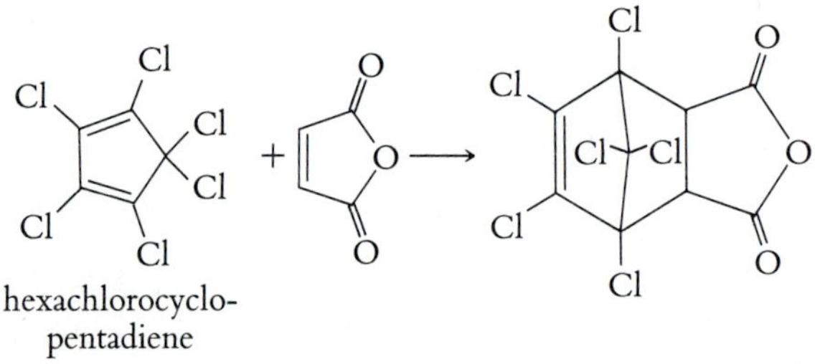

hexachlorocyclo-
pentadiene

The first three insecticides are prepared by Diels-Alder reactions. Mirex is obtained by the photochemical dimerization of hexachlorocyclopentadiene.

The same chlorinated diene has been used to prepare many broad-spectrum chlorinated insecticides, including dieldrin, aldrin, chlordane, and mirex. Chlordane is used in veterinary medicine to kill the fleas, ticks, and lice that torment domestic

Some chlorinated insecticides from hexachlorocyclopentadiene

dieldrin   aldrin   chlordane   mirex

animals, and other chlorinated insecticides are used against agricultural pests, mosquitoes, termites, and other destructive or annoying insects. Unfortunately, chlorinated insecticides don't discriminate between beneficial and harmful insects, and some insect pests may build up a resistance to them. Because chlorinated insecticides are nonbiodegradable, they tend to build up in the environment, where their toxic properties may threaten higher forms of animal life. For these and other reasons such insecticides have been supplanted in part by biodegradable "soft" insecticides and by alternative insect-control strategies.

The polybrominated and polychlorinated biphenyls (PBBs and PCBs) have caused serious environmental problems because they are toxic and probably carcinogenic, and because they persist in the environment. PBBs and PCBs are mixtures of various bromine or chlorine derivatives of biphenyl, such as 2,2′,4,4′,5,5′-hexabromobiphenyl, a major component of the fire retardant preparation called Firemaster BP-6. In 1973, Firemaster BP-6 was accidentally added to cattle feed in Michigan, resulting in the deaths of many farm animals; it has since been withdrawn from the market. Until recently, polychlorinated biphenyls had been used in a great number of applications, from cooling electrical transformers to manufacturing plastic liners for baby bottles. PCBs are probably the most widespread pollutants in the world; they occur in dangerously high concentrations in some fish from the Great Lakes and have been detected in the tissues of many other organisms, including humans and polar bears. Since 1979, the manufacture, processing, and distribution of PCBs has been prohibited in the United States.

A number of organic halides that are now in use appear to be quite safe, as long as they are handled with due care and used responsibly. But the reputation of these compounds as a group has been badly damaged by the misuse of a few of them, and this has stimulated a search for safer alternatives.

2,2′,4,4′,5,5′-hexabromobiphenyl

METHODOLOGY

Catalysts are substances that accelerate chemical reactions without themselves being consumed in the process. A catalyst can exert its effect by providing an alternative, low-energy pathway for a reaction or by increasing the concentration of some crucial intermediate. For example, an alcohol cannot be converted to an alkyl bromide by sodium bromide because OH^- is too poor a leaving group to be displaced by Br^-. A strong acid is needed to convert this poor leaving group to a better one, OH_2, by a protonation reaction:

$$R—OH + H^+ \longrightarrow R—OH_2^+$$

The protonated alcohol can then react with bromide ion by either an S_N1 or an S_N2 mechanism to form an alkyl bromide. Primary alcohols tend to react by the direct substitution (S_N2) mechanism; secondary and tertiary alcohols are more likely to form an intermediate carbenium ion, which then combines with halide ion.

Sulfuric acid catalyzes the reaction of alcohols with HBr by increasing the concentration of the protonated alcohol. Increasing the concentration of the catalyst should increase the reaction rate; we would expect this, in turn, to increase the amount of alkyl bromide produced in a given reaction period, but it might increase the rates of side reactions as well. Accelerating such side reactions as E1 elimination, polymerization, and ether formation might actually lower the product yield by converting an alcohol into undesired by-products.

In this experiment, you will prepare either 1- or 2-bromobutane from the corresponding alcohol. By using from 1 to 4 mL of sulfuric acid with each of the two alcohols, you and your co-workers will vary the ratio of catalyst to substrate from 0.25 to 1.00 (see Table 1). The optimum catalyst : substrate ratio is the *lowest* ratio that produces the maximum yield of alkyl halide. For example, if ratios 0.75 and 1.00 both yielded about the same amount of product, the ratio 0.75 would be preferred. Because variable product losses during the purification step may invalidate the results, the masses of the crude products will be used for the yield comparisons. The data from one or more laboratory sections can be combined for more reliable results.

You will carry out the reaction by refluxing the alcohol with about a 25% excess of 48% hydrobromic acid and the assigned volume of sulfuric acid. Since sulfur dioxide and HBr fumes are evolved during the reaction, a gas trap must be used to keep them from escaping into the laboratory (see Figure 1 of

See your lecture-course textbook for a more detailed discussion of nucleophilic substitution reactions.

Table 1.   Ratio of catalyst to alcohol in the eight runs

Run	Alcohol	mmol of Catalyst	Mole Ratio of Catalyst to Substrate
1	1-butanol (72 mmol)	18	0.25
2	1-butanol (72 mmol)	36	0.50
3	1-butanol (72 mmol)	54	0.75
4	1-butanol (72 mmol)	72	1.00
5	2-butanol (72 mmol)	18	0.25
6	2-butanol (72 mmol)	36	0.50
7	2-butanol (72 mmol)	54	0.75
8	2-butanol (72 mmol)	72	1.00

OP–22b). The alkyl bromide can be separated from the reaction mixture by codistilling it with the water present in the reaction flask. The product ordinarily forms the lower layer in the distillate and can be separated from the aqueous layer in a separatory funnel. With a low catalyst : substrate ratio, there may be enough unreacted alcohol in the distillate to make the organic layer less dense than water. If there is any question about the identity of the organic layer, test both layers by adding a drop of water to a drop of each. The organic layer may contain varying amounts of HBr, sulfuric acid, unreacted alcohol, and by-products; these impurities can be removed by washing with sulfuric acid, water, and aqueous sodium bicarbonate, followed by distillation. It is important to be aware that the alkyl bromides are less dense than sulfuric acid but more dense than water, so the alkyl bromide layer will be on the top in the first washing but on the bottom in the next two.

The infrared spectra of alkyl bromides are characterized by a strong C—Br stretching band at low frequency, from 690 to 515 cm^{-1}. Since the range of some infrared spectrometers does not extend below 600 cm^{-1}, the entire C—Br band is not observed in all cases. As shown on the NMR correlation chart (page 542), chemical shifts for protons on a carbon attached to a bromine atom (*alpha* protons) vary from about 2.7 to 4.1 ppm, depending on the classification of the proton. *Beta* protons are also shifted downfield, but to a much smaller extent.

The rate of a nucleophilic substitution reaction often depends on the structure of the substrate. If you do Minilab 17, you will compare the reaction rates of primary, secondary, and tertiary alcohols with HCl, using zinc chloride as a Lewis acid catalyst. Collateral Project 3 describes the synthesis of a tertiary alkyl chloride.

PRELAB ASSIGNMENT

Calculate the mass and volume of 72 mmol of 1-butanol and the theoretical yield of 1-bromobutane (both mass and volume). Repeat the calculations for 2-butanol and 2-bromobutane.

Reactions and Properties

$$CH_3CH_2CH_2CH_2OH + HBr \xrightarrow{H_2SO_4} CH_3CH_2CH_2CH_2Br + H_2O$$

or

$$\underset{\displaystyle CH_3CH_2\overset{\displaystyle |}{\underset{}{CH}}-CH_3}{\overset{\displaystyle OH}{}} + HBr \xrightarrow{H_2SO_4} \underset{\displaystyle CH_3CH_2\overset{\displaystyle |}{\underset{}{CH}}-CH_3}{\overset{\displaystyle Br}{}} + H_2O$$

Table 2.   Physical properties

	M.W.	b.p.	d.
1-butanol	74.1	117	0.810
2-butanol	74.1	99.5	0.808
hydrobromic acid (48%)	80.9		1.49
1-bromobutane	137.0	102	1.276
2-bromobutane	137.0	91	1.259

Figure 1.   IR spectra of the starting materials. (Reproduced from *The Aldrich Library of FT-IR Spectra,* by Charles J. Pouchert, with the permission of the Aldrich Chemical Company.)

1-Butanol

$$CH_3CH_2CH_2CH_2{-}OH$$

3335.4	1378.6	952.2
2959.7	1072.6	846.7
1466.0	1010.6	737.5

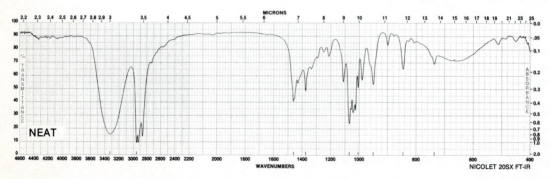

NEAT

Figure 1 *(continued)*

2-Butanol

$$CH_3CHCH_2CH_3$$
with OH above the second carbon

3343.1	1374.6	991.0
2967.2	1299.9	912.6
1457.7	1108.9	668.4

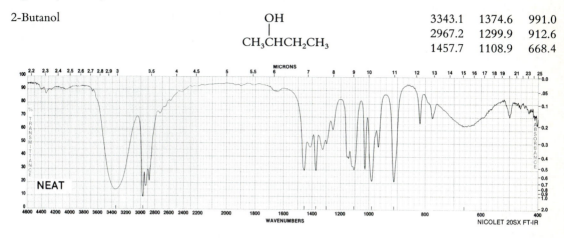

NEAT

NICOLET 20SX FT-IR

PROCEDURE

You will be assigned one of the two alcohols and either 1, 2, 3, or 4 mL of sulfuric acid.

Hydrobromic acid is toxic and very corrosive and can cause very serious damage to the skin, eyes, and respiratory tract. Wear gloves, and dispense under a hood; avoid contact with the acid, and do not inhale its vapors.

Sulfuric acid is very corrosive and can cause very serious damage to the skin and eyes; the concentrated acid reacts violently with water. Wear gloves, avoid contact, keep the acid away from water.

1-Butanol and 2-butanol are flammable and may cause eye or skin irritation. Avoid contact with both alcohols, do not breathe their vapors, and keep them away from flames.

1-Bromobutane and 2-bromobutane are flammable and may be harmful if swallowed, inhaled, or absorbed through the skin. Animal tests (by injection) suggest that 2-bromobutane may be carcinogenic, but no link to human cancer has yet been established. Avoid contact with the alkyl halides, do not breathe their vapors, and keep them away from flames.

SAFETY PRECAUTIONS

sulfuric acid

butanols

bromobutanes

Use acid-resistant boiling chips.

Reaction. Assemble a reflux apparatus equipped with a sodium hydroxide gas trap ⟨OP–22b⟩. Weigh 72 mmol of your assigned alcohol and put it in the boiling flask. *Under the hood,* cool the flask in ice, and carefully add 10 mL (89 mmol) of 48% aqueous hydrobromic acid **[contact and vapor hazard! wear gloves!],** followed by your assigned volume of concentrated sulfuric acid **[contact and reactive hazard!].** Reflux the mixture ⟨OP–7a⟩ for 1 hour from the time the solution starts to boil.

Separation. Codistill the alkyl bromide ⟨OP–15⟩ with the water present in the reaction mixture, using a simple distillation apparatus (no addition funnel) with a graduated cylinder as the receiver. Stop the distillation when the distillate is no longer cloudy and the organic layer no longer increases in volume over a period of 5–10 minutes. Add 10 mL of water to the distillate, shake the mixture in a separatory funnel, and separate the layers ⟨OP–13b⟩ cleanly. *Cautiously* wash the alkyl bromide layer ⟨OP–19⟩ with 5 mL of cold concentrated sulfuric acid **[possible violent reaction!],** then with a 10-mL portion of water and a 10-mL portion of saturated aqueous sodium bicarbonate (be sure to save the right layers!). Dry the alkyl bromide ⟨OP–20⟩ with a little calcium chloride. Decant and weigh the product at this point, and report the crude yield to your instructor or co-workers.

The residue is corrosive and produces very irritating fumes. Let it cool to room temperature, then dispose of it in a designated container under the hood.

Purification and Analysis. Purify the alkyl bromide by semimicro distillation ⟨OP–25a⟩ over a 3–4°C boiling range, and weigh it again. Record an infrared spectrum ⟨OP–33⟩ of the product, or obtain a spectrum from your instructor.

Report. Calculate the percentage yield of your crude and purified products. Using the data reported by other members of your class, determine the optimum ratio of catalyst to substrate for each reaction. Write a mechanism for each reaction, and try to explain any difference in the ratios. Interpret the infrared spectrum of your product as completely as you can, and use the appropriate starting-material spectrum in Figure 1 to show that the expected functional group conversion has taken place.

Collateral Projects

1. Obtain an NMR spectrum of your product in deuterochloroform, and interpret it as completely as you can.

2. Record gas chromatograms of the crude and purified alkyl bromide, and compare their purity. Spike the first chromatogram with the corresponding alcohol, and find out how much (if any) alcohol remains in the crude and purified product. Look for evidence of by-products in the crude product.

3. Prepare *t*-butyl chloride by cautiously adding 10 mL of concentrated hydrochloric acid **[contact and vapor hazard! wear gloves, and work under a hood!]** to 4 mL of *t*-butyl alcohol in a large test tube and shaking the mixture for about 5 minutes; use a rubber stopper and loosen it at intervals to release any pressure that has built up. Dissolve 1.5 g of calcium chloride in the reaction mixture, separate the layers, and wash the alkyl halide layer **[use care: pressure buildup!]** with 5 mL of saturated aqueous sodium bicarbonate. Dry the product over anhydrous calcium chloride. The product can be purified by semimicro distillation and characterized by IR or NMR spectrometry.

Nucleophilic Substitution Rates of Àlcohols

MINILAB 17

Avoid contact with the HCl/ZnCl$_2$ reagent, and do not breathe its vapors. Wear gloves, and work under a hood.

SAFETY
PRECAUTION

Label three clean, dry 10-cm test tubes, and measure 10 drops of (1) 1-butanol, (2) 2-butanol, and (3) 2-methyl-2-propanol into the respective test tube. Add 2 mL of ZnCl$_2$/HCl reagent **[contact and vapor hazard!]** to each test tube, stopper and shake each tube vigorously for 5 seconds, and let them stand at room temperature. Look for any evidence of a reaction, recording your observations during the first 15 minutes or so and at intervals throughout the lab period. Explain your observations, and conclude whether the results can best be explained by an S$_N$1 or S$_N$2 mechanism. Write balanced equations and propose mechanisms for any reactions for which you saw evidence.

The reagent contains 16 g of anhydrous zinc chloride dissolved in 10 mL of cold concentrated HCl.

Exercises

1. **(a)** Write structures for some by-products that might have formed in your reaction mixture by elimination or alternative nucleophilic substitution reactions. **(b)** Propose a mechanism for the formation of each by-product.

2. Construct a flow diagram for the synthesis carried out in this experiment. The flow diagram should show how each kind of by-product proposed in Exercise 1 is removed.

3. Write balanced equations showing how HBr and SO_2 were consumed in the gas trap.

4. **(a)** Propose a synthesis of aldrin starting only with cyclopentadiene and acetylene. **(b)** Show how dieldrin can be synthesized from aldrin.

Library Topic

Write a report describing the preparation and applications of PCBs and discussing the environmental consequences of their use.

Preparation of Phenacetin by the Williamson Ether Synthesis

Reactions of Alcohols. Reactions of Alkyl Halides. Preparation of Ethers. Nucleophilic Substitution.

Operations

OP–7a   Refluxing

OP–12   Vacuum Filtration

OP–21   Drying Solids

OP–23   Recrystallization

OP–23b Recrystallization from Mixed Solvents

OP–28   Melting-Point Determination

OP–34   Nuclear Magnetic Resonance Spectrometry

Objectives

To prepare phenacetin from acetaminophen by a Williamson ether synthesis.

To learn about the uses, limitations, and mechanism of the Williamson synthesis.

To learn more about nucleophilic substitution reactions.

To learn about some useful aryl ethers.

SITUATION

From the time when Carl Duisberg first made phenacetin (see Experiment 2), this analgesic drug has been synthesized by treating *p*-phenetidine (*p*-ethoxyaniline) with acetic anhydride. But the similarity of phenacetin to another analgesic drug, acetaminophen, suggests an alternative synthetic route. It should be possible to convert the hydroxy group of acetaminophen to the ethoxy function of phenacetin by means of a reaction that has been around for more than a century — the Williamson ether synthesis. In this experiment, you will prepare phenacetin from acetaminophen by the Williamson syn-

Acetaminophen is marketed under the trade names Tylenol and Datril.

Duisberg's synthesis of phenacetin:

Proposed synthesis of phenacetin:

nerolin

The phenolic OH group is the primary oxidation inhibitor in BHA, but the methoxy group has a synergistic effect.

thesis, reversing the biochemical process by which phenacetin is metabolized to acetaminophen in the human body.

BACKGROUND

Aromatic Ethers

Certain aromatic ethers live up to their name by exuding pleasant, distinctive odors, making them useful *aroma chemicals* for the formulation of perfumes. In the seventeenth century, Anna Maria de la Tremoïlle, the Italian princess of Neroli, introduced a perfume containing an oil distilled from orange blossoms. A major constituent of her *neroli oil* is the aromatic ether nerolin (2-methoxynaphthalene), which is still used to impart its characteristic odor of orange blossoms to perfumes. 2-Ethoxynaphthalene, which has the same aromatic qualities as its methyl homolog, is sometimes called "new nerolin" because of its more recent discovery and use by the perfume industry.

A number of anisole (methoxybenzene) derivatives are important aroma chemicals as well. *p*-Methoxyacetophenone (whose synthesis is described in Experiment 30) is a natural component of hawthorn flowers and an ingredient of some perfumes. *p*-Methylanisole occurs in the oil of the Malayan ylang-ylang tree and also contributes a floral odor to perfumes. Anisole itself has an agreeable odor but is better known for its antioxidant properties, which make it a good preservative for beer and a stabilizer for vinyl polymers. A more widely used antioxidant is butylated hydroxyanisole (BHA), a mixture of *t*-butyl-*p*-hydroxyanisoles that is used to preserve butter, meat, cereals, baked goods, and candies.

anisole   *p*-methylanisole   butylated hydroxyanisole

Unlike the aroma chemicals described here, phenacetin (*p*-ethoxyacetanilide) is virtually odorless. Because of its pain-killing and fever-reducing qualities, phenacetin was used for many years in drug preparations such as APC tablets, which contained aspirin and caffeine in combination with phenacetin. Recent studies suggest that phenacetin, when taken orally, may promote cancer in humans. It has also been known to induce hemolytic anemia in sensitive individuals and to cause kidney damage if taken in large amounts for a long period of time. Because of these adverse effects, phenacetin has been withdrawn from drug preparations in the United States, so its future as a therapeutic agent is in considerable doubt. Its synthesis from acetaminophen, which is still a widely used analgesic drug and generally considered a safe one, is the reverse of the process that takes place in the human body. When phenacetin is ingested, it is metabolized to acetaminophen, which is apparently the substance responsible for its analgesic and antipyretic effects.

METHODOLOGY

The traditional Williamson synthesis of ethers involves the reaction of an alkyl halide with the sodium salt of an alcohol or phenol. The alkoxide (or aryloxide) ion is strongly nucleophilic and displaces the halide by an S_N2 mechanism similar to that observed when preparing alcohols by the reaction of alkyl halides with hydroxide ion.

$$RO^- + \underset{}{\overset{}{C}}{-}X \longrightarrow \left[RO\cdots\overset{\delta-}{\underset{}{C}}\cdots\overset{\delta-}{X} \right] \longrightarrow ROC + X^-$$

Secondary and tertiary alkyl halides may undergo E2 elimination in the presence of alkoxide or aryloxide ions, producing alkenes as by-products and lowering the yield of ether. Most aryl halides are so unreactive that they cannot be used at all. As a rule, a Williamson synthesis is designed to avoid the use of such halides whenever possible. Thus, ethyl phenyl ether (phenetole) can be prepared from ethyl bromide and sodium phenoxide but not from bromobenzene and sodium ethoxide; and ethyl t-butyl ether is made by combining ethyl bromide and sodium t-butoxide rather than t-butyl bromide and sodium ethoxide.

Preparation of ethers by the Williamson synthesis:

$$CH_3CH_2Br + NaO-\!\!\left\langle\bigcirc\right\rangle \longrightarrow CH_3CH_2O-\!\!\left\langle\bigcirc\right\rangle + NaBr$$

phenetole

$$CH_3CH_2Br + NaO\overset{\overset{\displaystyle CH_3}{|}}{\underset{\underset{\displaystyle CH_3}{|}}{C}}CH_3 \longrightarrow CH_3CH_2O\overset{\overset{\displaystyle CH_3}{|}}{\underset{\underset{\displaystyle CH_3}{|}}{C}}CH_3 + NaBr$$

ethyl t-butyl ether

A notable exception to this rule is the synthesis of ethoxy-triphenylmethane, described in Minilab 18, which uses the tertiary halide triphenylmethyl (trityl) bromide as a reactant. Trityl bromide cannot undergo E1 or E2 elimination, and it reacts with the weak nucleophile ethanol, by an S_N1 mechanism (see Exercise 2).

Preparation of ethoxytriphenylmethane:

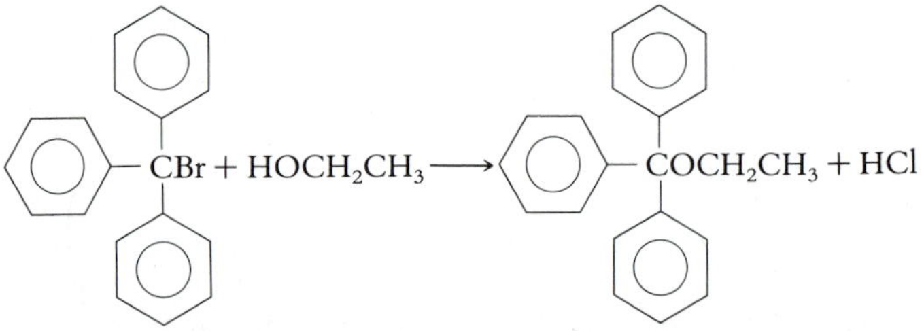

trityl bromide   ethoxytriphenylmethane
                                                    (trityl ethyl ether)

Because iodide ion is a very good leaving group, ethyl iodide is the most reactive of the ethyl halides.

You will prepare phenacetin by combining ethyl iodide (iodoethane) with acetaminophen in a basic solution. The base (sodium methoxide) is needed to convert acetaminophen to its more nucleophilic sodium salt, as shown in the Reactions and Properties section. Alcohols are usually converted to their salts by mixing them with bits of sodium metal. Since phenols are stronger acids than alcohols, weaker bases such as aqueous sodium hydroxide are sometimes used to produce aryloxide ions. Unfortunately, alkyl halides are virtually insoluble in water, and aryloxide ions are insoluble in nonpolar organic liquids. Therefore, if an alkyl halide is added to an aqueous solution containing aryloxide ions, the reactants remain in their separate phases and the reaction is extremely slow. There are several ways of getting around this difficulty: one is to add a phase-transfer catalyst to bring the species together across the phase boundary, and another is to use an organic solvent in which both reactants are soluble. You will use the second method, conducting the reaction in an alcoholic solution in which both ethyl iodide and the sodium salt of phenacetin are soluble. Sodium methoxide will be used to generate the aryl-oxide ion, because this strong base is considerably more soluble in alcohols than is sodium hydroxide. The product will be isolated by adding water to the reaction mixture to reduce its solubility, causing it to crystallize from the cooled solution. Although phenacetin is sometimes purified by recrystalliza-

Preparation of alkoxide and aryloxide ions

$$ROH + Na \rightarrow NaOR + \tfrac{1}{2}H_2$$

$$ArOH + NaOH \rightarrow NaOAr + H_2O$$

$$ArOH + NaOR \rightarrow NaOAr + ROH$$

tion from water, its solubility in that solvent is so low that an ethanol-water mixed solvent will be used here instead.

By obtaining the PMR spectrum of the product and comparing it to that of the starting material (Figure 1), you should be able to deduce whether or not the expected functional-group conversion took place. Information in the NMR Spectra section of Part III should help you interpret both spectra. You can also obtain an infrared spectrum of the product and compare it to that of the starting material (Collateral Project 1).

PRELAB ASSIGNMENT

Calculate the mass of 25 mmol of acetaminophen and the theoretical yield of phenacetin.

Reactions and Properties

Table 1.   Physical properties

	M.W.	m.p.	b.p.	d.
acetaminophen	151.2	170		
phenacetin	179.2	137–8		
ethyl iodide	156.0	−111	72	1.936
sodium methoxide	54.0	d		

Note: 25% sodium methoxide in methanol has a density of 0.945 g/mL and a concentration of about 4.4 M.

Figure 1.   PMR and IR spectra of acetaminophen. (PMR spectrum reproduced from *The Aldrich Library of NMR Spectra, Edition II,* by Charles J. Pouchert; IR spectrum reproduced from *The Aldrich Library of FT-IR Spectra,* by Charles J. Pouchert; both with the permission of the Aldrich Chemical Company.)

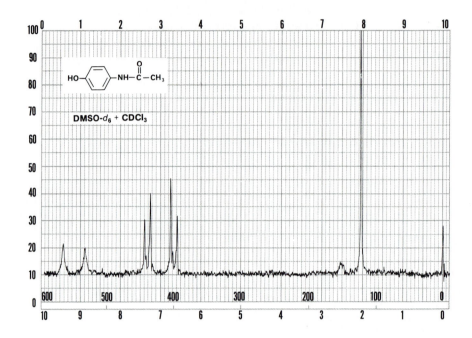

Acetaminophen

3161.9	1506.7	837.4
1654.6	1328.0	808.6
1565.3	1260.5	518.7

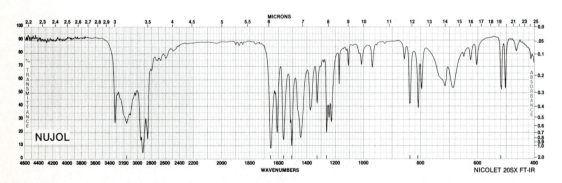

PROCEDURE

SAFETY
PRECAUTIONS

The solution of sodium methoxide in methanol is flammable, toxic, and very corrosive, capable of causing severe damage to skin and eyes. Wear gloves, and dispense the solution under a hood; avoid contact, do not breathe its vapors, and keep it away from flames and hot surfaces.

Ethyl iodide severely irritates the eyes, skin, and respiratory tract and can cause blisters on contact. Wear gloves, avoid contact with the liquid, and do not breathe the vapors.

Based on tests with laboratory animals, phenacetin has been classified as a carcinogen when taken orally. There is little likelihood that casual exposure will promote cancer, but it is only prudent to minimize your contact with the product.

Ethanol is flammable, so do not use flames during the recrystallization.

Reaction.   Under the hood, measure 6.0 mL (~26 mmol) of a 25% solution of sodium methoxide in methanol **[contact, vapor, and fire hazard! wear gloves!]** and 10 mL of absolute ethanol into a boiling flask; then mix in 25 mmol of acetaminophen. Insert a reflux condenser, and carefully add 3.0 mL (~37 mmol) of ethyl iodide **[contact and vapor hazard!]** through the condenser. Reflux the mixture ⟨OP–7a⟩ for an hour or more.

Separation.   Slowly add 40 mL of water down the condenser while the reaction mixture is still hot; if a precipitate forms, heat the flask until it dissolves. Pour the reaction mixture into a beaker, cool the beaker in ice until crystallization is complete, and collect the phenacetin by vacuum filtration ⟨OP–12⟩.

Purification.   Purify the phenacetin by recrystallization ⟨OP–23⟩ from an ethanol-water mixed solvent ⟨OP–23b⟩. Dry it to constant mass ⟨OP–21⟩, and measure its mass and melting point ⟨OP–28⟩. Record the PMR spectrum ⟨OP–34⟩ of the product in deuterochloroform, or obtain a PMR spectrum from your instructor.

Use 95% ethanol for the recrystallization.

Report. Calculate the percentage yield of phenacetin, and try to account for significant losses. Assign each signal in its PMR spectrum to the appropriate proton set. Compare your PMR spectrum with that of acetaminophen in Figure 1, and tell how the spectra confirm that the expected functional-group conversion took place. Write a mechanism for the reaction.

Collateral Projects

1. Record the infrared spectrum of the product in a KBr pellet or Nujol mull, and interpret it as completely as you can. Compare your spectrum to the IR spectrum of the starting material in Figure 1, and tell how the spectra confirm the expected functional-group conversion.

2. Perform a ferric chloride test ⟨Test C–13⟩ on your product. A positive test suggests the presence of unreacted acetaminophen.

MINILAB 18   Preparation of Ethoxytriphenylmethane

SAFETY
PRECAUTION

Since the reaction generates some gaseous hydrogen bromide, it should be conducted under a hood. Avoid contact with the reaction mixture, and do not breathe its vapors.

The product from Experiment 15 can be used.

Under the hood, mix 1.0 g of triphenylmethyl bromide with 10 mL of absolute ethanol in a test tube. Add a boiling chip, and heat the mixture gently over a steam bath or hot-water bath until no more HBr is evolved (test by holding moist blue litmus over the test tube); replace any ethanol that evaporates. Decant the hot solution into a small beaker. Let it cool until crystallization is complete; then collect the ethoxytriphenylmethane by vacuum filtration. Measure the mass and melting point of the dry product.

Exercises

1. Nerolin has been synthesized by treating a solution of 2-naphthol in aqueous sodium hydroxide with dimethyl sulfate, $(CH_3)_2SO_4$. Propose a mechanism for this reaction.

2. **(a)** Propose a mechanism for the reaction of trityl bromide with ethanol as described in Minilab 18. **(b)** Explain why the reaction takes place by this mechanism rather than the usual S_N2 route and why ethanol can be used instead of the stronger nucleophile sodium ethoxide. **(c)** Explain why neither E1 nor E2 elimination can occur during this reaction.

3. Write structures for some by-products that might form during the synthesis of phenacetin, and give equations and mechanisms for their formation.

4. Construct a flow diagram for the synthesis of phenacetin.

5. **(a)** The reaction of α-terpineol with mercury(II) acetate in tetrahydrofuran, followed by treatment with alkaline sodium borohyde, yields the bicyclic ether eucalyptol. Write a mechanism that explains this result. **(b)** Propose a systematic name for eucalyptol.

6. Outline a synthesis of each of the following ethers beginning with an alcohol or phenol of appropriate structure.

Synthesis of eucalyptol:

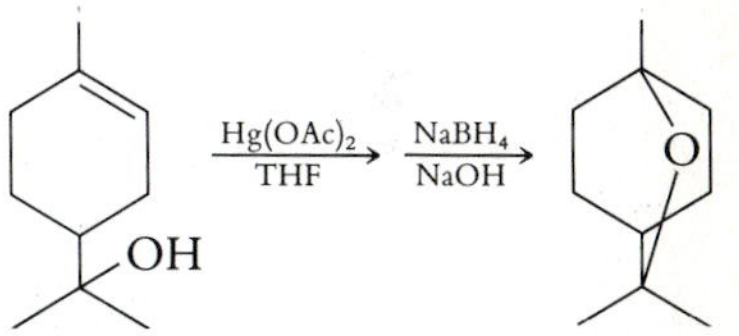

α-terpineol   eucalyptol

Eucalyptol, a major component of eucalyptus oil, is used in cough drops and mouthwashes.

(a) $OCH_2CH_2CH_3$ **(b)**

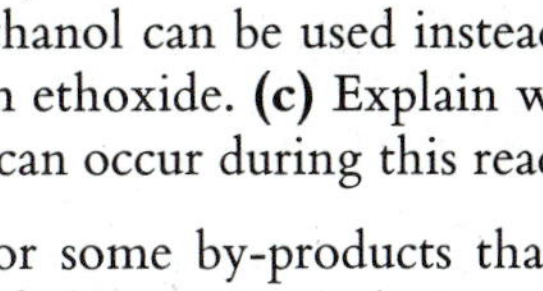

(c)

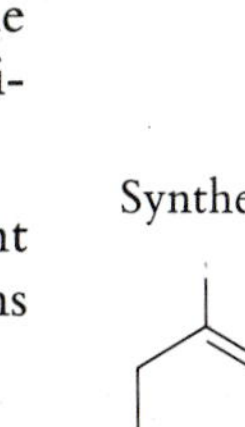

(d)

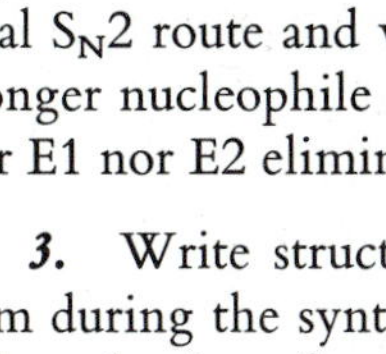

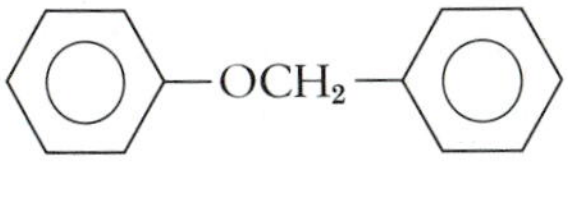

(e) OCH_3

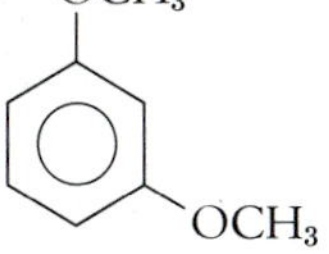

Library Topic

Report on the use of phenyl ethers as antioxidants, explaining how they function in this role and giving specific examples.

Preparation of Nonbenzenoid Aromatic Compounds

Reactions of Unsaturated Hydrocarbons. Preparation of Nonbenzenoid Aromatic Compounds. Hydride-Transfer Reactions. Carbenium Ions. Aromaticity.

Operations

OP–7a   Refluxing

OP–8   Cooling

OP–9   Mixing

OP–12   Vacuum Filtration

OP–12a   Semimicro Vacuum Filtration

OP–21   Drying Solids

OP–22b   Trapping Gases

OP–33   Infrared Spectrometry

OP–34   Nuclear Magnetic Resonance Spectrometry

Objectives

To prepare tropylium fluoborate from cycloheptatriene by a hydride-transfer reaction.

To prepare *meso*-tetraphenylporphin from pyrrole and benzaldehyde.

To verify the aromaticity of tropylium fluoborate using IR and NMR spectrometry.

To learn how to prepare stable carbenium ions by a hydride-transfer reaction.

To learn about some important nonbenzenoid aromatic compounds.

SITUATION

The German chemist Erich Hückel was a pioneer in applying the theories of quantum mechanics to organic molecules. One of his most valuable contributions was the *Hückel rule,* which predicts that certain cyclic, unsaturated systems having $4n + 2$ pi electrons will be aromatic. In 1931, he wrote, "One can expect that the seven-ring would have the tendency to give up its unpaired, non-bonding electron thus forming a positive ion. . . . Nothing is known about this." In other words, Hückel was predicting that a compound containing a cyclic $C_7H_7^+$ cation will show aromatic properties. At that time, aromaticity was associated with benzene and related six-membered ring compounds; no one knew that a *nonbenzenoid* aromatic compound of this type had already been synthesized.

In 1891, some other German chemists were trying to determine the structure of atropine, an important alkaloid from the plant *Atropa belladonna.* G. Merling had just brominated cycloheptatriene, a degradation product from atropine; but when he tried to distill the dibromide, a mass of yellow saltlike crystals collected in his distilling apparatus. He faithfully reported this observation in a German chemical journal, but it didn't appear to reveal anything about the structure of atropine, so it was filed away and forgotten for more than sixty years. When Merling's yellow salt was finally rediscovered in 1954, it was recognized as an aromatic compound of the kind that Hückel had foretold—but Hückel's "prediction" had been fulfilled forty years before he made it! Merling had prepared tropylium bromide from cycloheptatriene by the following reaction.

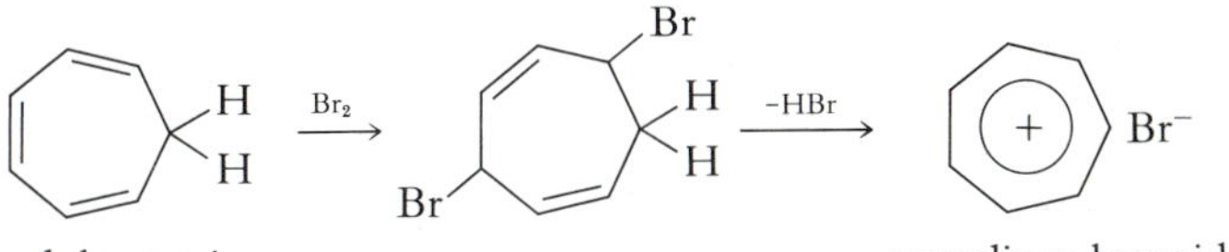

Hückel was alluding to a process by which a cyclic $\cdot C_7H_7$ radical with seven pi electrons loses one of them to form a $^+C_7H_7$ (tropylium) cation. In practice, tropylium compounds are seldom, if ever, prepared in this way.

cycloheptatriene   tropylium bromide

In this experiment, you will synthesize a similar tropylium salt, tropylium fluoborate, from cycloheptatriene by a simpler route, using a hydride-transfer reaction. You will also prepare *meso*-tetraphenylporphin, a representative of the porphyrins, a very important family of nonbenzenoid aromatics.

Tropylium is also called tropenium, tropenylium, and cycloheptatrienylium.

BACKGROUND

Molds, Blue Oils, and King George III

In 1945, Michael J. S. Dewar puzzled over a strange substance called stipitatic acid, which had been isolated from a culture of the mold *Penicillum stipitatum*. Because stipitatic acid showed

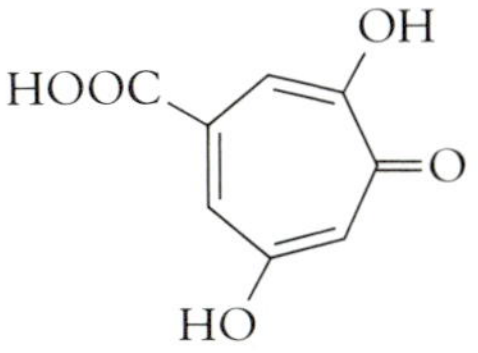

stipitatic acid

aromatic properties, such as undergoing substitution rather than addition reactions with bromine, other scientists had assigned it a benzene-type structure containing a six-membered aromatic ring. Dewar, however, disagreed with their reasoning. With little experimental evidence to go on, he proposed a seven-membered ring structure and declared that the compound was one of a previously unknown class of aromatic compounds, which he named *tropolones.* Dewar expected tropolone rings to show aromatic properties because one can draw resonance structures in which the six pi electrons are delocalized over the seven-membered ring. His guess turned out to be correct, and before long other investigators were proposing tropolone ring structures for compounds with similar properties. One sample of an oil had been distilled from the heartwood of a western red cedar *(Thuja plicata)* and set aside, nearly forgotten, for sixteen years while it slowly crystallized. The crystals yielded β-thujaplicin, an effective fungicide and antibiotic, which was shown to be an isopropyl derivative of tropolone. This compound and two other fungus-destroying thujaplicins are believed to be responsible for the great durability of red cedar, which was used by native Americans to build canoes long before settlers discovered its value in fence posts, shingles, and cedar chests.

tropolone   resonance structure of   β-thujaplicin
                        tropolone

Soon after the discovery that distillation of plant matter often yields valuable **essential oils,** it was found that some essential oils turn a beautiful azure blue on distillation. At first, it was believed that oxidation of the copper vessels used for distillation was responsible, but eventually the color was ascribed to certain decomposition products that were named *azulenes.* For example, guaiol, a natural sesquiterpene alcohol from guaiacum wood, can be decomposed to yield guaizulene; β-vetivone from vetiver oil gives rise to vetivazulene. Azulene precursors are found in everything from araucaria to zdravetz oil, as shown by the partial list in the margin. Sometimes the azulenes themselves occur naturally in small quantities. One group of scientists collected 160,000 liters of urine from pregnant mares in order to isolate 20 milligrams of vetivazulene!

Some essential oils that yield azulenes:

araucaria	lovage
cajeput	myrrh
camomile	niaouli
elemi	pimenta
galangal	rose
ginger	Siam wood
hops	valerian
juniper	ylang-ylang
lemongrass	zdravetz

guaiaol   guaiazulene   β-vetivone   vetivazulene

Unquestionably the most important nonbenzenoid aromatic compounds are the *porphyrins*—without them there could be no plant or animal life as we know it, since they play a key role in both photosynthesis and respiration. Porphyrins are built around the system of four linked pyrrole rings that constitutes the parent compound, porphin. The heavy line in the porphin molecule illustrated traces out an aromatic ring that has eighteen pi electrons. Hemoglobin, which is a conjugated protein containing an iron-complexed porphyrin known as heme, shuttles oxygen through the bloodstream and keeps the respiratory process going. Hemoglobin itself is blue in color; when it picks up oxygen in the lungs, it is converted to bright red oxyhemoglobin, which gradually gives up its oxygen to the cells and is reduced to hemoglobin again before it is carried back to the lungs.

Hemoglobin is an extremely complex protein whose structure differs slightly for different animal species; a typical empirical formula is $C_{738}H_{1166}O_{208}N_{203}S_2Fe$.

porphin   heme

Ordinarily, the excess porphyrins in the body are metabolized by the liver into iron-free substances such as biliverdin and bilirubin, which collectively make up the bile pigments. If this metabolic process breaks down, a disease known as porphyria results, giving rise to agonizing attacks that may resemble psychotic episodes. It has been speculated that King George III suffered from porphyria; medical records noted that his urine was red or discolored, a diagnostic symptom of the disease. This may have given rise to the episodes of apparent madness that characterized the later years of his reign.

See *Scientific American* (July 1969), p. 38, for more about George III's illness.

Proton transfer

$$HB_1 + B_2^- \longrightarrow$$

stronger   stronger
  acid   base

$$HB_2 + B_1^-$$

  weaker   weaker
   acid   base

Hydride transfer

$$R_1H + R_2^+ \longrightarrow$$

less stable
carbenium ion

$$R_2H + R_1^+$$

more stable
carbenium ion

trityl fluoborate

METHODOLOGY

Besides being the first representative of the class of non-benzenoid aromatics, tropylium was probably the first true carbenium ion synthesized. One way to prepare a stable carbenium ion is to remove a hydride ion from an appropriate hydrocarbon by a hydride-transfer reaction. Just as a strong acid can transfer a proton to a strong base, leaving behind a weaker (more stable) base, a hydrocarbon can transfer a hydride ion to a reactive carbenium ion, leaving behind a less reactive (more stable) carbenium ion. In this experiment, the reactive ion will be trityl (triphenylmethyl) and the stable one will be tropylium. The hydride transfer takes place when trityl fluoborate reacts with the hydrocarbon cycloheptatriene, yielding tropylium fluoborate and triphenylmethane. Trityl fluoborate itself is prepared as in Experiment 15, by treating the corresponding alcohol with 48% aqueous fluoboric acid in acetic anhydride. In the procedure used here, the trityl fluoborate will not be isolated but will be treated with cycloheptatriene while still in solution. Tropylium fluoborate precipitates as the dark color of the trityl carbenium ion disappears; adding ether completes the precipitation. Tropylium fluoborate can be used to prepare the aromatic salt tropylium iodide, as described in Minilab 19, and the nonaromatic "oxide" ditropyl ether, as described in Collateral Project 2.

The porphin derivative called *meso*-tetraphenylporphin will be prepared by a condensation reaction between benzaldehyde and pyrrole, using boiling propionic acid as the reaction solvent. The product precipitates from the cooled reaction mixture as lustrous purple crystals, which are collected by vacuum filtration. Although the yield of the preparation is quite low ($\sim 20\%$), the procedure is simple and the product is about 97% pure.

Aromatic compounds are characterized by an uninterrupted system of delocalized pi electrons that extends around the circumference of an unsaturated ring (see Figure 2). Evidence for a compound's aromaticity can be obtained from its spectra. For example, aromatic molecules generally absorb at longer wavelengths in the ultraviolet region than do comparable molecules with alternating single and double bonds. The infrared spectra of aromatic molecules show characteristic bands: aromatic C—H stretching vibrations generally occur between 3100 and 3000 cm^{-1}; C—H out-of-plane bending vibrations give rise to infrared bands in the 900–650 cm^{-1} region; "skeletal" vibrations involving carbon-carbon stretching within the ring result in one or more bands in the 1600–

Benzene

3035.7   1478.8
1960.4   1035.8
1815.0   673.3

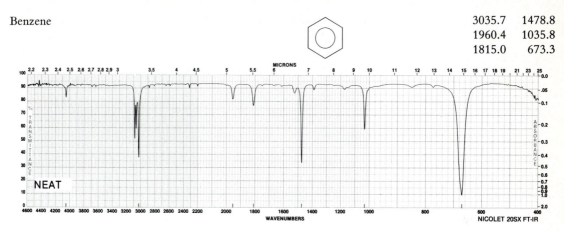

Figure 1.   IR spectrum of benzene. (Reproduced from *The Aldrich Library of FT-IR Spectra,* by Charles J. Pouchert, with the permission of the Aldrich Chemical Company.)

1400 cm^{-1} region. The infrared spectra of symmetrical unsubstituted aromatic compounds are particularly simple, as shown by the infrared spectrum of benzene in Figure 1.

The most unequivocal evidence for aromaticity is provided by NMR spectrometry. The circulation of electrons around an aromatic ring in a magnetic field results in a *ring current* (see Figure 2) that deshields ring protons, moving their PMR signals well downfield. As a rule, the larger the ring, the

Figure 2.   Ring-current effect in the tropylium ion

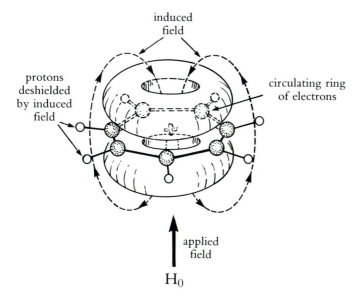

induced
field

protons
deshielded
by induced
field

circulating ring
of electrons

applied
field

H$_0$

stronger its ring current. The six-carbon benzene ring absorbs at about 7.3 δ, whereas the eighteen-carbon aromatic compound [18]annulene has protons that absorb near 8.9 δ. The protons of a tropylium ion are further deshielded because of its ion charge, which reduces the electron density at each ring atom by one-seventh. Most or all of the protons on an unsubstituted aromatic ring are equivalent, so the PMR spectra of unsubstituted aromatic compounds are very simple, usually consisting of one or two signals. By comparing the IR and PMR spectra of your product with those of the starting material (see Figure 3), you should find evidence for the aromaticity of tropylium fluoborate.

PRELAB ASSIGNMENTS

1. Calculate the mass of 12 mmol of triphenylmethanol and the theoretical yield of tropylium fluoborate.

2. Calculate the mass and volume of 10 mmol of benzaldehyde and the theoretical yield of *meso*-tetraphenylporphin.

Reactions and Properties

A. $Ph_3COH + HBF_4 \xrightarrow{Ac_2O} Ph_3C^+BF_4^- + H_2O$ $Ph_3C^+BF_4^- +$ $\longrightarrow Ph_3CH +$ BF_4^-

B. 4 (pyrrole) + 4 (benzaldehyde, CHO) + $\frac{3}{2}O_2 \xrightarrow{CH_3CH_2COOH}$ (*meso*-tetraphenylporphin) + $7H_2O$

pyrrole   benzaldehyde   *meso*-tetraphenylporphin

Table 1.   Physical properties

	M.W.	m.p.	b.p.	d.
acetic anhydride	102.1	−73	140	1.082
fluoboric acid	87.8		130 d	
triphenylmethanol	260.3	164	380	

Table 1.   *(continued)*

	M.W.	m.p.	b.p.	d.
cycloheptatriene	92.15	−80	117	0.887
benzaldehyde	106.1	−26	178	1.046
pyrrole	67.1		130–1	0.969
meso-tetraphenylporphin	614.8			

Figure 3.   IR and PMR spectra of cycloheptatriene. (IR spectrum reproduced from *The Aldrich Library of FT-IR Spectra,* by Charles J. Pouchert; PMR spectrum reproduced from *The Aldrich Library of NMR Spectra, Edition II,* by Charles J. Pouchert; both with the permission of the Aldrich Chemical Company.)

Cycloheptatriene

3024.0	907.2	708.9
1434.5	792.4	656.8
1294.8	741.8	588.2

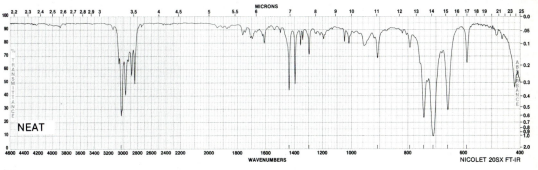

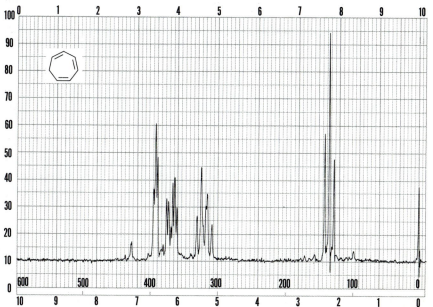

PROCEDURE

A. Preparation of Tropylium Fluoborate

Carry out the reaction under the hood and wear protective gloves.

Fluoboric acid is very toxic; the liquid can cause severe burns to skin and eyes, and the vapor irritates the respiratory system. Wear gloves, and work under a hood; avoid contact with the acid, and do not breathe its vapors.

Acetic anhydride is corrosive and lachrymatory, its fumes are highly irritating, and it reacts violently with water. Wear gloves, and work under a hood; avoid contact with the liquid, do not breathe its vapors, and keep it away from water.

Cycloheptatriene is harmful if swallowed, inhaled, or absorbed through the skin; avoid contact with the liquid, and do not breathe its vapors.

Ethyl ether is extremely flammable and its vapors are harmful; keep it away from flames and hot surfaces, and do not breathe its vapors.

Tropylium fluoborate is corrosive and can damage the eyes, skin, and respiratory tract; avoid contact with the solid, and do not breathe its dust.

Reaction. *Under the hood,* carefully measure 30 mL of acetic anhydride **[contact and vapor hazard! wear gloves!]** into a 125-mL Erlenmeyer flask, and cool the flask in an ice bath ⟨OP–8⟩. Add slowly, with swirling or magnetic stirring ⟨OP–9⟩, 2.0 mL of 48% aqueous fluoboric acid **[contact and vapor hazard!].** With continued cooling and shaking or stirring, add 12 mmol of pure triphenylmethanol in small portions to form a deeply colored solution of trityl fluoborate. Add 1.5 mL (∼ 14 mmol) of cycloheptatriene drop by drop, with cooling and shaking; if the solution remains dark, add more cycloheptatriene until the color disappears or fades to pale yellow. Then mix in 45 mL of anhydrous ethyl ether **[fire hazard!],** and let the mixture stand in the ice bath for at least 10 minutes, until precipitation is complete.

Separation and Analysis.   Collect the tropylium fluoborate by vacuum filtration ⟨OP–12⟩ or semimicro vacuum filtration ⟨OP–12a⟩, and wash it on the filter with anhydrous ethyl ether. Let the product air-dry thoroughly ⟨OP–21⟩, and weigh it. Record the PMR spectrum ⟨OP–34⟩ of tropylium fluoborate in dimethyl sulfoxide and its infrared spectrum ⟨OP–33⟩ using a KBr disc or Nujol mull, or obtain the spectra from your instructor.

Ordinary DMSO will give a strong signal near 2.5 δ, but this will not interfere with the tropylium fluoborate signal. DMSO-d₆ can be used if it is available.

Report.   Calculate the percentage yield of tropylium fluoborate. Interpret its IR and PMR spectra, and describe the spectral evidence supporting (or indicating) an aromatic structure for this salt. Compare the spectra of the product with those of cycloheptatriene in Figure 3, and discuss any significant similarities or differences.

B. *Preparation of* meso-*Tetraphenylporphin*

Work under the hood and wear protective gloves until the product has been filtered and washed. If the reflux cannot be conducted under the hood, use a gas trap. To reduce the likelihood of spillage, pyrrole may be dispensed from a buret under the hood.

SAFETY PRECAUTIONS

Propionic acid can burn the skin and eyes severely, and its vapors irritate the eyes and respiratory system. Use gloves, and dispense under a hood; avoid contact, do not breathe the vapors, and keep the liquid away from flames.

Pyrrole is poisonous and may be very harmful if inhaled, ingested, or allowed to contact the skin or eyes. Use gloves, and dispense under a hood; avoid contact, and do not breathe the vapors.

propionic acid

Under the hood, measure 35 mL of propionic acid **[contact and vapor hazard! wear gloves!]** into a boiling flask, and add 0.70 mL (~ 10.1 mmol) of freshly distilled pyrrole **[contact and vapor hazard!]** and 10 mmol of benzaldehyde. Assemble a reflux apparatus, using a sodium hydroxide gas trap ⟨OP–22b⟩ if the reaction cannot be carried out under the hood. Reflux the mixture ⟨OP–7a⟩ for 30 minutes. Cool the reaction mixture to room temperature, and carefully collect the product by semimicro vacuum filtration ⟨OP – 12*a*⟩ **[wear gloves! use a hood!].** Wash the filter cake thoroughly with methanol, then with hot water until the propionic acid odor is gone. Dry the *meso*-tetraphenylporphin in a desiccator ⟨OP – 21⟩, and weigh it.

Use a vacuum desiccator if one is available.

Report.   Calculate the percentage yield of *meso*-tetraphenylporphin, and describe the appearance of this product.

Collateral Projects

1.   Record and compare the ultraviolet spectra, from 200 – 400 nm, of ~ 10^{-4} solutions of tropylium fluoborate in 0.1 M HCl and of cycloheptatriene in absolute ethanol.

2.   Prepare ditropyl ether by dissolving about 1 g of tropylium fluoborate in 5 mL of water and slowly adding 4 mL of saturated aqueous sodium carbonate, with stirring. The product should be extracted with dichloromethane and the extracts washed with water, dried, and evaporated (no heat). Record and interpret the PMR spectrum of ditropyl ether, and write an equation for the reaction.

The liquid ether decomposes slowly in air.

MINILAB 19   Preparation of Tropylium Iodide

Dissolve 1 g of potassium iodide in 1 mL of water; then dissolve 1 g of tropylium fluoborate **[corrosive!]** in 5 mL of hot water (about 60°C) and immediately stir in the KI solution. Cool this mixture in an ice bath until precipitation is complete. Collect the deep red tropylium iodide by semimicro vacuum filtration, and wash the crystals on the filter with a little ice-cold water. Dry the product (use a vacuum desiccator, if possible), and weigh it.

Tropylium iodide is unstable in air, gradually changing to the red-black triiodide and various oxidation products.

Exercises

1. In water, tropylium fluoborate is a weak acid with a K_a of 1.8×10^{-5}, about the same as that of acetic acid. Write a balanced equation for an equilibrium reaction that explains the acidity of this salt.

2. **(a)** Trityl fluoborate from Experiment 24 could have been used to make tropylium fluoborate in this experiment. Write an equation for the reaction that would take place if you accidentally spilled a drop of water on the trityl fluoborate. **(b)** If you started with 4.0 g of trityl fluoborate, what percentage of it would be decomposed by the drop of water? (Assume that 1 drop $\cong$ 0.050 mL.)

3. Explain why the iodide (C_7H_7I) prepared from tropylium fluoborate is aromatic but the "oxide" [ditropyl ether, $(C_7H_7)_2O$] is not. Draw structures for both compounds, showing clearly the kind of bond between each ring and an I or O atom.

4. Predict which of the following compounds will show aromatic properties.

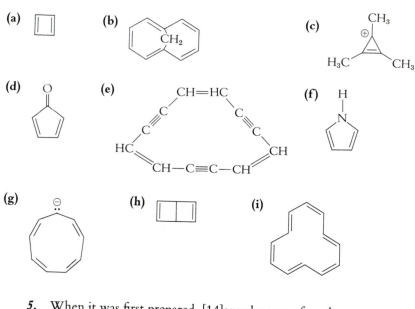

(a) (b) (c) CH₃

(d) (e) (f)

(g) (h) (i)

5. When it was first prepared, [14]annulene was found to be far less stable than expected; it was considered nonaromatic until NMR and x-ray diffraction studies showed otherwise. Build a molecular model of [14]annulene in the conformation shown, and attempt to explain its low stability.

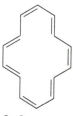

[14]annulene

cyclobutadiene

These metal complexes are "sandwich" compounds similar to ferrocene. One type is described as an "open-faced sandwich."

Library Topic

Cyclobutadiene, containing only four pi electrons on a ring, is not expected to be aromatic according to Hückel's $4n + 2$ rule. It is very unstable and has only recently been prepared in combination with certain metals. Describe the preparation of cyclobutadiene complexes, and discuss some recent speculations about the structure and properties of cyclobutadiene.

Mechanism of the Nitration of Arenes by Nitronium Fluoborate

Reactions of Arenes. Preparation of Nitro Compounds. Electrophilic Aromatic Substitution. Reaction Kinetics and Mechanisms

Operations

OP – 13*a* Semimicro Extraction

OP – 14 Evaporation

OP – 20 Drying Liquids

OP – 22 Drying Gases

OP – 32 Gas Chromatography

Objectives

To determine the *ortho* to *para* ratio for nitration of *t*-butylbenzene by nitronium fluoborate and evaluate the steric requirements of the reaction.

To determine the relative rates of nitration of mesitylene and toluene with nitronium fluoborate.

To identify the crucial intermediate in nitrations with nitronium fluoborate and propose a mechanism for the reaction.

To learn how to carry out a competitive rate study.

To learn about steric effects in organic reactions.

To learn about the structures of sigma and pi complexes and their participation in aromatic substitution reactions.

SITUATION

The electrophile in most aromatic nitrations is the nitronium ion (NO_2^+), which can be generated by mixing nitric acid with sulfuric acid or some other acid catalyst:

$$HNO_3 + 2H^+ \rightleftharpoons NO_2^+ + H_3O^+$$

The intermediate whose formation determines the rate and orientation of a reaction will be referred to here as the crucial intermediate.

Under the usual reaction conditions, the equilibrium concentration of NO_2^+ is so low that elevated temperatures and long reaction times are generally required for successful nitrations. Nitronium fluoborate (NO_2BF_4) supplies a much higher concentration of nitronium ions, allowing rapid nitration of most aromatic compounds at room temperature or below. George A. Olah and his co-workers have suggested that the "crucial" intermediate in nitrations with nitronium fluoborate is different from the intermediate proposed for other aromatic nitration reactions. You will try to find evidence indicating the probable structure of that intermediate, which should enable you to propose a reasonable mechanism for the reaction.

BACKGROUND

Intermediates in Nitration Reactions

The nitronium ion is an electron-hungry species—an *electrophile*—which can steal a pair of pi electrons from a benzene ring's aromatic sextet. According to a longstanding theory of aromatic substitution, such an electrophile attacks the aromatic ring to form a *sigma complex*, in which the electrophile (symbolized here by E^+) is connected to a ring carbon by a sigma bond.

The sigma complex (sometimes called an arenium ion) then loses a proton to a basic species in the reaction mixture to yield the corresponding substituted benzene.

The ring atoms *ortho* and *para* to the incoming electrophile bear most of the positive charge in the sigma complex. Thus, electron-donating groups located on those atoms stabilize the complex, favoring the formation of *ortho-* and *para*-substituted products. For example, only 4% of the product obtained from the nitration of toluene in mixed acid (nitric acid plus sulfuric acid) is *meta*-nitrotoluene.

Although the crucial intermediate in mixed-acid nitrations is apparently a sigma complex, a *pi complex* may be the crucial intermediate in some aromatic substitution reactions. A pi complex, in which the electrophile is loosely bonded to the pi electrons of an aromatic ring, would presumably rearrange to a sigma complex before giving rise to the product. The pi complex involving nitronium ion and benzene might be pictured as shown, with the electrophile sitting atop a "doughnut" of pi electrons, equidistant from all the ring atoms.

The pi cloud of an alkyl-substituted benzene is "lumpier" than that of benzene, with bulges at the positions *ortho* and *para* to the alkyl group. An electrophile that spends more time near these regions of higher electron density, will form an *oriented* pi complex rather than a symmetrical benzene-type complex. Both possibilities are illustrated.

Although alkyl groups alter the electron distribution in a pi cloud, they have only a small effect on its total electron density. Thus, increasing the number of methyl groups on a benzene ring should not increase the stability of its pi complexes very much. On the other hand, the stabilities of sigma complexes are very sensitive to the electronic effects of substituents. Mesitylene, with three methyl groups, forms a sigma complex with HBF_4 that is nearly 300,000 times more stable than the corresponding sigma complex involving toluene; but the pi complex that mesitylene forms with HCl is only about twice as stable as the complex that toluene forms. As a general rule, the more stable an intermediate is, the faster it will be formed. If the crucial intermediate in nitration by nitronium fluoborate is a pi complex, the nitration rates for mesitylene and toluene should be within a factor of 10 or so of one another. But if the crucial intermediate is a sigma complex, the nitration rate for mesitylene should be thousands of times greater than that for toluene.

When a substituent influences the outcome of a reaction as a result of its bulkiness, we say that a *steric effect* is operating. Steric effects in aromatic substitution can be detected by measuring the ratios of *ortho* products to *para* products for the same reaction with different substituents. There are twice as many *ortho* as *para* hydrogens in a monosubstituted arene, suggesting

Pi complex involving nitronium ion and benzene

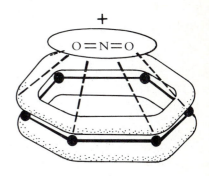

symmetrical
pi complex

oriented
pi complexes

HBF_4 sigma complex
of mesitylene

Table 1. *Ortho : para* ratios for the
nitration of arenes in mixed acid

Arene	*ortho : para* Ratio
toluene	1.57
ethylbenzene	0.93
isopropylbenzene	0.48
t-butylbenzene	0.22

sigma complex
for nitration of
t-butylbenzene

that the ratio of *ortho-* to *para*-substituted products should be about 2 : 1. As shown in Table 1, *ortho : para* ratios for the nitration of arenes in mixed acid are considerably lower than 2 : 1; for example, the ratio for *t*-butylbenzene is only 0.22 : 1. Apparently a bulky alkyl substituent hinders substitution at the *ortho* position by crowding the attacking electrophile in the transition state that leads to the sigma complex.

The steric requirements of pi complexes have not been as thoroughly studied as those of sigma complexes. If an incoming electrophile approaches the ring from the top to form a symmetrical pi complex, there should be little if any steric effect even with a bulky substituent such as the *t*-butyl group. On the other hand, the formation of an oriented pi complex might be markedly influenced by steric factors, giving product ratios comparable to those observed for mixed-acid nitrations.

METHODOLOGY

In this experiment, you will nitrate *t*-butylbenzene with nitronium fluoborate and determine the *ortho : para* ratio by gas chromatography. You will also nitrate a mixture of mesitylene and toluene to determine their relative reaction rates. From the results, you should be able to decide whether the crucial intermediate is a sigma complex, a symmetrical pi complex, or an oriented pi complex.

Nitronium fluoborate can be prepared by treating nitric acid with hydrofluoric acid and boron trifluoride:

$$HNO_3 + HF + 2BF_3 \longrightarrow NO_2^+BF_4^- + BF_3 \cdot H_2O$$

The stable crystalline salt ionizes in polar solvents to provide "ready-made" nitronium ions:

$$NO_2BF_4 \longrightarrow NO_2^+ + BF_4^-$$

Nitronium fluoborate was first used for aromatic nitrations by George A. Olah and his co-workers, who found it to nitrate aromatic hydrocarbons rapidly at room temperature or below, producing nearly quantitative (100%) yields of the products.

The relative rates of nitration for mesitylene and toluene will be measured by carrying out a competitive nitration reaction in which equimolar quantities of the two arenes compete for a limited amount of nitronium fluoborate. The arene that competes most successfully will form the most product, so the relative rates for the two arenes should be proportional to the relative amounts of nitroarene they produce:

$$\frac{\text{reaction rate for mesitylene}}{\text{reaction rate for toluene}} = \frac{\text{moles of nitromesitylene}}{\text{moles of nitrotoluenes}}$$

For a meaningful rate comparison, you should determine the relative rates *per reaction site;* otherwise, toluene, with five ring hydrogens, will have a statistical advantage over mesitylene, with only three. The rate per reaction site is proportional to the number of moles of product divided by the number of reaction sites, so the relative reactivity of a mesitylene site is given by this equation:

$$\frac{\text{reactivity of mesitylene site}}{\text{reactivity of toluene site}} = \frac{\text{moles of nitromesitylene}/3}{\text{moles of nitrotoluenes}/5}$$

Since the area of a peak on a gas chromatogram is roughly proportional to the mass of the component that produces it, you can estimate the relative number of moles of each product by dividing its peak area by its molecular weight. The peak areas for the three nitrotoluenes should be combined for this calculation.

The nitration reactions will be carried out at room temperature by adding 1 mmol of nitronium fluoborate in sulfolane to 5 or more mmol of each arene in the same solvent. Nitronium fluoborate absorbs moisture rapidly, so it should be dispensed from a buret protected by a drying tube. Sulfolane is an excellent solvent for the reaction because it dissolves both the nitronium salt and the arene, thus providing a homogeneous reaction mixture. It is also miscible with water, making it easy to separate the products from the reaction mixture. When water and ethyl ether are added to the reaction mixture, sulfolane and fluoboric acid (a by-product of the reaction) end up in the water layer, and the aromatic compounds are extracted into the ether layer. Evaporation of the ether leaves a mixture of unreacted arenes and nitrated products.

The reaction mixtures will be analyzed by gas chromatography on a column that has a silicone liquid phase or another suitable column that separates aromatic compounds in order of their boiling points. Unreacted arenes should elute from the column first, followed by *ortho-*, *meta-*, and *para-*nitroarenes in that order. The nitromesitylene peak should appear later than all the nitrotoluene peaks.

The component masses may not be exactly proportional to the peak areas, but they will be close enough for the purposes of this experiment.

The excess reactant prevents the formation of di- and trinitrated products, which would skew the results.

sulfolane

You will see an ether peak if you failed to evaporate the ether completely, and there may be additional peaks due to impurities in the commercial arenes used as starting materials. Compare your chromatograms with chromatograms of those arenes if such peaks make interpretation difficult.

Minilab 20 describes a simple mixed-acid nitration of naphthalene. The orientation of this nitration can be rationalized by drawing and comparing resonance structures for the sigma intermediates leading to the two possible products, 1-nitronaphthalene and 2-nitronaphthalene.

PRELAB ASSIGNMENT

Calculate the masses of 5.0 mmol of toluene and 5.0 mmol of mesitylene.

Reaction and Properties

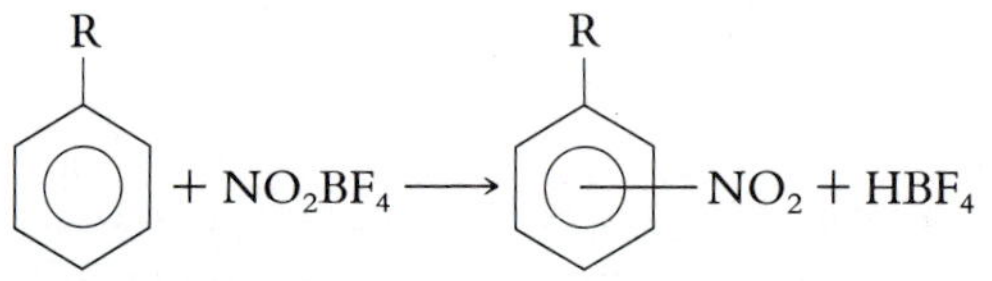

R = methyl or *t*-butyl

(NO_2 is predominantly *ortho* and *para*)

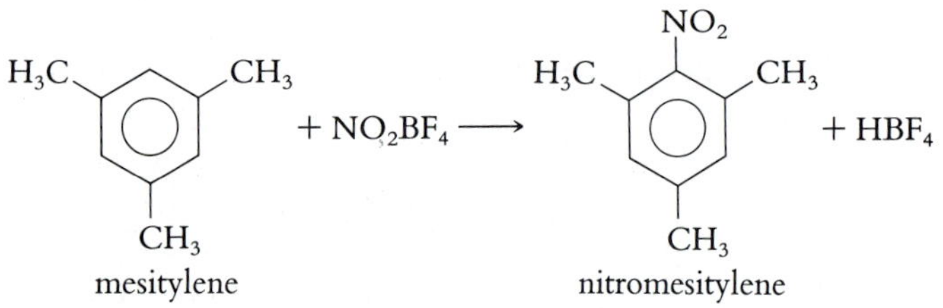

Table 2.   Physical properties

	M.W.	m.p.	b.p.	d.
toluene	92.1	-95	111	0.867
t-butylbenzene	134.2	-58	169	0.867
mesitylene	120.2	-45	165	0.862
nitronium fluoborate	132.8			
sulfolane	120.2	28	285	1.260
o-nitrotoluene	137.1	10	222	1.163
m-nitrotoluene	137.1	16	233	1.157
p-nitrotoluene	137.1	55	238	1.104
nitromesitylene	165.2	44	255	

PROCEDURE

Carry out the reactions under a hood, and wear protective gloves. If desired, students can work in pairs, with each student being responsible for one nitration reaction.

SAFETY PRECAUTIONS

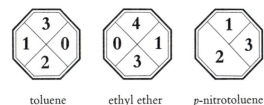

toluene   ethyl ether   *p*-nitrotoluene

Reactions. Accurately weigh 5.0 mmol of toluene **[vapor and fire hazard!]** and 5.0 mmol of mesitylene, and combine them in a clean, dry 15-cm test tube. Then mix in 2 mL of sulfolane with a clean, dry stirring rod. In another 15-cm test tube, mix 1.0 mL (6.5 mmol) of dry *t*-butylbenzene with 2 mL of sulfolane. Slowly add 2 mL of 0.5 M nitronium fluoborate **[contact hazard! wear gloves!]** in sulfolane to each test tube, swirling gently to mix the reactants. Cover the mouth of each test tube with Parafilm and shake gently; then let it stand at room temperature for 10 minutes, with occasional shaking.

Label the test tubes so they won't get mixed up.

The aqueous layer contains hydrofluoric acid and fluoboric acid; handle it carefully, and dispose of it properly.

Separation. Add 3 mL of ethyl ether and 6 mL of water to each test tube, and shake to extract ⟨OP–13a⟩ the products and unreacted arene into the ether layer **[wear gloves!].** Remove the aqueous layer with a capillary pipet. Wash the ether layer with 4 mL of water; then separate the ether layer and dry it ⟨OP–20⟩ with a little calcium chloride. Decant the solution, and evaporate the ether ⟨OP–14⟩ under the hood using a dry air stream ⟨OP–22⟩.

Analysis. Obtain a gas chromatogram ⟨OP–32⟩ of each product mixture, as directed by your instructor. Identify the peaks on the gas chromatograms, and measure the peak areas for all of the nitrated products.

Report. Calculate the *ortho : para* ratio for the nitration of *t*-butylbenzene and the reactivity (per reaction site) of mesitylene relative to toluene. Tell whether your results suggest a significant steric effect in the nitration by nitronium fluoborate, and cite the evidence supporting your conclusion. Decide whether the crucial intermediate in nitrations by nitronium fluoborate is a sigma complex, a symmetrical pi complex, or an oriented pi complex, and justify your conclusion. Based on your results, write a detailed mechanism for the reaction of *t*-butylbenzene with nitronium fluoborate to yield *p*-nitro-*t*-butylbenzene.

Collateral Projects

1. Carry out the same nitrations with mixed acid, and compare the results with those you obtained with nitronium fluoborate. Under the hood, cool 1 mL of concentrated nitric acid **[contact, vapor, and reactivity hazard!]** in an ice bath, and slowly add 1 mL of concentrated sulfuric acid **[contact and reactivity hazard!].** Measure 80 mmol of *t*-butylbenzene into a flask, and add the cold mixed acid drop by drop with cooling and stirring. Stir or shake for 30 minutes at room temperature. Then carefully wash the organic layer with ice cold water and saturated aqueous Na_2CO_3, and dry it with magnesium sulfate. Repeat using 80 mmol of mesitylene and 80 mmol of toluene combined. Analyze both mixtures by GC.

2. Nitrate any or all of the following hydrocarbons with nitronium fluoborate, and compare their *ortho : para* ratios with the ratio you obtained for *t*-butylbenzene: toluene, ethylbenzene, *n*-propylbenzene, isopropylbenzene, *n*-butylbenzene, isobutylbenzene. Explain your results.

Nitration of Naphthalene

Sulfuric acid and nitric acid can cause very serious burns, and they react violently with water and other chemicals. Nitric acid produces toxic nitrogen dioxide fumes during the reaction. Use gloves and a hood; avoid contact, and do not breathe the vapors.

SAFETY PRECAUTION

Under the hood, measure 1 mL of concentrated nitric acid **[contact, vapor, and reactivity hazard!]** into a large test tube, cool it in ice, and cautiously mix in 1 mL of concentrated sulfuric acid **[contact and reactivity hazard!]**. In small portions, add 1 g of powdered naphthalene, shaking after each addition and cooling as necessary to keep the temperature around 45 – 50°C. Then shake the reaction mixture in a 60°C water bath for 20 minutes, stir it into 50 mL of ice water, and decant the water when the yellow product has solidified. Boil the solid with 15 mL of water for 10 minutes or more **[in a hood!]**; then cool the mixture in ice, and collect the product by vacuum filtration. Weigh the crude 1-nitronaphthalene. Write a mechanism for the reaction, and explain its orientation.

If desired, the product can be purified by recrystallization from an ethanol-water mixed solvent ⟨OP – 23b⟩ and its melting point can be obtained.

Exercises

1. **(a)** It has been estimated that the *meta* product arising from direct nitration of *t*-butylbenzene makes up about 2.0% of the product mixture; the remaining *meta* product arises from isomerization of the *ortho* product. Use this estimate to calculate a more reliable value of the *ortho : para* ratio for *t*-butylbenzene. **(b)** Propose a mechanism for the isomerization reaction, which is apparently promoted by the HBF_4 formed during the reaction.

2. Mesitylene is brominated nearly 300 million times faster than benzene is in a solution of bromine in acetic acid. Do you think a sigma complex or a pi complex is formed in the rate-determining step of this reaction? Explain.

3. Construct a flow diagram for the nitration of *t*-butylbenzene as carried out in this experiment.

4. Predict the major product or products of the mononitration of **(a)** ethyl benzoate, **(b)** phenyl acetate, **(c)** phenyl

benzoate, **(d)** *m*-nitrotoluene, and **(e)** *p*-methoxybenzalde-
hyde.

Library Topic

Read the paper by George A. Olah and his co-workers
describing the nitration of arenes with nitronium fluoborate
[*J. Am. Chem. Soc. 83,* 4571 (1961)], and compare their results
and conclusions with your own.

The Bromination of Acetophenone by Potassium Bromate

Reactions of Aryl Ketones. Preparation of Aryl Halides. Electrophilic Aromatic Substitution. Reaction Mechanisms.

Operations

OP–7c	Temperature Monitoring
OP–8	Cooling
OP–9	Mixing
OP–13	Extraction
OP–14	Evaporation
OP–19	Washing Liquids
OP–20	Drying Liquids
OP–26a	Semimicro Vacuum Distillation
OP–32	Gas Chromatography
OP–33	Infrared Spectrometry

Objectives

To carry out the bromination of acetophenone using potassium bromate in aqueous sulfuric acid.

To determine the structure of the product using infrared spectrometry.

To propose a mechanism for the reaction of potassium bromate with acetophenone.

To learn about mechanisms for the halogenation of rings and side chains.

To learn about the physiological action of tear gas.

SITUATION

The acetyl (CH_3CO) group of acetophenone is a deactivating, *meta*-directing substituent. It therefore seems reasonable to assume that the bromination of acetophenone would

O
‖
C—CH₃

Br

m-bromoacetophenone

O
‖
C—CH₂Br

α-bromoacetophenone

A lachrymator is a substance that induces tears.

O
‖
CCH₂Cl

α-chloroacetophenone
(Agent CN, tear gas)

Cl

CH=C(CN)₂

o-chlorobenzylidene-
malonitrile
(Agent CS)

yield *m*-bromoacetophenone as the major product. If you carried out this reaction by the usual aromatic bromination procedure, you would learn very quickly — and tearfully — that the product was not *m*-bromoacetophenone but *α*-bromoaceto-phenone, a compound related to tear gas! This illustrates the pitfalls of trying to apply a general reaction procedure to a specific reaction without doing adequate research; a search of the chemical literature would have revealed that molecular bromine attacks the side chains of aromatic ketones much more rapidly than it does the benzene rings. In 1981, J. J. Harrison, J. P. Pellegrini, and C. M. Selwitz described a convenient procedure for the *meta*-bromination of deactivated aromatic compounds, using potassium bromate as the brominating agent. In this experiment, you will apply their method to the bromination of acetophenone and find out whether your product is *m*-bromoacetophenone or some other isomer.

BACKGROUND

Chemical Irritants

Known as tear gas or Agent CN, *α*-chloroacetophenone is a lachrymatory liquid used by police and military personnel to subdue lawbreakers and control riots. It is the active component of Mace (from *M*ethylchloroform-chloro*ace*tophenone), an aerosol mixture that also contains 1,1,1-trichloroethane (methylchloroform). In concentrations as low as a millionth of a gram per liter of air, *α*-chloroacetophenone causes a copious flow of tears and painful swelling of the eyes. Although exposure to the chemical may be extremely unpleasant, the effects are temporary and disappear rapidly in fresh air. An even more potent substance called Agent CS is now preferred for riot control because of its more rapid, aggressive action. The unfortunate victim of exposure to Agent CS is almost immediately incapacitated by severe burning of the eyes and upper respiratory tract, along with other physiological effects such as coughing, dizziness, runny nose, and a stinging sensation on moist skin.

The physiological action of tear gas is related to its ability to participate in nucleophilic substitution reactions. The chlorine on the side chain of *α*-chloroacetophenone is a highly reactive leaving group, and mucous membranes have nucleophilic sites that can displace such leaving groups. This results in reversible alkylation of the mucous membrane, which causes intense irritation to the affected tissue. The side-chain chlorine can also be displaced by moisture on the membrane, causing the release of hydrochloric acid—which makes its own contribution to the victim's misery.

Reactions of tear gas on a mucous membrane

$$\underset{\substack{\|\\ \text{PhCCH}_2\text{Cl}}}{\overset{\text{O}}{\|}} + \text{Membrane} \colon \rightleftharpoons \underset{\substack{\|\\ \text{PhCCH}_2}}{\overset{\text{O}}{\|}} \colon \text{Membrane} + \text{Cl}^-$$

$$\underset{\substack{\|\\ \text{PhCCH}_2\text{Cl}}}{\overset{\text{O}}{\|}} + \text{H}_2\text{O} \longrightarrow \underset{\substack{\|\\ \text{PhCCH}_2\text{OH}}}{\overset{\text{O}}{\|}} + \text{HCl}$$

α-Bromoacetophenone, the compound you hope *not* to synthesize in this experiment, is a potent lachrymator comparable to tear gas (*m*-bromoacetophenone is not, because halogen atoms on a benzene ring are notoriously poor leaving groups). Both α-bromoacetophenone and α-chloroacetophenone are synthesized by the reaction of acetophenone with the appropriate halogen. The side-chain bromination of acetophenone occurs by a mechanism that involves electrophilic addition of the halogen to the enolic form of acetophenone, followed by the transfer of a proton to a basic species.

Mechanism for the side-chain bromination of acetophenone

1. Acid-catalyzed enol formation:

$$\text{Ph}-\underset{\substack{\|\\ \text{O}}}{\overset{\text{O}}{\text{C}}}-\text{CH}_3 \underset{\text{acid}}{\rightleftharpoons} \text{PhC}\!\!\overset{\text{OH}}{\underset{\|}{}}\!\!=\!\text{CH}_2$$

2. Electrophilic addition:

$$\text{Ph}-\overset{\text{OH}}{\underset{|}{\text{C}}}=\text{CH}_2 + \text{Br}_2 \rightleftharpoons \text{Ph}\overset{\text{OH}}{\underset{\oplus}{\text{C}}}-\text{CH}_2\text{Br} + \text{Br}^-$$

3. Proton transfer:

$$\text{Ph}-\overset{\text{OH}}{\underset{\oplus}{\text{C}}}-\text{CH}_2\text{Br} + \text{B} \colon \rightleftharpoons \text{PhC}\overset{\text{O}}{\underset{\|}{}}-\text{CH}_2\text{Br} + \text{B} \colon \text{H}$$

METHODOLOGY

In this experiment, you will try to prepare *m*-bromoacetophenone from acetophenone and identify the product by infrared spectrometry. In a typical halogenation of an aromatic ring, a Lewis acid catalyst such as ferric bromide is used to convert the halogen to an electrophilic species that contains a positively charged halogen atom. This electrophile then attacks the aromatic ring to form a sigma complex, which loses a proton to yield the product.

The electrophilic species may also be represented as a polarized bromine molecule coordinated with the Lewis acid:

$$\overset{+}{Br}\!-\!\overset{-}{Br} : FeBr_3$$

Mechanism for the ring bromination of a monosubstituted benzene

1. $Br_2 + FeBr_3 \longrightarrow Br^+FeBr_4^-$

2.

sigma complex

3.

(Z = *meta*-directing substituent)

But molecular bromine attacks acetophenone at its side chain, so you must use another reagent, one that will attack only the benzene ring.

When acetophenone is treated with hypobromous acid (HOBr) in the presence of a strong mineral acid, *meta*-bromo and *ortho*-bromo products are obtained in a 6:1 ratio. The electrophilic species in this reaction has been represented as a hydrated bromine cation ($H_2O \cdot Br^+$). Harrison, Pellegrini, and Selwitz brominated several deactivated aromatic compounds with potassium bromate in aqueous sulfuric acid, obtaining the *meta*-bromo products in good yield. In the presence of acid, potassium bromate might be reduced to either elemen-

There was no evidence for the production of any *para* isomer.

tal bromine or hypobromous acid by one of the following reactions:

1. $H^+ + BrO_3^- \longrightarrow HOBr + O_2$
2. $4H^+ + 4BrO_3^- \longrightarrow 2Br_2 + 5O_2 + 2H_2O$

If the active brominating agent in the reaction with potassium bromate is hypobromous acid, you would expect to obtain *m*-bromoacetophenone as the major product, along with a little of the *ortho* product. If the active brominating agent is elemental bromine, you might obtain α-bromoacetophenone instead. Potassium bromate might even react by a different mechanism entirely, yielding *o*- or *p*-bromoacetophenone as the major product.

You will carry out the bromination reaction by adding potassium bromate to a cold solution of acetophenone in aqueous sulfuric acid. To keep the exothermic reaction under control and prevent unwanted side reactions, you will add the potassium bromate in small portions while stirring or swirling the reactants in a cooling bath. The sulfuric acid concentration is crucial — if it is too low, little or no product will be formed; if it is too high, potassium bromate may decompose violently, producing heat and elemental bromine.

The reaction mixture will be worked up via the usual operations and purified by distillation to remove unreacted acetophenone and high-boiling polybrominated products. Because the desired product tends to decompose at its normal boiling point, the distillation will be carried out under reduced pressure. With an efficient aspirator, the boiling point of the product can be reduced from $260°C$ to about $130-135°C$, but under typical laboratory conditions it may be somewhat higher than this. You can assess the purity of the product by gas chromatography and determine its identity by comparing its infrared spectrum with those in Figure 1. Once you know the structure of the major product, you should be able to propose a mechanism for its formation.

Because the theoretical yield is less than 5 mL, you should use the semimicro vacuum distillation procedure.

Hypohalous acids such as HOBr attack the aromatic ring of acetophenone, but hypohalite salts react quite differently. Sodium hypochlorite reacts with acetophenone (and other methyl ketones) to produce a carboxylic acid salt and chloroform:

$$PhCOCH_3 + 3NaOCl \longrightarrow PhCOONa + CHCl_3 + 2NaOH$$

The sodium hypochlorite first chlorinates the methyl group completely to produce $PhCOCCl_3$, which undergoes nucleophilic substitution in the alkaline solution, with hydroxide ion as the nucleophile. Although alkyl groups such as CH_3^- are very poor leaving groups, CCl_3^- is stabilized sufficiently by the three chlorine atoms to make it a reasonably good leaving group. If you do Minilab 21, you will use a commercial hypochlorite laundry bleach and a phase-transfer catalyst to synthesize benzoic acid from acetophenone.

PRELAB ASSIGNMENTS

1. Read or review OP–26 and read OP–26a; read or review the other operations as necessary.

2. Calculate the mass of 42 mmol of potassium bromate, the mass and volume of 40 mmol of acetophenone, and the theoretical yield of bromoacetophenone.

Reactions and Properties

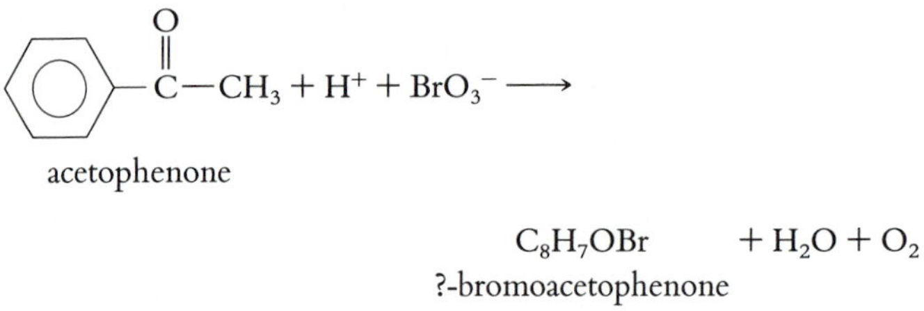

acetophenone

$$C_8H_7OBr \quad + H_2O + O_2$$
?-bromoacetophenone

Table 1. Physical properties

	M.W.	m.p.	b.p.	d
acetophenone	120.6	21	203	1.028
potassium bromate	167.0	350	370 d	3.27
α-bromoacetophenone	199.1	51	135[18]	1.647
o-bromoacetophenone	199.1		131[20]	
m-bromoacetophenone	199.1	8	131[16]	1.505
p-bromoacetophenone	199.1	54	130[15]	1.647

Note: A superscript on a boiling-point value indicates the pressure of the boiling-point determination, in torr.

Figure 1.   IR spectra of the four isomeric bromoacetophenones. (Reproduced from *The Aldrich Library of FT-IR Spectra,* by Charles J. Pouchert, with the permission of the Aldrich Chemical Company.)

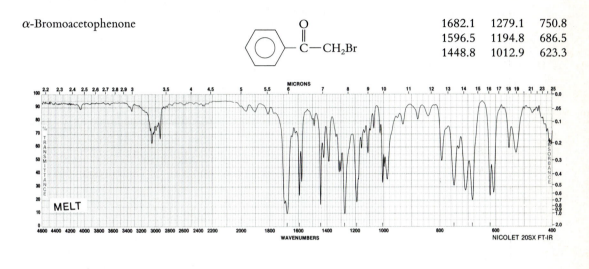

α-Bromoacetophenone		
1682.1	1279.1	750.8
1596.5	1194.8	686.5
1448.8	1012.9	623.3

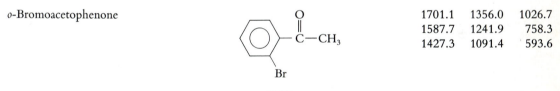

o-Bromoacetophenone		
1701.1	1356.0	1026.7
1587.7	1241.9	758.3
1427.3	1091.4	593.6

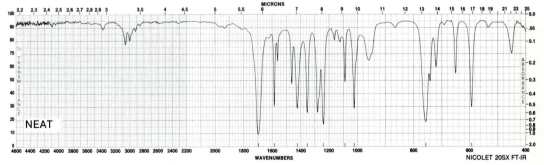

(continued)

Figure 1 *(continued)*

m-Bromoacetophenone

1688.6	1356.6	783.4
1567.9	1249.6	682.4
1420.5	1064.1	587.4

p-Bromoacetophenone

1687.8	1265.2	957.6
1587.8	1078.1	820.3
1395.7	1010.2	748.6

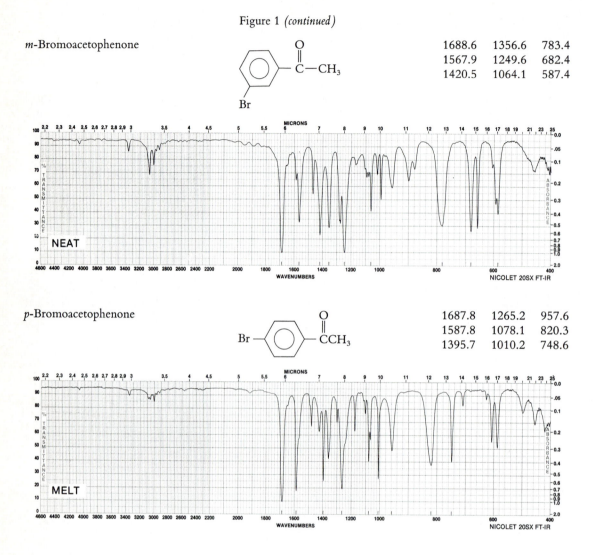

PROCEDURE

Make sure you use the right sulfuric acid solution! **If adding a little potassium bromate to the acidic solution results in a violent reaction, stop immediately and inform your instructor. Avoid contact with the sulfuric acid solution, which can irritate the skin and eyes.**

Potassium bromate is an oxidant and should be kept away from other chemicals or oxidizable materials.

The addition of saturated sodium chloride during the separation may release harmful fumes, so perform this operation under the hood and do not breathe the vapors.

Ethyl ether is extremely flammable, and its vapors are harmful. Keep it away from flames and hot surfaces, and do not breathe its vapors.

The product may be harmful if inhaled, ingested, or absorbed through the skin. Avoid contact, and do not breathe its vapors.

SAFETY PRECAUTIONS

acetophenone

ethyl ether

Reaction.   Weigh out 42 mmol of potassium bromate **[reactivity hazard!]** and divide it into ten roughly equal portions on a large square of weighing paper. In a 250-mL Erlenmeyer flask, dissolve 40 mmol of acetophenone in 40 mL of 9.6 M (62%) aqueous sulfuric acid **[see safety precautions!]**. Immerse the flask in a cooling bath ⟨OP–8⟩ containing cold tap water and a few pieces of ice. Add one portion of potassium bromate while mixing the reactants by swirling the flask in the cooling bath or by using a magnetic stirrer ⟨OP–9⟩. Monitor the temperature ⟨OP–7c⟩ of the reaction mixture as you continue to stir or swirl; let it drop to 28°C or lower after each addition, before adding the next portion of potassium bromate. Replenish the ice as necessary to maintain the temperature of the cooling bath during the addition. After adding the last portion of potassium bromate, make sure that all the solid has dissolved, and remove the flask from the cooling bath when the temperature drops below 25°C. Let the reaction mixture stand at room temperature for 5 minutes with stirring or occasional swirling.

You can place the solid in the "vee" of a folded weighing paper and use a flat-bladed spatula to divide it into tenths.

The temperature of the reaction mixture should remain between 25°C and 35°C during the addition, which should take about 30 minutes in all.

Use a flat-bottomed stirring rod to break up any lumps of potassium bromate.

Separation.   *Under the hood,* add 30 mL of saturated aqueous sodium chloride to the reaction mixture, and transfer it to a separatory funnel. Extract this solution ⟨OP–13⟩ with two 25-mL portions of ethyl ether, and combine the extracts. Wash ⟨OP–19⟩ the ether solution with water, followed by two 15-mL portions of saturated aqueous sodium bicarbonate **[be careful—possible violent reaction!]**. Dry the ether layer ⟨OP–20⟩ with magnesium sulfate, and evaporate the ether ⟨OP–14⟩.

Operations following the first washing do not have to be carried out under the hood.

A general-purpose silicone column (such as 10% OV-101/Chromosorb W) can be used for the gas chromatography, which is run at about 150°C.

Purification and Analysis.   Purify the product by semimicro vacuum distillation ⟨OP–26a⟩, and weigh it. Record a gas chromatogram ⟨OP–32⟩ of the purified product, and obtain the infrared spectrum ⟨OP–33⟩.

Report.   Interpret your infrared spectrum as completely as you can. Identify the major product, citing evidence to support your conclusion. Decide whether elemental bromine or hypobromous acid is more likely to be the brominating agent in this synthesis, and propose a detailed mechanism for the reaction. From your gas chromatogram, estimate the percentage purity of the major product and try to identify any significant impurities.

Collateral Projects

1.   Record a gas chromatogram of the crude product; indicate what compounds (or kinds of compounds) may be responsible for the extraneous peaks.

2.   Record the NMR spectrum of the product in deuterochloroform. Interpret it, and tell how it confirms your structure assignment.

Reaction of Acetophenone with Sodium Hypochlorite

MINILAB 21

SAFETY PRECAUTION   **The reaction evolves some chloroform, which has harmful vapors; laboratory tests on animals indicate that it is carcinogenic if taken orally. Carry out the reaction under the hood. Avoid contact with the sodium hypochlorite solution, and do not breathe its vapors.**

Under the hood, add 1 mL of acetophenone and 5 drops of Aliquat 336 (tricaprylmethylammonium chloride) to 40 mL of 5.25% aqueous sodium hypochlorite (Clorox, etc.) in a 125-mL Erlenmeyer flask. Shake or stir the mixture vigorously, and monitor the temperature occasionally. When the temperature reaches its maximum value and begins to fall, heat the mixture on a 65°C water bath **[in a hood!]** for about 10 minutes to drive off most of the chloroform. Add a few drops of acetone to the warm solution, with swirling. Test the solution with starch-iodide paper; if the test is positive (violet color), add more acetone drop by drop until the solution tests negative. Add 5 mL of 6 M hydrochloric acid, with stirring, and cool the mixture in ice water until precipitation is complete (the pH should be 2 or lower). Collect the benzoic acid by vacuum filtration, dry and weigh it, and measure its melting point. Write an equation for the reactions involved in the synthesis.

The reaction mixture can be heated to boiling with decolorizing carbon (just before adding HCl) and filtered to remove colored impurities.

Exercises

1. Why was saturated sodium chloride added to the reaction mixture before the extraction?

2. Construct a flow diagram for the synthesis of bromoacetophenone.

3. Outline a synthesis of α-chloroacetophenone (tear gas) starting with benzene and acetic acid.

4. Propose a synthetic route for preparing each of the following in high yield from benzene or toluene: **(a)** *m*-bromobenzoic acid, **(b)** *m*-bromotoluene, **(c)** α-bromotoluene (benzyl bromide), **(d)** *p*-dibromobenzene, and **(e)** *m*-dibromobenzene.

5. Propose a reasonable mechanism for the reaction of acetophenone with sodium hypochlorite as carried out in Minilab 21.

Library Topic

Write a report on the use of chemicals in law enforcement and warfare. Give structures and syntheses for representative compounds.

The Friedel-Crafts Acylation of Anisole

Reactions of Aromatic Ethers. Preparation of Carbonyl Compounds. Electrophilic Aromatic Substitution.

Operations

OP–7a	Refluxing
OP–9	Mixing
OP–10	Addition
OP–12	Vacuum Filtration
OP–13b	Separation of Liquids
OP–14	Evaporation
OP–19	Washing Liquids
OP–20	Drying Liquids
OP–22	Drying Gases
OP–22b	Trapping Gases
OP–25a	Semimicro Distillation
OP–28	Melting-Point Determination
OP–33	Infrared Spectrometry

Objectives

To carry out the Friedel-Crafts acylation of anisole using acetic anhydride.

To identify the product by analyzing its infrared spectrum.

To learn about the characteristics and applications of Friedel-Crafts reactions.

To learn about the properties and uses of some aromatic ketones.

SITUATION

A constituent of hawthorn flowers (genus *Crataegus*) known as crataegon has a pleasant odor that makes it useful for scenting soaps and perfumes. Spectral analysis of crataegon indicates that it is the *p*-methoxy derivative of acetophenone, which has itself been used in perfumes. Your goal in this experiment is to carry out a Friedel-Crafts reaction between anisole and acetic anhydride and find out whether the product is identical to natural crataegon.

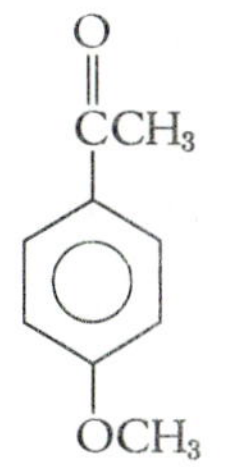

BACKGROUND

Friedel, Crafts, and Phenones

Aromatic ketones that have the carbonyl group adjacent to the benzene ring are frequently called phenones; the simplest member of this group is acetophenone. Acetophenone is a pleasant-smelling liquid that has been used to impart an odor of orange blossoms to perfumes and is also prescribed as a sleep-producing drug under the generic name hypnone. Acetophenone was first prepared by Friedel in 1857 — long before the discovery of the Friedel-Crafts reaction — by distilling a mixture of calcium benzoate and calcium acetate. Benzophenone is a white solid with a geranium-like odor that has been used as a fixative for perfumes and as a starting material for the manufacture of drugs and insecticides. It is the first aromatic ketone known to have been synthesized by a "Friedel-Crafts" reaction — Theodor Zincke prepared it accidentally, when he was trying to make something else, by heating benzoyl chloride with a metal in benzene.

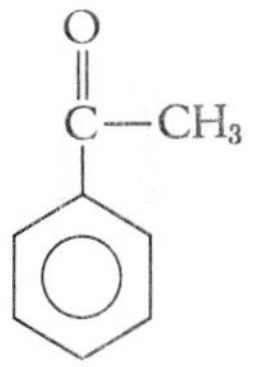

The natural and synthetic musks are powerfully odoriferous substances that supply the long-lasting, musky "end note" characteristic of some perfumes, deodorants, and after-shave lotions.

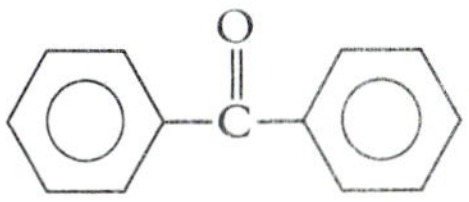

Natural and synthetic musks

crataegon

acetophenone

benzophenone

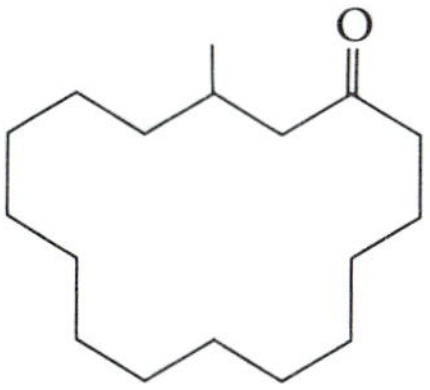

muscone 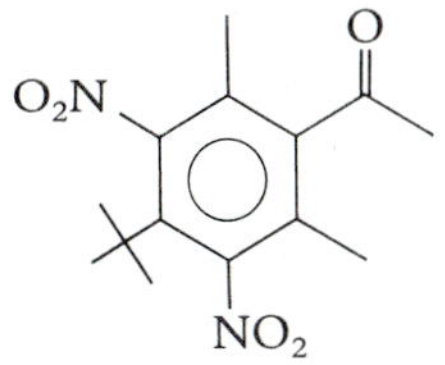   musk ketone 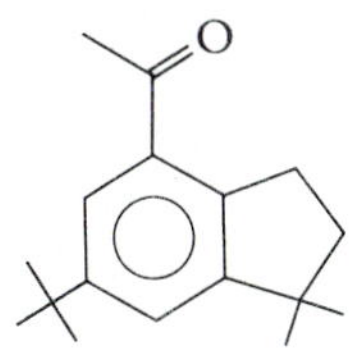   Celestolide

The large-ring ketone called muscone (3-methylcyclopentadecanone) is the major constituent of natural musk, which is a secretion from the musk pod of the male musk deer. Muscone is very costly and its use threatens the existence of the deer, so it has been largely replaced by synthetic musks. Among these are the phenones known as musk ketone and Celestolide. Musk ketone is prepared from *m*-xylene by two Friedel-Crafts reactions — alkylation with *t*-butyl chloride and acylation with acetyl chloride — followed by nitration of the aromatic ring.

Synthesis of musk ketone

m-xylene

Most phenones can be prepared by a Friedel-Crafts reaction of an aromatic compound with an appropriate acylating agent. The Friedel-Crafts reaction might well have been named the "Zincke reaction" if Theodor Zincke had understood the significance of an experiment that failed. In 1869, Zincke tried to synthesize 3-phenylpropanoic acid by combining benzyl chloride and chloroacetic acid in the presence of metallic silver (a variation of the Wurtz reaction). While carrying out the reaction with benzene as the solvent, Zincke observed, to his surprise, that a great deal of hydrogen chloride was evolved and that the major product was diphenylmethane instead of the expected carboxylic acid.

About four years later, a Frenchman named Charles Friedel was watching a student in Wurtz's laboratory perform a "Zincke reaction" using (appropriately) powdered zinc as the catalyst. When the reaction suddenly became violent, Friedel helped the student separate the solution from the zinc powder, thinking that removing the catalyst would moderate the reaction. To the astonishment of both, the reaction was just as violent in the absence of zinc! Although there is no record of his thought processes after this event, Friedel must have recognized its significance; in 1877, he and his collaborator, an American named Charles Mason Crafts, published a paper that marked the inception of the Friedel-Crafts reaction as one of the most important synthetic procedures in the history of or-

---

Zincke's attempted synthesis

$$PhCH_2Cl + ClCH_2COOH \xrightarrow[\text{benzene}]{Ag}$$

$$PhCH_2CH_2COOH$$

See Experiment 15 for the story of another Wurtz reaction that "failed," with momentous consequences.

The "Zincke reaction"

$$PhCH_2Cl + PhH \text{ (benzene)} \xrightarrow{Ag}$$

$$PhCH_2Ph + HCl$$

Crafts later returned to the United States, where he became president of the Massachusetts Institute of Technology.

ganic chemistry. Friedel and Crafts' basic discovery was a simple one — it was a chloride of the metal, and not the metal itself, that catalyzed the reaction of organic halides with aromatic compounds. They found that anhydrous aluminum chloride was the most effective catalyst of those then available: it is still the catalyst of choice for most Friedel-Crafts reactions.

In Zincke's experiment, traces of metal chloride were formed during the reaction as a result of the oxidation of the metal.

METHODOLOGY

The so-called Friedel-Crafts reaction is not a single reaction type, although the term has been most often applied to alkylations and acylations of aromatic compounds using aluminum chloride (or another Lewis acid catalyst) and a suitable alkylating or acylating agent.

A typical Friedel-Crafts acylation uses a carboxylic acid chloride or anhydride as the acylating agent and anhydrous aluminum chloride as the catalyst. Usually the acid anhydride offers better yields and a simpler workup than the acid chloride. The reaction requires more catalyst with the anhydride, because aluminum chloride is tied up by complexing with the carboxylic acid formed during the reaction, as well as by reacting with the acylating agent and aryl ketone. Since acetic anhydride is readily available and gives high yields of product, it is often used instead of acetyl chloride for synthesizing acetophenone and its derivatives.

Slightly more than 1 mole of AlCl₃ is used per mole of acyl chloride, but 2–3 moles may be needed per mole of anhydride.

Aluminum chloride complex with carbonyl compound

$$\underset{\underset{RCX}{\|}}{\overset{+}{O}}-\overset{-}{Al}Cl_3$$

In carrying out the acylation reaction, an excess of the aromatic hydrocarbon itself can be used as the solvent. Alternatively, another solvent such as methylene chloride, nitrobenzene, or carbon disulfide can be used. Since both of the latter have undesirable properties (nitrobenzene is high boiling and very toxic, carbon disulfide is toxic and extremely flammable), methylene chloride will be used as the solvent in this preparation. The acylation is highly exothermic, so it will be carried out by slowly adding acetic anhydride to the other reactants, then refluxing for a short time to complete the reaction. Pouring the product into ice water will decompose an aluminum chloride complex of the product and transfer inorganic salts to the aqueous phase. The product can then be recovered by evaporating the organic solvent and distilling the residue. The resulting methoxyacetophenone is a low-melting solid that should crystallize on cooling to room temperature. Its infrared spectrum can be obtained by the method for "melts" described in OP–33 or in a KBr disk or Nujol mull.

The reaction evolves gaseous HCl, so a gas trap is required.

In principle, the acylation of a monosubstituted benzene can yield any or all of three different disubstituted products.

Table 1.   Frequencies of C—H out-of-plane bending bands in aromatic hydrocarbons

Number of Adjacent Hydrogens	Frequency Range, cm^{-1}
1	900–860 (weak)
2	840–810
3	810–750
4	770–735
5	770–730

Only one isomer will be produced in significant quantity in this experiment, and you should be able to identify it by infrared spectrometry. Disubstituted benzenes can be distinguished by the location of their out-of-plane C—H bending bands, which occur at frequencies (wavenumbers) below 850 cm^{-1}. The frequency of such a band decreases with the number of adjacent hydrogens on the ring, as shown in Table 1. Thus, a *para*-disubstituted benzene, with its two sets of two adjacent hydrogens, should show an absorption band in the 840–810 cm^{-1} region; *meta* compounds, with three adjacent ring hydrogens, absorb in the 810–750 cm^{-1} region; and *ortho* compounds, with four adjacent ring hydrogens, absorb in the 770–735 cm^{-1} region. Absorption by the lone hydrogen of a *meta* compound is usually very weak, and its frequency may vary. Monosubstituted and *meta*-disubstituted benzenes have an additional band at 710–680 cm^{-1}, which arises from a bending vibration of the benzene ring.

Possible products from the Friedel-Crafts acylation of anisole

o-methoxyacetophenone   m-methoxyacetophenone

p-methoxyacetophenone

Your product is an ether as well as a ketone, so its infrared spectrum will contain bands characteristic of both functional groups. The carbonyl band of a phenone generally appears in the 1685–1665 cm^{-1} region, and a weak carbonyl overtone band may be observed at twice the frequency of the fundamental band. Aryl alkyl ethers display an asymmetrical C—O—C stretching band at 1275–1200 cm^{-1} and a symmetrical

C—O—C band near 1075–1020 cm^{-1}. These ether bands can be observed in the spectrum of anisole in Figure 1.

Minilab 22 demonstrates how a Friedel-Crafts reaction can be used to characterize aromatic hydrocarbons. Aromatic compounds undergo Friedel-Crafts alkylation reactions with chloroform to form triarylmethanes, which lose hydride ions to yield colored triarylmethyl cations:

$$3ArH + CHCl_3 \xrightarrow{AlCl_3} Ar_3CH + 3HCl$$

$$Ar_3CH + R^+ \longrightarrow Ar_3C^+ + RH \qquad (R = Ar_2CH^+, \text{ etc.})$$

Different aromatic hydrocarbons produce different colors, depending on their ring structures.

PRELAB ASSIGNMENT

Calculate the mass of 66 mmol of aluminum chloride, the mass and volume of 30 mmol of anisole, and the theoretical yield of methoxyacetophenone.

Reactions and Properties

| | anisole | acetic anhydride | ?-methoxy-acetophenone | |
| | | | | |

Table 2.   Physical properties

	M.W.	m.p.	b.p.	d.
anisole	108.2	−38	155	0.996
acetic anhydride	102.1	−73	140	1.082
methylene chloride	84.9	−95	40	1.327
o-methoxyacetophenone	150.2		245	1.090
m-methoxyacetophenone	150.2		240	1.034
p-methoxyacetophenone	150.2	38–9	258	1.082[41]

Figure 1.   IR spectrum of anisole. (Reproduced from *The Aldrich Library of FT-IR Spectra,* by Charles J. Pouchert, with the permission of the Aldrich Chemical Company.)

Anisole

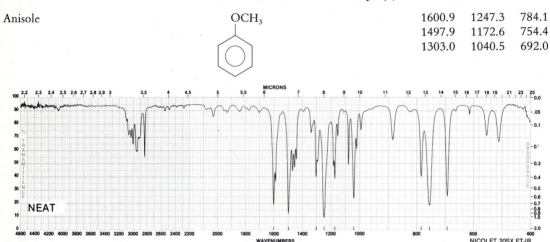

1600.9	1247.3	784.1
1497.9	1172.6	754.4
1303.0	1040.5	692.0

PROCEDURE

SAFETY PRECAUTIONS

aluminum chloride

acetic anhydride

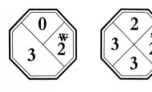

methylene chloride

anisole

Aluminum chloride reacts violently with water, generating hydrochloric acid; it can cause painful burns on moist skin and eyes, and inhaling the dust can irritate or burn the respiratory tract. Wear gloves and good safety goggles; avoid contact, do not breathe the dust, and keep it away from water.

Acetic anhydride is corrosive and lachrymatory, its fumes are highly irritating, and it reacts violently with water. Wear gloves, and dispense the liquid under a hood; avoid contact, do not breathe its vapors, and keep it away from water.

Methylene chloride may be harmful if ingested, inhaled, or absorbed through the skin; the vapors may be carcinogenic if inhaled. Minimize your contact with the liquid, and do not breathe its vapors.

Reaction.   Wear gloves and eye protection! Anhydrous aluminum chloride is deactivated by water, so protect it from atmospheric moisture and be sure your glassware is thoroughly dried. Assemble an apparatus for addition and reflux ⟨OP–10⟩, and attach a drying tube ⟨OP–22⟩ and gas trap ⟨OP–22b⟩ to the top of the reflux condenser, using dilute sodium hydroxide in the trap. Clamp the apparatus securely to a ringstand, so that you can lower the boiling flask to add the reactants without disturbing the rest of the apparatus. Carefully weigh 66 mmol of finely powdered *anhydrous* aluminum chloride **[contact, inhalation, and reactivity hazard!]** into a large dry vial and immediately stopper the vial. Add 25 mL of methylene chloride **[suspected carcinogen!]** and 30 mmol of anisole to the boiling flask; then cautiously add the aluminum chloride in small portions through a dry powder funnel, shaking the flask after each addition. Wash any adherent aluminum chloride into the flask with 5 mL of methylene chloride, and reassemble the apparatus. Under the hood, measure 3.0 mL (~32 mmol) of acetic anhydride **[contact and vapor hazard!]** into the addition funnel. Stopper the funnel, and put it in place on the reaction apparatus. Add the acetic anhydride slowly (about 20–30 drops a minute) so that the reaction mixture boils gently, while shaking or stirring ⟨OP–9⟩ to mix the reactants. Have a beaker of cold water handy so you can moderate the reaction if necessary. When the addition is complete, reflux the reaction mixture gently ⟨OP–7a⟩ for 30 minutes with a hot-water bath or steam bath.

Separation.   *Wear gloves and eye protection! Under the hood,* pour the warm reaction mixture *slowly* with vigorous stirring onto about 30 g of cracked ice in a large beaker **[take care: splattering is likely!]**. Use about 10 mL of ice water to rinse any residue out of the flask into the beaker. Separate ⟨OP–13b⟩ the organic layer, and wash it ⟨OP–19⟩ with one portion each of 3 M sodium hydroxide and saturated aqueous sodium chloride. Dry it ⟨OP–20⟩ with anhydrous magnesium sulfate, and evaporate the solvent ⟨OP–14⟩.

Purification and Analysis.   Purify the ?-methoxyacetophenone by semimicro distillation ⟨OP–25a⟩, using no cooling bath for the receiving flask. While it is still liquid, transfer the distillate to a watch glass or evaporating dish, and set it aside to crystallize. Wash the product on a Buchner funnel ⟨OP–12⟩ with a little cold low-boiling petroleum ether, and air-dry it. When it is completely dry, weigh the methoxyacetophenone, and measure its melting point ⟨OP–28⟩. Record its infrared spectrum ⟨OP–33⟩, or obtain a spectrum from your instructor.

Be sure to cap the aluminum chloride reagent bottle immediately.

Use a beaker of cold water to cool the mixture if it begins to boil.

The apparatus can be shaken by carefully moving the base of the ringstand back and forth. The addition should take about 10 minutes.

Evolution of HCl should stop during the reflux period.

If the product solidifies in the receiving flask, warm it over a steam bath.

Report. Interpret your infrared spectrum as completely as you can, and deduce the structure of the product. Cite all experimental evidence supporting your conclusion. Explain why this compound was formed rather than another of the isomers.

Collateral Projects

1. Record the PMR spectrum of the product in deuterochloroform. Interpret it as completely as you can, assigning the signals due to protons on the ring as well as those due to protons on the side chains.

2. Analyze the product by gas chromatography, and try to identify any extraneous peaks. Estimate the percentage purity of the product.

Friedel-Crafts Reactions of Aromatic Hydrocarbons

MINILAB 22

SAFETY
PRECAUTIONS

Wear gloves when handling the aluminum chloride and do not breathe its dust.

Chloroform is toxic and carcinogenic; avoid contact, and do not breathe its vapors. Be sure no flammable solvents are nearby when you light the burner.

Carefully measure about 0.1 g of anhydrous aluminum chloride **[contact, inhalation, and reactivity hazard!]** into a *dry* 10-cm test tube and heat it **[in a hood!]** over a burner flame until the aluminum chloride forms a thin coating of sublimed solid up the side of the tube. When it is cool, run a solution containing 1 drop of toluene in 8 drops of chloroform **[contact and vapor hazard! carcinogen!]** down the coated side of the tube and note any color change. Repeat the test using chloroform solutions of cyclohexane, biphenyl, and naphthalene in place of the toluene (use about 20 mg of each solid). Record and explain your observations, and write equations and a mechanism for the reaction of toluene with chloroform (assume *para* substitution).

Exercises

1. Write a mechanism for the Friedel-Crafts reaction of anisole with acetic anhydride.

2. Write equations for one or more reactions that might account for the production of HCl during the acylation reaction.

3. Construct a flow diagram for the synthesis carried out in this experiment.

4. Explain why Friedel-Crafts reactions are usually carried out by adding the alkylating or acylating agent *to* the aromatic compound rather than vice versa.

5. 2,5-Dichloro-2,5-dimethylhexane is an important starting material for the aroma chemicals called tetralin musks. Outline a synthesis of the tetralin musk versalide from this starting material and benzene, using any necessary inorganic or organic reagents.

versalide

Library Topic

Write a report on the chemistry of perfumery, telling how perfumes are formulated and what kinds of aroma chemicals are used to provide the different odor "notes." Give structures for some of the important aroma chemicals, and describe their syntheses.

Determination of the Structure of a Natural Product in Anise Oil

Reactions of Alkenylbenzenes. Preparation of Carboxylic Acids. Side-Chain Oxidation. Structure Determination.

Operations

OP–13 Extraction

OP–14 Evaporation

OP–20 Drying Liquids

OP–21 Drying Solids

OP–23*a* Semimicro Recrystallization

OP–28 Melting-Point Determination

OP–33 Infrared Spectrometry

Objectives

To determine the structure of the major component of anise oil using chemical degradation and spectrometric analysis.

To oxidize an alkenylbenzene to an aromatic carboxylic acid.

To learn how to carry out a potassium permanganate oxidation with a phase-transfer catalyst.

To learn about some methods for determining the structures of organic compounds.

To learn how infrared spectrometry can be used to determine the structures of unsaturated groups.

To learn about the sources and uses of anise oil.

SITUATION

Anise is an herb *(Pimpinella anisum)* similar to parsley that is indigenous to Greece, Egypt, and Asia Minor. When the dried fruits of anise or Chinese star anise are ground up and steam distilled, they yield an oily liquid with a strong odor of licorice. Anise seed contains about 1.5–3.5% anise oil. Most anise oil for commercial use comes from the brown, star-shaped seed clusters of the star anise *(Illicium verum),* which is a small evergreen tree of the magnolia family. Anise oil, or its synthetic equivalent, is widely used as a flavoring for licorice, cough drops, chewing gum, and liqueurs such as ouzo and anisette. The major constituent of anise oil is a compound having the molecular formula $C_{10}H_{12}O$, which we will call "anisene." Chemical and spectral analyses of anisene indicate that it contains a disubstituted benzene ring with an unsaturated side chain. Your goal in this experiment is to determine the structure of the side chain and its location on the benzene ring, from which you can deduce the structure of anisene.

BACKGROUND

The Structure Puzzle—Taking Molecules Apart and Putting Them Back Together

Today, when a chemist can run an NMR or mass spectrum of an organic compound and often determine its structure in a matter of minutes, it is hard to imagine how much time and effort were once required to determine the structures of even the simpler natural products. In a "classical" structure determination, the molecular formula is first determined by elemental analysis and molecular-weight measurement. Then the compound is degraded (broken down) into smaller structural units that are isolated and, if possible, identified. Finding how the smaller units fit together to form the original molecule is an intellectual challenge that might be compared to putting together a jigsaw puzzle with some pieces missing, others that don't belong, and still others that have been chewed up by the family dog and are no longer recognizable. Finally, when enough information has been amassed to allow a likely structure to be proposed, that structure must usually be proven by an independent synthesis in which the compound is built up again, from known compounds, by reactions whose outcome can be reliably predicted.

In many cases, structure determinations involved the efforts of dozens or even hundreds of chemists over many decades, with the generation of much irrelevant or misleading information and many synthetic dead ends. The advent of modern spectrometric methods has simplified the process enormously by providing detailed structural information that was not readily available to the chemists of earlier times.

As noted, the major component of anise oil, which can be called anisene until you have identified it, has the molecular formula $C_{10}H_{12}O$. Most open-chain saturated organic compounds (except those containing nitrogen or phosphorus) have $2n + 2$ hydrogen atoms for every n carbon atoms. If anisene were such a compound it would have $2(10) + 2 = 22$ hydrogen atoms, but since it has only 12 it is said to be "deficient" by 10 hydrogens. Every ring or unsaturated (pi) bond in a molecule represents a deficit of 2 hydrogens; that is, an open-chain compound must lose 2 hydrogen atoms to form a ring, and a saturated compound must lose 2 to form a double bond. Thus, its deficiency of 10 hydrogens indicates that there must be a total of 5 rings and/or unsaturated bonds in an anisene molecule. The number 5 is its *index of hydrogen deficiency (IHD),* which can be calculated from the following formula (for a compound with n carbon atoms and x hydrogen atoms):

$$\text{IHD} = \frac{(2n + 2) - x}{2}$$

> Each double bond (or double-bond equivalent in the Kekulé structure of an aromatic compound) counts as one unit of unsaturation, and a triple bond counts as two.

Catalytic hydrogenation of anisene yields a saturated compound with the formula $C_{10}H_{20}O$; the gain of eight hydrogens indicates that anisene has four unsaturated bonds, so it must contain only one ring. A high carbon:hydrogen ratio often indicates an aromatic structure, and we can account for the ring and three double bonds (or their equivalent in aromatic bonds) by supposing that anisene contains a benzene ring. This is confirmed by heating anisene with hydriodic acid, which produces a phenol with the molecular formula C_9H_9OH and a volatile compound identified as methyl iodide. This reaction is used to test for certain ether functions; methyl ethers yield methyl iodide, and the formation of a phenol indicates that anisene is an aryl methyl ether.

$$C_9H_9OCH_3 + HI \longrightarrow C_9H_9OH + CH_3I$$

At this point, we know that anisene contains a methoxyl (OCH_3) group and a benzene ring, which accounts for seven carbon atoms and three unsaturated bonds. That leaves three more carbons and a double bond to be accounted for. This remaining fragment could be a three-carbon unsaturated side chain; assuming that to be the case, its formula can be deter-

> The possibility that there are two side chains, a methyl group and a vinyl group, has been eliminated by chemical and spectroscopic evidence.

mined by subtracting the fragments already identified from the molecular formula of anisene:

molecular formula:	$C_{10}H_{12}O$
disubstituted benzene ring:	$-C_6H_4$
methoxyl group:	$-CH_3O$
side chain:	C_3H_5

Now we can write a partial structure for anisene:

All that remains is to determine the structure of the unsaturated side chain and its location on the benzene ring.

METHODOLOGY

The partial structure of anisene shows a C_3H_5 side chain on a benzene ring. Potassium permanganate is capable of oxidizing aliphatic side chains all the way down to the benzylic carbon atom, leaving a COOH group where the side chain was originally located. This kind of oxidation of anisene should yield one of three possible methoxybenzoic acids (whose melting points are given in Table 3). By identifying the oxidation product as one of these three, you will establish the position of anisene's side chain.

Recently it was discovered that many metal ions will dissolve in organic solvents if they are first complexed with an organic crown ether. This led to the use of "purple benzene" as an effective oxidizing agent. Potassium permanganate, when complexed with such an ether, will readily dissolve in organic solvents such as benzene; the resulting purple solution reacts rapidly with many oxidizable compounds, since the oxidant and the organic reactant are present in the same phase. Another way of obtaining purple benzene was reported in 1974 by Herriot and Picker, who added quaternary ammonium salts to dissolve permanganate ion (from an aqueous solution) in benzene, where it could oxidize water-insoluble organic compounds. The quaternary salt acts as a phase-transfer catalyst, transferring the permanganate ion from the aqueous to the

methoxybenzoic acid

a crown ether with
solubilized KMnO$_4$

See Experiment 19 for a discussion of phase-transfer catalysis.

organic phase where the reaction takes place. This catalyst causes the reaction to be carried out more rapidly (and under milder conditions) than would ordinarily be the case.

In this experiment, you will use a simplified procedure in which anisene and the phase-transfer catalyst are combined directly with aqueous potassium permanganate; thus, anisene itself is the second phase of the two-phase system, and no organic solvent is needed. Because you need only enough product for a melting point, you can start with a few drops of anisene, which can be obtained by steam-distilling anise seeds if desired (Collateral Project 1). As the reaction proceeds, permanganate ion is reduced to manganese dioxide, which forms a fine brown precipitate that is difficult to filter and wash. Fortunately, this precipitate can be dissolved during the workup by acidifying the solution and adding sodium bisulfite, which reduces manganese dioxide (and unreacted permanganate) to soluble manganese(II) sulfate. The methoxybenzoic acid can then be extracted with methylene chloride and purified by recrystallization from water.

Removal of manganese dioxide

$$MnO_2 + NaHSO_3 + H^+ \rightarrow$$
$$MnSO_4 + H_2O + Na^+$$

The out-of-plane bending vibrations of vinylic C—H bonds give rise to absorption bands in the $1000-650$ cm^{-1} region of an infrared spectrum. The frequencies of these bands may reveal the number and location of substituents on the carbon-carbon double bond, as shown in Table 1. There are four possible structures for an unsaturated C_3H_5 side chain, corresponding to the four structural types in Table 1. From the frequency (wavenumber) of anisene's vinylic C—H bending band (or bands), you should be able to deduce the structure of the side chain. But first you must locate the right absorption band(s), more easily said than done since aromatic C—H bonds give rise to strong bands in the same region (see Table 2).

Once you learn the position of the side chain on anisene's benzene ring, you should be able to locate the bands due to ring C—H bonds in its infrared spectrum, which will help you pick out the vinylic C—H bands from the remaining strong bands in the $1000-650$ cm^{-1} region.

Table 1.   Out-of-plane bending vibrations of vinylic C—H bonds

Structural type	Frequency range (cm^{-1})
RCH=CH$_2$	995–985 and 915–905
RCH=CHR (cis)	730–665
RCH=CHR (trans)	980–960
R$_2$CH=CH$_2$	895–885

R = alkyl or aryl

There may be some weak bands in that region that arise from minor components of anise oil.

Table 2.   Out-of-plane bending vibrations of aromatic C—H bonds

Ring substitution	Frequency range (cm^{-1})
ortho	770–735
meta	810–750 and 710–690
para	840–810

A different kind of side-chain oxidation is described in Minilab 23. In the presence of aqueous base, the side-chain methylene of fluorene is slowly oxidized by air to a carbonyl group, yielding the ketone fluorenone.

fluorene   fluorenone

A phase-transfer catalyst and vigorous magnetic stirring will be used to speed up the reaction so that it can be completed in an hour or so.

Reactions and Properties

$Q^+ \approx (CH_3(CH_2)_7)_3\overset{\oplus}{N}CH_3$

Table 3.   Physical properties

	M.W.	m.p.	b.p.	d.
potassium permanganate	158.0			
o-methoxybenzoic acid	152.2	101		
m-methoxybenzoic acid	152.2	110		
p-methoxybenzoic acid	152.2	185		
toluene	92.2	−95	111	0.867

PROCEDURE

Obtain some anise oil (or pure anisene) from your instructor, or isolate it from anise seeds as described in Collateral Project 1. Record its infrared spectrum ⟨OP–33⟩, or obtain a spectrum from your instructor.

Potassium permanganate can react violently with oxidizable materials; keep it away from other chemicals and combustibles.

Sodium bisulfite produces harmful vapors when it reacts with acids; do not breathe them.

1–2 mL of HCl and 0.5–1.0 g of NaHSO₃ should be sufficient.

Reaction.   Dissolve 0.5 g of crystalline potassium permanganate **[reactivity hazard!]** in 10 mL of water in a 125-mL Erlenmeyer flask by heating the mixture on a steam bath for several minutes. Add 2 drops of tricaprylmethylammonium chloride (Aliquat 336) and 5 drops of anisene or anise oil, and stopper the flask loosely. Using gloves or paper towels to protect your hands from the steam, swirl and shake the flask vigorously over a steam bath for 15 minutes or more, loosening the stopper occasionally to prevent pressure buildup. Cool the reactants to room temperature and acidify the reaction mixture to blue litmus paper with 6 M hydrochloric acid. Add just enough solid sodium bisulfite in small portions, with vigorous shaking, to reduce any excess permanganate and dissolve the brown precipitate of manganese dioxide. Test the solution with litmus after each bisulfite addition, and add 6 M HCl drop by drop as needed to keep it acidic. When the brown color has disappeared and only a white precipitate remains, test the solution with pH paper; if necessary, reduce the pH to 2 with additional HCl.

Separation.   Add 10 mL of saturated aqueous sodium chloride, and extract the reaction mixture ⟨OP–13⟩ with two 10-mL portions of methylene chloride. Dry the combined extracts ⟨OP–20⟩ with magnesium sulfate, and evaporate the solvent ⟨OP–14⟩.

Purification and Analysis.   Recrystallize the solid residue ⟨OP–23a⟩ from boiling water, and dry it ⟨OP–21⟩ to constant mass. Measure the melting point ⟨OP–28⟩ of the methoxybenzoic acid.

Report. Interpret the infrared spectrum as completely as you can. Deduce the location and structure of the side chain, and draw the structure of anisene. Justify your conclusion, citing all the experimental evidence that supports it. Derive a systematic name for anisene, and find its traditional common name in the *Merck Index* or another reference book.

Collateral Projects

1. Isolate anise oil from anise seeds as follows. Weigh out 10 g or more of fresh anise seeds; grind them finely using a spice grinder or a mortar and pestle. Isolate the anise oil by steam distillation and extraction of the distillate with methylene chloride, as described for clove oil in Experiment 10. Dry the methylene chloride layer with magnesium sulfate, and evaporate the solvent completely. You can obtain a gas chromatogram of the oil and estimate the percentage of anisene it contains.

Do not extract the methylene chloride layer with NaOH.

2. Record the PMR spectrum of anisene in deuterochloroform, and interpret it as completely as you can.

Air Oxidation of Fluorene to Fluorenone

MINILAB 23

Mix 1.0 g of fluorene with 10 mL of heptane **[fire hazard!]** in a 125-mL Erlenmeyer flask. Add 5 mL of 10 M (~30%) aqueous NaOH **[contact hazard!]**. Add 10 drops of tricaprylmethylammonium chloride (Aliquat 336), and stir the mixture *vigorously* for an hour or more, using a 1-inch or larger magnetic stirbar. The stirring rate should be high enough to produce a froth on the surface of the reaction mixture. Transfer the mixture to a 15-cm test tube, cool it in ice, and carefully remove the aqueous (lower) layer with a capillary pipet. Collect the crude fluorenone by semimicro vacuum filtration, washing it on the filter with a little 1 M HCl, then with water. Dry it at room temperature or in an oven at 60°C. Recrystallize the dry fluorenone from cyclohexane **[fire hazard!]**, and measure its mass and melting point.

Exercises

1. **(a)** Write a balanced equation for the reaction of anisene with potassium permanganate, assuming that acetic acid is a by-product of the reaction. (*Note:* the reaction mixture is alkaline.) **(b)** Assuming that 5 drops of anisene is about 1.0 mmol, calculate the mass of potassium permanganate required to oxidize that much anisene, and the percentage in excess that was actually used.

Excess permanganate is used because some of it may decompose spontaneously during the reaction.

2. Describe the probable role of the phase-transfer catalyst in this reaction, giving equations for the relevant reactions.

3. **(a)** Draw the structure of the compound, $C_{10}H_{20}O$, that is obtained by the catalytic hydrogenation of anisene. **(b)** Draw the structure of the compound, C_9H_9OH, that is obtained when anisene is treated with hydriodic acid.

4. The pK_a of the methoxybenzoic acid you prepared is 4.47. What percentage of this product would you lose if the pH of the reaction mixture were 4.00 when you extracted it with methylene chloride?

5. You could confirm the structure of anisene by synthesizing it from known starting materials. Outline a synthesis of anisene from benzene and alcohols that have four carbon atoms or fewer.

6. Construct a flow diagram for the preparation of methoxybenzoic acid as carried out in this experiment.

7. The structure shown has been proposed for coniferyl alcohol, which can be obtained by the hydrolysis of coniferin, a natural product found in the sap of conifer trees. Assuming that coniferyl alcohol has not been reported in the literature, tell how you would go about proving its structure. Indicate what chemical tests and degradations might be carried out, describing the expected results and conclusions. Summarize the information that could be derived from spectral analysis. Then show how the alcohol could be synthesized from readily available starting materials.

proposed structure for
coniferyl alcohol

Library Topic

Capsaicin is a pungent substance found in Tabasco sauce, cayenne, and red pepper. Describe how its structure was determined, outlining the reaction steps involved.

The Borohydride Reduction of Vanillin to Vanillyl Alcohol

Preparation of Alcohols. Reactions of Carbonyl Compounds. Nucleophilic Addition. Reduction.

Operations

The operations you use will depend on the procedure you develop (see the Prelab Assignment).

Objectives

To develop a procedure for the reduction of vanillin to vanillyl alcohol and carry it out in the laboratory.

To learn about the use of complex metal hydrides in syntheses.

To learn about the properties and uses of vanillin and related aromatic compounds.

SITUATION

Lignin is a complex polymer that gives rigidity to trees and other woody plants; after cellulose, it is the second most abundant organic material on earth. Nearly 80% of the "building blocks" of coniferous lignin are guiacylpropane units, and such units can be broken down chemically to yield vanillin, the main component of vanilla flavorings. If there were enough demand to justify its large-scale manufacture from lignin, vanillin could easily become one of the most abundant chemical raw materials. One way of creating such a demand might be to find ways of converting vanillin to other compounds with commercial potential. Vanillin can be reduced by sodium borohydride to vanillyl alcohol, which could perhaps be used to develop new flavoring or perfume ingredients. Unfortunately, the procedure for this synthesis was published in an obscure Swedish journal, *Acta Universitatis Lundensis,* which ceased publication in the 1960s. Since you are not likely to come across that journal in your library, you will have to develop your own procedure for the preparation of vanillyl alcohol.

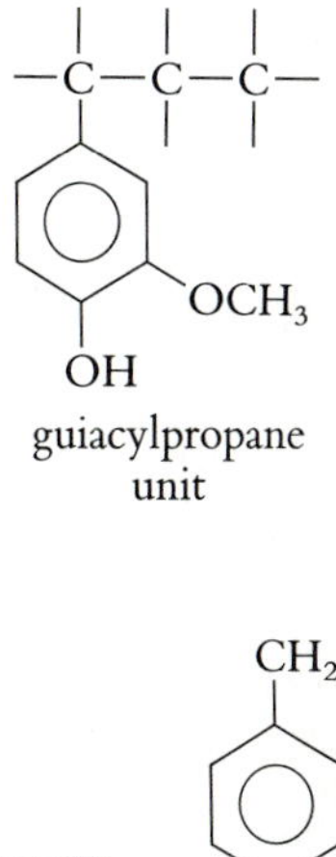

guiacylpropane unit

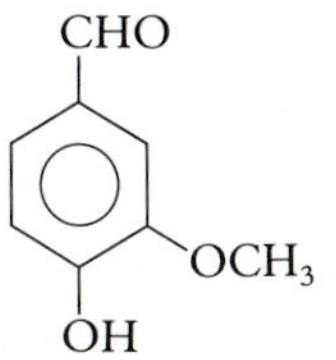

vanillin

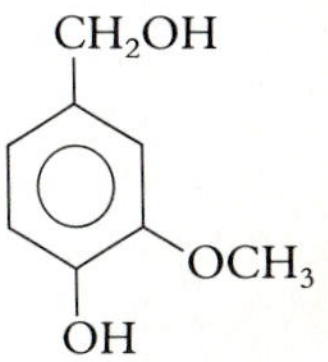

vanillyl alcohol

BACKGROUND

The Fragrant Aromatics

Chemists recognized at an early date that certain compounds obtained from natural sources showed a higher ratio of carbon to hydrogen than did typical aliphatic compounds. These compounds also had distinctly different chemical properties. Because many of them came from such pleasant-smelling sources as the essential oils of cloves, sassafras, cinnamon, anise, bitter almonds, and vanilla, they were called *aromatic* compounds. The name has stuck, although it is no longer associated with the odor of such compounds but rather with their structure and properties. Many aromatic compounds do justify the name, however, and among the most interesting and important of these are the chemical relatives of vanillin.

In 1520, the Spanish conquistador Hernando Cortez was served an exotic new drink by Montezuma II, emperor of the Aztecs. Cortez enjoyed the drink, a combination of chocolate and vanilla, and it soon found its way back to Europe. The vanilla plant (which is actually a climbing orchid, *Vanilla planifolia*) was also shipped back to the Old World in the hope that vanilla could be produced there successfully. The plant grew well but, mysteriously, would not fruit outside its native country. This mystery remained unsolved for more than 300 years, until someone discovered that the plant was pollinated by a native Mexican bee with an exceptionally long proboscis. A method of hand pollination was soon developed.

Vanilla flavoring comes from the fruit of the vanilla plant, a long, narrow pod that, after curing, looks somewhat like a dark brown string bean. The principal component, vanillin, does not exist as such in the fresh "bean" but is formed by the enzymatic breakdown of a glucoside during the curing process.

Cortez later repaid Montezuma's hospitality by razing his capital.

isoeugenol   vanillin

Although the finest vanilla flavoring is still obtained from natural vanilla, synthetic vanillin is far less costly. It is widely used as a component of flavorings, perfumes, and pharmaceutical products and as a raw material for the synthesis of such drugs as L-dopa. At one time, most synthetic vanillin was made from isoeugenol, a naturally occurring and widely used perfume ingredient. Most vanillin is now synthesized using lignin derived from wood pulp.

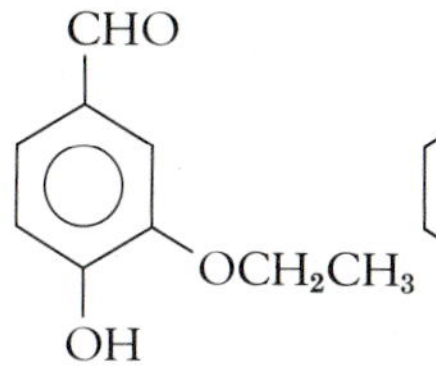

Many other natural aromatic compounds are related to vanillin. Bourbonal (3-ethoxy-4-hydroxybenzaldehyde) derives its trivial name from the island of Bourbon (now called Reunion), which produces a fine quality of vanilla. Although generally prepared from vanillin, bourbonal has been isolated in small quantities from vanilla beans and other natural sources. Safrole is a fragrant compound derived from sassafras and camphor oils. Once widely used as a flavoring in root beer, toothpaste, and chewing gum, it has since been banned for such uses because of its toxic, irritant qualities and because it produces liver tumors in rats and mice.

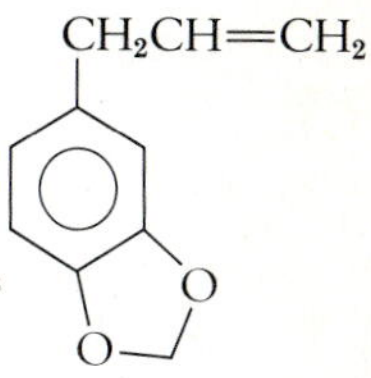

Piperonal can be derived from safrole by isomerization to isosafrole (which is analogous to isoeugenol) and oxidation. Because of its sweet "cherry pie" odor of heliotrope, piperonal is used widely in perfumes, cosmetics, and soaps.

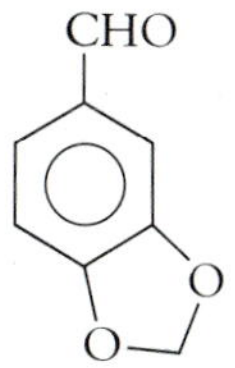

The most important of all spices, black pepper, gets its "bite" from piperine, which contains the same methylenedioxy ($-OCH_2O-$) unit found in safrole and piperonal.

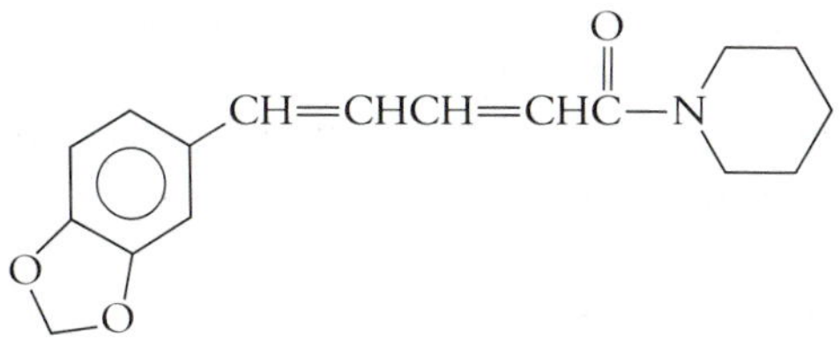

bourbonal   safrole

Bourbonal has a strong vanilla odor that is three or four times as powerful as that of vanillin itself.

piperonal

piperine (*trans* double bonds)

Hydrolysis of piperine followed by oxidation yields both piperonal and the cyclic amine piperidine.

Capsaicin, an even more pungent compound related to vanillin, is the fiery component of Tabasco sauce, hot paprika, and cayenne peppers. Hydrolysis of capsaicin yields vanillylamine, which can easily be synthesized from vanillin.

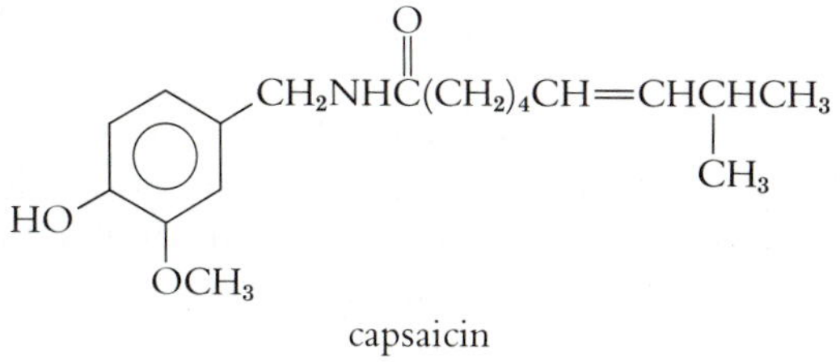

capsaicin

Another hot compound with the 3-methoxy-4-hydroxy grouping is zingerone, the pungent principle of ginger.

$$
\underset{\text{zingerone}}{
\begin{array}{c}
\text{HO} \\
\text{OCH}_3
\end{array}
\;\;
\text{CH}_2\text{CH}_2\overset{\displaystyle\text{O}}{\overset{\|}{\text{C}}}\text{CH}_3
}
$$

zingerone

Apparently the same structural features that contribute to the pleasant odors of vanillin and piperonal produce quite a different effect in these fiery flavoring agents.

METHODOLOGY

When lithium aluminum hydride (LiAlH$_4$) was first introduced as a reducing agent in the late 1940s, it brought about a revolution in the preparation of alcohols by **reduction.** The two most popular methods for making alcohols from carbonyl compounds had been reduction by sodium metal in a hydroxylic solvent and catalytic reduction with gaseous hydrogen under pressure. The simplicity and convenience of the hydride reaction, however, soon made it the preferred method for a broad spectrum of chemical reductions. Lithium aluminum hydride is a powerful reducing agent with the ability to reduce aldehydes, ketones, acyl chlorides, lactones, epoxides, carboxylate esters, carboxylic acids, nitriles, and nitro compounds to alcohols or amines. Unlike catalytic hydrogenation, hydride reduction does not reduce carbon-carbon multiple bonds (except in some α,β-unsaturated compounds). However, the high reactivity of lithium aluminum hydride is a disadvantage in some applications. Since it reacts violently with water and other hydroxylic solvents to release hydrogen gas, it can only be used in aprotic solvents, such as ethyl ether or diglyme, under strictly anhydrous conditions. It is also expensive and somewhat hazardous to use — even grinding it in a mortar can cause a fire.

By contrast, sodium borohydride (NaBH$_4$) is a much milder reducing agent that is comparatively safe to handle in the solid form; it can even be used in aqueous or alcoholic solutions. In such solutions, it is used primarily for reducing aldehydes and ketones, since it is unreactive toward many

reduction: the conversion of a compound in a higher oxidation state to a compound in a lower one, usually accompanied by a gain of hydrogen atoms or a loss of oxygen atoms, or both

functional groups affected by lithium aluminum hydride. Because of its selectivity, sodium borohydride is widely used for reducing aldehydes and ketones that contain other functional groups. For example, a 3-keto bile acid ester *(1)* can be reduced to the corresponding steroid alcohol without disturbing the ester function or the bromine atom.

Sodium borohydride reductions are usually carried out in water or in an alcohol such as methanol, ethanol, or 2-propanol. The reagent is not stable at low pH values, and it decomposes to the extent of about 4.5% per hour at 25°C in neutral solution. Thus, aqueous reactions are usually run in dilute sodium hydroxide. When compounds having acidic functional groups are reduced, enough of the aqueous NaOH must be used to neutralize them and maintain a pH of 10 or higher. Sodium borohydride reacts slowly with ethanol and methanol, but these solvents are usually suitable when the reaction time is no more than 30 minutes at 25°C; for longer reaction times or reactions at higher temperatures, isopropyl alcohol is a better solvent.

Sodium borohydride reductions involve nucleophilic addition of hydride ion (H : ⁻) to the carbonyl carbon, but apparently no free hydride ions are generated. Kinetic evidence suggests that one solvent molecule bonds to the boron atom while it is transferring hydride to the carbonyl compound, and another solvent molecule provides a proton to the carbonyl oxygen:

1 M sodium hydroxide is suitable for most reductions.

Unless they are neutralized, acidic functional groups may cause rapid decomposition of the sodium borohydride.

The rate of NaBH₄ decomposition in alcohols is in the order CH₃OH > C₂H₅OH > i-C₃H₇OH. Isopropyl alcohol is more difficult to remove during the workup, however.

(R = alkyl or H)

From the stoichiometry of the reaction (see page 322), it can be seen that 1 mole of sodium borohydride will reduce 4

Some textbooks show the borohydride reaction producing a tetraalkoxyborate salt of the product alcohol but recent studies indicate that it yields the product directly.

moles of aldehyde or ketone. In practice, it is wise to use a 50–100% excess of borohydride to compensate for any that reacts with the solvent or other hydroxylic materials. Since the reaction is first-order in sodium borohydride (as well as the carbonyl compound), using an excess will also increase the reaction rate.

In most reactions with sodium borohydride, the aldehyde or ketone is dissolved in the reaction solvent and added *to* a solution of sodium borohydride, with external cooling if necessary, at a rate slow enough to keep the reaction temperature below 25°C. The amount of solvent is not crucial, but enough should be used to readily dissolve each reactant and facilitate the workup of the reaction mixture. The solubility of sodium borohydride per 100 g of solvent is reported to be 55 g in water at 25°C, 16.4 g in methanol at 20°C, and 4.0 g in ethanol at 20°C.

Higher temperatures may decompose the hydride, especially in methanol or ethanol.

The time required to complete the reaction depends on the reaction temperature and the reactivity of the substrate. Kinetic studies of borohydride reduction in isopropyl alcohol have shown that aldehydes are considerably more reactive than ketones and that aliphatic carbonyl compounds are more reactive than aromatic ones. For example, the comparatively reactive ketone 4-*t*-butylcyclohexanone is completely reduced at room temperature in 20 minutes, but benzophenone is reduced only by heating it at the boiling point of isopropyl alcohol for 30 minutes. Most reactions of aldehydes and aliphatic ketones are complete in 30 minutes at room temperature, but those of aromatic ketones or particularly hindered ketones may require more time or higher reaction temperatures.

Reaction conditions in borohydride reductions

Et = Ethyl
i-Pr = isopropyl
t-Bu = *t*-butyl
r.t. = room temperature

After the reaction is complete, the excess sodium borohydride is decomposed by acidifying the reaction mixture to about pH 6 (slowly and with stirring) with 3–6 M HCl. Hydrogen gas is evolved during this process as the excess sodium borohydride decomposes, so there must be no flames in the vicinity. Depending on the properties of the product and the

Addition of acid may also generate some diborane (B_2H_6), which can cause side reactions if other reducible groups (COOH, COOR, C = C) are present.

solvent used, the product can be isolated by filtration or extraction, or by the evaporation of excess solvent. If the product is a solid that crystallizes from the reaction mixture, it can be collected by vacuum filtration, followed by extraction of the aqueous solution to recover dissolved product. Liquids or water-soluble products are generally isolated from an aqueous solution by ether extraction and recovered by evaporating the ether. If the reaction solvent is an alcohol, the reaction mixture is usually concentrated before the extraction step by evaporating most of the alcohol. Water can then be added to facilitate the extraction. The product can be purified by any appropriate method.

Vanillyl alcohol tends to form supersaturated solutions in water; it may be necessary to scratch the flask to cause the alcohol to crystallize.

Minilab 24 gives a procedure for an interesting photochemical reduction of benzophenone by a very unlikely reducing agent, isopropyl alcohol (2-propanol). Absorption of light can convert a benzophenone molecule to a high-energy "triplet-state" molecule, which behaves like a diradical. A triplet-state benzophenone molecule can strip a hydrogen atom from a 2-propanol molecule, which soon loses a second hydrogen atom to ground-state benzophenone, forming a molecule of acetone. This leaves two diphenylhydroxymethyl radicals that can combine to yield a molecule of benzpinacol. When a solution of benzophenone in 2-propanol is allowed to stand in strong light, benzpinacol starts to precipitate in a few days.

Photochemical reduction of benzophenone to benzpinacol

PRELAB ASSIGNMENT

After reading the Methodology section, develop a procedure for the sodium borohydride reduction of 25 mmol of vanillin to vanillyl alcohol. Calculate or estimate the quantities of reactants and solvents you will need, describe the reaction conditions, and tell how you intend to separate and purify the product. Your procedure should be clear and detailed enough so that anyone with sufficient background could carry it out successfully.

Alternatively, several students can get together to work out a set of procedures in which some experimental parameters are varied, to see how such variations affect the yield and purity of the product.

Reactions and Properties

$$\underset{\substack{\text{CHO} \\ \\ \\ \text{OH}}}{4\text{ }\bigcirc\text{OCH}_3} + \text{NaBH}_4 + 4\text{H}_2\text{O} \longrightarrow \underset{\substack{\text{CH}_2\text{OH} \\ \\ \\ \text{OH}}}{4\text{ }\bigcirc\text{OCH}_3} + \text{H}_3\text{BO}_3 + 4\text{NaOH}$$

Table 1. Physical properties

	M.W.	m.p.	b.p.
vanillin	152.2	79	285
sodium borohydride	37.83		400d
vanillyl alcohol	154.2	115	d

Vanillyl alcohol is reported to be soluble *cold* in alcohol and ether and soluble *hot* in water, alcohol, ether, and benzene. It is relatively insoluble in cold water and benzene.

Figure 1.   IR spectrum of vanillin. (Reproduced from *The Aldrich Library of FT-IR Spectra,* by Charles J. Pouchert, with the permission of the Aldrich Chemical Company.)

Vanillin

3171.4	1509.6	859.6
1665.7	1266.6	733.5
1588.0	1154.9	633.2

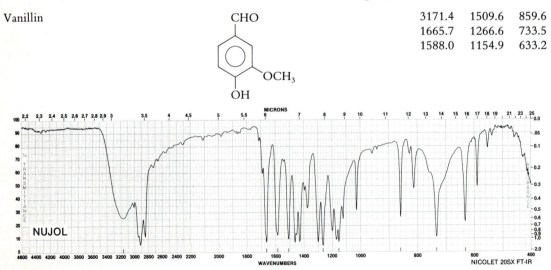

PROCEDURE

SAFETY PRECAUTIONS

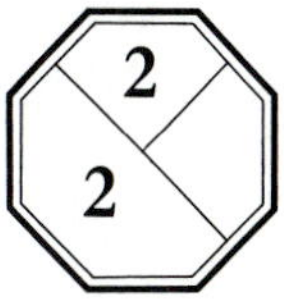

Sodium borohydride is toxic and corrosive, and it can react violently with concentrated acids, oxidizing agents, and other substances. Aqueous borohydride solutions with pH values below 10.5 have been known to decompose violently, so be sure your reaction mixture (if aqueous) is sufficiently alkaline. Avoid contact with sodium borohydride, do not breathe its dust, and keep it away from other chemicals.

Hydrogen is generated when the reaction mixture is acidified, so be sure there are no flames nearby during this step.

sodium borohydride

Develop your own procedure for this experiment as instructed in the Prelab Assignment, and submit it to your instructor for approval. Carry out the synthesis in the laboratory, measure the yield and melting point of the purified vanillyl alcohol, and turn it in.

Remember that phenols are acidic.

Keep the container of $NaBH_4$ tightly capped when it is not in use.

Report. Calculate your percentage yield, and compare your results with those of other students in your laboratory section. Suggest ways of improving the yield and/or purity of your product.

Collateral Projects

1. Record the infrared spectrum of the product, and interpret it as completely as you can. Compare it with the spectrum of vanillin in Figure 1, and describe any evidence suggesting that the expected reaction has taken place.

2. Reduce another carbonyl compound (such as benzaldehyde or camphor) with sodium borohydride after developing an appropriate procedure, and turn in a pure sample of the resulting alcohol. Your instructor must approve the procedure before you start.

3. Carry out the following tests from Part III on vanillin and on your product: 2,4-dinitrophenylhydrazine ⟨Test C–11⟩, ferric chloride ⟨Test C–13⟩, and Tollens' test ⟨Test C–23⟩. Interpret the results, and comment on the purity of your product.

Photoreduction of Benzophenone to Benzpinacol

MINILAB 24

SAFETY PRECAUTIONS   **Avoid contact with acetic acid, and do not breathe its vapors.**

Keep 2-propanol away from heat or flames.

If the laboratory is very cold, some benzophenone may precipitate before it reacts.

Combine 1.0 g of benzophenone with 5 mL of 2-propanol (isopropyl alcohol) in a test tube, and warm the mixture over a steam bath to dissolve the benzophenone. Add a drop of glacial acetic acid, stopper the test tube tightly, and shake it. Then set it on a windowsill where it will receive direct sunlight. After a week or more, collect the product by vacuum filtration, dry it, and measure its melting point. Leave the filtrate in sunlight for another week or so to see if any additional product forms, and cool it in ice before filtering. Weigh the benzpinacol, and turn it in.

Exercises

1. Write a mechanism for the reduction of vanillin by sodium borohydride under the reaction conditions you used.

2. Write a balanced equation for the acid-catalyzed decomposition of sodium borohydride in water.

3. Construct a flow diagram for the procedure you used in preparing vanillyl alcohol.

4. Sodium borohydride is a strong base as well as a reducing agent. In the reduction of base-sensitive compounds, should the substrate be added to the borohydride solution or the borohydride to the substrate? Explain.

5.   **(a)** Draw the structure of the product you would have isolated if you had used $NaBD_4$ in D_2O in this experiment, and write a mechanism explaining the result. **(b)** Give the product structure for the same reaction using $NaBD_4$ in H_2O. Assume that there is no hydrogen transfer between $NaBD_4$ and the solvent.

6.   A student dissolved vanillin in 10 M sodium hydroxide and let the mixture stand overnight before adding sodium borohydride. Although the product was a white solid, it melted over a broad temperature range that was much lower than the melting point of vanillyl alcohol. The instructor suggested washing the product with dilute sodium bicarbonate; when the student did so, about half of the product dissolved and the remainder melted at 115°C. Explain what happened, and write an equation for the reaction.

Library Topic

Write a short report on the sources, nature, and uses of lignin, including a description of a commercial process for producing vanillin from the by-products of papermaking.

The Effect of Reaction Conditions on the Condensation of Furfural with Cyclopentanone

Reactions of Carbonyl Compounds. Preparation of α,β-Unsaturated Carbonyl Compounds. Nucleophilic Addition. Condensation Reactions. Carbanions.

Operations

OP–8	Cooling
OP–9	Mixing
OP–12	Vacuum Filtration
OP–13	Extraction
OP–13b	Separation of Liquids
OP–14	Evaporation
OP–19	Washing Liquids
OP–20	Drying Liquids
OP–21	Drying Solids
OP–23	Recrystallization
OP–23b	Recrystallization from Mixed Solvents
OP–25b	**Distillation of Solids**
OP–26a	Semimicro Vacuum Distillation
OP–28	Melting-Point Determination
OP–34	Nuclear Magnetic Resonance Spectrometry

Objectives

To carry out the condensation of cyclopentanone with furfural under two different sets of reaction conditions.

To identify the major product of each reaction from its PMR spectrum and explain its formation.

To learn more about aldol-type condensation reactions.

To learn more about phase-transfer catalysis and its applications.

To gain experience in analyzing PMR spectra.

SITUATION

A paper published in the *Journal of Organic Chemistry* in 1957 describes a Claisen-Schmidt reaction (a type of aldol condensation) between cyclopentanone and furfural, in which a low-melting yellow solid was obtained in about 60% yield. If the same reaction is carried out with excess furfural and an effective phase-transfer catalyst, a high-melting, golden-orange solid is recovered in nearly 100% yield. In this experiment, you will carry out the condensation reaction under the two sets of reaction conditions and try to identify the products by NMR spectrometry.

BACKGROUND

Furfural — The Oat-Hull Aldehyde

Furfural, also known as 2-furaldehyde, is the most important member of the furan series of aromatic compounds. The furan ring is aromatic because it has six pi electrons (two from the oxygen atom) distributed about a five-membered ring (see Figure 1). However, its aromatic sextet is less stable than that of benzene. The furan ring can therefore undergo reactions such as electrophilic addition, cycloaddition, and cleavage more readily than the corresponding benzene compounds do.

Furfural has been known since its isolation by Dobereiner in 1832. It can be prepared in large quantities by treating such "waste" materials as bran, oat hulls, corncobs, and peanut shells with dilute acids. These materials contain polysaccharides known as pentosans that are hydrolyzed to pentoses (simple five-carbon sugars). These in turn are converted to furfural by acid-catalyzed dehydration.

furfural

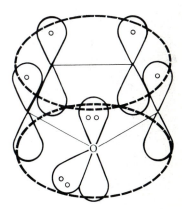

Figure 1.   Aromatic furan ring

Conversion of a pentose to furfural (pentoses also exist in a cyclic hemiacetal form)

$$\xrightarrow{-3H_2O}$$

The first commercial process for producing furfural was developed in 1922 by the Quaker Oats Company, which was trying to convert oat hulls into a better cattle feed at the time. Instead,

Furfural selectively removes alkenes and arenes from petroleum fractions, giving them better viscosity and greater resistance to oxidation.

it came up with a valuable commercial product that could be produced cheaply on a large scale. Among the many applications of furfural are in the purification of lubricating oils, the extractive distillation of butadiene (used in the manufacture of rubber), the synthesis of phenolic resins, and the manufacture of a large number of chemical intermediates.

Furfural behaves like a typical aromatic aldehyde in many of its reactions; it can be oxidized and reduced to the corresponding carboxylic acid and alcohol and, like benzaldehyde, undergoes the Cannizzaro reaction and the benzoin condensation. It also reacts with compounds having active methylene groups, such as aldehydes and ketones, to yield condensation products. For example, the reaction of furfural with acetone yields furfurylideneacetone by a Claisen-Schmidt condensation, and its condensation with acetic anhydride and sodium acetate forms furylacrylic acid by a Perkin reaction.

Claisen-Schmidt condensation of furfural and acetone

furylacrylic acid

furfurylideneacetone

METHODOLOGY

The Claisen-Schmidt condensation is actually a kind of crossed aldol condensation; it was discovered by Schmidt in 1880 and later improved by Claisen. It is regarded as a reaction between an aromatic aldehyde and an aliphatic aldehyde or ketone that yields an α,β-unsaturated aldehyde or ketone under basic conditions. The mechanism is essentially the same as that for a typical aldol condensation (for example, the self-condensation of acetaldehyde) up to the formation of a β-hydroxy carbonyl compound. In the Claisen-Schmidt condensation, this intermediate is then dehydrated by an E1cb mechanism involving the formation of a resonance-stabilized carbanion, which loses hydroxide ion to form the unsaturated product. The overall mechanism is illustrated on page 329 for the condensation of acetone and benzaldehyde.

E1cb refers to a unimolecular elimination reaction that proceeds via a carbanion.

In the literature procedure mentioned in the Situation section, cyclopentanone and furfural are dissolved in ethyl ether and shaken with an aqueous solution of dilute sodium hydroxide. Because the organic reactants are not very soluble in water, they tend to stay in the ether phase, making the reaction rather slow. Extraction of the reaction mixture with ethyl ether yields an impure yellow liquid, which is vacuum-distilled to produce a low-melting yellow solid that we will call product **A**. Even though the product usually does not solidify until the distillate is cooled, the distillation should be monitored closely to make sure that crystals do not form in the vacuum adapter and plug it up. Product **A** is further purified by recrystallization from ethanol.

When the same reactants are shaken or stirred in the presence of an effective phase-transfer catalyst, such as tricaprylmethylammonium chloride (Aliquat 336), golden-orange crystals of product **B** begin to crystallize from the reaction mixture almost immediately. The exothermic reaction generates enough heat to vaporize the ether; thus, it is necessary to cool the reactants in an ice bath before and during the reaction period. Product **B** can be separated by filtration at the end of the reaction period and purified by recrystallization from 2-butanone (methyl ethyl ketone). The role played by the phase-transfer catalyst in aldol-type condensations is not entirely clear, but presumably the catalyst carries hydroxide ions into the organic phase where they can generate enolate ions, which then react with furfural molecules.

From the structures of the reactants and your knowledge of aldol-type condensation reactions, you should be able to propose likely structures for the two products. The PMR spectra will show clearly which product you have actually prepared in each case. You should pay particular attention to the number and multiplicity of signals produced by the methylene protons of the cyclopentanone ring. Comparing your spectra with the PMR spectra of the reactants in Figure 2 should help you identify the signals from particular proton sets.

Although the stereochemistry of the products has not been reported in the literature, most reactions of this type yield the E isomers, as in the condensation of benzaldehyde with 4,4-dimethyl-1-tetralone:

Mechanism of a Claisen-Schmidt condensation

$$CH_3\overset{O}{\overset{\|}{C}}CH_3 \xrightarrow{OH^-} \overset{\ominus}{CH_2}\overset{O}{\overset{\|}{C}}CH_3 \xrightarrow{PhCHO}$$

$$Ph\overset{O\ominus}{\overset{|}{C}}H-CH_2\overset{O}{\overset{\|}{C}}CH_3 \xrightarrow{H_2O}$$

$$Ph\overset{OH}{\overset{|}{C}}HCH_2\overset{O}{\overset{\|}{C}}CH_3 \xrightarrow{OH^-}$$

$$Ph\overset{OH}{\overset{|}{C}}H\overset{\ominus}{C}H\overset{O}{\overset{\|}{C}}CH_3 \xrightarrow{-OH^-}$$

$$PhCH=CH\overset{O}{\overset{\|}{C}}CH_3$$

Read the Methodology section in Experiment 19 for further information about phase-transfer catalysis.

4,4-dimethyl-
1-tetralone

E-2-benzal-4,4-dimethyl-
1-tetralone

The nearby carbonyl group has a deshielding effect on the vinylic proton of the *E* isomer, raising its chemical shift to 7.7 ppm compared to a value of 6.6 ppm in the *Z* isomer. The infrared spectra of these compounds also differ in that the C=C stretching band for the *E* isomer (at 1610 cm^{-1}) is nearly as strong as the carbonyl band, but for the *Z* isomer appears to be absent.

If you do Minilab 25, you will carry out a Claisen-Schmidt condensation of benzaldehyde and acetone to prepare dibenzalacetone (also called dibenzylideneacetone).

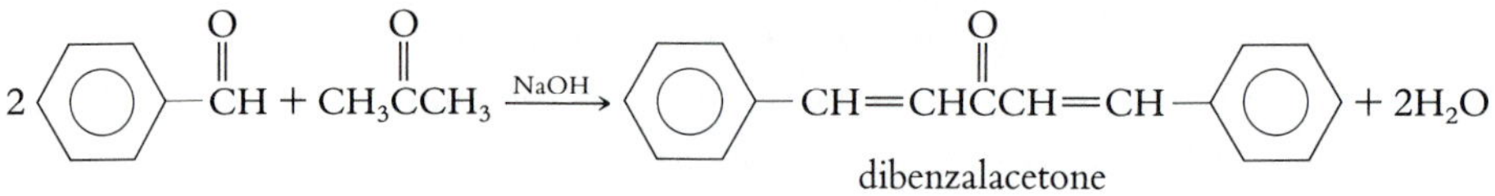

dibenzalacetone

PRELAB ASSIGNMENTS

1. Read OP–25*b*, and read or review the other operations as necessary. Read OP–26 if you have not previously carried out a vacuum distillation.

2. For part A, calculate the mass and volume of 50 mmol of cyclopentanone and of 50 mmol of furfural.

3. For part B, calculate the mass and volume of 10 mmol of cyclopentanone.

Reactions and Properties

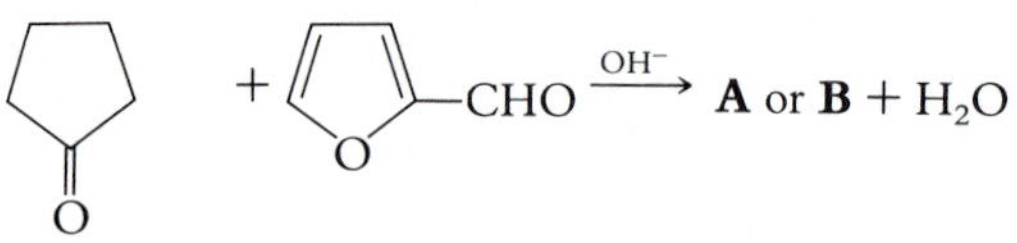

cyclopentanone   furfural   (equation not balanced)

Table 1.   Physical properties

	M.W.	m.p.	b.p.	d.
furfural	96.1	−39	162	1.159
cyclopentanone	84.1	−51	131	0.949
2-butanone	72.1	−86	80	0.805
tricaprylmethylammonium chloride	404.2			0.884
product **A**	162.2	60.5	154^{15}	
product **B**	240.3	162		

Figure 2.   PMR spectra of the starting materials. (Reproduced from *The Aldrich Library of NMR Spectra, Edition II,* by Charles J. Pouchert, with the permission of the Aldrich Chemical Company).

Cyclopentanone

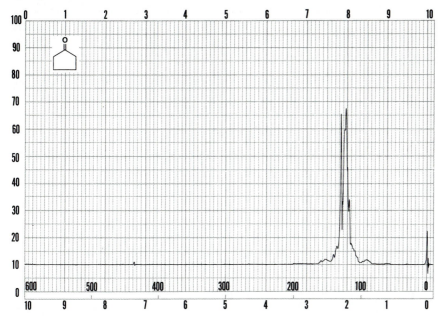

2-Furaldehyde (furfural)

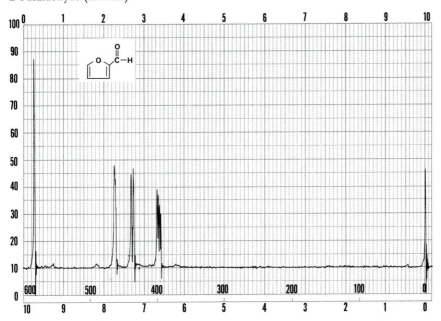

PROCEDURE

Furfural irritates the skin, eyes, and respiratory tract and may cause allergic skin or respiratory reactions. Wear gloves while handling furfural and the reaction mixture (which will stain your hands yellow), and do not breathe the vapors.

Cyclopentanone is flammable, and inhalation, ingestion, or skin absorption may be harmful. Avoid contact with the liquid, and do not breathe its vapors.

Ethyl ether is extremely flammable and may be harmful if inhaled. Do not breathe its vapors, and keep it away from flames and hot surfaces.

A. Claisen-Schmidt Condensation of Cyclopentanone and Furfural

Reaction. In a 125-mL Erlenmeyer flask, dissolve 50 mmol of cyclopentanone in 25 mL of ethyl ether **[fire and vapor hazard!]**, and add 45 mL of 0.1 M aqueous sodium hydroxide. Cool the mixture to 5°C in an ice bath ⟨OP–8⟩, and add 50 mmol of freshly distilled furfural **[contact and vapor hazard! wear gloves!]**. Cork the flask loosely, and stir the reaction mixture vigorously ⟨OP–9⟩ in a cold-water bath (10–15°C) ⟨OP–8⟩ for 45 minutes. Use a magnetic stirrer if available; otherwise, the flask should be shaken continuously **[wear gloves!]** with cooling throughout the reaction period.

If necessary, replace any ether that evaporates during the reaction period.

Separation. Filter the reaction mixture by vacuum filtration ⟨OP–12⟩, and thoroughly wash any solid on the filter with 10 mL of ethyl ether, combining the wash liquid with the filtrate. Save this solid and weigh it when it is dry. Shake the filtrate (aqueous and ether layers) in a separatory funnel to extract product from the aqueous layer, and separate the layers ⟨OP–13b⟩. Extract the aqueous layer ⟨OP–13⟩ with 15 mL of ethyl ether, and combine the ether extract with the initial ether layer. Wash this solution ⟨OP–19⟩ twice with saturated aqueous sodium chloride, dry it over magnesium sulfate ⟨OP–20⟩, and evaporate the solvent ⟨OP–14⟩.

If a solid continues to form in the reaction mixture after the initial filtration, remove it by filtration before you evaporate the solvent.

Purification and Analysis. Purify the residue by semimicro vacuum distillation ⟨OP–26a⟩ with *no* cooling bath, following the precautions described for distillation of solids ⟨OP–25a⟩.

Any unreacted starting materials and residual ether should distill below 100°C; this forerun should be disposed of as directed by your instructor. Cool the distillate (if necessary) until it completely solidifies; then purify it by recrystallization from ethanol-water ⟨OP–23*b*⟩, and dry it ⟨OP–21⟩. Measure the mass and melting point ⟨OP–28⟩ of product **A**. Record its PMR spectrum in deuterochloroform ⟨OP–34⟩, or obtain a spectrum from your instructor.

The low-boiling forerun, which includes unreacted starting materials, can be distilled using a steam bath.

B. Claisen-Schmidt Condensation Using a Phase-Transfer Catalyst

See part A for precautions regarding furfural, cyclopentanone, and ethyl ether.

2-Butanone is flammable, and ingestion, inhalation, or skin absorption may be harmful. Avoid contact, do not breathe its vapors, and keep it away from flames or hot surfaces.

2-butanone

Reaction. In a 125-mL Erlenmeyer flask, dissolve 10 mmol of cyclopentanone in 10 mL of ethyl ether **[fire and vapor hazard!]**. Then add 12 mL of 0.1 M aqueous sodium hydroxide and 6 drops of tricaprylmethylammonium chloride (Aliquat 336) or 0.2 g of another suitable phase-transfer catalyst. Cool the mixture to 5°C in an ice bath ⟨OP–8⟩, and add 2.0 mL of furfural **[contact and vapor hazard!]** with swirling. Stopper the flask and shake it vigorously ⟨OP–9⟩ for 15 minutes **[wear gloves!]**, swirling it frequently in the ice bath to prevent pressure buildup from vaporizing ether. Let the reaction mixture stand at room temperature, with occasional shaking, for 10 minutes or more.

Use a metal water or steam bath with the rings removed for the cooling bath.

Use a magnetic stirrer if available.

Separation. Collect the product by vacuum filtration ⟨OP–12⟩, and wash it on the filter with two portions of ethyl ether. Partially dry the crude product by blotting it between filter papers.

Purification and Analysis. Recrystallize product **B** ⟨OP–23⟩ from 2-butanone **[fire and vapor hazard!]**, using about 12 mL of the solvent per gram of crude product. Wash the product on the filter with ethyl ether, dry it ⟨OP–21⟩, and weigh it. Measure the melting point ⟨OP–28⟩ of product **B**.

Record its PMR spectrum in deuterochloroform ⟨OP – 34⟩, or obtain a spectrum from your instructor.

Report. Interpret the PMR spectra as completely as you can. Draw the structures of products **A** and **B**, name them, and cite all the evidence supporting your conclusions. Write balanced equations and mechanisms for the formation of the products. Discuss the effect of reaction conditions on the outcome of the Claisen-Schmidt reaction, and tell what conditions promote the formation of each product and why.

Collateral Projects

1. Students can carry out part B using different phase-transfer catalysts and compare their crude yields to find out which catalysts are most effective. Suggested catalysts are tetrabutylammonium bromide, tetrabutylphosphonium bromide, cetyltrimethylammonium bromide, and 1-hexadecyl-pyridinium chloride.

2. Record, compare, and interpret the infrared spectra of the two products.

MINILAB 25   Preparation of Dibenzalacetone

Dissolve 2.0 mL of benzaldehyde in 20 mL of 95% ethanol in a small Erlenmeyer flask, and stir in 20 mL of 0.5 M aqueous sodium hydroxide, followed by 0.25 mL of acetone (use a measuring pipet). Shake the mixture frequently over a 30-minute period; then scratch the sides of the flask to initiate crystallization (if necessary). When crystallization is complete, cool the flask in an ice bath and collect the product by vacuum filtration, washing it on the filter with cold water. Purify the dibenzalacetone by recrystallization from ethanol, and dry it. Weigh the dry product, and measure its melting point. Propose a mechanism for the reaction.

Exercises

1. **(a)** What is the probable identity of the solid that was filtered from the reaction mixture in part A? How could you have confirmed its identity? **(b)** What percentage of the cyclopentanone you started with in part A was converted to products **A** and **B**?

2. Diagram a reasonable phase-transfer process for the formation of product **B**, using the format illustrated in Experiment 19.

3. Construct a flow diagram for the synthesis in part A.

4. From your PMR spectra, is it more likely that your products are *Z* or *E* stereoisomers? What additional information would you need to be reasonably certain?

5. A student forgot to add the furfural in part A, but recovered a small amount of liquid that distilled at 139–142°C at 20 torr and did not solidify on cooling. The PMR spectrum of the liquid showed no signals from vinylic or hydroxylic protons. Propose a structure for this product and write a mechanism for its formation.

6. **(a)** Determine the limiting reactant in Minilab 25, and calculate the theoretical yield of dibenzalacetone. If you completed the Minilab, calculate your percentage yield. **(b)** Would a phase-transfer catalyst speed up this reaction? Why or why not? **(c)** How might the reaction conditions be altered to prepare benzalacetone instead?

Library Topic

Write a report on the production and uses of furan and furfural, and some of their derivatives.

The Electronic Effect of a *para*-Iodo Substituent

Reactions of Aromatic Amines. Preparation of Aryl Halides. Aromatic Nucleophilic Substitution. Carboxylic Acids. Diazonium Salts. Linear Free-Energy Relationships.

Operations

OP–6   Measuring Volume

OP–8   Cooling

OP–9   Mixing

OP–12   Vacuum Filtration

OP–21   Drying Solids

OP–23   Recrystallization

Objectives

To prepare *p*-iodobenzoic acid from *p*-aminobenzoic acid via a diazonium salt.

To measure the acid ionization constant of *p*-iodobenzoic acid.

To determine the sigma value for the *para*-iodo substituent and the rho value for the ionization of benzoic acids in 95% ethanol.

To find out whether a *para*-iodo substituent acts as an electron-withdrawing or an electron-donating substituent on a benzene ring.

To learn about linear free-energy relationships.

SITUATION

In general chemistry, you learned that the electrons in a covalent bond tend to migrate toward the more electronegative atom, building up its electron density at the expense of the less electronegative atom. The situation is not always so simple in conjugated organic compounds, where the electrons of unshared pairs or in pi bonds may migrate *away* from the more electronegative atom into the "electron sink" of a delocalized pi system. For example, as illustrated for anisole (methoxybenzene), lone-pair electrons from an oxygen-containing substituent such as methoxyl (—OCH₃) tend to migrate toward and overlap with the pi-electron cloud of the benzene ring. Iodine is less electronegative than oxygen and an iodo substituent (—I) has three lone pairs of electrons to share, so it might also act as an electron donor to a benzene ring if there is sufficient orbital overlap. The electronic effect of a substituent can be measured by a parameter called its *sigma value;* electron-withdrawing groups have positive sigma values, and electron-donating groups have negative ones. In this experiment, you will measure the sigma value of the *para*-iodo substituent to find out whether or not it donates electrons to a benzene ring.

Overlap of lone-pair electrons with pi system in anisole

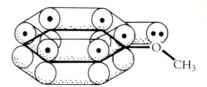

BACKGROUND

Linear Free-Energy Relationships

Consider a general substituent Z at the *para* position of benzoic acid. If Z is more electronegative than carbon, it will tend to withdraw electrons through the sigma-bond framework of the benzene ring by an *inductive effect,* and this should increase the strength of the acid by stabilizing its conjugate base. Conversely, a substituent that is more electropositive than carbon can donate electrons inductively and weaken an acid. But inductive effects decrease rapidly with distance, and because in this example the *para* substituent is remote from the reaction site (the COOH group), its inductive effect may be quite small.

 Substituents that have unpaired electrons and are adjacent to a conjugated system tend to donate electrons by a *resonance effect.* Electron donation by resonance can cause a buildup of electron density at locations within the ring and on conjugated substituents, as illustrated by the following resonance structures for the conjugate base of a substituted benzoic acid:

Z—⟨◯⟩—COOH

stabilized by
electron withdrawal
through sigma bonds

Z—⟨◯⟩—COOH ⇌ H⁺ +

$$\left[\ddot{Z} - \left\langle \bigcirc \right\rangle - C \diagdown_{O^{\ominus}}^{O} \longleftrightarrow {}^{\oplus}Z = \left\langle \bigcirc \right\rangle^{\ominus} - C \diagdown_{O^{\ominus}}^{O} \longleftrightarrow {}^{\oplus}Z = \left\langle \bigcirc \right\rangle = C \diagdown_{O^{\ominus}}^{O^{\ominus}} \right] \text{etc.}$$

In this example, the conjugate base is destabilized by the concentration of negative charge near the reaction site; destabilization of the conjugate base shifts the ionization equilibrium to the left and weakens the acid. Resonance effects are transmitted freely throughout a conjugated system and do not decrease significantly with distance. Substituents such as the methoxyl group can donate electrons by resonance but withdraw them inductively, and thus may either weaken or strengthen an acid depending on the relative importance of the two effects. Thus, a *meta*-methoxyl substituent increases the acidity of benzoic acid by its inductive effect, but a *para*-methoxyl substituent decreases the acidity by resonance.

For example, the pK_a of phosphorous acid, which has a K_a value of 1.0×10^{-2}, is 2.00.

The pK_a of an acid is defined as the negative logarithm of its ionization constant:

$$pK_a = -\log K_a$$

At a given temperature, the pK value for any equilibrium reaction is directly proportional to its standard free-energy change, ΔG^0. This can be shown by using the above definition to rewrite the thermodynamic equation that relates ΔG^0 to the equilibrium constant:

$$\Delta G^0 = -(2.303 \cdot RT) \cdot \log K = (2.303 \cdot RT) \cdot pK$$

The value of ΔG^0 for a reaction is a measure of its tendency to go to completion, so pK must measure the same tendency. Negative pK_a and ΔG^0 values are observed for strong acids that dissociate completely, or nearly so. Most organic acids have positive pK_a values because they are only partly dissociated in ionizing solvents; a high positive pK_a value indicates a weak acid and a low one a stronger acid.

Suppose we compare the pK_a values for a pair of acids that differ in only one respect — that one acid has a substituent at a site where the other has only a hydrogen. Since each acid's pK_a is a measure of its acid strength, the *difference* between their pK_a values must measure the substituent's effect on acid strength. For example, pK_a for the ionization of benzoic acid in water at 25°C is 4.19, whereas pK_a for *p*-nitrobenzoic acid is 3.41. The difference between these pK_a values, or $4.19 - 3.41 = 0.78$, is a quantitative measure of the *para*-nitro substituent's acid-strengthening effect. Similarly, *p*-toluic acid (p-CH_3—C_6H_4COOH) has a pK_a value of 4.36; subtracting this value from the pK_a for benzoic acid gives a difference of

−0.17, indicating that the methyl substituent has a small acid-weakening effect. The difference between the pK_a value for the unsubstituted acid and that for the substituted acid, which we shall call ΔpK, is a measure of the substituent's ability to donate or withdraw electrons.

$$\Delta pK = pK_a \text{ of unsubstituted acid} - pK_a \text{ of substituted acid} \quad \textbf{(1)}$$

Because substituents that give negative ΔpK values weaken acids, they must be electron donors, whereas those that give positive ΔpK values are electron acceptors. The numerical value of ΔpK in either case measures the magnitude of the substituent's effect.

Don't confuse a negative ΔpK value, which is associated with an acid-weakening substituent, with a negative pK_a value, which indicates a strong acid. The sign of ΔpK results from the way this parameter is defined in Equation 1.

The ionization of benzoic acid in water at 25°C has been chosen as the reference reaction for measuring the effects of substituents. The ΔpK values for the benzoic acid system are called *substituent constants* and are symbolized by the Greek letter sigma (σ):

$$\sigma = \Delta pK(bz) = pK_a(bz) - pK_a(Zbz) \quad \textbf{(2)}$$
$$(bz = \text{benzoic acid; Zbz} = \text{substituted benzoic acid})$$

Table 1 (on page 342) lists the reported sigma values for some common *para* substituents.

The effect of a substituent will usually change if the substrate or the reaction type is changed, or even if the reaction conditions are varied. For example, the pK_a values for phenylacetic acid and *p*-nitrophenylacetic acid in water are 4.28 and 3.85, respectively; thus, ΔpK for the *para*-nitro group on this substrate is 0.43, which is considerably less than its ΔpK of 0.78 in the benzoic acid system. We can use the ratio of these ΔpK values to estimate the relative effect of a substituent on the ionization of phenylacetic acid, keeping in mind that σ is ΔpK(benzoic acid):

p-nitrobenzoic acid

p-nitrophenylacetic acid

$$\frac{\Delta pK(\text{phenylacetic acid})}{\sigma} = \frac{0.43}{0.78} = 0.55$$

In other words, the nitro group has only about half as much acid-strengthening effect on phenylacetic acid as it does on benzoic acid, undoubtedly because it is farther from the reaction site in phenylacetic acid. Comparing the effects of a large number of substituents on the same reaction yields an "averaged" $\Delta pK/\sigma$ value of 0.49 for the ionization of phenylacetic acid. We call this value the *reaction constant,* which is symbolized by the Greek letter rho (ρ). Rearranging the equality $\rho = \Delta pK / \sigma$ leads us to the *Hammett equation,*

$$\Delta pK = \rho\sigma \qquad (3)$$

which can also be written as

$$pK_a \text{ (unsubstituted acid)} - pK_a \text{ (substituted acid)} = \log \frac{K_a \text{ (substituted acid)}}{K_a \text{ (unsubstituted acid)}} = \rho\sigma$$

According to the Hammett equation, the effect that any substituent (Z) will have on the pK of a particular reaction can be estimated by multiplying the sigma value for Z (a parameter that measures the electronic effect of Z on a reference reaction) by the rho value for the reaction in question (a parameter that measures the sensitivity of the reaction to the electronic effects of substituents). Because of the proportionality between pK and ΔG^0 mentioned previously, the Hammett equation is called a *linear free-energy relationship.*

By definition, the reaction constant, or rho value, for the ionization of benzoic acids in water is 1; reactions that are more sensitive to substituent effects than the reference reaction have rho values greater than 1, and less sensitive reactions have rho values between 0 and 1. Negative rho values are observed for reactions in which electron donors increase the extent of reaction and electron acceptors decrease it, rather than the other way around. The rho value for a particular reaction may change if the solvent or other reaction conditions are changed. Just as a car's power is needed more on a steep hill than on the level, substituents have the greatest effect when a reaction is most difficult and the least when it is easiest.

METHODOLOGY

In this experiment, you will first prepare *p*-iodobenzoic acid from *p*-aminobenzoic acid (PABA). By measuring the pK_a value of *p*-iodobenzoic acid, you can then determine the sigma value for the *para*-iodo substituent. An aryl iodide can be prepared by treating the corresponding aromatic amine with nitrous acid in the presence of a mineral acid, then heating the resulting diazonium salt with aqueous potassium iodide.

$$ArNH_2 + HONO + H^+ \longrightarrow ArN_2^+ + 2H_2O$$
$$ArN_2^+ + I^- \longrightarrow ArI + N_2$$

Most displacement reactions of diazonium salts appear to be S_N1 reactions in which the diazonium salt loses nitrogen (as N_2) to form an intermediate aryl carbocation (Ar^+). A nucleophilic species then combines with the carbocation to form the product — or another intermediate that yields the product after a proton exchange.

The diazonium salt of a primary aromatic amine can be prepared by dissolving or suspending the amine in dilute hydrochloric or sulfuric acid, cooling the solution to 5°C or below, and adding aqueous sodium nitrite. Sodium nitrite reacts with some of the acid to generate nitrous acid according to the equation

$$NaNO_2 + H^+ \longrightarrow HONO + Na^+$$

An excess of nitrous acid can bring about undesirable side reactions, so it is important to weigh the amine and sodium nitrite accurately. The solution must be kept cold until it is added to the potassium iodide solution, because diazonium salts can react with the solvent at elevated temperatures. *p*-Iodobenzoic acid separates as a fine precipitate that may be difficult to filter; using a reasonably large Buchner funnel and an efficient aspirator will help speed up the process.

You will determine the pK_a of the *p*-iodobenzoic acid in 95% ethanol by measuring the pH of a solution that contains roughly equimolar amounts of the acid and its conjugate base. (For an alternative method, see Collateral Project 1.) The solution will be prepared by adding 1 mL of ~0.5 M aqueous sodium hydroxide to a solution containing about 1 mmol of *p*-iodobenzoic acid dissolved in 19 mL of absolute (100%) ethanol. That much NaOH should convert about half of the *p*-iodobenzoic acid to its conjugate base according to the following reaction equation, leaving the remaining half unreacted:

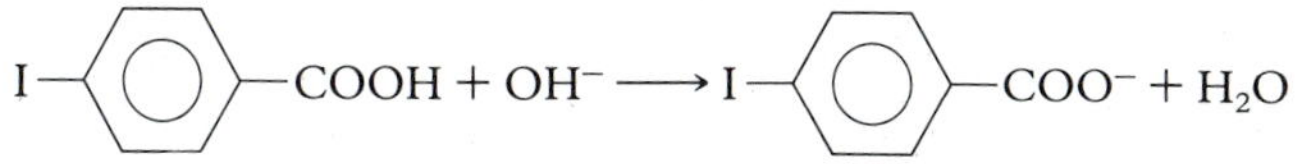

From the molar amounts of acid and conjugate base and the pH of the solution, you can calculate the pK_a of the acid using Equation 4 (where n_{HA} = moles of acid and n_A = moles of conjugate base):

$$pK_a = pH + \log \frac{n_{HA}}{n_{A^-}} \qquad (4)$$

Reactions of diazonium salts with CuCl and some other reagents are believed to involve free-radical intermediates.

The attacking species in the iodination reaction may be I_3^- (triiodide) rather than I^-, or a mixture of both ions.

For example, combining 1 mL of 0.490 M NaOH with 1.005 mmol of acid would produce 0.490 mmol of the conjugate base and leave 0.515 mmol of unreacted acid.

COOH

Z

Table 1.   Molecular weights and sigma values for *para*-substituted benzoic acids

Acid	M.W.	Substituent (Z)	Sigma Value
p-aminobenzoic acid	137.1	$-NH_2$	−0.66
p-hydroxybenzoic acid	138.1	$-OH$	−0.37
p-anisic acid	152.2	$-OCH_3$	−0.27
p-toluic acid	136.2	$-CH_3$	−0.17
benzoic acid	122.1	none	0.00
p-chlorobenzoic acid	156.6	$-Cl$	0.23
terephthalic acid	166.1	$-COOH$	0.45
p-nitrobenzoic acid	167.1	$-NO_2$	0.78

By the same procedure, you and your co-workers will measure pK_a values for most or all of the acids listed in Table 1, from which you can calculate their ΔpK values. Graphing the ΔpK values for the acids against the sigma values for the substituents yields a *Hammett plot* (see Equation 3) from which you should be able to determine the sigma value for the *para*-iodo substituent and the rho value for the ionization of benzoic acids in 95% ethanol.

Minilab 26 describes an unusual synthesis of benzoic acid. When benzaldehyde is mixed with sand and a pipe cleaner or two is stuck into the mixture, the liquid climbs up the pipe cleaner by capillary action, where it slowly oxidizes to benzoic acid. You may wish to exercise your creativity by twisting some pipe cleaners together to make a benzoic acid "tree" or some other object.

PRELAB ASSIGNMENTS

1.   Calculate the mass of 10 mmol of *p*-aminobenzoic acid, of 10 mmol of sodium nitrite, and of 15 mmol of potassium iodide. Calculate the theoretical yield of *p*-iodobenzoic acid.

2.   Be prepared to calculate the mass of 1 mmol of any acid listed in Table 1.

Reactions and Properties

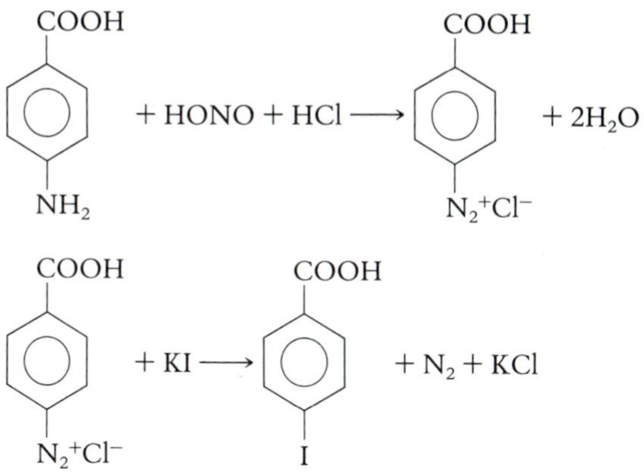

COOH + HONO + HCl ⟶ COOH + 2H$_2$O
NH$_2$ N$_2^+$Cl$^-$

COOH + KI ⟶ COOH + N$_2$ + KCl
N$_2^+$Cl$^-$ I

Table 2.   Physical properties

	M.W.	m.p.
p-aminobenzoic acid	137.1	189
p-iodobenzoic acid	248.0	270
potassium iodide	166.0	681
sodium nitrite	69.0	271

PROCEDURE

A. Preparation of p-Iodobenzoic Acid

**SAFETY
PRECAUTIONS**

benzoic acid

Reaction.   Mix 10 mmol of *p*-aminobenzoic acid (PABA) with 10 mL of 3 M hydrochloric acid in a small beaker; then cool the suspension in an ice bath ⟨OP–8⟩ until its temperature falls below 5°C. Dissolve 10 mmol of sodium nitrite in 10 mL of deionized water in another beaker, and cool it to 5°C or below in the ice bath. Add the sodium nitrite solution to the PABA mixture in small portions, with stirring ⟨OP–9⟩. Carry out the addition slowly enough so that the temperature of the reaction mixture does not rise above 10°C. In a 400-mL (or larger) beaker, dissolve 15 mmol of potassium iodide in 100 mL of water. Pour the diazonium salt solution, with stirring, into the potassium iodide solution. Heat the mixture on a steam bath **[foaming may occur!]** for 15–20 minutes, stirring occasionally; then cool it to room temperature or below.

When the reaction is complete, the foam should subside.

Separation and Purification.   Separate the precipitated *p*-iodobenzoic acid from the reaction mixture by vacuum filtration ⟨OP–12⟩, washing it on the filter with cold water. Purify the product by recrystallization ⟨OP–23⟩ from 95% ethanol. Dry the product thoroughly ⟨OP–21⟩, and weigh it. (At your instructor's request, measure its melting point.)

Use a thermometer with a range extending to 280°C or higher for determining the melting point.

B. Measurement of pK_a Values

Each member of a team of two to four students should carry out pH measurements on his or her *p*-iodobenzoic acid and two or more acids from Table 1. Alternatively, the instructor may assign one or more acids from Table 1 to each student and make all the data obtained available to everyone.

Accurately measure 19.00 mL of *absolute* ethanol from a buret or graduated pipet into a clean, dry 50-mL beaker, and cover the beaker with a watch glass. Weigh about 1 mmol of the dry *p*-iodobenzoic acid to the nearest milligram, and dissolve it in the ethanol. Using a 1-mL volumetric pipet ⟨OP–6⟩, add exactly 1 mL of standardized aqueous 0.5 M sodium hydroxide to this solution and mix thoroughly. From its label, record the concentration of the NaOH solution in your lab notebook. Measure the pH of the mixture with a pH meter that has been standardized against a pH 7 buffer solution. Repeat the pH measurement with two or more of the carboxylic acids listed in Table 1. Use your data to calculate pK_a values for *p*-iodobenzoic acid and the other acids you worked with. Re-

If necessary, your instructor will show you how to operate the pH meter. Your group should measure pH values for at least six of the acids, including benzoic acid.

port your pK_a values for the acids from Table 1 to your co-workers (or instructor), and record the values for the remaining acids.

It is a good idea to obtain the raw data (pH values, etc.) from your partners so that you can verify their calculations.

Report. Calculate ΔpK values for all the acids, and construct a Hammett plot. From your plot, determine the sigma value for the *para*-iodo substituent and the rho value for the reaction in 95% ethanol. Describe the electronic effect of the *para*-iodo substituent, and tell whether an inductive or a resonance effect can best account for your results. Tell whether the ionization of benzoic acid in ethanol is more or less sensitive to substituent effects than its ionization in water. Explain how you arrived at your conclusions in each case.

Collateral Projects

1.   As an alternative to the method described in part B, you can measure the pK_a of *p*-iodobenzoic acid (or the other acids) by carrying out a pH titration. Dissolve approximately 1 mmol of the acid in 20 mL of 95% ethanol. Titrate this solution with ~0.05 M KOH in 95% ethanol, adding about 1 mL at a time and measuring the pH after each addition. When the pH begins to rise rapidly, add the KOH in 0.5-mL increments and continue to titrate until the pH begins to level off. Plot pH versus volume of base; mark the inflection point of the curve, and determine the volume of base at that point. The pH reading at exactly half that volume (the half-neutralization point) should equal the pK_a of the acid. Note that you do not have to know the exact mass of pure acid or the exact base concentration with this method.

The inflection point of a curve is the point where it changes direction; in this case, it is the point where the slope of the curve reaches its maximum and begins to decrease.

2.   Record, compare, and interpret the infrared spectra (KBr pellet or Nujol mull) of *p*-aminobenzoic acid and your product.

3.   Prepare *p*-iodonitrobenzene from *p*-nitroaniline or *o*-iodobenzoic acid from anthranilic acid by the procedure described in part A. The first product can be recrystallized from ethanol and the second from water.

Synthesis of Benzoic Acid from Benzaldehyde

You may also be able to recover some benzoic acid from the sand by boiling the hard crust in water and filtering by gravity.

Put about 15 mL of clean dry sand into a small beaker. Stir in just enough benzaldehyde to moisten all the sand — there should be no standing liquid on top of the sand. Push one or two pipe cleaners into the sand so that one end of each touches the bottom of the beaker and the other is exposed to the air. After a week or so, scrape off the white solid and purify it by recrystallization from boiling water. Dry the benzoic acid, measure its melting point, and write an equation for the reaction.

Exercises

1.   Compare the rho value for the ionization of benzoic acid in 95% ethanol with that in water and try to explain the difference, based on the ionizing power of each solvent.

2.   Derive Equation 4 in the Methodology section from the equilibrium constant expression for the ionization of an acid, HA.

3.   Construct a flow diagram for the preparation of *p*-iodobenzoic acid.

4.   **(a)** A careless student heated the diazonium salt solution to boiling before adding potassium iodide and isolated a white solid with a melting point of 215°C. What product did he obtain instead of *p*-iodobenzoic acid? **(b)** Propose a mechanism for the reaction that formed it.

5.   Using resonance structures, explain why *m*-methoxybenzoic acid is more acidic than benzoic acid, whereas *p*-methoxybenzoic acid is less acidic than benzoic acid.

6.   Predict the acid ionization constant (K_a) of *p*-trifluoromethylbenzoic acid in water and in 95% ethanol, given that a *para*-trifluoromethyl substituent ($-CF_3$) has a sigma value of 0.54.

7.   **(a)** Would you expect the rho value for the ionization of substituted phenols (Z—C₆H₄OH) in water to be
greater than or less than 1? Explain. **(b)** Give examples of
reactions that you would expect to have negative rho values.

Library Topic

Write a report describing some natural sources, physiological functions, and commercial uses of *p*-aminobenzoic acid
(PABA).

Shifting the Equilibrium in an Esterification Reaction

Reactions of Carboxylic Acids. Preparation of Esters. Nucleophilic Acyl Substitution.

Operations

OP–6   Measuring Volume

OP–7a Refluxing

OP–25 Simple Distillation

Objectives

To carry out the Fischer esterification of acetic acid with ethanol.

To measure the equilibrium constant for the esterification reaction.

To observe the effect on the yield of ethyl acetate of varying the ratio of reactants and removing the products.

To learn about some methods for shifting a chemical equilibrium.

SITUATION

Ethyl acetate is an important commercial solvent used in the manufacture of lacquers, varnishes, smokeless powder, perfumes, and artificial fruit flavorings. It is usually manufactured by a Fischer esterification process, in which acetic acid and ethanol are heated in the presence of an acidic catalyst. Such esterification reactions are reversible, so the yield of ethyl acetate is low unless the equilibrium is shifted to favor the products. In a typical batch-process industrial esterification, the reactants are heated together until the reaction comes to equilibrium, after which the ester is distilled from the reaction vessel. Because the mole ratio of the reactants determines how much product will be present at equilibrium, this ratio is carefully controlled to bring the reaction to a predetermined degree of completion before the distillation. In this experiment, you and your co-workers will carry out an esterification reaction using different mole ratios of ethanol to acetic acid. Your primary goals are to determine the mole ratio needed to bring the reaction to 85% completion at equilibrium and to find out if the reaction can be brought to 100% completion — or nearly so — by distilling the products from the equilibrium mixture.

BACKGROUND

How Fast and How Far?

When a certain reaction is being considered for use in a chemical process, investigators must provide answers for two vital questions: *will it go,* and *how long will it take?* The first is a question of thermodynamics (reaction equilibria) and the second of kinetics (reaction rates). If the first question can be answered affirmatively, it may then be necessary to find out how *far* the reaction will go to completion — that is, to measure its equilibrium constant. This is usually done by monitoring the concentrations of reactants or products while the reaction proceeds. When they no longer change with time, the reaction has reached equilibrium. The equilibrium concentration of one or more species can then be used to determine the extent of the reaction, from which the equilibrium constant can be calculated.

The value of the equilibrium constant is important because it provides a measure of how far a reaction will proceed under standard conditions. Table 1 illustrates this effect for a reaction of the type $A + B \rightleftharpoons C + D$, showing the extent of reaction for various K values. For example, the maximum attainable yield from a reaction with $K = 2.25$ is 60% if equimolar concentrations of reactants are used and no attempt is made to shift the equilibrium. Since 60% yields are not acceptable in most industrial preparations, it is necessary to shift the equilibrium in favor of the products by using an excess of one reactant or by removing one of the products, or both. Many ingenious ways of removing a product have been devised. Water can be removed by distillation — with or without an azeotroping agent — or by use of a drying agent such as sulfuric acid or alumina. Some esters that boil at a lower temperature than their component alcohols can be removed by distillation; others can be removed by extracting them into a water-immiscible solvent such as ethylene chloride during the reaction.

The actual thermodynamic expression for the equilibrium constant involves activities rather than concentrations, but the latter can be used as a first approximation.

Table 1.   Extent of reaction as a function of K

K	Max. % Yield
1.00	50
2.25	60
5.44	70
16.0	80
81.0	90
361	95
9800	99
1.0×10^6	99.9

Note: Assume a reaction of the type $A + B \rightleftharpoons C + D$, starting with equimolar quantities of reactants.

Esterification provides some excellent examples of the principles of reaction equilibria and kinetics. The K values of many esterification reactions are close to 1, so the effect of structure on equilibria can be observed without difficulty. Reaction rates also vary considerably in response to changes in the structures of both the carboxylic and the hydroxylic reactant. Highly branched acids and alcohols react slowly because of steric interference in the transition states leading to products. Electronic effects can be observed by comparing the effects of different electron-withdrawing and electron-donating substituents on the reaction rate.

METHODOLOGY

The reaction of acetic acid and ethanol to form ethyl acetate and water is a typical esterification reaction that can be catalyzed by a strong acid such as sulfuric acid. The role of the catalyst is apparent from the reaction mechanism shown; protonation of the carbonyl group of the acid makes it more susceptible to nucleophilic attack by the alcohol and thus increases the reaction rate.

Esterification reactions involving aliphatic reactants generally have equilibrium constants ranging between 1 and 10. If equimolar quantities of reactants are used, such reactions will be only about 50–75% complete at equilibrium. According to Le Châtelier's principle, an equilibrium can be shifted to favor the products by adding more of a reactant or by removing one or more of the products. You and your co-workers will use both approaches: varying the amount of ethanol in the reaction mixture, and distilling ethyl acetate and water from the mixture at equilibrium.

You will monitor the progress of the reaction by withdrawing aliquots from the reaction mixture at intervals and titrating them with aqueous sodium hydroxide. As the reaction proceeds, the amount of acetic acid in the reaction mixture will decrease, until, at equilibrium, it should remain constant. The volume of titrant (aqueous NaOH) necessary to neutralize the acetic acid at any stage of the reaction can be used to determine the reaction's degree of completion at that point. The degree of completion (α) at any time t is simply the amount of ester present at that time divided by the theoretical yield of ester; thus an α value of 0.60 means that 60% of the ester that *could* be formed has been formed at that time.

Acetic acid is the limiting reactant, so the theoretical yield of ester in moles is equal to the initial number of moles of acetic

Esterification mechanism

acid, which can be calculated by multiplying the volume of aqueous NaOH consumed in the initial titration, V_0, by the molar concentration of the NaOH solution, M_b.

$$\text{moles of ester at 100\% completion} = M_b V_0 \qquad (1)$$

Because every mole of acetic acid that reacts forms a mole of ester, the amount of ester (in moles) present in the reaction mixture at time t is equal to the amount of acetic acid that has reacted by that time. This in turn is equal to the difference between the volume of aqueous NaOH consumed by acetic acid before the reaction begins and the volume consumed at time t, multiplied by the base concentration. Because some base also will be consumed by the sulfuric acid catalyst, you will measure this difference by titrating the reaction mixture just after the catalyst has been added (but before the reaction begins) and again at time t; since the same concentration of sulfuric acid is present at both times, its effect will cancel out. Letting V_1 represent the volume of titrant just after adding the catalyst and V_t represent the volume of titrant at time t yields the following equation:

$$\text{moles of ester at time } t = M_b(V_1 - V_t) \qquad (2)$$

The degree of completion of the reaction at time t is the ratio between the amount of ester at time t and the amount at 100% completion and is obtained by dividing Equation 2 by Equation 1:

$$\alpha = \frac{\text{moles of ester at } t}{\text{moles of ester at completion}} = \frac{V_1 - V_t}{V_0} \qquad (3)$$

The equilibrium constant of the reaction (K_e) can be calculated from its degree of completion at equilibrium, α_e, as shown in Equation 4, where n is the initial mole ratio of ethanol to acetic acid.

The α_e value is calculated by using the equilibrium titrant volume, V_e, in Equation 3.

$$K_e = \frac{[\text{EtOAc}][\text{H}_2\text{O}]}{[\text{HOAc}][\text{EtOH}]} = \frac{(\alpha_e)^2}{(1 - \alpha_e)(n - \alpha_e)} \qquad (4)$$

$$(\text{Et} = \text{CH}_3\text{CH}_2\!\!-\!\!; \ \text{Ac} = \text{CH}_3\text{CO}\!\!-\!\!)$$

You will use your experimental results and those of your co-workers to estimate the mole ratio of reactants that should bring the esterification reaction to 85% completion at equilibrium ($\alpha_e = 0.85$). The equilibrium constant calculated using Equation 4 could be used for this estimate if it were truly constant; unfortunately, K_e values based on molar concentrations are constant only in dilute solutions, a condition seldom realized in organic syntheses. By comparing your equilibrium "constant" with those of your co-workers, you will be able to observe how K_e varies with reactant ratios in esterification reactions.

Le Châtelier's principle tells us that removing one or more products from a reaction at equilibrium should shift the equilibrium to favor the formation of more product. You will see whether this holds true for the esterification reaction by distilling ethyl acetate and water from the reaction mixture after it reaches equilibrium. Ethyl acetate forms a ternary azeotrope with water and ethanol (83% ethyl acetate, 9% water, 8% ethanol) that boils at 70.2°C, so the two products should distill at a temperature substantially below their normal boiling points. After all the ethyl acetate has distilled, the temperature should rise to about 78°C as a binary azeotrope of ethanol and water (95.5% ethanol) begins to come over. The amount of unreacted acetic acid remaining in the product mixture can then be determined by titration as before.

An ester can be converted to its component alcohol and acid by a hydrolysis reaction that is essentially the reverse of the esterification reaction that forms it. The pH of the reaction mixture decreases as the acidic component is formed, providing a means of monitoring the reaction rate. If you do Minilab 27, you will compare the hydrolysis rates of several esters by monitoring the pH values of their aqueous solutions as a function of time.

PRELAB ASSIGNMENT

Calculate the mole ratios of ethanol to acetic acid for 10, 15, 20, 25, and 30 mL of ethanol (assume these volumes are accurate to three significant digits).

Reaction and Properties

$$CH_3\overset{\overset{\displaystyle O}{\displaystyle \|}}{C}OH + CH_3CH_2OH \underset{}{\overset{H_2SO_4}{\rightleftharpoons}} CH_3\overset{\overset{\displaystyle O}{\displaystyle \|}}{C}OCH_2CH_3 + H_2O$$

Table 2. Physical properties

	M.W.	m.p.	b.p.	d.
acetic acid	60.1	17	118	1.049
ethanol	46.1	−117	78.5	0.789
ethyl acetate	88.1	−84	77	0.900
sulfuric acid (98%)	98.1	10	338	1.84

PROCEDURE

Your instructor will assign you one of the following quantities of ethanol to work with: 10, 15, 20, 25, *or* 30 mL. You should measure the ethanol accurately from a buret. If you are assigned 25 or 30 mL, use 2-mL aliquots for the titrations; otherwise, use 1-mL aliquots.

SAFETY PRECAUTIONS

Acetic acid can cause serious damage to skin and eyes; its vapors are highly irritating to the eyes, mucous membranes, and upper respiratory tract. Wear gloves, and dispense under a hood; avoid contact, and do not breathe its vapors.

Ethanol is very flammable, so keep it away from flames and hot surfaces.

Sulfuric acid is very corrosive and can cause very serious damage to the skin and eyes; the concentrated acid reacts violently with water. Wear gloves, avoid contact, and keep the acid away from water.

Ethyl acetate is flammable, and inhalation, ingestion, or skin absorption may be harmful. Avoid contact, do not breathe its vapors, and keep it away from flames.

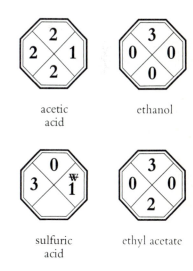

acetic acid

ethanol

sulfuric acid

ethyl acetate

A. Attaining Equilibrium

Clean and number two or more 125-mL Erlenmeyer flasks, and add 25 mL of distilled or deionized water to each. (You will need at least four flasks, but the first two can be cleaned and reused if necessary.) Assemble the reaction apparatus by inserting a Claisen connecting tube into a 50-mL or 100-mL boiling flask, stoppering the straight arm, and inserting a reflux condenser into the bent arm. Under the hood, accurately

Use a 100-mL flask if you started with 25 or 30 mL of ethanol. Clamp the apparatus securely so that you can remove the reaction flask.

measure 5 mL of glacial acetic acid **[contact and vapor hazard!]** into the reaction flask, and add your assigned volume of absolute ethanol. Mix the reactants thoroughly. Use a 1-mL or 2-mL volumetric pipet to transfer an aliquot ⟨OP–6⟩ into the first Erlenmeyer flask **[do not pipet by mouth!]**. Cool the reactants in ice. Then add 3 drops of concentrated sulfuric acid **[contact hazard!]**, swirl to mix thoroughly, and *immediately* pipet an aliquot into the second flask. Swirl both flasks to mix the contents, stopper them, and set them aside.

Use acid-resistant boiling chips.

Replace the reaction flask on the apparatus, and reflux the reaction mixture ⟨OP–7a⟩ gently for 30 minutes or more. During the reflux period, add 2 drops of phenolphthalein solution to the first Erlenmeyer flask, and titrate its contents with 0.25 M aqueous sodium hydroxide until the mixture reaches the pale pink endpoint, to get V_0. Titrate the sample in the second flask by the same procedure, to get V_1. After the reaction mixture has refluxed for about 30 minutes, remove the heat source and let it cool for a few minutes. Then withdraw an aliquot through the straight arm of the Claisen head, and transfer it to another flask containing 25 mL of water. Resume the reflux, and titrate the solution as before. Collect and titrate additional aliquots from the boiling flask every 10 minutes, letting the reactants cool before you collect each aliquot, until the volume of titrant no longer decreases significantly between two successive titrations. Record the final titrant volume as V_e. Calculate α_e for the reaction. Report this value and your mole ratio to your co-workers, and obtain their values for the same parameters.

Don't forget to add the phenolphthalein!

If the titrant volume increases between two titrations, repeat the titration.

B. Shifting the Equilibrium

Assemble an apparatus for simple distillation ⟨OP–25⟩, using your reaction flask as the boiling flask and a graduated cylinder as the receiver. Slowly distill the reaction mixture until the temperature rises to 78°C *or* until you have collected 10 mL of liquid, whichever comes first. Cool the boiling flask to room temperature or below. Add the distillate to it, mix thoroughly, and titrate an aliquot of this solution as before.

Report. Using your group's results, construct a graph of α_e versus n (the plot will not necessarily be a straight line), and estimate the mole ratio of reactants that should result in 85% completion. Calculate the degree of completion after the distillation and discuss its effect on the yield. Calculate the equilibrium constant for each mole ratio and describe the effect of the reactant ratio on the product yield and the value of K_e.

Collateral Project

You can measure equilibrium constants for other aliphatic esters (such as methyl acetate, isopropyl acetate, and butyl acetate) by the same general procedure, using the appropriate alcohols. Some of the esters may form a separate layer during the reaction, in which case aliquots should be withdrawn from the lower aqueous layer.

Hydrolysis Rates of Esters   MINILAB 27

Place 2 mL of 50% aqueous ethanol, 2 drops of a universal indicator (see Table 3) and 2 drops of 1 M aqueous sodium hydroxide in each of four numbered test tubes. Stopper and shake the tubes. Add 5 drops of one of the following esters to each of the four test tubes: ethyl acetate, ethyl benzoate, ethyl formate, ethyl butyrate. Shake each test tube until the ester has dissolved, and record the color of each solution. Immerse the test tubes in a 50°C water bath, and keep them at 50°C for 30 minutes or more, recording the colors every 5 minutes or so. Estimate the pH of each solution at each time, tabulate your results, and arrange the esters in order of reactivity. Write balanced equations for all the hydrolysis reactions, and propose an explanation for the reactivity order you observed.

Table 3.   Colors and approximate pH values for Gramercy universal indicator

pH	Color
4	red
5	red-orange
5.5	orange
6	yellow-orange
6.5	yellow
7	yellow-green
7.5	green
8	dark green
8.5	blue-green
9	blue
9.5	violet
10	red-violet

Exercises

1. Assuming that all the ester that had been formed could be recovered, calculate the mass of ethyl acetate you would have obtained **(a)** at equilibrium and **(b)** after the distillation.

2. If all of the ethyl acetate calculated in Exercise 1(b) distilled as the 70.2°C azeotrope, about how much water would remain in the boiling flask after the distillation?

3. Use your K_e value to calculate the mole ratio of reactants that should result in 85% completion, and compare the result with the value you determined graphically.

4. **(a)** If, after adding sulfuric acid to the reaction mixture, you did not withdraw an aliquot immediately but waited several minutes, would your calculated equilibrium constant have been too high or too low? Explain. **(b)** How would your calculated K_e be affected if you withdrew the aliquot for the equilibrium titration before the reaction actually reached equilibrium? Explain.

5. Derive the expression for the equilibrium constant in terms of α_e as given in Equation 4.

6. **(a)** Use your K_e value to estimate the equilibrium constant for the hydrolysis of ethyl acetate. **(b)** How could the reaction conditions be changed to favor this hydrolysis, rather than esterification? **(c)** Write a mechanism for the acid-catalyzed hydrolysis of ethyl acetate.

7. Besides the direct esterification procedure used in this experiment, there are many other synthetic methods used in preparing esters. Propose a high-yield synthesis for each of the following esters. Do not use the Fischer procedure; begin with any appropriate carboxylic acids, alcohols, or phenols.

(a)

(b)

(c)

methyl
mesitoate

Library Topic

The Fischer esterification proceeds by a mechanism that has been classified as $A_{AC}2$. Certain other esterification and hydrolysis reactions proceed by different mechanisms, such as the hydrolysis of methyl mesitoate in concentrated sulfuric acid ($A_{AC}1$), the hydrolysis of methyl salicylate in aqueous sodium hydroxide ($B_{AC}2$), and the transesterification of t-butyl benzoate with methanol ($A_{AL}1$). After consulting references on organic reaction mechanisms, explain what is meant by each symbol used in this classification system, and write mechanisms for these reactions.

Synthesis and Identification of an Unknown Carboxylic Acid

Reactions of Acid Anhydrides. Reactions of Carbon-Carbon Double Bonds. Preparation of Carboxylic Acids. Qualitative Analysis.

Operations

OP–11 Gravity Filtration

OP–12 Vacuum Filtration

OP–21 Drying Solids

OP–28 Melting-Point Determination

Objectives

To prepare a carboxylic acid by the reaction of maleic anhydride with zinc and hydrochloric acid.

To identify the carboxylic acid from its neutralization equivalent and its melting point.

To learn how to determine the neutralization equivalent of an acid.

To learn about the properties and uses of maleic anhydride.

SITUATION

Maleic anhydride is manufactured by air oxidation of benzene in the presence of a vanadium oxide catalyst:

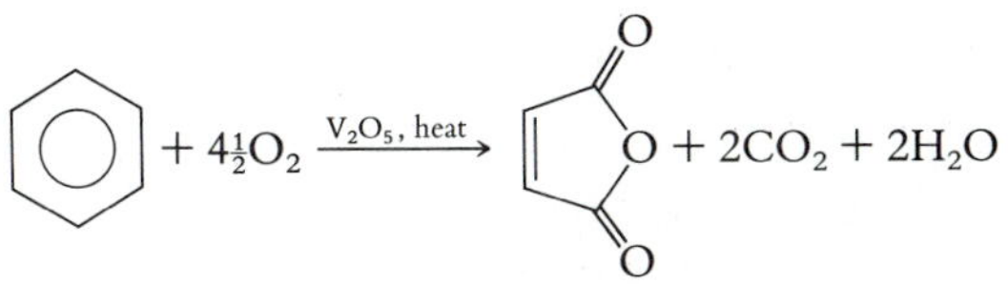

maleic
anhydride

Because both air and benzene are cheap and abundant, maleic anhydride is an attractive starting material for many other marketable chemicals. For example, tetrahydrofuran, an excellent solvent and reaction medium, has been synthesized by the catalytic hydrogenation of maleic anhydride:

$$
\begin{array}{c}
\underset{\text{CH}-\text{C}}{\overset{\text{CH}-\text{C}}{\Bigg\|}}\overset{\displaystyle O}{\underset{\displaystyle O}{}}\!\!\diagdown\!\!O + 5H_2 \xrightarrow{\text{catalyst}} \underset{\text{CH}_2\text{CH}_2}{\overset{\text{CH}_2\text{CH}_2}{\Big|}}\!\!\diagdown\!\!O + 2H_2O
\end{array}
$$

tetrahydrofuran

As a research project, an inexperienced graduate student decided to develop a safer and more convenient synthesis of tetrahydrofuran. He reasoned that using zinc and hydrochloric acid to generate hydrogen *in situ* would allow him to avoid most of the hazards associated with the use of hydrogen gas, as well as the need for expensive apparatus and catalysts.

$$Zn + 2HCl \longrightarrow H_2 + ZnCl_2$$

So he stirred some zinc and maleic acid into boiling water, added a little concentrated HCl, and was pleased when the reaction mixture started fizzing vigorously, as expected. But when the reaction mixture cooled down, a white solid crystallized from solution. Since tetrahydrofuran is a low-boiling liquid, the graduate student realized that something had gone wrong and threw away the product in disgust, thinking it was unreacted maleic anhydride. In fact, he had synthesized a different compound by a little-known reaction that was first reported by E. Emmett Reid in 1972. In this experiment, you will carry out Reid's reaction with maleic anhydride and try to identify the white solid it produces.

BACKGROUND

Cyclic Anhydrides and Reid's Reaction

Besides being used in the production of tetrahydrofuran, maleic anhydride is an important raw material for manufacturing agricultural chemicals, polyester resins, surface coatings, dye intermediates, pharmaceuticals, lubricant additives, and many other commercially significant products. As mentioned in Experiment 23, maleic anhydride is also an excellent dienophile that forms Diels-Alder adducts with most conjugated dienes.

Maleic acid, which is formed by a reaction of maleic anhydride with water, helps keep fats and oils from becoming rancid and has other commercial uses similar to those of the anhydride. Maleic acid's geometric isomer, fumaric acid, occurs naturally in many plants and is sometimes used to replace citric acid or tartaric acid in beverages and baking powder.

When Dr. E. Emmett Reid added maleic anhydride to a mixture of granular zinc and boiling water, he noticed an immediate reaction and the separation of a white precipitate. After acidifying the reaction mixture with hydrochloric acid, he recovered the product and identified it as a carboxylic acid that was not maleic acid. This result was not published until 1972, when it was mentioned—almost as an afterthought— in a "chemical autobiography" describing Reid's seventy-six years of research. Reid's reaction works only with maleic anhydride and certain other unsaturated carboxylic acids and anhydrides, including maleic acid, fumaric acid, cinnamic acid, and butynedioic acid. Although a detailed mechanism for the reaction has not yet been proposed, it seems to involve a direct reaction between zinc and the substrate. The function of the hydrochloric acid is probably not to generate hydrogen but to convert an intermediate zinc salt into the final product.

Cyclic anhydrides are often prepared by heating an appropriate dicarboxylic acid with acetic anhydride, which removes water from the acid to form its anhydride and produce acetic acid. Like other reactive functional derivatives of carboxylic acids, anhydrides react readily with nucleophilic reagents. If you do Minilab 28, you will prepare succinic anhydride from the corresponding acid and observe its reaction with p-toluidine.

maleic acid

fumaric acid

cinnamic acid

$$HOOCC{\equiv}CCOOH$$
butynedioic acid

The reaction with cinnamic acid requires a zinc-mercury amalgam rather than pure zinc.

succinic acid + acetic anhydride → succinic anhydride + acetic acid

METHODOLOGY

You will synthesize the unknown acid by adding granular zinc to a hot aqueous solution of maleic anhydride, whose infrared spectrum is shown in Figure 1 of Experiment 23. Because powdered maleic anhydride reacts quite rapidly with atmospheric moisture, it is usually manufactured in the form of briquettes, which must be pulverized before use. If maleic anhydride is provided in powdered form, you should open the reagent jar only momentarily, and replace the cap immediately after you have removed the amount needed. After the initial reaction is complete, concentrated hydrochloric acid is added to liberate the final product and dissolve any excess zinc. Because the product is somewhat soluble even in cold water, the solution should be concentrated to increase the yield. The product should require no further purification after it crystallizes from the cold solution.

The *equivalent weight* of an acid, sometimes called its *neutralization equivalent,* can be defined as the mass of the acid that provides a mole of protons (H^+) in a neutralization reaction. For example, 1 mole of acetic acid (CH_3COOH, M.W. = 60) and $\frac{1}{2}$ mole of oxalic acid (HOOCCOOH, M.W. = 90) both contain 1 mole of acidic hydrogens that can be removed as H^+ ions by a strong base. Thus, the equivalent weight of acetic acid is 60, and the equivalent weight of oxalic acid is 45 (half of its molecular weight). If the structure of an acid is known, its equivalent weight can be calculated by dividing its molecular weight by the number of acidic hydrogens per molecule.

The equivalent weight of an unknown acid is usually determined by titrating an accurately weighed sample of the acid with a standardized solution of sodium hydroxide. The volume of the base used in the titration and its molar concentration can be used to calculate the number of moles of base neutralized by the acid, which is equal to the number of moles of protons that the acid provided. For example, suppose 0.222 g of an unknown carboxylic acid with a melting point of 210°C was titrated with 23.2 mL of 0.115 M NaOH. Multiplying the molarity of the base by the volume used (in liters) gives the number of moles of base and protons. Dividing the mass of the acid by the number of moles of protons then yields the equivalent weight of the acid, which is about 83.

Note that equivalent weight, like molecular weight, is dimensionless.

$$\text{moles of } H^+ = \text{moles of NaOH} = \frac{0.115 \text{ mol}}{1 \text{ L}} \times 0.0232 \text{ L}$$

$$= 2.67 \times 10^{-3} \text{ mol}$$

$$\text{equivalent weight} = \frac{\text{mass of acid}}{\text{moles of } H^+} = \frac{0.222}{2.67 \times 10^{-3}} = 83.2$$

The melting point of this unknown compound is impossibly high for a carboxylic acid with a molecular weight near 83.

The unknown acid must have two or more carboxyl groups and a molecular weight that is some multiple of 83. A glance at Table 7 in Appendix VI indicates that the unknown may be phthalic acid, whose molecular weight is 166.

The unknown acid you will synthesize in this experiment is also listed in Table 7 in Appendix VI, so you should be able to identify it easily after you have measured its equivalent weight and melting point. With your instructor's permission, you can also obtain the IR or PMR spectrum of the acid to help confirm its identity (Collateral Projects 1 and 2).

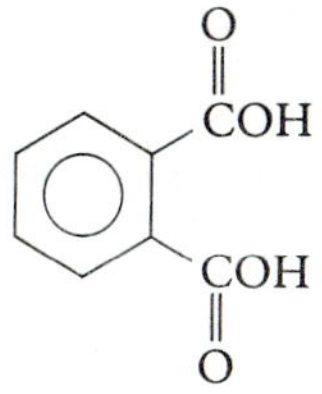

phthalic acid

PRELAB ASSIGNMENT

Calculate the mass of 25 mmol of maleic anhydride and of 27.5 mmol of zinc.

Properties

Table 1.   Physical properties

	M.W.	m.p.	b.p.
maleic anhydride	98.1	53	202
zinc	65.4	419	907 d

PROCEDURE

Maleic anhydride is corrosive and toxic and can cause severe damage to the eyes, skin, and upper respiratory tract. Wear gloves, avoid contact, and do not breathe the dust. If you must powder maleic anhydride briquettes, do it under the hood and wear safety goggles and protective clothing.

Hydrochloric acid is poisonous and corrosive; contact or inhalation can cause severe damage to the eyes, skin, and respiratory tract. Wear gloves, and dispense under a hood; avoid contact, and do not breathe the vapors.

SAFETY PRECAUTIONS

maleic anhydride    hydrochloric acid

SAFETY
PRECAUTIONS

The reaction of zinc with hydrochloric acid produces hydrogen gas, which is highly flammable; keep the reaction mixture away from flames.

The product may irritate the eyes, skin, or respiratory tract; avoid contact with it, and do not breathe its dust.

Put the cap back on the maleic anhydride bottle!

Reaction.   Dissolve 25 mmol of maleic anhydride **[contact and inhalation hazard!]** in 15 mL of boiling deionized water in an Erlenmeyer flask. Remove the flask from the heat source, and immediately add 27.5 mmol of 40-mesh zinc in three or four portions, swirling after each addition. Let the flask stand for 10–15 minutes with occasional swirling. *Under the hood,* slowly add 5 mL of concentrated HCl **[contact and vapor hazard!]** with swirling.

Separation.   When the zinc (or most of it) has dissolved, heat the mixture to boiling under the hood; any white solid that forms during the reaction should dissolve. Filter the hot solution through fluted filter paper ⟨OP–11⟩. Boil the filtrate under the hood until its volume has been reduced to about 5 mL (stop boiling if it becomes cloudy). Cover the beaker with a watch glass, and set it aside to cool to room temperature. Then cool the beaker in an ice bath until crystallization is complete. Collect the product by vacuum filtration ⟨OP–12⟩, washing it on the filter with a little cold acetone. Air-dry the product on the filter for a few minutes, then dry it ⟨OP–21⟩ to constant mass.

If any solid forms on the sides of the beaker, scrape it into the solution before cooling.

Analysis.   Measure the mass and melting point ⟨OP–28⟩ of the thoroughly dried product. Then weigh about 0.2 g of product to the nearest milligram and dissolve it in 25 mL of water in a 125-mL Erlenmeyer flask. Add 2 drops of phenolphthalein solution, and titrate with a standardized ~0.2 M NaOH solution.

Record the concentration of the solution in your lab notebook.

Report.   Calculate the equivalent weight of your unknown acid, and identify it by referring to Table 7 in Appendix VI. Write the structure of the product and a balanced equation for its formation, and give its systematic and common names. Calculate the theoretical yield and the percentage yield of the product.

Collateral Projects

1. Record the IR spectrum of the product (KBr pellet or Nujol mull). Compare it with the spectrum of the starting material in Figure 1 of Experiment 23. Interpret both spectra as completely as you can.

2. Predict the PMR spectrum you would expect from the product (see Exercise 6); then record its spectrum in DMSO-d_6 or another suitable solvent and see whether your prediction was accurate.

Preparation of Succinic Anhydride

MINILAB 28

All apparatus must be dry, and the product should be protected from moisture.

Acetic anhydride can cause serious burns, and its vapors are highly irritating and corrosive; wear gloves, and use the anhydride only under a fume hood.

Avoid contact with succinic acid and succinic anhydride, and do not breathe their dust.

Avoid contact with toluene, do not breathe its vapors, and keep it away from flames.

SAFETY PRECAUTIONS

Under the hood, combine 1.0 g of succinic acid with 2.0 mL of acetic anhydride **[contact and vapor hazard!]** in a test tube fitted with a cold-finger condenser ⟨OP – 7b⟩. Add an acid-resistant boiling chip, heat the mixture to boiling, and reflux it gently for 15 minutes. Stopper the tube to keep out moisture, and let the solution cool to room temperature. If necessary, scratch the sides of the test tube with a stirring rod to induce crystallization; then cool the mixture in ice. Carefully **[wear gloves! work under a hood!]** collect the succinic anhydride **[contact hazard!]** by semimicro vacuum filtration, washing it on the filter with two portions of ice-cold anhydrous ethyl ether. Weigh the dry product, and measure its melting point.

You can also carry out the reaction of succinic acid with *p*-toluidine. Mix about 25 mg of succinic anhydride with

Before the product crystallizes, remove the boiling chip by decantation or with an acid-resistant spatula.

1 mL of toluene [**fire and vapor hazard!**] in a small test tube, and heat it over a steam bath until all the solid dissolves. Add about 25 mg of *p*-toluidine, stopper and shake the tube, and observe the result. Propose a structure for the product, and write a balanced equation for its formation.

Exercises

1.   Estimate the amount of product that was lost in the filtrate, assuming that the volume of the reaction mixture was 5 mL when you filtered it. Compare this with the amount of product you would have lost if you had not boiled off most of the water. The solubility of the product is 6.8 g/100 mL at 20°C.

2.   Construct a flow diagram for the synthesis carried out in this experiment.

3.   Propose structures for the products you would obtain by treating fumaric acid, cinnamic acid, and butynedioic acid with excess zinc and HCl. Write balanced equations for the reactions.

4.   What was wrong with the graduate student's idea (see the Situation section) that the reaction of maleic acid with zinc and aqueous HCl would yield tetrahydrofuran?

5.   Suppose that 0.196 g of an unknown carboxylic acid is titrated using 19.3 mL of 0.196 M aqueous NaOH. What is the probable identity of the unknown if its melting point is 132 ± 5°C? (Use Table 7 in Appendix VI.)

6.   Describe the PMR spectrum you would expect from the product of this experiment, giving peak areas, multiplicities, and approximate chemical shifts for all signals.

7.   Write balanced equations for the reactions of maleic anhydride with water, cyclohexanol, aniline, and isoprene (2-methyl-1,3-butadiene).

8.   Write a mechanism for the reaction of succinic anhydride with *p*-toluidine as carried out in Minilab 28.

Library Topic

Write a short report about the manufacture and uses of maleic anhydride, maleic acid, and fumaric acid.

Preparation of the Insect Repellent
N,N-Diethyl-*m*-Toluamide

Reactions of Carboxylic Acids. Preparation of Amides. Nucleophilic
  Acyl Substitution. Acid Chlorides.

Operations

OP–7a	Refluxing	
OP–8	Cooling	
OP–9	Mixing	
OP–10	Addition	
OP–13	Extraction	
OP–14	Evaporation	
OP–19	Washing Liquids	
OP–20	Drying Liquids	
OP–22b	Trapping Gases	
OP–26a	Semimicro Vacuum Distillation	
OP–32	Gas Chromatography	
OP–33	Infrared Spectrometry	

Objectives

To prepare N,N-diethyl-*m*-toluamide from *m*-toluic acid.

To assess the purity and analyze the infrared spectrum of the product.

To learn how to carry out a Schotten-Baumann reaction.

To learn about some insect repellents and how they work.

SITUATION

There is hardly a soul who hasn't cringed on hearing the high-pitched whine of a hungry mosquito. Besides their capacity to torment us, these bloodthirsty insects have a well-earned reputation for spreading diseases among humans, including malaria, yellow fever, dengue, and viral encephalitis.

O
‖
C—N(CH₂CH₃)₂

CH₃

deet
(*N,N*-diethyl-*m*-toluamide)

See R. H. Wright, "Why Mosquito Repellents Repel," *Scientific American,* July 1975, p. 104.

The phenomenon of insect repellence is undoubtedly more complicated than the explanation here might suggest. There must be individual variations in skin chemistry or physical characteristics to explain why some unlucky individuals are eaten alive by mosquitoes and others escape unscathed.

The diol 2-butyl-2-ethyl-1,3-propanediol is even more effective than deet. In laboratory tests, one application has protected against mosquito bites for up to 196 days!

Numerous methods of controlling mosquitoes have been developed, but the hardy creatures can withstand a variety of adverse conditions. Mosquitoes have been known to breed in the hot alkaline volcanic pools of Uganda, and even in a tank of hydrochloric acid in India! Since we are not likely to eradicate mosquitoes from the earth anytime soon, we must resign ourselves to living with them. This fate is made more tolerable by the availability of effective insect repellents. One of the best of these is *N,N*-diethyl-*m*-toluamide (deet), now an ingredient of nearly all commercial mosquito repellents in the United States and Canada. In this experiment, you will synthesize deet from *m*-toluic acid and characterize it by infrared spectrometry.

BACKGROUND

Chemical Mosquito Evasion

Mosquito "repellents" don't really repel mosquitoes in the same way that a disagreeable odor might repel a human from its source. Instead, they appear to jam the insect's sensors so that it can't find its victim. Warm objects generate convection currents in the air around them; warm living objects also evolve carbon dioxide, which alerts a mosquito to the presence of a blood source and starts it on its flight. This flight is initially random, but when the insect encounters a warm, moist stream of air it moves toward the source, which is generally a living object. Unless the object takes rapid evasive action, the mosquito is usually able to follow the convection current until it makes contact (unless it gets squashed first).

When you (the intended victim) are protected by an effective insect repellent, the mosquito still knows you're around but is unable to find you. This is because the repellent prevents the insect's moisture sensors from responding normally to the high humidity of your convection current. Ordinarily, when a mosquito passes from warm, moist air into drier air, its moisture sensors send fewer signals to its central nervous system, causing it to turn back into the air stream. By blocking these sensors, the repellent reduces the signal frequency and convinces the mosquito that it is heading into drier rather than moister air; so it turns away just before landing.

The molecular features that make a compound a good insect repellent are not well understood at this time. Repellents occur in nearly every chemical family and exhibit a large variety of molecular shapes. A number of *N,N*-disubstituted

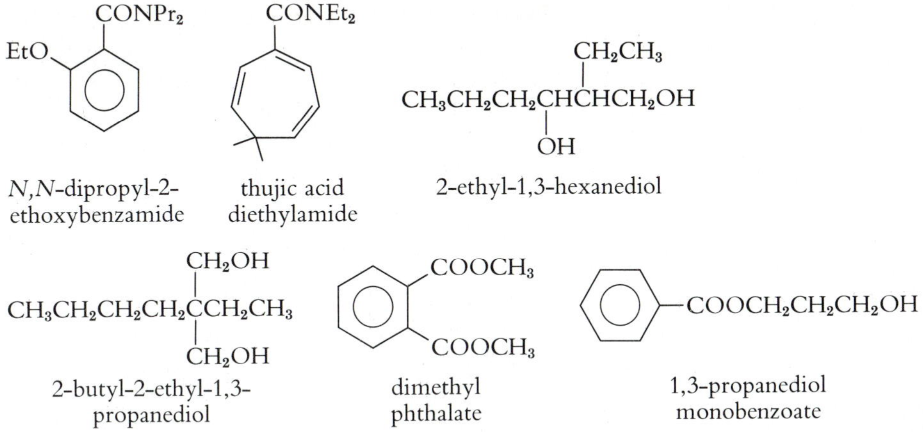

Figure 1.   Some typical insect repellents

amides similar to deet are represented (see Figure 1), including the diethylamide of thujic acid, a constituent of the western red cedar that may be partly responsible for that tree's resistance to insect attack. Some of the more effective repellents are esters such as dimethyl phthalate and diols such as 2-ethyl-1,3-hexanediol. "Bug pills" that can be taken orally are already on the market. These provide protection by causing a repellent chemical to be released through the skin. When more is learned about the structural features that make a substance act as a repellent, it should be possible to develop even more effective and convenient repellents for long-term protection against insect bites.

METHODOLOGY

Amides are usually prepared by treating a carboxylic acid derivative with ammonia or with a primary or secondary amine. Acid anhydrides or esters are sometimes used as the acid derivative, but acyl chlorides are the most useful for preparing the widest variety of amides. Because of the high reactivity of acyl chlorides, the reactions are usually rapid and exothermic —so much so that, in many cases, the rate must be controlled by cooling or by using an appropriate solvent. When the reaction is carried out in an inert solvent such as ethyl ether or

General reactions for preparing amides

$$R\overset{\overset{\displaystyle O}{\|}}{-}C-Z + NH_3 \longrightarrow R\overset{\overset{\displaystyle O}{\|}}{C}NH_2 + HZ$$

$$R\overset{\overset{\displaystyle O}{\|}}{-}C-Z + R'NH_2 \longrightarrow R\overset{\overset{\displaystyle O}{\|}}{C}NHR' + HZ$$

$$R\overset{\overset{\displaystyle O}{\|}}{-}C-Z + R'\underset{\underset{\displaystyle R''}{|}}{-}NH \longrightarrow R\overset{\overset{\displaystyle O}{\|}}{C}N\underset{\underset{\displaystyle R''}{|}}{R'} + HZ$$

Z = Cl, OR, OCOR, etc.

Example of reaction in inert solvent

$$RCCl + 2R'NH_2 \longrightarrow RCNHR'$$
$$+ R'NH_2{}^+Cl^-$$

Schotten-Baumann method for amides

$$RCCl + R'NH_2 + NaOH \longrightarrow$$

$$RCNHR' + NaCl + H_2O$$

Reaction of an amine salt with a base

$$RNH_3{}^+Cl^- + NaOH \rightarrow$$
$$RNH_2 + NaCl + H_2O$$

benzene, it is necessary to use at least a 100% excess of the nitrogen compound, because the reaction produces HCl that reacts with one equivalent of amine (or ammonia) to form the amine salt. This salt may also be difficult to separate from the product. One way to avoid these disadvantages is to add a base such as pyridine, triethylamine, or sodium hydroxide to react with the evolved HCl.

The synthetic method that uses aqueous sodium hydroxide (or potassium hydroxide) as the base is known as the Schotten-Baumann reaction. It has been used to prepare esters and other acyl derivatives as well as amides. Some acyl chlorides, particularly low-molecular-weight aliphatic ones, hydrolyze rapidly in water to form carboxylic acids and thus are unsuitable reactants for this procedure. However, aromatic and long-chain aliphatic acyl chlorides are nearly insoluble in water and therefore hydrolyze much more slowly; the small amount of acyl chloride lost by hydrolysis can be compensated for by using an excess of this reactant. The presence of aqueous alkali in the reaction mixture also makes it possible to use an amine hydrochloride rather than the free amine, as the salt reacts with base to liberate the amine *in situ*. Since many amines are volatile, corrosive, and quite unpleasant to handle, the use of the comparatively well-behaved amine salt can be a definite plus.

The acyl chlorides used for preparing amides are generally themselves prepared from the corresponding carboxylic acids. Several reagents are commonly used for this transformation; each has certain advantages and disadvantages. Thionyl chloride ($SOCl_2$) has the great advantage that all of the inorganic reaction products (HCl and SO_2) are gases and can be easily removed from the acyl chloride, which is often used for further reactions without isolation.

Methods for preparing acyl chlorides

$$\underset{RCOH}{\overset{O}{\parallel}} + PCl_5 \longrightarrow \underset{RCCl}{\overset{O}{\parallel}} + POCl_3 + HCl$$

$$3\underset{RCOH}{\overset{O}{\parallel}} + 2PCl_3 \longrightarrow 3\underset{RCCl}{\overset{O}{\parallel}} + 3HCl + P_2O_3$$

$$\underset{RCOH}{\overset{O}{\parallel}} + SOCl_2 \longrightarrow \underset{RCCl}{\overset{O}{\parallel}} + SO_2 + HCl$$

In this experiment, you will prepare *m*-toluoyl chloride by treating *m*-toluic acid with excess thionyl chloride and will then run a Schotten-Baumann reaction with diethylamine (from the hydrochloride) to prepare *N,N*-diethyl-*m*-toluamide. Because corrosive acidic gases are released during the reactions, a sodium hydroxide gas trap will be used to keep them out of the atmosphere. In the Schotten-Baumann reaction, efficient mixing is necessary to provide adequate contact between the water-insoluble acyl chloride and the amine. Use of the detergent sodium lauryl sulfate helps disperse the acyl chloride into smaller particles to increase the reaction rate. After the product is isolated by being extracted with ether, it can be purified by vacuum distillation and analyzed by gas chromatography.

The excess thionyl chloride from the first step is decomposed by its reaction with sodium hydroxide in the second.

By comparing the infrared spectrum of your product with those of the starting materials in Figure 2, you should see evidence for the functional group conversions that have taken place. The PMR spectrum of the product (see Collateral Project 1) consists of well-separated signals that are easy to assign and can be used to confirm its structure.

You may also find it instructive to compare the infrared spectrum and properties of *N,N*-diethyl-*m*-toluamide with those of another aromatic amide, benzamide, whose preparation is described in Minilab 29.

PRELAB ASSIGNMENTS

1. Calculate the mass of 30 mmol of *m*-toluic acid, the mass of 25 mmol of diethylamine hydrochloride, and the theoretical yield of *N,N*-diethyl-*m*-toluamide.

2. Read OP–26 if you have not previously carried out a vacuum distillation.

Reactions and Properties

A.

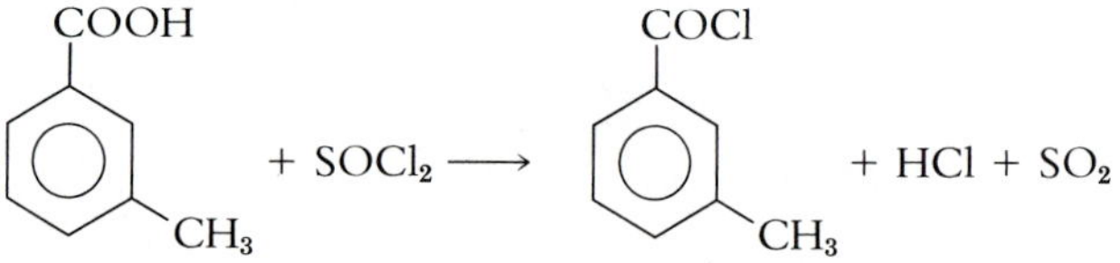

COOH
[m-toluic acid structure] + SOCl₂ ⟶ COCl
[m-toluoyl chloride structure] + HCl + SO₂

$$\text{COOH (on ring with CH}_3\text{)} + SOCl_2 \longrightarrow \text{COCl (on ring with CH}_3\text{)} + HCl + SO_2$$

m-toluic acid   m-toluoyl chloride

B. $(CH_3CH_2)_2NH_2^+Cl^- + NaOH$
diethylamine
hydrochloride

$$\longrightarrow (CH_3CH_2)_2NH + NaCl + H_2O$$
diethylamine

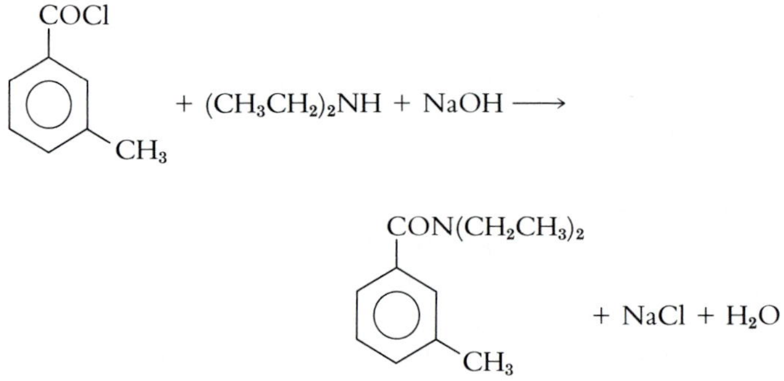

COCl
[ring with CH₃] $+ (CH_3CH_2)_2NH + NaOH \longrightarrow$

CON(CH₂CH₃)₂
[ring with CH₃] $+ NaCl + H_2O$

deet

Table 1.   Physical properties

	M.W.	m.p.	b.p.	d.
m-toluic acid	136.2	111–13		
thionyl chloride	119.0		79^{746}	1.65
diethylamine hydrochloride	109.6	227–30		
N,N-diethyl-m-toluamide	191.3		160^{19}	0.996

Figure 2.   IR spectra of the starting materials. (Reproduced from *The Aldrich Library of FT-IR Spectra,* by Charles J. Pouchert, with the permission of the Aldrich Chemical Company.)

m-Toluic acid

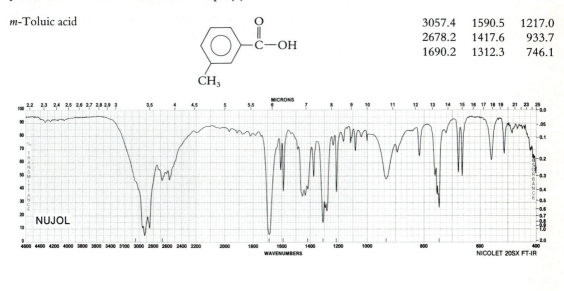

3057.4	1590.5	1217.0
2678.2	1417.6	933.7
1690.2	1312.3	746.1

Diethylamine   CH$_3$CH$_2$NHCH$_2$CH$_3$

2964.2	1377.6	1046.3
2742.8	1326.6	776.2
1451.5	1138.2	727.5

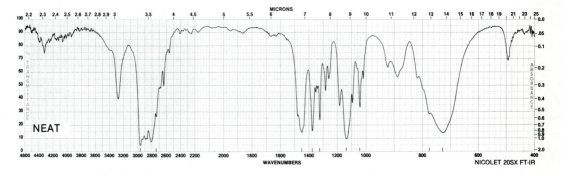

PROCEDURE

A. *Preparation of* m-*Toluoyl Chloride*

All glassware must be thoroughly dried for this procedure. If the reaction cannot be carried out under a hood, equip the reflux condenser with a gas trap ⟨OP–22b⟩ containing dilute NaOH.

<table>
<tr><td align="right">SAFETY
PRECAUTION</td><td>**Thionyl chloride and *m*-toluoyl chloride are both corrosive and lachrymatory and can cause severe damage to the eyes, skin, and respiratory system. Thionyl chloride decomposes violently on contact with water to produce poisonous gases (HCl and SO₂). Wear gloves, and work under a hood; avoid contact with thionyl chloride and the reaction mixture, keep them away from water, and do not breathe their vapors.**</td></tr>
</table>

Thionyl chloride is in excess, so it can be measured with a small graduated cylinder.

Under the hood, combine 30 mmol of *m*-toluic acid with 2.6 mL (~36 mmol) of thionyl chloride **[contact and vapor hazard! wear gloves!]** in a boiling flask. Reflux ⟨OP–7a⟩ the mixture *gently* for at least 20 minutes, until the evolution of gases stops. Cool the reaction mixture to 10°C or below in an ice bath. Leave the *m*-toluoyl chloride under the hood in the stoppered reaction flask (supported in a small beaker) until you are ready to use it in the next step.

B. *Preparation of* N,N-*Diethyl-m-toluamide*

<table>
<tr><td align="right">SAFETY
PRECAUTIONS</td><td>**Diethylamine hydrochloride irritates the skin, eyes, and respiratory tract, so avoid contact and inhalation. In the reaction mixture, it forms diethylamine, which is corrosive and has toxic vapors.**

Ethyl ether is extremely flammable and may be harmful if inhaled. Do not breathe its vapors, and keep it away from flames and hot surfaces.</td></tr>
</table>

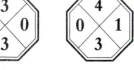

diethylamine   ethyl ether

Reaction. If magnetic stirrers are available, use the alternative procedure below. Otherwise, measure 35 mL of 3.0 M aqueous sodium hydroxide into an Erlenmeyer flask, and cool the flask in an ice bath ⟨OP–8⟩. Add, in small portions and with swirling, 25 mmol of diethylamine hydrochloride [contact hazard!]; then mix in 0.1 g of sodium lauryl sulfate. *Under the hood,* transfer the cold *m*-toluoyl chloride [vapor and contact hazard! wear gloves!] from part A to a *dry* addition funnel supported on a ringstand, and stopper the funnel immediately. Remain under the hood and wear protective gloves for the remainder of the reaction. Add ⟨OP–10⟩ the *m*-toluoyl chloride in small portions (about 1 mL each) to the reaction flask, swirling the flask in the ice bath after each addition until the reaction subsides. When the addition is complete, remove the flask from the ice bath, stopper it, and shake and swirl it vigorously for several minutes. Then loosen (but do not remove) the stopper, and heat the mixture on a steam bath for about 15 minutes, with continued shaking. At this point, the odor of the acid chloride should be gone and the reaction mixture should be basic to litmus. (Go on to the Separation step.)

Warning: Do not pour the m-toluoyl chloride into the reaction flask by mistake! If you do so, a violent reaction will result.

Reaction with Magnetic Stirring. Assemble an apparatus for addition and reflux ⟨OP–10⟩, using a 100-mL boiling flask equipped with a magnetic stirbar. Unless the reaction will be carried out under a hood, leave the gas trap attached to the reflux condenser. Lower the boiling flask, add 35 mL of 3.0 M aqueous sodium hydroxide, and cool the flask in an ice bath ⟨OP–8⟩. While stirring the NaOH solution ⟨OP–9⟩, add 25 mmol of diethylamine hydrochloride [contact hazard!] in small portions, followed by 0.1 g of sodium lauryl sulfate. *Under the hood,* transfer the cold *m*-toluoyl chloride [vapor and contact hazard! wear gloves!] from part A to the *dry* addition funnel; stopper the funnel immediately, and replace it on the reassembled reaction apparatus. Keeping an ice bath handy for moderating the reaction if necessary, slowly add the acid chloride ⟨OP–10⟩ to the reaction mixture while stirring vigorously ⟨OP–9⟩. When the addition is complete, heat and stir the reaction mixture over a steam bath for about 15 minutes. At this point, the acid chloride odor should be gone, and the reaction mixture should be basic to litmus.

Use enough clamps so that you will be able to lower the reaction flask without leaving any components unsupported.

Separation. Cool the reaction mixture to room temperature or below, and extract it ⟨OP–13⟩ with three 20-mL portions of ethyl ether. Wash the combined ether extracts ⟨OP–19⟩ with 30 mL of saturated aqueous sodium chloride, dry the ether layer ⟨OP–20⟩ with magnesium sulfate, and evaporate the solvent ⟨OP–14⟩.

You can also purify the product by column chromatography (see Collateral Project 3).

Purification and Analysis. Purify the *N,N*-diethyl-*m*-toluamide by semimicro vacuum distillation ⟨OP–26a⟩. Weigh it, and analyze it by gas chromatography ⟨OP–32⟩. Record the IR spectrum of your product ⟨OP–33⟩, or obtain a spectrum from your instructor.

Report. Compare your spectrum with the IR spectra of the reactants in Figure 2. Point out important bands that have appeared, disappeared, or changed location in going from reactants to product, and tell how these bands provide evidence that the expected reaction has taken place. Estimate the percentage purity of the product from its gas chromatogram.

Collateral Projects

1. Record the PMR spectrum of your product in deuterochloroform (or obtain a spectrum from your instructor), and assign all signals to the appropriate sets of protons. Compare the PMR spectrum of the product with those of the reactants in Figure 3 and try to account for any similarities or differences.

2. With your instructor's permission, you can test the effectiveness of your product as a mosquito repellent. Deet itself is a mild irritant, and the deet you synthesized might contain harmful impurities, so do not apply it directly to your skin. Instead, prepare a 15% solution of the deet in isopropyl alcohol and use it to saturate a piece of absorbent cheesecloth. When the treated cloth is dry, take it to a mosquito-infested area, and drape it over one arm. Place an untreated piece of cheesecloth on the other arm to serve as a control, and see which arm is targeted by more mosquitoes.

3. Purify your product by column chromatography rather than vacuum distillation, using an activated alumina column and petroleum ether as the eluent. Deet should be the first substance to come off the column in quantity.

Figure 3.   PMR spectra of the starting materials. Note that the offset PMR signal has been shifted 10 ppm upfield of its true position. (Reproduced from *The Aldrich Library of NMR Spectra, Edition II,* by Charles J. Pouchert, with the permission of the Aldrich Chemical Company.)

m-Toluic acid

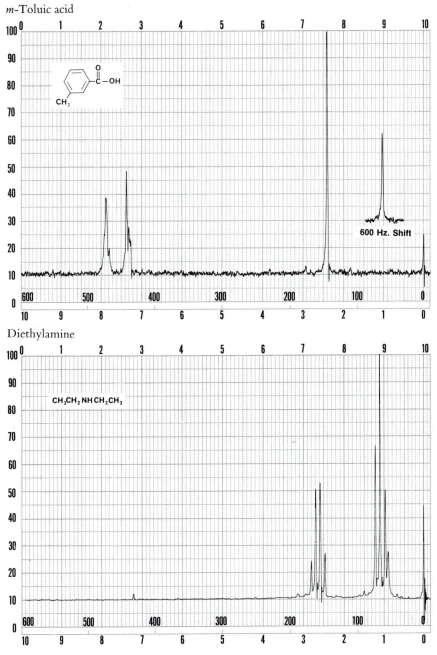

600 Hz. Shift

Diethylamine

CH₃CH₂ NH CH₂CH₃

MINILAB 29   Preparation of Benzamide

Both concentrated ammonia and benzoyl chloride can cause serious burns to skin and eyes, and their fumes irritate the eyes and respiratory system. Wear gloves, use them only under the hood, and do not breathe their vapors.

Under the hood, measure 2 mL of concentrated aqueous ammonia **[contact and vapor hazard!]** into a 15-cm test tube, and cool it well in an ice bath. Wearing protective gloves, *slowly* add 1 mL of benzoyl chloride **[contact and vapor hazard! violent reaction!]** to the cold ammonia with stirring. Let the solution stand in the ice bath for 5 minutes; then collect the benzamide by vacuum filtration, washing it on the filter with cold water. Recrystallize the benzamide from boiling water, dry and weigh it, and measure its melting point. Write a balanced equation and a mechanism for the reaction.

You may want to record the IR spectrum of your benzamide using a KBr disc or Nujol mull. Compare the spectrum with the one you obtained for *N,N*-diethyl-*m*-toluamide, and explain any significant similarities or differences.

Exercises

1. **(a)** Write balanced equations for the reactions that took place in the gas trap. **(b)** Write balanced equations for the reactions of excess *m*-toluoyl chloride and thionyl chloride with aqueous sodium hydroxide.

2. Construct a flow diagram for the preparation of deet, including any by-products formed by the reactions in Exercise 1(b).

3. Write reasonable mechanisms for **(a)** the reaction of *m*-toluic acid with thionyl chloride, and **(b)** the reaction of *m*-toluoyl chloride with diethylamine.

4. The insect repellent *N*-butylacetanilide is used to treat clothing for fleas and ticks. Outline a synthesis of this compound, starting with aniline.

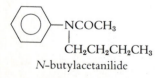

N-butylacetanilide

5. **(a)** Calculate the total volume of 3 M NaOH required for part B, including the amount used up in combining with excess reactants. **(b)** Calculate the percentage excess of NaOH that was actually used.

6. A student misread the label on a reagent bottle and used a 50% NaOH solution in this experiment instead of the 3 M solution. No deet was obtained from the ether layer after extraction, but acidification of the aqueous layer yielded a white solid that melted at 112°C. Identify the solid and tell how it might have formed, giving balanced equations for the reactions.

7. Hair is made up of the protein keratin, which is a polyamide containing many α-amino-acid residues. Hair is often responsible for clogged drains, which can be unclogged by drain cleaners that contain lye (sodium hydroxide). Explain how such drain cleaners get rid of hair.

$$\underset{\text{acetamide}}{CH_3\overset{\displaystyle O}{\overset{\|}{C}}NH_2}$$

$$\underset{\substack{N,N\text{-dimethyl-}\\\text{formamide}}}{H\overset{\displaystyle O}{\overset{\|}{C}}\underset{\displaystyle CH_3}{\overset{|}{N}}CH_3}$$

8. Acetamide (M.W. = 59) is a solid that melts at 82°C; *N,N*-dimethylformamide (M.W. = 73) is a liquid with a melting point of −61°C. Explain.

Library Topic

Deet repels insects, but there are many other chemicals that attract them. Report on some of the chemicals (called pheromones) that function as insect attractants, and tell how they can be used for insect control.

The Reaction of Phthalimide with Sodium Hypochlorite

Reactions of Amides and Imides. Nucleophilic Acyl Substitution. Functional Derivatives of Carboxylic Acids. Molecular Rearrangements.

Operations

OP–7   Heating

OP–8   Cooling

OP–12 Vacuum Filtration

OP–21 Drying Solids

OP–23 Recrystallization

OP–28 Melting-Point Determination

Objectives

To carry out the reaction of phthalimide with alkaline sodium hypochlorite.

To deduce the identity of the product and write a mechanism for its formation.

To learn about the chemistry of phthalimide and some related functional derivatives of carboxylic acids.

To learn about molecular rearrangements involving electron-deficient nitrogen atoms.

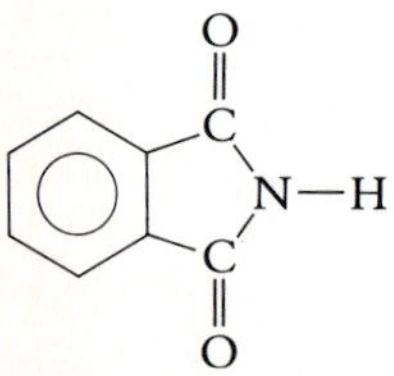

phthalimide

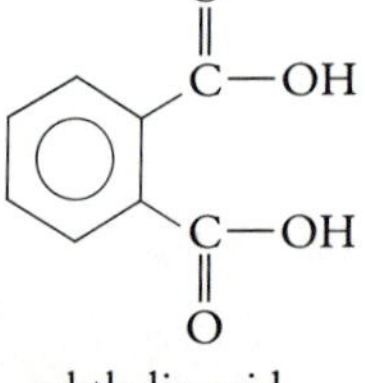

phthalic acid

SITUATION

While attempting to hydrolyze some phthalimide to phthalic acid, a nearsighted chemistry professor mistook a plastic bottle of chlorine bleach for a similar bottle containing aqueous sodium hydroxide, and added some to the reaction mixture before he noticed the error. Thinking he might still salvage the reaction, he then added the designated amount of sodium hydroxide and continued to heat the reactants. But the crystalline solid that precipitated from the acidified reaction mixture had the wrong melting point for phthalic acid. Curious about this unexpected result, the professor sent a sample of the product to a commercial laboratory for analysis. The analytical report showed that the product contained nitrogen but no chlorine. Based on the product's elemental composition and a determination of its molecular weight (see Exercise 4) the professor came up with the molecular formula $C_7H_7NO_2$. Unfortunately, in the *CRC Handbook* alone are listed eighteen compounds with that formula. By analogy with some related reactions, it seems probable that the reaction of phthalamide with sodium hypochlorite involves some kind of molecular rearrangement. With that in mind, you should be able to prepare the product, deduce one or more reasonable structures for it, and use its melting point to determine the correct structure.

Analysis of Sample

Carbon:	61.30%
Hydrogen:	5.15%
Nitrogen:	10.23%
Chlorine:	0.00%

BACKGROUND

Reactions of Imides

Imides are compounds containing a —CO—N—CO— functional group, and their chemistry resembles that of amides and other carboxylic acid functional derivatives. For example, imides can undergo nucleophilic acyl substitution reactions with good nucleophiles. The amide formed in such a reaction may react further under the appropriate conditions. Thus phthalimide is hydrolyzed by cold aqueous sodium hydroxide to phthalamic acid (as the sodium salt); if the reaction mixture is heated, phthalic acid is obtained:

Nucleophilic acyl substitution with an imide (H^+ may be obtained from water or another solvent)

$$\overset{O}{\overset{\|}{R}}\overset{O}{\overset{\|}{C}}NHCR' \xrightarrow{Nu:} \xrightarrow{H^+}$$

$$\overset{O}{\overset{\|}{R}}CNu + R'\overset{O}{\overset{\|}{C}}NH_2$$

phthalimide   phthalamic acid   phthalic acid

Because of the two carbonyl groups adjacent to nitrogen, the N—H hydrogen of an imide is acidic enough to be removed by moderately strong bases. Thus, phthalimide reacts with potassium hydroxide or potassium carbonate to form potassium phthalimide (as shown on page 380).

Sodium hypochlorite is a good oxidizing and chlorinating agent that is very stable in its aqueous solutions. In such solutions, it can chlorinate most nitrogen compounds having

Mixing chlorine bleach with aqueous ammonia produces chlorinated ammonia derivatives such as nitrogen trichloride (NCl_3), a very poisonous gas. Because of such potentially dangerous reactions, household chemicals should never be mixed together.

Reaction of phthalimide with potassium carbonate

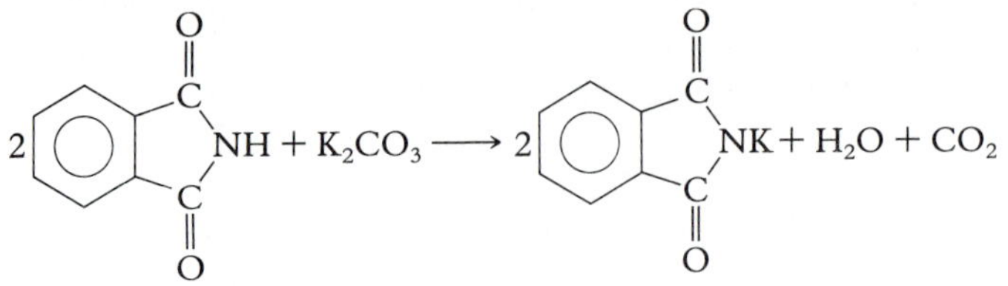

$$2 \quad \underset{\substack{O \\ \| \\ C}}{\overset{\substack{O \\ \| \\ C}}{\bigcirc}}NH + K_2CO_3 \longrightarrow 2 \quad \underset{\substack{O \\ \| \\ C}}{\overset{\substack{O \\ \| \\ C}}{\bigcirc}}NK + H_2O + CO_2$$

potassium
phthalimide

N—H bonds, including imides, amides, amines, ammonia, and sulfonamides, as well as their basic salts. For example, sodium hypochlorite converts indole to N-chloroindole by the following reaction:

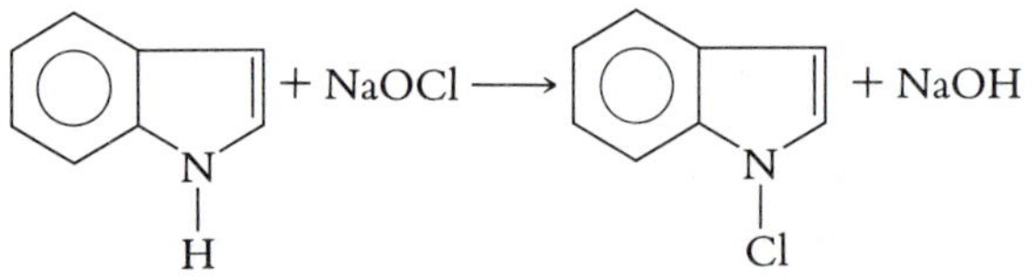

$$\underset{\substack{N \\ H}}{\bigcirc} + \text{NaOCl} \longrightarrow \underset{\substack{N \\ Cl}}{\bigcirc} + \text{NaOH}$$

N-chloroindole

Rearrangements to Nitrogen

Nitrene mechanism for the Curtius rearrangement

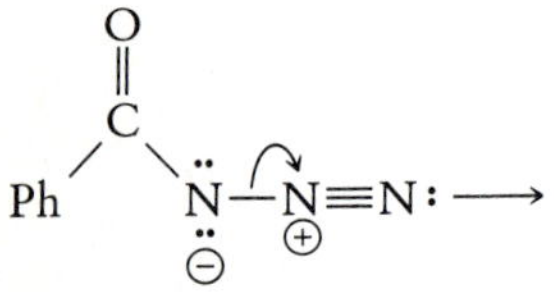

benzoyl
azide

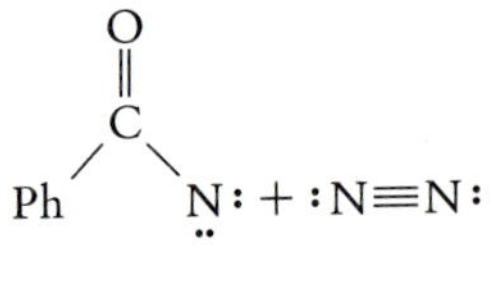

a nitrene

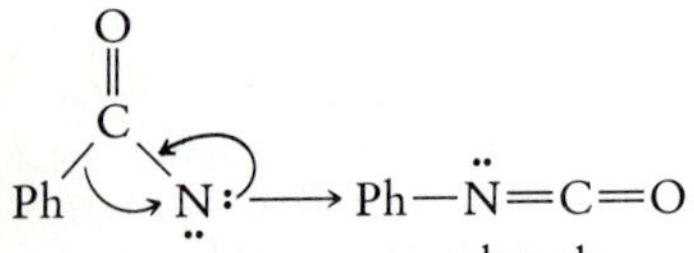

phenyl
isocyanate

Acyl azides are compounds containing the azide ($-N_3$) functional group on a carbonyl carbon. They can be prepared by treating acid chlorides with sodium azide (NaN_3). When an acyl azide is heated, it evolves gaseous nitrogen to yield an isocyanate by a reaction called the Curtius rearrangement. At one time it was thought that the rearrangement took place via an electron-deficient intermediate called a *nitrene*, as illustrated for benzoyl azide. This reaction resembles a carbenium-ion rearrangement, except that the migrating group moves to an electron-deficient nitrogen atom instead of to a carbon atom. In this example, migration of a phenyl group to the nitrogen atom would restore its missing pair of electrons and yield a stable product.

Recent studies of such rearrangements suggest that free nitrenes are probably not involved. The rearrangement step is thought to proceed by a concerted mechanism such as the following, where Z is a good leaving group and R an alkyl or aryl group:

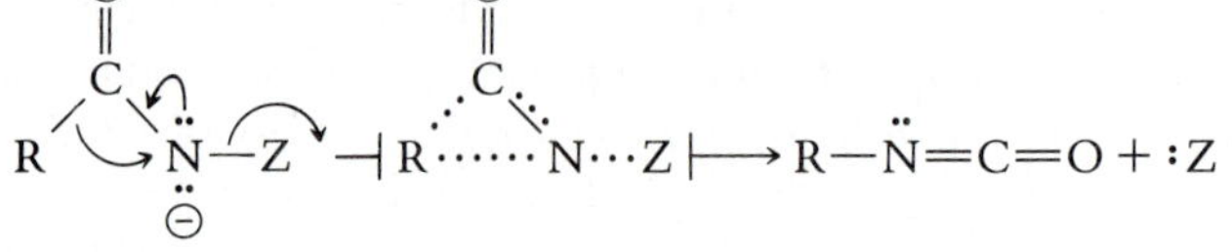

Thus, the migration to a *potentially* electron-deficient nitrogen occurs *as* the leaving group is being ejected, not after it has already left to form a nitrene. Depending on the reaction conditions, the resulting isocyanate may be isolated as such, or it may react further with the solvent. If the Curtius rearrangement is run in an alcoholic solution, for example, the alcohol adds to the C=N bond of the isocyanate to form a urethane. In water, a carbamic acid is formed at first, but it spontaneously decarboxylates to yield an amine as the final product.

The Curtius rearrangement is only one of a number of reactions involving migration to an electron-deficient nitrogen atom. Some of these reactions, such as the Beckmann rearrangement, involve reactants or intermediates with an RC=NZ grouping. The transition state for the rearrangement step is similar to that for a Curtius-type rearrangement:

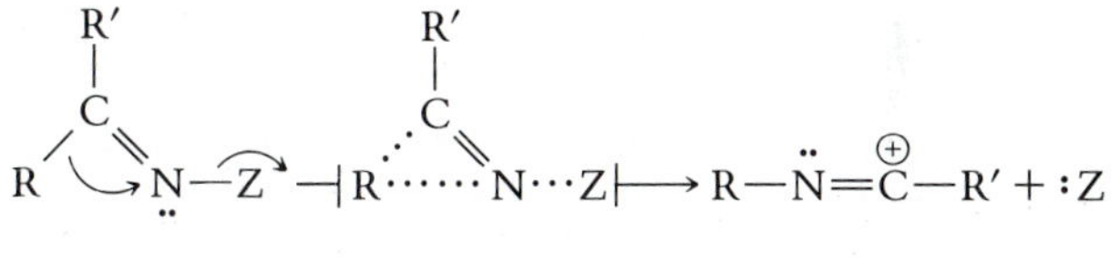

The carbenium-ion intermediate formed in the rearrangement step can subsequently react with a nucleophile, such as water or another solvent, to yield a stable product. Minilab 30 involves a reaction of this type, the Beckmann rearrangement of benzophenone oxime to benzanilide.

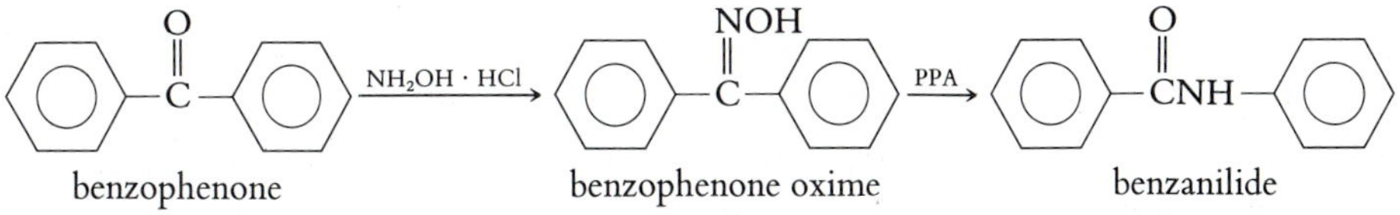

Reactions of isocyanates with hydroxylic solvents

$$RN{=}C{=}O + R'OH \longrightarrow RNHCOR'$$
a urethane

$$RN{=}C{=}O + HOH \longrightarrow RNHCOH$$
a carbamic acid

$$\longrightarrow RNH_2 + CO_2$$

Under the alkaline reaction conditions used in some rearrangements of this type, the carbon dioxide is converted to carbonate ion.

benzophenone   benzophenone oxime   benzanilide

METHODOLOGY

Most chlorine-type laundry bleaches are aqueous sodium hypochlorite solutions with a concentration of about 5.25% NaOCl by mass. Thus, a commercial product like Clorox or Javex is a convenient, inexpensive source of this reagent in solution. Phthalimide can be prepared by heating phthalic anhydride with urea; near the end of the reaction period, the reaction mixture suddenly expands to about three times its initial volume, due to the rapid evolution of carbon dioxide.

You will carry out the reaction of phthalimide with sodium hypochlorite by heating the reactants together with aqueous sodium hydroxide. The resulting basic solution should contain the sodium salt of the product, which is precipitated with acetic acid after most of the sodium hydroxide has been neutralized with hydrochloric acid. Considerable foaming may occur during the acidification step, but it can be reduced by keeping the reactants cool and adding the acid slowly. The solid product is separated from the reaction mixture by vacuum filtration and purified by recrystallization.

The product will be one of the compounds listed in Table 1, some of whose structures are shown in Figure 1. After having read the Background section carefully, you should be able to propose one or more reasonable structures for the product and a mechanism for its formation. The melting point of the product should then lead you to the correct structure. With your instructor's permission, you can also use the infrared spectrum of the product to help verify its structure (see the Collateral Project).

Table 1.   Compounds with the molecular formula $C_7H_7NO_2$ listed in the *CRC Handbook of Chemistry and Physics*

Compound	m.p.
2-hydroxybenzaldoxime*	63
3-hydroxybenzaldoxime	90
benzohydroxamic acid*	131–2
2-aminobenzoic acid*	146–7
3-aminobenzoic acid	174
4-aminobenzoic acid	188–9
2-hydroxybenzamide*	142
3-hydroxybenzamide	170.5
4-hydroxybenzamide	162
methyl 3-pyridinecarboxylate*	42–3
methyl 4-pyridinecarboxylate	8.5
1-methyl-3-pyridinecarboxylic acid*	218
2-hydroxy-5-nitrosotoluene	134–5
5-hydroxy-2-nitrosotoluene*	165
α-nitrotoluene	(b.p. 225–7)
2-nitrotoluene*	−10
3-nitrotoluene	16
4-nitrotoluene	54.5

Note: The structures of the compounds designated by asterisks are shown in Figure 1.

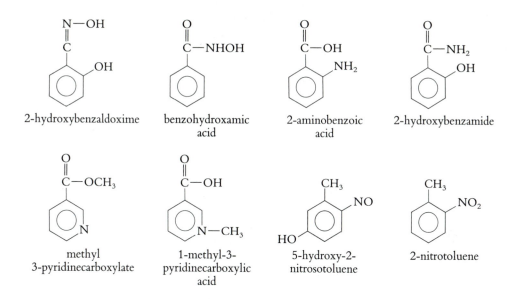

Figure 1.   Structures of some representative compounds from Table 1

PRELAB ASSIGNMENT

Calculate the mass of 50 mmol of phthalic anhydride, of 25 mmol of urea, and of 30 mmol of phthalimide. Calculate the theoretical yield of the final product.

Reactions and Properties

A. (phthalic anhydride) + H_2NCNH_2 (urea) $\longrightarrow$ 2 (phthalimide) + CO_2 + H_2O

B. (phthalimide) $\xrightarrow{\text{NaOCl, NaOH}} \xrightarrow{\text{H}^+} C_7H_7NO_2$

Table 2.   Physical properties

	M.W.	m.p.
phthalic anhydride	148.1	132
urea	60.1	135
phthalimide	147.1	238
sodium hydroxide	40.0	322
sodium hypochlorite	74.4	

Note: 5.25% sodium hypochlorite has a concentration of about 0.74 M.

Figure 2.   IR spectrum of phthalimide. (Reproduced from *The Aldrich Library of FT-IR Spectra,* by Charles J. Pouchert, with the permission of the Aldrich Chemical Company.)

Phthalimide

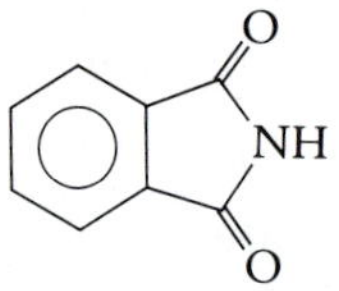

3194.9	1605.1	717.4
1753.3	1308.0	648.1
1690.3	1053.6	535.2

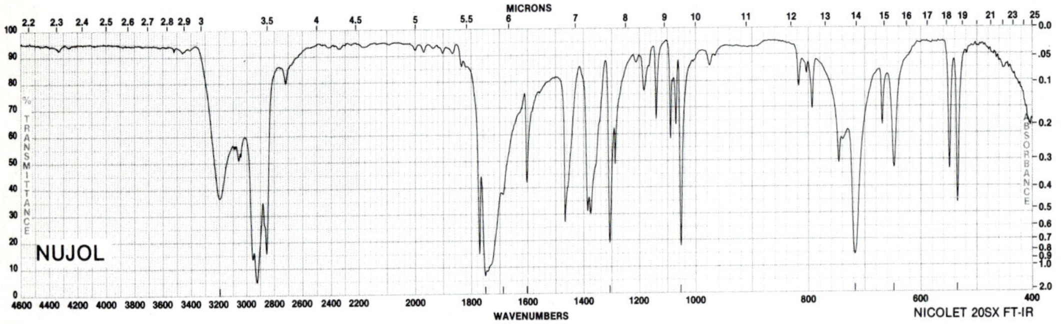

PROCEDURE

A. Preparation of Phthalimide

(This part can be omitted if commercial phthalimide is provided.)

SAFETY PRECAUTION

Phthalic anhydride irritates the skin and eyes, and its dust can irritate the respiratory system if it is inhaled. Avoid contact and do not breathe the dust.

Hot oil causes serious burns and may ignite above its flash point. Clamp the oil bath securely, keep it away from flames, and monitor its temperature closely.

Combine 50 mmol of pure phthalic anhydride **[contact and inhalation hazard!]** with 25 mmol of urea in a 100-mL boiling flask equipped with an air condenser. Heat the reactants in a 130–135°C oil bath ⟨OP–7⟩ **[fire hazard!]**; within about 20 minutes, the mixture should suddenly froth up and become nearly solid. At this point, stop heating the oil bath but leave the reaction flask in the oil while it cools down. Add about 10 mL of cold water to break up the solid; then collect the product by vacuum filtration ⟨OP–12⟩, washing it with a little cold water. Dry the phthalimide ⟨OP–21⟩, and measure its mass and melting point ⟨OP–28⟩.

The air condenser should be a wide-bore condenser that is not *connected to a water line; the distilling column from a lab kit works well.*

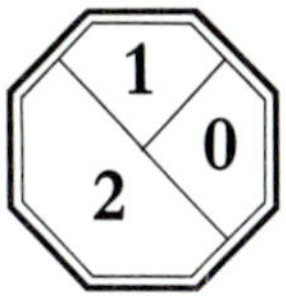

phthalic anhydride

B. Reaction of Phthalimide with Sodium Hypochlorite

SAFETY PRECAUTIONS

Sodium hydroxide is very corrosive, and its solutions can cause severe damage to the skin, eyes, and mucous membranes. Wear gloves, avoid contact, and wash up any spills immediately.

Phthalimide and the product may be harmful if inhaled, ingested, or absorbed through the skin. Avoid contact and inhalation.

Hydrochloric acid and acetic acid can cause severe damage to the skin and eyes, and their vapors are highly irritating to the eyes and respiratory tract. Wear gloves, and dispense them under a hood; avoid contact, and do not breathe their vapors.

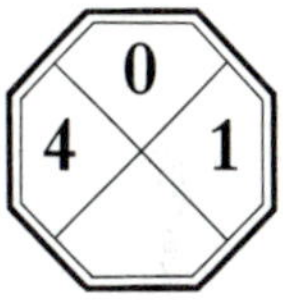

sodium hydroxide

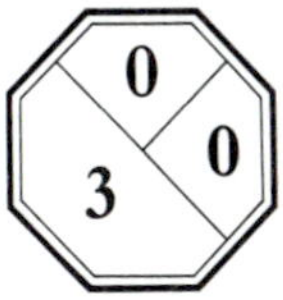

hydrochloric acid

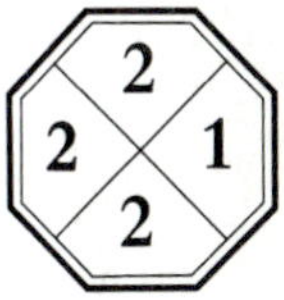

acetic acid

Reaction. Measure 45 mL (~33 mmol) of 5.25% aqueous sodium hypochlorite into a small Erlenmeyer flask, and add 15 mL of 8 M (25%) aqueous NaOH **[contact hazard!]**. Cool the solution below 5°C in an ice or ice-salt bath ⟨OP–8⟩. Add 30 mmol of finely powdered phthalimide and swirl vigorously; then add 10 mL more of the 8 M NaOH solution and again swirl to mix the reactants. The temperature of the reaction mixture should rise slowly as the solid dissolves, then more rapidly. When the temperature stops rising, heat the reaction mixture on a steam bath to 80°C. Keep it at that temperature for several minutes; then let it stand for about 10 minutes at room temperature.

If necessary, transfer the product mixture to a larger container to keep it from foaming over.

The product tends to char when heated, so it is best to dry it at room temperature.

Under the hood, cool the reaction mixture in an ice bath, and carefully stir in 15 mL of concentrated hydrochloric acid **[contact and vapor hazard! wear gloves!].** Test the solution with pH paper, and slowly add *just* enough additional concentrated HCl to bring its pH down to 10. Add 5 mL of glacial acetic acid **[contact and vapor hazard! foaming may occur!]** *slowly* with continuous swirling or stirring to precipitate the product. Let the reaction mixture stand in the ice bath until precipitation is complete.

Separation and Purification.   Collect the precipitate by vacuum filtration ⟨OP–12⟩, and wash it on the filter with ice water until the odor of acetic acid is gone. Purify the product by recrystallization from boiling water ⟨OP–23⟩, using decolorizing carbon as necessary to remove colored impurities. Dry the product ⟨OP–21⟩, and measure its melting point ⟨OP–28⟩.

Report.   Calculate the percentage yield of phthalimide and of the final product. Give the name and structure of the product, and tell how you arrived at your conclusion. Propose a detailed mechanism that explains how phthalimide is converted to the product, showing the transition state for any rearrangement step.

Collateral Project

With your instructor's permission, you can record the infrared spectrum (KBr disc or Nujol mull) of the product and use it to assist you in identifying the product. Compare its spectrum with that of phthalimide in Figure 2, and interpret it as completely as you can.

MINILAB 30

The Beckmann Rearrangement of Benzophenone Oxime

SAFETY PRECAUTIONS

Polyphosphoric acid is corrosive; avoid contact.

Hydroxylamine hydrochloride is corrosive and very toxic; wear gloves, and avoid contact.

Measure 5.0 mL of polyphosphoric acid (PPA) **[contact hazard!]**, 1.0 g of benzophenone, and 1.2 g of hydroxylamine hydrochloride **[contact hazard! wear gloves!]** into a small Erlenmeyer flask. Swirl the flask in a boiling water bath until the reactants are well mixed. Continue heating the flask in the bath for 30 – 45 minutes (more if necessary), with occasional swirling, until frothing has stopped. Add about 30 g of crushed ice to the warm reaction mixture, and shake or swirl it until the ice has melted. Collect the product by vacuum filtration, washing it on the filter with ice water. Purify the benzanilide by recrystallization from 95% ethanol. Dry and weigh it, and measure its melting point. Propose a detailed mechanism for the reaction, showing the transition state of the rearrangement step.

The polyphosphoric acid will be easier to pour if it is kept in a warm place. It is hygroscopic, so keep it away from water.

Exercises

1. What gas was responsible for the foaming when you acidified the reaction mixture in part B? Write a balanced equation for the reaction that liberated the gas.

2. **(a)** A student misread the directions and acidified the reaction mixture to a pH of 2 with HCl, and was surprised when there was no precipitate to filter. Explain what went wrong and write an equation for the reaction that caused the problem. **(b)** How could the student have recovered the product and salvaged the experiment?

3. Construct a flow diagram for the synthesis carried out in part B.

4. **(a)** When 35 mg of the product from this experiment was mixed with 0.46 g of camphor, the melting point of the mixture was found to be 157°C. Calculate the approximate molecular weight of the product if the melting point of pure camphor is 179°C and its freezing-point depression constant (K_f) is 40 K·kg·mol^{-1}. **(b)** Show how the molecular formula $C_7H_7NO_2$ can be derived using this result and the analytical data given in the Situation section.

5. When 9-fluorenone hydrazone is treated with sodium nitrite in aqueous sulfuric acid, it rearranges to form phenanthridone. Propose a detailed mechanism that explains

Remember that compounds with —NH₂ groups are diazotized by nitrous acid.

this reaction, showing the transition state for the rearrangement step.

9-fluorenone hydrazone   phenanthridone

6. **(a)** When a famous German chemist warmed a solution of *N*-bromoacetamide in base, he recognized an unmistakable pungent odor that slowly faded and was replaced by the strong ammonialike odor of an escaping gas. What were the two substances that his nose told him were there? Write a mechanism explaining the formation of both. **(b)** Who was the chemist, and what reaction had he discovered?

Library Topic

Report on the use of potassium phthalimide in the synthesis of amines and amino acids, including a discussion of the phthalimidomalonic ester method and the Gabriel synthesis. Illustrate synthetic routes to specific products using each method.

Synthesis of Sulfa Drugs from Acetanilide

EXPERIMENT 39

Reactions of Amines. Preparation of Sulfonamides. Electrophilic Aromatic Substitution. Nucleophilic Acyl Substitution. Sulfonic Acid Functional Derivatives.

Operations

OP–7a   Refluxing

OP–8   Cooling

OP–11   Gravity Filtration

OP–12   Vacuum Filtration

OP–14   Evaporation

OP–20   Drying Liquids

OP–21   Drying Solids

OP–22b   Trapping Gases

OP–23   Recrystallization

OP–28   Melting-Point Determination

Objectives

To prepare a sulfanilamidopyridine from acetanilide.

To learn how to carry out the chlorosulfonation of an aromatic ring and the amidation of a sulfonyl chloride.

To learn about the discovery, preparation, and applications of the sulfa drugs.

SITUATION

The discovery of Prontosil in the early 1930s led to worldwide research centered around the sulfanilamide derivatives known as sulfa drugs. An all-out effort was made to develop effective, broad-spectrum antibacterials; more than 5000 derivatives of sulfanilamide were synthesized and tested in the ten years between 1935 and 1945. These efforts culminated in the development and widespread use of such drugs as sulfadiazine, sulfathiazole, and sulfapyridine. These "wonder

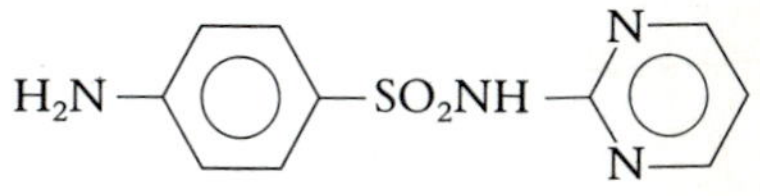

sulfadiazine

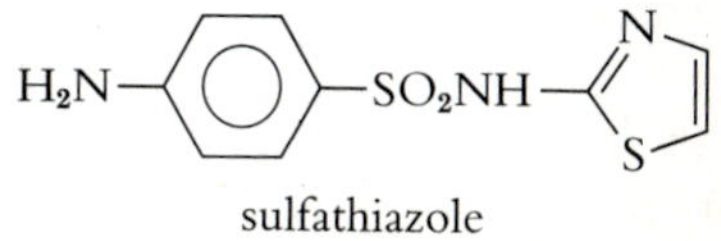

sulfathiazole

sulfapyridine
(2-sulfanilamidopyridine)

3-sulfanilamidopyridine

Domagk was later awarded a Nobel prize for his discovery, but Hitler did not permit him to accept it.

drugs" are credited with saving many lives during World War II and thereafter, until they were eventually supplanted by penicillin and other modern antibiotics. Even today sulfa drugs are used to treat urinary infections, bacillary dysentery, and meningitis, and there are numerous applications for sulfa drugs in veterinary medicine. Most of the sulfanilamides that were prepared between 1935 and 1945 have long since been forgotten, but an examination of the early literature reveals that an isomer of sulfapyridine was more effective than sulfapyridine itself against certain types of bacteria. In this experiment, you will prepare either sulfapyridine (2-sulfanilamidopyridine) or its lesser-known isomer, 3-sulfanilamidopyridine.

BACKGROUND

The First Wonder Drugs

The history of the sulfa drugs began in 1932 when a German dye factory (I.G. Farbenindustrie) patented a rather ordinary azo dye called Prontosil, which incorporated a sulfamoyl ($-SO_2NH_2$) group to improve its resistance to fading.

Prontosil

Because a German chemist named Paul Ehrlich believed that some dyes might be useful as therapeutic agents, Prontosil was routinely tested for effectiveness against bacterial cultures, but it failed to show any antibacterial activity. Then Gerhard Domagk, a pharmacologist working for I. G. Farben, decided to test Prontosil on mice infected with hemolytic streptococci. When the mice got well, Domagk published his results, and the world had its first important antibacterial drug — or nearly so. It remained for a group of French chemists at the Pasteur Institute to prove that Prontosil breaks down in biological systems to form *sulfanilamide,* which is the true antibacterial agent. Prontosil is effective only *in vivo* (in the body), whereas sulfanilamide is biologically active both *in vivo* and *in vitro* (in cultures outside a living host).

Sulfanilamide does not actually kill bacteria; rather, it inhibits their growth by depriving them of the essential nutrient folic acid and its derivatives. Mammals get their folic acid ready-made in the food they eat. Bacteria, however, must make their own in order to function, and sulfanilamide interrupts this process because of its close resemblance to *p*-aminobenzoic acid (PABA). A bacterium synthesizes folic acid by combining a molecule of pteridine with one each of PABA and glutamic acid. This process is directed by certain enzymes, including one called dihydropteroate synthetase that is responsible for attaching a PABA residue onto the pteridine. When a molecule of sulfanilamide is present, the enzyme apparently mistakes it for a molecule of PABA and uses it instead, thus keeping the pteridine out of circulation and reducing folic acid synthesis to a critical level.

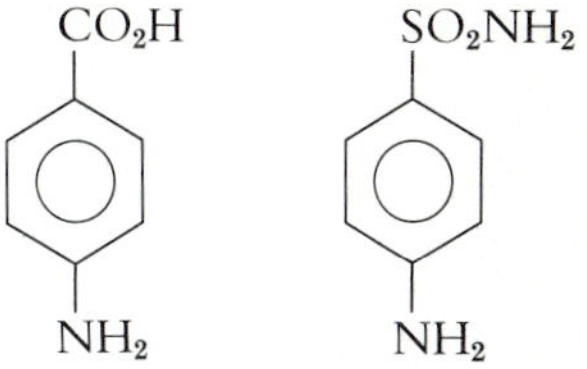

$$CO_2H \qquad\qquad SO_2NH_2$$

PABA   sulfanilamide

Another theory of sulfa drug action holds that sulfanilamide merely binds to the active site on the enzyme and blocks out the PABA.

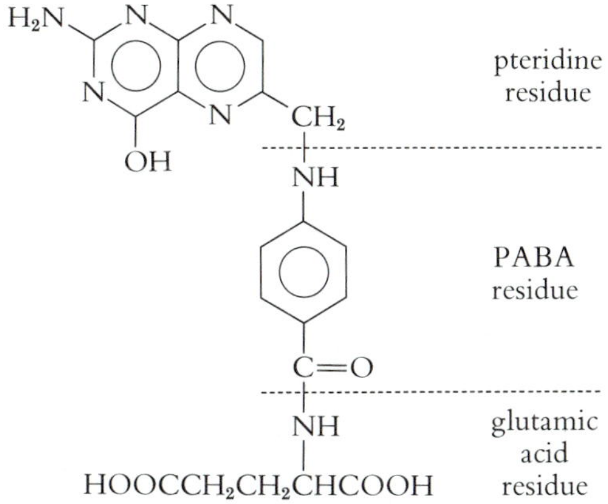

folic acid

The discovery of the antibacterial effects of sulfanilamide led to an explosion of interest in these compounds. Derivatives such as sulfapyridine and sulfathiazole were found to have many times the activity of sulfanilamide itself. However, these and other sulfa drugs have some undesirable characteristics, such as a tendency to crystallize out in the kidneys and block the renal passages, with sometimes fatal results. The "sulfonamide era" came to an abrupt end when penicillin was developed—production of sulfa drugs dropped from an all-time high of 10 million pounds in 1943 to less than half that amount the following year.

Although sulfa drugs are now used only in veterinary medicine and to combat a few bacterial infections in humans, the sulfonamide era led to improvements in therapeutic techniques, stimulated research on antibiotic drugs, advanced the theory of drug design, and resulted in the development of many related drugs for specific illnesses.

METHODOLOGY

The usual synthesis of sulfanilamide and its derivatives is carried out by combining *p*-acetamidobenzenesulfonyl chloride (also called acetylsulfanilyl chloride or ASC) with ammonia or an amine, followed by hydrolysis to remove the acetyl group. Acetylsulfanilyl chloride itself is prepared by chlorosulfonating acetanilide, which is easily obtained from aniline.

Synthesis of sulfonamides

At your instructor's request, you may prepare sulfanilamide (Collateral Project 2) in addition to or in place of a sulfanilamidopyridine.

2-aminopyridine   3-aminopyridine

In this experiment, you will prepare a sulfanilamide derivative using 2-aminopyridine or 3-aminopyridine as the amine. Combination of the 2-amino compound with ASC yields (after hydrolysis) 2-sulfanilamidopyridine, which was the first sulfa drug to incorporate a heterocyclic ring into its structure; it is still one of the most effective of the sulfa drugs. 3-Aminopyridine reacts with ASC to yield the isomeric 3-sulfanilamidopyridine. This compound was never used extensively as an antibacterial, perhaps because of the higher cost of 3-aminopyridine.

Chlorosulfonation is an electrophilic aromatic substitution reaction that results in the transformation of ArH to $ArSO_2Cl$, an arenesulfonyl chloride. It is usually carried out by treating the aromatic compound with an excess of chlorosulfonic acid ($ClSO_3H$). Because gaseous HCl is evolved during the reaction, a gas trap must be used. After the reaction is complete, the reaction mixture is poured onto crushed ice to destroy the excess sulfonic acid and dissolve the inorganic products of the reaction. It is essential to remove all water from the crude acetylsulfanilyl chloride because this compound readily hydrolyzes to the corresponding sulfonic acid, which inhibits its reaction with the aminopyridines. The water is removed by dissolving the partly dried ASC in acetone, filtering off the insoluble sulfonic acid, and drying the solution thoroughly with magnesium sulfate.

Sulfonamides are synthesized by methods similar to those used in preparing carboxamides from carboxylic acid chlorides and amines. Unsubstituted sulfonamides are formed by adding the arenesulfonyl chloride solution to an excess of ammonia, as described in Collateral Project 2. The extra NH_3 ties up HCl (generated during the reaction) that would otherwise hydrolyze the sulfonyl chloride. Since the aminopyridines used in this experiment are relatively expensive, keeping them in excess is not practical; instead, the reaction will be carried out in dry acetone (as the solvent) with a small amount of pyridine to react with the evolved HCl. The product, a 4-acetamido-*N*-pyridylbenzenesulfonamide, crystallizes out of solution and can be collected by vacuum filtration.

The acetyl group is removed by refluxing this intermediate product in dilute sodium hydroxide. Although both the sulfonamide and carboxamide linkages could, in theory, be hydrolyzed by base, sulfonamide groups are much more resistant to hydrolysis and are unaffected under the reaction conditions. After the reaction mixture is neutralized, the final product can be isolated by filtration and recrystallized from aqueous acetone.

Biological testing of the drugs (see Collateral Project 1) can be carried out on nutrient agar plates inoculated with bacteria by introducing paper discs soaked in different sulfa drugs, incubating the plates, and observing the effect of each drug on the growth of the bacterial colonies. Their relative effectiveness can be roughly evaluated by measuring the size of the *zone of inhibition* surrounding the antibacterial agent. Since the testing requires sterile conditions and special equipment, it should be performed in a separate section of the laboratory or, if possible, in a biology lab equipped with the necessary materials and apparatus.

pyridine

sulfonamide
linkage

carboxamide
linkage

Many sulfonamides are unaffected even by fusion at 250°C with 80% sodium hydroxide.

PRELAB ASSIGNMENT

Calculate the mass of 25 mmol of acetanilide, the mass of 20 mmol of an aminopyridine, and the theoretical yield of the expected sulfanilamidopyridine.

Reactions and Properties

Reactions are illustrated for 2-aminopyridine; the reaction of 3-aminopyridine is analogous.

A.

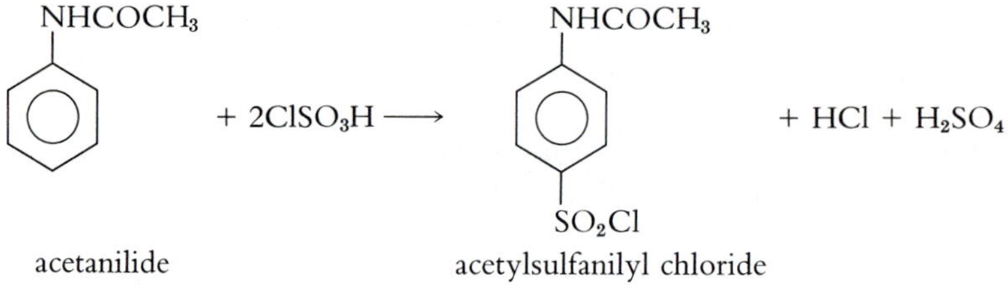

NHCOCH$_3$

$+ 2ClSO_3H \longrightarrow$

NHCOCH$_3$

SO$_2$Cl

$+ HCl + H_2SO_4$

acetanilide   acetylsulfanilyl chloride

B.

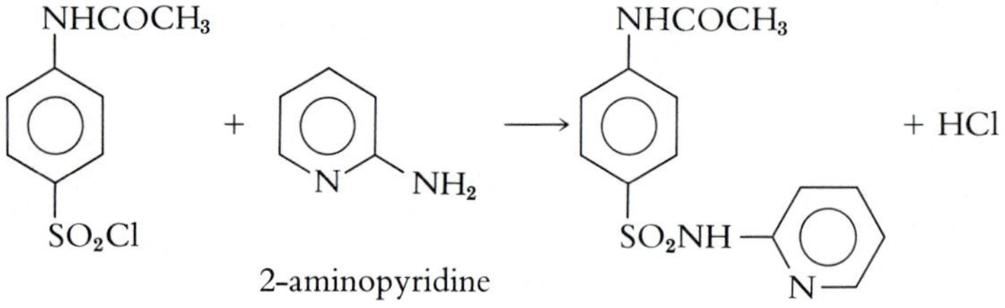

NHCOCH$_3$

SO$_2$Cl

$+$

2-aminopyridine

$\longrightarrow$

NHCOCH$_3$

SO$_2$NH —

$+ HCl$

C.

NHCOCH$_3$

SO$_2$NH —

4-acetamido-N-(2-pyridyl)-benzenesulfonamide

$+ H_2O \xrightarrow{OH^-} H_2N$ — — SO$_2$NH — $+ CH_3COOH$

sulfapyridine

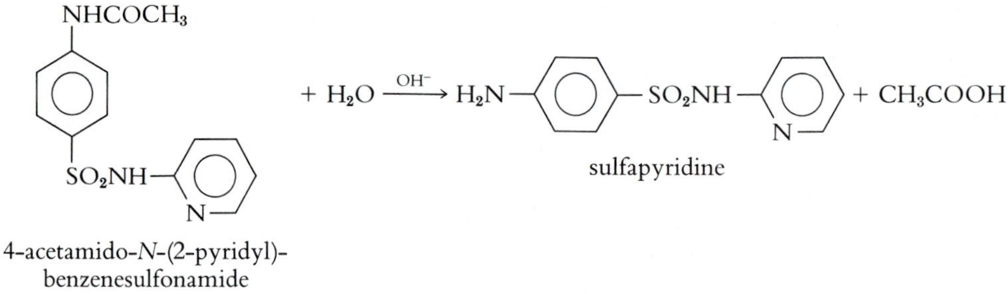

Table 1.   Physical properties

	M.W.	m.p.	b.p.	d.
acetanilide	135.2	114	304	
chlorosulfonic acid	116.5	−80	158	1.766[18]
acetylsulfanilyl chloride	233.7	149		
2-aminopyridine	94.1	59−60	204	
3-aminopyridine	94.1	54−55	252	
pyridine	79.1	−42	115.5	0.982
2-sulfanilamido-pyridine	249.3	190−92		
3-sulfanilamido-pyridine	249.3	258−59		

PROCEDURE

Wear protective gloves during this experiment, particularly while handling chlorosulfonic acid, the aminopyridines, and acetylsulfanilyl chloride. You must use an efficient fume hood during part A and the first part of part B. The glassware for part A must be thoroughly dried.

A. Preparation of Acetylsulfanilyl Chloride (ASC) from Acetanilide

Measure 25 mmol of acetanilide into a dry 250-mL round-bottomed flask, and melt it with a suitable heat source. Then swirl the flask to spread the acetanilide evenly around the bottom half of the flask as it solidifies. Cool the flask in an ice bath ⟨OP – 8⟩ for 10 minutes. Assemble a gas trap ⟨OP – 22*b*⟩ containing dilute sodium hydroxide, using a thermometer adapter or rubber stopper to fit the flask. *Under the hood,* with the flask still in the ice bath, *carefully* add 10 mL of chlorosulfonic acid **[contact, vapor, and reactivity hazard! wear gloves!]** to the flask through a dry funnel, and immediately connect the gas trap. Remove the flask from the ice bath, and swirl it until the acetanilide has completely dissolved; then heat it on a steam bath for 10 minutes.

Cool the reaction mixture in an ice bath for 10 minutes. Then *slowly* and *carefully* **[violent reaction! work under a hood!]** pour the liquid in a thin stream, with constant stirring, into a 150-mL beaker half-filled with cracked ice and water. Carefully rinse the flask with a little ice water, and add the rinsings to the beaker. Break up the precipitate with a stirring rod, and collect it by vacuum filtration ⟨OP – 12⟩. Wash the crude ASC **[contact hazard!]** with ice water, and let it air-dry

A friable solid is one that can readily be crumbled to a powder.

on the filter for at least 10 minutes. Press the solid between large filter papers [wear gloves!] until it is friable. Stir and shake the ASC with 20 mL of dry acetone in a small Erlenmeyer flask until no more dissolves. Add 1 g of anhydrous magnesium chloride, and swirl for 2–3 minutes; then filter by gravity ⟨OP–11⟩, and wash the residue with a little dry acetone, adding the wash to the filtrate. Treat the ASC solution with additional 1-g portions of magnesium sulfate until it is thoroughly dry ⟨OP–20⟩; at this point, the drying agent should remain as a fine white powder in the bottom of the flask and not clump up into large crystals. Use enough acetone for the washing to keep the total volume of the ASC solution close to 20 mL. Use this (filtered) solution promptly in the next step, since ASC does not keep well.

B. Reaction of Acetylsulfanilyl Chloride with 2- or 3-Aminopyridine

The aminopyridines are poisonous and can irritate the skin, eyes, and respiratory tract. 2-Aminopyridine is readily absorbed through the skin and can be fatal if inhaled or swallowed. Wear gloves, avoid contact and inhalation, and wash your hands thoroughly after using these chemicals.

Acetone is very flammable, so keep the liquid away from flames and hot surfaces.

Pyridine has a strong, unpleasant odor, and inhalation, ingestion, or skin absorption may be harmful. Use gloves and a fume hood; avoid contact, and do not inhale its vapors.

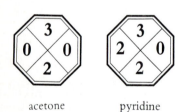

acetone   pyridine

The pyridine should be dried over KOH pellets.

Under a hood, dissolve 20 mmol of 2- or 3-aminopyridine [**poison! contact hazard!**] in 5 mL of dry acetone in a small Erlenmeyer flask, and add 2.5 mL of dry pyridine. Add the ASC solution from part A, swirl to mix, and watch the mixture for evidence of reaction. If a precipitate forms immediately, break up any lumps with a stirring rod. After the reaction has ceased, stopper the flask and set it aside for at least 24 hours; any oil that formed initially should completely crystallize by this time. Cool the reaction flask thoroughly in an ice bath, and collect the product by vacuum filtration ⟨OP–12⟩, washing it on the filter with water. Collect a second crop of crystals by concentrating the filtrate to half its volume by evaporation, cooling it in ice, and repeating the vacuum filtration.

If the product is to be tested for antibacterial activity (see Collateral Project 1), purify the first crop of crystals by recrystallization from acetone, and use that batch in subsequent steps. The second crop can then be hydrolyzed separately or weighed and turned in.

C. Preparation of 2- or 3-Sulfanilamidopyridine

Mix the product from part B with 15 mL of 3 M aqueous sodium hydroxide, and reflux the mixture ⟨OP–7a⟩ for 30 minutes. Transfer the cooled reaction mixture to a beaker, and carefully make it *just* acidic to blue litmus paper with 6 M HCl. Cool the mixture in an ice bath, and when crystallization is complete, collect the precipitate by vacuum filtration ⟨OP–12⟩. Recrystallize the product from 85% aqueous acetone ⟨OP–23⟩, filtering the hot solution through a fluted filter paper. Dry the purified product ⟨OP–21⟩, weigh it, and measure its melting point ⟨OP–28⟩.

If you add too much HCl and the precipitate dissolves, add enough dilute NaOH to neutralize the solution.

Use decolorizing carbon if necessary. A rather large volume of recrystallization solvent may be needed, but be careful not to add too much.

Report. Calculate the percentage yield of your product, and try to account for significant losses. Explain why it was important to avoid adding too much 6 M HCl when you precipitated the final product, and write balanced equations for the reaction of the product with HCl and NaOH.

Collateral Projects

1. If the necessary facilities are available and your instructor permits, test the antibacterial activity of 2-sulfapyridine and 3-sulfapyridine, using sulfanilamide as a reference compound. Work in pairs, using a nutrient agar plate previously inoculated with a specific type of bacterium (*E. coli* or another type that can be handled safely). Dissolve 0.10 g each of 2-sulfanilamidopyridine, 3-sulfanilamidopyridine, and sulfanilamide (as a reference compound) in separate flasks containing 30 mL of boiling 25% ethanol. Mark four quadrants on the bottom of the Petri dish with a grease pencil and number the quadrants 1–4; then write your names on the cover.

Sterilize a metal-tipped forceps by dipping the ends in 70% ethanol and then placing them in a burner flame. Use the forceps to dip a sterile paper disc into one of the solutions while it is still boiling. Transfer the disc to a piece of filter paper, and

For successful results, the product must be made from the purest available starting materials and scrupulously purified in part C. It may be necessary to recrystallize the product a second time, using decolorizing carbon.

"Quadrant plates" with the quadrants already marked are available.

Throughout this process, care must be taken to protect the sample discs and the agar plate from contamination by undesirable bacteria. The filter paper and watch glass should be as sterile as possible.

The Petri dish should be opened by lifting the cover straight up, just high enough to allow room for the forceps, to keep out airborne bacteria.

The cloudiness on the plate is due to bacterial growth, so a clear area indicates inhibition.

Use a small metric ruler and measure the diameter in millimeters.

It is a good idea to compare your results with those of other students before drawing any conclusions about the relative effectiveness of the sulfapyridines, since the action of the drug is affected by impurities.

The ASC should be used just after the vacuum filtration; it can still be somewhat moist.

If a precipitate forms when the reaction mixture is cooled, the reaction is incomplete; continue to reflux until no solid forms on cooling.

cover it immediately with a watch glass. Repeat the process with the other solutions, sterilizing the forceps after each transfer. A fourth disc should be dipped into boiling distilled water for use as a control. When the discs are dry, again sterilize the forceps and transfer each disc to the appropriate quadrant inside the Petri dish: 1 for sulfanilamide, 2 for 2-sulfanilamidopyridine, etc.

Incubate the agar plate at 28 – 30°C in an incubator oven or temperature-controlled room and observe it after 24 and 48 hours of incubation. Holding the Petri dish up so that you can observe it from the bottom, measure the diameter of the clear circle surrounding each sample disc and that surrounding the control. Calculate the area of each circle and subtract the area for the control from that for each sample.

Compare **(a)** the relative effectiveness of the three sulfa drugs, as indicated by the areas of inhibition after 24 hours; and **(b)** their long-term persistence, as indicated by the decrease in size of the zone of inhibition between 24 and 48 hours. If the size of the zone does not decrease over that period, your product may contain bactericidal impurities that kill the bacteria rather than just inhibiting their growth. Compare your results with those of other students in your lab section, and rank the three compounds in order of their apparent antibacterial effectiveness.

2. You can synthesize sulfanilamide from acetylsulfanilyl chloride prepared as described in part A. *Under the hood*, combine 12 mL of concentrated aqueous ammonia **[contact and vapor hazard!]** with 12 mL of water, and add the mixture to the ASC in an Erlenmeyer flask. Stir and rub the mixture with a stirring rod until a smooth paste is obtained; then heat it on a steam bath **[in a hood!]** for 20 – 30 minutes with occasional swirling. Cool the flask in an ice bath, acidify the solution to a pH of 3 with 6 M sulfuric acid, and collect the solid by vacuum filtration after further cooling. Combine it with 10 mL of 3 M HCl, and reflux the mixture for 30 – 45 minutes (heat gently at first to prevent charring). Cool the reaction mixture, and add 10 mL of water. Make the solution just basic to litmus by cautiously adding solid sodium bicarbonate **[foaming may occur!]** in small portions, with swirling. Cool in ice, collect the crude sulfanilamide by vacuum filtration, and purify it by recrystallization from water, using decolorizing carbon if necessary. Dry the sulfanilamide, and measure its mass and melting point.

3. Record the infrared spectrum of your sulfanilamido-pyridine (or of sulfanilamide) and interpret it as completely as you can.

Exercises

1. (a) What was the precipitate that remained when you dissolved the ASC in acetone? Write a balanced equation for its formation. **(b)** Write a balanced equation for the violent reaction of excess chlorosulfonic acid with water in part A.

2. In the reaction sequence outlined in the Methodology section, an acetyl group is added in the first step, only to be removed in the last, which adds two steps to the synthesis. Why are those extra steps necessary?

3. What type of reaction is the chlorosulfonation of acetanilide? Propose a mechanism for this reaction.

4. (a) What type of reaction is the combination of ASC with 2- or 3-aminopyridine? What type of reaction is the hydrolysis of this product to a sulfanilamidopyridine? **(b)** Propose mechanisms for both reactions.

5. (a) Sulfapyridine has also been synthesized by the base-catalyzed reaction of 2-chloropyridine with sulfanilamide. Write a balanced equation for this reaction and designate the reaction type. **(b)** Propose a mechanism for this reaction.

6. Construct a flow diagram for the synthesis you carried out.

Library Topic

Some modern drugs such as salicylazosulfapyridine, probenecid, tolbutamide, and dapsone bear structural relationships to the sulfa drugs. Give the structures of these drugs, and report on their physiological action and applications.

The Preparation of Para Red and Related Azo Dyes

Reactions of Amines. Reactions of Diazonium Salts. Preparation of Azo Compounds. Electrophilic Aromatic Substitution.

Operations

OP–8   Cooling

OP–12 Vacuum Filtration

OP–21 Drying Solids

Objectives

To prepare Para Red and one or more other azo dyes.

To dye cloth by the ingrain process.

To learn how to carry out diazotization and coupling reactions.

To learn about dyes and dyeing.

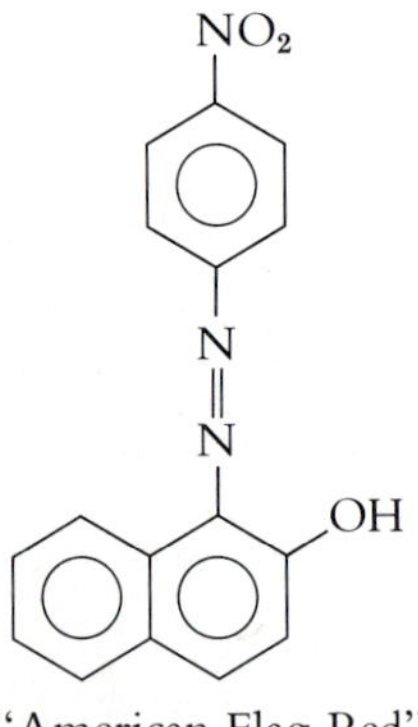

"American Flag Red"
(Para Red)

In the Persian fairy tale "The Three Princes of Serendip," the title characters were forever discovering things they were not looking for at the time—thus, serendipity is the aptitude for making happy discoveries by accident.

SITUATION

A synthetic dye made from *p*-nitroaniline and 2-naphthol was once called "American Flag Red" because it was used to dye the cloth used for the stripes in the American flag. Also known as Para Red, this compound is a member of the large family of dyestuffs called the azo dyes. Many azo dyes can be easily prepared by diazotizing an aromatic amine and mixing the resulting diazonium salt with an activated aromatic compound, such as a phenol or another aromatic amine. In this experiment, you will prepare Para Red and some other azo dyes and learn how to dye cloth by a technique known as the ingrain process.

BACKGROUND

Dyes and Serendipity

The ability of certain plant materials to dye cloth in different colors was probably known long before people began to keep written records. The rapid advancement of organic chemistry as a science is due, in large part, to the discovery that chemists could prepare synthetic colors that were in many ways superior to the natural ones. The first commercially important synthetic dye was made almost by accident when an eighteen-year-old research assistant, William Henry Perkin, naively tried to synthesize quinine by oxidizing allyltoluidine.

See Bib–L43 for the fascinating story of the preparation of the first synthetic dye, pittacal.

Perkin's idea

allyltoluidine   quinine

We now know that he had attempted the impossible — the structure of allyltoluidine bears hardly any resemblance to that of quinine. In 1856, the structural theory of organic chemistry was in a primitive stage of development and Perkin had only the molecular formulas to go on. Allyltoluidine's formula is $C_{10}H_{13}N$, and that of quinine is $C_{20}H_{24}N_2O_2$, so by adding a little oxygen and eliminating a molecule of water from two of allyltoluidine . . . at least it seemed like a good idea at the time! So Perkin painstakingly oxidized allyltoluidine with potassium dichromate and came up with a reddish-brown precipitate that was definitely not quinine. Most chemists would have thrown out the stuff and started over, but it had properties that interested Perkin, so he decided to try the same reaction with a simpler base, aniline. This time he obtained a black precipitate that, when extracted by ethanol, formed a beautiful purple solution that greatly impressed some of the local dyers. Perkin knew a good thing when he saw it, so he promptly gave up his study of chemistry and went into the business of manufacturing "aniline purple," or mauve as the dye soon came to be known.

While Perkin was getting the synthetic-dye industry under way, other chemists were experimenting with aniline and the vast array of other compounds that could be extracted from coal tar. One of these was a brewery chemist named Peter

aniline

Perkin's dyestuffs plant was so successful that he was able to retire at the age of 36 and devote the rest of his life to pure research. Read an account of his remarkable career in *Scientific American*, Feb. 1957, p. 110.

mauve

An account of Griess's work can be found in *J. Chem. Educ. 35,* 187 (1958).

Griess, who took time off from the brewing of Alsopps' Pale Ale to discover the azo dyes. Undiscouraged by the fact that many of the diazo compounds he prepared had an unfortunate tendency to explode, Griess did some fundamental research into the diazotization of aromatic amines and went on to discover the coupling reaction by which virtually all azo colors are now synthesized. Aniline Yellow and Bismark Brown were developed in the 1860s; since then, the production of azo dyes has grown very rapidly. The number of azo dye formulations now exceeds that of all other dyes put together.

Bismark Brown

Aniline Yellow

Azo dyes are used to dye cloth by several different processes. In the *direct process,* the acidic or basic form of the dye is dissolved in water, the solution is heated, and the cloth that is to be dyed is immersed in the hot solution. The dye molecules attach themselves to the cloth fibers by direct chemical interactions. In the *disperse process,* a water-insoluble dye is suspended in water and a small amount of a carrier substance is added. The carrier dissolves the dye and carries it into the fibers, where it becomes trapped because of its hydrophobic nature. *Ingrain* (or *developed) dyes* are synthesized right inside the fiber. Rather than mixing the diazonium salt and the coupling component directly, one immerses the cloth in separate solutions of each. The relatively small molecules of the separate components can diffuse into the spaces between the fibers, but once they have combined within the fiber to form the dye, the larger dye molecules are trapped there.

The color of a dye depends on the wavelengths of the light it absorbs. If a dye absorbs light at a certain wavelength, the color perceived by the human eye consists of the remaining visible wavelengths that are reflected. The colors of the dyes you will prepare in this experiment should be some shades of yellow, orange, or red; a yellow color is usually associated with absorption of light of shorter wavelengths and a deep red color with absorption of light of longer wavelengths. The light-absorbing portion of a dye molecule, called a *chromophore,* is a conjugated system of delocalized pi electrons. The chromophoric system of Para Red includes the benzene and naphthalene rings, the two doubly bonded nitrogen atoms that connect them, and the unsaturated nitro group. In general, the more extended the chromophore, the longer the wavelength of the light it absorbs. Thus, Para Red, with 21 atoms in its chromophore, absorbs light of considerably longer wavelengths than does Aniline Yellow, with only 14 atoms in its chromophore. Certain saturated substituents called *auxochromes* can, in effect, extend a conjugated system by resonance. Auxochromes such as the OH, OCH_3, and NR_2 groups have one or more pairs of nonbonded electrons that they can share with a chromophore, thereby increasing the wavelength of the light it absorbs. Such groups are likely to have a significant effect only if they are *ortho* or *para* to the N=N group so that they can participate in resonance with the rest of the chromophore.

METHODOLOGY

The preparation of an azo compound involves two stages, known as *diazotization* and *coupling*. In the first stage, a primary aromatic amine reacts with nitrous acid (HONO), forming a diazonium salt. The nitrous acid is generated *in situ* from sodium nitrite and a mineral acid; the reaction is carried out at a low temperature so as not to decompose the diazonium salt. In the second stage, the diazonium salt is added to a weakly acidic or basic solution of the coupling component, which is usually a phenol or another amine. Phenols couple most readily in mildly alkaline solutions, whereas amines react best in acid. However, too low a pH will prevent an amine from reacting by causing protonation of the amino group, whereas too high a pH will cause the diazonium salt to change to a diazotate ion, which is incapable of coupling. The coupling reaction is an electrophilic aromatic substitution reaction with the diazonium salt acting as the electrophile. Because the diazo ($-\overset{+}{N}\equiv N$) group is only weakly electrophilic, the coupling component must contain strongly activating groups, such as OH or NR_2, in order for the reaction to occur.

A diazonium salt can be prepared by dissolving a primary amine in about 2.5 equivalents of dilute hydrochloric acid (or other suitable acid) in a flask, cooling the solution to 5°C or below in an ice-salt bath, and adding an equivalent amount of an aqueous solution of sodium nitrite while keeping the temperature at or below 10°C. Toward the end of the addition, the solution can be tested for excess nitrous acid using starch-iodide paper.

The coupling reaction is carried out by adding the diazo compound, with cooling and stirring, to a solution of a coupling component in dilute acid or base. If the coupling compo-

Diazotization

$$ArNH_2 \xrightarrow{\text{HONO}} Ar-N_2^+$$
$$\text{diazonium salt}$$

Coupling

$$Ar-N_2^+ + H-Ar' \xrightarrow{-H^+}$$
$$\text{coupling}$$
$$\text{component}$$

$$Ar-N=N-Ar'$$
$$\text{azo compound}$$

(Ar' must contain an activating group such as —OH or —NR_2.)

See the next section for the reactions used in preparing Para Red.

If the amine is insoluble, the diazotization is carried out in suspension, with stirring. To obtain as fine a suspension as possible, the amine is dissolved with heating and the solution is cooled rapidly with stirring.

Formation of unreactive species at low and high pH

Low pH: $\text{ArNR}_2 \underset{}{\overset{\text{H}^+}{\rightleftharpoons}} \text{ArNHR}_2{}^+$

High pH: $\text{ArN}_2{}^+ \underset{}{\overset{\text{OH}^-}{\rightleftharpoons}}$

$$\text{ArN}{=}\text{N}{-}\text{O}^-$$
diazotate
ion

The sodium carbonate solution should be added slowly to reduce foaming.

COOH
NH₂

anthranilic acid
(*o*-aminobenzoic acid)

nent is a phenol, it can be dissolved in about 2 equivalents of 1 M sodium hydroxide and cooled before adding the diazonium salt solution. The pH should be adjusted, if necessary, after the addition; the azo dye crystallizes out on cooling. If the coupling component is an amine, it should be dissolved in 1 equivalent of 1 M HCl; after the diazo component is added and coupling is complete, the solution is neutralized to litmus paper by adding 3 M aqueous sodium carbonate. In some cases, the yield and appearance of the product may be improved by adding solid sodium chloride (about 2 grams) to the mixture containing the precipitated dye and heating the mixture on a steam bath for a few minutes. The dye is then cooled in an ice bath before filtering. Some of the dyes decompose on heating, so they should be dried at room temperature.

You will dye a piece of cloth by soaking it in the solution containing the diazonium salt and then in the solution of the coupling component, so that the color will develop on the fabric itself. You may also wish to experiment with some other dyeing methods, as described in Collateral Projects 1 and 2.

When preparing an azo dye, it is important to keep the diazonium salt solution cold before adding it to the coupling component, because heating the solution brings about another kind of reaction. If you do Minilab 31, you will observe the effect of heating the diazonium salt of anthranilic acid in solution and identify the product that forms.

PRELAB ASSIGNMENT

Calculate the mass of 10 mmol of *p*-nitroaniline, the mass of 10 mmol of *2*-naphthol, and the theoretical yield of Para Red. Be prepared to calculate the mass of 10 mmol of any other coupling component or diazo component and the theoretical yield of any other azo dye.

Reactions and Properties

Equations are given for the preparation of Para Red only.

Diazotization

NH₂

$+ \ 2\text{HCl} + \text{NaNO}_2 \longrightarrow$

$\overset{\oplus}{\text{N}}{\equiv}\text{N}: \ \overset{\ominus}{\text{Cl}}$

$+ \ 2\text{H}_2\text{O} + \text{NaCl}$

NO₂
p-nitroaniline

NO₂
p-nitrobenzenediazonium chloride

Coupling

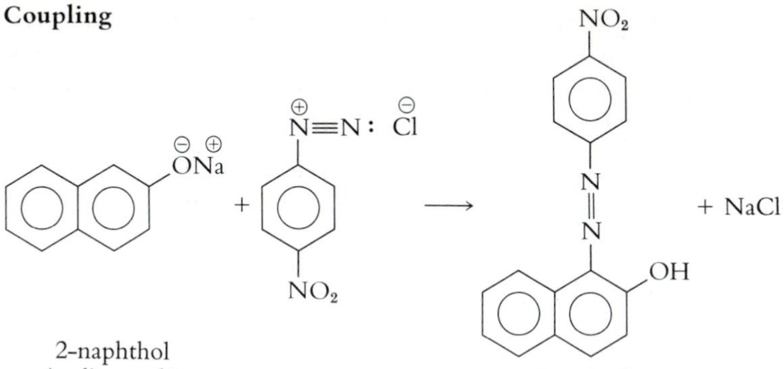

2-naphthol
(sodium salt)

Para Red

+ NaCl

Table 1.   Diazo and coupling components

Diazo component	M.W.	Coupling component	M.W.
aniline	93.1	aniline	93.1
m-anisidine	123.2	*N*-methylaniline	107.2
m-nitroaniline	138.1	*N,N*-dimethylaniline	121.2
m-toluidine	107.2	*m*-phenylenediamine	108.1
p-anisidine	123.2	phenol	94.1
p-nitroaniline	138.1	1-naphthol	144.2
p-toluidine	107.2	2-naphthol	144.2
		resorcinol	110.1

PROCEDURE

Each student should prepare Para Red and at least one other dye. With the instructor's permission, students can work in small groups, with each person preparing a different set of dyes. Any diazo component from Table 1 can be used in combination with any coupling component.

Aromatic amines are very harmful if swallowed, inhaled, or absorbed through the skin; they may also cause allergic skin reactions. *p*-Toluidine has induced cancer in laboratory animals when administered orally, and aniline is a suspected carcinogen. Wear gloves, and dispense the amines under a hood; avoid contact, and do not inhale their vapors.

SAFETY PRECAUTION

aromatic amines

Sodium nitrite may cause a fire if mixed with combustible materials, so keep it away from other chemicals and combustibles.

Most phenols are toxic and irritant and some are very corrosive, causing severe irritation or damage to skin and eyes. Phenol can cause poisoning by ingestion, inhalation, or skin absorption, and has been reported to cause cancer in laboratory animals when applied to the skin. Wear protective gloves; avoid contact and do not inhale dust or vapors.

Some azo dyes such as *p*-phenylazoaniline (Aniline Yellow) and 1-phenylazo-2-naphthol have been reported to cause cancer in laboratory animals when ingested or applied to the skin. Although it is unlikely that momentary exposure to an azo dye will cause cancer, you should minimize your contact with the dyes.

A. Preparation of an Azo Dye

1. Diazotization of an Aromatic Amine. Dissolve 10 mmol of the diazo component [contact and vapor hazard! wear gloves!] in 8 mL of 3 M HCl; heat the solution gently and add up to 10 mL of water if necessary to get most of the solid in solution. Cool this solution to 5°C in an ice bath ⟨OP–8⟩ with stirring. Continue to stir as you add 10 mL of 1 M sodium nitrite (freshly prepared) slowly enough so that the temperature remains below 10°C during the addition. Test the solution with starch-iodide paper and add a little more sodium nitrite drop by drop, if necessary, to give a positive test (blue-violet color). Leave your diazonium salt solution in the ice bath and go to step 2a or 2b, depending on the nature of your coupling component.

The amine salt may precipitate as you cool the solution, but it will diazotize satisfactorily if the reaction mixture is well stirred.

Test the starch-iodide paper by moistening it with dilute HCl and adding a drop of sodium nitrite solution; if it does not turn blue-violet, obtain fresh test paper.

If you add too much sodium nitrite, destroy the excess by stirring in just enough urea to give a negative starch-iodide test.

2a. Coupling with a Phenol. If your coupling component is a phenol, dissolve or suspend 10 mmol of the phenol [contact hazard! wear gloves!] in 20 mL of 1 M NaOH (use 40 mL for a dihydroxyphenol), and cool the solution in an ice bath. Slowly add the diazonium salt solution to the coupling component solution, with stirring, and leave the mixture in the ice bath for 15 minutes or more until crystallization is complete. If little or no colored solid appears, adjust the pH with dilute HCl or NaOH to induce precipitation. Collect the azo dye [may be

The phenol may not be completely dissolved.

carcinogenic!] by vacuum filtration ⟨OP–12⟩, washing it on the filter with water. Dry your azo dye ⟨OP–21⟩, and weigh it.

2b. Coupling with an Amine. If your coupling component is an aromatic amine, dissolve or suspend 10 mmol of the amine **[contact and vapor hazard! wear gloves!]** in 10 mL of 1 M HCl (use 20 mL for a diamine), and cool the solution in an ice bath. Slowly add the diazonium salt solution, with stirring, to the coupling component solution, and leave the mixture in the ice bath for 15 minutes or more. Neutralize the solution to litmus with 3 M aqueous sodium carbonate, then allow it to stand in the ice bath until crystallization is complete. Collect the azo dye **[may be carcinogenic!]** by vacuum filtration ⟨OP–12⟩, washing it on the filter with water. Dry your azo dye ⟨OP–21⟩, and weigh it.

B. Dyeing a Cloth by the Ingrain Process

Select one or more pairs of diazo components and coupling components. Prepare separate solutions of the diazonium salt and the coupling component as described in part A, and keep them cold. Dilute the solution of the coupling component to 100 mL with water, and soak a piece of clean white cloth in it for 2–3 minutes. Remove the cloth with a forceps or a pair of stirring rods, and blot it between towels to remove most of the water; then hang it up to dry. Dilute the diazonium salt solution to 100 mL with ice water, insert the dry cloth, and agitate the solution with a stirring rod for a few minutes to dye the cloth uniformly. Remove the cloth, and dry it as before. Turn in the dyed cloth with your report.

Report. Write balanced equations for all the syntheses of azo dyes you carried out. Calculate the percentage yield of each dye, and describe the color of both the dye itself and the dyed cloth. Tabulate the colors of the dyes prepared by your group or lab section, and discuss the effects of structural features (such as substituents and chromophore size) on color.

Collateral Projects

1.   Use your dye crystals for direct dyeing by suspending 0.5 g of the dye in 100 mL of hot water and acidifying the mixture with a few drops of concentrated sulfuric acid [contact hazard!]. Immerse pieces of wool or cotton cloth in the mixture for 5 minutes or more. Remove the cloth, rinse it with water, and let it dry. Adjusting the pH of the dyeing mixture with dilute HCl or NaOH may give better results in some cases.

2.   Use your dye crystals for disperse dyeing by suspending 0.5 g of the dye in 100 mL of hot water and stirring in 0.1 g of biphenyl (the carrier) and 2–3 drops of liquid detergent. Then immerse a piece of cloth made of Dacron or another polyester in the mixture, and heat the solution on a steam bath for 15–20 minutes. Remove the cloth and let it dry.

If any solid remains undissolved, filter the solution.

3.   Dissolve 3–5 mg of an azo dye in 10 mL of 95% ethanol. Record the ultraviolet-visible spectrum of the dye ⟨OP–35⟩ over the tungsten lamp range (~800–350 nm), diluting the solution with more ethanol if necessary. Compare the λ_{max} values of different dyes, and try to explain some of the differences you observe. Note that such comparisons are meaningful only if the bands you are comparing arise from the same kind of electronic transition; such bands should be similar in appearance and intensity.

MINILAB 31

Diazotization and Hydrolysis of Anthranilic Acid

With your instructor's permission, you can also record and interpret its IR or PMR spectrum.

Diazotize 1.37 g (10 mmol) of anthranilic acid as described in step 1 of part A of the Procedure section. Heat the solution to boiling, and boil it gently until no more gas is evolved. Cool it in an ice bath until crystallization is complete. Collect the product by vacuum filtration, and purify it by recrystallization from boiling water, using a little decolorizing carbon. Dry the product, and measure its mass and melting point. Test the compound with 2.5% ferric chloride as described in Test C–13 of Part III. Deduce the structure of the product, and verify your conclusion by checking its melting point in an appropriate reference book. Write balanced equations for the reactions involved, and identify the gas that was evolved.

Exercises

1. Write a detailed mechanism for the coupling reaction of *p*-nitrobenzenediazonium chloride with 2-naphthol.

2. A student was attempting to prepare chrysoidine by coupling 10 mmol of *m*-phenylenediamine with an equimolar amount of aniline. To her surprise, she had to add 20 mL of 1 M sodium nitrite in the diazotization step before the solution turned starch-iodide paper blue. After pouring the diazonium salt solution into the solution of the other component, she recovered a dark-colored precipitate that was not chrysoidine. The following day the filtrate contained another precipitate that she identified as resorcinol. **(a)** What did she do wrong, and what was the structure of the azo dye she synthesized? **(b)** Write balanced equations for its synthesis and for the reaction that formed resorcinol.

3. Another student was trying to prepare *p*-dimethyl-aminoazobenzene (Butter Yellow) by coupling 10 mmol of *N,N*-dimethylaniline with an equimolar amount of aniline. He added 20 mL of 1 M sodium nitrite to the diazo component; mixing this solution with the coupling component yielded some Butter Yellow along with a pale yellow oil. **(a)** What did he do wrong, and what was the yellow oil? **(b)** Write a balanced equation for its formation.

4. Why is it important to keep the temperature low during diazotization and coupling? Give the structure of the product that might form if the solution is heated during the diazotization of *p*-nitroaniline. Write an equation for its formation.

5. Why does coupling of *p*-nitrobenzenediazonium chloride occur mainly *para* to the $-N(CH_3)_2$ group of *N,N*-dimethylaniline but *ortho* to the $-OH$ group of 2-naphthol?

6. Outline syntheses of Aniline Yellow and Bismark Brown starting with benzene.

chrysoidine

Library Topic

Write a short report on the chemistry and uses of food colorings. Outline syntheses for some azo dyes that have been used as food colorings and discuss the controversy surrounding such dyes as FD&C Red No. 2.

The Synthesis and Spectral Analysis of Aspirin

Reactions of Phenols. Preparation of Esters. Nucleophilic Acyl Substitution.

Operations

OP–11 Gravity Filtration

OP–12 Vacuum Filtration

OP–21 Drying Solids

OP–23 Recrystallization

OP–23b Recrystallization from Mixed Solvents

OP–33 Infrared Spectrometry

OP–34 Nuclear Magnetic Resonance Spectrometry

Objectives

To synthesize aspirin from salicylic acid and assess its purity.

To analyze the IR and PMR spectra of aspirin.

To learn about the history and uses of aspirin.

To learn about the IR and PMR spectra of carboxylic acids and esters and about long-range NMR coupling effects.

SITUATION

Aspirin is so familiar to us that we tend to take it for granted, but it is truly one of the world's wonder drugs. No other drug can do so many things so well — from reducing fever, pain, and inflammation to preventing hypertension and coronary disease. In this experiment, you will learn a little more about this remarkable substance by preparing it from salicylic acid and exploring its molecular structure with infrared and nuclear magnetic resonance spectrometry.

BACKGROUND

Hermann Kolbe's Blunder

Although aspirin is the most important pain reliever of modern times, it was considered so unremarkable by its discoverer that it was forgotten for forty years before being resurrected by chemists working for the Bayer Company. The aspirin story began even before its discovery, when, in 1763, clergyman Edward Stone read a paper on the uses of willow bark to the Royal Society of London. Stone had found the bark of the white willow, *Salix alba,* effective in reducing the intermittent fever (ague) of malaria. Because of its derivation from *Salix* species, the active principle of willow bark was named salicin.

Salicin can be broken down in the presence of water and an oxidant to form the simpler molecules of glucose (simple sugar) and a sweet-smelling liquid named salicylaldehyde. In 1838, some German chemists discovered that salicylaldehyde obtained from another source, the meadowsweet plant (*Spiraea* species), reacts with strong alkali to yield a white solid on neutralization. The new compound was named spirsäure (*Spiraea* + *säure,* the German name for an acid) by its discoverers but was called salicylic acid by the English, who traced its lineage back to the white willow tree.

Salicylic acid and its derivatives, known collectively as the salicylates, comprise an important family of medicinally useful compounds. The properties of salicylic acid were discovered as the result of a "brilliant mistake" by Hermann Kolbe, one of the most eminent chemists of his day. Around 1870, Kolbe was approached by a physician who was looking for a safe substitute for carbolic acid (phenol). This acid had long been known to be an effective germicide for external use, but it was much too caustic to be taken internally. Twenty years earlier, Kolbe had invented a synthesis of salicylic acid from phenol and carbon dioxide. Perhaps, he reasoned, salicylic acid would break down in the human body to regenerate phenol and thus act as a safe, *internal* source of this germicide. He carried out a number of experiments that "proved" to his satisfaction that salicylic acid was indeed an effective germ-killer, and soon recommended its use on patients suffering from a variety of bacterial diseases and infections. The first reports seemed promising — patients were still dying after salicylic acid treatments, but they felt much better while doing so! Pretty soon doctors began to suspect that the salicylic acid "cured" only those patients who would have survived without any medi-

The aspirin family tree

cine. The blunder, of course, was Kolbe's — salicylic acid does not break down into phenol in the body, and his test results were later discarded as invalid.

Still, it was apparent that the acid provided some benefits: it made the patients feel more comfortable and reduced fever, and its curative effect on rheumatism was remarkable. Salicin was known to be mildly effective against pain and fevers, so salicylic acid was soon prescribed for those symptoms. The acid was found to relieve the pain of neuralgia, sciatica, neuritis, and headaches, as well as rheumatism. Salicylic acid might not be worth much as a germicide, but it was unsurpassed as a fever-fighter and pain-killer.

Aspirin Rediscovered

Salicylic acid has one serious drawback as a pain reliever — it irritates the mucous membranes lining the mouth, esophagus, and stomach. The irritation can be so severe that some patients treated with it almost preferred to live with their rheumatism rather than take the cure. One of these disgruntled patients was the father of Felix Hoffman, a staff chemist working for the Bayer Company. By a happy coincidence, Hoffman's company was interested in finding a substitute for salicylic acid at about the same time his father was suffering from its side effects, and this convergence of Hoffmann's research and personal interests provided the motivation that led to aspirin's rediscovery in 1893.

Hoffman knew that phenolic compounds were corrosive because of the presence of a free OH group, and he reasoned that "masking" the OH with some easily removed substituent might provide the benefits of salicylic acid without the irritation. While studying some of the known derivatives of salicylic acid, Hoffman came across one in which the hydrogen of the OH group was replaced by an acetyl group; this was acetylsalicylic acid, which had been prepared some forty years earlier by Charles Gerhardt. Tests proved that acetylsalicylic acid was superior to all known pain-killers, both in its effectiveness against pain and fever and its freedom from serious side effects. It was soon being marketed by Bayer under the trade name "aspirin."

Hoffman's intuition about the function of the acetyl group in aspirin turned out to be quite accurate. Aspirin apparently passes through the acidic environment of the stomach intact, but loses its acetyl group in the alkaline medium of the small intestine to regenerate salicylic acid. The salicylic acid is

aspirin

The name *aspirin* was derived from *acetylspirsäure*, the German name for acetylsalicylic acid.

then absorbed into the bloodstream and carried throughout the body.

METHODOLOGY

Acetic anhydride is an effective acetylating agent that reacts rapidly with alcohols or phenols in the presence of a catalyst to yield acetate esters. In this experiment, acetic anhydride will be used to convert salicylic acid to aspirin, and also to serve as a solvent for the reaction. Since acetic anhydride is so reactive, it will not be necessary to heat the reactants to reflux temperatures; merely warming them in a water bath will be sufficient. When the reaction is complete, water is added to destroy the excess acetic anhydride, converting it to water-soluble acetic acid.

The most likely impurities in the crude product are salicylic acid itself and a polymeric by-product formed by intramolecular reactions of salicylic acid. The polymer can be removed by dissolving the product in aqueous sodium bicarbonate solution. Aspirin reacts with sodium bicarbonate to form a water-soluble sodium salt, whereas the polymeric by-product is insoluble and is removed by filtration. Aspirin is reprecipitated from the solution of its salt by acidifying the filtrate with hydrochloric acid. Unreacted salicylic acid and other impurities are then removed by recrystallization from an ethanol-water mixture.

You will assess the purity of your aspirin by testing it for the presence of salicylic acid with ferric chloride, which forms highly colored complexes with phenolic compounds. By testing the crude and partly purified product as well as the final product, you may be able to determine whether the impurity (if there is any) resulted from incomplete reaction of the starting materials or was formed during the workup of the product.

Aspirin is both an ester and a carboxylic acid, and its infrared spectrum shows characteristics of both kinds of compounds. In the solid or liquid state, most carboxylic acids exist as dimers held together by strong hydrogen bonding. The O—H stretching band of a monomeric carboxylic acid appears as a rather sharp band near 3520 cm^{-1}, but hydrogen bonding in the dimer moves the band to lower frequencies and causes it to spread out over much of the region between 3300 cm^{-1} and 2500 cm^{-1}. The infrared spectrum of a typical carboxylic acid also features a C=O stretching band around 1725–1680 cm^{-1}, a C—O stretching band in the 1315–1280 cm^{-1} region, and an out-of-plane O—H bending band near 920 cm^{-1}. The

$$
\underset{\text{acetic anhydride}}{CH_3\overset{\overset{\displaystyle O}{\|}}{C}-O-\overset{\overset{\displaystyle O}{\|}}{C}CH_3}
$$

The interconversion of aspirin with its sodium salt is described in Experiment 2.

The melting point of aspirin is not a reliable indicator of its purity since it decomposes at high temperatures.

The "C—O" band vibrations for both carboxylic acids and esters are actually coupled vibrations involving some adjacent atoms.

Stretching vibrations of carboxylic acids and esters

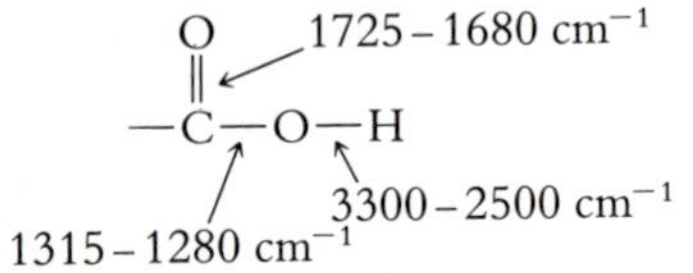

1725 – 1680 cm^{-1}

3300 – 2500 cm^{-1}

1315 – 1280 cm^{-1}

carboxylic acid

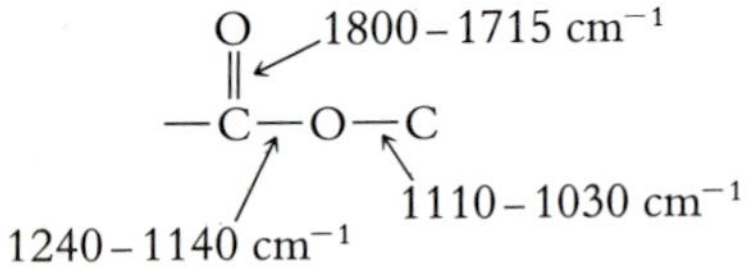

1800 – 1715 cm^{-1}

1110 – 1030 cm^{-1}

1240 – 1140 cm^{-1}

carboxylic ester

carbonyl stretching band of an ester usually occurs at higher frequency than that of a carboxylic acid, and the acyl C—O stretching band of an ester ranges from about 1240 cm^{-1} to 1140 cm^{-1}. The aromatic ring of an aryl ester increases the C=O frequency and decreases the acyl C—O frequency of the ester function. These bands occur at 1770 cm^{-1} and 1190 cm^{-1}, respectively, for phenyl acetate, compared to about 1740 cm^{-1} and 1240 cm^{-1} for ethyl acetate.

The PMR spectrum of aspirin shows a complex pattern of aromatic proton signals characteristic of *ortho* substitution by groups of differing electronegativity. The —COOH substituent withdraws electrons from the ring, deshielding nearby ring protons, and the —OCOCH$_3$ substituent donates electrons by resonance, shielding the ring protons. This effect is particularly noticeable for proton H$_a$ in Figure 1, whose signal occurs well downfield of the rest because of its proximity to the —COOH group. Although H$_a$ has only one "nearest neighbor" proton, its PMR signal has four peaks, because the delocalized pi cloud of the ring allows long-range coupling between nonadjacent protons. Thus, the signal due to H$_a$ is split into a doublet by proton H$_b$, and each peak of that doublet is split into another closely spaced doublet by proton H$_c$, as illustrated in Figure 1. The long-range coupling constant for protons *meta* to one another is only about 1–3 Hz, compared to typical coupling constants for *ortho* protons of 6–10 Hz.

Figure 1.   The effect of long-range coupling on an aromatic proton signal

aspirin

$J_{ac} = \sim$ 1–3 Hz

$J_{ab} = \sim$ 6–10 Hz

coupling between ring protons

idealized PMR signal of H$_a$

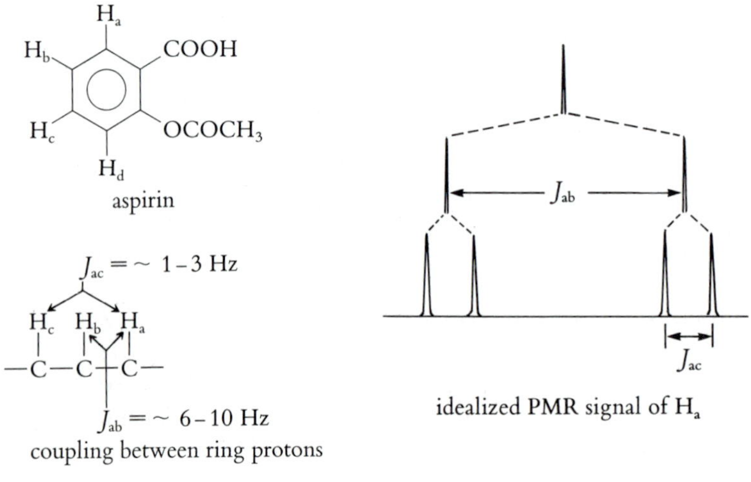

PRELAB ASSIGNMENT

Calculate the mass of 15 mmol of salicylic acid and the theoretical yield of aspirin.

Reactions and Properties

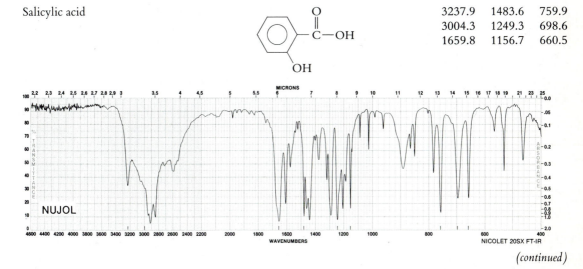

salicylic acid   acetic anhydride   aspirin   acetic acid

Table 1.   Physical properties

	M.W.	m.p.	b.p.	d.
salicylic acid	138.1	159		
acetic anhydride	102.1	−73	140	1.082
aspirin	180.2	135 d		

Figure 2.   IR and PMR spectra of salicylic acid. (IR spectrum reproduced from *The Aldrich Library of FT-IR Spectra*, by Charles J. Pouchert; PMR spectrum reproduced from *The Aldrich Library of NMR Spectra, Edition II*, by Charles J. Pouchert; both with the permission of the Aldrich Chemical Company.)

Salicylic acid

3237.9	1483.6	759.9
3004.3	1249.3	698.6
1659.8	1156.7	660.5

(continued)

Figure 2 *(continued)*

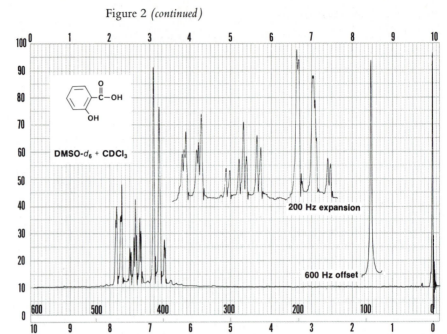

Note: The offset signal has been moved 600 Hz (10 ppm) upfield; the other signals have been expanded by a factor of 3 along the horizontal axis and moved upfield.

PROCEDURE

SAFETY PRECAUTIONS

acetic anhydride

sulfuric acid

Acetic anhydride is corrosive and lachrymatory, its fumes are highly irritating, and it reacts violently with water. Use gloves and a hood; avoid contact, do not breathe its vapors, and keep it away from water.

Sulfuric acid is very corrosive and can cause very serious damage to the skin and eyes; concentrated sulfuric acid reacts violently with water. Wear protective gloves, and keep the concentrated acid away from water.

Reaction.   *Under the hood,* add 5.0 mL (~42 mmol) of acetic anhydride **[contact, vapor and reactivity hazard! wear gloves!]** to 15 mmol of salicylic acid in a dry 125-mL Erlenmeyer flask. Add 3–4 drops of concentrated sulfuric acid **[contact and reactivity hazard!],** and swirl the resulting white slurry. Heat the mixture in a 45–50°C water bath with frequent swirling or stirring for 5–7 minutes. By this time, the salicylic acid should have dissolved; if it has not, continue heating until the solution is clear. Let the flask stand at room temperature until crystallization begins. If no product has precipitated when the solution is near room temperature, induce crystallization by vigorously scratching the wall of the flask at the surface of the solution with a glass stirring rod, or by adding a few seed crystals of pure aspirin. When a heavy precipitate has formed, stir in 30 mL of water, and break up any lumps with a flat-bottomed stirring rod. Cool the mixture in an ice bath until crystallization is complete.

The reaction mixture should be a thick, semisolid slurry at this point.

Separation. Separate the aspirin from the reaction mixture by vacuum filtration ⟨OP–12⟩, washing it with cold water. Save a small amount (about 50 mg) of the crude product in a labeled 10-cm test tube for analysis, and transfer the rest to a beaker.

If some aspirin remains in the reaction flask, use a little of the filtrate to wash it into the funnel.

Purification. Slowly stir in 25 mL of 5% aqueous sodium bicarbonate, using a flat-bottomed stirring rod or a spatula to break up the solid and dissolve as much of it as possible. When the fizzing sound of the reaction has stopped, remove any undissolved solid by gravity filtration ⟨OP–11⟩. Slowly add 8 mL of 3 M hydrochloric acid, with continuous stirring, to precipitate the aspirin. Test the mixture with pH paper to see that it is strongly acidic; if it is not, add enough 3 M HCl to bring the pH below 3. Cool the mixture in an ice bath, and collect the aspirin by vacuum filtration. Save another small sample of the partly purified aspirin in a test tube for analysis.

Recrystallize the aspirin ⟨OP–23⟩ from an ethanol-water mixture ⟨OP–23*b*⟩, using the following procedure to reduce the likelihood of hydrolysis. Dissolve the aspirin in the minimum volume of boiling 95% ethanol, and add another 1 mL of ethanol, measuring the total volume of ethanol used. Add twice that volume of warm (~60°C) water to the solution while it is still at the boiling point, and swirl to mix. If some precipitate forms, heat the solution *gently* until it is clear; then let it cool until crystallization is complete. Collect the aspirin by vacuum filtration, washing it on the filter with ice cold water. Save another small sample of the aspirin for analysis. Dry ⟨OP–21⟩ and weigh the remaining aspirin.

Do not boil the solution after you add water; if necessary, add a little more ethanol to dissolve any precipitate.

Analysis. Place a small quantity of salicylic acid in a labeled 10-cm test tube, and dissolve each reserved aspirin sample and the salicylic acid in 1 mL of 95% ethanol. Add a drop of 1% ferric chloride to each test tube, and record your observations. Record the infrared spectrum ⟨OP–33⟩ of aspirin in a KBr

The crystals need not be completely dry for the analysis.

disc or Nujol mull and its PMR spectrum ⟨OP – 34⟩ in deuter-ochloroform, or obtain the spectra from your instructor. If possible, use a sweep offset for the COOH signal and a scale expansion for the aromatic proton signals when you record the PMR spectrum.

Report. Discuss the purity of your aspirin samples, and account for the presence of salicylic acid in any of them. Locate and clearly label both carbonyl bands, both acyl C—O bands, and the O—H bending and stretching bands on the infrared spectrum, and assign as many of the other bands as you can. Locate the PMR signal for the proton designated H_a in Figure 1, determine the values of J_{ab} and J_{ac}, and assign as many of the other PMR signals as you can. Compare your aspirin spectra with those of salicylic acid in Figure 2, and explain any significant similarities and differences.

Collateral Projects

1.   Test commercial aspirin tablets for salicylic acid by the ferric chloride test described above. Try the test on freshly purchased aspirin and on aspirin that has been on the shelf for some time. Indicate whether there is any evidence of decomposition, and write an equation for the most likely decomposition reaction.

What kind of odor would you expect if the aspirin had decomposed?

2.   Test commercial aspirin tablets for starch (often used in them as a binder) by boiling 2 mg of ground-up tablets in 2 mL of water and adding a drop of a solution of iodine in potassium iodide. Starch forms a blue-violet complex with iodine.

Exercises

1.   (a) Write an equation for a reaction that might form salicylic acid during the workup of the product. (b) Tell how you could reduce or prevent the contamination of your product due to this reaction.

2.   Write a detailed mechanism for the reaction of salicylic acid with acetic anhydride, showing clearly the function of the catalyst, sulfuric acid.

3.   Propose a structure for the polymer (draw several repeating units) formed during the synthesis of aspirin.

4.   Construct a flow diagram for the synthesis of aspirin.

5.   Calculate the volume of 3 M HCl needed to just neutralize the aqueous sodium bicarbonate used in the initial purification of aspirin. The density of 5% $NaHCO_3$ is 1.034 g/mL.

6.   A small bottle of 5-grain aspirin tablets holds 100 tablets, each containing about 0.325 g of aspirin. Calculate the cost of the acetic anhydride and salicylic acid required to prepare the aspirin in such a bottle, using the market prices given in the margin and assuming equimolar quantities of the reactants. (Your instructor may suggest that you use the current market prices listed in *Chemical Marketing Reporter* instead.)

Market prices, December, 1987:
acetic anhydride: $0.435/lb. (r.r. tank cars)
salicylic acid: $1.33/lb. (U.S.P. crystal, drums)

7.   Which would you expect to be the stronger acid, aspirin or salicylic acid? Explain.

8.   Construct a flow diagram outlining a method for separating the components of an analgesic drug preparation that contains aspirin, acetaminophen, and phenacetin. (See Experiment 2 for the structures.)

Library Topic

Acetic anhydride is one of the top fifty chemicals produced in the United States; it is manufactured at the rate of about two billion pounds per year. Report on some commercial uses for acetic anhydride other than in aspirin production.

Nucleophilic Strength and Reactivity in S_NAr Reactions

Reactions of Aryl Halides. Nucleophilic Aromatic Substitution. Reaction Kinetics.

Operations

OP – 6   Measuring Volume

OP – 35a Colorimetry

Objectives

To measure the rate of the reaction between 2,4-dinitrochlorobenzene and a secondary amine and calculate the second-order rate constant for the reaction.

To determine the nucleophilic strengths of morpholine and piperidine in an S_NAr reaction.

To learn how to measure reaction rates by spectrophotometry and how to evaluate kinetic data for a second-order reaction.

To learn about nucleophilic substitution reactions involving activated aryl halides.

SITUATION

The ability of a nucleophile to displace a leaving group from a particular type of substrate can be measured by a parameter called the nucleophilic constant. Although nucleophilic constants have been determined for many nucleophiles in reactions with aliphatic substrates, these values do not apply to reactions of the same nucleophiles with aromatic substrates. In this experiment, you and your co-workers will measure rate constants for the reactions of 2,4-dinitrochlorobenzene with two nucleophiles, morpholine and piperidine, and use the rate constants to determine the corresponding aromatic nucleophilic constants for the nucleophiles.

BACKGROUND

The S$_N$Ar Reaction

Nucleophilic aromatic substitution can occur (1) through an S$_N$1 mechanism, as in some substitution reactions of aryldiazonium salts; (2) by an elimination-addition mechanism involving an aryne intermediate; or (3) by a bimolecular addition-elimination route, often called the S$_N$Ar reaction. In the S$_N$Ar reaction, the nucleophile attacks an activated aromatic nucleus to form an intermediate complex, which then loses a leaving group to yield the product.

The nature of the intermediate complex in the S$_N$Ar reaction has long been the subject of speculation. The first evidence of its structure was obtained in 1898 when Jackson and Boos mixed picryl chloride (2,4,6-trinitrochlorobenzene) with sodium methoxide in methanol and isolated a red salt, which was converted by ethanol to another red salt. J. Meisenheimer prepared the second salt by two different methods, treating either 2,4,6-trinitroanisole with potassium ethoxide or 2,4,6-trinitrophenetole with potassium methoxide. He was then able to assign the salt the structure labeled **1**.

General S_NAr mechanism

1. $Ar—Z + Nu: \longrightarrow$ Ar with $\overset{\ominus}{\underset{Nu}{\overset{Z}{\diagdown}}}$

2. Ar $\overset{\ominus}{\underset{Nu}{\overset{Z}{\diagup}}} \longrightarrow Ar—Nu + Z^-$

(Ar must be activated by electron-withdrawing groups; Z is the leaving group and Nu: the nucleophile.)

Table 1.   Nucleophilic constants for various nucleophiles

Nucleophile	n
CH_3OH	0.00
F^-	2.7
Cl^-	4.37
pyridine	5.23
NH_3	5.50
aniline	5.70
Br^-	5.79
CH_3O^-	6.29
$(CH_3CH_2)_3N$	6.66
$(CH_3CH_2)_2NH$	7.0
pyrrolidine	7.23
piperidine	7.30
I^-	7.42

Note: n values are measured relative to methanol, with methyl iodide as the substrate.

Many other Meisenheimer complexes have since been prepared and characterized, and a large body of experimental evidence indicates that the intermediates in S_NAr reactions are in fact sigma complexes of the Meisenheimer type. Their role in such reactions explains why only activated aromatic compounds undergo bimolecular substitution easily: nitro groups and similar electron-withdrawing substituents remove excess electron density from the aromatic nucleus, stabilizing the intermediate complex.

The mechanism of most S_NAr reactions thus appears to be a simple two-step process involving the initial addition of the nucleophile to form a Meisenheimer complex, followed by the departure of the leaving group. If the first step of the reaction is rate-limiting, the nucleophilic strength of the reactant — its ability to donate its electron pair to the substrate and form a sigma bond — should affect the reaction rate. The *Swain-Scott equation,*

$$\log \frac{k}{k_0} = ns \tag{1}$$

relates reaction rates to the strength of the nucleophile (n) and the sensitivity of the substrate to nucleophilic substitution (s). It has been applied widely to aliphatic systems, and some values of the nucleophilic constant n for reactions of various nucleophiles with methyl iodide are given in Table 1. Attempts to use such correlations for S_NAr reactions have met with less success, because changing the substrate or solvent often changes the order of nucleophilic strength. However, an equation of this kind can be useful in comparing the nucleophilic strengths of various reactants with reference to the same class of substrates. For this experiment, we define a nucleophilic constant for the S_NAr reaction as follows:

$$n_{Ar} = \frac{1}{s} \log \frac{k}{k_0} \tag{2}$$

The reaction of 2,4-dinitrochlorobenzene with ammonia in absolute ethanol, for which the rate constant (k_0) is 4.0×10^{-6} L mol^{-1} s^{-1} at 25°C, is used as the reference reaction for Equation 2. For this substrate and nucleophile, s is 1 and n_{Ar} is zero by definition. The rate constant k is for a reaction involving the nucleophile whose n_{Ar} value is being determined.

METHODOLOGY

In this experiment, you will measure the rates of the reactions of piperidine and morpholine with 2,4-dinitrochlorobenzene and use the rate constants to calculate their nucleophilic constants. The reaction of 2,4-dinitrochlorobenzene with an amine is second-order in both substrate and nucleophile and follows the general rate equation

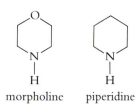

morpholine   piperidine

$$\frac{dx}{dt} = k(S_0 - x)(N_0 - 2x) \qquad (3)$$

where x is the concentration of the product at time t and S_0 and N_0 are the initial concentrations in the reaction mixture of substrate and nucleophile, respectively. The computations can be simplified considerably if the experiment is carried out with the initial concentration of nucleophile being just twice that of 2,4-dinitrochlorobenzene. Equation 3 then becomes $dx/dt = 2k(S_0 - x)^2$, and the integrated rate equation is

$$\frac{1}{(S_0 - x)} = 2kt + \frac{1}{S_0} \qquad (4)$$

By measuring the concentration of the product, x, at regular intervals during the reaction, one can calculate the term on the left side of Equation 4, which, when plotted versus time, should yield a straight line with slope $2k$.

The time should be measured in seconds.

The concentration of the product will be determined indirectly by measuring the absorbance of 380-nm light by aliquots that are removed from the reaction mixture at various times. Each aliquot must first be quenched by adding dilute acid to stop the reaction; this is done so that the concentration will remain constant until you are ready to take the absorbance readings. The concentration term in Equation 4, $S_0 - x$, can be shown to be proportional to $A_\infty - A$; therefore,

The N-(2,4-dinitrophenyl) amines are yellow-orange and absorb strongly in the visible region around 380 nm.

$$\frac{1}{(S_0 - x)} = \frac{q}{(A_\infty - A)} \qquad (5)$$

A_∞ is the absorbance when the reaction is 100% complete, and A is the absorbance at time t.

The "infinity" value of the absorbance, A_∞, can be obtained by warming the reaction mixture to complete the reaction and measuring the absorbance of the resulting solution. The proportionality constant q can then be calculated from the relationship $q = A_\infty / S_0$. Because of their different reaction rates,

S_0 is the initial substrate concentration in the reaction mixture, not in the stock solution.

the amines will be used in different initial concentrations so that both reactions will be about half complete after 30 minutes. The absorbance values of the products at 380 nm are so high that to obtain readings in a convenient range the solutions must first be diluted.

If you do Minilab 32, you will isolate an aryl chloride from moth flakes and identify it.

PRELAB ASSIGNMENT

Read OP – 35*a* on the use of spectrophotometry for quantitative analysis.

Reaction and Properties

2,4-dinitrochlorobenzene   piperidine (Y = CH₂) or morpholine (Y = O)   dinitrophenylated amine

The extra mole of amine combines with HCl liberated during the reaction.

Table 2.   Physical properties

	M.W.	m.p.	b.p.	d.
2,4-dinitrochlorobenzene	202.6	53	315	
morpholine	87.1	−5	128	1.000
piperidine	85.2	−9	106	0.861

PROCEDURE

If equipment is limited, you may be asked to work in pairs or larger groups. Each student or group should have a timer or a watch that measures in seconds. All glassware must be clean and dry. It is recommended that the stock solutions be dispensed from burets.

2,4-Dinitrochlorobenzene presents a serious risk of poisoning if it is ingested or inhaled. If the compound or its solutions come in contact with the skin, very unpleasant and persistent dermatitis may result. Although the 2,4-dinitrochlorobenzene solution is very dilute, you should avoid contact with it.

Morpholine and piperidine are toxic and corrosive, capable of causing severe damage to the skin, eyes, and respiratory system. Avoid contact with their solutions.

Ethanol is very flammable, so keep the ethanolic solutions away from flames or hot surfaces.

SAFETY
PRECAUTIONS

2,4-dinitro-
chlorobenzene

morpholine   piperidine

A. Kinetic Runs

Do not pipet any solutions by mouth! Using a volumetric pipet, measure ⟨OP–6⟩ 25 mL of the quenching solution, 0.2 M sulfuric acid in 50% ethanol, into each of seven small numbered flasks with stoppers. Measure about 18 mL of absolute ethanol into a 25-mL volumetric flask, and pipet in 4 mL of the 0.10 M stock solution of 2,4-dinitrochlorobenzene in ethanol **[contact hazard!].** Pipet 2 mL of a 0.40 M solution of piperidine in ethanol **[contact hazard!]** into the volumetric flask, and start your timer (or record the starting time) immediately. Promptly fill the flask to the mark with absolute ethanol, stopper and shake it, and transfer the contents to a small Erlenmeyer flask. Using a clean, dry volumetric pipet, withdraw a 1-mL aliquot of this solution, and transfer it to the first quenching flask as you record the quenching time; then stopper both the quenching flask and the reaction flask. Swirl the contents of the reaction flask occasionally, withdrawing another 1-mL aliquot every 5 minutes or so and transferring it to a quenching flask, until a total of six aliquots have been removed and quenched. Then cork the reaction flask loosely, and warm it in a 50°C water bath for 2 hours or more (or let it stand for at least 48 hours at room temperature) to drive the reaction to completion. Pipet a final 1-mL aliquot into the last quenching flask to prepare the A_∞ solution.

Repeat the kinetic run with morpholine as the nucleophile. Prepare the reaction mixture by measuring about 12 mL of absolute ethanol into a 25-mL volumetric flask, pipetting in 10 mL of 0.10 M 2,4-dinitrochlorobenzene in ethanol, followed by 2 mL of 1.0 M morpholine in ethanol (start timer),

For greatest accuracy, the times should be measured when about half the solution has drained from the pipet.

Rinse the pipet with the reaction mixture just before delivering each aliquot. Record the exact time of the transfer to the quenching flask in each case.

and adding absolute ethanol to the mark. Then continue as described above.

B. Absorbance Measurements

Clean, dry, and number as many 15-cm test tubes as you have solutions to analyze. Using a volumetric pipet, measure 10 mL of 95% ethanol into each one. Pipet 1 mL of the solution from each quenching flask into the corresponding test tube, shake to mix the solutions, and stopper the test tubes. Zero the spectrophotometer ⟨OP – 35a⟩ at 380 nm and set the 100% transmittance control using 95% ethanol. Make sure the sample cuvette is positioned correctly, and record the transmittance of each solution. Use the same instrument to analyze all your samples, including the A_∞ solution.

If you have access to a calculator or computer that will do linear regressions, you can use it to compute the slope more accurately.

Report.   Calculate the absorbance of each solution, and use these data to compute $1/(S_0 - x)$ for each aliquot. For each amine, plot these values versus time. Determine the second-order rate constant from the slope of the line. Use the rate constants to calculate the n_{Ar} value for each amine. Arrange morpholine, piperidine, and ammonia ($n_{Ar} = 0$) in order by their n_{Ar} values and try to explain their relative nucleophilicities. Write a detailed mechanism for the reaction of piperidine with 2,4-dinitrochlorobenzene.

Collateral Projects

1.   Measure the reaction rates for one of the amines at different temperatures (0°C, 20°C, and 40°C, for example). Then plot ln k versus $1/T$ to determine the activation energy of the reaction, based on the following form of the Arrhenius equation:

$$\ln k = -\frac{E_{act}}{RT} + \ln A \qquad (R = 8.31 \text{ J mol}^{-1} \text{ K}^{-1})$$

You can also use the data to calculate the enthalpy and entropy of activation.

2.   Carry out one of the reactions described in the Procedure section in other solvents, such as 2-propanol and aqueous ethanol, to measure the effect of solvent polarity on the rate constants. If you make up your own solutions, it is *essential* that you wear protective gloves and avoid contact with 2,4-dinitrochlorobenzene and the amines.

Identification of an Aryl Chloride in Moth Flakes

Vapors from the moth repellent may irritate the eyes and respiratory system. Carry out the sublimation under a hood, and avoid breathing the vapors.

SAFETY PRECAUTION

Grind up 0.5 g of moth flakes with a mortar and pestle, and sublime ⟨OP–24⟩ the finely powdered solid from a 45°C hot-water bath. Carry out a Beilstein test ⟨Test C–5⟩ on the sublimate to confirm the presence of halogen. Measure the melting point of the sublimate and record its PMR spectrum in deuterochloroform (or obtain a spectrum from your instructor). Identify the aryl chloride from its spectrum and by referring to Table 10 in Appendix VI.

"Old-style" moth repellents contain naphthalene, which will not sublime at 45°C.

Exercises

1. The rate constants for the reactions of piperidine with 2,4-dinitrochlorobenzene and with 2,4-dinitrobromobenzene are virtually identical. Identify the rate-determining step in the mechanism for the reaction with the chloro compound, and explain your reasoning.

2. Explain why the reaction of 2,4-dinitrochlorobenzene with piperidine or morpholine is second-order, even though there are three reactant species in the overall equation for the reaction.

3. Could you substitute chlorobenzene for 2,4-dinitrochlorobenzene in this experiment and still obtain useful results? Explain your answer with reference to the mechanism.

4. **(a)** Derive Equation 5 using Beer's law and evaluate the constant q in terms of Beer's law parameters. **(b)** Derive Equation 4 from Equation 3.

5. Outline a synthesis of the spiro Meisenheimer complex shown, using 1-chloro-2,4,6-trinitrobenzene and any appropriate inorganic and organic reactants.

6. **(a)** Calculate the concentrations of the dinitrophenylamines in the solutions used for the spectrophotometry infinity reading. **(b)** If you know the path length of the cuvette

a spiro Meisenheimer complex

A *spiro system* contains two rings having one carbon atom in common.

you used, calculate the molar absorptivities of these products at
380 nm.

7. Explain why "quenching" the reaction mixture with
H_2SO_4 stops the reaction.

Library Topic

Read the paper by Bunnett, Garbisch, and Pruitt, *J. Am.
Chem. Soc. 79,* 385 (1957). Then tell what is meant by the
"element effect" in reactions of 1-substituted 2,4-dinitroben-
zenes, and explain how it was used to elucidate the mechanism
of the S_NAr reaction. Explain why aryl fluorides showed
anomalous behavior in the reactions studied.

The Enamine Synthesis of 2-Propionylcyclohexanone

Reactions of Carbonyl Compounds. Preparation of Dicarbonyl Compounds. Nucleophilic Acyl Substitution. Enamines.

Operations

OP – 7a   Refluxing

OP – 10   Addition

OP – 13   Extraction

OP – 14   Evaporation

OP – 20   Drying Liquids

OP – 22a   Excluding Moisture

OP – 25c   Water Separation

OP – 26   Vacuum Distillation

OP – 33   Infrared Spectrometry

Objectives

To prepare 2-propionylcyclohexanone from cyclohexanone and propionic anhydride.

To learn how to identify enols from their infrared spectra.

To learn about the characteristics of enamines and their applications in organic syntheses.

SITUATION

In 1954, Gilbert Stork and his co-workers sent a letter to the editor of the *Journal of the American Chemical Society* that began with this sentence: "We have discovered a new method for the alkylation and acylation of ketones." Their brief communication caused quite a stir in the world of organic chemistry because it announced a synthetic tool of exceptional utility, which came to be known as the Stork enamine reaction. A few years later, Siegfried Hunig and his associates described how Stork's reaction can be used to lengthen the chain of a carbox-

See Hunig, Lucke, and Benzing, *"Kettenverlangerung von Carbonsauren um 6C-Atome," Berichte 91,* 129 (1958).

ylic acid by six carbon atoms. By this route, these German chemists were able to convert propionic acid to pelargonic acid (nonanoic acid) at an overall yield of 65%.

Chain-lengthening of propionic acid

$$CH_3CH_2\overset{\overset{\displaystyle O}{\|}}{C}OH \xrightarrow{\text{enamine synthesis}} CH_3(CH_2)_7\overset{\overset{\displaystyle O}{\|}}{C}OH$$
propionic acid   pelargonic acid

In this experiment, you will use the Stork enamine reaction to prepare 2-propionylcyclohexanone, an intermediate in the pelargonic acid synthesis.

BACKGROUND

Enamines and The Stork Reaction

Enamines are 1-aminoalkenes, the nitrogen analogs of enols. Just as enols tend to isomerize to the corresponding ketones, enamines with N—H bonds are in equilibrium with the corresponding imines. Enamines without N—H bonds, however, cannot tautomerize to imines, and these are the kind of enamines used in the Stork reaction.

Ketones can be alkylated by a base-catalyzed reaction in which an alpha proton is removed to form an enolate ion, which acts as a carbon nucleophile toward alkyl halides. However, multiple alkylation can also occur, with the second alkyl group usually ending up on the same alpha carbon atom as the first. The requirement for a strongly basic catalyst (such as sodium methoxide) precludes the use of reactants having base-sensitive functional groups and can result in side reactions, such as aldol condensations.

If a ketone such as cyclohexanone is condensed with a secondary amine such as pyrrolidine, it will, under the proper reaction conditions, yield the enamine 1-pyrrolidinocyclohexene *(1)*. This enamine can then be alkylated under neutral reaction conditions to yield a monoalkylated iminium salt, which can be hydrolyzed to the corresponding 2-alkyl ketone. Since one of the enamine's two resonance structures has a negative charge on the alpha carbon atom, the alkylation step can be viewed as a nucleophilic substitution reaction with a carbanion acting as the nucleophile. Enamines thus provide an

Tautomeric equilibria of enols and enamines

Direct alkylation of a carbonyl compound

Formation of an enamine

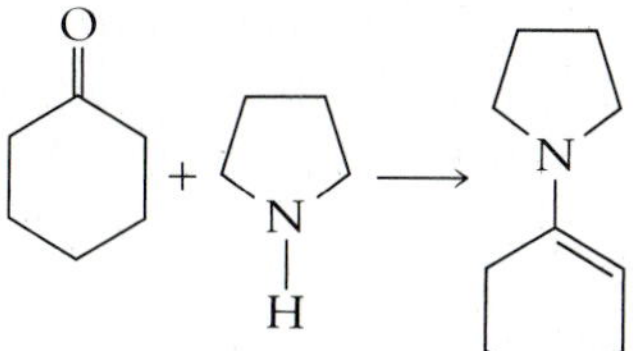
pyrrol-
idine
1

Alkylation and hydrolysis of an enamine

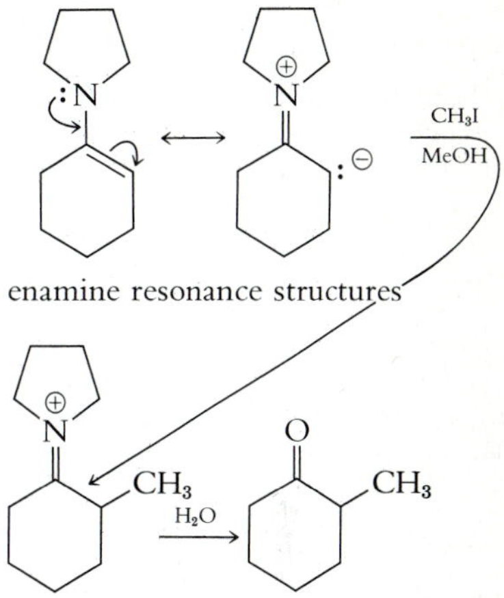
CH_3I
MeOH
enamine resonance structures
CH_3
H_2O
iminium salt

indirect method for preparing alkyl (and acyl) derivatives of carbonyl compounds; the enamine functional group both activates the molecule and prevents unwanted side reactions.

Enamines can be acylated by acid chlorides and anhydrides; this provides a unique method for lengthening a carbon chain. For example, acetyl chloride reacts readily with 1-morpholinocyclohexene ***(2)*** to yield, after hydrolysis of the acylated enamine, 2-acetylcyclohexanone. Treatment of this diketone with strong base results in ring cleavage to form 7-oxooctanoic acid, which can be reduced by a Wolff-Kischner reaction to octanoic acid. Other carboxylic acids can also undergo the same chain-lengthening process via their acyl chlorides or anhydrides.

Enamine synthesis of octanoic acid

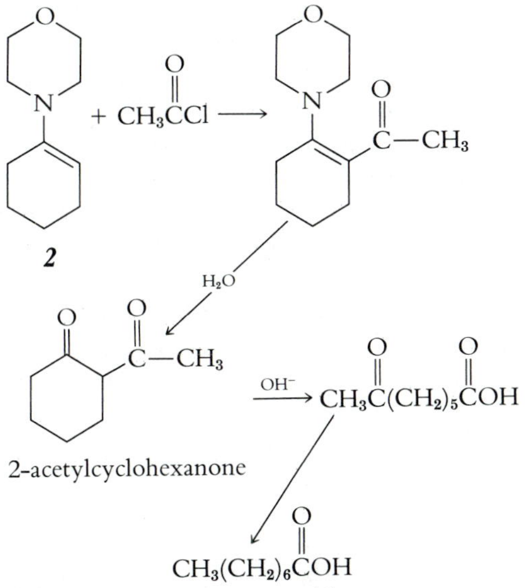
CH_3CCl
$C-CH_3$
2
H_2O
$C-CH_3$
2-acetylcyclohexanone
OH^-
$CH_3C(CH_2)_5COH$
$CH_3(CH_2)_6COH$
octanoic acid

SO₃H

CH₃

p-toluenesulfonic acid

A small excess of the amine is generally used, as some may be distilled off with the water.

Triethylamine is added to neutralize the propionic acid evolved during the reaction.

METHODOLOGY

Because this experiment involves three reflux periods and an overnight reaction period, it is important that you plan your lab time carefully. You should be ready to start the overnight reaction by the end of the first lab period, if at all possible.

The formation of a ketone enamine is usually carried out in the presence of a catalyst such as *p*-toluenesulfonic acid; water is removed during the reaction to prevent the premature hydrolysis of the enamine. If toluene is used as a reaction solvent and the reaction is run under reflux, the water can be removed as an azeotrope through distillation into a water separator. Although pyrrolidine enamines form more rapidly than morpholine or piperidine enamines do, morpholine compounds give higher yields in acylation reactions. Therefore, the morpholine enamine of cyclohexanone will be used in this preparation. The acylation of 1-morpholinocyclohexene will be carried out, using propionic anhydride as the acylating agent, by simply heating the reactants in methylene chloride and allowing the solution to stand. The acylated enamine is then hydrolyzed by refluxing it with water, and the product is purified by vacuum distillation.

You can analyze the product by recording its infrared spectrum; the infrared spectra of cyclohexanone and propionic acid are shown in Figure 1 for comparison. Many α-diketones exist partly or entirely in the enol form, as illustrated for 2,4-pentanedione (acetylacetone). Hydrogen bonding between the enol OH group and the carbonyl oxygen generally causes the O—H stretching band to broaden and move into the $3200-2500$ cm^{-1} region; sometimes this band is so broad and low that it may be overlooked. At the same time, the carbonyl band broadens and moves from its normal ketonic position at about 1715 cm^{-1} to the $1650-1580$ cm^{-1} region. If both keto and enol forms exist in equilibrium, the keto carbonyl band will absorb in the usual frequency range for a ketone and may appear as a doublet. The infrared spectrum of 2-propionylcyclohexanone should tell you whether this compound exists in the keto form, the enol form, or as a mixture of both.

Keto and enol forms for 2,4-pentanedione

$$CH_3CCH_2CCH_3 \rightleftharpoons CH_3C=CHCCH_3$$

keto form   enol form

The PMR spectra (see Collateral Project 2) of α-diketones that exist in enolic forms are characterized by an enolic OH signal at very low field (~15 δ for 2,4-pentanedione). This peak is often quite broad as a result of proton exchange.

PRELAB ASSIGNMENTS

1. Read OP–25c and read or review the other operations as necessary.

2. Calculate the mass and volume of 0.10 mol of cyclohexanone, of 0.13 mol of morpholine, of 0.12 mol of triethylamine, and of 0.11 mol of propionic anhydride. Calculate the theoretical yield of 2-propionylcyclohexanone.

Reactions and Properties

A.

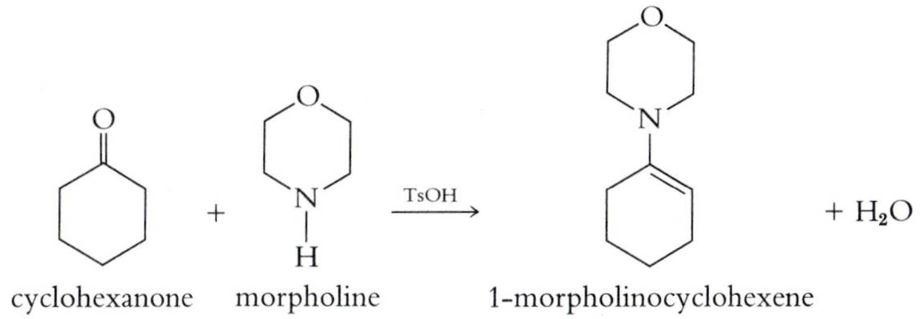

cyclohexanone   morpholine   1-morpholinocyclohexene

B.

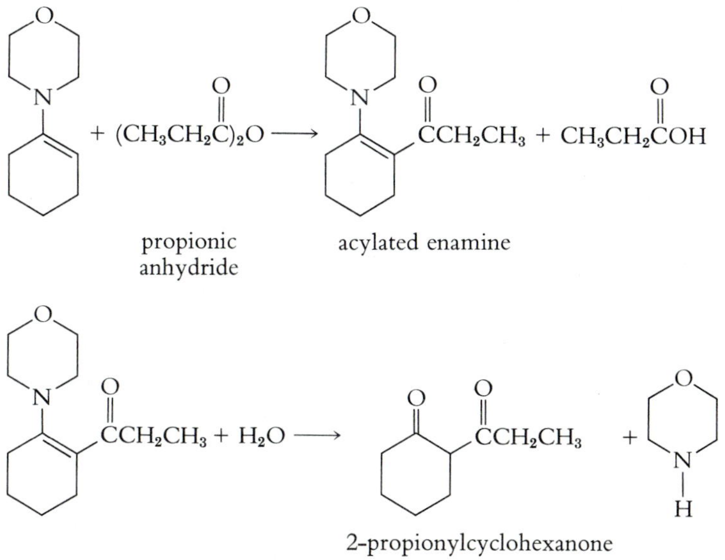

propionic anhydride   acylated enamine

2-propionylcyclohexanone

Table 1.   Physical properties

	M.W.	m.p.	b.p.	d.
cyclohexanone	98.2	−16	156	0.948
morpholine	87.1	−5	128	1.000
toluene	92.15	−95	111	0.867
p-toluenesulfonic acid	172.2	104–5	140^{20}	
1-morpholino-cyclohexene	167.3		119^{10}	
propionic anhydride	130.15	−45	168^{712}	1.011
triethylamine	101.2	−115	89	0.728
2-propionylcyclo-hexanone	154.2		125^{20}	

Figure 1.   IR spectra of cyclohexanone and propionic acid. (Reproduced from *The Aldrich Library of FT-IR Spectra,* by Charles J. Pouchert, with the permission of the Aldrich Chemical Company.)

Cyclohexanone

2937.8	1311.2	908.4
1714.0	1221.7	749.8
1449.5	1118.8	489.7

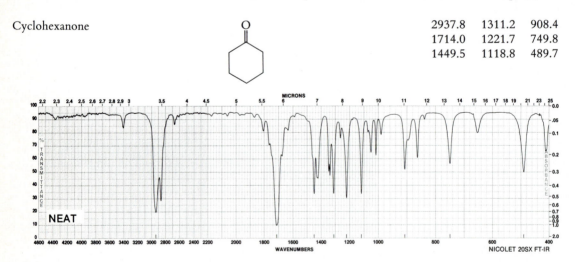

NEAT

Figure 1 *(continued)*

Propionic acid

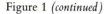

$$CH_3CH_2-\overset{\overset{\displaystyle O}{\|}}{C}-OH$$

2648.0	1415.5	1079.2
1715.2	1290.5	932.0
1466.8	1240.4	847.7

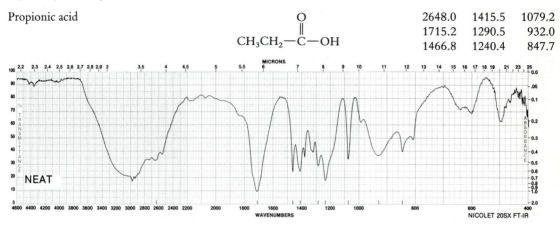

NEAT

NICOLET 20SX FT-IR

PROCEDURE

The apparatus must be thoroughly dried, preferably in an oven. Wear protective gloves while handling the chemicals, and dispense all liquids under a hood.

A. Preparation of 1-Morpholinocyclohexene

Toluene is flammable, and inhalation, ingestion, or skin absorption may be harmful. Avoid contact, do not breathe its vapors, and keep it away from flames.

Morpholine is flammable, toxic, and corrosive, capable of causing severe damage to the skin, eyes, and respiratory tract. Wear gloves, and dispense from a hood; avoid contact, and do not breathe its vapors.

p-Toluenesulfonic acid is toxic and corrosive, capable of causing serious damage to the skin, eyes, and respiratory tract. Avoid contact, and do not breathe the dust.

SAFETY PRECAUTIONS

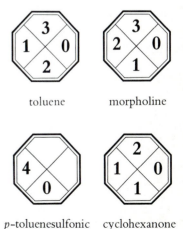

toluene

morpholine

p-toluenesulfonic acid

cyclohexanone

Assemble a reflux apparatus equipped with a water separator ⟨OP–25c⟩ and a drying tube ⟨OP–22a⟩. Fill the water separator to the side arm with toluene **[vapor and fire hazard!]**. In the reaction flask, combine 0.10 mol of cyclohexanone, 0.13 mol of morpholine **[contact and vapor hazard!]**,

Place the toluene from the water separator in a designated solvent recovery container.

Toluene boils at about 20°C and morpholine at 35°C at 20 torr.

0.10 g of *p*-toluenesulfonic acid **[contact hazard!]**, and 20 mL of dry toluene. Reflux the reaction mixture ⟨OP–7a⟩ gently for at least an hour, until the volume of water in the trap appears to remain constant throughout a period of 10–15 minutes. Remove the toluene and excess morpholine by vacuum distillation ⟨OP–26⟩ from a hot-water bath. If the morpholinocyclohexene will not be used the same day, keep it in a tightly stoppered flask in the refrigerator.

B. Preparation of 2-Propionylcyclohexanone

SAFETY PRECAUTIONS

propionic
anhydride

triethylamine

methylene
chloride

Propionic anhydride is corrosive and lachrymatory and can cause severe damage to the eyes, skin, and respiratory tract. Wear gloves, and dispense in a hood; avoid contact, and do not breathe its vapors.

Triethylamine is toxic, flammable, corrosive, and lachrymatory and can cause severe damage to the eyes, skin, and respiratory tract. Wear gloves, and dispense in a hood; avoid contact, do not breathe its vapors, and keep it away from flames or hot surfaces.

Methylene chloride may be harmful if ingested, inhaled, or absorbed through the skin, and it is a suspected carcinogen. Minimize your contact with the liquid, and do not breathe the vapors.

Reactions. Using the boiling flask that contains the 1-morpholinocyclohexene as the reaction flask, assemble an apparatus for addition and reflux ⟨OP–10⟩, including a drying tube ⟨OP–22a⟩. *Under the hood,* prepare a solution containing 0.12 mol of triethylamine **[contact and vapor hazard! wear gloves!]** in 100 mL of dry methylene chloride **[suspected carcinogen!]**, and another solution containing 0.11 mol of propionic anhydride **[contact and vapor hazard!]** in 50 mL of dry methylene chloride. Add the triethylamine solution to the reaction flask and place the propionic anhydride solution in the addition funnel. Reflux the mixture using a 50°C water bath while adding the propionic anhydride solution drop by drop over a period of 30 minutes or more. Disassemble the apparatus, insert the drying tube in the reaction flask, and let it stand in a warm place for 24 hours or longer. Evaporate the methylene chloride ⟨OP–14⟩ on a water bath using a cold

Place the recovered methylene chloride in a designated solvent recovery container.

trap. Cool the residue to room temperature or below, add 25 mL of water, and reflux the mixture for 30 minutes to hydrolyze the acylated enamine.

Separation.   Extract the reaction mixture ⟨OP–13⟩ with three portions of methylene chloride, dry the combined organic layers ⟨OP–20⟩, and evaporate the solvent.

Purification and Analysis.   Purify the 2-propionylcyclohexanone by vacuum distillation ⟨OP–26⟩, weigh it, and record its infrared spectrum ⟨OP–33⟩.

Report.   Calculate the percentage yield of 2-propionylcyclohexanone. Interpret the infrared spectrum of the product as completely as you can. Determine whether it exists in the keto or enol form (or as a mixture of both), and draw the structure of the most probable enol form.

Collateral Projects

1.   The product can be converted to 7-oxononanoic acid by an alkaline cleavage reaction. A general procedure is described on p. 505 of *Organicum* ⟨Bib–B10⟩.

2.   Record the PMR spectrum of 2-propionylcyclohexanone in deuterochloroform, and interpret it as completely as you can. Use the sweep offset control to locate any enolic proton signal.

3.   Assess the purity of your 2-propionylcyclohexanone by gas chromatography.

Exercises

1.   Why is it important to keep the reaction apparatus free of moisture in part A? Write an equation for the reaction that would take place if water were present.

2.   Acylation of enamines differs from their alkylation reaction in that an acylated iminium salt loses a proton to form an enamine before hydrolysis. Propose a detailed mechanism for the acylation of 1-morpholinocyclohexene by propionic anhydride and the hydrolysis of the resulting enamine.

3. **(a)** Write an equation and a mechanism for the alkaline cleavage of 2-propionylcyclohexanone to the salt of 7-oxononanoic acid. **(b)** Outline a complete synthesis of pelargonic acid from propanoic acid.

4. Construct a flow diagram for the synthesis of 2-propionylcyclohexanone.

5. Outline enamine syntheses for the following compounds, starting with cyclohexanone and using any other needed reagents.

(a) (2-propionylcyclohexanone structure: $CH_2CCH_2CH_3$)

(b) (bicyclic structure with CH_3 and $=O$)

(c) $HOC(CH_2)_5C(CH_2)_5COH$

6. Polyol esters of pelargonic acid (nonanoic acid) are used in the manufacture of synthetic lubricants. Write the structure of the ester made by combining 4 mol of pelargonic acid with 1 mol of pentaerythritol, $C(CH_2OH)_4$.

Library Topic

Write a report giving the structure of "queen substance" and telling how it is produced and used by the queen honeybee. Outline a laboratory synthesis of queen substance starting with cyclohexanone and having 7-oxooctanoic acid as an intermediate product.

Synthesis of Dimedone and Measurement of Its Tautomeric Equilibrium Constant

Reactions of α,β-Unsaturated Carbonyl Compounds. Preparation of Dicarbonyl Compounds. Nucleophilic Addition to Carbon-Carbon Double Bonds. Condensation Reactions. Decarboxylation. Reaction Equilibria.

Operations

OP–7a	Refluxing
OP–12	Vacuum Filtration
OP–14	Evaporation
OP–21	Drying Solids
OP–22a	Excluding Moisture
OP–23	Recrystallization
OP–28	Melting-Point Determination
OP–34	Nuclear Magnetic Resonance Spectrometry

Objectives

To prepare dimedone from mesityl oxide and dimethyl malonate.

To determine the equilibrium constant for the keto-enol equilibrium of dimedone using NMR spectrometry.

To learn more about conjugate addition, condensation, and decarboxylation reactions and their mechanisms.

To learn more about keto-enol equilibria and stable enolic compounds.

SITUATION

5,5-Dimethyl-1,3-cyclohexanedione, also called dimedone or methone, is represented as a diketone in most textbooks of organic chemistry. Some "diketones," such as 2,4-butanedione, are known to exist mainly in the enolic form.

Keto-enol equilibrium for dimedone

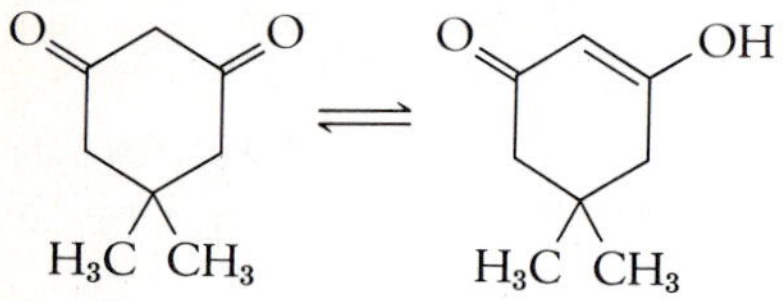

diketone

enolic
ketone

Thus, the diketone structure given for dimedone may be inaccurate, or at least incomplete. In this experiment, you will synthesize dimedone from dimethyl malonate and mesityl oxide and use NMR spectrometry to find out whether the diketone or its enolic form is the predominant species in solution.

BACKGROUND

Stable and Unstable Enols

Enols have long been known as unstable intermediates in reactions involving carbonyl compounds, such as the bromination of acetone. Although acetone exists overwhelmingly in the keto form, it is the minute amount of the enolic form present that actually reacts with bromine under acidic conditions.

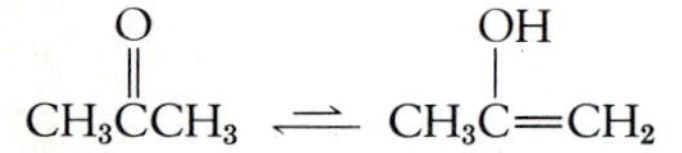

$$CH_3CCH_3 \rightleftharpoons CH_3C=CH_2$$
~100% keto   0.00025% enol
acetone

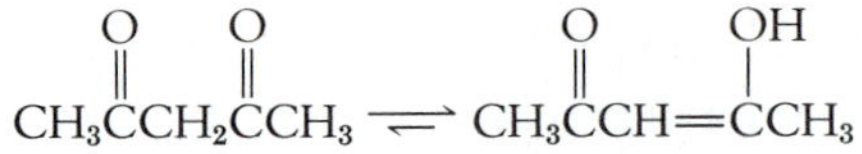

$$CH_3CCH_2CCH_3 \rightleftharpoons CH_3CCH=CCH_3$$
20% keto   80% enol
acetylacetone

Other enols are considerably more stable than that of acetone — ethyl acetoacetate contains about 7.5% enol at equilibrium, and acetylacetone (2,4-pentanedione) contains about 80% enol. Certain natural compounds such as Vitamin C (an enediol) exist almost entirely in the enolic form. In a sense, even phenols are enolic tautomers that could, in theory, exist in equilibrium with keto forms. For example, phlorglucinol forms carbonyl-type derivatives with hydroxylamine and similar derivatizing agents, which suggests that it exists in equilibrium with a triketone.

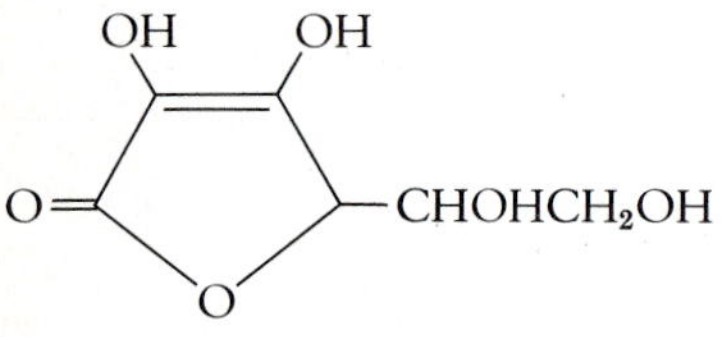

Vitamin C
(ascorbic acid)

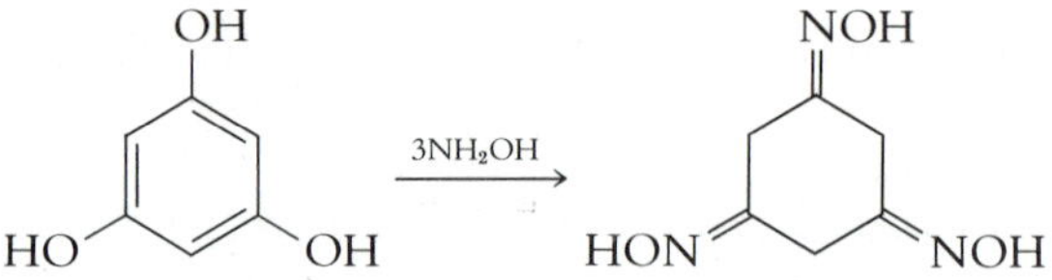

phlorglucinol

phlorglucinol
oxime

Ordinarily the keto form of a phenolic compound is the less stable tautomer, since its formation involves a loss of resonance energy. However, in the early 1950s, chemists realized that bases such as guanine and thymine should exist mainly in

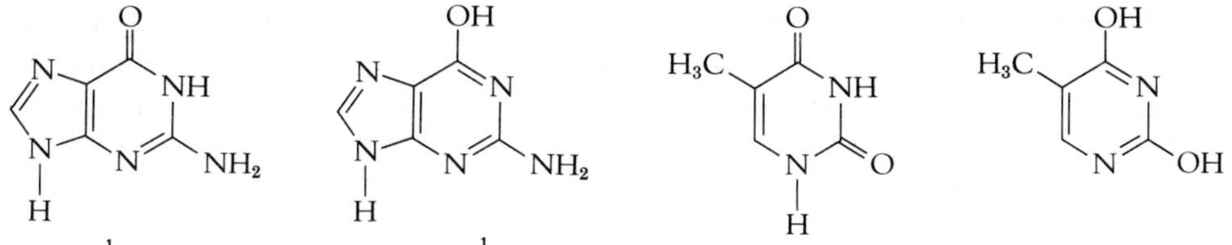

keto   enol

guanine

keto   enol

thymine

the keto forms at the pH of physiological systems — previously most textbooks had illustrated only the enol forms. This fact provided the key to the structure of DNA and thus to the genetic code. The bases, which are attached to the polyester backbone of a nucleic acid molecule, are responsible for the transmission of genetic traits (by DNA) and the synthesis of protein (by RNA) in living beings. Only the keto form allows certain base pairs (such as adenine-thymine and guanine-cytosine) to form — a process that is essential to the operation of nucleic acid molecules and thus to life itself.

METHODOLOGY

The synthesis of dimedone from mesityl oxide involves four distinct reactions that take place in sequence (see page 443). The first is a *Michael addition* of dimethyl malonate to mesityl oxide. The combination of sodium methoxide with dimethyl malonate generates a carbanion (sodiomalonic ester), which undergoes nucleophilic addition to the conjugated system of mesityl oxide. The next reaction is a *cyclization* that resembles the Dieckmann condensation except that the attacking nucleophile is a carbanion adjacent to a ketonic carbonyl group rather than to an ester carbonyl group. Both of these reaction steps occur spontaneously when dimethyl malonate and mesityl oxide are combined in a solution of the base in methanol. The resulting ester is then *hydrolyzed* to yield a carboxylic acid, which is easily *decarboxylated* to form dimedone.

A standard synthesis of dimedone requires the preparation *in situ* of sodium ethoxide by adding metallic sodium to absolute ethanol. Since this operation is quite hazardous, commercial sodium methoxide in methanol will be used instead.

A sodiomalonic ester forms immediately when sodium methoxide is added to dimethyl malonate; it reacts readily with

$$Na^+[CH(CO_2CH_3)_2]^-$$

sodiomalonic ester

The Dieckmann condensation is an intramolecular Claisen condensation of a diester that results in ring formation.

See *Organic Syntheses* Coll. Vol. II, p. 200.

The mesityl oxide should be freshly opened or recently distilled (see Safety Precautions).

mesityl oxide, which should be added in small portions. The Michael addition and the cyclization should go to completion during the subsequent reflux period. The cyclic ester is hydrolyzed with boiling sodium hydroxide, and the solution is acidified to form the carboxylic acid, which loses carbon dioxide on heating. The resulting dimedone is then isolated by vacuum filtration and recrystallized from aqueous acetone.

One of the best ways to detect and analyze enol content is by NMR spectrometry. The enolic OH protons of β-diketones absorb far downfield, with chemical shift values of about 10–16 δ. Enols also show vinyl proton (H—C=C) absorption in the usual range for these signals, about 4.5–6.0 δ. The keto form of a diketone can usually be recognized by the signal of the protons alpha to both carbonyl groups [H—C(C=O)$_2$]; these protons absorb near 3.5 δ. Since there are two such protons for every enolic vinyl proton in the enol form of dimedone, the equilibrium constant for dimedone's keto-enol equilibrium can be determined from the integrated areas of these two signals using the following formula:

$$K = \frac{[\text{enol}]}{[\text{keto}]} = \frac{2 \times \text{area of the H—C=C signal}}{\text{area of the H—C(C=O)}_2 \text{ signal}}$$

Enols are stabilized by hydrogen-bonding interactions with other polar species, so the equilibrium constant might vary depending on the solvent used (see Collateral Project 1). Infrared spectrometry can also be used to detect the existence of enols, as described in Experiment 43. You may want to compare the infrared spectra of dimedone in the condensed (solid) state and dimedone in solution to see if the proportion of enol differs (see Collateral Project 2).

In the presence of a secondary amine, dimedone reacts with an aldehyde to form a *bis*-dimedone derivative by a mechanism that apparently involves an aldol condensation and a Michael addition. *Bis*-dimedone derivatives cyclize in acidic solutions to form octahydroxanthenediones; both kinds of derivatives are used to identify unknown aldehydes. If you do Minilab 33, you will prepare the dimedone derivatives of benzaldehyde and test your mechanism-writing skills.

Bis-dimedone derivatives are also called dimethones, and octahydroxanthenediones are also called dimethone anhydrides.

2 dimedone + PhCH benzaldehyde →(base)→ *bis*-dimedone derivative →(H⁺)→ octahydroxanthenedione derivative

PRELAB ASSIGNMENT

Calculate the mass and volume of 25 mmol of dimethyl malonate and the theoretical yield of dimedone.

Reactions and Properties

All of the cyclohexanedione derivatives can also be represented by enol forms.

Table 1.   Physical properties

	M.W.	m.p.	b.p.	d.
mesityl oxide	98.2	−52	129	0.858
dimethyl malonate	132.1	−62	181	1.154
sodium methoxide	54.0			
methanol	32.0	−94	65	0.791
dimedone	140.2	149–51		

A 25% solution of sodium methoxide in methanol has a density of 0.945 g/mL.

Figure 1.   PMR and IR spectra of mesityl oxide. (PMR spectrum from *The Aldrich Library of NMR Spectra, Edition II,* by Charles J. Pouchert; IR spectrum from *The Aldrich Library of FT-IR Spectra,* by Charles J. Pouchert; both with the permission of the Aldrich Chemical Company.)

Mesityl oxide (4-methyl-3-penten-2-one)

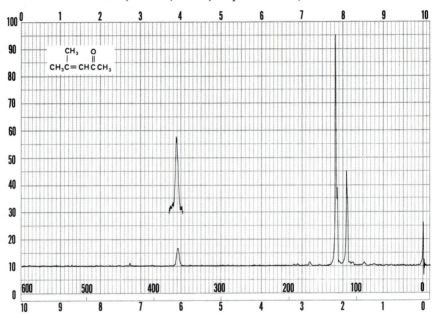

Mesityl oxide

2914.1	1448.7	1165.5
1689.7	1357.1	964.8
1620.3	1219.9	621.8

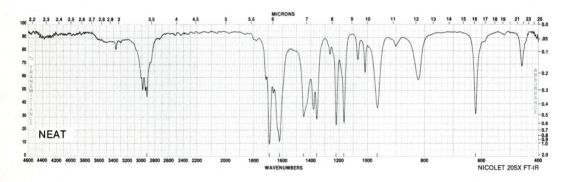

PROCEDURE

The solution of sodium methoxide in methanol is corrosive, toxic, and flammable; contact or inhalation can cause severe damage to the skin, eyes, and respiratory tract. Wear gloves, and avoid contact; do not inhale its vapors, and keep it away from flames.

Mesityl oxide is a lachrymator and strong irritant; inhalation, ingestion, and skin absorption are harmful. Avoid contact, and do not breathe its vapors. Mesityl oxide forms explosive peroxides on standing, so it should never be distilled to dryness.

SAFETY PRECAUTIONS

mesityl oxide

methanol

Reaction. Equip a dry reflux apparatus ⟨OP–7a⟩ with a drying tube ⟨OP–22a⟩. Place 25 mmol of dimethyl malonate in the reaction flask. Mix in 6.0 mL (~26 mmol) of a 25% solution of sodium methoxide in methanol **[contact and vapor hazard!].** Attach the reflux condenser, and heat the solution just to boiling on a steam bath. Remove the heat source, and add 3.0 mL (~26 mmol) of recently distilled mesityl oxide **[contact and vapor hazard!]** through the condenser in several small portions over a period of 2–3 minutes, swirling the reaction flask after each addition. Reflux ⟨OP–7a⟩ the reaction mixture for 1 hour.

The solid that forms when sodium methoxide is added should dissolve on heating.

Remove the reflux condenser, connect the reaction flask to a trap and aspirator, and evaporate the methanol ⟨OP–14⟩ using a steam bath or hot-water bath. Stop the evaporation when the solid residue is still barely moist but all standing liquid has been removed. Add 20 mL of 3 M aqueous sodium hydroxide to the reaction flask and reflux the mixture for another hour to hydrolyze the intermediate ester. Pour the warm reaction mixture, with stirring, into a 150-mL beaker containing 15 mL of 6 M hydrochloric acid. Carefully boil the mixture for 10 minutes or more **[foaming may occur!],** until no more carbon dioxide is evolved. Cool the beaker in an ice bath until precipitation is complete.

There may be considerable bumping during the evaporation, so use fresh boiling chips and gentle heating.

Boiling the solution should remove any remaining methanol as well as carbon dioxide. Any oil that forms during this step should crystallize on cooling.

Separation, Purification, and Analysis. Collect the dimedone by vacuum filtration ⟨OP–12⟩, and recrystallize it ⟨OP–23⟩ from 50% aqueous acetone. Dry ⟨OP–21⟩ and weigh the dimedone, and measure its melting point ⟨OP–28⟩. Record

the PMR spectrum ⟨OP – 34⟩ of dimedone in deuterochloroform, or obtain a spectrum from your instructor.

Report.   On the PMR spectrum, identify the enol, H—C=C,
and the keto, H—C(C=O)$_2$, signals. Calculate the keto-enol
equilibrium constant, and indicate which species predominates
in the solution. Compare the PMR spectrum of your product
to that of mesityl oxide shown in Figure 1, and interpret both
spectra as completely as you can.

Collateral Projects

1.   Record PMR spectra of dimedone in solvents other
than deuterochloroform, such as DMSO-d$_6$, and compare the
keto-enol equilibrium constants for the different solvents. Try
to explain any differences.

2.   Record the infrared spectrum of dimedone in a KBr
disc or Nujol mull and then in a chloroform solution ⟨OP –
33⟩. Identify any absorption bands that give evidence for the
presence of the enol form or the keto form in both spectra, and
compare the relative amounts of enol. Try to explain any differences.

MINILAB 33

Synthesis of Dimedone Derivatives of Benzaldehyde

**SAFETY
PRECAUTION**

**Piperidine is very toxic and corrosive. Avoid contact,
and do not breathe its vapors.**

Dissolve 0.1 g of benzaldehyde in 4 mL of 50% ethanol.
Stir in 0.3 g of dimedone and a drop of piperidine **[contact
and vapor hazard!],** and boil the mixture gently for 5 minutes. If the hot solution is clear, add water drop by drop until it
just becomes cloudy. Cool the solution in an ice bath until
crystallization is complete. Collect the product by semimicro
vacuum filtration, wash it with a little cold 50% ethanol, and
dry and weigh it.

Dissolve 0.1 g of the *bis*-dimedone derivative in 4 mL of absolute ethanol with gentle heating. Stir in 1 mL of water and 2 drops of 6 M hydrochloric acid, and boil the mixture gently for 5 minutes. Add water drop by drop to the hot solution until it just becomes cloudy, and then cool it until crystallization is complete. Collect the octahydroxanthene derivative by semi-micro vacuum filtration, wash it with cold 50% ethanol, and dry and weigh it.

Measure the melting points of both derivatives. Write balanced equations for the reactions and propose detailed mechanisms for the formation of both products.

Exercises

1. **(a)** Write a mechanism for the addition of dimethyl malonate to mesityl oxide in the presence of sodium methoxide in methanol. **(b)** Write a mechanism for the cyclization step of the synthesis of dimedone. **(c)** Write a mechanism for the decarboxylation step of this synthesis, illustrating the activated complex involved.

2. Give IUPAC names for all the organic compounds (except methanol) illustrated in the Reactions and Properties section.

3. Construct a flow diagram for the synthesis of dimedone.

4. Interpret the infrared spectrum of mesityl oxide (Figure 1) as completely as you can. Compare the frequency of its carbonyl band to the "normal" value of 1715 cm^{-1} for aliphatic ketones, and explain any significant difference.

5. What inexpensive starting material do you think is used to prepare mesityl oxide commercially? Outline the synthesis of mesityl oxide from this starting material.

6. A very confused student hydrolyzed the intermediate ester with 6 M HCl instead of 3 M NaOH and boiled the reaction mixture with 50% NaOH instead of 6 M HCl. Filtration of the cooled reaction mixture yielded only a little dimedone. Finally realizing his mistake, the student acidified the filtrate and another solid precipitated; unfortunately, its melting point was different from that of dimedone or any of the expected intermediates. What was the second product, and how did it form?

7.   Draw structures of some by-products that might be present in your product if you neglected to use a drying tube in the first step of the synthesis of dimedone. Write reaction pathways for their formation.

8.   When heated with sodium ethoxide in ethanol, the diethyl esters of butanedioic acid and heptanedioic acid both yield products having six-membered rings. Write structures for both products and mechanisms for their formation.

9.   Write a mechanism for the following reaction of dimedone with 3-buten-2-one.

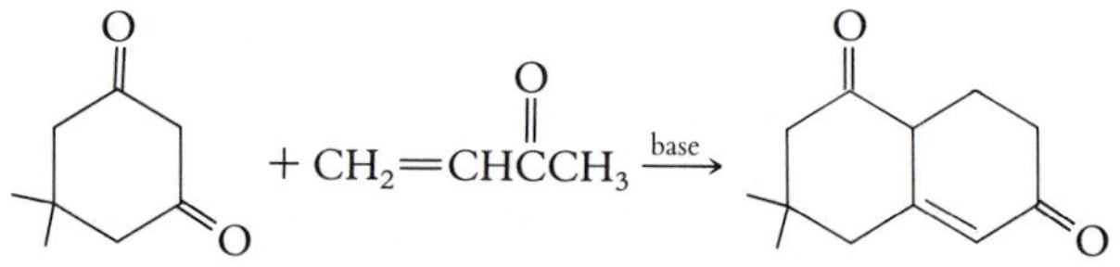

10.   The reaction of dimedone with formaldehyde has been used as a qualitative and quantitative test for the aldehyde. Calculate the percentage by weight of formaldehyde in a pickling solution (density = 1.08 g/mL) if 1.00 mL of the solution reacts with excess dimedone to produce 3.90 g of the *bis*-dimedone derivative.

Library Topic

Like the synthesis of dimedone carried out in this experiment, the Robinson annulation (or annelation) reaction involves a Michael addition followed by a ring-forming condensation reaction. Describe some applications of this reaction in the synthesis of steroids and terpenoids, and write a detailed mechanism for a typical Robinson annulation.

Analysis of Fatty-Acid Content of Commercial Cooking Oils and Fats

Reactions and Preparation of Carboxylic
  Esters. Transesterification. Fats and Oils.

Operations

OP – 11   Gravity Filtration

OP – 13   Extraction

OP – 14   Evaporation

OP – 32   Gas Chromatography

Objectives

To convert a mixture of triglycerides to the corresponding methyl esters.

To analyze the fatty-acid content of a commercial cooking oil or fat.

To learn how to identify fatty-acid esters from their equivalent-chain-length values.

To learn about the history and significance of fats and oils.

SITUATION

Fats and oils are glyceryl esters of long-chain carboxylic acids called fatty acids. The dietary quality of a fat or oil is related to the kinds and proportions of the fatty acids that constitute those esters. Animal fats and coconut oil contain a relatively high proportion of saturated fatty acids (SFA), which apparently raise serum cholesterol levels in humans and increase the risk of heart disease. Many vegetable oils and fish oils are high in polyunsaturated fatty acids (PFA), which apparently have the opposite effect, lowering the amount of cholesterol in the blood by hastening its excretion. Recent research suggests that monounsaturated fatty acids (MFA), which occur in some vegetable oils, also reduce the incidence of heart disease. A high degree of unsaturation has become a major selling point for certain cooking oils and fats because of these apparent health benefits. In this experiment, you will analyze a commercial product and use its ratio of unsaturated to saturated fatty acids as a measure of its dietary quality.

BACKGROUND

Fatty Acids

$$CH_3(CH_2)_{14}COOH$$

palmitic acid

$$CH_3(CH_2)_{16}COOH$$

stearic acid

Olive oil was so important to the medieval alchemists that they devised a special symbol, a circle within a cross, to represent it.

Fats and oils have been put to many uses throughout human history. Excavation of an Egyptian tomb more than 5000 years old yielded several earthenware vessels containing substances identified as palmitic acid and stearic acid, most likely formed by the breakdown of palm oil and beef or mutton tallow, which were placed there as provisions for the deceased. The Egyptians used olive oil as a lubricant when moving huge stones for their building projects and a mixture of fat and lime as axle grease for their war chariots. The Romans used candles made of beeswax and tallow for illumination, and the Phoenicians of 600 B.C. were trading soap to the Gauls, who undoubtedly needed it. Pliny the Elder described a process for making soap by boiling goat fat with wood ashes, then treating the pasty mass with sea water to harden it. Soap factories have been excavated at Pompeii, which was buried by a volcano in 79 A.D. The art of painting with oils or waxes goes back at least to the ancient Egyptians, who used the encaustic technique (pigments mixed with natural waxes) to paint portraits on their mummy cases.

Modern research has revealed the importance of fats and oils in nutrition and is just beginning to clarify their role in human disease. One fatty acid, linoleic acid, is known to be an essential component of the human diet, since it is not synthesized in the body. Polyunsaturated fatty acids (PFA) such as linoleic acid are believed to be involved in the biosynthesis of the prostaglandins, which help control blood pressure and muscle contraction. They play a vital role in the functioning of biological membranes and have been implicated in the occurrence or prevention of such illnesses as atherosclerosis, cancer, and multiple sclerosis. Recent research seems to indicate that

$$CH_3(CH_2)_4\overset{\overset{\displaystyle H}{|}}{C}=\overset{\overset{\displaystyle H}{|}}{C}CH_2\overset{\overset{\displaystyle H}{|}}{C}=\overset{\overset{\displaystyle H}{|}}{C}(CH_2)_7COOH$$

linoleic acid

PFAs of the $\omega 3$ and $\omega 6$ families are especially effective in preventing the buildup of cholesterol in arteries, which is a major cause of heart disease. Certain $\omega 3$ PFAs found in fish oils, called eicosapentaenoic acid and docosahexaenoic acid, appear to be even more beneficial than the PFAs in vegetable oils. Besides lowering serum cholesterol levels, they also reduce the tendency of the blood to clot.

Fatty acids are named or symbolized by several different systems. Most of the common fatty acids have trivial names that have been used for years, partly because of their simplicity compared to the systematic names. Trivial names do not reveal the structures of the corresponding acids however, so several shorthand notations have been adopted. Oleic acid can be represented by c-9-18 : 1, where c means $cis,$ 9 refers to the position of the double bond (numbering from the COOH group), 18 is the total number of carbon atoms in the chain, and 1 is the number of double bonds. Linoleic acid would be represented by c,c-9, 12-18 : 2 using this system. Since all of the acids to be studied in this experiment are $cis,$ the c prefix will be omitted here.

$$CH_3CH_2CH{=}CH \ldots$$

$$\underset{\omega 1}{\bigcirc}\underset{\omega 2}{\bigcirc}\underset{\omega 3}{\bigcirc}$$

The ω designation refers to the distance of the last double bond from the methyl end of the fatty acid chain, as illustrated.

Linoleic acid, for example, is named cis-9-cis-12-octadecadienoic acid under the IUPAC system.

$$CH_3(CH_2)_7{-}\overset{H}{\underset{|}{C}}{=}\overset{H}{\underset{|}{C}}{-}(CH)_7COOH$$
oleic acid (9-18:1)

METHODOLOGY

The fats and oils to be analyzed in this experiment are triglycerides, that is, esters of the trihydroxy alcohol glycerol containing three fatty-acid residues. The fatty acids can be obtained by hydrolyzing a triglyceride, usually in the presence of a base, to yield the corresponding acid salts. This is the process used in soap formation, with the acid salts comprising the "soap." Alternatively, a triglyceride can be transesterified with an alcohol such as methanol to yield a mixture of methyl esters. Since the esters are much more suitable for GLC analysis than the acids themselves, the second method will be used in this experiment. The fatty acid composition of each fat can then be derived from the percentages of the corresponding esters.

When saturated methyl esters of fatty acids are chromatographed on a suitable GLC column, their retention times (T_r) increase with the length of the carbon chain according to the relationship log $T_r \propto$ number of carbons. The retention times of unsaturated fatty esters do not coincide with those of saturated esters; on a nonpolar liquid phase such as Apiezon their retention times are shorter than those of the saturated esters of the same chain length; whereas on a polar liquid phase such as

$$CH_3(CH_2)_{16}COOCH_2 \qquad CH_2OH$$
$$CH_3(CH_2)_{16}COOCH \qquad\ CHOH$$
$$CH_3(CH_2)_{16}COOCH_2 \qquad CH_2OH$$
a triglyceride,   glycerol
tristearin

$$CH_3(CH_2)_{16}COO^-Na^+$$

a soap, sodium stearate

$$CH_3(CH_2)_{16}COOCH_3$$

methyl ester of stearic acid
(methyl stearate)

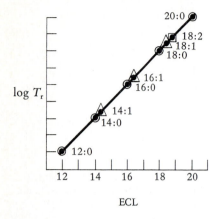

Figure 1. Plot of log T_r versus equivalent chain length for a DEGS liquid phase

diethylene glycol succinate (DEGS) they are longer. The actual value of T_r depends on the number and positions of the double bonds. By comparing the retention times of a large number of such esters, a series of equivalent-chain-length (ECL) values has been worked out for various esters on different liquid phases. The ECL for a *saturated* ester is the same as its actual chain length — for instance, 14.0 for myristic acid (as the methyl ester), 18.0 for stearic acid, etc. The ECL for an *unsaturated* ester on a polar liquid phase is greater than its actual chain length — for instance, 18.43 for methyl oleate (9-18:1) and 19.22 for methyl linoleate (9,12-18:2), both measured on DEGS. If the retention times for two or more saturated esters are known, a calibration curve can be constructed by plotting the logarithms of their retention times against their chain lengths, as shown in Figure 1. To identify an unknown fatty-acid ester, one can calculate the logarithm of its retention time, read its ECL value off the graph, and compare that value with the ECL values of known methyl esters (see Table 1).

In order for ECL values to provide an accurate means of identifying the esters, the same type of column and carrier gas should be used in all cases. Even then, the experimental values will not usually correspond exactly to literature values because of differences in such factors as the kind of column support, the amount of liquid phase, the age of the column, and the experimental conditions. In addition, some peaks may overlap in

Table 1. Shorthand notation and ECL values (on DEGS liquid phase) for representative fatty acids

Acid (as methyl ester)	Shorthand notation	ECL value	Type
lauric	12:0	12.00	SFA
myristic	14:0	14.00	SFA
myristoleic	9-14:1	14.71	MFA
palmitic	16:0	16.00	SFA
palmitoleic	9-16:1	16.55	MFA
stearic	18:0	18.00	SFA
oleic	9-18:1	18.43	MFA
linoleic	9,12-18:2	19.22	PFA
linolenic	9,12,15-18:3	20.12	PFA
behenic	22:0	22.00	SFA

complex mixtures, so that two or more columns must be used for their separation, and deviations from linearity in the log T_r plot may occur when short-chain fatty acids are analyzed. The relatively simple mixtures of fatty acids that you will encounter in this experiment should not present any serious experimental difficulties, however.

The methanolysis (transesterification) of the oil will be carried out by saponifying the triglycerides in methanolic sodium hydroxide and esterifying the resulting fatty acids in methanol with a boron trifluoride catalyst. After the methyl esters have been isolated from the reaction mixture, they will be mixed with an internal standard, methyl heptadecanoate, and analyzed by gas chromatography. To minimize the effect of retention time drift due to changes in operating parameters, relative retention values should be calculated by dividing the retention time (or distance) for each peak by that of the $17:0$ peak. Once the $17:0$ peak has been located, those of the $16:0$ and the $18:0$ esters can be identified by the fact that the log T_r interval between their peaks and $17:0$ is the same. Then a log T_r versus ECL plot can be prepared, from which the other esters can be identified. Since the peak area for a component should be nearly proportional to its mass, the percentage composition of the mixture can be determined from peak areas with good accuracy. Assuming that both monounsaturated and polyunsaturated fatty acids are beneficial and saturated fatty acids are harmful, you can evaluate the dietary quality of the fat or oil you have analyzed in terms of the ratio (MFA + PFA)/ SFA, and then compare your results with those for the other fats and oils analyzed in your laboratory section.

Heptadecanoic acid ($17:0$) does not occur naturally, so the $17:0$ peak will not interfere with other component peaks.

The $16:0$ ester often gives the first strong peak on the chromatogram.

The ECL graph can be refined after several more esters have been identified by using their reported ECL values to plot a more accurate straight line.

Soap has long been made by heating fats or oils with lye (sodium hydroxide) and other alkaline substances, including potash (potassium carbonate) from wood ashes. Alcohol is sometimes used to speed up the saponification reaction by bringing the base and fat together in solution; however, the soap must then be salted out or the alcohol evaporated. A simpler procedure using a phase-transfer catalyst to saponify a vegetable shortening (containing mainly palmitic, stearic, oleic, and linoleic acid residues) is described in Minilab 34.

Although the triglycerides found in plants are usually mixtures containing several different fatty-acid residues, the fat derived from nutmeg contains only one, myristic acid. If you do Minilab 35, you will isolate trimyristin from ground nutmeg.

Reactions and Properties

$$R'COOCH_2 \qquad\qquad\qquad R'COOCH_3$$

R'COOCH₂   R'COOCH₃
| +
R''COOCH + 3CH₃OH $\xrightarrow[\text{BF}_3]{\text{NaOH}}$ R''COOCH₃ + CH₂CHCH₂
| +
R'''COOCH₂   R'''COOCH₃

(R groups can be alike or different.)

Table 2.   Physical properties

	M.W.	m.p.	b.p.	d.
methanol	32.0	−94	65	0.791
boron trifluoride– methanol complex	131.9		59[4]	
boron trifluoride	67.8	−127	−100	

The complex has the composition $BF_3 \cdot 2CH_3OH$.

Figure 2.   IR spectra of saturated and polyunsaturated eighteen-carbon methyl esters. (Reproduced from *The Aldrich Library of FT-IR Spectra*, by Charles J. Pouchert, with the permission of the Aldrich Chemical Company.)

Methyl stearate

2925.2	1360.7	1169.8
1744.9	1302.8	1117.3
1465.8	1247.5	721.3

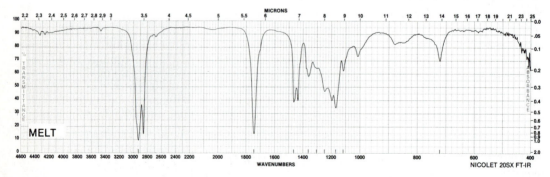

Figure 2 *(continued)*

Methyl linolenate

$$CH_3(CH_2CH=CH)_3(CH_2)_7\overset{\overset{\textstyle O}{\textstyle \|}}{C}OCH_3$$

2930.3	1361.7	1171.7
1742.7	1308.2	1103.8
1435.7	1245.5	722.4

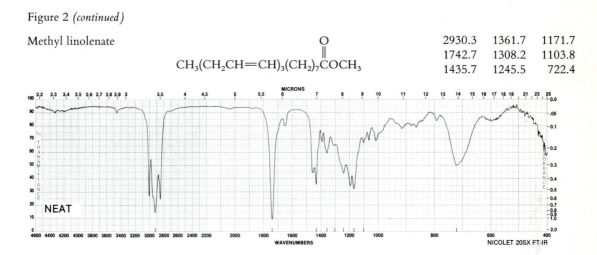

NEAT

WAVENUMBERS

NICOLET 20SX FT-IR

PROCEDURE

[Adapted from a procedure in the *Journal of Chemical Education 51,* 406 (1974) with permission.]

A selection of commercial products may be provided, or students may be asked to bring samples of various fats and oils. It is suggested that several different classes of fats and oils be analyzed, such as soybean or corn oil, safflower or sunflower oil, peanut oil, olive oil, vegetable shortening, and lard. It is important that the glassware and reagents be dry for this experiment.

SAFETY PRECAUTIONS

The methanolic sodium hydroxide solution is flammable and toxic. Avoid contact with the liquid, do not breathe its vapors, and keep it away from flames.

The solution of boron trifluoride in methanol is flammable, corrosive, and toxic and can cause severe damage to the eyes, skin, and respiratory tract. The fumes are highly irritating. Use gloves and a hood; avoid contact with the solution, and do not breathe its vapors.

Petroleum ether is extremely flammable, so keep it away from flames and hot surfaces.

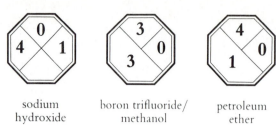

sodium   boron trifluoride/   petroleum
hydroxide   methanol   ether

Reaction.   In a *dry* 15-cm test tube, combine 0.15 g of a commercial cooking oil or fat with 5 mL of 0.5 M methanolic sodium hydroxide solution **[contact and vapor hazard!]**. Heat the reaction tube over a steam bath until the oil or fat has completely dissolved (about 3–5 minutes). *Under the hood,* add 6 mL of a 12.5% (weight/volume) solution of boron trifluoride in methanol **[contact and vapor hazard! wear gloves!]**, and boil the mixture over the steam bath for 2 minutes.

The BF₃ solution can be prepared by mixing 4 mL of methanol with 1 mL of 50% BF₃-methanol complex. If you prepare this solution yourself, work under the hood, and wear gloves!

Separation.   Transfer the reaction mixture to a separatory funnel using 30 mL of low-boiling (~ 35–$60°$C) petroleum ether and rinsing out the test tube with some of the solvent. Add 20 mL of saturated aqueous sodium chloride, and shake to extract $\langle OP$–$13 \rangle$ the methyl esters into the petroleum ether layer. Filter the petroleum ether layer by gravity $\langle OP$–$11 \rangle$, and evaporate the solvent completely $\langle OP$–$14 \rangle$.

The inorganic reactants and most of the methanol should end up in the aqueous layer.

Analysis.   Obtain a gas chromatogram $\langle OP$–$32 \rangle$ of a 1–2 μL sample of the methyl alkanoate mixture on a DEGS/Chromosorb W or similar column. Add a drop of a 50% solution of methyl heptadecanoate in methylene chloride to the remaining mixture, and obtain its gas chromatogram also. Measure the areas of all peaks on the first gas chromatogram. Identify the methyl heptadecanoate peak, and measure the retention times of all peaks on the second gas chromatogram.

If your data points don't fit the plot, you probably misidentified the esters. Make another guess and try again.

Report.   Calculate relative retention times and log T_r values for all the components, and use these data to identify tentatively several of the methyl esters. For these esters, plot log T_r versus their ECL values, and try to fit the rest of your data to the plot, redrawing it as necessary to get the best straight line. Identify the esters, and calculate the percentage of each fatty acid in your fat or oil. Calculate the total percentages of saturated, monounsaturated, and polyunsaturated fatty acids, and the ratio of unsaturated to saturated fatty acids. Tabulate the class results, and rank the different fats and oils according to their apparent degree of effectiveness in preventing heart disease.

Collateral Projects

1. Test your cooking oil or fat for unsaturation using Test C – 7 or C – 19 of Part III.

2. Record an infrared spectrum of your ester mixture, and compare it with the spectra of the eighteen-carbon methyl esters in Figure 2. Try to account for significant similarities and differences in the spectra, and point out any bands that show evidence of unsaturation.

Preparation of a Soap
Using a Phase-Transfer Catalyst

MINILAB 34

The sodium hydroxide solution can cause serious burns; wear gloves, and avoid contact.

SAFETY PRECAUTION

Measure 10 g of vegetable shortening (for example, Crisco) into a small beaker, and heat it over a steam bath until it melts. Stir in 4.5 mL of 8 M ($\sim$25%) aqueous sodium hydroxide [**contact hazard!**] and 2 drops of tricaprylmethylammonium chloride (Aliquat 336). Heat the mixture on the steam bath, with stirring, for 20 minutes or more, until all the oily globules have disappeared. Cool the reaction mixture; then add 3 mL of water, and use a flat-bottomed stirring rod to break up and wash the soap particles. Separate the soap by vacuum filtration, washing it on the filter with several small portions of ice water. Dry the soap in a desiccator, and weigh it. Draw the structures of the major components of the soap you prepared, and write an equation for the reaction of glyceryl tristearate (tristearin) with N₂OH.

MINILAB 35   # Extraction of Trimyristin from Nutmeg

Accurately weigh 2 g of finely ground nutmeg. Combine it with 10 mL of ethyl ether [fire hazard!], and reflux the mixture ⟨OP–7a or OP–7b⟩ gently for 45 minutes or more over a hot-water or steam bath. Filter the mixture by gravity, washing the nutmeg residue on the filter with a little ether; then evaporate the ether from the filtrate. Recrystallize the trimyristin from 95% ethanol. Dry the trimyristin, weigh it, and measure its melting point. Draw the structure of trimyristin (glyceryl trimyristate), and calculate the percentage of trimyristin in the nutmeg.

Exercises

1. Write the structures of the fatty acids whose triglycerides were in the fat or oil you analyzed, and give their IUPAC names.

2. On the gas chromatogram of a mixture of methyl esters, methyl palmitate had a retention time of 150 seconds and methyl heptadecanoate a retention time of 198 seconds. What is the probable identity of a methyl ester with a retention time of 363 seconds on the same chromatogram?

3. Explain why there was no glycerol peak on the gas chromatogram of the methyl alkanoate mixture, even though glycerol was a product of the hydrolysis.

4. Why was the Lewis acid boron trifluoride used to catalyze the transesterification reaction rather than a protic acid such as hydrochloric acid?

5. Calculate the (MFA + PFA)/SFA ratio for palm oil, whose fatty-acid content is approximately 9% linoleic, 2% myristic, 40% oleic, 45% palmitic, and 4% stearic acid.

6. **(a)** Neat's-foot oil consists almost entirely of the triglycerides of oleic and palmitic acids. How many different triglycerides of these two acids can it contain? **(b)** Draw structures for the triglycerides that contain two oleic acid units and one palmitic acid unit.

$$CH_3(CH_2)_{10}CH_2OSO_2O^-Na^+$$
sodium lauryl sulfate

7. Outline a synthesis of the detergent sodium lauryl sulfate from glyceryl trilaurate (trilaurin).

Library Topic

Write a report on the health implications of the various types of fats and oils. Include a description of the HDL and LDL forms of cholesterol and a discussion of the role they play in heart disease.

Determination of the Structure of an Unknown Aldohexose

Reactions of Monosaccharides. Preparation of Aldaric Acids. Preparation of Phenylosazones. Structure Determination.

Operations

OP–11 Gravity Filtration

OP–12 Vacuum Filtration

OP–21 Drying Solids

OP–23*b* Recrystallization from Mixed Solvents

OP–28 Melting-Point Determination

OP–31 Measuring Optical Rotation

Objectives

To determine the structure of an unknown aldohexose.

To prepare an aldaric acid and a phenylosazone from an aldohexose.

To gain additional experience with polarimetric measurements.

To learn about some methods for the structure analysis of carbohydrates.

To learn about simple sugars and the sweet-taste sensation.

SITUATION

Two simple sugars can be obtained from the complex lipids called cerebrosides, which occur in nerve and brain tissues. One of these "brain sugars" is glucose; the other is an aldohexose that has the same molecular formula as glucose but is only half as sweet. In this experiment, you will attempt to establish the structure of the unknown aldohexose by converting it to an aldaric acid and to a phenylosazone.

BACKGROUND

Sweet Molecules

One of the most apparent properties of the sugars is their sweetness. Fructose is the sweetest known sugar — 1.8 times sweeter (in its crystalline form) than sucrose — but other substances are sweeter than any of the sugars. Cyclamates such as sodium cyclohexylsulfamate are approximately 30 times sweeter than sucrose, saccharin is about 350 times sweeter, and 1-*n*-propoxy-2-amino-4-nitrobenzene (P-4000) is estimated to be 4000 times sweeter (see Figure 1).

Many explanations have been advanced to explain the sweet taste sensation. One theory asserts that all sweet compounds contain a so-called AH,B couple that is necessary to bind the compound to the taste-bud receptor site. According to the theory, A and B are electronegative atoms such as oxygen or nitrogen (but sometimes halogen or even carbon) that must be 2.5 – 4.0 Å apart in order to interact (by hydrogen bonding) with the receptor site, which possesses a similar AH,B couple. In the sugars, AH and B are assumed to be an OH group and the oxygen atom of an adjacent OH group, respectively — ideally in a *gauche* conformation. In an *anti* conformation they would be too far apart to interact, and in a *syn* conformation they could hydrogen-bond intramolecularly rather than with the receptor site.

Fructose crystallizes as β-D-fructo-pyranose, which is the sweetest form. When dissolved in water, some of it changes to the furanose form, causing its sweetness to decrease soon after solution.

Proposed interaction of a sweet molecule with the receptor site

Gauche conformation in a sugar molecule

Another possible requirement for the sweet taste sensation is a hydrophobic or "greasy" site, γ, on the sweet molecule, which is approximately 3.5 Å from A and 5.5 Å from B in the triangular grouping illustrated. The dissymmetry of this arrangement could explain why enantiomers sometimes have different tastes. In β-D-fructopyranose, the AH, the B, and the γ sites are considered to be the C-1 OH, the CH$_2$OH oxygen atom, and the ring methylene group, respectively. Proposed AH,B groupings for some other sweet compounds are illustrated in Figure 1.

Proposed location of hydrophobic site on sweet molecules

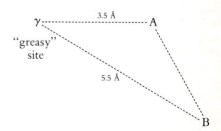

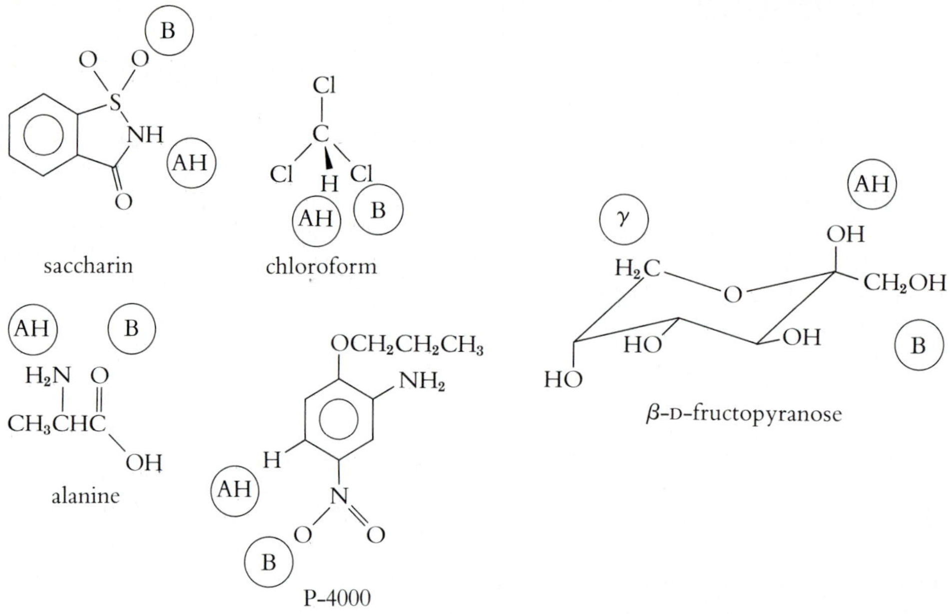

Figure 1.   AH,B groupings in sweet molecules

METHODOLOGY

Monosaccharides can be characterized by a number of different methods, such as converting them to osazones and oxidizing them to aldaric acids. The formation of osazones from monosaccharides in effect "homogenizes" the sugar molecules at the first two carbon atoms of the chain by converting both $-CO-CH_2OH$ and $-CHOH-CHO$ to identical groupings:

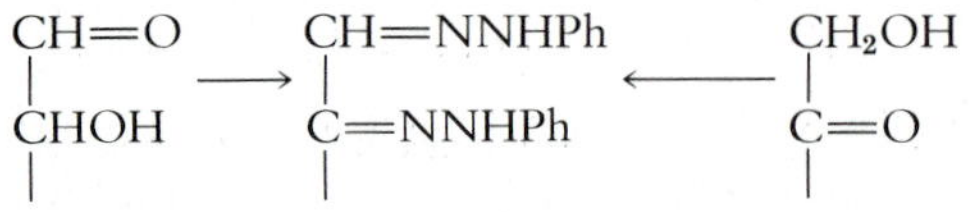

$$\begin{array}{ccc}
CH{=}O & CH{=}NNHPh & CH_2OH \\
| & | & | \\
CHOH & C{=}NNHPh & C{=}O \\
| & | & |
\end{array}$$

Thus aldoses and ketoses that are identical in configuration from C-3 on down the chain, but different at C-1 or C-2, yield identical osazones. Considering only the D-hexoses (in their open-chain forms, for convenience) we can identify four C-3 to C-6 structural units that yield different osazones. For example, allose, altrose, and the ketohexose allulose, illustrated by Rosanoff symbols, all contain the structural unit **A** (Figure 2)

D-allose   D-altrose   D-allulose

Rosanoff symbols for functional groups

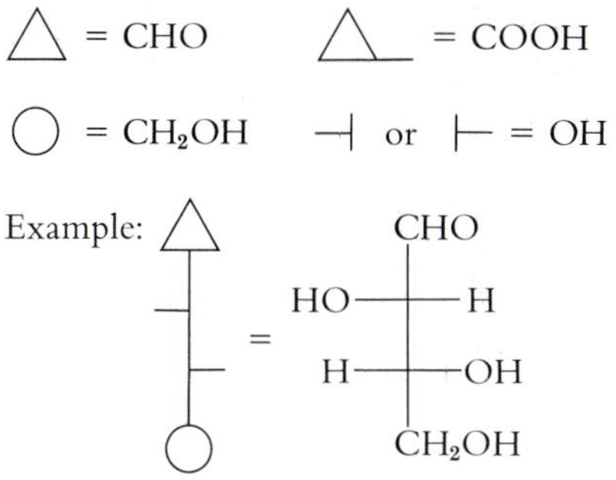

△ = CHO   △̲ = COOH

◯ = CH₂OH   ┤ or ├ = OH

Example:

CHO
HO——H
H——OH
CH₂OH

Figure 2.   C-3 to C-6 structural units of D-hexoses

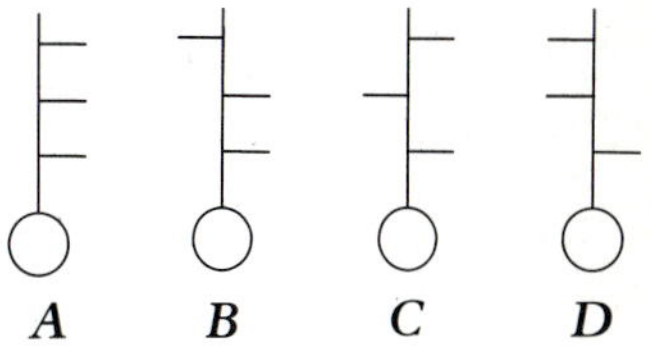

A   *B*   *C*   *D*

and should yield the same osazone. You will prepare the osazone of your unknown monosaccharide by a simple test-tube reaction and measure its melting point. Because osazones decompose on melting, you will have to determine the melting point by a special technique that requires rapid heating.

Nitric acid oxidation of a monosaccharide converts both aldehyde and primary alcohol functional groups to carboxyl groups, forming an aldaric acid. Thus, glucose is oxidized to D-glucaric acid, which has COOH groups at each end of the molecule. (This acid actually exists as a dilactone.) Since glucaric acid, like glucose, has no plane of symmetry, it is optically active and will rotate the plane of polarized light. The acid obtained by oxidizing D-ribose *does* have a plane of symmetry and should therefore be an optically inactive *meso* compound. Thus, the optical activity or inactivity of the aldaric acid formed from oxidation of a monosaccharide can yield valuable structural information.

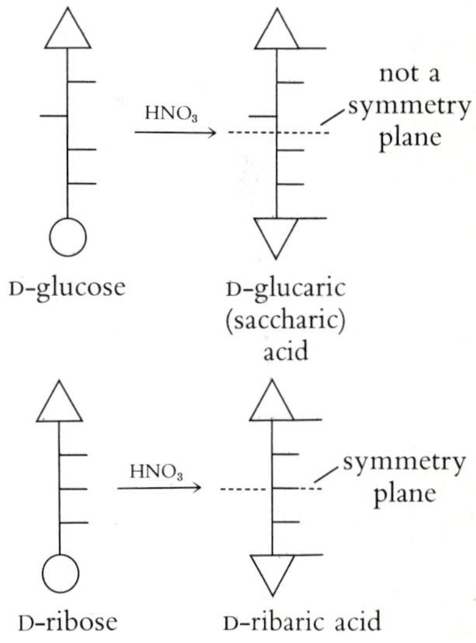

You may wish to confirm the identity of the unknown aldohexose in this experiment by determining its specific rotation, as described in Collateral Project 1. Different anomeric forms of carbohydrates have different optical rotations, so a little ammonia is added to catalyze a mutarotation reaction that yields an equilibrium mixture of anomers. Other tests used to characterize carbohydrates are described in Collateral Projects 2 and 3.

Ketones such as acetone and cyclohexanone react with adjacent *cis*-hydroxyl groups of monosaccharides to form cyclic ketals. Ketal formation can be used to protect certain hydroxyl groups, allowing selective reactions at the unprotected sites. For example, the reaction of D-fructose (shown in the β-pyranose ring form) with excess cyclohexanone yields a di-

Optically active and inactive aldaric acids

D-glucose   D-glucaric (saccharic) acid   not a symmetry plane

D-ribose   D-ribaric acid   symmetry plane

Use of protective groups in a carbohydrate synthesis

β-D-fructopyranose

1,2:4,5-di-*O*-cyclohexylidene-
D-fructopyranose

3-*O*-benzyl-
D-fructopyranose

ketal having a single unprotected hydroxyl group at C-3. Alkylation of this hydroxyl group by benzyl chloride, followed
by hydrolysis of the ketal protective groups, gives 3-*O*-ben-
zyl-D-fructose. If you do Minilab 36, you will prepare the
intermediate diketal 1,2:4,5-di-*O*-cyclohexylidene-D-fructo-
pyranose.

Reactions and Properties

Table 1.   Osazone melting points

Structural unit	Osazone m.p.
A	178
B	205
C	173
D	201

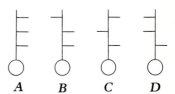

PROCEDURE

Concentrated nitric acid is poisonous and very corrosive, capable of causing severe damage to the skin, eyes, and respiratory tract. Use gloves and a hood; avoid contact, do not breathe its vapors, and keep it away from other chemicals.

Methanol is very flammable, and ingestion, inhalation, and skin absorption are harmful. Avoid contact, do not breathe its vapors, and keep it away from flames and hot surfaces.

The brown vapors of nitrogen dioxide gas that are evolved during the reaction are poisonous. Do not breathe them!

SAFETY PRECAUTIONS

nitric acid

A. Preparation of the Aldaric Acid

Reaction.   *The reaction must be carried out under an efficient fume hood.* Dissolve 2.0 g of the unknown aldohexose in 20 mL of water in an evaporating dish. Cautiously add 8 mL of concentrated nitric acid **[contact hazard! wear gloves!]** with stirring. Heat the mixture over a boiling-water bath, but reduce the heat immediately when a vigorous reaction begins, accompanied by the evolution of red-brown fumes of NO_2 **[vapor hazard!].** When the reaction has subsided, heat the mixture on the boiling-water bath, with occasional stirring, until nearly all of the water has evaporated and a moist, pasty mass remains.

Do not evaporate to dryness or the product may char.

Separation and Purification.   Cool the residue, add 10 mL of ice water, and collect the aldaric acid by vacuum filtration, washing it several times with cold water. Dissolve the product in 12 mL of ice-cold 2 M (∼7.5%) aqueous sodium hydroxide with stirring. Quickly filter the solution through a coarse fluted filter paper ⟨OP–11⟩ to remove undissolved impurities. Cool the solution in an ice bath, and add 5 mL of 6 M hydrochloric acid with constant stirring. Let the solution stand until crystallization is complete. Collect the product by vacuum filtration, washing it with water followed by cold methanol **[contact, vapor, and fire hazard!].** Dry ⟨OP–21⟩ and weigh the aldaric acid.

Allow sufficient time for complete crystallization—the product may not precipitate immediately. Be sure the solution is strongly acidic (pH 2 or lower).

Analysis.   Make up a solution containing about 0.5 g of the aldaric acid per 25 mL of solution, using 0.25 M (1%) aqueous sodium hydroxide as the solvent. If the solution is cloudy or contains undissolved solid, vacuum filter it ⟨OP–12⟩ through a "slow" filter paper. Measure the optical rotation of this solution ⟨OP–31⟩, using water as the solvent blank.

If the filtrate is still cloudy, it can be refiltered through a bed of Celite (a filtering aid), which is prepared by pouring an aqueous slurry of Celite onto the filter paper while the aspirator is running.

B.   Preparation of the Phenylosazone

phenylhydrazine

Dissolve 0.2 g of the unknown aldohexose in 4 mL of water in a test tube, and stir in 0.4 g of phenylhydrazine hydrochloride **[contact hazard! suspected carcinogen! wear gloves!]**, 0.6 g of sodium acetate trihydrate (or 0.36 g of anhydrous sodium acetate), and 0.4 mL of saturated aqueous sodium bisulfite. Stopper the test tube loosely with a cork, and leave it in a boiling-water bath for 30 minutes, with occasional shaking. Add 5 mL of water, and cool the reaction mixture in an ice bath. Collect the phenylosazone by vacuum filtration ⟨OP–12⟩, washing it on the filter with a little ice-cold methanol. Recrystallize the product from an ethanol-water mixture ⟨OP–23b⟩, dissolving it in the ethanol first. Dry the phenylosazone thoroughly ⟨OP–21⟩ at room temperature and weigh it. Pulverize enough of the solid for two melting-point

Washing the product on the filter with ice-cold methanol will speed up the drying.

capillaries ⟨OP–28⟩, and measure the temperature, T_1, at which the first sample melts with rapid heating (a rise of 10–20°C per minute). After the melting-point apparatus has cooled below T_1, heat it at a rate of 3–6°C per minute; insert the second capillary just as the temperature reaches T_1. Record the temperature at which this sample becomes completely liquid; this is the melting point of the phenylosazone.

Report. If the optical rotation of the aldaric acid is essentially equal to that of the blank, assume the acid is optically inactive; otherwise, calculate its specific rotation. Deduce the structure of the unknown aldohexose, and explain your reasoning. Write equations for the reactions you carried out, showing the stereochemical structures of the products. Calculate the percentage yields. Draw the Haworth (flat-ring) structure of your aldohexose, and use the formula index of the *Merck Index* or another reference work to find its trivial name.

Collateral Projects

1.   Measure the optical rotation of a solution containing about 1 g (accurately weighed) of the unknown aldohexose in 25 mL of aqueous solution. Mix 2 drops of concentrated aqueous ammonia **[contact and vapor hazard!]** into the solution, and again measure its rotation. If the optical rotation changes with time, wait until it equilibrates before taking a final reading. Write an equation for the equilibration reaction. Calculate the specific rotation of the equilibrium mixture, and compare it to the appropriate literature value from the *Merck Index* or another reference work.

2.   Use the Seliwanoff test to confirm that the unknown monosaccharide is an aldose. Place 4-mL portions of Seliwanoff's reagent (0.5 g of resorcinol per liter of 4 M HCl) in each of two test tubes. Add 1 mL of a 1% aqueous solution of the unknown monosaccharide to one test tube and 1 mL of a 1% aqueous solution of fructose to the other. Simultaneously place both test tubes in a boiling-water bath. Ketoses form a colored solution in a few minutes; aldoses react much more slowly.

3. Compare the rates of osazone formation for as many of the following carbohydrates as possible: arabinose, fructose, galactose, glucose, mannose, sucrose, and xylose. Use the procedure in part B but omit the saturated sodium bisulfite. Insert test tubes containing the reaction mixtures into a boiling-water bath simultaneously, and shake them occasionally to prevent supersaturation. Measure the time required for a visible precipitate to form in each case. Compare the rate for sucrose with the rates for its constituents, glucose and fructose, and explain any significant difference. If a microscope is available, examine the crystals, and sketch their characteristic shapes.

MINILAB 36

Preparation of 1,2 : 4,5-Di-O-cyclohexylidene-D-fructopyranose

SAFETY PRECAUTIONS

Avoid contact with cyclohexanone and methanol, and do not breathe their vapors.

Avoid contact with sulfuric acid.

Measure 2.2 mL of cyclohexanone into a 10-cm test tube, cool it in an ice bath, and add 5 drops of concentrated sulfuric acid **[contact hazard!]** with stirring. Add 1.0 g of *dry* finely powdered D-fructose, and stir vigorously to dissolve it. When the fructose has dissolved and a waxy solid begins to form in quantity, let the test tube stand in an ice bath (replenish the ice as necessary) with occasional stirring throughout the lab period (or until it completely solidifies). Then let the test tube stand overnight at room temperature. Collect the product by semimicro vacuum filtration, washing it on the filter with two portions of ice-cold methanol **[vapor and fire hazard!]**. If the product is still gummy, stir it thoroughly with a few milliliters of high-boiling petroleum ether **[fire hazard!]**, pressing it against the sides of the test tube with your stirring rod, and refilter. Recrystallize the product from high-boiling petroleum ether or cyclohexane **[fire hazard!]**, dry it at room temperature, weigh it, and measure its melting point.

Exercises

1. **(a)** Draw structures of all the monosaccharides that would yield the same phenylosazone as the aldohexose you analyzed. **(b)** Draw structures of the aldaric acids that would be obtained by oxidation of all possible D-aldohexoses, and indicate whether each one is optically active or optically inactive.

2. It is believed that the OH group on C-4 and the oxygen atom on C-3 of an aldopyranose molecule can interact as an AH,B couple with the sweet taste receptor sites. Using molecular models of the pyranose ring forms of the aldohexose you identified and glucose, suggest a reason why the former is not as sweet as the latter.

3. **(a)** The α and β anomers of the aldohexose you identified have specific rotations in water of $+150.7°$ and $+52.8°$, respectively. From the specific rotation of the equilibrium mixture (see Collateral Project 1), calculate the percentage of each anomer in the mixture. **(b)** Draw chair-form rings for both anomers, indicate which one should be more stable, and explain how the composition of the equilibrium mixture verifies your conclusion (if it does).

4. (a) α-D Glucose reacts with excess cyclohexanone to form a diketal that has a furanose ring. Draw the structure of this diketal, and explain why it is more stable than the corresponding pyranose-form diketal. (Models will be helpful.) **(b)** Outline a synthesis of 3-O-acetyl-D-glucose from D-glucose using this diketal as an intermediate.

5. Explain why the diketal formed in the reaction of D-fructose with cyclohexanone in Minilab 36 has a pyranose ring and not a furanose ring.

6. **(a)** Draw the structure of the aldohexose that would yield the same phenylosazone as D-glucose. **(b)** Draw the structure of the aldohexose that would yield the same aldaric acid as L-glucose.

Library Topic

Write a report on the chemical basis of taste, giving examples of compounds that produce sweet, sour, salty, and bitter tastes. Describe the chemical species or structural features that appear to contribute to each kind of taste sensation, and give examples illustrating the effect of structural variations on taste quality or intensity.

EXPERIMENT 47

Determination of the Structure of an Unknown Dipeptide

Reactions of Peptides. Nucleophilic Aromatic Substitution. Amino Acids. Structure Determination.

Operations

OP – 4	Basic Glass Working
OP – 13a	Semimicro Extraction
OP – 14	Evaporation
OP – 17	Thin-Layer Chromatography
OP – 18	Paper Chromatography
OP – 22	Drying Gases

Objectives

To determine the structure of an unknown dipeptide.

To learn how to identify amino acids and their DNP derivatives using chromatography.

To learn how to carry out reactions on a micro scale.

To learn how to identify the *N*-terminal residue of a peptide.

To learn about some naturally occurring peptides.

SITUATION

The structures of long-chain polypeptides, such as those found in the venom of poisonous snakes, can be determined by breaking them down into shorter chains of amino acids and analyzing these fragments. For example, an enzyme called cathepsin C hydrolyzes a polypeptide into dipeptide units, and the enzyme trypsin breaks a polypeptide chain after each lysine or arginine residue. By analyzing each fragment from these and other reactions, it is possible to identify overlapping sequences of amino acids, from which the entire primary structure of the polypeptide can be deduced. In this experiment, you

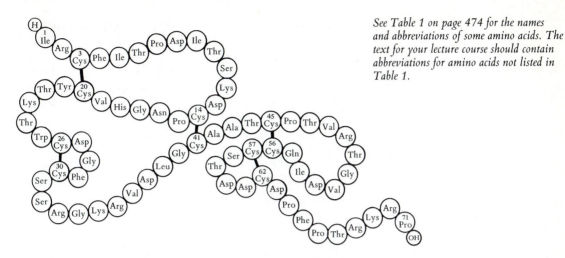

See Table 1 on page 474 for the names and abbreviations of some amino acids. The text for your lecture course should contain abbreviations for amino acids not listed in Table 1.

Figure 1.   Venom toxin of the Indian cobra *(Naja naja)*. [Reprinted from *Biochemical and Biophysical Research Communications, 55,* 435 (1973), by permission.]

will determine the structure of an unknown dipeptide and find out whether it is one of the dipeptides that could be obtained from the polypeptide in cobra venom (shown in Figure 1).

BACKGROUND

Mushrooms, Black Mambas, and Memory Molecules

The naturally occurring polypeptides and proteins range in size from tripeptides such as glutathione to complex proteins having molecular weights on the order of 1 million; their variations in structure and function cover just as broad a range. Some of them, such as the keratins of hair, horns, nails, and claws, act as biological building materials and have little or no biological activity; others function as hormones, enzymes, antibiotics, or toxins, exerting a profound effect on biochemical reactions.

Amino acids and most polypeptides actually exist as dipolar ions (zwitterions), such as $H_3\overset{+}{N}CH_2CO\overset{-}{O}$ for glycine. For clarity, the neutral forms are used throughout this experiment.

$$H_2NCHCH_2CH_2CONHCHCONHCH_2COOH$$
$$\underset{COOH}{|} \qquad \underset{CH_2SH}{|}$$
$$\text{glutathione}$$

Bradykinin, sometimes known as the "pain molecule," is released by enzymatic cleavage of plasma glycoprotein when-

Cys—Tyr—Ile
| |
S
|
S
|
Cys—Asn—Gln
|
Pro—Leu—GlyNH$_2$
oxytocin

Cys—Tyr—Phe
| |
S
|
S
|
Cys—Asn—Gln
|
Pro—Arg—GlyNH$_2$
vasopressin

As little as one mushroom of the *Amanita phalloides* type can cause death in an adult. Some *Amanita* mushrooms (particularly *A. muscaria,* known as "soma" or "fly agaric") have been eaten deliberately for their reported hallucinogenic properties, occasionally with fatal consequences.

Arg—Pro—Pro—Gly—Phe—Ser—Pro—Phe—Arg
bradykinin

ever tissues are damaged. It causes the sensation of pain by bonding to certain receptors on nerve endings. Peptides similar to bradykinin are present in wasp venom; other kinins act as hormones, stimulating a variety of physiological responses such as the contraction or relaxation of smooth muscles and the dilation of blood-vessel walls.

Oxytocin and vasopressin are both secreted by the posterior pituitary gland. Their structures are identical except for two amino-acid units, but their functions are entirely different. Vasopressin increases retention of water in the kidneys and is used as an antidiuretic in treating a form of diabetes that is characterized by excessive urine flow. Oxytocin intensifies uterine contractions during childbirth and is used clinically to induce labor. Both are cyclic compounds (cyclopeptides) having a tripeptide side chain.

Most of the poisonous mushrooms of the genus *Amanita* contain the cyclopeptides known as phallotoxins and amatoxins. These mushrooms are very deceptive, since the victim is usually not aware that he or she has been poisoned for ten hours or more after eating them. After a day or so of violent cramps, nausea, vomiting, and other symptoms, the patient appears to recover and may be sent home from the hospital. Death from kidney and liver failure often follows in several days. α-Amanitin and the other amatoxins are believed to attack the nuclei of liver and kidney cells, depleting their nuclear RNA and inhibiting the synthesis of more RNA, so that protein synthesis stops and the cells die.

α-amanitin

Larger polypeptides, containing from 60 to 74 amino-acid residues, are found in snake venoms such as those of the African black mamba and the Indian cobra (Figure 1). Most of these venom toxins have long chains that are cross-linked by four or more cystine disulfide bridges.

The venom toxin of the black mamba is probably the quickest—it can kill a mouse in less than five minutes.

One of the most exciting new areas of polypeptide biochemistry is the study of "memory molecules." Researchers have discovered that an animal's learned behavior can be forgotten when a substance interfering with peptide synthesis is injected into the animal's brain at a certain time. This suggests that peptide synthesis is involved in the consolidation of long-term memory. A peptide called scotophobin has been isolated from the brains of rats that had been trained to fear the dark.

Ser—Asp—Asn—Asn—Gln—Gln—Gly—Lys—Ser—Ala—Gln—Gln—Gly—Gly—TyrNH$_2$

scotophobin

(from the Greek *scotos*, dark; and *phobos*, fear)

When scotophobin was injected into untrained rats, they also became afraid of the dark! This raises the intriguing possibility that breaking the chemical memory code could enable chemists to synthesize the peptides corresponding to any kind of learning experience. Perhaps someday it will be possible to get an injection of Biology 101 or Philosophy 416 rather than absorbing the subject matter in the classroom!

METHODOLOGY

In this experiment, you will determine the structure of an unknown dipeptide, which will contain one or more of the amino acids in Table 1 (on page 474). To do so, you will first identify the amino acids it contains, by hydrolyzing it and analyzing the resulting amino-acid mixture by paper chromatography. To determine its complete structure, however, you must determine the sequence in which those amino acids appear in the dipeptide. This can be done by the dinitrophenylation procedure developed by Frederick Sanger. 2,4-Dinitrofluorobenzene (DNFB) is an unusually reactive aryl halide that undergoes nucleophilic substitution reactions with unprotonated amino groups. Because most peptides exist as zwitterions in neutral solutions, sodium bicarbonate is used to raise the pH of the peptide solution enough to deprotonate the amino group, allowing it to react with the DNFB. This reaction yields a dipeptide having a dinitrophenyl (DNP) substituent attached to the *N*-terminal amino-acid residue, as shown in the Reactions and Properties section.

Sanger won the Nobel Prize for Chemistry in 1956 for determining the structure of insulin and again in 1980 for his work on nucleic acid structures.

NO$_2$

O$_2$N—〈 〉—

2,4-dinitrophenyl group
(DNP)

The *N*-terminal amino acid residue is the one at the end of the chain that has the free amino group.

Table 1.   Names and abbreviations of selected amino acids

Name	Abbreviation
histidine	His
lysine	Lys
serine	Ser
aspartic acid	Asp
glycine	Gly
threonine	Thr
alanine	Ala
proline	Pro
tyrosine	Tyr
valine	Val
phenylalanine	Phe
leucine	Leu

The solvent ratios are volume ratios. Thus, a 7:3 1-propanol:ammonia solvent system can be prepared by mixing 70 mL of 1-propanol with 30 mL of concentrated aqueous ammonia.

ninhydrin

In the alkaline solution, the DNP-substituted dipeptide exists in the anionic form, with a polar —COO⁻ group. Thus, extracting this solution with ether removes the less polar DNFB but leaves the DNP dipeptide in the aqueous layer. Lowering the pH with hydrochloric acid protonates the carboxyl group, allowing the DNP dipeptide to be extracted from the aqueous layer by ether and then isolated by evaporating the ether. It is important to remove the last traces of ether before hydrolysis, so the residue is dissolved in acetone, which is also evaporated. The DNP dipeptide is then hydrolyzed in aqueous hydrochloric acid, yielding the free *C*-terminal amino acid and a dinitrophenyl derivative of the *N*-terminal amino acid. The DNP amino acid can be identified by thin-layer chromatography on silica gel, using known DNP amino acids as standards for comparison. The *C*-terminal amino acid can also be analyzed if desired (see the Collateral Project).

The dipeptides and DNP amino acids are quite expensive; therefore, this experiment will be carried out on a micro scale, using only a milligram of the dipeptide for each reaction. You will use a capillary pipet for volume measurements and solvent extractions and sealed capillary tubes for the hydrolysis reactions. It takes some practice to fill and seal a capillary tube properly, so it is recommended that you practice the technique with water before using the dipeptide and DNP dipeptide solutions. The capillary method is very reliable when performed correctly, but there is always the chance that an improperly sealed tube will break or leak during the overnight hydrolysis period. To prevent the considerable loss of time that would result from a ruined sample, you should prepare at least three tubes for each hydrolysis reaction.

The paper chromatography of amino acids can be carried out in a multitude of developing solvents; no one solvent system is the best under all circumstances. The 16:1:4 2-propanol:formic acid:water solvent system used in this procedure is suitable for most amino acids, but certain combinations of amino acids may be separated more readily with a different solvent system, such as 7:3 1-propanol:ammonia or 15:5:6 2-butanone:propionic acid:water. The amino acids are colorless, so their spots will be made visible using a solution of ninhydrin, which reacts with most amino acids (proline is an exception) to form a blue-violet product.

The DNP amino acids are more easily separated on silica gel than on paper, so they will be analyzed by thin-layer chromatography. Because the DNP derivatives are light-sensitive, their chromatograms should be developed in a location away from strong light. The yellow spots formed by the DNP amino acids are easy to locate but they fade with time, so you should circle them soon after the TLC plate is developed. Dinitrophenol is a possible by-product of the dinitrophenylation procedure; however, its yellow spot is usually evident only when an alkaline developing solvent is used. If TLC analysis of your unknown produces two or more distinct spots, exposing the chromatogram to vapors of hydrochloric acid should bleach out the dinitrophenol spot.

Casein is a milk protein that contains all the common amino acids and is particularly rich in the essential ones. Cheese production is based on the precipitation of casein from milk by an acid, often the lactic acid formed during the natural souring of milk. If you do Minilab 37, you will isolate casein from skim milk by curdling the milk with acetic acid.

PRELAB ASSIGNMENT

Read OP – 18 and the parts of OP – 4 and OP – 13a relating to this experiment.

Reactions and Properties

A. O_2N—⟨benzene ring, NO$_2$⟩—F + H_2NCHC(=O)—$NHCHC$(=O)—OH $\longrightarrow$
                                          |R |R′

O_2N—⟨benzene ring, NO$_2$⟩—NHCHC(=O)—NHCHC(=O)—OH + HF
                              |R |R′

B. O_2N—⟨benzene ring, NO$_2$⟩—NHCHC(=O)—NHCHC(=O)—OH + H_2O $\xrightarrow{\text{HCl}}$
                                   |R |R′

O_2N—⟨benzene ring, NO$_2$⟩—NHCHC(=O)—OH + H_2NCHC(=O)—OH
                              |R |R′

(continued)

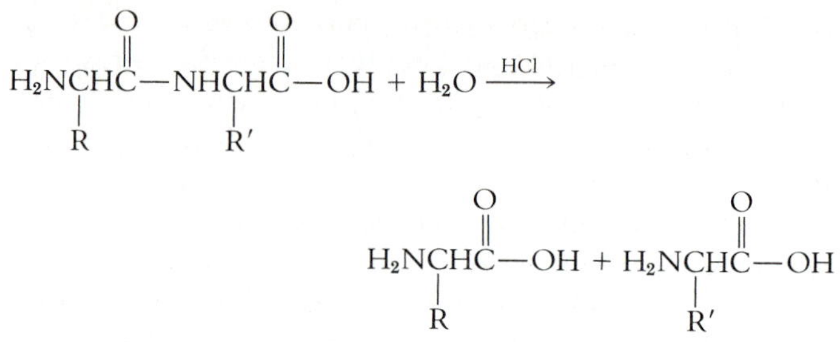

$$\underset{\underset{R}{|}}{H_2NCHC}\overset{\overset{O}{\|}}{—}NHCHC\overset{\overset{O}{\|}}{—}\underset{\underset{R'}{|}}{OH} + H_2O \xrightarrow{\ HCl\ }$$

$$\underset{\underset{R}{|}}{H_2NCHC}\overset{\overset{O}{\|}}{—}OH + \underset{\underset{R'}{|}}{H_2NCHC}\overset{\overset{O}{\|}}{—}OH$$

R, R′ = amino acid side chains

Table 2.　Physical properties

	M.W.	m.p.	b.p.
dinitrofluorobenzene	186.1	27.5–30	178[15]

PROCEDURE

Use a Pasteur (capillary) pipet for all volume measurements; the pipet should deliver about 35–45 drops per milliliter. Organize your time efficiently, so that you can have your hydrolysis tubes prepared and in the oven by the end of the first laboratory period.

SAFETY PRECAUTIONS

2,4-Dinitrofluorobenzene is highly toxic if ingested, inhaled, or absorbed through the skin. It can cause severe dermatitis in sensitive individuals, and it is suspected of causing cancer in laboratory animals. Wear gloves and avoid contact with the DNFB reagent, and do not breathe its vapors.

Ethyl ether is extremely flammable and may be harmful if inhaled. Do not breathe its vapors, and keep it away from flames and hot surfaces.

A.　Dinitrophenylation of the Dipeptide

In a small conical centrifuge tube or a 7.5-cm test tube, combine about 1 mg of your unknown dipeptide with 4 drops of distilled or deionized water, 1 drop of 4% aqueous sodium bicarbonate, and 8 drops of a 5% solution of dinitrofluorobenzene in ethanol **[contact and vapor hazard! wear gloves!]**. Cover the tube with Parafilm, and shake it occasionally over a

1-hour period. Every 10 minutes, check the pH by dipping the closed end of a melting-point capillary into the solution and touching it to a strip of narrow-range pH paper. Add a drop or two of 4% sodium bicarbonate solution as necessary to keep the pH between 8 and 9.

Add 10 drops of water and 10 drops of 4% sodium bicarbonate solution to the reaction mixture. Extract the mixture 2 or 3 times with ethyl ether ⟨OP–13a⟩, each time using a volume of ether approximately equal to the total volume of solution. Carry out each extraction by (1) drawing the mixture into the capillary tube and ejecting it rapidly, repeating the operation a half-dozen times or more; (2) cooling the mixture in ice water and waiting for any emulsion to settle; (3) withdrawing the ether layer with the capillary pipet. Add enough 6 M hydrochloric acid (2 or 3 drops are usually sufficient) to the aqueous layer to bring its pH down to 1 or 2. Then extract the solution with two portions of ethyl ether as before, and combine and save the extracts. Be careful not to include any of the aqueous layer when you withdraw the ether layers.

Perform more extractions if necessary; the ether layer should be colorless after the last extraction.

The ether layers may be discarded.

Use a dry capillary pipet to distribute the ether solution equally among three wells on a porcelain spot plate, and let the ether evaporate *completely* under the hood. Dissolve each residue in 2 drops of acetone, and let the acetone evaporate under the hood. Then dissolve each residue in 4 drops of 6 M hydrochloric acid, and prepare at least three hydrolysis tubes by one of the following methods.

Use the closed end of a capillary melting-point tube for stirring.

Method 1: Warm a capillary melting-point tube by holding it near its open end and moving it in and out of the "cool" part of a Bunsen burner flame for several seconds; then immediately insert the open end into the liquid in one of the wells [**be careful not to burn yourself!**]. As the tube cools (blowing on it may help), it should begin to fill with liquid. When 2–3 cm of liquid is inside the tube, take the open end out of the liquid in the well, and cool the *closed* end under the water tap. Holding the tube upright near the top, carefully strike it at a point below the liquid level with your fingernail until nearly all the liquid is in the bottom of the tube. Holding the closed end so that your fingers just cover the liquid, rotate the tube as you move it rapidly in and out of the cool part of the burner flame for several seconds. (The tube should be inserted lengthwise into the flame, so that as much of it is heated as possible.) Immediately seal the open end ⟨OP–4⟩ by rotating it in the inner cone of the flame, near its apex. Because of the vacuum created by preheating the tube, its end should collapse and bend over as it seals.

Capillary tubes are fragile—don't tap the tube too hard or it will break.

Method 2: Insert one end of an open capillary tube into one of the wells and draw in (by capillary action) about 2–3 cm of liquid. Hold the tube horizontally and tap it so that the liquid is in the middle, then seal one end by rotating it in the outermost edge of the flame. Hold the open end up and strike the lower part of the tube with your fingernail until nearly all the liquid is in the bottom, then seal the other end as described above.

Put the capillary tubes in a small test tube labeled with your name and "DNP amino acid."

B. Hydrolysis of the Dipeptide and the DNP Dipeptide

Dissolve about 1 mg of your unknown dipeptide in 10 drops of 6 M hydrochloric acid. Use this solution to prepare three or more capillary tubes by one of the methods described in part A. Put the capillary tubes in a small test tube labeled with your name and "amino acids." Leave this test tube and the one from part A in a 100–110°C oven overnight or longer.

C. Separation and Analysis of the Amino Acids and the DNP Amino Acid

SAFETY PRECAUTION Both chromatographic developing solvents contain flammable or toxic liquids with harmful vapors, and chloroform is a suspected carcinogen. Use a hood. Avoid contact with the solvents, do not breathe their vapors, and keep them away from flames and hot surfaces.

2-propanol   formic acid   chloroform   acetic acid

Under a hood, prepare a large beaker (or another suitable developing chamber) for paper chromatography ⟨OP–18⟩ by adding enough of the 16:1:4 2-propanol:formic acid:water solvent system **[contact and vapor hazard!]** to form a 1-cm layer on the bottom. Cover the beaker with plastic wrap, and let the system equilibrate for at least an hour (preferably over-night) before using it. Prepare a similar developing chamber for the TLC analysis ⟨OP–17⟩ using the 23:1:10 chloro-form:acetic acid:*t*-pentyl alcohol solvent system **[contact and vapor hazard! suspected carcinogen!].** Cover the beaker with plastic wrap, and let the system equilibrate for at least a half-hour before using it.

Use a file to open one of the (cooled) amino acid capillary tubes just above the liquid level, and empty it into one well of a spot plate. Carefully evaporate the solvent ⟨OP–14⟩ under a dry air stream ⟨OP–22⟩, and dissolve the residue in 1 drop of water. Obtain a sheet of Whatman #1 chromatography paper that is at least 12 cm long (in the direction of development) and wide enough to accommodate the standard amino-acid solutions as well as your unknown solution. Spot the paper in two (or more) places with the amino-acid solution; then spot it with the amino-acid standards. Develop the chromatogram under a hood in the 16:1:4 2-propanol:formic acid:water solvent system. Under the hood, spray the paper lightly with the ninhydrin spray reagent. Develop the color by heating the paper in a 100–110°C oven for about 10 minutes. Measure the R_f values of all the spots.

Handle the paper by the edges so that you don't get fingerprints in the developing region. Ninhydrin will reveal your fingerprints as well as the spots!

Open one of the DNP amino acid capillary tubes, and empty the contents into a small test tube or centrifuge tube, washing it down with 10–15 drops of water. Extract the aqueous solution twice with ethyl ether ⟨OP–13a⟩, using a volume of ether equal to the volume of the solution for each extraction. Combine the extracts. Transfer the solution to a well on a spot plate, and let the ether evaporate under the hood. Dissolve the residue in 3 drops of acetone, and immediately use this solution to spot a silica-gel TLC plate in two places. Spot the plate with the standard solutions of DNP amino acids, and develop it under a hood with the 23:1:10 chloroform:acetic acid:*t*-pentyl alcohol solvent system, in a location away from strong light. Circle the yellow spots as soon as the developed plate is dry; you may use a "black light" to make them more clearly visible. Measure the R_f values of all the spots.

Save the aqueous layer, which contains the C-terminal amino acid (see the Collateral Project).

If your unknown mixture forms more than one yellow spot, hold the chromatogram over an open bottle of concentrated HCl [in a hood!] momentarily.

Report.   Tabulate the R_f values from your chromatograms, and identify the amino acids and the DNP amino acid. Draw the structure of the unknown dipeptide, name it, and tell how you arrived at your conclusion. Write equations for the reactions undergone by the dipeptide during the dinitrophenylation and hydrolysis steps. Tell whether the dipeptide you identified could be obtained by using cathepsin C to break down the polypeptide from cobra venom illustrated in Figure 1.

Collateral Project

If the identity of the dipeptide is not obvious from your results, you can evaporate the reserved aqueous layer from the DNP-dipeptide hydrolysate, add a drop of water to the residue, and identify the *C*-terminal amino acid by paper chromatography as described above.

MINILAB 37

Isolation of Casein from Milk

Measure 20 mL of *skim* milk into a beaker and heat it to 40°C. Add 10% (~ 1.7 M) aqueous acetic acid drop by drop with stirring until no more precipitate forms. Stir until the casein coagulates into an amorphous mass. Then decant the liquid, and transfer the wet solid to a vacuum filtration apparatus. With the aspirator running, press the casein with a clean cork to squeeze out as much water as possible; then dry it further between paper towels. Transfer the casein to a small beaker, cover it with 95% ethanol, and use a flat-bottomed stirring rod to stir and crush it until it is finely divided. Collect the casein by vacuum filtration, wash it with two small portions of acetone, and let it dry at room temperature. Weigh the casein and calculate the mass percentage of casein in the skim milk (density ~ 1.03 g/mL). Add a drop of concentrated nitric acid **[contact and reactivity hazard!]** to a small piece of casein in a test tube, and record your observations. Explain why this test is positive for proteins containing aromatic amino-acid residues. Write an equation for the reaction of a tyrosine residue with nitric acid (casein contains about 6% tyrosine).

Exercises

1.   When you extracted the DNP-dipeptide hydrolysate with ether, why wasn't the *C*-terminal amino acid extracted along with the DNP amino acid? Write equations explaining your answer.

2. **(a)** Propose a detailed mechanism for the reaction of 2,4-dinitrofluorobenzene with the dipeptide you identified. **(b)** With reference to the mechanism, explain why the reaction would not take place at low pH.

3. A student accidentally added aqueous sodium bisulfate instead of sodium bicarbonate to the dinitrophenylation reaction mixture just before the first ether extraction. Her DNP-dipeptide hydrolysate produced no spots when analyzed by TLC. What happened to her product, and why?

4. Another student decided to save time by adding a large excess of sodium bicarbonate at the beginning of the dinitrophenylation reaction instead of adding smaller amounts during the reaction. When he developed his TLC plate with a solvent system containing butanol and ammonia, a large yellow spot showed up that didn't match any of the DNP amino acid standards. What compound formed the yellow spot? Write an equation for its formation.

5. **(a)** Some amino acids, such as lysine and tyrosine, yield dinitrophenylated derivatives even when they are not at the end of a peptide chain. Explain, giving structures for the DNP derivatives. **(b)** Such derivatives do not usually interfere with the identification of the DNP *N*-terminal amino acids, since they are not extracted from the aqueous hydrolysis mixture at pH 1 by ethyl ether. Explain.

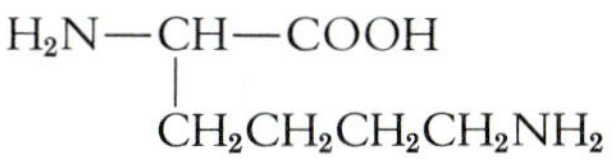

6. Construct a flow diagram for the preparation and hydrolysis of the DNP dipeptide.

Library Topic

Report on the methods used to determine the structures of polypeptides and proteins, describing the applications of such reagents and enzymes as phenyl isothiocyanate, dansyl chloride, trypsin, chymotrypsin, cathepsin C, and cyanogen bromide.

$$H_2N—CH—COOH$$
$$|$$
$$CH_2CH_2CH_2CH_2NH_2$$

lysine

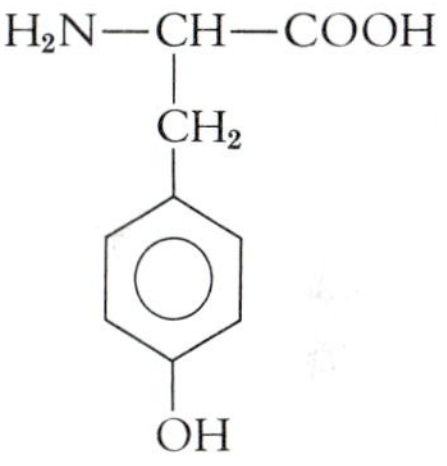

$$H_2N—CH—COOH$$
$$|$$
$$CH_2$$

tyrosine

The Preparation of 2-Phenylindole

Reactions of Carbonyl Compounds.   Preparation of Heterocyclic Amines.   Fischer Indole Synthesis.   Phenylhydrazones.

Operations

OP–12   Vacuum Filtration

OP–21   Drying Solids

OP–23*b*   Recrystallization from Mixed Solvents

OP–24*a* Vacuum Sublimation

OP–28   Melting-Point Determination

OP–36   Mass Spectrometry

Objectives

To prepare 2-phenylindole from acetophenone and phenylhydrazine.

To learn how to perform a vacuum sublimation.

To learn how to interpret a simple mass spectrum.

To learn about the Fischer indole synthesis.

To learn about some natural indoles.

SITUATION

The indole ring system occurs in many compounds that have profound effects on biological systems; these range from plant growth hormones to mind-altering drugs such as psilocybin and LSD. Nearly all of the physiologically active indoles have substituents at carbon-3 of the indole ring system and none at carbon-2. It seems likely that a nonsubstituted carbon-2 is necessary for an indole molecule to be able to bind to a receptor site where it can exert its physiological effects. In this experiment, you will prepare a physiologically inactive 2-substituted indole and analyze its mass spectrum.

BACKGROUND

Of Toads and Toadstools

The indoles constitute a large family of natural and synthetic compounds with extraordinary properties and functions. Indoles as a group display an ambivalent nature, as illustrated by the parent compound; indole has an odor of fine jasmine in dilute solutions but smells like feces when undiluted. 3-Methylindole is a major product of the digestive putrefaction of proteins and is mainly responsible for the repulsive odor of feces, giving rise to its common name, skatole (from *scat,* meaning animal droppings). However, it is also found in cabbage sprouts, tea, and even lilies!

In the sixteenth century, in Mexico, the Spanish conquistadors observed the Aztecs using some small brown mushrooms called *teonanacatl* ("flesh of the gods") in their religious ceremonies. According to the Spanish friar Bernardino de Sahagun, "They ate these little mushrooms with honey, and when they began to be excited by them, they began to dance, some singing, others weeping. . . . Some saw themselves dying in a vision and wept; others saw themselves being eaten by a wild beast; others imagined that they were capturing prisoners in battle, that they were rich, that they possessed many slaves, that they had committed adultery and were to have their heads crushed for the offense." Naturally the Spanish friars disapproved of these ceremonies, which were thereafter practiced only by secret cults until rediscovered in the 1930s and brought to public attention by R. Gordon Wasson. The little mushrooms were from several different species of the genus *Psilocybe;* their hallucinogenic constituents are two indoles of the tryptamine group, psilocin and psilocybin. A similar tryptamine derivative, bufotenin, is the active principle of cohoba snuff, which is inhaled by certain Indians of South America and the Caribbean area because it produces hallucinations and intoxication. There is an indole ring in the complex molecule of lysergic acid, which can be isolated from the ergot fungus on rye and is the source of LSD.

indole

skatole

tryptamine

psilocybin

psilocin

bufotenin

Bufotenin has also been isolated from both toads (genus *Bufo*) and toadstools (*Amanita citrina,* etc.).

tryptophan

3-indoleacetic acid

indigo

serotonin

Other indole derivatives are more beneficial, such as the essential amino acid tryptophan, the plant growth hormone 3-indoleacetic acid (also called heteroauxin), the beautiful dye indigo, and the bioregulator serotonin. Tryptophan is important as a source of serotonin and the B-vitamin nicotinamide, both of which are formed during its metabolism. Because the human body cannot ordinarily biosynthesize aromatic compounds, tryptophan and the other aromatic amino acids (phenylalanine and tyrosine) must be obtained from food. Indoleacetic acid promotes the enlargement of plant cells and is the principal natural growth regulator for many plants. Indigo was one of the first natural dyes to be prepared synthetically. Its manufacture resulted from the brilliant efforts of Adolph von Baeyer, who confirmed its structure after eighteen years of research. Although the role of serotonin in the human body is not fully understood, it appears to play an important role in mental processes. Some researchers see it as a mind stabilizer that helps to preserve sanity. The resemblance between serotonin and many mind-altering drugs is striking; it seems likely that the psychological activity of these drugs is caused by their interference with the action of serotonin.

METHODOLOGY

The preparation of 2-phenylindole by the Fischer indole synthesis involves the acid-catalyzed cyclization of a phenylhydrazone with the loss of a molecule of ammonia.

a phenylhydrazone

The phenylhydrazone can be prepared by combining phenylhydrazine (or a substituted phenylhydrazine) with an aldehyde or ketone of the structure RCH_2COR', where the R groups can be alkyl, aryl, or hydrogen. The generally accepted mechanism for the *indolization reaction* is one proposed by Robinson and Robinson in 1918. It includes the following steps:

1. A tautomeric shift of a proton from carbon to the β-nitrogen

2. Protonation of that nitrogen

Robinson mechanism

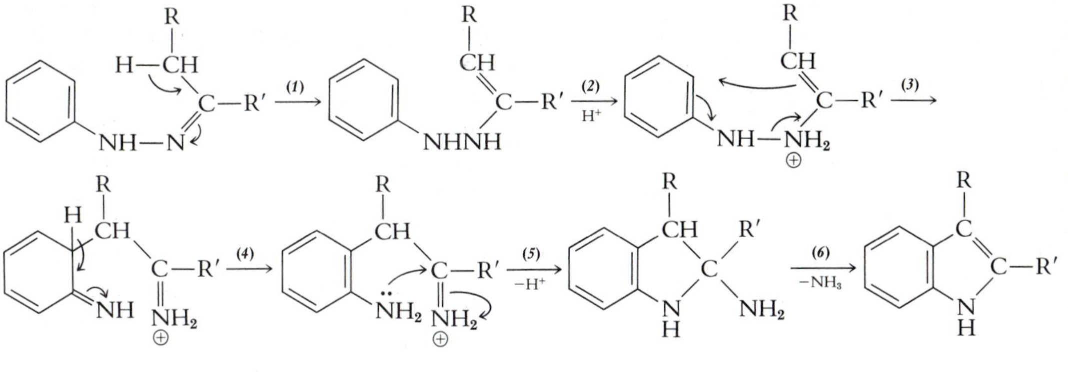

3. A concerted electron shift that forms a C—C bond to the ring and breaks the N—N bond

4. Another tautomeric shift of a proton from carbon to nitrogen

5. Nucleophilic attack by nitrogen on a doubly bonded carbon atom, followed by the loss of a proton

6. Loss of ammonia to yield an aromatic pyrrole ring

2-Phenylindole has been prepared from acetophenone by a variety of methods, including the following:

1. A direct reaction of acetophenone with phenylhydrazine and polyphosphoric acid in which acetophenone phenylhydrazone is formed *in situ*

See *Organicum,* p. 605 ⟨Bib – B10⟩.

2. Prior preparation of acetophenone phenylhydrazone, which is then stirred at 170°C with anhydrous zinc chloride

See *Organic Syntheses,* Coll. Vol. III, p. 125.

3. Heating acetophenone phenylhydrazone in 100% phosphoric acid

Reported by Grineva, Sadovskaya, and Ufimtsev, in *J. Gen. Chem. USSR 33,* 545 (1963) (English translation).

In this experiment, you will use polyphosphoric acid as the catalyst, but will prepare the acetophenone phenylhydrazone beforehand to avoid the highly exothermic reaction that results from procedure 1.

The crude acetophenone phenylhydrazone, which results from heating the reactants in ethanol with a little acetic acid to catalyze the reaction, can be used directly in the second step of the synthesis. The reaction in polyphosphoric acid is exothermic and requires only a few minutes to go to completion, after which the reaction mixture is poured into water to dissolve the acid and precipitate crude 2-phenylindole. You can purify

Acetophenone phenylhydrazone is light- and heat-sensitive. Do not use heat to dry it; store it in the dark.

Polyphosphoric acid (PPA) is made by heating H_3PO_4 with P_2O_5. It contains about 55% triphosphoric acid along with other polyphosphoric acids.

2-phenylindole by recrystallization from a mixed solvent —
either ethanol : water or toluene : hexane — followed by vacuum sublimation. If you use toluene : hexane, the product
should be well dried before recrystallization. The recrystallization from ethanol : water is complicated by the fact that the
crystals dissolve quite slowly in ethanol; it is advisable to use a
reflux condenser (such as a cold finger) to keep the solvent
from boiling away. About 25 mL of absolute ethanol should be
sufficient to dissolve the product.

Read OP–36 if you are not familiar with the principles and terminology of mass spectrometry.

The mass spectrum of indole (Figure 1) is characterized by
an intense molecular ion peak ($m/e = 117$), which is also the
base peak. The molecular ion is probably formed by loss of a
nonbonded electron from the nitrogen atom:

Most substituted indoles also have strong molecular ion peaks.
Neutral fragments lost from the molecular ions of most indoles

Figure 1.   Mass spectrum of indole. (Reproduced from *Mass Spectrometry of Heterocyclic Compounds,* by Q. N. Porter and J. Baldas.
Copyright © 1971 by John Wiley & Sons, Inc. Reprinted by permission of John Wiley & Sons, Inc.)

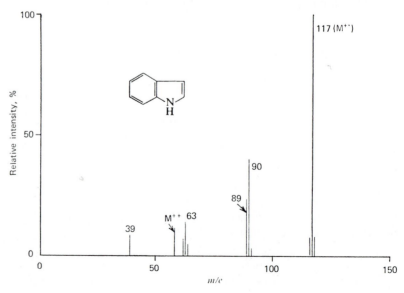

include HCN and CH_2N, forming ions with m/e values of $M - 27$ and $M - 28$, respectively. The loss of CH_2N from 1-, 2-, and 3-phenylindoles leaves a $C_{13}H_9^+$ ion, which could be a fluorenyl cation; if so, the molecular ion must undergo considerable "scrambling" (intramolecular rearrangement) before it ejects CH_2N. Phenylindoles may lose the phenyl-substituted analogs of HCN and CH_2N (C_6H_5CN and C_6H_5CHN) to form additional daughter ions. Because the molecular ion peak of 2-phenylindole is quite strong, its $M + 1$ and $M + 2$ peaks are also prominent, making it easy to compare their intensities with the expected values. The mass spectra of some indoles show "half-mass" peaks that result from the formation of ions having a 2^+ charge. For example, the doubly charged molecular ion of indole itself gives an M^{++} peak with an m/e value of 58.5.

The infrared spectra of heteroaromatic compounds (see the Collateral Project) are similar to those of analogous aromatic compounds, with some additional bands arising from the heteroatoms. Thus, pyrroles and indoles, like other aromatic compounds, show ring-stretching bands in the 1650–1300 cm^{-1} region and C—H out-of-plane bending bands in the 910–665 cm^{-1} region. Most five-membered pyrrole rings are distinguished by a characteristically strong, broad C—H bending band near 740 cm^{-1}. Heteroaromatic amines with N—H bonds absorb in the 3500–3220 cm^{-1} region; the N—H stretching band of indoles is particularly distinct and occurs near 3450 cm^{-1}.

Indigo is an indole derivative that has been synthesized by many different pathways. If you do Minilab 38, you will prepare indigo by a very simple synthesis involving the condensation of o-nitrobenzaldehyde with acetone in an alkaline solution. In the presence of a reducing agent, indigo changes to a colorless base-soluble compound called leucoindigo. When a piece of cloth is dipped into a solution of leucoindigo and exposed to air, indigo is regenerated, dyeing the cloth the color of blue jeans.

fluorenyl cation

See the Infrared Spectra section in Part III for a discussion of the effect of substitution on aromatic C—H bending bands.

Indigo is not colorfast, but the popularity of faded blue jeans has turned this deficiency into an advantage.

The interconversion of indigo and leucoindigo

PRELAB ASSIGNMENTS

1.   Read OP–24*a* and OP–36, and read or review the other operations as necessary.

2.   Calculate the mass and volume of 10 mmol of acetophenone and the theoretical yield of 2-phenylindole.

Reactions and Properties

A.

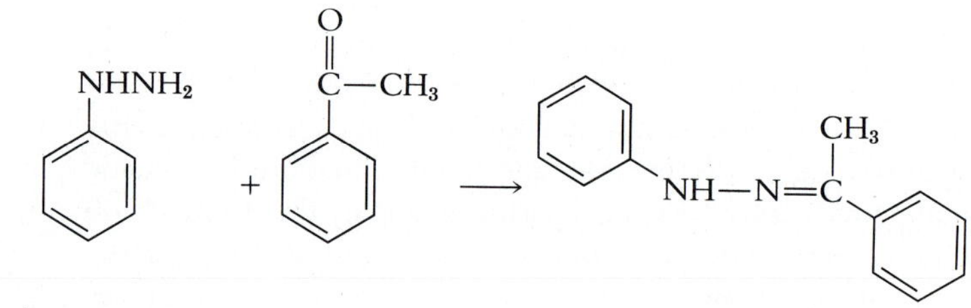

phenylhydrazine   acetophenone   acetophenone phenylhydrazone

B.

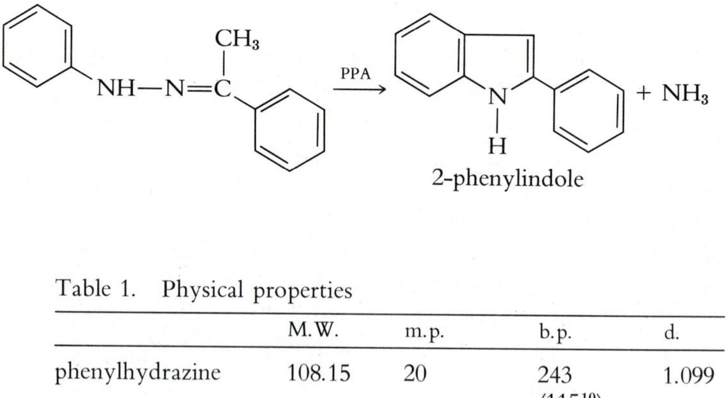

2-phenylindole

Table 1.   Physical properties

	M.W.	m.p.	b.p.	d.
phenylhydrazine	108.15	20	243 (115^{10})	1.099
acetophenone	120.2	20.5	202	1.028
2-phenylindole	193.25	188–89	250^{10}	
acetophenone phenylhydrazone	210.35	105–06		

Figure 2.   IR spectra of the starting materials. (Reproduced from *The Aldrich Library of FT-IR Spectra,* by Charles J. Pouchert, with the permission of the Aldrich Chemical Company.)

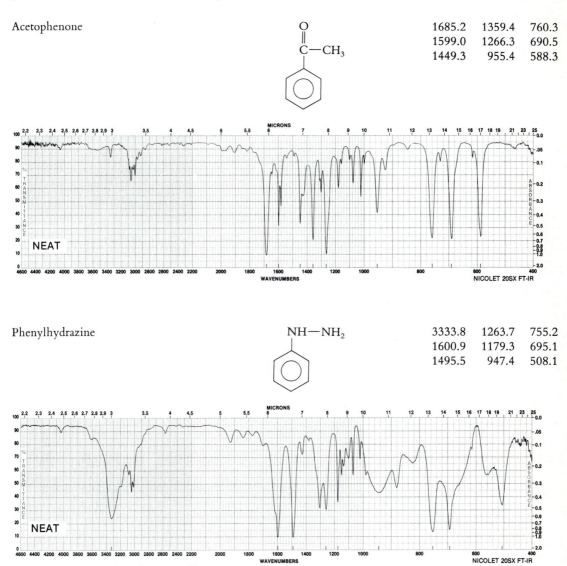

Acetophenone

1685.2	1359.4	760.3
1599.0	1266.3	690.5
1449.3	955.4	588.3

Phenylhydrazine

3333.8	1263.7	755.2
1600.9	1179.3	695.1
1495.5	947.4	508.1

PROCEDURE

Because acetophenone phenylhydrazone is heat- and light-sensitive, it should be used soon after it is prepared, or else stored in a cool, dark place.

A. Preparation of Acetophenone Phenylhydrazone

SAFETY PRECAUTION

phenylhydrazine   acetophenonone

Impure phenylhydrazine is dark reddish-brown and will give an inferior product.

If necessary, scratch the sides of the test tube to promote crystallization.

Phenylhydrazine is corrosive and very toxic if inhaled, ingested, or absorbed through the skin; contact may cause painful and persistent skin eruptions in sensitive individuals. It is also a suspected carcinogen. Use gloves, and dispense it under a hood; avoid contact, and do not breathe its vapors.

In a test tube, dissolve 10 mmol of acetophenone in 5 mL of 95% ethanol. Carefully stir in 1.0 mL (~10 mmol) of freshly distilled phenylhydrazine [**contact and vapor hazard! suspected carcinogen! wear gloves!**], followed by 2 drops of glacial acetic acid [**contact and vapor hazard!**]. Add a boiling chip, and heat the reactants gently over a steam bath for 15 minutes, taking care not to boil away the ethanol (add more if necessary). Cool the mixture in ice, and collect the acetophenone phenylhydrazone by vacuum filtration when crystallization is complete. Wash the product on the filter with 1 M hydrochloric acid followed by 1–2 mL of ice-cold 95% ethanol, and let it air-dry. Wearing protective gloves, blot the phenylhydrazone as dry as possible between large filter papers.

B. Preparation of 2-Phenylindole

SAFETY PRECAUTIONS

Polyphosphoric acid can be warmed gently on a steam bath to make it easier to pour.

Polyphosphoric acid can cause serious burns, particularly to the eyes, so avoid contact with skin and eyes.

Some indoles are suspected of causing cancer, so you should minimize your contact with the product.

Reaction. Measure about 10 mL of polyphosphoric acid [**contact hazard!**] into a dry 20-cm test tube. Clamp the tube so that the portion containing the liquid extends inside a steam bath (remove enough rings to let the test tube just slip through). Clamp a thermometer in the liquid, near one wall of the test tube (but not touching it) to allow room for stirring. Warm the polyphosphoric acid to 50°C; then mix in the ace- tophenone phenylhydrazone with a stirring rod, and heat the reactants, with stirring. When the temperature of the solution reaches 100°C, turn off the steam. Use a cold-water bath to keep the temperature below 125°C if necessary. When the temperature begins to drop, replace the steam bath and heat the mixture for 10–15 minutes with stirring.

Don't use the thermometer as a stirring rod!

Separation.   Cautiously pour the warm reaction mixture into 30 mL of ice water, using more water for the transfer, and stir until all the polyphosphoric acid has dissolved. Collect the crude 2-phenylindole by vacuum filtration ⟨OP–12⟩, wash- ing it on the filter with cold methanol and letting it air-dry.

Purification and Analysis.   Recrystallize the product from either ethanol : water or toluene : hexane following the procedure for recrystallization from mixed solvents ⟨OP–23*b*⟩. Use about 0.1 g of decolorizing carbon. Dry the 2-phenylindole at room temperature ⟨OP–21⟩, and weigh it. Purify the product by vacuum sublimation ⟨OP–24*a*⟩, and measure its mass and melting point ⟨OP–28⟩. Obtain a mass spectrum ⟨OP–36⟩ of the product from your instructor.

Your instructor may suggest that you sublime only a small amount for a melting-point determination.

Report.   Calculate the percentage yield of 2-phenylindole after recrystallization. On the mass spectrum, locate the molecular ion peak, the $M + 1$ and $M + 2$ peaks, and all other peaks whose intensities are 10% or more of the base-peak intensity. Derive the molecular formula of 2-phenylindole from its structure, use the formula to calculate the expected intensities of the $M + 1$ and $M + 2$ peaks, and compare the experimental and theoretical values. Characterize as many of the other peaks as you can, in each instance giving the formula of the species lost from the molecular ion and the formula of the resulting daughter ion.

Collateral Project

Record an infrared spectrum of 2-phenylindole in a KBr disc or Nujol mull. Interpret this spectrum as completely as you can, compare it with the spectra of the starting materials shown in Figure 2, and discuss the evidence suggesting that the expected reaction has taken place.

MINILAB 38   Dyeing with Indigo

You may be able to recover more indigo by adding a few drops of NaOH to the filtrate.

Dissolve 0.5 g of *o*-nitrobenzaldehyde in 2 mL of acetone. Add water drop by drop until the solution becomes cloudy; then add a drop or so of acetone to clear it up. Add 20 drops of 1 M sodium hydroxide slowly, with stirring; the solution should warm up and turn dark brown. Let the mixture stand for 15 minutes or more. Then cool it in ice, and collect the indigo by semimicro vacuum filtration. Wash the dye with a little ethanol, then with ethyl ether, and let it air-dry.

Transfer the indigo to a 15-cm test tube, add 0.15 g of sodium dithionite, and grind the solids together with the end of a stirring rod until they are well mixed. Add 10 mL of 1 M aqueous sodium hydroxide, and heat the mixture on a steam bath with stirring. When the indigo has dissolved, stopper and shake the tube until the indigo-blue color disappears (if the color persists, add a little more sodium dithionite). Immerse a small piece of cotton cloth in the solution with a stirring rod, stopper the tube immediately, and shake it for about 30 seconds. Remove the cloth, blot it dry between paper towels, and hang it up to dry. Note what happens when the test tube is left open for a time and then stoppered and shaken. Describe and explain your observations.

Exercises

1. **(a)** Rewrite the general mechanism given in the Methodology section so that it applies to the synthesis of 2-phenylindole, and sketch the activated complexes for steps 3 and 5. **(b)** Write a mechanism for the reaction that forms acetophenone phenylhydrazone, showing the role of acetic acid.

2. Draw structures for the indoles that would be obtained from the phenylhydrazones of the following carbonyl compounds: **(a)** acetone, **(b)** phenylacetaldehyde (2-phenyl-ethanal), **(c)** cyclopentanone, **(d)** pyruvic acid (CH_3CO-COOH), and **(e)** camphor.

3. Outline syntheses of 1-phenylindole and 3-phenyl-indole from appropriate starting materials.

4. Construct a flow diagram for the synthesis of 2-phenylindole.

5. A student who was running behind schedule discovered that all the phenylhydrazine in the lab had been used up, so she used 10 mmol of phenylhydrazine hydrochloride instead. When she cooled the reaction mixture from part A, no product crystallized from the solution. Explain.

6. When the phenylhydrazone of isobutyrophenone (2-methyl-1-phenyl-1-propanone) is heated to 150°C with polyphosphoric acid, it yields a mixture of the two compounds shown. Propose a mechanism that explains the formation of each product.

7. R. B. Woodward's total synthesis of strychnine began with the preparation of 2-(3,4-dimethoxyphenyl) indole. Show how this compound can be prepared from catechol and phenylhydrazine.

2-(3,4-dimethoxyphenyl)
indole

Library Topic

Write a report on natural and synthetic plant growth hormones (auxins), describing their sources and chemical structures and giving examples of their applications.

The Multistep Synthesis of Benzilic Acid from Benzaldehyde

Reactions of Carbonyl Compounds. Preparation of Carboxylic Acids. Nucleophilic Addition. Oxidation. Molecular Rearrangements.

Operations

OP–7a   Refluxing

OP–11   Gravity Filtration

OP–12   Vacuum Filtration

OP–21   Drying Solids

OP–23   Recrystallization

OP–28   Melting-Point Determination

OP–33   Infrared Spectrometry

Objectives

To prepare benzilic acid by a multistep synthesis from benzaldehyde.

To learn about the benzoin condensation and benzilic acid rearrangement.

To obtain additional experience in writing reaction mechanisms.

To learn about the role of mandelonitrile and its glycosides in cyanogenetic plants.

SITUATION

The oil pressed from bitter almonds contains amygdalin, which can be hydrolyzed to yield benzaldehyde and hydrogen cyanide. Benzaldehyde contributes the familiar almond flavor to marzipan and maraschino cherries. When almond oil is purified by washing with a basic solution, a small amount of white solid is often produced. This observation led to the discovery (more than a century and a half ago) of the benzoin condensation, which is a cyanide-catalyzed combination of benzaldehyde molecules. For many years, it was thought that only cyanide ion could catalyze the benzoin condensation, but recently Vitamin B_1 was found to be an effective catalyst as well. Benzoin is an important starting material for the synthesis of numerous organic compounds. For example, it can be reduced to benzil, which undergoes a molecular rearrangement in the presence of strong base to form a salt of benzilic acid. In this experiment, you will carry out the benzoin condensation using thiamine hydrochloride (a form of Vitamin B_1) as the catalyst and then convert the benzoin to benzilic acid.

BACKGROUND

Justus von Liebig and the Bitter Almond Tree

Chemical warfare is generally thought of as being a recent and uniquely human invention, but, as with many of our "discoveries," nature beat us to it. An otherwise unexceptional insect, the millipede *Apheloria corrugata,* discourages predators with a dose of poison gas powerful enough to kill a mouse. Many trees in the rose family (such as the cherry, apple, peach, plum, and apricot) insure their continued survival by protecting their seeds and foliage with cyanide-containing substances. There are cases on record of human fatalities from eating the seeds of these species. The culprit in both of these examples is the cyanohydrin mandelonitrile, which can be decomposed by enzymes or stomach acids into benzaldehyde and lethal hydrogen cyanide.

One man who considered apple seeds a delicacy ate a cup of them at one sitting and died from cyanide poisoning.

mandelonitrile   benzaldehyde   hydrogen cyanide

amygdalin

In most cyanogenetic (cyanide-forming) plants, mandelonitrile is present as a carbohydrate derivative called a glycoside. The most common of these glycosides is amygdalin, which is an acetal of mandelonitrile and the carbohydrate gentiobiose. Although amygdalin itself is not toxic, it can be bro-

Laetrile

benzoin

benzil

benzilic acid

ken down enzymatically under certain conditions to mandelonitrile and its decomposition product, hydrogen cyanide. Closely related to amygdalin is the controversial cancer drug Laetrile, which allegedly kills malignant cells by releasing hydrogen cyanide (or possibly mandelonitrile) at the cancer site.

One of the most prolific sources of amygdalin is the bitter almond, which (unlike the sweet varieties used for human consumption) is grown for the oil that can be pressed from its seed kernels. The characteristic odor of this oil comes from benzaldehyde formed in the breakdown of amygdalin. In the early 1800s, workers often purified almond oil by washing it with aqueous base to extract acids, a process that resulted in the formation of a white solid later identified as benzoin. Friedrich Wöhler and Justus von Liebig studied this reaction in 1832 and found that benzoin formation is due to the catalytic action of sodium cyanide (formed when the HCN from amygdalin reacted with the base) on benzaldehyde. Surprisingly, cyanide ion is almost the only substance that catalyzes this reaction; so the early discovery of the benzoin condensation resulted from the coincidental presence of both hydrogen cyanide and benzaldehyde when the bitter almond oil was washed with base.

Benzoin itself can be oxidized by a variety of oxidizing agents to the diketone benzil. It was not long after von Liebig's work on the benzoin condensation that he discovered yet another unusual reaction, the benzilic acid rearrangement. When benzil is treated with hydroxide ion, it is converted to the benzilate ion, which, on acidification, forms benzilic acid. This reaction, the oldest known molecular rearrangement, is the prototype of a general class of rearrangements to electrophilic carbon atoms.

METHODOLOGY

The benzoin condensation has traditionally been carried out with sodium cyanide as a catalyst, but thiamine hydrochloride (Vitamin B_1) has proven to be an effective catalyst as well.

thiamine hydrochloride

It has the considerable advantage of being relatively hazard-free, unlike the very poisonous cyanide. In the presence of sodium hydroxide, thiamine hydrochloride loses two protons to form a nucleophilic species that can attack the carbonyl carbon of benzaldehyde. The thiamine residue is sufficiently electron-withdrawing to increase the acidity of the adjacent hydrogen atom, allowing its removal by the base to yield a carbanion, which attacks another molecule of benzaldehyde. Loss of the thiamine residue from the product then yields benzoin.

Major steps in the mechanism of the thiamine-catalyzed benzoin condensation

$$PhCH \atop \displaystyle\mathop{}^{O} + Th^- \xrightarrow{H_2O} PhCHTh \atop \displaystyle\mathop{}^{OH} \xrightarrow{OH^-} PhCTh \atop \displaystyle\mathop{}^{OH} \xrightarrow{PhCHO} \underset{Th\ H}{PhC-CPh} \atop \displaystyle\mathop{}^{OH\ O^{\ominus}} \longrightarrow \underset{H}{PhC-CPh} \atop \displaystyle\mathop{}^{O\ OH} + Th^-$$

$$Th^- = $$

Benzoin can be oxidized to benzil by mild oxidizing agents, including ammonium nitrate in the presence of a catalytic amount of copper(II) acetate. The benzoin is probably oxidized by the direct action of cupric ion, which is reduced to cuprous ion in the process; but because Cu^{2+} is regenerated by the reaction of Cu^+ with ammonium nitrate, the copper(II) acetate functions as a catalyst. When treated with alcoholic potassium hydroxide, benzil undergoes the molecular rearrangement that was discovered by von Liebig. This rearrangement involves nucleophilic attack by hydroxide ion on one carbonyl carbon, followed by migration of the adjacent phenyl group to the other carbonyl carbon. The resulting salt of benzilic acid can be converted to the free acid by treatment with hydrochloric acid.

Because this is a multistep synthesis, you should try to keep material losses at a minimum in each step—otherwise, your overall yield will be quite low. It is important that the benzaldehyde be pure; impure benzaldehyde generally contains benzoic acid, which will inhibit the condensation reaction. If there is any doubt about the purity of the benzaldehyde, it should be distilled before it is used. The benzoin condensa-

The 1-hour reaction time specified in the procedure is a bare minimum. If time permits, you should be able to improve your yield by increasing the reaction time and allowing plenty of time for crystallization.

tion is carried out by simply refluxing the reactants for at least an hour; benzoin precipitates from solution when the reaction mixture is cooled. Two additional reflux periods are needed for the preparation of benzil and benzilic acid. Reasonably pure benzil is required for good results in the benzilic acid rearrangement, so the crude benzil should be recrystallized from ethanol. The reaction of benzil with potassium hydroxide should yield a suspension of potassium benzilate, which dissolves when the reaction mixture is heated. Benzilic acid is then precipitated from the filtered solution and purified by recrystallization from water.

You can gain additional experience in spectral interpretation by analyzing and comparing the infrared spectra of benzaldehyde and the products. These spectra should show bands from several different kinds of O—H, C—O, and C=O bonds, and in most of them you can observe aromatic C—H vibrations with no interference from aliphatic C—H bands.

Benzil is a diketone that undergoes the characteristic reactions of carbonyl compounds. It can be reduced to the corresponding diol by sodium borohydride (see Collateral Project 2), and it reacts with derivatives of ammonia to form compounds with C=N double bonds. As described in Minilab 39, the reaction of benzil with 1,2-diaminobenzene (o-phenylenediamine) results in the formation of a heterocyclic compound, 2,3-diphenylquinoxaline.

PRELAB ASSIGNMENT

Calculate the mass and volume of 150 mmol of benzaldehyde and the theoretical yields of benzoin, benzil, and benzilic acid expected from that much benzaldehyde.

Reactions and Properties

A. Benzoin condensation

benzaldehyde   benzoin

B.

benzil

C. Benzilic acid rearrangement

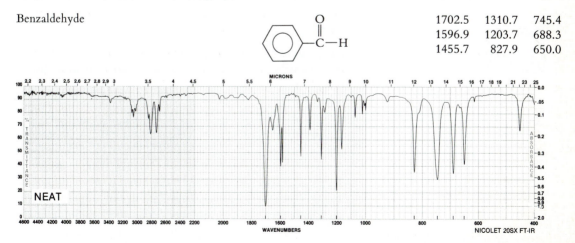

potassium benzilate

benzilic acid

Table 1.   Physical properties

	M.W.	m.p.	b.p.	d.
benzaldehyde	106.1	−26	178	1.042
thiamine hydrochloride	337.3	248 d		
benzoin	212.2	137		
benzil	210.2	95–6		
benzilic acid	228.2	151		
ammonium nitrate	80.0	170		
potassium hydroxide	56.1			

Potassium hydroxide pellets are about 85% KOH.

Figure 1.   IR spectrum of benzaldehyde. (Reproduced from *The Aldrich Library of FT-IR Spectra,* by Charles J. Pouchert, with the permission of the Aldrich Chemical Company.)

Benzaldehyde

1702.5	1310.7	745.4
1596.9	1203.7	688.3
1455.7	827.9	650.0

PROCEDURE

A. Preparation of Benzoin from Benzaldehyde

benzaldehyde

methanol

*Alternatively, you can let the reaction
mixture stand at room temperature for 24
hours or more.*

**Methanol is flammable and toxic, and inhalation, inges-
tion, or skin absorption may be harmful. Avoid contact,
do not breathe its vapors, and keep it away from flames
and hot surfaces.**

Dissolve 2.6 g of thiamine hydrochloride in 8 mL of water in a
125-mL Erlenmeyer flask. Add 30 mL of 95% ethanol, and
cool the solution in an ice bath. Add 5 mL of 3 M aqueous
sodium hydroxide solution drop by drop, with stirring, slowly
enough so that the temperature does not exceed 20°C. Add
150 mmol of *pure* benzaldehyde to this solution, and heat the
mixture at 60°C for 1 hour or more. Cool the reaction mixture
in an ice bath, and scratch the inside of the flask to induce
crystallization if necessary. When crystallization is complete,
collect the benzoin by vacuum filtration ⟨OP–12⟩, and wash
it on the filter with cold water, then with two 10-mL portions
of ice-cold methanol. Dry ⟨OP–21⟩ and weigh the product
before proceeding to the next part of this procedure. Save
enough benzoin to do a melting-point determination and
record an infrared spectrum.

B. Preparation of Benzil from Benzoin

ammonium
nitrate

acetic acid

**Ammonium nitrate may cause a fire or an explosion if it
is heated or allowed to contact combustible materials;
keep it cool and away from other chemicals and combustibles.**

**Acetic acid is corrosive, and its vapors are highly irritat-
ing. Avoid contact with the acetic acid solution, and do
not breathe its vapors.**

Multiply the mass of the dry benzoin by 0.5 and combine
that mass of ammonium nitrate with 35 mL of 80% (volume/
volume) aqueous acetic acid in a boiling flask. Mix in the
benzoin and 0.15 g of copper(II) acetate. Attach a reflux condenser ⟨OP–7a⟩, and heat the flask gently with occasional
shaking to start the reaction, which should be accompanied by

a vigorous evolution of nitrogen. When the gas evolution subsides, heat the solution slowly to the boiling point, and reflux it gently for an hour or more.

Cool the reaction mixture to 50°C and pour it, with stirring, onto 75–100 mL of crushed ice in a beaker. Collect the benzil by vacuum filtration, wash it twice with water, and recrystallize it ⟨OP–23⟩ from 95% ethanol. Dry the benzil at 75°C or below, and weigh it. Save enough benzil to do a melting-point determination and record an infrared spectrum.

Be sure to dispose of the filtrate properly; ammonium nitrate solutions can explode when concentrated.

C. Preparation of Benzilic Acid from Benzil

The solution of potassium hydroxide in ethanol is flammable, toxic, and corrosive; it can cause severe damage to the eyes, skin, and respiratory tract. Wear gloves, and avoid contact; do not breathe the vapors, and keep it away from flames.

Hydrochloric acid is very corrosive; contact or inhalation can cause severe damage to the skin, eyes, and respiratory tract. Wear gloves, and dispense under a hood; avoid contact, and do not breathe its vapors.

SAFETY PRECAUTIONS

potassium   hydrochloric
hydroxide   acid

Reaction.   For every gram of dry benzil, measure out approximately 2.5 mL of aqueous 6 M potassium hydroxide and 3 mL of 95% ethanol. Combine the benzil with the liquids, and reflux the mixture gently for 15 minutes on a steam bath.

Separation.   Pour the hot reaction mixture, with stirring, into 100 mL of water and let it stand a few minutes; then warm the mixture to 50°C, with stirring to dissolve the potassium benzilate. There may be a colloidal suspension of unreacted benzil or by-products at this point, but most of the solid should dissolve; if it does not, add more warm water. Add about a gram of decolorizing carbon and 0.5 g of filtering aid (Celite), stir or swirl the mixture at 50°C for 2 minutes, and filter the hot solution through fluted filter paper ⟨OP–11⟩. Mix 15 mL of concentrated hydrochloric acid **[contact and vapor hazard! wear gloves, and work in a hood!]** with 100 mL of crushed ice in a beaker. Add 10–15 mL of the potassium benzilate solution with stirring, and scratch the sides of the beaker until crystals begin to form. Slowly add the rest of the solution, with continuous stirring. When the addition is complete, test the solution with pH paper; if the pH is higher than 3, add more HCl. Cool the solution in ice, and collect the benzilic acid by vacuum filtration, washing it on the filter with cold water.

Purification and Analysis.   Recrystallize the benzilic acid ⟨OP– 23⟩ from boiling water; then dry and weigh it. Measure the melting points ⟨OP–28⟩ of the benzoin, benzil, and benzilic acid. Record the infrared spectra ⟨OP–33⟩ of the three products, or obtain their spectra from your instructor.

Report.   Calculate the percentage yield for each step of the synthesis and the overall yield of benzilic acid; try to account for significant losses. Interpret the infrared spectra of the products as completely as you can, and discuss the evidence suggesting that each reaction step has taken place as expected.

Collateral Projects

1.   Test both your crude and purified benzil for the presence of unreacted benzoin by dissolving a few crystals of each in 1 mL of ethanol and adding a drop of dilute sodium hydroxide solution. Test a little benzoin in the same way and compare the results. Shake each solution with air, let them stand, and record your observations.

2.   Benzil can be reduced to hydrobenzoin (PhCHOH-CHOHPh, a mixture of stereoisomers) by the general procedure described in Experiment 32, using 95% ethanol as the solvent. The reaction time is about 10 minutes.

Synthesis of 2,3-Diphenylquinoxaline from Benzil

MINILAB 39

SAFETY PRECAUTION

Avoid contact with *o*-phenylenediamine; it can cause dermatitis and serious eye damage.

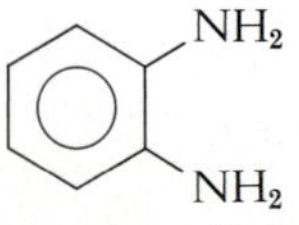

o-phenylenediamine

Dissolve 1 g of benzil in 4 mL of warm 95% ethanol, and add a solution of 0.5 g of *o*-phenylenediamine in 4 mL of 95% ethanol. Warm the solution in a 50°C water bath for 20–30 minutes. Add water to just saturate the warm solution (watch

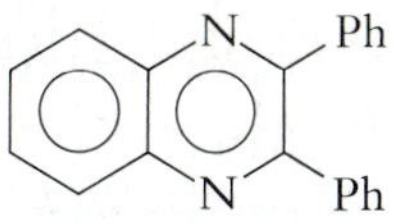

2,3-diphenylquinoxaline

for a slight cloudiness), and then cool it in ice until the crystallization is complete. Recover the 2,3-diphenylquinoxaline by vacuum filtration. The product can be recrystallized ⟨OP–23b⟩ from a mixed solvent of ethanol and water and a melting point obtained, if desired. Write a balanced equation, and propose a mechanism for the reaction.

The pure compound should melt at 125–126°C.

Exercises

1. Propose a mechanism for the rearrangement of benzil to potassium benzilate in the presence of potassium hydroxide.

2. Write a detailed mechanism for the cyanide-catalyzed condensation of benzaldehyde, and explain the role played by the cyanide.

3. Write a synthetic pathway illustrating the preparation of compound **1** from a suitable five-carbon starting material, using the reactions described in this experiment.

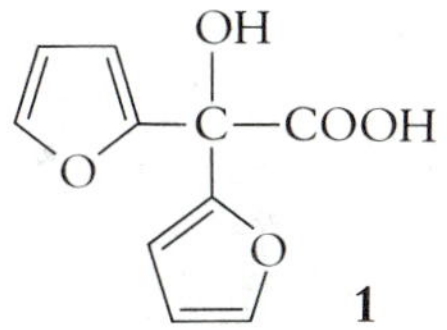

4. Construct a flow diagram for the synthesis of benzilic acid from benzaldehyde.

5. Benzil reacts with urea in the presence of sodium hydroxide to form the sodium salt of 5,5-diphenylhydantoin (Dilantin Sodium), a powerful anticonvulsant used to treat epilepsy. By analogy with the benzilic acid rearrangement, propose a mechanism for this reaction.

6. In the presence of a little sulfuric acid, benzilic acid reacts with acetone to form a compound with the molecular formula $C_{17}H_{16}O_3$. Propose a structure for this product, and write a balanced equation for the reaction.

1

5,5-diphenylhydantoin

Library Topic

Write a short report on the Laetrile controversy, describing the production of Laetrile, how it is used in treating cancer, and some of the arguments for and against its use.

Using the Chemical Literature in an Organic Synthesis

Multistep Synthesis. Searching the Chemical Literature.

Objectives

To search the literature for published information about a target compound.

To design a synthesis of the target compound from available starting materials.

To search the literature for information about the methods and materials to be used in the synthesis.

To carry out the synthesis in the laboratory and purify the product.

To verify the identity of the product and assess its purity.

SITUATION

In the previous synthetic experiments, you could rely on a procedure printed in this book. For this experiment, you are on your own. You will have to operate like a research chemist — outlining a reasonable synthetic path to a target compound and carrying it out in the laboratory, using what you already know about organic synthesis and what you can learn from the chemical literature.

BACKGROUND

Searching the Chemical Literature

The literature of chemistry can be subdivided into three general classes: primary, secondary, and tertiary sources. Most primary sources in chemistry contain descriptions of original research carried out by qualified chemists. Primary sources include scientific periodicals such as the *Journal of Organic Chemistry,* patents, dissertations, technical reports, and government bulletins. Secondary sources contain material from the primary literature that has been systematically organized, condensed, or restated so as to make it more accessible and understandable to users. Secondary sources include most monographs, textbooks, dictionaries, encyclopedias, reference works, review publications, and abstracting journals concerned with chemistry. Tertiary sources are intended to aid users of the primary and secondary sources or to provide facts about chemists and their work. Tertiary sources include guides to the chemical literature, directories of scientists and scientific organizations, bibliographies, trade catalogs, and publications devoted to the financial and professional aspects of chemistry. Some sources may combine several different functions; *Chemical & Engineering News* prints articles about chemical research as well as financial and professional information and thus serves as both a secondary and a tertiary source.

Primary sources are used when it is important to obtain the most accurate and detailed information available on a topic; errors can always occur when the material reappears in a secondary source, and important information may be left out. Because the primary sources are not organized in any systematic way, it is generally necessary to refer to other sources to determine where the desired information can be found. The Bibliography following Appendix VII lists a number of secondary and tertiary sources that can be used either to obtain information directly or to gain access to the primary literature —and sometimes both. For example, *Beilstein's Handbook of Organic Chemistry* (described below) gives detailed, reliable information about organic compounds and also provides citations to the literature in which the information was first reported.

A few secondary sources (such as *Beilstein*) critically evaluate primary material and correct errors before printing it; such secondary sources may be very reliable.

Most of the reference works cited in the Bibliography are limited in scope; they make no attempt to cover the entire field of chemistry or to list all of the known organic compounds. The two major works that do attempt that kind of coverage are *Chemical Abstracts* (abbreviated as *CA*) and *Beilstein's Handbook of Organic Chemistry* (generally referred to as simply *Beilstein*). *Chemical Abstracts* prints brief summaries (abstracts) of scientific papers, patents, and other printed material related to chemistry shortly after their publication. *Beilstein* summarizes, in German, all the important information published about specific compounds, but it is more than 25 years behind the current literature in most areas. To locate all the important original work dealing with a particular compound, it is often necessary to consult *Chemical Abstracts* and *Beilstein* as well.

Before using *Chemical Abstracts* for the first time, be sure to read the introduction that appears in issue 1 of each volume (a new volume is published every six months), which describes the layout of the abstracts. Each abstract of a scientific article, or paper, contains an abstract number, the title and author of the paper, a citation that tells where the original paper can be located, and a concise summary of the important information in the paper. The contents of *Chemical Abstracts* can now be searched by computer (see Bib–M1), but the best way to locate abstracts of interest in the printed volumes is to consult the indexes. Although each weekly issue has its own indexes, the semiannual and collective indexes are far more useful for literature searches. A Collective Index is now published every five years; prior to 1957, such indexes came out every ten years. In 1972, the Subject Index was divided into two parts: the Chemical Substances Index and the General Subject Index. Author, formula, and patent indexes are also published, along with ancillary materials such as the Ring Systems Handbook, Registry Handbook, Index Guide, and Service Source Index, which are updated periodically.

The *Chemical Abstracts* nomenclature rules for the 8th and 9th Collective Indexes are described in detail in the Subject Index of volume 66 and the Index Guide for volume 76, respectively.

The first thing you *must* do before searching *Chemical Abstracts* for information about a particular compound or subject is to find the *CA index name* of the compound or the *CA index heading* for the subject. In some cases, it may be possible to derive (or guess) the index name, but with complex compounds this may be difficult without a thorough knowledge of the nomenclature rules for the appropriate index of *Chemical Abstracts*. The Index Guide, which now appears with each Collective Index and at intervals in between, gives cross-references from alternative names of substances to the *CA index name*. Thus, the entry under "aniline" lists the index name benzeneamine, followed by the *CA registry number* [62-53-3]. The Index Guide does not list every compound indexed or give every synonym for the compounds it does list, so you may have to try different approaches to find what you are looking for. If you can't locate a specific compound, try looking up a possible parent (unsubstituted) compound under its trivial name; the index name of this parent compound should begin with a root name under which you will find your compound listed in the Chemical Substances Index (or the Subject Index prior to 1972). For example, suppose you are searching for the following compound:

$$CH_3O \text{---} \langle \bigcirc \rangle \text{---} CH = CH\overset{\displaystyle O}{\overset{\|}{C}} \text{---} \langle \bigcirc \rangle$$

The unsubstituted compound (PhCH=CHCOPh) is known by such names as chalcone and benzalacetophenone. Looking up "chalcone" in a recent Index Guide provides the *CA* index name "2-propen-1-one, 1,3-diphenyl," so you will find the substituted compound listed in the Chemical Substances Index under the same root name, as "2-propen-1-one, 3-(4-methoxy)phenyl-1-phenyl." Keep in mind that the index name of a compound may change from time to time. This compound was listed under "chalcone, 4-methoxy" during the 8th Collective Index period (1967–1971) and before. Between the 8th and 9th Collective Index periods, some major changes were made in the *Chemical Abstracts* nomenclature rules; rigorously systematic names are now used for most chemical substances.

If you have trouble finding the *CA* index name for a compound using the Index Guide, you might look for it in another source such as the *Merck Index* or the *Dictionary of Organic Compounds*. The Registry Handbook gives index names for over 6 million compounds indexed by *Chemical Abstracts*, so if you can locate the registry number for a compound in another source ⟨for example, Bib–A3, Bib–A4, Bib–A5, Bib–A6, Bib–A8⟩ you can easily find its index name in the Registry Handbook. If you have a fairly good idea of what the index name for a compound might be, you may also be able to locate it in the formula indexes.

Once you locate the index name used during a particular index period, you will find abstracts listed under that name in the Chemical Substances Index or Subject Index for that period. Abstract citations in indexes from 1967 on are given in the form **80:**12175e, where the first number is the *Chemical Abstracts* volume number and the second is the abstract number. In an index from prior to 1967, an abstract citation such as **51:**4321^e refers to the volume and column number (there are two columns on each page) in which the abstract appears; the superscript (either a letter from a through i or a number from 1 through 9) indicates the location of the abstract in that column. The information you are looking for may appear in the abstract itself, or you may have to read the original article cited in the abstract. Citations of such articles now appear in the form *J. Am. Chem. Soc.* **1983,** 105(25), 7473–4 (Eng.), where the abbreviated name of the publication appears first, followed by the date, volume and issue number, page numbers, and language in which the paper is written. (Earlier citations are given with the volume number first, followed by the pages and

The letter e in this citation is a check-letter. If you looked up abstract number 12715 by mistake, you would find that its check-letter is b.

year.) The full name of the publication will be found in the *Chemical Abstracts Service Source Index* (CASSI), which also provides a brief publication history of each source and a list of libraries that carry it.

To carry out a thorough search of *Chemical Abstracts,* it is best to start with the most recent Collective Index and all semiannual indexes published since then and work your way back through the previous collective indexes, using the index guides or other sources to locate the appropriate index names. Trivial names are used more frequently in the earlier indexes, and these can often be recognized in the formula indexes or located in other sources ⟨for example, Bib–A3 or Bib–A8⟩. To bring your search up to date, you may also want to search indexes of current issues published since the last semiannual index. For further information about the use of *Chemical Abstracts,* refer to one or more of the literature guides in category M of the Bibliography. For instructions on conducting a search of *Chemical Abstracts* by computer, see Bib–M1 or other appropriate sources.

Because *Chemical Abstracts* began publication in 1907, it does not cover some of the early literature that *Beilstein* does; even after 1907, *Beilstein* and *Chemisches Zentralblatt* (a now-defunct German abstracting journal) may offer better coverage or more detailed information on some topics. If you are looking for information about a specific compound, you can conduct a very adequate literature search (for most purposes) by searching *Chemical Abstracts* from about 1950 to the present date and *Beilstein* for earlier information.

Beilstein's Handbook of Organic Chemistry (Handbuch der Organischen Chemie) ⟨Bib–A9⟩ is by far the most comprehensive source of *organized* information about organic compounds. *Beilstein* provides information on the structure, characterization, natural occurrence, preparation, purification, energy parameters, physical properties, and chemical properties of organic compounds. It also cites the primary sources from which the information was obtained. *Beilstein,* unlike *Chemical Abstracts,* evaluates its sources critically and doesn't hesitate to correct errors from previous series. The fifth supplemental series is now being published in an English-language edition; when it is complete (only a few volumes are now in print), *Beilstein* will cover the chemical literature through 1979. The basic series *(Hauptwerke)* and four previous supplemental series *(Erganzungswerke* I–IV, which is abbreviated E I–IV) are published only in German and cover the literature through 1959.

To obtain all the information about a particular compound in *Beilstein,* you must search the basic series and all the
available supplemental series. The enormous size of this
"handbook"—along with the language barrier—could make
that a fearsome prospect, but there is really no reason to be
intimidated by *Beilstein.* The amount of German you need to
know is quite limited and can be learned quickly with the help
of a slim dictionary written specifically for *Beilstein* users
⟨Bib–A21⟩. And *Beilstein* is so well organized that it is not
difficult to find the information you seek. If a compound has
been around for some time, you can locate its *Beilstein* entries
by the following procedure:

1.   Write the molecular formula of the compound with
C and H first, followed by other elements in alphabetical
order.

2.   Locate the formula in volume *(Band)* 29 of the second supplemental series *(General-Formelregister, Zweites Erganzungswerke)* and look for the name of the compound among
those listed under that formula. Although the names are in
German, many are similar or identical to the English names. (If
you need help, use a German-English dictionary.)

3.   Write down the volume number and pages on which
information about the compound appears in the basic series
(H) and the first and second supplemental series (E I and E II),
and look up the appropriate entries in those volumes. (Each
"volume" may include several individually bound subvol-
umes.)

4.   Once you know the index name of the compound
and its page number in the basic series, you can locate its entry
in the corresponding volume of any later series. Alternatively,
you can look it up in the cumulative subject index for that
volume.

For example, the notation for indigo ($C_{16}H_{10}N_2O_2$) in volume
29 of E II reads "Indigo **24,** 417, I 370, II 233." So you will
find entries for indigo on page 417 of volume 24 in the basic
series and pages 370 and 233 of volume 24 in the first and
second supplemental series, respectively. The page number in
the basic series, written as "**H,** 417," is called its *coordinating
reference;* to find indigo's entry in volume 24 of a later series,
you can locate the pages with **H,** 417 printed at the top and leaf
through them until you find the entry for indigo. Knowing
the E II index name of a compound may also help you locate it

indigo

*Each compound is also assigned a system
number, which can be used to locate its
entries in the same way.*

in the current cumulative subject index *(Sachsregister)* for the appropriate volume; these indexes list entries in all the series through E IV. The cumulative indexes for some volumes are combined; thus, the listing for indigo is found in the volume 23–25 subject index, and it reads "**24** 417 d, I 370 d, II 233 e, IV 469." The letters refer to the location of an entry on the page; thus, "II 233 e" means that information about indigo will be found under the fifth entry on page 233 of E II. There is no listing for E III because supplementary series III and IV were issued jointly for volumes 17–27. Entries in the joint series are designated by E IV rather than by E III/IV.

Locating entries for a compound such as adamantane, which does not appear in the E II formula indexes, may take a little more time. All *Beilstein* entries are organized according to a detailed system, and learning that system is the best way to get complete access to the information contained in this source. However, you can usually locate such entries by either (1) finding the volume number in which a structurally similar compound appears and searching the cumulative indexes of that volume or (2) locating a *Beilstein* reference from another source. The E II indexes indicate that cyclohexane appears in volume 5, which contains all cyclic compounds lacking functional groups, so you will find adamantane listed in the cumulative indexes for that volume. You can also find *Beilstein* references for many compounds in certain reference books ⟨Bib–A1, Bib–A2, Bib–A4, Bib–A5, Bib–A6, for example⟩. The entry for adamantane in the *CRC Handbook* gives the notation "B5[4], 469," referring to a *Beilstein* entry on page 469 in volume 5 of E IV. If you look up that entry you will find a back-reference (E III 393) and a coordinating reference (**H**, 165) that will help you find information about adamantane in the other series.

Another important reference work that you should be familiar with is *Reagents for Organic Synthesis* (known as *Fieser*) ⟨Bib–B20⟩, which provides information about the preparation, purification, handling, and hazards of many chemical reagents, as well as examples of their use, with literature citations. *Fieser* adopts a critical approach to some of the information it reports (particularly in the first volume) and often gives helpful suggestions. Only the individual volumes are indexed, but if you locate the entry for a reagent in a recent volume, it will provide back-references to the previous volumes. The *Aldrich Catalog* ⟨Bib–A6⟩ also lists *Fieser* references for many chemicals sold by the Aldrich Chemical Company.

adamantane

The *Beilstein* system is described in Bib–M7 and in the "Notes for Users" (in English) at the beginning of each volume in the recent supplemental series.

Note that adamantane is assigned a back-reference even though it didn't appear in the basic series; if it had been known at the time, it would have appeared on page 165.

METHODOLOGY

Your instructor will either suggest a target compound for you to synthesize or allow you to choose your own (subject to approval). Your instructor will also inform you whether you will work individually or in teams of two or more students.

If you are preparing a compound that has been synthesized previously, you should carry out a thorough search of the primary literature using *Chemical Abstracts* (and *Beilstein* as well, if appropriate). If you locate a specific procedure for the synthesis of your target compound in the literature, you should not assume that it will be suitable for your purposes as written. It may call for unavailable or excessively hazardous chemicals or suggest techniques and apparatus that are not appropriate to your laboratory situation. Literature procedures tend to be rather sketchy in any case, so you will probably have to "fill in the blanks" from your own experience or by consulting other sources. It is often possible to adapt a literature procedure (preferably a tested procedure from a source such as *Organic Syntheses*) that has been used to prepare a similar compound. If the target compound is fairly complex, you will probably have to synthesize one or more of the starting materials for its preparation from simpler compounds, so it may be necessary to work out a detailed synthetic pathway outlining all the steps in your synthesis.

Before planning your synthesis in detail, you should read the appropriate sections of Appendix VII that describe specific sources in the Bibliography. Category B contains a number of books on organic synthesis, and several other categories list sources that you should find useful as well. To work out a reaction path for the synthesis, you can refer to your lecture text or a source such as B26 or B29. Works such as B8–B19, J7, and J8 should help you locate literature procedures for the specific reactions you will be using. It may also be worthwhile to consult such works as A3, A8, A9, and K4 for selected references to literature procedures, or B1, B3, and B4–B7 for tested procedures that are more detailed and reliable than most procedures from the primary literature. *Organic Reactions* ⟨Bib–B2⟩ describes the applications and limitations of many synthetic reactions and might help you tailor a synthetic procedure to fit the particular starting materials you will be working with. It is always a good idea to consult *Fieser* ⟨Bib–B20⟩ for information about the reagents you plan to use and ideas about how best to use them.

To find out about potential health and safety hazards associated with the synthesis, you should refer to sources from category C in the Bibliography. To learn more about the laboratory techniques you will be using, see category D and the appropriate sources from category E. Refer to sources listed in category E for information about chromatographic analysis of reaction mixtures and to those in category F for information about the characterization and identification of pure compounds by spectrometric methods. Additional information about "wet" chemical and spectrometric methods of identification is provided in sources from category G. For additional help in carrying out a literature search, consult appropriate sources listed in category M.

PRELAB ASSIGNMENTS

1. Read or review Appendix VII, and familiarize yourself with the Bibliography.

2. Find the *CA* index names of your target compound for the appropriate collective index periods and its *CA* registry number. Conduct a thorough search of the chemical literature for information about the target compound, and outline a synthesis of the target compound from readily available starting materials. (Check to see whether the materials are available in your laboratory.) For the chemicals you will be using, find information concerning physical properties, purification methods, safety and health hazards, safe handling procedures, and disposal procedures. If the synthesis requires laboratory methods you have not encountered before, find information about these methods in the literature, and sketch any special apparatus required for them. Locate any published spectra or other data that will help you characterize the target compound.

3. Write a complete description of your plan for the synthesis, including a detailed procedure for each preparation and a description of the methods you propose to use to analyze the purity of and verify the structure of the product. The procedure should be designed to yield about 2–5 g of the final product, unless your instructor indicates otherwise. Your writeup should also include a tabulation or description of all relevant properties of the chemicals you will be using, a clear statement of what you intend to accomplish, and a concise description of the experimental methodology. Submit your writeup to the instructor for approval.

PROCEDURE

If necessary, purify any starting materials, reagents, or solvents that appear to be insufficiently pure. Carry out the synthesis in the laboratory, purify the product, weigh the thoroughly dried product, and compute the percentage yield. Keep detailed notes of your work and observations as described in Appendix II. Measure the physical constants of the product, assess its purity by an appropriate chromatographic method, and confirm its identity by one or more spectrometric methods. With your instructor's permission, you can use a derivative melting point instead of (or in addition to) the spectrometric analysis.

Report.   Write up your report as if it were a scientific paper being submitted to a professional publication such as the *Journal of Organic Chemistry*. Such papers are traditionally written in an impersonal, third-person style (for instance, "the solution was stirred" rather than "I stirred the solution"). Your report should include the following:

The ACS Style Guide ⟨Bib–L18⟩ provides detailed information about writing a scientific paper. You should study papers from some professional journals for additional guidance.

1.   A brief but descriptive title

2.   Your name(s) and affiliation

3.   A brief abstract that summarizes the principal results of the work

4.   An introductory section (usually untitled) that provides a concise statement of the purpose and possible applications of the work, supported by descriptions of related work from the literature. If your synthesis differs significantly from those reported in the literature, tell how and why.

5.   A section titled "Experimental Methods" that gives enough detail about your materials and methods so that another experienced worker could repeat your work. Give current *CA* index names and registry numbers for important starting materials and the product and provide information about their purity where possible. Note any significant hazards and safety precautions in a separate paragraph labeled "Caution."

6.   A section titled "Results and Discussion" that summarizes the important experimental results and points out any special features, limitations, or implications of your work. You should include any data (including spectral parameters) that will help justify your conclusion. You may also wish to suggest

You may use separate sections titled "Results" and "Discussion" instead.

different approaches to the problem or areas that require further study.

7. A section titled "Conclusions" that states any conclusions you can draw from your work, based on the evidence presented

8. A section titled "References" that gives complete citations for the literature sources referred to in the report

Your instructor may suggest additions to or modifications of the above items and may request that your laboratory notes be submitted with the report.

Exercises

1. The recent synthesis of dodecahedrane has been described as the "Mount Everest of alicyclic chemistry." **(a)** Give the *CA* index name and registry number and the structure of dodecahedrane. **(b)** Locate the paper in which the synthesis of dodecahedrane was first reported. Give its title, the authors and their affiliation, and a standard literature citation showing where and when the paper appeared. **(c)** Summarize the salient points of the synthesis in your own words, specifying the starting material, the number of synthetic steps required, the yield of dodecahedrane, and the purification method. Tell how dodecahedrane was characterized, and report any spectral parameters. **(d)** Find another paper that predicts the heat of formation for dodecahedrane, cite it as described in (b), and give the predicted value. **(e)** Find a systematic name for dodecahedrane that is different from the *CA* index name. **(f)** Quote the reference to an ancient Greek philosopher that appears in one of the above papers, and explain it.

2. Herbert C. Brown was awarded the Nobel Prize for Chemistry in 1979 for his work with reagents for organic syntheses, principally organoboranes. **(a)** Give the structure of Brown's reagent 9-BBN (9-borabicyclo[3.3.1]nonane) and describe some of its applications and characteristics, giving equations where appropriate. Cite several recent papers reporting the use of 9-BBN. **(b)** Cite a paper published in 1985 in which Brown and his co-workers claim to have proven that the norbornyl cation does not have the nonclassical stabilization energy that had long been predicted. Summarize the evidence that led to this conclusion.

3. **(a)** Cite the paper in which the use of pyridinium chlorochromate to oxidize alcohols to carbonyl compounds was first reported. **(b)** Tell how pyridinium chlorochromate is prepared, describe a typical experimental procedure for oxidation of a primary alcohol to an aldehyde, and discuss the stoichiometry of the reaction. **(c)** Describe the oxidative cyclization of (−)-citronellol by pyridinium chlorochromate, giving structures for the product and any apparent intermediates. **(d)** Report on any hazards associated with the use of pyridinium chlorochromate, and describe safe disposal procedures for the reagent.

4. **(a)** Give a concise definition of the Knoevenagel condensation. **(b)** Describe typical experimental conditions for conducting a Knoevenagel condensation between an aldehyde and diethyl malonate. **(c)** Find and summarize a detailed procedure for the synthesis of ethyl coumarin-3-carboxylate using this reaction. **(d)** What is the Doebner modification of the Knoevenagel condensation?

5. **(a)** Describe any hazards associated with the use of thionyl chloride, and describe proper handling precautions and disposal procedures for this reagent. **(b)** Tell how thionyl chloride can be purified for use as a chemical reagent. **(c)** Cite a paper in which thionyl chloride was used to convert an amino acid to an ester in one step, and briefly describe the experimental conditions.

6. **(a)** Give the current *CA* index name and registry number for (+)-camphor. **(b)** Draw a structure for (+)-camphor that shows its absolute configuration. **(c)** What is the melting point of the oxime of (+)-camphor? **(d)** Find an infrared spectrum for camphor and give the wavenumbers of the major absorption bands. **(e)** Tell where you can find information about camphor in *Beilstein* (through E IV). **(f)** Tell how most synthetic camphor is currently produced, giving equations for the reactions.

7. **(a)** Give the *CA* index name of the compound whose *CA* registry number is [5543-57-7]. **(b)** Give the trivial name of this compound, and describe its major application. **(c)** Tell where information about this compound can be found in *Beilstein*. **(d)** Cite an article in which the resolution of the racemic compound was described, and give the absolute configuration of the (−) form.

8. For each of the following abbreviated names, provide the full name of the journal, any names under which it was previously published, the year in which it was first published (as volume 1, under a current or previous name), the language or languages in which it is published, and a nearby library that carries it. **(a)** *Helv. Chim. Acta,* **(b)** *Dokl. Akad. Nauk SSSR,* **(c)** *Chem. Ber.,* and **(d)** *Uzb. Khim. Zh.*

9. Find detailed synthetic procedures for the following compounds, briefly describe the experimental conditions, and write equations for the relevant reactions: **(a)** hexaphenylbenzene, **(b)** vanillic acid, **(c)** 1,2-cyclononadiene, and **(d)** octadecanedioic acid.

10. For each of the following compounds, find and reproduce as many different kinds of published spectra (or spectral parameters) as you can: **(a)** mandelic acid, **(b)** resorcinol, **(c)** exaltone, **(d)** bourbonal, and **(e)** testosterone.

SYSTEMATIC ORGANIC QUALITATIVE ANALYSIS

Part III describes how to classify and identify unknown organic compounds using chemical and spectrometric methods.

Background: The Chemist as Detective

Professional organic chemists are often confronted with unknown compounds whose identity must be established. If a chemist carries out a reaction that is expected to produce a certain compound, he or she must prove that the product is, in fact, the expected one. This may require no more than a melting or boiling point and some spectral data. However, if the outcome of the reaction is *not* known, the problem of characterizing the products is a more difficult one, and the chemist will ordinarily have to go through some kind of systematic procedure for identifying them. The same is true of a chemist who has isolated unknown components from natural or synthetic products.

The process of identifying an unknown can be compared to the approach used by a detective in identifying the perpetrator of a crime. The detective first looks for clues that help to characterize the criminal, narrowing down the range of suspects and perhaps suggesting the most productive areas of investigation. When one or more likely suspects has been tracked down, the detective has to find and analyze evidence to help eliminate some suspects from consideration and to establish the probable guilt of the prime suspect so that an indictment can be obtained. Finally, the detective has to amass enough additional evidence to convince a jury that the accused is, in fact, guilty of the crime.

In carrying out the identification of an organic compound you, like the detective, should be constantly on the lookout for clues to its identity. Spectral and chemical data should allow you to confine your search to a particular chemical family. Additional physical and chemical evidence will help you narrow down the list of "suspects" and focus your attention on a few of the most probable compounds. Finally, the preparation of one or more derivatives should lead you to a definite conclusion and provide you with sufficient evidence to convince the "jury" (your instructor) that your compound is, in fact, what you believe it to be.

As in the solution of any other problem, you must first ask yourself the right questions before you can arrive at the correct answer. Some important questions to be answered for an unknown compound are these: (1) Is it pure? (2) What functional group(s) does it contain? (3) Are there any other significant structural features that might aid in its identification? Each bit of evidence that you obtain should, if interpreted correctly, help reveal the answer to one or more of these questions; and all of them combined should provide you with an answer to the ultimate question, "What is it?"

A detective trying to solve a case will almost invariably come upon clues that lead nowhere or, even worse, to false conclusions. The same is true in chemical problem-solving, so it is important to keep an open mind throughout your investigation and to avoid jumping to conclusions before all the evidence is in. You may formulate tentative conclusions based on your initial observations (for example, if it turns chromic anhydride reagent green, it may be an alcohol), but you should be ready to revise or discard such conclusions if they are not supported by subsequent observations (for example, its IR spectrum shows a C=O stretching band but no O—H band, so it may be an aldehyde instead).

Chemical and physical evidence can be misleading for a variety of reasons:

1. Some compounds of a given family may undergo an atypical reaction with a given reagent and yield either a false positive or a false negative result.

2. Some reagents give positive tests with more than one functional group.

3. Impurities may complicate or invalidate a test.

4. Spectral bands may occur outside the expected frequency ranges or may be incorrectly assigned.

Because of these and other possible sources of error, it is wise not to rely on a single piece of evidence in formulating a conclusion. For example, the classification of an unknown as a secondary alcohol may be established by a slow reaction with Lucas's reagent, a positive chromic anhydride test, *and* an infrared band in the $1100 \ cm^{-1}$ region, but not by any one of these alone.

There is no single best way to identify an organic compound and no one path to follow. Although a useful approach is described in the Methodology section, it is not necessary (or even desirable, in some cases) to follow it inflexibly. You must use your own judgment and initiative in choosing which tests to perform, which physical properties to measure, which derivatives to prepare, and which spectra to record. By keeping your eyes and mind open at all times, you may find clues to the structure or identity of your compound that suggest a way to bypass some of the usual intermediate steps.

A qualitative analysis problem can provide an exciting and challenging experience for those who apply all their skill and ingenuity to its solution. Besides testing your mastery of the operations learned previously in the laboratory, it furnishes many practical applications of the concepts learned during the lecture course in organic chemistry. Throughout the analysis of an unknown compound you may apply your knowledge of functional group chemistry, nomenclature, acid-base equilibria, structure-property relationships, spectral analysis, organic synthesis, and many other areas of organic chemistry.

Methodology

This section describes the classical "wet chemistry" approach to qualitative analysis and the use of infrared and nuclear magnetic resonance spectrometry for the identification of unknown organic compounds. It is possible to identify most unknown compounds by chemical methods alone, but you may be allowed to use spectral methods as well, if the necessary instruments are available. It is assumed that the unknowns will be limited to the following ten classes: *alcohols, aldehydes, ketones, amides, amines, carboxylic acids, esters, halides, aromatic hydrocarbons,* and *phenols.* An unknown may contain an additional subsidiary functional group (such as the nitro group in *p*-nitrophenol or the ether group in *p*-anisic acid), but it will be classified as a member of one of the above families. The analysis scheme is divided into four parts, as outlined in the margin. The *preliminary work* includes a gross physical examination of the unknown, an ignition test (which can provide useful clues to its structure), and an evaluation of its purity. The *functional class determination* includes solubility tests and classification tests that should allow you to assign the unknown to one of the ten chemical families. *Spectral analysis* (optional) can greatly facilitate the identification of the unknown by providing information about its functional groups and other structural features. The *identification* phase requires a literature search, the preparation of derivatives, and the accumulation of additional physical and chemical evidence to establish the identity of the unknown.

Qualitative analysis scheme

1. Preliminary work
   a. Gross physical examination
   b. Ignition
   c. Estimation of purity
   d. Purification (optional)

2. Functional class determination
   a. Solubility tests
   b. Functional class tests

3. Spectral analysis
   a. Infrared spectra
   b. NMR spectra

4. Identification
   a. Examination of literature
   b. Additional tests and data
   c. Preparation of derivatives

If the unknowns provided are all in a single family or a small group of families, part 2 can be omitted or modified.

PRELIMINARY WORK

Observation of the physical state, color, and odor of a compound may provide some clues to its identity. For instance, the fact that an unknown is a solid eliminates all organic compounds that are liquids at room temperature, and an intrinsic color suggests that chromophoric groups having conjugated double bonds or rings are present.

Organic compounds that are nonflammable in the ignition test may contain a high ratio of halogen to hydrogen or have a very high molecular weight. A yellow, sooty flame often indicates an aromatic compound. Aliphatic hydrocar-

Light or dull colors (light yellow, tan, brown, black, etc.) may suggest that the unknown is impure, whereas intense yellows, oranges, reds, etc., are usually intrinsic colors.

For example, methanol burns with a bluer flame than 1-octanol because it has a higher oxygen/hydrogen ratio.

OP–23c tells how to choose a recrystallization solvent.

The purity of liquid unknowns can also be determined by GLC ⟨OP–32⟩.

bons burn with a clean yellow flame, whereas compounds with a high oxygen content tend to burn bluer.

The melting or boiling point of an unknown gives a rough estimate of its purity. Solids having a melting point range greater than about 3–4°C should ordinarily be purified by recrystallization. A liquid with a boiling-point range greater than a few degrees (or a nonreproducible micro boiling point) should be purified by distillation and the high-boiling or low-boiling fractions discarded. It is very important to determine the melting or boiling point accurately, since compounds are listed in order of melting and boiling points in the tables used to identify them.

FUNCTIONAL-CLASS DETERMINATION

The traditional procedure for classifying unknowns by functional group involves the use of preliminary screening tests that categorize the unknowns into broad groups according to solubility or other properties, followed by functional class tests that place them in chemical families. The use of infrared spectroscopy and other spectrometric methods can simplify this process considerably; a skilled analyst can often classify a compound by its infrared spectrum alone. However, the classical tests still have their uses, particularly for students, who usually do not have extensive experience in spectral analysis. It is recommended that the solubility tests and *at least* one functional class test be performed on each unknown (except for amines and carboxylic acids, for which the solubility tests may suffice) even when an infrared spectrum is obtained.

Solubility Tests

In these tests, compounds that dissolve to the extent of about 30–35 mg per mL of water will be considered soluble.

Cyclic and branched compounds are usually more soluble than straight-chain compounds of the same carbon number; a phenyl group has about the same effect on solubility as an *n*-butyl group.

Neutral water-soluble compounds are placed in Class S_n, whereas acidic and basic water-soluble compounds are placed in Classes S_a and S_b respectively (see Figure 1, page 524).

Water solubility usually indicates the presence of at least one oxygen or nitrogen atom and a relatively low molecular weight. The borderline for water solubility in the case of monofunctional oxygen or nitrogen compounds is usually around five carbon atoms. For example, 1-butanol (with four carbon atoms) is soluble, whereas 1-pentanol (with five carbons) is not. Of the ten classes of compounds considered here, water solubility can be expected for low-molecular-weight alcohols, aldehydes, ketones, amides, amines, carboxylic acids, esters, and phenols. Most amines and carboxylic acids can be distinguished from the rest by testing their aqueous solution with litmus paper. Phenols and aromatic amines may be too weakly acidic or basic to turn litmus red or blue.

Water-insoluble compounds that are soluble in 5% hydrochloric acid (Class B) contain basic functional groups, usually incorporating nitrogen atoms. Of the families considered here, only amines fall into this category. Some amines may form insoluble hydrochloride salts as they dissolve, so solubility behavior should be observed carefully to detect any change in the appearance of the unknown when it is shaken with the solvent.

Compounds that are insoluble in water *and* 5% hydrochloric acid but soluble in 5% sodium hydroxide contain acidic functional groups that give them pK_a values of approximately 12 or less. Carboxylic acids and phenols fall into this category; the latter can be recognized by their insolubility in 5% sodium bicarbonate. Care should be taken in interpreting this test. Some compounds, such as reactive esters, may react with the solvent to form soluble reaction products; long-chain carboxylic acids may form relatively insoluble salts that yield a soapy foam when shaken.

Acids with pK_a values of about 6 or less will dissolve in 5% sodium bicarbonate. This class includes most carboxylic acids but not phenols, so the solubility test with sodium bicarbonate can distinguish between these two types of compounds.

Compounds that are insoluble in all the previous solvents but dissolve in, or react with, cold concentrated sulfuric acid (Class N) include high-molecular-weight alcohols, aldehydes, ketones, amides, and esters. Unsaturated compounds and some aromatic hydrocarbons (those containing several alkyl groups on the benzene ring) are also soluble in this reagent. Solution in sulfuric acid is accompanied by protonation of a basic atom (nitrogen or oxygen) or by some other reaction such as sulfonation, dehydration, addition to multiple bonds, or polymerization.

Compounds that are insoluble in all of the solvents (Class X) include most aromatic hydrocarbons and the halogen derivatives of aliphatic and aromatic hydrocarbons.

An outline of the solubility scheme is given in Figure 1 (on page 524). It is seldom necessary to test an unknown with every solvent, since if it dissolves in water, for example, it will dissolve in aqueous solutions of HCl, NaOH, and NaHCO$_3$ as well. In most cases, an unknown is not tested further once it is found to dissolve in a given solvent, since its solubility classification is established at that point. Compounds that are soluble in 5% NaOH, however, are also tested with 5% NaHCO$_3$ to determine whether they are carboxylic acids or phenols. It should be emphasized that solubility classifications can be misleading; there are exceptions and borderline cases. For in-

Amines containing two or more aryl groups and certain hindered amines may be insoluble in 5% HCl.

Compounds soluble in both solvents are placed in Class A$_1$; those soluble in 5% NaOH but insoluble in 5% NaHCO$_3$ are assigned to Class A$_2$.

Some reactions in cold concentrated sulfuric acid

$$RCHO \xrightarrow{H_2SO_4} R\overset{\oplus}{C}HOH$$

$$RCH_2OH \xrightarrow{H_2SO_4} RCH_2OSO_3H$$

$$RCH_2\underset{\underset{R''}{|}}{\overset{\overset{OH}{|}}{C}}{-}R' \xrightarrow{H_2SO_4} RCH{=}\underset{\underset{R''}{|}}{C}{-}R'$$

$$RCOOH \xrightarrow{H_2SO_4} R\overset{\oplus}{C}(OH)_2$$

$$RCH{=}CHR' \xrightarrow{H_2SO_4} RCH_2\underset{\underset{}{}}{\overset{\overset{OSO_3H}{|}}{C}}HR'$$

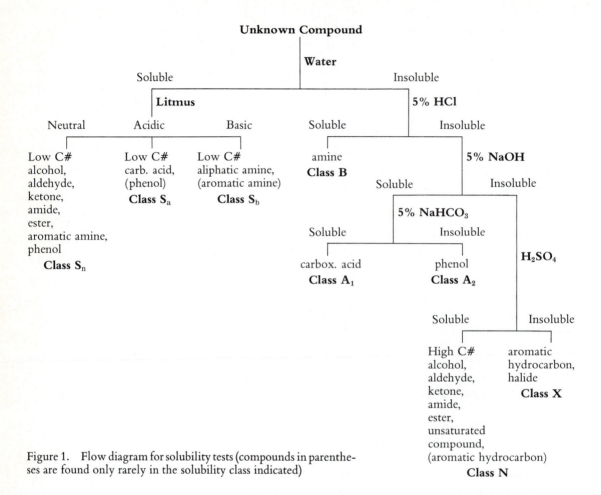

Figure 1.   Flow diagram for solubility tests (compounds in parentheses are found only rarely in the solubility class indicated)

stance, some phenols are acidic enough to dissolve in 5% $NaHCO_3$, and some weakly basic amines do not dissolve in 5% HCl. Therefore, the solubility tests should always be supported by other chemical or spectral evidence before a definite conclusion is drawn.

Functional-Class Tests

The results of the solubility tests should eliminate some functional classes and support others, allowing you to determine the functional class of your unknown compound with only a few chemical tests. An infrared spectrum of the unknown can further limit the possibilities, so that it may be necessary to use a test only to confirm the presence of a functional group suggested by the spectrum. In any case, you should choose the tests

Table 1.   Functional-class tests

Family	Number	Test	Comments
Alcohols	C–9	Chromic acid	Negative for 3° alcohols
	C–17	Lucas's test	Negative for 1° and high M.W. alcohols
	C–1	Acetyl chloride	Useful if other tests inconclusive
Aldehydes	C–9	Chromic acid	Alcohols also react
	C–23	Tollens's test	
Aldehydes and ketones	C–11	2,4-Dinitrophenyl-hydrazine	Ketones react with C–11, not with C–9 or C–23
Amides	C–2	Alkaline hydrolysis	Best for amides of ammonia or low M.W. amines
Amines	C–12	Elemental analysis	Detects N; used with solubility test results
		Solubility tests	Soluble in 5% HCl
Aromatic hydrocarbons	C–3	Aluminum chloride–chloroform	Should be substantiated by tests indicating absence of functional groups (such as halogens)
Carboxylic acids		Solubility tests	Soluble in 5% NaOH and 5% NaHCO$_3$
Esters	C–14	Ferric hydroxamate	
	C–2	Alkaline hydrolysis	Can be used to prepare derivatives
Halogenated hydrocarbons	C–5	Beilstein's test	Simple, not always reliable
	C–21	Silver nitrate/ethanol	Negative for vinyl and aryl halides
	C–10	Density test	Usually negative for monochloroalkanes
	C–12	Elemental analysis	Can distinguish Cl, Br, and I
Phenols	C–13	Ferric chloride	Positive for most (but not all) phenols
	C–8	Bromine water	Aromatic amines also react

carefully so that they will provide the information you need in a minimum number of steps. The tests used to detect functional groups are listed in Table 1. Within families, functional-class tests are listed in approximate order of simplicity and utility. It may be necessary to perform two (or more) tests for the same functional group in case the first test is inconclusive. Other chemical tests that can be used to provide additional structural information about the unknown are described in Table 4 (on page 547). Procedures for the tests are listed by test number (such as C–8 for bromine water) and in alphabetical order beginning on page 551.

SPECTRAL ANALYSIS

Infrared Spectra

Because infrared absorption bands are associated with specific chemical bonds, it is usually possible to deduce the functional class of an organic compound from its infrared spectrum. Infrared "frequencies" are often measured in *wavenumbers;* the

The principles and experimental methods of infrared spectrometry are discussed in OP–33. This section will be concerned with the interpretation of infrared spectra.

wavenumber in cm^{-1} of a vibration is simply the number of waves per centimeter. The wavenumber is thus the inverse of the wavelength, which is generally measured in micrometers (μm); the two can be interconverted as shown:

$$\text{wavenumber in cm}^{-1} = \frac{10^4 \, \mu\text{m/cm}}{\text{wavelength in } \mu\text{m}}$$

Thus, a C=O bond vibration at 1710 cm^{-1} has a wavelength of 5.85 μm (10,000/1710). On a typical infrared spectrum, wavelengths increase and wavenumbers decrease going from left to right. For that reason, wavenumber ranges will be specified here with the higher value first.

The stretching vibrations of chemical bonds resemble the vibrations of springs, in that stronger bonds have higher vibrational energies and frequencies than weaker ones. Thus, triple bonds generally absorb at higher frequencies than double bonds do, and double bonds absorb at higher frequencies than single bonds do. Because of the exceptionally low mass of the hydrogen atom, however, single bonds involving hydrogen (C—H, O—H, N—H, etc.) have higher vibrational frequencies than most double and triple bonds. Stretching bands involving single bonds to hydrogen occur at the high-frequency (left) end of an infrared spectrum, in the region between 3700 and 2700 cm^{-1} (2.7–3.7 μm). Triple bonds usually absorb between 2700 and 1850 cm^{-1} (3.7–5.4 μm), and double bonds and aromatic bonds absorb between 1950 and 1450 cm^{-1} (5.1–6.9 μm). Most absorption bands between 1500 and 600 cm^{-1} (6.7–16.7 μm) are produced by bending vibrations or single-bond stretching vibrations. An absorption band in this region may be associated with more than one bond; for example, the so-called C—O stretching bands of esters arise from the vibration of C—C—O units rather than of isolated carbon-oxygen bonds.

Examining the following regions of the infrared spectrum will help you locate the most useful infrared bands quickly.

It takes less energy to bend a bond than to stretch it, so most bending vibrations have comparatively low frequencies.

Region 1: 3600–3200 cm^{-1} (2.8–3.1 μm). Bands in this region can arise from O—H and N—H stretching vibrations of alcohols, phenols, amines, and amides. O—H bands are generally very strong and broad; N—H bands are somewhat weaker, and in the case of primary amines and amides, they have two peaks.

Region 2: 3100–2500 cm⁻¹ (3.2–4.0 µm). This region contains most of the C—H stretching vibrations. A strong band in the $3000-2850$ cm⁻¹ region, arising from C—H bonds to sp^3 carbon atoms, is present for most organic compounds. The sp^2 C—H bonds associated with aromatic hydrocarbons and unsaturated compounds absorb at higher frequencies ($3100-3000$ cm⁻¹), and the C—H bonds of aldehyde (CHO) groups absorb at lower frequencies. The O—H bond of a carboxylic acid gives rise to a very broad absorption band in this region.

Region 3: 1750–1630 cm⁻¹ (5.7–6.1 µm). This region contains most of the carbonyl (C=O) stretching bands of aldehydes, ketones, carboxylic acids, amides, esters, and other carbonyl compounds. The carbonyl band is usually strong and quite unmistakable. Unsaturated compounds may have a C=C stretching band in the $1670-1640$ cm⁻¹ region, but this band is nearly always weaker and narrower than a carbonyl band.

Region 4: 1350–1000 cm⁻¹ (7.4–10.0 µm). This region is usually cluttered with many absorption bands, but it is often possible to identify the C—O stretching bands of alcohols, phenols, carboxylic acids and esters, and some C—N stretching bands of amines and amides. Some bending bands in this region are useful for confirming the identities of certain functional groups.

Shading is used on the infrared spectrum in Figure 2 (on page 528) to show the locations of the four regions. In this example, the absence of any band in Region 1 (or a broad band in Region 2) eliminates from consideration all compounds containing O—H and N—H bonds, such as alcohols, phenols, and primary or secondary amines and amides. In Region 2, the appearance of a weak "shoulder" on the C—H band at 3050 cm⁻¹ indicates an sp^2 C—H bond associated with either an aromatic ring or a carbon-carbon double bond. Region 3 shows a strong carbonyl band near 1720 cm⁻¹, and Region 4 shows a strong C—O band at 1276 cm⁻¹ as well as a weaker one at 1109 cm⁻¹. The absence of an O—H or N—H band and the presence of the C=O and two C—O bands suggest that the compound responsible for this spectrum is an ester. The compound is, in fact, the aromatic ester ethyl benzoate.

Ethyl benzoate

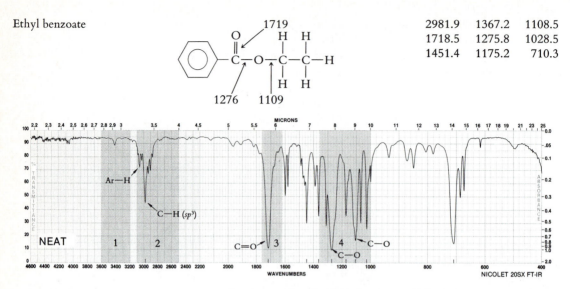

2981.9	1367.2	1108.5
1718.5	1275.8	1028.5
1451.4	1175.2	710.3

Figure 2.   Classification of a compound from its IR spectrum. (Spectrum from *The Aldrich Library of FT-IR Spectra,* by Charles J. Pouchert, used with the permission of the Aldrich Chemical Company.)

Table 2 summarizes the locations and characteristics of important absorption bands that occur in the four regions described above. Additional information is provided on the infrared correlation chart on the back cover of this book. When you interpret the infrared spectrum of an unknown organic compound, it is suggested that you first search the four infrared regions for spectral bands that will help you determine what kind of compound you have, or at least eliminate most of the possibilities. Then you should study the spectral characteristics of the appropriate classes of compounds to confirm your tentative classification and to obtain additional structural information about the unknown. For example, if the initial search suggests that your unknown is an alcohol, you should then (1) confirm that fact by locating the bands described in the section on the IR bands of alcohols; (2) find out whether the alcohol is aromatic from the information given in the section on aromatic hydrocarbons (which applies to other aromatic compounds as well); and (3) attempt to classify the alcohol as primary, secondary, or tertiary from the frequency of its C—O band.

It is always a good idea to study published spectra such as those in the Aldrich collections ⟨Bib – F32 and Bib – F33⟩ or in

Table 2.   Important bands in regions 1–4 of IR spectra

Region	Frequency range (cm⁻¹)	Bond type	Family	Comments
1	3500–3200	N—H	amine, amide	Weak-medium. 1°: 2 bands; 2°: 1 band; 3°: no bands. See also Region 3.
	3600–3200	O—H	alcohol, phenol	Broad, strong. Sharp band around 3600 cm⁻¹ in dilute solution. See Region 4.
2	3300–2500	O—H	carboxylic acid	Broad, strong, centered around 3000. See Regions 3 and 4.
	3100–3000	C—H	aromatic hydrocarbon	May be shoulder on stronger sp^3 C—H band. See discussion of aromatic hydrocarbons in text.
	2850–2700	$\overset{\overset{\text{(O)}}{\|\|}}{\text{C}}$—H	aldehyde	Weak-medium, usually sharp. See Region 3.
3	1750–1630	C=O	aldehyde, ketone, carboxylic acid, ester, amide	Strong. Position depends on nature of functional group and other structural features. See Region 2 for aldehydes; Regions 2 and 4 for acids; Region 4 for esters; Region 1 for amides.
4	1350–1210	C—O	carboxylic acid	Medium-strong.
	1300–1180	C—O	phenol	Strong.
	1200–1000	C—O	alcohol	Strong. Frequency in order 3° > 2° > 1°.
	1310–1160	C—O	ester	Strong. Should be accompanied by weaker band in the same region as the alcohol C—O stretch.

Note: Tentative classifications as to family must be confirmed by referring to the descriptions in the text.

your lecture textbook. Although the following descriptions will tell you the approximate locations of absorption bands and some of their characteristics, only a thorough study of the actual spectra will make you proficient in spectral interpretation. In many cases, the shape of an absorption band will help you identify it. For example, an absorption band near 3300 cm⁻¹ could arise from an O—H bond, an N—H bond, or the ≡C—H bond of a terminal alkyne. But most O—H stretching bands are broad and blunt, whereas N—H bands are broad and sharp (often with two prongs), and ≡C—H bands are narrow and sharp. With a little practice, you should be able to recognize certain bands on sight, so that you will not mistake them for other bands that appear in the same frequency region.

The most useful infrared bands associated with the ten classes of compounds we will consider are described below. A summary of the main spectral features that characterize each

See Experiment 14 for a description of the C—H vibrations that appear in the IR spectra of most organic compounds.

functional class is followed by a description of individual bond vibrations and a representative infrared spectrum in each case. The wavenumber ranges given are for solids (in Nujol mulls or KBr discs) or neat liquids; values for solutions may differ. Although the wavenumber ranges apply to most organic compounds in each class, compounds with certain structural features (such as highly strained rings) may absorb outside of the ranges. Absorption bands on the infrared spectra are designated either as stretching (ν) or bending (δ) bands; only those bands that are most useful for functional group classification are labeled. Note that the exact wavenumbers of the strongest bands (designated by tick marks along the lower edge of the spectra) are listed with each spectrum.

Characteristic Infrared Bands of Alcohols. The presence of a strong, broad band centered around 3300 cm^{-1} and a strong C—O band in the 1200–1000 cm^{-1} region is usually good evidence for an alcohol. A C—O band above 1200 cm^{-1} may suggest a phenol, particularly when it is accompanied by bands indicating an aromatic ring structure. (See Figure 3.)

O—H *stretch:* 3600–3200 cm^{-1} (strong, broad). Usually centered near 3300 cm^{-1}.

C—O *stretch:* 1200–1000 cm^{-1} (medium to strong). Most saturated aliphatic alcohols absorb near 1050 cm^{-1} if they are primary, near 1110 cm^{-1} if they are secondary, and near 1175 cm^{-1} if they are tertiary. Alicyclic alcohols and alcohols with aromatic rings or vinyl groups on the carbinol carbon absorb at frequencies about 25–50 cm^{-1} lower, for each structural class, than the values specified.

Characteristic Infrared Bands of Phenols. Phenols are characterized by a strong, broad band centered around 3300 cm^{-1} and a strong band near 1230 cm^{-1}, accompanied by bands indicating an aromatic structure (see the section on aromatic hydrocarbons below). (See Figure 4.)

O—H *stretch:* 3600–3200 cm^{-1} (strong, broad).

O—H *bend:* 1390–1315 cm^{-1} (medium).

C—O *stretch:* 1300–1180 cm^{-1} (strong). Usually close to 1230 cm^{-1}. This band may be split, with several distinct maxima.

Figure 3. IR spectrum of a primary alcohol, 2-methyl-1-propanol. (Spectrum from *The Aldrich Library of FT-IR Spectra,* by Charles J. Pouchert, used with the permission of the Aldrich Chemical Company.)

2-Methyl-1-propanol

$$CH_3CHCH_2OH$$
$$|$$
$$CH_3$$

3328.1	1387.5	940.3
2957.5	1247.4	818.5
1470.9	1041.8	669.8

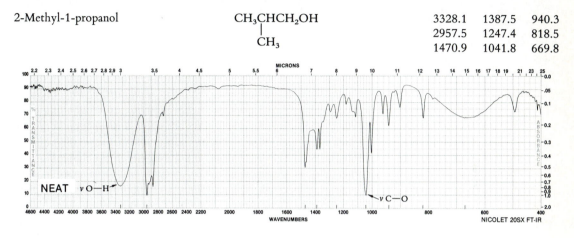

Figure 4. IR spectrum of phenol. (Spectrum from *The Aldrich Library of FT-IR Spectra,* by Charles J. Pouchert, used with the permission of the Aldrich Chemical Company.)

Phenol

OH

3372.6	1224.4	751.8
1595.3	1168.0	689.8
1498.9	809.8	506.0

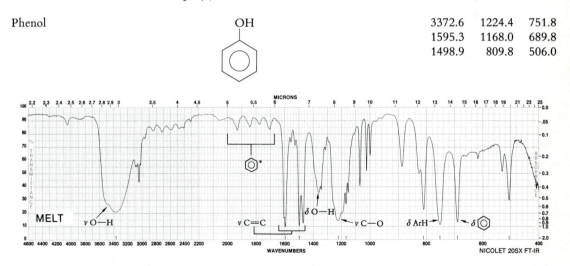

Note: The bands marked ⬡ * are aromatic overtone-combination bands.

3-Methylpentanal

$$\underset{\text{CH}_3\text{CHCH}_2\text{CH}}{\overset{\text{CH}_3 \quad \text{O}}{\vert \qquad \vert\vert}}$$

2960.1	1468.3	1016.6
2718.7	1368.8	898.9
1727.6	1170.8	524.1

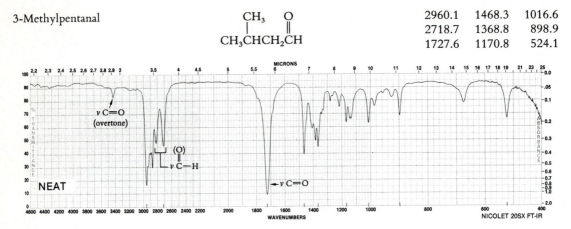

Figure 5. IR spectrum of an aldehyde, 3-methylpentanal. (Spectrum from *The Aldrich Library of FT-IR Spectra,* by Charles J. Pouchert, used with the permission of the Aldrich Chemical Company.)

Characteristic Infrared Bands of Aldehydes. The presence of a sharp, medium-intensity band near 2720 cm^{-1} and a strong carbonyl band around 1700 cm^{-1} is good evidence for an aldehyde. (See Figure 5.)

$$\overset{(\text{O})}{\vert\vert}$$
C—H *stretch:* 2850–2700 cm^{-1} (two bands, weak to medium). From the carbonyl C—H bond. Most aldehydes have two bands near 2850 and 2720 cm^{-1}, with the low-frequency band well separated from other aliphatic C—H bands.

C=O *stretch:* 1740–1685 cm^{-1} (strong). Most unconjugated aldehydes absorb near 1725 cm^{-1}; conjugation of the carbonyl group with an aromatic ring or other unsaturated system shifts the band to the 1700–1665 cm^{-1} region. A weak overtone of this band may appear near 3400 cm^{-1}.

Characteristic Infrared Bands of Ketones. The presence of a strong carbonyl band around 1700 cm^{-1} is good evidence for a ketone if other bands described in Table 2 (O—H, N—H, C—O, and aldehyde C—H) are absent. Bands in the 1300–1100 cm^{-1} region, which may be mistaken for C—O bands, arise from C—C—C vibrations involving the carbonyl carbon. (See Figure 6.)

2-Pentanone

$$CH_3-\overset{\overset{\displaystyle O}{\|}}{C}-CH_2CH_2CH_3$$

2963.9	1366.0	1170.7
1717.4	1295.5	727.0
1422.9	1235.5	591.9

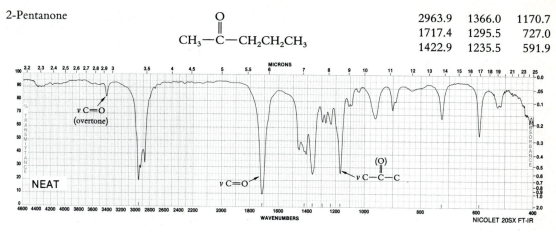

Figure 6. IR spectrum of a ketone, 2-pentanone. (Spectrum from *The Aldrich Library of FT-IR Spectra,* by Charles J. Pouchert, used with the permission of the Aldrich Chemical Company.)

C=O *stretch:* 1750–1660 cm^{-1} (strong). Most unconjugated aliphatic ketones absorb around 1710 cm^{-1}, and conjugated ketones absorb near 1670 cm^{-1}. A weak C=O overtone band is usually evident near 3400 cm^{-1}.

$\overset{\overset{\displaystyle (O)}{\|}}{C}$—C—C *stretch/bend:* 1300–1100 cm^{-1} (medium). Often multiple bands; unconjugated ketones absorb around 1230–1100 cm^{-1}, and conjugated ketones around 1300–1230 cm^{-1}.

Characteristic Infrared Bands of Carboxylic Acids. The presence of a very broad band centered near 3000 cm^{-1} and a carbonyl band around 1700 cm^{-1} is good evidence for a carboxylic acid. (See Figure 7 on page 534.)

O—H *stretch:* 3300–2500 cm^{-1} (strong, very broad). C—H stretching bands are generally superimposed on this band.

C=O *stretch:* 1725–1665 cm^{-1} (strong). Unconjugated acids absorb around 1725–1700 cm^{-1}, and conjugated acids around 1700–1665 cm^{-1}.

C—O *stretch:* 1350–1210 cm^{-1} (strong). May be broken up into a number of sharp peaks for long-chain acids.

Figure 7.   IR spectrum of a carboxylic acid, hexanoic acid. (Spectrum from *The Aldrich Library of FT-IR Spectra,* by Charles J. Pouchert, used with the permission of the Aldrich Chemical Company.)

Hexanoic acid

$$CH_3(CH_2)_3CH_2{-}\overset{\displaystyle O}{\overset{\displaystyle \|}{C}}{-}OH$$

3191.2	1710.7	1293.4
2959.4	1467.5	1213.2
2669.9	1413.8	939.2

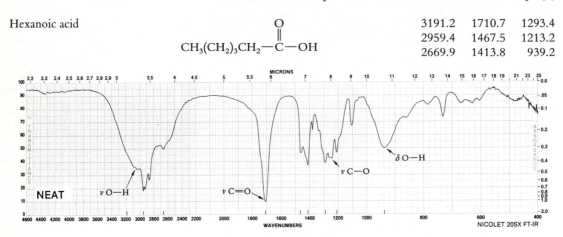

Figure 8.   IR spectrum of a carboxylic ester, *sec*-butyl acetate. (Spectrum from *The Aldrich Library of FT-IR Spectra,* by Charles J. Pouchert, used with the permission of the Aldrich Chemical Company.)

sec-Butyl acetate

$$CH_3{-}\overset{\displaystyle O}{\overset{\displaystyle \|}{C}}{-}O{-}\overset{\displaystyle CH_3}{\overset{\displaystyle |}{\underset{\displaystyle |}{\underset{\displaystyle CH_2CH_3}{CH}}}}$$

2975.6	1373.2	1031.5
1737.8	1244.6	996.5
1457.6	1096.7	944.8

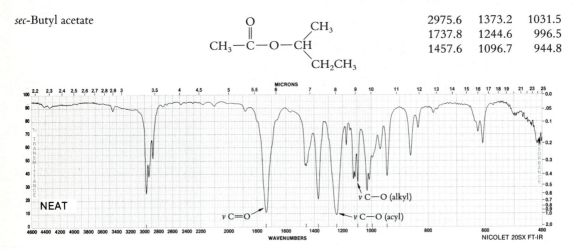

O—H *bend:* 950–870 cm^{-1} (medium, broad). Another O—H bending band near 1430 cm^{-1} is often obscured by C—H bending bands.

Characteristic Infrared Bands of Esters. The presence of a strong carbonyl band around 1740 cm^{-1} and an unusually strong C—O band in the 1310–1160 cm^{-1} region is good evidence for an ester, if there is no O—H band. (See Figure 8.)

C=O *stretch:* 1775–1715 cm^{-1} (strong). Near 1770 cm^{-1} for phenyl esters (RCOOAr) and vinyl esters, 1740 cm^{-1} for most unconjugated esters, and 1730–1695 cm^{-1} for formates and conjugated esters.

C—O *stretch:* 1310–1160 cm^{-1} (strong, broad). Occurs near 1310–1250 cm^{-1} for conjugated esters, 1240 cm^{-1} for acetates, and 1175 cm^{-1} for other unconjugated esters. This vibration involves the acyl-oxygen linkage and is often broader and stronger than the carbonyl band. A weaker "alkyl-oxygen" C—O stretching band occurs at lower frequencies, usually around 1040 and 1100 cm^{-1} for aliphatic esters of primary and secondary alcohols, respectively; aromatic esters absorb at somewhat higher frequencies.

Characteristic Infrared Bands of Amines. Primary amines are characterized by a medium-intensity, two-pronged band near 3350 cm^{-1} and two medium-strong bands near 1615 and 800 cm^{-1}, the latter very broad. Secondary amines have a single weak band near 3300 cm^{-1} and a broad band near 715 cm^{-1}. Tertiary amines may be distinguished by a shift of the methylene stretching band to ~2700 cm^{-1} and the presence of a C—N band. (See Figure 9 on page 536.)

N—H *stretch:* 3500–3200 cm^{-1} (weak to medium, broad). Primary aliphatic amines give rise to a two-pronged band centered near 3350 cm^{-1}, secondary amines have one weak band near 3300 cm^{-1}, and tertiary amines have none. Aromatic primary and secondary amines absorb near 3400 and 3450 cm^{-1}, respectively.

N—H *bend (scissoring):* 1650–1500 cm^{-1} (medium to strong). Usually near 1615 cm^{-1} for primary amines. Seldom observed for secondary aliphatic amines; secondary aromatic amines absorb near 1515 cm^{-1}.

3-Methylpentylamine	CH₃	3366.4	1384.1	847.7
		2955.0	1066.7	815.5
	CH₃CHCH₂CH₂NH₂	1467.7	919.2	770.5

3-Methylpentylamine

$$CH_3CHCH_2CH_2NH_2$$

with CH_3 attached above

3366.4	1384.1	847.7
2955.0	1066.7	815.5
1467.7	919.2	770.5

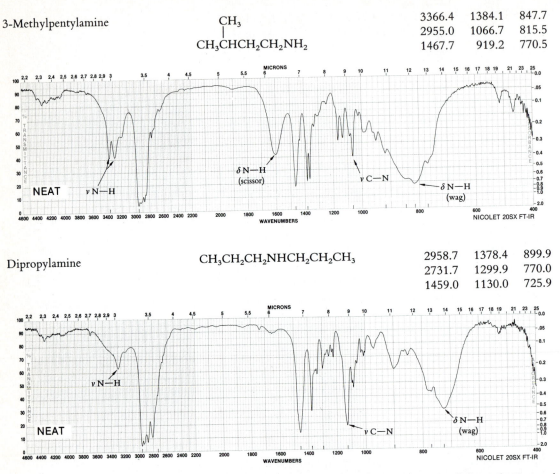

Dipropylamine $CH_3CH_2CH_2NHCH_2CH_2CH_3$

2958.7	1378.4	899.9
2731.7	1299.9	770.0
1459.0	1130.0	725.9

Figure 9.   IR spectra of a primary amine, 3-methylpentylamine, and a secondary amine, dipropylamine. (Spectra from *The Aldrich Library of FT-IR Spectra,* by Charles J. Pouchert, used with the permission of the Aldrich Chemical Company.)

N—H *bend (wagging):* 910–660 cm⁻¹ (medium to strong, broad). Usually strong and very broad, around 910–770 cm⁻¹ for primary amines. Closer to 715 cm⁻¹ for secondary amines.

C—N *stretch:* 1340–1020 cm⁻¹ (medium to strong). Around 1340–1250 cm⁻¹ for aromatic amines, 1250–1020 cm⁻¹ for aliphatic amines. Like that of the C—O band, the frequency of an aliphatic C—N band varies with changes in the structure of the attached alkyl group.

2-Methylpropanamide

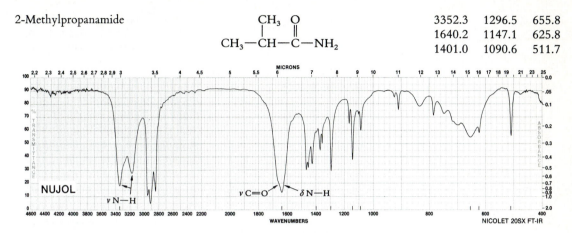

3352.3	1296.5	655.8
1640.2	1147.1	625.8
1401.0	1090.6	511.7

Figure 10.   IR spectrum of an amide, 2-methylpropanamide. (Spectrum from *The Aldrich Library of FT-IR Spectra,* by Charles J. Pouchert, used with the permission of the Aldrich Chemical Company.)

Characteristic Infrared Bands of Amides.   The presence of a carbonyl band near 1640 cm^{-1} and two bands (or peaks) in the 3400–3000 cm^{-1} region is good evidence for an amide. (See Figure 10.)

N—H *stretch:* 3450–3300 cm^{-1} and 3225–3180 cm^{-1} (one or two bands, medium to strong). Primary amides have two bands (or a two-pronged band) near 3400 and 3200 cm^{-1}. Secondary amides have a single N—H stretching band near 3340 cm^{-1}, with an N—H bending overtone near 3080 cm^{-1}.

C=O *stretch:* 1695–1615 cm^{-1} (strong). Usually centered near 1640 cm^{-1}.

N—H *bend:* 1655–1615 cm^{-1} (primary) *or* 1570–1515 cm^{-1} (secondary) (medium to strong). This band usually overlaps with the carbonyl band on spectra of primary amides obtained using KBr discs or mulls; it appears at lower frequencies on spectra obtained in solution. The band is near 1540 cm^{-1} for most secondary amides, and an overtone can sometimes be seen at about 3080 cm^{-1}.

Characteristic Infrared Bands of Organic Halides.   Alkyl chlorides and bromides show fairly strong absorption between 800 and

1-Chloropentane $CH_3(CH_2)_4Cl$

2959.3	1282.1	789.3
1467.0	1037.2	730.8
1380.3	925.7	653.7

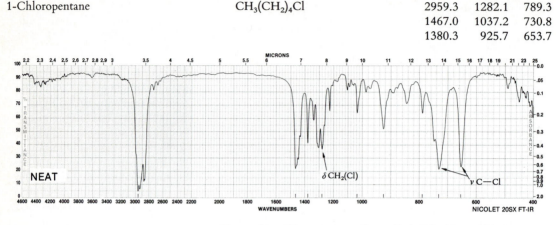

Figure 11. IR spectrum of an alkyl chloride, 1-chloropentane. (Spectrum from *The Aldrich Library of FT-IR Spectra,* by Charles J. Pouchert, used with the permission of the Aldrich Chemical Company.)

500 cm^{-1}. Additional chemical evidence is usually needed to characterize organic halides. (See Figure 11.)

$$\underset{\mid}{(X)}$$

C—H *bend:* 1300 – 1150 cm^{-1} (medium). Observed only for halides with terminal halogen atoms (—CH$_2$X).

C—Cl *stretch:* 850 – 550 cm^{-1} (medium to strong). Two bands near 725 and 645 cm^{-1} when the chlorine is terminal; below 625 cm^{-1} otherwise, unless several chlorine atoms are on the same or adjacent carbons. Ar—Cl bonds absorb around 1175 – 1000 cm^{-1}.

C—Br *stretch:* 760 – 500 cm^{-1} (medium to strong). Near 645 cm^{-1} when the bromine is terminal. Ar—Br bonds absorb around 1175 – 1000 cm^{-1}.

Characteristic Infrared Bands of Aromatic Hydrocarbons. Most compounds containing benzene rings are characterized by (1) aromatic C—H (Ar—H) stretching bands near 3070 cm^{-1}, (2) a distinctive pattern of weak bands in the 2000 – 1650 cm^{-1} region, (3) two sets of bands near 1600 cm^{-1} and 1515 – 1400 cm^{-1}, and (4) one or more strong absorption bands in the 900 – 675 cm^{-1} region. The presence of such bands and the absence of absorption bands characteristic of functional groups

Isopropylbenzene

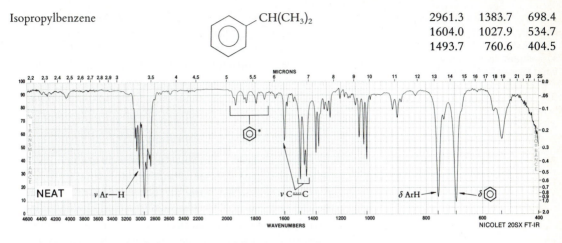

2961.3	1383.7	698.4
1604.0	1027.9	534.7
1493.7	760.6	404.5

Figure 12. IR spectrum of an arene, isopropylbenzene. The bands marked ⊚* are aromatic overtone-combination bands. (Spectrum from *The Aldrich Library of FT-IR Spectra,* by Charles J. Pouchert, used with the permission of the Aldrich Chemical Company.)

suggest an aromatic hydrocarbon. Alkenes and other compounds with carbon-carbon double bonds also show C—H stretching bands above 3000 cm⁻¹ and strong C—H bending bands below 900 cm⁻¹, but the absorption patterns of aromatic compounds in the 2000–1650 cm⁻¹ and 1600–1400 cm⁻¹ regions are quite distinctive. (See Figure 12.)

Ar—H *stretch:* 3100–3000 cm⁻¹ (weak to medium). May appear as a shoulder on the left side of a stronger sp^3 C—H band.

Overtone-combination vibrations: 2000–1650 cm⁻¹ (multiple bands, weak). The band pattern in this region is related to the kind of ring substitution, as illustrated in Figure 13.

C⋯C *stretch:* 1615–1585 cm⁻¹ and 1515–1400 cm⁻¹ (variable). Bands near 1600 cm⁻¹ are often split, and bands in the low-frequency region are usually centered near 1500 cm⁻¹ and 1430 cm⁻¹.

Ar—H *out-of-plane bend:* 910–730 cm⁻¹ (strong). The band frequency varies with the number of adjacent ring hydrogens:

two adjacent hydrogens: 855–800 cm⁻¹

three adjacent hydrogens: 800–765 cm⁻¹

four or five adjacent hydrogens: 770–730 cm⁻¹

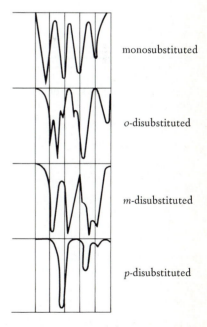

monosubstituted

o-disubstituted

m-disubstituted

p-disubstituted

Figure 13. Typical absorption patterns of substituted aromatic compounds in the 2000–1670 cm⁻¹ region

For example, a *meta*-disubstituted benzene has three adjacent ring hydrogens and one isolated hydrogen, so it should have a strong C—H bending band in the 800–765 cm^{-1} region and possibly a weak one around 910–835 cm^{-1}, along with a ring-bending band around 715–680 cm^{-1}.
MC - for msp. 540 - guide

The principles and experimental methods of NMR spectrometry are discussed in OP–34. This section will be concerned with the interpretation of NMR spectra.

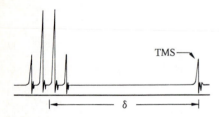

Figure 14.   Chemical shift of a PMR signal

A weak band in the 910–835 cm^{-1} region may also be observed for rings with isolated hydrogens. Monosubstituted, *meta*-disubstituted, and some trisubstituted benzenes show an additional ring-bending band around 715–680 cm^{-1}.

Nuclear Magnetic Resonance Spectra

Like infrared spectra, nuclear magnetic resonance (NMR) spectra can be used to detect certain functional groups, but they are more often used to provide detailed information about molecular structures. The proton NMR (PMR) spectrum of a compound can help you determine what structural features are present or absent and, in some cases, may allow you to reconstruct its entire molecular structure.

A PMR spectrum provides data in the form of chemical shifts, signal areas, signal multiplicities, and coupling constants. The use of these parameters to work out a molecular structure is a fascinating mental exercise, comparable to the work of a cryptographer who reconstructs messages from coded symbols. The *chemical shift* is the distance, measured in hertz or parts per million, from the center of a signal to some reference signal, usually that of tetramethylsilane (TMS). The TMS signal occurs farther upfield (to the right) than nearly all other signals, so the chemical shift of a signal is usually measured as its distance downfield from (to the left of) that of TMS, as shown in Figure 14.

The *signal area*, which is the sum of the areas under all the peaks in a proton signal, is proportional to the number of protons giving rise to the signal. This area is usually determined by an electronic integrator that traces a line across each proton signal after it is recorded; the area of the signal is proportional to the vertical rise of the integrator pen as it crosses the signal. Integrated signal areas can be converted to proton numbers using the following relationship:

number of protons responsible for signal =

$$\text{total number of protons} \times \frac{\text{area under signal}}{\text{area under all signals}}$$

For example, suppose a compound with the molecular formula $C_{10}H_{14}$ has four signals with relative areas of 42, 7, 14, and 35.

The sum of the areas is 98, so the number of protons responsible for the first signal is

$$14 \times \frac{42}{98} = 6$$

By a similar calculation, it can be shown that the other signals arise from 1, 2, and 5 protons, respectively. If the molecular formula of a compound is not known, relative proton numbers can be obtained by reducing the signal areas to the lowest possible ratio of integers.

The signal generated by a given set of protons may be split into several peaks as a result of coupling interactions with nearby proton sets. The *multiplicity* of a signal is simply the number of separate peaks it contains; its *coupling constant* is the distance between two adjacent peaks in the signal, measured in hertz (Hz). Figure 15 shows the signals of two sets of protons that are interacting with one another; the protons of set *a* have split the signal of the protons of set *b* into four peaks (a quartet) and the *b* protons have split the signal of the *a* protons into three peaks (a triplet). The coupling constant, which is equal for the two signals, is represented by J_{ab}. In the simplest case, the number of protons responsible for splitting the signal of a neighboring set of protons can be determined by subtracting 1 from the number of peaks in that signal; thus, the three peaks in the *a* signal could have been produced by two neighboring *b* protons, and the four peaks in the *b* signal by three neighboring *a* protons. An interacting triplet-quartet grouping of this kind is usually produced by an ethyl (CH_3CH_2-) group. Note that the two signals are not perfectly symmetrical but appear to "lean" toward one another, with the peaks on the side facing the other signal being higher. This and the fact that their coupling constants are equal are additional evidence that the protons responsible for the two signals are, in fact, coupling with one another and not with some other proton sets in the molecule.

You can use the following methods to derive structural information from a PMR spectrum.

1. If the spectrum is integrated, measure the integrated area of each signal and use it to determine the number of protons responsible for the signal. Each set of equivalent protons (protons in the same molecular environment) gives rise to a signal, and the relative signal areas can tell you how many protons are in each set. For example, the spectrum of 3,3-di-

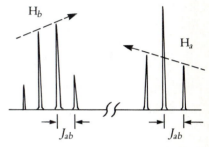

(*J* values equal, high sides facing)

Figure 15.   Signals of nearest-neighbor protons

The peaks in a signal are generally of unequal height. "Ideal" triplets and quartets have relative peak area ratios of 1:2:1 and 1:3:3:1, respectively. As shown in Figure 15, these signal patterns may be distorted in an actual spectrum.

If the spectrum is not integrated, add the heights of the peaks in each signal to get a rough estimate of its relative area.

Figure 16. Correlation chart relating PMR chemical shifts to proton environments

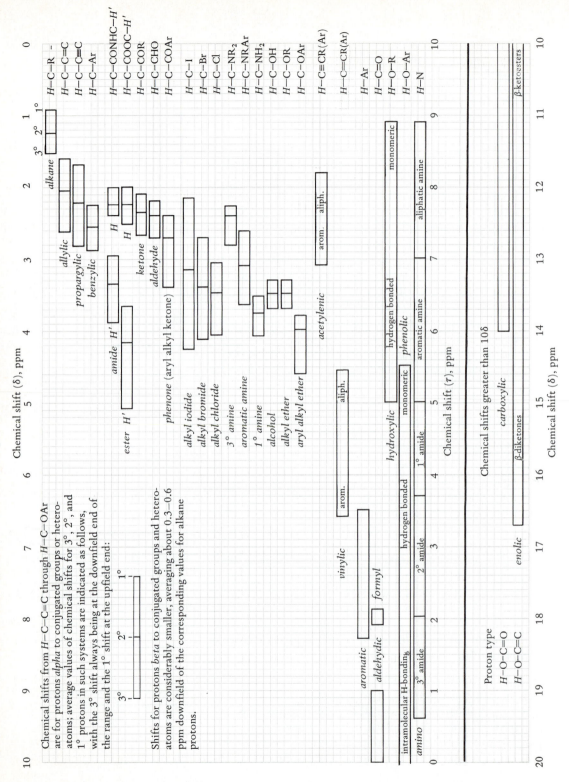

methyl-2-butanone has two signals with an area ratio of 3:1, because there are nine hydrogen atoms on the three equivalent methyl groups to one side of the carbonyl group and only three on the other methyl group.

$$CH_3 - \underset{\underset{\displaystyle CH_3}{\displaystyle |}}{\overset{\overset{\displaystyle CH_3}{\displaystyle |}}{C}} - \overset{\overset{\displaystyle O}{\displaystyle \|}}{C} - CH_3$$

3,3-dimethyl-2-butanone

2. Determine the chemical shift of each signal on the delta scale by measuring the distance, in parts per million, from the center of the signal to the TMS reference peak. The chemical shift of a signal may indicate what kind of protons are responsible for the signal or may suggest their relative locations in the molecule. For example, alkyl hydrogen atoms that are remote from electron-withdrawing substituents should have a chemical shift of approximately 0.9 ppm if they are primary, 1.3 ppm if they are secondary, and 1.5 ppm if they are tertiary. Electron-withdrawing groups containing oxygen, nitrogen, or halogens tend to move PMR signals downfield, increasing their chemical shifts. Benzene rings give rise to large downfield shifts, making it quite easy to recognize aromatic compounds from their PMR spectra. The correlation chart in Figure 16 summarizes chemical shift data for a number of proton types, and some selected chemical shift values for compounds in the common functional classes are given in Table 3 (on page 544).

3. Determine the multiplicity of each signal by counting the number of distinguishable peaks in the signal. (Very small peaks may be obscured by background noise.) If the signal of a proton set is reasonably symmetrical and contains evenly spaced peaks, it should be possible to estimate how many nearby protons are coupled with the protons in that set by subtracting 1 from the multiplicity of the signal. Irregular signals and signals that have been split by several dissimilar proton sets can be analyzed by more advanced methods.

See Bib – F7 for an introduction to the interpretation of complex splitting patterns.

4. Measure the coupling constant of each signal, when applicable, and try to determine from the J values and the way a signal "leans" what other signals might be coupled with it.

All of this information can be used to build up the structure of a molecule piece by piece. For example, consider the spectrum in Figure 17 (on page 545) of a ketone with the molecular formula $C_7H_{14}O$. Signals a and b have a relative area ratio of 6:1; since the compound contains 14 protons, 6/7 of them, or 12, must be responsible for signal a, and 1/7 of them, or 2, for signal b. Signal a has only two peaks, indicating that the a protons have only one neighboring proton. The seven peaks in signal b indicate that the b protons have six neigh-

A neighboring, or nearest-neighbor, proton is one on an adjacent carbon atom. Protons that are not nearest neighbors (usually in conjugated systems) sometimes couple weakly with one another, but we will not consider such long-range coupling effects here.

Table 3. Approximate PMR chemical shift values for different types of protons

Type of compound	Type of proton	Chemical shift (ppm or δ)
Alcohol	H—O—R	1–5.5
	H—C—OH	3.4–4
Phenol	H—O—Ar	4–12
	H—Ar	6–8.5
Aldehyde	H—C=O	9–10
	H—C—CHO	2.2–2.7
Ketone	H—C—COR	2–2.5
Carboxylic acid	H—O—C=O	10.5–12
	H—C—COOH	2–2.6
Ester	H—C—COOR	2–2.5
	H—C—OC=O	3.5–5
Amine	H—N—R (aliphatic)	1–3
	H—N—R (aromatic)*	3–5
	H—C—N	2.2–4
Amide	H—N—C=O*	5–9.5
	H—C—CON	2–2.4
	H—C—NHC=O	3–4
Halide	H—C—Br	2.5–4
	H—C—Cl	3–4
Aromatic hydrocarbon	H—Ar	6–8.5
	H—C—Ar	2.2–3

Note: Signals of proton types marked with an asterisk are often very broad.

H₃C—C(H)(b,CH₃ a)—C(=O)—C(H)(CH₃ a)—CH₃

2,4-dimethyl-3-pentanone

bors, and the higher chemical shift of this signal suggests that these protons are next to the electron-withdrawing carbonyl group. The only alkyl group in which a single proton has six equivalent protons for neighbors is the isopropyl group, $(CH_3)_2CH—$. Two such groups provide the required total of 12 *a* and 2 *b* protons, and attaching them both to a carbonyl group gives the complete structure of the ketone, 2,4-dimethyl-3-pentanone. Note that this structure accounts for all the features of the PMR spectrum: the 6 : 1 area ratio for the *a* and *b* protons; the higher chemical shift of the signal for the *b*

Figure 17. PMR spectrum of compound with molecular formula $C_7H_{14}O$. (Spectrum from *The Aldrich Library of NMR Spectra, Edition II*, by Charles J. Pouchert, used with the permission of the Aldrich Chemical Company.)

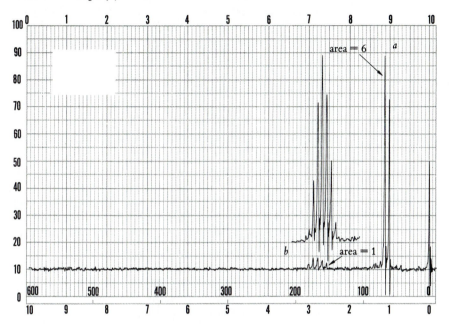

Note: Signal of *b* protons has been enlarged for clarity.

protons resulting from their proximity to the carbonyl group; the splitting of the *a* signal into two peaks by each neighboring *b* proton; the splitting of the *b* signal into seven peaks by each group of six neighboring *a* protons; and the equal coupling constants for the two signals.

For further information about the interpretation of PMR spectra, refer to the textbook for your lecture course or to appropriate sources in category F of the Bibliography.

IDENTIFICATION

After you have identified the main functional group present in your unknown, you are ready to identify the compound itself by comparing its properties with those of other compounds listed in the literature and obtaining sufficient evidence to eliminate all of the possibilities but one.

Examination of the Literature

At your instructor's option, you may either use the tables of selected organic compounds provided in Appendix VI or consult a more complete listing such as the *CRC Handbook of Tables for Organic Compound Identification* ⟨Bib – A10⟩.

The dividing line between liquids and solids is usually near a melting point of 25°. Some compounds with melting points over 25° (such as t-butyl alcohol) may appear to be liquids at room temperature, particularly when impure. When in doubt, try freezing the compound in an ice bath and estimating its melting point.

Most literature tabulations separate the compounds into solids and liquids, listing the solids in order of increasing melting point and the liquids in order of increasing boiling point. Tertiary amines are often listed separately from primary and secondary amines because their derivatives are different; the various classes of halides may also be listed separately. If your compound has a narrow melting- or boiling-point range (suggesting that it is relatively pure) around a temperature below 200°C, you should be safe in considering only those compounds with melting or boiling points within about ± 5°C of your observed value. However, if your boiling or melting point is 200°C or higher, particularly if you are using an uncorrected thermometer, this range should be extended to as much as ± 10°C. A table listing each compound within the chosen range should then be prepared. It should include the boiling (or melting) point, any other appropriate physical constants, and the melting points of useful derivatives. The chemical formulas for the compounds should be drawn, with the aid of the *CRC Handbook of Chemistry and Physics,* the *Merck Index,* the *Dictionary of Organic Compounds,* or other appropriate sources (see Appendix VII) when needed. At this point, you should be able to tell what physical properties and chemical tests or spectra will be most useful for narrowing down the list further, or which compounds can be eliminated on the basis of spectral data and tests already performed. In some cases, it may be possible to prepare a derivative immediately. However, if the list is a long one, you will probably need to reduce its length before you can decide on the most suitable derivative.

Your instructor may wish to check your melting- or boiling-point value and tell you whether it is within the ±5° range.

Melting points for some derivatives may not be listed, either because the compound does not form the derivative or because its melting point has not yet been reported in the literature.

Additional Tests and Data

The refractive index of a liquid can be very useful for purposes of identification. With an accurate refractometer that measures to the fourth decimal place, for example, it may be possible to eliminate compounds with refractive index values that deviate by more than ± 0.001 or so from the observed value.

This assumes that the unknown is pure and that a thermostatted refractometer is used.

Density cannot be measured as accurately as refractive index, but it can be useful in distinguishing between compounds having different structural features. Aromatic com-

Table 4.   Chemical tests providing structural information

Family or Structural Feature	Number	Test	Use
Alcohols	C–17	Lucas's test	To classify alcohols as 1°, 2°, 3°, etc.
	C–16	Iodoform test	To detect —CH(OH)CH$_3$ groupings
Aldehydes	C–6	Benedict's test	To distinguish aliphatic from aromatic aldehydes
Aldehydes and Ketones	C–16	Iodoform test	To detect —COCH$_3$ groupings
Amines	C–15	Hinsberg's test	To classify amines as 1°, 2°, or 3°
	C–4	Basicity test	To distinguish alkyl from aryl amines
	C–20	Quinhydrone	Complements C–15
Carboxylic acids	C–18	Neutralization equivalent	To determine the equivalent weight of an acid
Halogenated Hydrocarbons	C–10	Density test	To distinguish aliphatic and aromatic Cl, Br, and I compounds
	C–22	Sodium iodide/acetone	To classify halides as 1°, 2°, and 3°, etc.
	C–21	Silver nitrate/ethanol	Complements C–22
Unsaturation	C–7	Bromine/carbon tetra-chloride	To detect C=C and C≡C bonds
	C–19	Potassium permanganate	Complements C–7

pounds are usually more dense than comparable aliphatics, for example, and density varies among halogen compounds in the order RI > RBr > RCl.

It will usually be helpful to go back and re-interpret an infrared spectrum at this point, since it may provide information that can eliminate some compounds having (or lacking) subsidiary functional groups and certain structural features. PMR spectra can be extremely valuable if interpreted skillfully; a correct identification can often be made with very little additional evidence.

The tests listed in Table 4 can be used to provide structural information about the unknown. Procedures are listed alphabetically beginning on page 551.

Preparation of Derivatives

The preparation of at least one derivative is necessary to confirm the identity of your unknown. In some cases, it may be

Table 5. Derivatives of selected ketones

Compound	b.p.	Oxime	Semicarbazone	2,4-Dinitro-phenylhydrazone
2-methylcyclohexanone	163	43	195	137
2,6-dimethyl-4-heptanone	168	210	121	92
3-methylcyclohexanone	169	· · ·	180	155
4-methylcyclohexanone	169	37	199	130
2-octanone	172	· · ·	122	58

A good derivative should have a melting point above 50°C and below 250°C. Its melting point should be significantly different from the melting point of the unknown itself.

See, for example, Bib–A8, Bib–A9, Bib–A10, Bib–B4, Bib–G1, Bib–G4.

necessary to prepare two or more derivatives to be certain of your identification. Because a considerable amount of time can be wasted in preparing the wrong derivative — one that doesn't distinguish sufficiently between the possibilities — it is important to choose your derivatives very carefully. For example, suppose your compound is a ketone boiling at 168°C and you prepare the list of possibilities shown in Table 5.

The oxime would not be a good derivative, because melting points have not been reported for two of the possibilities. Also, two of the oximes melt below 50°C, which is undesirable since low-melting solids are usually difficult to recrystallize. Semicarbazones are reported for all the unknowns, but those for 2,6-dimethyl-4-heptanone and 2-octanone melt within a degree of each other. The 2-octanone might be distinguished by an iodoform test, but that would still leave 2-methyl- and 4-methylcyclohexanone, whose semicarbazones melt within 4°C of each other — a little too close to distinguish them with certainty. (Your derivative might melt at 197°C, for example.) The 2,4-dinitrophenylhydrazones, on the other hand, are well spread out: only the 2-methyl- and 4-methylcyclohexanone derivatives melt anywhere close to each other. This would be the best derivative in this case.

It should be pointed out that (since even professional chemists occasionally make mistakes) there are some conflicts in the literature with regard to physical constants, including the melting points of derivatives. If you suspect that a literature value may be incorrect, it is wise to check it against the values reported for the same derivative in some of the other sources listed in the Bibliography.

Procedures for preparing derivatives are listed by chemical family beginning on page 566. Melting points of the derivatives of selected organic compounds are listed in Appendix VI.

Procedures

PRELIMINARY WORK

Observe and describe the physical state and color of the unknown.

Carry out an ignition test by placing a drop of a liquid or about 25 mg of a solid in a small evaporating dish under the hood and igniting it with a burning wood splint **[be sure there are no flammable liquids nearby!]**. If it burns, observe the color of the flame and whether it is clean or sooty.

Obtain the melting point of a solid unknown ⟨OP–28⟩ or the boiling point of a liquid ⟨OP–29a⟩, and purify it, if necessary, by recrystallization ⟨OP–23a⟩, distillation ⟨OP–25a⟩, or another suitable method. If the compound boils at a temperature higher than 200°C, it may be advisable to use vacuum distillation ⟨OP–26a⟩ for purification.

A thermometer correction should be applied for temperatures over 200°C ⟨OP–28⟩.

SOLUBILITY TESTS

If the unknown is a liquid, measure the number of drops in 1.0 mL, using a medicine dropper and a small graduated cylinder. Divide this number by 5 to get the number of drops in 0.2 mL. Measure 0.2 mL of the liquid (by drops) into a 10-cm test tube, and add 3.0 mL of water; then shake vigorously for a minute or so, or until the liquid appears to dissolve. If the liquid is soluble, the mixture should be homogeneous with no separate layer or suspended droplets of the unknown evident. If it is insoluble or only partly soluble, the mixture should look cloudy or show suspended droplets when shaken.

If the unknown is a solid, accurately weigh out 0.10 g of the solid, and estimate all subsequent 0.1-g portions with reference to this quantity. Grind the solid to a fine powder, and mix it with 3.0 mL of water in a small test tube, shaking or stirring it for a few minutes, or until it dissolves. Grinding the solid

If the quantity of unknown is limited, the amounts of solvent and solute can be scaled down appropriately.

Always stopper the test tube before shaking it (don't use your thumb for a stopper!).

Do not mistake air bubbles (which may form on shaking) for suspended liquid droplets.

A small mortar and pestle or a flat-bottomed stirring rod and watch glass can be used for grinding. Be certain they are clean.

against the sides of the tube with a stirring rod may accelerate the process.

    If the unknown is water-soluble, test its aqueous solution with red and blue litmus paper. If there is no reaction to litmus and you suspect that the unknown is an aromatic amine or a phenol, dissolve a little of it in 5% HCl (for an amine) or 5% NaOH (for a phenol) and see if its odor disappears.

Salts of amines and phenols are generally odorless.

    If the unknown is insoluble in water, test its solubility in 3.0 mL of 5% HCl by the same procedure as described above. If it does not dissolve completely in the HCl, separate it from the solvent by filtering it, removing it with a capillary pipet, or decanting the solvent; then carefully neutralize the solvent to red litmus with dilute NaOH solution. The formation of a precipitate, a separate liquid phase, or even a cloudy solution upon neutralization classifies the unknown as "soluble" in HCl.

    If the unknown is insoluble in 5% HCl, test it with 5% NaOH and with 5% $NaHCO_3$ by the same procedure as before. This time neutralize the separated solvent to blue litmus with dilute HCl if the unknown does not dissolve completely. If it is insoluble in all of the foregoing, *carefully* test its solubility in 3.0 mL of cold (room temperature or below) concentrated sulfuric acid **[contact hazard!]**. Shake this mixture vigorously, and look for any evidence of reaction, such as the generation of heat, a distinct change in color, the formation of a precipitate, or the evolution of a gas. Solubility or a definite reaction is considered a positive test. The nature of any reaction that takes place may provide some clues to the identity of the unknown.

When testing with NaOH, notice whether the mixture foams on shaking. This could indicate a long-chain carboxylic acid.

A slight color change may be caused by impurities.

    Based on the solubility behavior, decide which solubility class your unknown belongs to (see Figure 1 on page 524) and what families it may belong to.

SPECTRAL ANALYSIS (OPTIONAL)

    Obtain an infrared spectrum of the unknown ⟨OP–33⟩ using the neat liquid or (if the sample is a solid) a Nujol mull or KBr disc. If the spectrum must be obtained when the sample is in solution, some of the important bands may be shifted by the solvent, so it may be necessary to consult the literature for interpretation. Decide what functional group (or groups) may be present, and choose one or more classification tests to confirm it. A PMR spectrum ⟨OP–34⟩ may also be obtained and examined, if permitted.

CLASSIFICATION TESTS

Procedures for all classification tests, both functional class tests and those providing additional structural information, are given below. Always read any Safety Precautions before starting a test. When each test is performed for the first time, it is advisable to run a *control* and (sometimes) a *blank* at the same time. Only by doing so will you know exactly what to look for in deciding whether the test with the unknown is positive or negative. A control is a known compound that is expected to give a predictable result with the test reagent. A blank is run by combining all reagents as in the actual test but omitting the unknown.

The volumes of most liquid reactants are given by drops in the procedures; it may be convenient to dispense reagents for classification tests from dropper bottles. Droppers should be of the same kind so that they will deliver roughly the same volume per drop. If you use a graduated pipet rather than a dropper, assume that 10 drops is about 0.3 mL of an organic liquid or 0.5 mL of an aqueous solution.

Unless otherwise indicated, solutions for the classification tests are aqueous.

Unless otherwise indicated, all compounds suggested as controls in the following procedures should give a positive test.

Always use different droppers for the reagent and the unknown. Clean the droppers thoroughly after using them. Do not insert your medicine droppers into reagent bottles—you may contaminate the reagents.

C–1. *Acetyl Chloride.* **(a)** *Under the hood,* carefully add about 10 drops of acetyl chloride **[hazard!]** to 10 drops (or 0.4 g) of the unknown in a test tube. Observe any evolution of heat; carefully exhale over the mouth of the test tube to see if a cloud of HCl gas is revealed by the moisture in your breath. After a minute or two, pour the mixture into about 2 mL of water, shake it, and note any phase separation. Carefully smell the mixture for evidence of an ester aroma, which is usually pleasant and fruity.

Reaction

$$CH_3COCl + ROH$$

$$\longrightarrow CH_3COOR + HCl$$

Control: 1-butanol

Acetyl chloride is corrosive and toxic. Use gloves and a hood; do not breathe the vapors.

***N,N*-Dimethylaniline and ammonia are toxic and have harmful vapors; avoid contact, and do not breathe the vapors.**

SAFETY PRECAUTIONS

Positive test: Evidence of reaction (heat, HCl), especially if accompanied by phase separation or an ester-like odor. Amines and phenols also react, but the former will not yield pleasant odors. If the test is inconclusive or if you suspect a tertiary alcohol, carry out variation **(b)**.

*Tertiary alcohols will not form esters by procedure **(a)** but should in the presence of the base used in **(b)**.*

(b) *Under the hood,* mix 5 drops of acetyl chloride [**hazard!**] with 10 drops of *N,N*-dimethylaniline [**hazard!**], and carefully add 5 drops (or 0.2 g) of the unknown. Allow the solution to stand for a few minutes, and look for evidence of a reaction (no HCl will be evolved since it reacts with the base). If heat is not evolved, warm the mixture on a 50°C water bath for 15 minutes. Cool the mixture, add 1 g of ice and 1 mL of concentrated ammonia [**hazard!**], mix, and let stand. If a layer separates, remove it with a dropper or pipet and test it for ester using the ferric hydroxamate test ⟨C–14⟩.

C–2. *Alkaline Hydrolysis.* **(a)** *Amides.* Place 0.1 g of the unknown (3 drops, if liquid) in a test tube containing 4 mL of 6 M sodium hydroxide [**contact hazard!**]. Secure a small piece of filter paper over the top of the tube, and moisten it with 2 drops of 10% cupric sulfate. Boil the mixture for a minute or two, and note any color change on the filter paper. Remove the paper and note the odor of the vapors while the solution is boiling. Acidify the solution with 6 M HCl; if a carboxylic acid precipitates, save it for use as a derivative.

Positive test: Paper turning blue, ammonia or amine-like odor. Amides of higher amines that do not turn the paper blue may nevertheless give an amine-like odor. Some amides will yield a precipitate or a separate liquid phase (the carboxylic acid) when the hydrolysis mixture is acidified. The characteristic odor of a carboxylic acid may also be observed.

If the test is inconclusive, try one of the following:

1. Increase the reaction time.

2. Repeat the hydrolysis at 200°C using 20% KOH in glycerine.

3. Distill off the amine and characterize it using one of the amine classification tests.

(b) *Esters.* Mix 1 mL of the unknown (1 g, if solid) with 10 mL of 6 M sodium hydroxide [**hazard!**], and reflux the mixture for half an hour, or until the solution is homogeneous. Note whether the organic layer has disappeared (if the unknown was water-insoluble) and whether the odor of the unknown is gone. If a separate organic layer or residue remains, reflux the mixture longer until it disappears or until it is apparent that no reaction is taking place. Cool the mixture, remove any organic layer if necessary, and acidify the aqueous solution with 6 M sulfuric acid. If a carboxylic acid precipitates on acidification, save it for use as a derivative.

Reactions

$RCONR'_2 + NaOH \longrightarrow$
$\qquad RCOONa + R'_2NH$

$RCOONa + H^+ \longrightarrow$
$\qquad RCOOH + Na^+$

R' = H, alkyl, or aryl

Controls: benzamide, acetanilide

Distill a volatile amine into a receiver containing dilute HCl, then neutralize the distillate with base before testing.

Reactions

$RCOOR' + NaOH \longrightarrow$
$\qquad RCOONa + R'OH$

$RCOONa + H^+ \longrightarrow$
$\qquad RCOOH + Na^+$

Most esters boiling under 110°C will hydrolyze in half an hour; higher-boiling esters may take up to 2 hours, or longer.

Positive test: Disappearance of the organic layer (for a water-insoluble ester) and of the odor (usually pleasant) of the unknown; the formation of a precipitate (if the carboxylic acid is a solid) or the appearance of the odor of a carboxylic acid upon acidification. If evidence for the formation of an acid is inconclusive, make the solution basic again with sodium hydroxide and saturate it with potassium carbonate to see if an organic layer (the alcohol) separates. Note the odor of this layer, which should be different from that of the original ester. Esters with boiling points higher than 200°C are usually unreactive in aqueous NaOH and may be hydrolyzed by KOH in diethylene glycol; see the work by Shriner et al. ⟨Bib – G4⟩ for a procedure.

Controls: ethyl benzoate, butyl acetate

C–3. *Aluminum Chloride and Chloroform.* Under the hood, make up a solution containing 2 – 3 drops (or 0.1 g) of a solubility class X unknown in 2 mL of dry chloroform **[hazard!].** Place about 0.2 g of anhydrous aluminum chloride **[hazard!]** in a dry test tube and heat it over a flame, angling the test tube so the $AlCl_3$ sublimes onto the inner wall of the tube a few centimeters above the bottom. Allow the tube to cool until it can be held comfortably in the hand; then let a few drops of the solution flow down the side of the tube so that it contacts the aluminum chloride. Note any color change at the point of contact.

Reaction

$$ArH \xrightarrow{\text{CHCl}_3,\ \text{AlCl}_3} Ar_3C^+AlCl_4^-$$

(and other carbenium ion species)

Controls: toluene, biphenyl

Positive test: Bright yellow-orange, red, blue, green, or purple color, depending on the type of aromatic compound. A light yellow color is inconclusive or negative.

Chloroform has been found to cause cancer in laboratory animals when ingested, and its vapors are harmful.

Aluminum chloride can cause serious damage if inhaled or allowed to contact the skin or eyes, and it reacts violently with water and other chemicals. Use gloves and a hood; avoid contact with the reagents, and do not breathe the vapors or dust.

SAFETY PRECAUTIONS

C– 4. *Basicity Test.* If the unknown is water-soluble, dissolve 4 drops (0.10 g of a solid) in 3 mL of water and measure its pH using pH paper or a universal indicator. If the unknown is insoluble in water, shake about 4 drops (or 0.10 g) with 3 mL of an acetate – acetic acid buffer (pH 5.5).

Reaction (in buffer)

$$RNH_2 + H^+ \xrightarrow{\text{pH 5.5}} RNH_3^+$$

(and similar reactions for 2° and 3° amines)

Controls: aniline and *n*-butylamine; *p*-toluidine and dibutylamine

Positive test: Most water-soluble aliphatic amines will give pH values above 11, and water-insoluble aliphatic amines should dissolve in the buffer. Most water-soluble aromatic amines give pH values below 10, and water-insoluble aromatic amines do not dissolve in the buffer. Test C–8 can also be used to test for aromatic amines.

Control: chlorobenzene

C–5. *Beilstein's Test.* Make a small loop in the end of a length of copper wire (10 cm or longer), and heat the loop to redness in a flame. Place a small amount of unknown on the loop, and heat it in the nonluminous (blue) flame of a burner, near the lower edge.

Positive test: A distinct green or blue-green flame, due to the presence of copper halide.

Benedict's reagent contains cupric sulfate (toxic!), sodium citrate, and sodium carbonate.

C–6. *Benedict's Test.* Add 5 drops or 0.2 g of the unknown to 5 mL of water and mix in 5 mL of Benedict's reagent. Heat the mixture to boiling. Observe whether a precipitate forms, and if one forms, what its color is.

Control: butyraldehyde

Positive test: Aliphatic aldehydes generally produce a yellow to orange suspension or precipitate; it may appear greenish in the blue solution. Some other compounds, such as α-hydroxyketones and reducing sugars, also react. Most ketones and aromatic aldehydes do not react.

Reaction

$$-\underset{|}{\overset{|}{C}}=\underset{|}{\overset{|}{C}}- + Br_2 \longrightarrow -\underset{|}{\overset{Br}{\overset{|}{C}}}-\underset{|}{\overset{Br}{\overset{|}{C}}}-$$

C–7. *Bromine in Freon TF.* *Under the hood,* dissolve 2 drops or 50 mg of the unknown in 1 mL of Freon TF (1,1,2-trichlorotrifluoroethane). Add drop by drop, with shaking, a dilute solution (~0.2 M – 1.0 M) of bromine in Freon TF **[hazard!].** Continue adding until the red color persists or until about 20 drops have been added. Immediately after the addition, exhale over the mouth of the test tube and observe whether a cloud of HBr gas is apparent.

Controls: cyclohexene, ethyl acetoacetate (reacts with HBr formation)

SAFETY PRECAUTION

Bromine is corrosive and toxic. Avoid contact, and do not breathe its vapors.

Positive test: Decolorization of more than 1 drop of the bromine solution, *without* the evolution of HBr, is characteristic of olefinic unsaturation. Aldehydes, ketones, amines, phenols, and many other compounds react by substitution to evolve HBr.

C–8. *Bromine Water.* Dissolve 3 drops or 0.1 g of the unknown in 10 mL of water, and check the pH of the solution with pH paper. *Under the hood,* add saturated bromine water **[hazard!]** drop by drop until the bromine color persists; watch for evidence of a precipitate.

If the unknown is insoluble in the water, try adding just enough ethanol to bring it into solution.

Controls: phenol, aniline

Reaction for phenol (aromatic amines and substituted phenols undergo similar reactions)

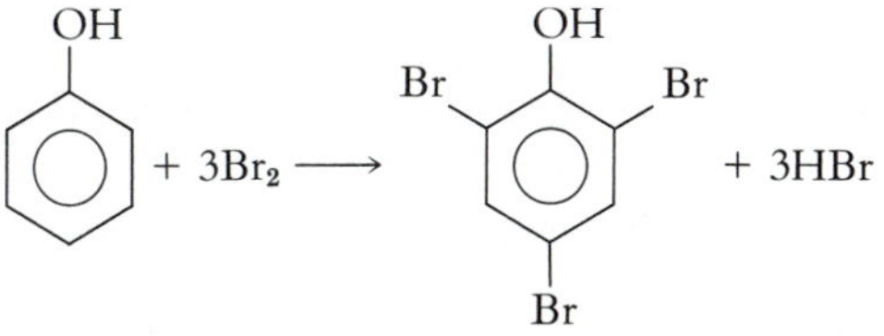

Positive test: Decolorization of the bromine, accompanied by simultaneous formation of a white (or nearly white) precipitate. The pH of the initial solution should be less than 7 if the unknown is a phenol. Aromatic amines also react.

C–9. *Chromic Acid.* Dissolve 1 drop of a liquid or ~30 mg of a solid unknown in 1 mL of reagent-grade acetone. (If there is any doubt about the purity of the acetone, test it with a drop of the reagent beforehand.) Add 1 drop of the chromic acid reagent **[hazard!]** and swirl, noting the time required for a color change.

Controls: 1-butanol, butyraldehyde, benzaldehyde

The chromic acid reagent contains chromium (VI) oxide and concentrated sulfuric acid.

The chromic acid reagent is corrosive and may be carcinogenic; avoid contact.

SAFETY PRECAUTION

Reactions

$$3RCH_2OH + 4CrO_3 + 6H_2SO_4 \longrightarrow 3RCOOH + 2Cr_2(SO_4)_3 + 9H_2O$$

$$3R_2CHOH + 2CrO_3 + 3H_2SO_4 \longrightarrow 3R_2CO + Cr_2(SO_4)_3 + 6H_2O$$

$$3RCHO + 2CrO_3 + 3H_2SO_4 \longrightarrow 3RCOOH + Cr_2(SO_4)_3 + 3H_2O$$

Positive test: Formation of an opaque blue-green suspension or emulsion within 2 seconds for a primary or secondary alcohol. With aliphatic aldehydes, the solution generally turns cloudy in 5 seconds and a green precipitate forms within 30 seconds, whereas aromatic aldehydes require 30–120 seconds or longer. The generation of some other dark color, particularly with the color of the liquid remaining orange, should be considered a negative test.

Controls: toluene, chlorobenzene, 1-chlorobutane, 1-bromobutane

C–10. *Density Test.* Add a few drops of a solubility class X unknown to 1 mL of deionized water, stir gently, and note whether the unknown floats or sinks.

Approximate density ranges for halogenated hydrocarbons:

alkyl chloride (mono)	0.85–1.0
alkyl bromide (mono)	1.1–1.5
alkyl iodide (mono)	over 1.4
alkyl chloride (poly)	1.1–1.7
alkyl bromide (poly)	1.5–3.0
aryl chloride	1.1–1.3
aryl bromide	1.3–2.0
aryl iodide	over 1.8

Interpretation: Of the compounds that are insoluble in cold, concentrated sulfuric acid (see Solubility Tests, page 549), most aromatic hydrocarbons and monochloroalkanes will float, whereas aryl chlorides and all bromides and iodides, along with polychloroalkanes, will sink. For further differentiation, measure the density of the unknown by carefully pipetting exactly 1 mL of the liquid (use a volumetric pipet) into a vial and weighing the liquid on an accurate balance to at least two decimal places. Density ranges for most common organic halides are listed in the margin.

Controls: cyclohexanone, benzaldehyde

C–11. *2,4-Dinitrophenylhydrazine.* Dissolve 1 drop or about 40 mg of the unknown in 1 mL (more, if necessary) of 95% ethanol. Add the solution to 2 mL of 2,4-dinitrophenylhydrazine–sulfuric acid reagent **[hazard!].** Shake and let the mixture stand for 15 minutes, or until a precipitate forms. If no precipitate has formed at the end of 15 minutes, scratch the inside of the test tube.

SAFETY PRECAUTION

2,4-dinitrophenylhydrazine is toxic, and sulfuric acid is corrosive; avoid contact with the reagent.

Reaction

$$R-\underset{\underset{R'}{|}}{C}=O + H_2N-NH-\underset{\underset{O_2N}{}}{\langle\bigcirc\rangle}-NO_2 \longrightarrow R-\underset{\underset{R'}{|}}{C}=N-NH-\underset{\underset{O_2N}{}}{\langle\bigcirc\rangle}-NO_2 + H_2O$$

R, R′ = alkyl, aryl, or H

Positive test: Formation of a crystalline yellow or orange-red precipitate. Some carbonyl compounds initially form oils that may or may not become crystalline; a few may require *gentle* heating, but overheating can cause oxidation of allylic or other reactive alcohols, resulting in a false positive test. Some aromatic compounds (hydrocarbons, halides, phenols, and phenyl esters) may form slightly soluble complexes with the reagent; some alcohols may be contaminated with small amounts of the corresponding aldehyde or ketone. In these cases, the amount of precipitate should be quite small, and comparison with a control will show the difference between a positive test and a doubtful one. The color of the precipitate may give a clue to the structure of the carbonyl compound, since unconjugated aliphatic aldehydes and ketones usually yield yellow 2,4-dinitrophenylhydrazones, whereas aromatic and α,β-unsaturated aldehydes and ketones yield orange-red precipitates.

An orange-red color can also be caused by impurities, including 2,4-dinitrophenylhy-drazine itself.

C–12. *Elemental Analysis.* *Under the hood,* place about 0.5 g of "Dri-Na" sodium-lead alloy **[hazard!]** in a small *dry* test tube held vertically by an asbestos-lined clamp. Melt the alloy with a burner flame, and continue heating until the sodium vapor rises about 1 cm up the tube. Add 2 drops of the unknown (about 20 mg of a solid) directly onto the molten alloy so that it does not touch the sides of the tube. Heat gently to start the reaction, remove the flame until the reaction subsides, then heat the tube strongly for a minute or two, keeping the bottom a dull red color. Let the tube cool to room temperature; then carefully add 3 mL of water, and heat gently for a minute or so until the excess sodium has been decomposed and gas evolution ceases. If necessary, filter the solution, wash the filter paper with 2 mL of water, and combine the wash water with the filtrate. The filtrate should be colorless or just slightly yellow; if it is darker, repeat the fusion with stronger heating or more of the alloy.

The sodium-lead alloy is much less hazardous than elemental sodium; it reacts with water to give a comparatively gentle reaction with no fire hazard.

If the unknown is quite volatile, it may be advisable to place it in the test tube before adding the alloy and to heat them together gently until charring occurs. Then heat the tube more strongly and proceed as directed.

Controls: acetamide (N), bromobenzene (Br)

Reactions

$$RX \xrightarrow{\text{Na(Pb)}} NaX$$

$$\text{Nitrogen Compound} \xrightarrow{\text{Na(Pb)}} NaCN$$

$$AgNO_3 + NaX \longrightarrow \mathbf{AgX} + NaNO_3$$

$$Ag\mathbf{Cl} + 2NH_3 \longrightarrow Ag(NH_3)_2Cl$$

$$2\mathbf{Br}^- + Cl_2 \xrightarrow{\text{CCl}_4} 2Cl^- + \underset{\text{(red-brown)}}{Br_2}$$

$$2I^- + Cl_2 \xrightarrow{\text{CCl}_4} 2Cl^- + \underset{\text{(purple)}}{I_2}$$

$$(X = Cl, Br, or I)$$

SAFETY PRECAUTIONS

Sodium can cause serious burns, and the sodium-lead alloy may react violently with some substances; avoid contact, and keep it away from other chemicals. Do *not* use this procedure with sodium metal; it will react violently when water is added.

The PNB reagent (*p*-nitrobenzaldehyde in dimethyl sulfoxide) is harmful if it is inhaled or allowed to contact the skin; wear gloves, avoid contact, and do not breathe its vapors.

To test for *nitrogen,* put 10 drops of the fusion solution into a small test tube, and add enough solid sodium bicarbonate, with stirring, to saturate it (excess solid should be present). Add 1–2 drops of this solution to a test tube containing 20 drops of PNB reagent. A purple color is a positive test for nitrogen (green indicates sulfur).

The boiling removes (as HCN) the cyanide ion that is formed if nitrogen is present and that would react to yield AgCN in this test.

To test for the *halogens,* acidify 10 drops of the fusion solution with dilute nitric acid, boil gently under the hood for a few minutes, and add a drop or two of 0.3 M aqueous silver nitrate. The formation of a heavy white precipitate of silver chloride indicates the presence of chlorine, whereas bromine yields a pale yellow and iodine a yellow precipitate, respectively. If only a faint turbidity is produced, it may be caused by traces of impurities or by incomplete sodium fusion. To confirm the presence of chlorine, remove the solvent by filtration (or by centrifugation and decanting), add 2 mL of 3 M aqueous ammonia, and shake. If the precipitate is silver chloride, it will dissolve. Silver bromide will be only slightly soluble, and silver iodide will be insoluble. To test further for bromine and iodine, acidify 3 mL of the original stock solution with dilute sulfuric acid, boil for a few minutes, and add 1 mL of Freon TF and a drop or two of freshly prepared chlorine water. Shake and look for a color in the lower layer (purple for iodine, reddish-brown for bromine).

Controls: phenol, salicylic acid

C–13. *Ferric Chloride.* Dissolve 1 drop or about 40 mg of the unknown in 2 mL of water (or in a water-alcohol mixture if it does not dissolve in water) and add 1–3 drops of 2.5% ferric chloride solution.

Positive test: Formation of an intense red, green, blue, or purple color suggests a phenol or an easily enolizable compound. Some phenols do not react under these conditions.

Many aromatic carboxylic acids form tan precipitates; aliphatic hydroxyacids yield yellow solutions.

C–14. *Ferric Hydroxamate Test.*   Before carrying out the ferric hydroxamate test, dissolve 1 drop or about 40 mg of the unknown in 1 mL of 95% ethanol. Add 1 mL of 1 M hydrochloric acid; then add 2 drops of 2.5% ferric chloride. If a definite color other than yellow results, the ferric hydroxamate test cannot be used. Save the solution for comparison with the following test.

The preliminary test eliminates those phenols and enols that give colors with ferric chloride in acidic solution and that would therefore give a false positive result in the ferric hydroxamate test.

Hydroxylamine hydrochloride is toxic and can cause a form of anemia; avoid contact.

SAFETY PRECAUTION

Mix 1 mL of 0.5 M ethanolic hydroxylamine hydrochloride **[hazard!]** with 5 drops of 6 M sodium hydroxide **[hazard!],** add 1 drop (or about 40 mg) of the unknown, and heat the solution to boiling. Allow it to cool slightly and add 2 mL of 1 M hydrochloric acid. If the solution is cloudy at this point, add enough 95% ethanol to clarify it. Add 2 drops of 2.5% ferric chloride solution, and observe any color produced. If the color does not persist, continue to add the ferric chloride solution until the color becomes permanent.

Control: butyl acetate

Reactions

$$\underset{}{\overset{O}{\underset{\|}{R\,C}}}{-}OR' + H_2NOH \longrightarrow \underset{}{\overset{O}{\underset{\|}{R\,C}}}{-}NHOH + R'OH$$

$$3RCONHOH + FeCl_3 \longrightarrow \underset{\text{ferric hydroxamate}}{(RCONHO)_3Fe} + 3HCl$$

Positive test: A burgundy or magenta color that is distinctly different from the color obtained in the preliminary test.

C–15. *Hinsberg's Test.*   *Under the hood,* mix 3 drops of the unknown (0.1 g if a solid) with 5 mL of 3 M sodium hydroxide in a test tube and add 6 drops of benzenesulfonyl chloride **[hazard!]** or 0.2 g of *p*-toluenesulfonyl chloride **[hazard!].** Stopper the tube and shake it intermittently for 3–5 minutes; then remove the stopper and warm the solution over a steam bath for a minute. The solution should be basic at this point (if

Controls: aniline, *N*-methylaniline, *N,N*-dimethylaniline

Some N,N-dialkylanilines form a purple dye if the reaction mixture gets too hot. If this occurs, repeat the reaction in a 15–20°C water bath.

not, add more NaOH). If there is a liquid or solid residue in the tube, separate it from the solution (by filtration, use of a pipet, or some other means), and test its solubility in 5 mL of water and in dilute hydrochloric acid. Acidify the original solution with 6 M hydrochloric acid and, if no precipitate forms immediately, scratch the sides of the test tube and cool.

SAFETY PRECAUTION   **Arenesulfonyl chlorides are toxic and corrosive; avoid contact, do not breathe the vapors, and *use them under a hood.***

Reactions

1° $RNH_2 + ArSO_2Cl + 2NaOH \rightarrow ArSO_2NR^-Na^+ + NaCl + 2H_2O$
                              (soluble)

   $ArSO_2NR^-Na^+ + HCl \rightarrow ArSO_2NHR + NaCl$
                              (insoluble)

2° $R_2NH + ArSO_2Cl + NaOH \rightarrow ArSO_2NR_2 + NaCl$
                              (insoluble)

3° $R_3N + ArSO_2Cl \rightarrow$ no reaction

   $R_3N + HCl \rightarrow R_3NH^+Cl^-$
                              (soluble)

Interpretation: Primary amines usually yield no appreciable amount of liquid or solid residue after the initial reaction; a white solid (an arenesulfonamide) should precipitate when the solution is acidified. Most *secondary amines* yield a white solid that does not dissolve in water or dilute HCl. *Tertiary amines* should not react; the residue will be the original liquid or solid amine, which should dissolve in dilute HCl. If the test yields an oil, it should be possible to tell whether it is the original (liquid) amine or an arenesulfonamide that has failed to crystallize — the latter are more dense than water.

Some high-molecular-weight and cyclic primary alkylamines form sodium salts that are insoluble in the alkaline solution. These salts should dissolve in the 5 mL of water used to test the solubility of the precipitate; if not, the test will falsely suggest a secondary amine. Other primary amines may yield disulfonyl derivatives that are also insoluble in the original solution, so it is wise to reserve judgment on any test suggesting a secondary amine unless it can be confirmed independently. Water-soluble tertiary amines should yield a clear solution that does not form a separate phase on acidification.

If a pure amine yields a solid residue after the original reaction and forms a precipitate upon acidification of the filtrate, it is probably a primary amine that has produced some of the disulfonyl derivative.

C–16. *Iodoform Test.* Dissolve 3 drops (or 0.1 g) of the unknown in 1 mL of water, and add 1 mL of 3 M sodium hydroxide solution. Add 0.5 M iodine–potassium iodide reagent drop by drop until the iodine color persists after shaking. Place the test tube in a 60°C water bath, and add more of the iodine–potassium iodide solution until the brown color persists for 2 minutes; then add 3 M NaOH drop by drop until the color just disappears (a light yellow color may remain). Remove the test tube from the water bath, and add 10 mL of cold water. If a precipitate does not form immediately, let the test tube stand for 15 minutes. If there is any doubt about the identity of the precipitate, obtain its melting point, which should be about 119–121°C.

Dissolve the unknown in 3 mL or more of pure methanol if it is not soluble in water.

Controls: isopropyl alcohol, 2-butanone

Reactions

1. Methyl carbinols

$$\underset{\overset{|}{\text{OH}}}{\text{RCH}}-\text{CH}_3 + 4\text{I}_2 + 5\text{NaOH} \longrightarrow \underset{\overset{\|}{\text{O}}}{\text{RC}}-\text{CI}_3 + 5\text{NaI} + 5\text{H}_2\text{O}$$

$$\underset{\overset{\|}{\text{O}}}{\text{RC}}-\text{CI}_3 + \text{NaOH} \longrightarrow \underset{\overset{\|}{\text{O}}}{\text{RC}}-\text{ONa} + \underset{\text{(iodoform)}}{\text{CHI}_3}$$

2. Methyl ketones and acetaldehyde

$$\underset{\overset{\|}{\text{O}}}{\text{RC}}-\text{CH}_3 + 3\text{I}_2 + 3\text{NaOH} \longrightarrow \underset{\overset{\|}{\text{O}}}{\text{RC}}-\text{CI}_3 + 3\text{NaI} + 3\text{H}_2\text{O}$$

$$\underset{\overset{\|}{\text{O}}}{\text{RC}}-\text{CI}_3 + \text{NaOH} \longrightarrow \underset{\overset{\|}{\text{O}}}{\text{RC}}-\text{ONa} + \text{CHI}_3$$

Positive test: Formation of a yellow precipitate with the characteristic odor of iodoform. Methyl carbinols must have at least one hydrogen atom on the carbinol carbon in order to react, since they are oxidized to methyl ketones initially. Other compounds that can also yield iodoform in this test include some conjugated aldehydes, such as acrolein and furfural, and certain 1,3-dicarbonyl or dihydroxy compounds.

C–17. *Lucas's Test.* *Under the hood,* place 2 mL of the Lucas reagent **[hazard!]** in a small test tube. Add 4 drops of the unknown, stopper the tube immediately, and shake vigorously; then allow the mixture to stand for 15 minutes or more.

Reaction

$$\text{ROH} + \text{HCl} \xrightarrow{\text{ZnCl}_2} \text{RCl} + \text{H}_2\text{O}$$

SAFETY
PRECAUTION

Lucas's reagent ($ZnCl_2$ in concentrated HCl) can cause serious burns; avoid contact.

Controls: n-butyl, sec-butyl, and t-butyl alcohols

If there is any question whether the alcohol is secondary or tertiary, repeat the test using concentrated HCl instead of Lucas's reagent. A tertiary alcohol should react within minutes, whereas a secondary alcohol should not react at all.

Polyhydroxy alcohols boiling higher than 140°C are often soluble in the reagent and can be tested.

Interpretation: Tertiary alcohols that are soluble in the reagent should turn the reagent cloudy almost immediately and soon form a separate layer of alkyl chloride. *Secondary alcohols* usually turn the clear solution cloudy in 3–5 minutes and form a distinct layer within 15 minutes. *Primary alcohols* do not react under these conditions. Most allylic and benzylic alcohols give the same result as tertiary alcohols, except that the chloride formed from allyl alcohol is itself soluble in the reagent and separates out only upon addition of ice water. The test is not valid for alcohols that are not soluble in the reagent. This includes most alcohols with six or more carbons and boiling points higher than 140–150°C. You should interpret with caution any apparent positive result (such as immediate phase separation) that has been obtained using a water-insoluble alcohol having a boiling point above 140°C.

Reaction

RCOOH + NaOH $\longrightarrow$
                    RCOONa + H_2O

Control: adipic acid

Use bromothymol blue as the indicator if the solvent is ethanol.

N.E. =

$$\frac{\text{mass of sample in mg}}{\text{mL of NaOH} \times \text{molarity NaOH}}$$

C-18. *Neutralization Equivalent.* Accurately weigh (to 3 decimal places) about 0.2 g of the unknown carboxylic acid and dissolve it in 50–100 mL of water, ethanol, or a mixture of the two, depending on its solubility. Titrate the solution with a standardized solution of 0.1 M sodium hydroxide using phenolphthalein as the indicator. Calculate the neutralization equivalent (N.E.) of the acid using the formula shown.

Interpretation: The neutralization equivalent (equivalent weight) of a carboxylic acid is equal to its molecular weight divided by the number of carboxyl groups. For instance, the N.E. of adipic acid [$HOOC(CH_2)_4COOH$; M.W. = 146] is 73. An acid that has an unusually low neutralization equivalent for its boiling or melting point probably contains more than one carboxyl group.

If alcohol is the solvent, it is best to run a blank using the pure solvent and compare the result with that for the unknown.

The reaction rate depends on the solubility of the unknown in the solvent. A solid that is only sparingly soluble should be finely powdered, and the reaction mixture should be shaken vigorously.

C-19. *Potassium Permanganate.* Dissolve 1 drop (or 30 mg) of the unknown in 2 mL of water or 95% ethanol and add 0.1 M potassium permanganate drop by drop until the purple color of the permanganate persists, or until about 20 drops have been added. If a reaction does not take place immediately, shake the mixture and let it stand for up to 5 minutes. Disregard any decolorization that takes place after 5 minutes have elapsed.

Positive test: Decolorization of more than 1 drop of the purple permanganate solution, accompanied by the formation of a brown precipitate (or reddish-brown suspension) of manganese dioxide. The test is positive for most compounds containing double and triple bonds, including some alkenes that do not react with bromine in Freon TF. Easily oxidizable compounds such as aldehydes, aromatic amines, phenols, formic acid, and formate esters also give positive tests. Alcohols that contain oxidizable impurities may react with a small amount of permanganate, so decolorization of only the first drop should not be considered a positive test; most pure alcohols will not react in less than 5 minutes. Aromatic hydrocarbons and conjugated alkadienes do not, as a rule, give positive tests with cold, neutral potassium permanganate.

C–20. *Quinhydrone.* This test should be run in conjunction with Test C–4 or with an infrared spectrum that shows whether the amine is aliphatic or aromatic. Shake 1 drop (or 30 mg) of the unknown amine with 6 mL of water in a 15-cm test tube. If the amine dissolves, add 6 mL more of water; otherwise, add 6 mL of ethanol. Shake the mixture, add 1 drop of 2.5% quinhydrone in methanol **[hazard!],** and let it stand for 2 minutes. If no distinct color develops, add 5 more drops of the reagent, shake the mixture again, and allow it to stand for another 2 minutes.

The components of quinhydrone (hydroquinone and benzoquinone) are toxic and irritants. Avoid contact; wash hands after using.

Interpretation: Most alkyl amines produce strong colors with 1 drop of the reagent; aryl amines require 6 drops. Controls should be run for comparison, since the colors produced (especially by aromatic amines) may be difficult to characterize accurately. The test is not applicable to nitro-substituted aromatic amines and phenylenediamines.

C–21. *Silver Nitrate in Ethanol.* Add 1 drop or 50 mg of the unknown (dissolved in a little ethanol, if it is a solid) to 2 mL of 0.1 M ethanolic silver nitrate **[hazard!].** Shake the mixture and let it stand. If no precipitate has formed after 5 minutes, heat the solution to boiling and boil it for 30 seconds. If a precipitate forms, note its color, and see if it dissolves when the mixture is shaken with 2 drops of 1 M nitric acid.

Reaction

$$3\!-\!\overset{|}{\underset{|}{C}}\!=\!\overset{|}{\underset{|}{C}}\!-\ +\ 2KMnO_4\ +\ 4H_2O\ \longrightarrow$$

$$3\!-\!\overset{HO}{\underset{|}{\underset{|}{C}}}\!-\!\overset{OH}{\underset{|}{\underset{|}{C}}}\!-\ +\ 2MnO_2\ +\ 2KOH$$

Control: cyclohexene

Controls: butylamine, dibutylamine, tributylamine, aniline, *N*-methylaniline, *N,N*-dimethylaniline

SAFETY PRECAUTION

Colors of quinhydrone test:
  1° aliphatic: violet
  2° aliphatic: rose
  3° aliphatic: yellow
  1° aromatic: rose
  2° aromatic: amber
  3° aromatic: yellow

Reaction

$$RX + AgNO_3 \longrightarrow RONO_2 + AgX$$

(Other organic products, such as ROH, ROEt, and alkenes, may also be formed.)

**SAFETY
PRECAUTION**

**Silver nitrate is caustic and toxic; avoid contact with
skin or eyes.**

Controls: n-butyl chloride, *t*-butyl
chloride, *n*-butyl bromide, bromobenzene, ethyl iodide

Halides that react at room temperature:
  Chlorides: 3°, allyl, benzyl.
  Bromides: 1°, 2°, 3° alkyl (except
  geminal di- and tribromides); allyl,
  benzyl, CBr₄.
  Iodides: All aliphatic and alicyclic
  except vinyl.

Halides that react upon heating:
  Chlorides: 1°, 2° alkyl.
  Bromides: alkyl *gem*-di- and
  tribromides.
  Some activated aryl halides, such as
  2,4-dinitrohalobenzenes.

Halides that do not react:
  Chlorides: alkyl *gem*-di- and
  trichlorides, CCl₄.
  Most aryl and vinyl halides.

Reactions

$$RCl + NaI \xrightarrow{\text{acetone}} RI + NaCl$$

$$RBr + NaI \xrightarrow{\text{acetone}} RI + NaBr$$

Controls: n-butyl chloride, *t*-butyl
chloride, *n*-butyl bromide, bromobenzene, 1,2-dichloroethane

Halides that react at room temperature
(X = Br, Cl only; starred compounds
liberate iodine, turning the solution
red-brown.)
  1° alkyl bromides; benzyl and allyl
  halides; α-halo ketones, esters, and
  amides; CBr₄*; *vic*-dihalides*;
  triaryl halides*.

Interpretation: The rate of reaction of a halide with silver
nitrate depends on its structure. Tertiary, allylic, and benzylic
halides react fastest, followed by secondary and primary halides. Most aryl and vinyl halides are unreactive, except for aryl
halides activated by two or more nitro groups. Alkyl halides
with more than one halogen atom on the same carbon (geminal halides) are generally less reactive than the corresponding monohaloalkanes. This is not true, however, if the halogen
atoms are on an allylic or benzylic carbon atom — such compounds react at room temperature. Alkyl iodides react faster
than the corresponding bromides, which are more reactive
than alkyl chlorides. The types of compounds that react at
room temperature, after boiling the solution, or not at all, are
listed.

The color of the silver salt formed may indicate the type
of halide responsible. Silver chloride is white, silver bromide is
pale yellow or cream-colored, and silver iodide is yellow.
These salts will not dissolve when the mixture is acidified.
Some carboxylic acids and alkynes yield silver salts, but these
should dissolve in the acidic solution.

C–22. Sodium Iodide in Acetone. Place 1 mL of the sodium
iodide – acetone reagent into a small test tube. Add 2 drops of
the halogen compound (or 0.1 g of a solid dissolved in the
minimum volume of acetone). Shake the mixture and allow it
to stand for 3 minutes, noting whether a precipitate or a red-
brown color forms. If there is no reaction at the end of this
time, place the test tube in a 50°C water bath (replenish the
acetone if necessary), and leave it there for an additional 6
minutes; then cool the mixture to room temperature.

Interpretation: Certain alkyl chlorides and bromides react
to precipitate sodium chloride or bromide; others may also
liberate elemental iodine, which is red-brown in acetone.
Since the reaction is, in most cases, a bimolecular displacement
by iodide ion, the order of reactivity is generally methyl >
primary > secondary > tertiary > aryl, vinyl. Many cycloal-
kyl halides (cyclopentyl is an exception) react more slowly
than the corresponding open-chain compounds and may give
no precipitate after heating. Bromides react faster than chlo-

rides, and the test is not applicable to iodides. Vicinal dihalides (with two halogen atoms on adjacent carbon atoms) and some geminal halides (with two or more halogen atoms on the same carbon atom), along with triarylmethyl halides, undergo redox reactions to liberate iodine while precipitating the sodium halide. A summary of the reactivities of various halides is given.

Halides that react at 50°C: most 1°, 2° alkyl chlorides; most 2°, 3° alkyl bromides; *gem*-di- and tribromides*.

Halides that do not react: 3° alkyl chlorides; aryl, vinyl halides; cyclopropyl, cyclobutyl, cyclohexyl halides; *gem*-polychloro compounds (except benzyl and allyl).

If a precipitate forms upon mixing but does not persist, the test should be considered negative. Some halides may be contaminated with isomeric impurities that could yield a little precipitate of sodium halide, so it is advisable to run a control and compare the amount of precipitate obtained in each case.

C-23. Tollens's Test. Prepare the reagent immediately before use as follows. Measure 2 mL of 0.3 M aqueous silver nitrate **[hazard!]** into a thoroughly cleaned test tube, and add 1 drop of 3 M sodium hydroxide. Then add 2 M aqueous ammonia drop by drop, with shaking, until the precipitate of silver oxide just dissolves (avoid an excess of ammonia). Add 1 drop or 40 mg of the unknown to this solution, shake the mixture, and let it stand for 10 minutes. If no reaction has occurred by the end of this time, heat the mixture in a 35°C water bath for 5 minutes. Immediately after the test is completed, wash the contents of the test tube down the drain with plenty of water and rinse the tube with dilute nitric acid *(important!)*.

Reaction

$$RCHO + 2Ag(NH_3)_2OH \longrightarrow$$

$$2Ag + RCOONH_4 + H_2O + 3NH_3$$

Control: benzaldehyde

The reagent and test solution must never be stored—explosive silver fulminate forms on standing. Avoid contact with silver nitrate solutions.

SAFETY PRECAUTION

Positive test: Formation of a silver mirror on the inside of the test tube. If the tube is not sufficiently clean, a black precipitate or a suspension of metallic silver may form instead. In addition to aldehydes of all kinds, certain aromatic amines, phenols, and α-alkoxy- or dialkylaminoketones may give positive tests.

Some cyclic ketones (such as cyclopentanone) may also give a positive Tollens's test.

DERIVATIVES

The following procedures are suitable for preparing derivatives of most common organic compounds in the specified class. In some cases, a variation of the procedure (the use of different reaction conditions or recrystallization solvents, etc.)

may be required for satisfactory results. If you do not obtain the expected derivative, consult one of the references listed in the Bibliography ⟨Bib – B4 or Bib – G1 through Bib – G4⟩ for alternative procedures. You may also wish to prepare a derivative that is not described in this section but that will better differentiate between the possibilities. If so, you should consult your instructor about the feasibility of preparing such derivatives (the reagents may not be available in your stockroom) and find an appropriate procedure in one of the sources listed.

In most cases, the relative quantities of unknown and reagent are not crucial, since any excess is removed during the isolation and purification of the derivative. In cases where the quantities could affect the results, the quantity of the unknown is given in millimoles so that the required mass can be calculated using an estimate of its molecular weight.

In preparing a derivative, you will be using very small quantities of reactants and reagents, so you should use small-scale apparatus whenever feasible. Reactions are usually run in large test tubes and only a few require refluxing. Filtrations, distillations, extractions, and other operations are carried out on a semimicro scale as described in OP – 12a, OP – 25a, OP – 13a, etc. Whenever the recrystallization solvent is specified as an alcohol-water mixture or some other mixture, the procedure for recrystallization from mixed solvents ⟨OP – 23b⟩ with semimicro apparatus ⟨OP – 23a⟩ should be used. The derivative will usually be dissolved in the less polar solvent and the solution saturated by adding the more polar solvent drop by drop.

As always, appropriate precautions should be taken to avoid contacting or inhaling the reagents and unknown. Many of the reagents are highly reactive and may react violently with water or other substances. All should be considered toxic, and many of them are corrosive, lachrymatory, or have other unpleasant properties. Always read the Safety Precautions before beginning any derivative preparation and heed their warnings.

Read OP – 7b for methods of carrying out small-scale reactions under reflux.

Use 95% ethanol when recrystallizing from ethanol-water mixtures.

Alcohols

D – 1. *3,5-Dinitrobenzoates, p-Nitrobenzoates.* Mix 0.2 g of pure 3,5-dinitrobenzoyl chloride *or* p-nitrobenzoyl chloride **[hazard!]** with 0.1 g of the alcohol. Heat the mixture over a *small* flame so that it is just maintained in the liquid state (do not overheat — decomposition will result). If the alcohol boils

Reaction

ArCOCl + ROH $\longrightarrow$ ArCOOR + HCl

under 160°C, heat the mixture for 5 minutes; otherwise heat it for 10–15 minutes. Allow the melt to cool and solidify. Break it up with a stirring rod, and stir in 4 mL of 0.2 M sodium carbonate. Heat the mixture to 50–60°C on a steam bath, and stir it at that temperature for half a minute. Then cool it and collect the precipitate by small-scale vacuum filtration. Wash the precipitate several times with cold water, and recrystallize it from ethanol or an ethanol-water mixture.

If a microburner is not available, remove the barrel of your burner and light it with the gas turned down.

In general, the higher-boiling (or melting) alcohols will require a higher eth-anol:water ratio in the recrystallization solvent.

The acid chlorides are corrosive and lachrymatory. Avoid contact with skin, eyes, and clothing.

D–2. *α-Naphthylurethanes, Phenylurethanes.* Unless you are certain that your unknown alcohol is anhydrous, dry it with magnesium sulfate or another suitable drying agent, and dry a small test tube in an oven or over a flame. Stopper the tube and allow it to cool. *Under the hood,* mix 5 drops (or 0.2 g) of the alcohol with 5 drops of phenyl isocyanate **[hazard!]** *or* α-naphthyl isocyanate **[hazard!].** If no reaction takes place immediately, heat the mixture in a 60–70°C water bath for 5–15 minutes. Cool the test tube in ice and scratch its sides, if necessary, to induce crystallization. Collect the precipitate by decantation or vacuum filtration. Recrystallize it from about 5 mL of high-boiling petroleum ether or heptane (filter the hot solution to remove high-melting impurities).

The presence of moisture in the alcohol will result in the formation of diphenyl- or di-α-naphthylurea, which melt at 241° C and 297°C, respectively.

Tertiary alcohols do not form ure-thanes readily.

Reaction

$$ArN{=}C{=}O + ROH \longrightarrow$$

$$\overset{\displaystyle O}{\underset{\displaystyle \|}{ArNHC}}{-}OR$$

The isocyanates are irritants and lachrymators. Avoid contact with liquid and vapors; use a hood.

Aldehydes and Ketones

D–3. *2,4-Dinitrophenylhydrazones.* Dissolve 0.10 g of the unknown in 4 mL of 95% ethanol. Add 3 mL of the 2,4-dinitrophenylhydrazine–sulfuric acid reagent **[hazard!],** and allow the solution to stand at room temperature until crystallization is complete. If necessary, warm the solution gently for a minute on a steam bath. If no precipitate appears after 15 minutes (or if precipitation does not seem complete), add water drop by drop to the warm solution until it is cloudy, heat to clarify, and cool. Recrystallize the derivative by dissolving it in

Reaction

$$R-\underset{\underset{R'}{|}}{C}=O + H_2N-NH-\underset{\underset{O_2N}{}}{\bigcirc}-NO_2 \longrightarrow R-\underset{\underset{R'}{|}}{C}=N-NH-\underset{\underset{O_2N}{}}{\bigcirc}-NO_2 + H_2O$$

(R, R' = alkyl, aryl, or H)

6 mL or less of boiling 95% ethanol and adding water (to a maximum of 2 mL) drop by drop. If the derivative does not dissolve in 6 mL of boiling ethanol, add ethyl acetate drop by drop until it goes into solution.

SAFETY PRECAUTION

2,4-Dinitrophenylhydrazine is toxic, and sulfuric acid is corrosive; avoid contact with the reagent.

If the unknown is water-soluble, omit the ethanol.

Reaction

$$-\underset{|}{\overset{O}{\overset{\|}{C}}} + H_2NNHCNH_2 \longrightarrow$$

$$-\underset{|}{\overset{O}{\overset{\|}{C}}}=NNHCNH_2 + H_2O$$

D–4. *Semicarbazones.* Mix together 0.2 g of semicarbazide hydrochloride **[hazard!],** 0.3 g of sodium acetate, 2 mL of water, and 2 mL of 95% ethanol in a test tube. Add 0.2 g of the aldehyde or ketone and stir. (If the solution is cloudy, add more ethanol until it clears up.) Shake the mixture for a minute or two, and then let it stand, cooling it in ice if necessary to induce crystallization. If no crystals form, place the test tube in a boiling-water bath for a few minutes, and then allow it to cool. Collect the crystals, wash them with cold water, and recrystallize the product from alcohol or an alcohol-water mixture.

SAFETY PRECAUTION

Semicarbazide hydrochloride has been found to cause cancer in mice when ingested; avoid contact with the reagent.

Reaction

$$-\underset{|}{\overset{O}{\overset{\|}{C}}} + H_2N-OH \longrightarrow$$

$$-\underset{|}{\overset{}{C}}=N-OH + H_2O$$

D–5. *Oximes.* Oximes are suitable derivatives for most ketones and for some (though not all) aldehydes. They can be prepared by the method given for semicarbazones ⟨D–4⟩, using hydroxylamine hydrochloride **[hazard!]** in place of semicarbazide hydrochloride. It is usually necessary to warm the reactants on a steam bath for 10 minutes or more. Adding a few milliliters of cold water to the reaction mixture may hasten precipitation.

SAFETY PRECAUTION

Hydroxylamine hydrochloride is toxic and can cause a form of anemia. Avoid contact.

Amides

D–6. *Hydrolysis Products.* The acid and amine portions of an amide can be obtained by hydrolysis and one or both of them characterized. Reflux 0.6 g of the amide with 10 mL of 3 M sodium hydroxide for 15 minutes or more, then reassemble the apparatus for semimicro simple distillation. Place 4 mL of 3 M hydrochloric acid in the receiver (a 25-mL round-bottomed flask is recommended). Since the amine may be a gas, the receiving flask should be attached directly to the vacuum-adapter outlet and the vacuum side arm should be connected to a gas trap containing dilute HCl. Distill the reaction mixture until about 6 mL of distillate has been collected.

If the alkaline hydrolysis classification test ⟨*C–2*⟩ *suggests that the amide is difficult to hydrolyze, use 6 M NaOH or a longer reaction time, or both.*

To characterize the amine, add 10 mL of 3 M sodium hydroxide to the distillate; then add 0.4 mL of benzenesul-fonyl chloride **[hazard! see Procedure D–9]** or 0.6 g of *p*-toluenesulfonyl chloride **[hazard!].** Proceed according to the directions given in Procedure D–9 for preparing arenesul-fonamides.

If the amide is known to be unsubstituted on nitrogen, omit this procedure.

Arenesulfonyl chlorides are toxic and corrosive; avoid contact, and do not breathe the vapors.

SAFETY PRECAUTION

To characterize the carboxylic acid, carefully acidify the residue in the boiling flask with 6 M hydrochloric acid. If a precipitate forms, collect it by vacuum filtration, wash it, re-crystallize it (from water, alcohol-water, or another suitable solvent), and obtain its melting point. If no precipitate forms, prepare a *p*-nitrobenzyl derivative according to the directions in Procedure D–15.

D–7. *N-Xanthylamides.* This procedure is suitable only for primary amides. *Under the hood,* dissolve 0.4 g of xanthydrol in 5 mL of glacial acetic acid **[hazard!],** and shake until the solid dissolves. Decant the solution from any undissolved residue, add 0.2 g of the amide (or as close to 1.5 mmol as you can estimate), and warm the mixture in an 85°C water bath for 15–20 minutes. If the amide is not soluble in acetic acid, dissolve it in a minimum volume of ethanol before adding it to the reaction mixture; then add 1 mL of water after the reaction is complete (just before cooling). Let the solution cool until precipitation is complete. Collect the solid by vacuum filtra-tion, and recrystallize the derivative from an ethanol-water mixture.

Reaction

SAFETY
PRECAUTION

Acetic acid can cause burns, and its vapors are very irritating. Use under a hood; avoid contact.

Amines (Primary and Secondary)

D‑8. *Benzamides.*   *Under the hood,* combine 0.2 g of the primary or secondary amine (or as close to 2 mmol as you can estimate) with 2 mL of 3 M sodium hydroxide. Then add 0.8 mL of benzoyl chloride **[hazard!]** drop by drop, with vigorous shaking. Continue to shake the stoppered test tube for about 5 minutes, then carefully neutralize the solution to a pH of 8 with 3 M HCl (use pH paper). Break up the solid mass with a stirring rod, if necessary, and collect the derivative by vacuum filtration. After washing the product thoroughly with cold water, recrystallize it from an alcohol-water mixture.

Reaction

$$PhC\overset{\underset{\parallel}{O}}{} \!-\! Cl + R\!-\!\underset{\underset{R'}{|}}{N}H \longrightarrow PhC\overset{\underset{\parallel}{O}}{} \!-\! \underset{\underset{R'}{|}}{N}\!-\!R + HCl$$

(R′ can be alkyl, aryl, or H.)

SAFETY
PRECAUTION

Benzoyl chloride is a lachrymator and a strong irritant. Avoid contact with the liquid or its vapors; use under a hood.

Reactions

$$RNH_2 + ArSO_2Cl \xrightarrow{\text{NaOH}} \xrightarrow{\text{HCl}}$$
$$ArSO_2NHR$$

$$R_2NH + ArSO_2Cl \xrightarrow{\text{NaOH}}$$
$$ArSO_2NR_2$$

D‑9. *Benzenesulfonamides and* p-*Toluenesulfonamides.*   *Under the hood,* combine in a test tube 0.2 g (or as close to 2 mmol as you can estimate) of the primary or secondary amine, 0.4 mL of benzenesulfonyl chloride **[hazard!]** *or* 0.6 g of p-toluenesulfonyl chloride **[hazard!],** and 10 mL of 3 M sodium hydroxide. Stopper the tube and shake it frequently over a period of 3–5 minutes. Remove the stopper and warm the tube gently over a steam bath for a minute or two. Let the solution cool.

SAFETY
PRECAUTION

Arenesulfonyl chlorides are toxic and corrosive. Avoid contact, do not breathe the vapors, and use under a hood.

(a) If a precipitate forms, collect it by vacuum filtration, wash it with water, and recrystallize it from an ethanol-water mixture.

(b) If no precipitate forms on cooling, acidify the solution to pH 6 with 6 M HCl, cool the mixture to complete crystallization, and collect and purify the product as described above.

A precipitate formed in the Hinsberg test ⟨C–15⟩ can also be purified and used as a derivative.

D–10. *α-Naphthylthioureas and Phenylthioureas.* *Under the hood,* dissolve 0.2 g (or as close to 2 mmol as you can estimate) of the primary or secondary amine in 2 mL of absolute ethanol and add 0.2 mL of phenylisothiocyanate **[hazard!]** *or* 0.3 g of α-naphthylisothiocyanate **[hazard!].** Reflux the mixture for 5 minutes (using a water or steam bath) so that the alcohol boils gently. Cool the mixture and scratch the sides of the test tube to induce crystallization. If no crystals form, reflux the mixture for another 20 minutes or so. Add 1 mL each of ethanol and water to facilitate transfer. Collect the derivative by vacuum filtration, wash it with dilute ethanol (one washing with 50% and one with 95% ethanol are recommended), and recrystallize the product from ethanol or from an alcohol-water mixture. If the derivative appears impure, it may be necessary to extract the impurities by boiling the solid with 4 mL of petroleum ether before recrystallizing it, or to carry out several recrystallization steps.

Protect the reactants and the reaction mixture from water.

Adding water drop by drop until the solution becomes cloudy may facilitate crystallization. If an oil forms, scratching the sides of the tube will usually cause it to crystallize.

Reaction

$$\text{ArN}=\text{C}=\text{S} + \underset{\underset{\displaystyle R'}{|}}{\text{RNH}} \longrightarrow \text{ArNH}\overset{\overset{\displaystyle S}{\|}}{\text{C}}\underset{\underset{\displaystyle R'}{|}}{\text{NR}}$$

(R′ may be alkyl, aryl, or H.)

The isothiocyanates are toxic and irritant. Avoid contact; do not inhale the vapors; use gloves and a hood.

SAFETY PRECAUTION

Amines (Tertiary)

D–11. *Methiodides.* *Under the hood,* mix 0.2 g of the tertiary amine with 0.2 mL of methyl iodide **[hazard!]** in a test tube, and warm the mixture in a 50–60°C water bath for 5 minutes. Cool the mixture in ice, collect the product by vacuum filtration, and recrystallize it from alcohol, ethyl acetate, or a mixture of the two.

Reaction

$$\text{R}_3\text{N} + \text{CH}_3\text{I} \longrightarrow \text{R}_3\text{NCH}_3^+\text{I}^-$$

(The R groups may be alkyl, aryl, or a combination of the two.)

Methyl iodide is toxic and corrosive, and it has been found to cause cancer in mice by injection. Use gloves and a hood, avoid contact, and do not breathe the vapors.

SAFETY PRECAUTION

D–12. *Picrates.* Dissolve 0.2 g of the unknown in 5 mL of 95% ethanol, and add 5 mL of a saturated solution of picric acid in ethanol **[hazard!]**. Heat the solution to boiling over a steam bath, and let it cool to room temperature. Collect the crystals by vacuum filtration, and wash them with cold ethanol or methanol. (The product may be recrystallized from alcohol, if necessary.)

Reaction

Picrates of primary amines, secondary amines, and aromatic hydrocarbons can be prepared by essentially the same method. The picrates of many aromatic hydrocarbons dissociate when heated and cannot be recrystallized. Some picrates explode when heated, so use caution when obtaining the melting point.

SAFETY PRECAUTION

In solid form, picric acid is unstable and can explode when subjected to heat or shock. It is also toxic and an irritant. Avoid contact with solutions containing picric acid.

Carboxylic Acids

D–13. *Amides.* This derivative is suitable for aromatic acids and for most aliphatic acids with six or more carbon atoms. Since Procedures D–13 and D–14 both involve preparation of the acid chloride, it is convenient to prepare one of the D–14 amides with the same acid chloride solution.

Reactions

$$RCOOH + SOCl_2 \longrightarrow$$
$$RCOCl + HCl + SO_2$$

$$RCOCl + NH_3 \longrightarrow$$
$$RCONH_2 + HCl$$

The reaction apparatus must be dry.

If you do not intend to prepare an anilide or p-toluidide, add the entire reaction mixture to 15 mL of ammonia.

Under the hood, reflux 1 g of the carboxylic acid with 5 mL of thionyl chloride **[hazard!]** for 20–30 minutes, using a steam or hot-water bath. Cool the acid chloride solution in an ice bath, and (if desired) save part of it in a stoppered container to prepare an anilide or *p*-toluidide. Still under the hood, *slowly* add 2 mL of this solution to a beaker containing 5 mL of ice-cold concentrated aqueous ammonia **[hazard!],** with constant stirring. Let the solution stand for 5 minutes, separate the derivative by vacuum filtration, and recrystallize it from water or an alcohol-water mixture.

Thionyl chloride and ammonia can both irritate or cause burns to the skin and eyes; their vapors are highly irritating and toxic. Use gloves and a hood.

SAFETY PRECAUTION

D–14. *Anilides and* p-*Toluidides.* Prepare the acid chloride **[hazard!]** as directed in Procedure D–13, or use the reaction mixture saved from the amide preparation. *Under the hood,* dilute about 2 mL of the acid chloride reaction mixture with 2 mL of anhydrous ethyl ether, and add in portions (shaking after each addition) a solution containing 0.8 g of *p*-toluidine **[hazard!]** *or* 0.7 mL of aniline **[hazard!]** dissolved in 10 mL of anhydrous ethyl ether. Continue the addition until the odor of the acid chloride has disappeared. If the odor persists after all the arylamine solution has been added, heat the mixture gently on a steam bath for a minute or two. Wash the ether solution in a large test tube with 5 mL of 1.5 M HCl followed by 5 mL of water, separating the aqueous layers from the ether layer with a capillary pipet. Evaporate the solvent under aspirator vacuum. Recrystallize the derivative from water or an alcohol-water mixture.

Reactions

$$RCOOH + SOCl_2 \longrightarrow RCOCl + HCl + SO_2$$

$$RCOCl + ArNH_2 \longrightarrow RCONHAr + HCl$$

The aromatic amines are toxic if inhaled, ingested, or absorbed through the skin. Avoid contact with skin, eyes, and clothing; do not breathe the vapors.

Avoid contact with the acid chloride; do not breathe its vapors.

SAFETY PRECAUTIONS

D–15. p-*Nitrobenzyl Esters.* Preparation of this derivative is recommended only when the previous derivatives are not suitable or when the carboxylic acid is obtained in aqueous solution (for instance, after the hydrolysis of an ester or amide). If the acid is obtained by hydrolysis, the alcohol or amine portion should be identified first so that the molecular weight of the acid portion can be estimated from a list of possibilities. Then you can estimate the amount of acid in the hydrolysis solution, and measure out enough of it to provide 2 mmol of the acid.

It is important that p-nitrobenzyl chloride not be present in excess, as it is difficult to remove from the product.

Reaction

$$RCOONa + O_2N-\langle\bigcirc\rangle-CH_2Cl \longrightarrow RCOOCH_2-\langle\bigcirc\rangle-NO_2 + NaCl$$

Mix an estimated 2 mmol of the carboxylic acid with 3 mL of water (or use the measured solution from the amide or ester hydrolysis), and add a drop of phenolphthalein solution. Add 3 M NaOH drop by drop until the solution turns pink. Heat (if necessary) to put the acid salt into solution; then add ~2 drops of 1.5 M HCl to discharge the pink color. *Under the hood,* add 0.30 g (1.75 mmol) of *p*-nitrobenzyl chloride **[hazard!]** dissolved in 10 mL of 95% ethanol, and reflux the mixture gently for about 1½ hours. Allow the solution to cool to room temperature. If precipitation has not occurred by then, add 1 mL of water and scratch the sides of the reaction vessel. When crystallization is complete (it may take 20 minutes or so), collect the product by vacuum filtration and wash it twice with small portions of 5% sodium carbonate, then once with water. Recrystallize the derivative from alcohol or an alcohol-water mixture. If the melting point of the product is close to 71°C (the melting point of *p*-nitrobenzyl chloride), the reaction may not have gone to completion, and a longer reflux time may be required.

Di- and triprotic acids should be refluxed for 2–3 hours.

SAFETY PRECAUTIONS

p-**Nitrobenzyl chloride is lachrymatory and can cause blisters. Avoid contact with the reagent or its vapors.**

Avoid contact between the product and the skin— blistering may result.

Esters

D–16. Hydrolysis Products. Carry out the hydrolysis of the ester according to the procedure in Test C–2, using 2–3 times the amounts given there. If a solid carboxylic acid precipitates on acidification of the solution (use concentrated HCl, *under the hood,* with care) it can be recrystallized from water or an alcohol-water mixture and used as a derivative itself. It can also, if necessary, be converted to one of the carboxylic acid derivatives. If no precipitate forms, make the solution alkaline with dilute sodium hydroxide and distill it until about 5 mL of distillate has been collected. Acidify the residue in the boiling flask and extract it with ether, then evaporate the ether to

Reactions

$$RCOOR' + NaOH \longrightarrow$$
$$RCOONa + R'OH$$

$$RCOONa + H^+ \longrightarrow$$
$$RCOOH + Na^+$$

The acidified residue can be used to prepare a p-nitrobenzyl ester directly.

isolate the carboxylic acid for a derivative preparation. The alcohol portion of the ester often separates out when the distillate is saturated with potassium carbonate. It can then be isolated with a separatory funnel or (if necessary) by extraction with ether and used to prepare a derivative as well.

It may be necessary to purify the acid or alcohol before attempting to prepare a derivative.

D–17. N-*Benzylamides.* This procedure works well only with methyl and ethyl esters. To prepare the *N*-benzylamide of an ester of a higher alcohol, first reflux 1 g of the ester with 5 mL of 5% sodium methoxide in methanol for 30 minutes, and then evaporate the methanol. Use the resulting ester in the following procedure.

Reaction

$$RCOOR' + CH_3OH \xrightarrow{CH_3ONa}$$
$$RCOOCH_3 + R'OH$$

Under the hood, combine 1 g of a methyl or ethyl ester with 3 mL of benzylamine **[hazard!]** and 0.1 g of powdered ammonium chloride. Reflux the mixture for 1 hour. Cool the reaction mixture and wash it with a few milliliters of water, using a capillary pipet to remove the water. If no precipitate forms, add a drop or two of 3 M HCl, and scratch the sides of the test tube. If a precipitate still does not form, transfer the mixture to an evaporating dish (use 1 mL of water in the transfer), and boil it for a few minutes to remove the excess ester. Collect the derivative by vacuum filtration, wash it with petroleum ether, and recrystallize it from an alcohol-water mixture or ethyl acetate.

Procedure D–17 forms a derivative of the carboxylic acid portion of the ester; Procedure D–18 forms a derivative of the alcohol portion.

Reaction

$$RCOOR' + PhCH_2NH_2 \xrightarrow{NH_4Cl}$$
$$RCONHCH_2Ph + R'OH$$

(R′ should be ethyl or methyl.)

Benzylamine is highly irritating to the skin, eyes, and mucous membranes. Avoid contact, do not breathe its vapors, and *use a hood.*

SAFETY PRECAUTION

D–18. *3,5-Dinitrobenzoates.* Combine 1 g of the ester with 0.8 g of 3,5-dinitrobenzoic acid **[hazard!]**, and add a drop of concentrated sulfuric acid **[hazard!]**. Reflux the mixture until the 3,5-dinitrobenzoic acid dissolves; then continue refluxing for an additional 30 minutes. Dissolve the cooled reaction mixture in 20 mL of anhydrous ethyl ether. Wash the ether solution twice with 10-mL portions of 0.5 M sodium carbon-

This procedure is not satisfactory for esters of tertiary and some unsaturated alcohols.

Reaction

ate [*caution:* **foaming**], then with water. Evaporate the ether and dissolve the residue (often an oil) in 2–3 mL of boiling ethanol. Filter the hot solution, add water until the mixture becomes cloudy, and cool the solution to crystallize the product. Recrystallize the derivative from an alcohol-water mixture, if necessary.

SAFETY PRECAUTION   **Avoid contact with sulfuric acid and 3,5-dinitrobenzoic acid.**

Alkyl Halides

In addition to the following derivative, density values ⟨Test C–10⟩ and refractive index values are often very useful in characterizing alkyl and aryl halides.

Tertiary alkyl halides will not form this derivative. If the unknown has been shown to be an alkyl chloride, use the alternative procedure with ethylene glycol.

D–19. S-*Alkylthiuronium Picrates.*   Mix together 0.2 g of the alkyl halide, 0.2 g of powdered thiourea [**hazard!**], and 2 mL of 95% ethanol. Boil the mixture on a steam bath for a few minutes, add 2 mL of a saturated solution of picric acid [**hazard!**] in ethanol, and boil the mixture for a few minutes longer. Allow the solution to cool, collect the derivative by vacuum filtration, and recrystallize it from ethanol or an ethanol-water mixture.

Reaction

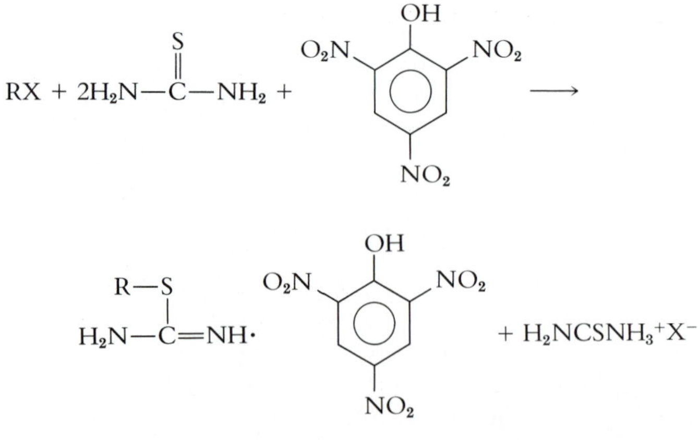

$$X = Cl, Br, I \quad R = alkyl$$

If the derivative fails to form under these conditions (alkyl chlorides are often unreactive), follow the same procedure using 6 mL of ethylene glycol as the solvent. Heat the solution in an oil bath at 120°C for 30 minutes before adding the picric acid solution; then continue heating it for 15 minutes. Add 6 mL of water, and cool the mixture in an ice bath to assist crystallization.

Aryl Halides

In addition to preparing the following derivative, you can characterize aryl halides with alkyl side chains by oxidizing the side chain ⟨Procedure D–21⟩. Density values ⟨Test C–10⟩ and refractive index values are also very useful.

D–20. *Nitro Compounds.* *(a)* *Under the hood,* mix 0.5 g of the unknown with 2 mL of concentrated sulfuric acid **[hazard!].** Cautiously add 2 mL of concentrated nitric acid **[hazard!]** drop by drop with shaking or stirring; then heat the mixture on a 60°C water bath for 10 minutes, shaking frequently. Pour the product onto 15 mL of cracked ice, with stirring. After the ice has melted, collect the precipitate by vacuum filtration, wash it with water, and recrystallize it from an ethanol-water mixture. Additional recrystallizations may be necessary if the product melts at a temperature lower than expected or over a wide range.

Reaction

$$ArH + HNO_3 \xrightarrow{H_2SO_4} ArNO_2 + H_2O$$

(and di- or trinitro derivatives in some cases)

If the product separates as an oil, it is probably a mixture of compounds with different numbers of nitro groups. See the work by Cheronis, Entrikin, and Hodnett ⟨Bib–G1⟩ for a more detailed discussion of nitration, as well as additional procedures.

(b) Follow procedure *(a)*, but use 2 mL of fuming nitric acid **[hazard!]** instead of concentrated nitric acid, and heat the solution for 10 minutes on a steam bath. Add the acid slowly to minimize the generation of brown (NO_2) fumes.

Procedure *(a)* will yield mononitro derivatives of most comparatively reactive aryl halides and hydrocarbons, whereas procedure *(b)* should produce dinitro or trinitro derivatives of reactive compounds and mononitro derivatives of unreactive ones. Di- and trialkylbenzenes (such as the xylenes and mesitylene) usually yield trinitro derivatives by procedure *(b)*, whereas monoalkylbenzenes such as toluene yield dinitro compounds. Since the di- and trinitro derivatives are often easier to purify than the mononitro derivatives, procedure *(b)* is usually preferred.

Aromatic Hydrocarbons

In addition to preparing the following derivative, you can characterize aromatic hydrocarbons by the preparation of picrates ⟨D–12⟩ and nitro compounds ⟨D–20⟩.

Reaction

$$Ar\!-\!R \xrightarrow{\ \text{KMnO}_4\ } ArCOOH$$

D–21. *Aromatic Carboxylic Acids.* The following procedure is for an aromatic hydrocarbon or aryl halide with one alkyl side chain. If the unknown may have more than one side chain, increase the quantities of the reagents in proportion to the anticipated number of side chains.

Dissolve 1.5 g of potassium permanganate **[hazard!]** in 25 mL of water, and add 0.5 g of the aromatic hydrocarbon or halide followed by 0.5 mL of 6 M aqueous sodium hydroxide **[hazard!].** Reflux the mixture for 1 hour or until the permanganate color has disappeared. Cool the reaction mixture to room temperature, acidify it with 6 M sulfuric acid, and boil it for a few minutes. If necessary, stir a little solid sodium bisulfite into the hot solution to destroy any unreacted manganese dioxide, keeping the solution acidic during the treatment. Cool the mixture, collect the product by vacuum filtration, and recrystallize the acid from water or an ethanol-water mixture.

SAFETY PRECAUTION

Avoid contact with the caustic NaOH solution and with potassium permanganate. Keep the latter away from oxidizable materials.

Phenols

D–22. *Aryloxyacetic Acids.* *Under the hood,* combine 0.5 g of the unknown with 0.7 g of chloroacetic acid **[hazard!]** and 3 mL of 8 M sodium hydroxide **[hazard!]**. Heat the mixture on a steam bath for an hour; then cool it to room temperature and add 6 mL of water. Acidify the solution to pH 3 with 6 M hydrochloric acid, extract it with 20 mL of ethyl ether, and wash the ether extract with 5 mL of cold water. Extract the derivative from the ether using about 10 mL of 0.5 M sodium carbonate, then acidify the sodium carbonate solution with 6 M hydrochloric acid **[caution: foaming]**. Collect the precipitate by vacuum filtration, and recrystallize it from water.

Reaction

$$ArOH + ClCH_2COOH \longrightarrow$$
$$ArOCH_2COOH + HCl$$

You can add a drop of a solution of Congo Red indicator, which turns blue around pH 3.

Chloroacetic acid is corrosive and toxic. Avoid contact with the skin, eyes, and clothing; do not breathe the vapors.

Avoid contact with the highly caustic NaOH solution.

SAFETY PRECAUTIONS

The equivalent weight of the aryloxyacetic acid (and thus of the phenol as well) can be determined by obtaining its neutralization equivalent ⟨Test C–18⟩.

D–23. *Bromo Derivatives.* *Under the hood,* prepare a brominating solution by dissolving 4.5 g of potassium bromide in 30 mL of water and *carefully* adding 1 mL (about 3 g) of pure bromine **[hazard!]**. Dissolve 0.3 g of the phenol in 6 mL of 50% ethanol (try 95% ethanol or acetone if it doesn't dissolve). Add the brominating solution drop by drop, with stirring, until the bromine color persists after shaking. Add 15 mL of water and shake the mixture; then collect the derivative by vacuum filtration. Wash the product with 1 M sodium bisulfite to remove excess bromine, and recrystallize it from ethanol or an alcohol-water mixture.

Reaction for phenol (other phenols undergo similar reactions)

Bromine is very poisonous and causes severe burns. Handle it with gloves under the hood; do not breathe its vapors.

SAFETY PRECAUTION

D–24. *α-Naphthylurethanes.*   *Under a hood,* carry out the procedure for preparing *α*-naphthylurethanes of alcohols ⟨D–2⟩, but add a drop of pyridine **[hazard!]** to catalyze the reaction. For particularly unreactive phenols, it may be helpful to add 1 mL of pyridine and a drop of 10% triethylamine in petroleum ether to the reaction mixture and to heat the mixture at 70°C for about 30 minutes. If the urethane does not precipitate on cooling, add 1 mL of 0.5 M sulfuric acid.

Reaction

$$\text{ArN=C=O} + \text{Ar'OH} \longrightarrow \text{ArNH}\overset{\displaystyle O}{\overset{\|}{\text{C}}}\text{—OAr'}$$

SAFETY PRECAUTIONS

Pyridine is toxic and an irritant and has an unpleasant odor. Avoid contact, and use a hood.

α-Naphthyl isocyanate is lachrymatory and a strong irritant. Avoid contact and use a hood.

Report

Your report should include information in each of the following categories, as appropriate:

Unknown number

Results of preliminary examination

Physical constants determined

Results of solubility tests

Spectrometric results

Results of functional class tests

Interpretation of spectra and tests

Results of literature examination

Additional tests and data

Probable compounds

Derivatives prepared

Discussion and conclusion

Where convenient, you should tabulate your data and results so that they are presented clearly and concisely. Some of the categories may be combined — for example, your interpretation of the functional class tests might be included in a table describing the test results. Insofar as possible, the report should reveal the thought processes leading to your conclusion as well as the physical and chemical data supporting it.

Exercises

1. Write balanced equations for the chemical reactions undergone by your unknown, including those involved in classification tests and the preparation of derivatives.

2. Specify the functional class (chemical family) of each of the following unknowns and give any additional structural information suggested by the tests. Explain the reasoning behind your deductions.

(a) Unknown #22 is insoluble in all of the solubility-test solvents except cold, concentrated sulfuric acid. It gives a blue-green suspension with the chromic acid reagent, yields no precipitate with Benedict's solution, and immediately forms a separate organic layer when shaken with concentrated HCl and $ZnCl_2$ (Lucas's test). The C—H stretching band in the infrared spectrum extends from about 2750 to 3050 cm^{-1}, but there is no strong absorption at a higher frequency.

(b) Unknown #7B is insoluble in water but soluble in dilute sodium hydroxide. It decolorizes potassium permanganate and bromine in Freon TF solutions. Addition of a saturated solution of bromine in water causes decolorization of the bromine and yields a white precipitate.

(c) Unknown #128 is in solubility class B and yields an infrared spectrum with a strong absorption band centered around 3400 cm^{-1}. The unknown is insoluble in a pH 5.5 buffer and forms an insoluble precipitate when shaken with *p*-toluenesulfonyl chloride in aqueous sodium hydroxide.

(d) Unknown #33C has a strong absorption band in the 1750 – 1650 cm^{-1} region of its infrared spectrum, but it forms no precipitate with 2,4-dinitrophenylhydrazine reagent. When it is refluxed with 6 M sodium hydroxide, a vapor is generated that turns red litmus paper blue. Acidification of the alkaline hydrolysis solution yields a white precipitate.

(e) Unknown #2B is insoluble in cold concentrated sulfuric acid, and a drop of the unknown in water sinks to the bottom. The unknown gives an orange-red color with aluminum chloride and chloroform. It reacts immediately to form precipitates with both ethanolic silver nitrate and sodium iodide in acetone; the colors of the precipitates are pale yellow and white, respectively.

(f) Unknown #44 is water-soluble and reacts with acetyl chloride to yield a liquid that smells like nail polish. It dissolves in the Lucas reagent but gives no separate phase or cloudiness within an hour, even when ice water is added. Treating the unknown with iodine in sodium hydroxide forms a yellow precipitate with a medicinal odor.

3. Indicate the solubility class to which each of the following is most likely to belong.

(a) propanoic acid

(b) toluene

(c) p-anisidine

(d) p-cresol

(e) 3-methylpentanal

(f) p-toluic acid

(g) 2-chlorobutane

(h) 2-aminobutane

(i) ethylene glycol

(j) acetophenone

4. Give reasonable mechanisms for the reactions involved in each of the following classification tests or derivative preparations.

(a) the alkaline hydrolysis of N-ethylbenzamide

(b) the reaction of p-cresol with bromine water

(c) the preparation of the 2,4-dinitrophenylhydrazone of 2-butanone

(d) the reaction of ethyl acetate with hydroxylamine in alkaline solution

(e) the reaction of diethylamine with p-toluenesulfonyl chloride in aqueous sodium hydroxide

(f) the iodoform reaction of 2-butanone (give a mechanism for each reaction step involved)

(g) the Lucas reaction of t-butyl alcohol

(h) the reaction of 2-chloro-2-methylbutane with ethanolic silver nitrate

(i) the reaction of 1-bromobutane with sodium iodide in acetone

(j) the formation of the 3,5-dinitrobenzoate of 2-butanol

(k) the formation of the methiodide of triethylamine

(l) the formation of the aryloxyacetic acid of p-cresol

ADVANCED
PROJECTS

The open-ended projects in Part IV are intended to be carried out by advanced or honors students who have mastered the major operations described in Part V. Most of the projects in Part IV require some familiarity with the qualitative analysis methods described in Part III as well.

The following research-type projects require advanced laboratory skills and a comprehensive knowledge of organic chemistry. Many of the projects are open-ended and can lead to further investigations not suggested in the outlines. You may have to make extensive use of the library; you must exercise independent judgment in planning the best course of action. Although some of the projects will not lead directly to new or unique contributions to the sum of chemical knowledge, others could (at least potentially) yield hitherto unknown facts and new discoveries.

Each project is outlined only briefly so that you will have considerable latitude in developing and pursuing a course of action. Each project includes one or more key references — additional references can be found in Appendix VII. A record of all laboratory work should be kept in a bound notebook, and each project should be written up in a form acceptable for publication in a professional journal such as the *Journal of Organic Chemistry*.

See Experiment 50 for information about carrying out a literature search.

PROJECT 1

Isolation and Testing of a Natural Growth Inhibitor

juglone

Key References:

J. Chem. Educ. 49, 436 (1972). Synthesis of juglone.

J. Chem. Educ. 50, 782 (1973). General information, isolation of juglone.

J. Chem. Educ. 54, 156 (1977). Isolation of juglone from walnut hulls.

K. Paech and M. V. Tracey, *Modern Methods of Plant Analysis,* Vol. 3 (Berlin: Springer-Verlag, 1955), pp. 565–625. Cultivation and biological testing of plants.

Juglone is a natural substance occurring in walnut trees that is known to "poison" some kinds of plants and inhibit their development. It has been speculated that juglone acts as a kind of chemical defense agent for walnut seedlings, favoring their growth at the expense of competing species.

Your major objectives are as follows:

1. To isolate and purify a quantity of juglone from black-walnut hulls.

2. To prepare synthetic juglone and establish its identity with the natural compound using spectrometric methods.

3. To test the inhibitory activity of purified juglone on the growth of various species or varieties of plants.

Juglone is an effective growth inhibitor for many plants of the heath family (genus *Ericaceae*); both wild and domesticated species of such plants could be tested. The use of wild plants may require special cultivation methods. Different varieties of tomatoes or other domestic plants could be tested to determine whether certain varieties are more resistant than others. Biological testing could be carried out on a cross section of typical plants from other families to determine what characteristics the juglone-sensitive plants may have in common. The juglone molecule could also be modified chemically to see what effect, if any, the modifications have on its inhibitory properties. Possible modifications include introducing substituents on the aromatic ring and converting the hydroxyl group to an ester linkage.

Isolation and Analysis of an Essential Oil   PROJECT 2

The leaves, stems, and other parts of a plant can usually be steam-distilled to yield appreciable quantities of an essential oil, which contains the plant's more volatile components and is largely responsible for its odor. The number and relative abundance of the components of such an oil can be determined by gas chromatography. The major components can be isolated by fractional distillation, column chromatography, and other methods. Individual components can often be identified by spectrometric methods (using collections of standard spectra, when available) along with traditional qualitative analysis procedures such as the determination of physical properties and the preparation of derivatives.

Vacuum fractional distillation is recommended for oils that may decompose at atmospheric pressure.

Your major objectives are as follows:

1.   To obtain an essential oil from a natural plant product by steam distillation. If possible, the plant should be identified and collected from your local area.

2.   To analyze the essential oil by gas chromatography.

3.   To isolate and identify at least one (preferably more) component of the essential oil.

In selecting a plant species to analyze, keep two things in mind. First, the plant must be abundant enough to provide the large amounts of plant material that may be needed. Second, plants containing appreciable quantities of essential oil will ordinarily have a pronounced odor. You may find it especially interesting to select a plant whose volatile oil is known to possess medicinal properties and to attempt to identify the active component or components of the oil. Many such plants are described in the references by Grieve and Angier. If the essential oil is too complex or does not have any very abundant

Key References:

E. Guenther, *Essential Oils* (3 Volumes) (New York: Van Nostrand, 1948). Isolation methods, sources, and properties of essential oils.

K. Paech and M. V. Tracey, *Modern Methods of Plant Analysis* (Berlin: Springer-Verlag, 1955). Methods for the isolation and analysis of plant constituents.

M. Grieve, *A Modern Herbal* (New York: Dover, 1971). Species-by-species description of medicinal plants and their uses.

B. Angier, *Field Guide to Medicinal Wild Plants* (Harrisburg, Penna.: Stackpole Books, 1978) Guide to medicinal plants in the United States (see also the field guides listed in Project 3).

constituents, it may be extremely difficult to separate and identify its components unless special analytical tools (GC/IR, GC/MS, HPLC, etc.) are available. In any event, it may be wise to screen a number of different species by gas chromatography before selecting one that has a comparatively simple mixture of constituents.

Most of the commonly encountered essential oils are described in the reference by Guenther; however, this comprehensive work is more than forty years old and some of its information is out of date. You should plan on searching through *Chemical Abstracts* to determine whether the chemical makeup of any given plant has already been reported in the chemical literature. It is clearly more interesting and challenging to analyze a new plant than to repeat a previous analysis, but your project can also be designed to verify the presence of previously characterized constituents in a species under study.

Isolation and Identification of Pigments from Wild Plants

PROJECT 3

Key References:

T. W. Goodwin, *Chemistry and Biochemistry of Plant Pigments* (London: Academic Press, 2nd ed., 1976). Isolation and analysis methods.

E. Gibbons, *Stalking the Healthful Herbs* (New York: David McKay, 1966). Nutritional aspects of wild plant usage.

E. Gibbons and G. Tucker, *Euell Gibbons' Handbook of Edible Wild Plants* (Virginia Beach, Va.: Donning, 1979). Comprehensive listing of edible wild plants of the United States.

Lee Peterson, *A Field Guide to Edible Wild Plants of Eastern and Central North America* (Boston: Houghton Mifflin, 1978). Guide to the identification and uses of wild plants.

D. R. Kirk, *Wild Edible Plants of the Western United States* (Healdsburg, Calif.: Naturegraph Publishers, 1970). Field guide for plants of the Western United States.

Carotenoids and other natural plant pigments can be isolated from plant material by solvent extraction and column chromatography and identified by spectrometric methods. Although the pigments in tomatoes, carrots, spinach leaves, and many other domesticated plants have been analyzed thoroughly, the same cannot be said for many common wild plants. Carotene content, in particular, has nutritional significance because the carotenes are Vitamin A precursors. Some edible wild plants such as curled dock *(Rumex crispa)* are said to contain more Vitamin A than carrots.

Your major objectives are as follows:

1. To identify and collect specimens of one or more edible wild plants.

2. To isolate and identify as many as possible of the pigments present in the leaves or other edible parts of the plant.

3. To determine the amount of β-carotene and other Vitamin A precursors as a fraction of the dry weight of the plant matter and to convert this to equivalent Vitamin A content.

General methods for isolating and analyzing plant pigments are described in the reference by Goodwin. Carotenoid pigments are always found in the leaves of plants; they may also occur in brightly colored fleshy fruits or in the roots of some plants. For example, the very common "Queen Anne's lace" *(Daucus carota)* is simply a wild version of the domestic carrot. It might be of interest to compare the carotene content of its root with that of a domestic carrot at the same stage of growth.

Many other possibilities should suggest themselves after you have scanned one or both of the references by Gibbons. The field guides by Peterson and Kirk will help you identify the relevant species in your area.

The Effect of
Molecular Modification on Odor

Slight changes in molecular structure can have drastic (and sometimes unpredictable) effects on the odor of a substance. *p*-Hydroxybenzaldehyde is described in the *Merck Index* as having a "slight, agreeable, aromatic odor." Its two functional groups and benzene ring make possible a number of interesting molecular modifications. Aldehyde functions can be oxidized to yield carboxylic acids, reduced to form alcohols, condensed with active hydrogen compounds, treated with Grignard reagents, etc. The resulting functional groups can, in turn, be altered to form esters and many other derivatives. Phenolic hydroxyl groups can be converted to ester and ether linkages. The benzene ring itself is subject to a wide variety of electrophilic substitution reactions. For example, nitration of the ring followed by reduction and diazotization should allow the synthesis of a number of 1,2,4-trisubstituted benzene derivatives.

Your major objectives are as follows:

1. To prepare as many related compounds as you can, starting with *p*-hydroxybenzaldehyde and using the reaction types indicated above or any other reactions that suggest themselves.

2. To obtain sufficient analytical data, including spectra and GLC chromatograms where feasible, to establish the identity and purity of each product.

3. To observe and characterize the odor of each product that is prepared.

p-hydroxybenzaldehyde

Since some products may be toxic when inhaled, be careful not to inhale an appreciable amount.

Key References:

R. W. Moncrieff, *The Chemical Senses,* 3rd ed. (Cleveland, Ohio: CRC Press, 1967). Treatise on the senses of taste, smell, and vision.

R. B. Wagner and H. D. Zook, *Synthetic Organic Chemistry* (New York: Wiley, 1953). Guide to traditional synthetic methods.

See Appendix VII and category B of the Bibliography.

Bear in mind that the presence of two functional groups may cause complications in certain reactions unless one of them is "masked" or otherwise altered. In a sodium borohydride reduction of the aldehyde functional group, for example, the hydroxyl group will decompose the reducing agent unless —OH is converted to —O^-Na^+ by running the reaction in basic solution. Oxidation of the carbonyl group will be accompanied by oxidation of the hydroxyl group unless the latter is first converted to an ester or ether functional group. Accordingly, you must plan each synthesis carefully in order to optimize the yield of the desired product — or, in some cases, to get any product at all. General synthetic procedures can be found in many of the references on organic synthesis listed in the Bibliography. A search of the chemical literature, as described in Experiment 50, may yield procedures for specific conversions. The reference by Moncrieff provides considerable information, with specific examples, about the effect of molecular structure on odor.

Stereochemistry of a Hydride Reduction of 4-*t*-Butylcyclohexanone

PROJECT 5

Key References:

E. J. Goller, "Stereochemistry of Carbonyl Addition Reactions," *J. Chem. Educ. 51,* 182 (1974).

A. S. Kushner and T. Vaccariello, "Use of a New Chemical Reducing Agent in the Undergraduate Organic Laboratory," *J. Chem. Educ. 50,* 154 (1973).

$NaAlH_2(OCH_2CH_2OCH_3)_2$

sodium *bis*(2-methoxyethoxy)aluminum hydride

[also called Vitride, Red-al, and sodium dihydrido-*bis*(2-methoxyethoxy)aluminate]

Reductions and Grignard reactions of 4-*t*-butylcyclohexanone have been studied extensively to elucidate the stereochemistry of carbonyl addition reactions. Since the bulky *t*-butyl group effectively locks the molecule into one conformation (that in which *t*-butyl is equatorial), the conformation of the hydroxyl group in the product should be the same as its developing conformation in the transition state leading to the product. Both "steric approach control" and "product development control" concepts have been used to rationalize the results of such reactions as lithium aluminum hydride reduction, which yields mostly *equatorial* alcohol, and Grignard additions, which result in predominantly *axial* OH groups.

A comparatively new reducing agent, sodium *bis*(2-methoxyethoxy)aluminum hydride, is considerably safer and easier to handle than lithium aluminum hydride. It also contains two organic side chains that could conceivably affect the stereochemical outcome of a reduction reaction. Other new hydride reducing agents are discussed in the work by Fieser and Fieser listed in the Bibliography ⟨B–20⟩.

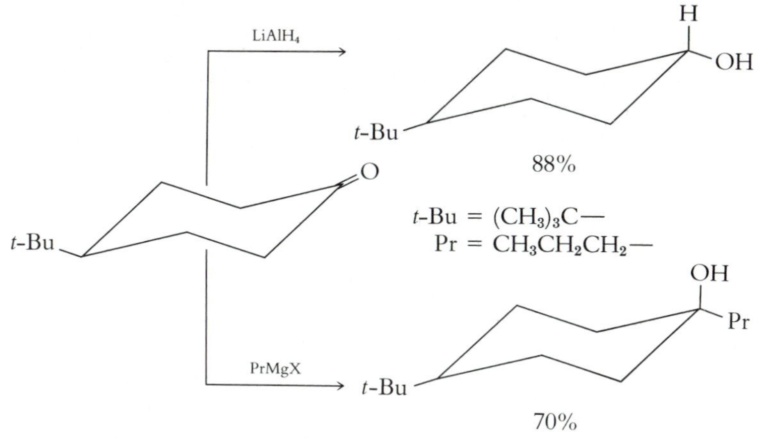

t-Bu = (CH$_3$)$_3$C—
Pr = CH$_3$CH$_2$CH$_2$—

Your major objectives are as follows:

1.   To carry out the reduction of 4-*t*-butylcyclohexa-none using sodium bis(2-methoxyethoxy)aluminum hydride or some other "new" hydride reducing agent (not LiAlH$_4$ or NaBH$_4$) and analyze the product mixture by GLC.

2.   To separate the axial and equatorial products by an appropriate separation method and identify each product by a spectroscopic method, so that the stereoisomer responsible for each GLC peak can be determined.

3.   To compare the results with those obtained in other carbonyl addition reactions and attempt to explain them on the basis of the concepts discussed in the article by Goller.

A study of the literature should reveal the most productive methods of separation and analysis. If time permits, additional reducing agents (or some Grignard reagents not mentioned in the literature) may be used as well.

Comparison of the Migratory Aptitudes of Aryl Groups

PROJECT 6

The pinacol rearrangement is a well-known reaction that involves the migration of an alkyl or aryl group, presumably by means of a bridged intermediate of the kind illustrated in the margin. Since the intermediate contains a positive charge distributed over the migrating group, migration rates should be

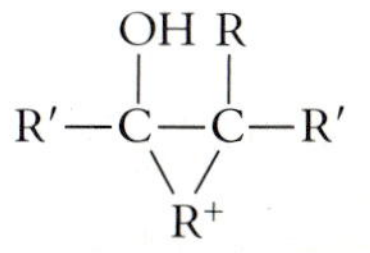

$$2\,Ph\overset{\displaystyle O}{\overset{\|}{-C-}}Ar \longrightarrow Ph\overset{\displaystyle HO}{\underset{\underset{\displaystyle Ar}{|}}{\overset{|}{-C-}}}\overset{\displaystyle OH}{\underset{\underset{\displaystyle Ar}{|}}{\overset{|}{C-}}}Ph$$

enhanced by substituents that stabilize such a charge and retarded by substituents that destabilize it. By preparing different pinacols from aryl phenyl ketones and analyzing the products of their rearrangement, it should be possible to compare the migratory aptitudes of different aryl groups with that of an unsubstituted benzene ring.

Your major objectives are as follows:

1. To prepare different aryl phenyl ketones by Friedel-Crafts reactions and convert them to pinacols.

2. To induce a pinacol rearrangement of each pinacol and analyze the product mixture by NMR or by cleavage and GLC (or both).

3. To compare the migratory aptitudes of the aryl groups in terms of some appropriate quantitative measure.

Key Reference:
N. M. Zaczek et al., "Migratory Aptitudes," *J. Chem. Educ.* **48**, 257 (1971).

The reference by Zaczek et al. outlines the procedures used with *p*-chlorophenyl and *p*-toluyl migrating groups. You should extend the scope of that study by testing a wider range of substituents, perhaps using *ortho*- and *meta*-substituted benzene rings, different *para*-substituents, polynuclear aryl groups such as naphthyl, etc.

PROJECT 7

Investigation of the Relative Stability of Endocyclic and Exocyclic Double Bonds

A reaction exhibits thermodynamic control when the composition of the product mixture is determined by the relative stabilities of the products, rather than by their relative rates of formation.

1-Benzylcycloalkanols can be dehydrated to yield alkenes containing either an *endocyclic* or an *exocyclic* double bond. This is illustrated by the dehydration of 1-benzylcyclopentanol, which yields about 75% of the *endo* product. Since the reaction is believed to involve thermodynamic control of the product mixture, it is evident that the alkene containing a double bond inside the ring is more stable, even though the *exo* double bond is stabilized by a phenyl group.

(major product)

Changing the ring size can alter the relative amounts of *endo* and *exo* products, but it does not appear to increase the amount of *exo* alkene to much over 25%. The use of different substituents on the starred carbon atom in structure **1** will presumably alter the *exo : endo* ratio. Some substituents might even make the *exo* product the predominant one.

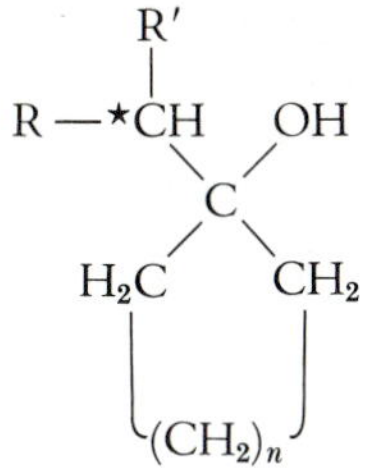

Your major objectives are as follows:

1. To prepare several 1-alkylcyclopentanols or 1-alkylcyclohexanols from the appropriate cyclic ketones and Grignard reagents.

2. To dehydrate each of the cyclic alcohols using the same reaction conditions.

3. To determine the percentage of *endo* and *exo* products in each reaction mixture using one or more appropriate analytical techniques.

4. To attempt to explain the observed variations in the *exo : endo* ratio on the basis of ring size and substituents.

In choosing the groups R and R' for incorporation into your cycloalkanols, consider the factors involved in stabilizing double bonds and also the availability of starting materials for the corresponding Grignard reagents. The method used for analyzing the product mixture should establish the identity as well as the quantity of each component. Thus, a GLC chromatogram alone is not sufficient unless the identity of the alkene peaks can be established by an independent method. The study can be extended by starting with small- or large-ring cyclic ketones other than cyclopentanone and cyclohexanone.

1

$n = 2,3$

Key Reference:
G. R. Newkome, J. W. Allen, and G. M. Anderson, "The Preparation and Dehydration of 1-Benzylcycloalkanols," *J. Chem. Educ. 50,* 372 (1973).

The R and R' groups can be alkyl, aryl, or hydrogen. They can also contain functional groups.

Synthesis and Activity of Cockroach Sex Pheromone Mimics

PROJECT 8

Scientists working for the U.S. Department of Agriculture have discovered that bornyl acetate stimulates sexual excitement in the male American cockroach, *Periplaneta americana*. Although bornyl acetate is not identical to the natural sex pheromone secreted by the female, the males of the species apparently mistake it for the real thing. The activity of this pheromone mimic depends on its stereochemistry; the D form of bornyl acetate is much more active than the L enantiomer.

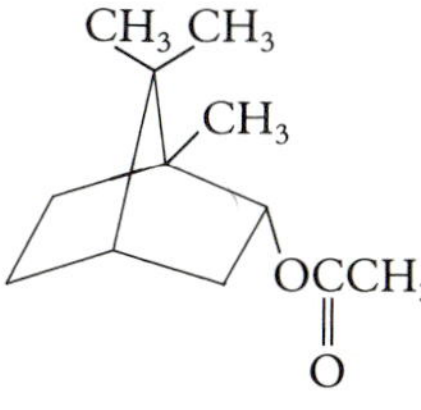

D-(−)-bornyl acetate

Key Reference:
W. S. Bowers and W. G. Bodenstein, "Sex Pheromone Mimics of the American Cockroach," *Nature* 232, 259 (1971).

The researchers did not investigate the activity of other stereoisomers such as isobornyl acetate. The reduction of camphor usually yields a mixture of borneol and isoborneol, from which the isomers can be separated and resolved. These alcohols can be converted by standard esterification methods to acetates and other esters.

Your major objectives are as follows:

1. To prepare a mixture of borneol and isoborneol by the reduction of camphor, and to separate and purify the alcohols. (As an alternative you can use special reducing agents that produce one isomer predominantly.)

2. To convert the pure alcohols to acetates or other esters for testing.

3. To measure the apparent effectiveness (as sex pheromone mimics) of the esters you have prepared, by testing them on *Periplanata americana.*

If you start with racemic camphor, you may also wish to resolve each of the products and test their enantiomers separately. You will be expected to carry out a literature search (as described in Experiment 50) to find information about the reactions and processes you will be using. Your project can be designed to demonstrate the effect of stereochemistry on activity or to explore the effect of structural modifications such as varying the acid component of a particular ester. You might also try to observe the effect of your esters on other cockroach species. The procedure for carrying out bioassays with cockroaches is described by Bowers and Bodenstein — capturing and identifying the cockroaches is up to you.

PROJECT 9

Preparation of a New Organic Compound

To prove the structure of α-pinene, Adolf von Baeyer started by oxidizing it to pinonic acid with potassium permanganate. This reaction has been shown to proceed in very high yield with "purple benzene," a solution of potassium permanganate in benzene containing a crown ether. Pinonic acid is a relatively uncommon compound containing two reactive functional groups. It should therefore be possible, using it as the starting material, to prepare a totally new organic compound — one that has never been reported in the chemical literature.

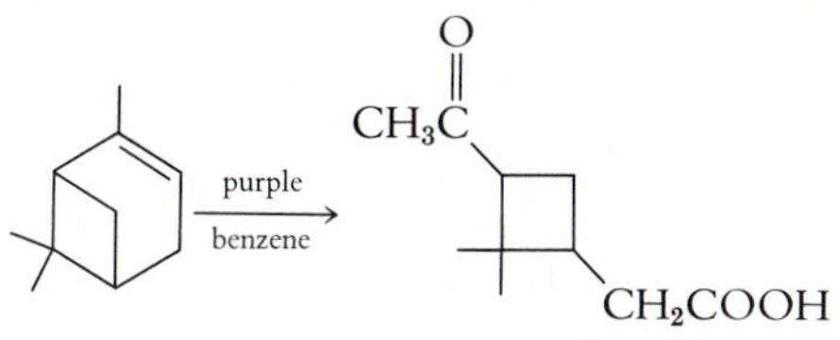

pinonic acid

SAFETY PRECAUTION

Crown ethers are toxic and have been reported to cause testicular atrophy in males.

Your major objectives are as follows:

1. To synthesize pinonic acid from α-pinene using "purple toluene."

2. To plan the synthesis of a new organic compound from pinonic acid and conduct a literature search to establish that the compound has not been previously reported.

The new compound should not be a simple functional group derivative (such as a phenylhydrazone or ester) of pinonic and itself. Its preparation should require two or more synthetic steps and should be approved by your instructor beforehand.

3. To carry out the synthesis of the compound and establish its structure by spectrometric and chemical methods.

4. To characterize the new compound as thoroughly as possible by measuring its physical properties, obtaining spectra, observing its chemical properties, etc.

Your literature search should encompass both *Beilstein* and *Chemical Abstracts* (see Experiment 50). Since crown ethers are quite expensive, you should try to make do with only a few grams of pinonic acid. This means that every reaction step in your synthesis must be carefully planned to provide the maximum yield of pure product. Keep in mind that some reagents may react with both functional groups and may even cause ring cleavage or other unanticipated reactions. It may be necessary to use a protective group for one function (such as ethylene ketal for the ketone function) while carrying out a reaction on the other. The final product should be obtained in as pure a form as possible, so that the physical and spectral data obtained would be suitable for submission to a professional journal.

Key Reference:
D. J. Sam and H. F. Simmons, "Crown Polyether Chemistry. Potassium Permanganate Oxidations in Benzene," *J. Am. Chem. Soc. 94,* 4024 (1974).

See Appendix VII and category B of the Bibliography for sources of synthetic methods.

PART
V

THE OPERATIONS

Part V contains descriptions of all the laboratory operations
employed in the experiments and projects in this book.

Elementary Operations

Cleaning and Drying Glassware

Cleaning Glassware

Clean glassware is essential for satisfactory results in the organic chemistry laboratory, since small amounts of impurities can inhibit chemical reactions, catalyze undesirable side-reactions, and invalidate the results of chemical tests or rate studies. It is important to clean glassware as soon as possible after it is used. Otherwise, residues may harden and become more resistant to cleaning agents; they may also attack the glass itself, weakening it and making future cleaning more difficult. It is particularly important to wash out strong bases such as sodium hydroxide or sodium methoxide quickly, since these substances can etch the glass permanently and cause "frozen" joints after prolonged contact. When glassware has been thoroughly cleaned, water applied to its inner surface should wet the whole surface and not form droplets or leave dry patches.

Most glassware can be cleaned adequately by vigorous scrubbing with hot water and soap (or a good laboratory detergent), using a brush of appropriate size and shape to reach otherwise inaccessible spots. A plastic trough or another suitable container can serve as a "dishpan." Organic residues that cannot be removed by detergent and water will often dissolve in organic solvents such as acetone or methylene chloride. An acetone rinse should be used to remove excess methylene chloride (or other chlorinated solvent) from glassware before scrubbing it with soap and water. After washing, always rinse the glassware thoroughly with water (a final distilled-water rinse is a good idea) and check it to make sure that the water wets its surface evenly. If it doesn't pass the wettability test, scrub it some more, or use a cleaning solution.

Very troublesome residues can usually be removed with a special commercial cleaning solution such as Nochromix.

Drying Glassware

The easiest (and cheapest) way to dry glassware is to let it stand overnight in a position that allows easy drainage. The outer

Always clean any dirty glassware at (or before) the end of each laboratory period.

See OP–2 for methods for dealing with frozen joints.

Used glassware that has been scratched or etched may not wet evenly.

A nylon mesh scrubbing cloth is useful for cleaning spatulas, stirring rods, large beakers, and the outer surfaces of other glassware. Pipe cleaners are ideal for cleaning narrow funnel stems, eyedroppers, etc.

Use organic solvents sparingly and recycle them, as they are much more costly than water. Never use reagent-grade acetone for washing; keep wash acetone separate from other grades.

surfaces of glassware may be dried with a soft cloth, if necessary. However, the surfaces that will be in contact with chemicals should *not* be dried in this way, because of the possibility of contamination. If a piece of glassware is needed shortly after washing, it can be drained of excess water and rinsed with one or two *small* portions of technical-grade acetone, then dried in a gentle stream of *dry* air ⟨OP–22⟩. Keep in mind that thorough drying is not necessary if the glassware is to be used with aqueous solutions. In that case, it is usually sufficient to drain the glassware for a few minutes after rinsing.

Glassware that is to be used under anhydrous conditions, for a Grignard reaction, for example, must be dried thoroughly before use. If possible, the glassware should be cleaned during the previous lab period, allowed to dry overnight or longer, and then dried in an oven set at about 110°C for 15 minutes or longer. **[Caution: Use asbestos gloves or tongs when handling hot glass!]** The apparatus should be assembled and fitted with one or more drying tubes ⟨OP–22a⟩ as soon as possible after the glassware is removed from the oven; otherwise, moisture will condense inside it as it cools. If the glassware must be cleaned the same day as it is used, the wet glassware should be drained to remove excess water, and then rinsed with acetone and flushed thoroughly with dry air before oven-drying. Glassware can also be dried by playing a "cool" Bunsen-burner flame over the surface of the assembled apparatus, but this practice should *never* be used in laboratories where volatile solvents such as ether are in use. It should be done only with the instructor's permission and according to his or her directions.

The acetone should then be recycled for washing.

Compressed air from an air line often contains pump oil and moisture.

Using Standard-Taper Glassware

OPERATION 2

Most ground-joint glassware used in organic chemistry is of the straight, standard-taper type with rigid joints. The size of a tapered joint is designated by two numbers (e.g., 19/22), in which the first number is the diameter, in millimeters, of the large end of the joint, and the second is the length of the taper. Straight adapters are available that make it possible to use different joint sizes in the same assembly. Organic lab kits that can be assembled to perform many different operations, from refluxing to fractional distillation, are available commercially.

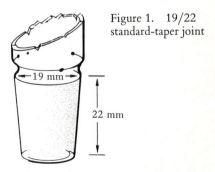

Figure 1.   19/22 standard-taper joint

19 mm

22 mm

The setups for the various operations are usually illustrated in the instruction booklet that comes with each kit.

Lubricating Joints

A thin layer of stopcock grease can be applied to the joints of standard-taper glassware to provide an air-tight seal and prevent freezing. Hydrocarbon greases are easily removed by acetone and other solvents, but because they are soluble in many organic liquids they may introduce undesired contaminants during such operations as distilling or refluxing. Silicone lubricants are suitable for most operations, particularly under conditions of high vacuum or high temperature. Unfortunately, silicones are difficult to remove. When applied improperly, they tend to form a film that can make it almost impossible to get the glass clean. A thorough scrubbing with methylene chloride may remove most, if not all, of the residue. Because of the likelihood of contamination, some chemists prefer not to lubricate ground joints except when an airtight seal is required (as in vacuum distillation) or when caustic solutions are being used.

Stopcock grease is applied only to the inner (male) joint of each component, by spreading a *very thin* coating on the top half of the joint. When the components are assembled, the joints are pressed firmly together with a slight twist (if necessary) to form a seal around the entire joint. In no case should excess grease extend beyond the joint inside the apparatus.

Each joint should be examined to see that the grease extends completely around the joint with no breaks.

Assembling Glassware

Equipment and supplies:
  organic chemistry lab kit (or
    equivalent)
  stopcock grease (optional)
  ringstands
  clamps
  rings and wire-gauze squares, as
    needed

Because standard-taper joints are rigid, an apparatus must be assembled carefully to avoid excess strain that can result in breakage. First the necessary clamps and rings are placed at appropriate locations on the ringstands. The apparatus is then assembled *from the bottom up, starting at the heat source.* If a heating bath or mantle is to be used, it should be placed at a location from which it can be lowered and removed when the heating period is over. The reaction flask or boiling flask is then clamped securely at the proper distance from the heat source ⟨OP–7⟩. As other components are added, they are clamped to the ringstands, but the clamp jaws are not tightened completely until all of the components are in place and aligned properly. When the clamps have been positioned so that all glass joints slide together without applying excessive force, the joints are seated and the clamps tightened carefully. Each joint is then examined for possible leaks or loose connections.

You should use as many clamps as necessary to provide adequate support for all parts of the apparatus. A vertical setup such as the one for addition and reflux ⟨OP–10⟩ requires at least two clamps for security, because if the setup is bumped, the clamp holding the reaction flask may rotate and deposit your glassware on the lab bench — usually with very expensive consequences. Some vertical components, such as Claisen adapters, need not be clamped if they are adequately supported by the component below. Nonvertical components such as distilling condensers should be clamped; if left unsupported, they may fall if the ringstands are accidentally moved apart.

If it will be necessary to lower a receiving flask (or other container) to add or remove substances, the flask can be supported by a ring and wire gauze or by a lab jack, blocks of wood, etc.

For clamping condensers and other components at an angle to a ringstand, an adjustable clamp with a wing nut on the shaft is required. This wing nut must be tightened after the apparatus is aligned. Occasionally a strong rubber band can substitute for a clamp when there will be little force on a joint; the vacuum adapter, for example, can be fastened to a condenser by passing a rubber band around the tubulation on both. A vacuum adapter should never be allowed to hang unsupported, since stopcock grease cannot be relied on to hold joints together against the force of gravity.

Spring clamps can also be used to hold glass joints together.

Figure 2 on page 602 illustrates the steps to be followed in assembling a typical ground-glass apparatus.

Disassembling and Cleaning Glassware

Ground-joint glassware should be disassembled promptly after it is used, as joints that are left coupled for extended periods of time may freeze together and become difficult or impossible to separate without breakage. Frozen joints can sometimes be loosened by tapping the outer joint gently with the wooden end of a spatula or by applying steam to the outer joint while rotating it slowly. The components are then separated by pulling them apart with a twisting motion. If these procedures do not work, the instructor should be consulted.

Strong bases can cause ground-glass joints to fuse together permanently.

If stopcock grease was used, each joint should be wiped free of excess grease and cleaned with methylene chloride or another appropriate solvent. The glassware should always be cleaned thoroughly ⟨OP–1⟩ after use; otherwise, residues may dry and harden, becoming more difficult to remove and sometimes attacking the surface of the glass.

Steps

1. Position clamps, rings.

2. Position heat source.

3. Secure boiling flask (clamp tightly).

4, 5. Add Claisen, 3-way adapters.

6. Clamp condenser in place.

7. Attach vacuum adapter with rubber band or spring clamp.

8. Attach receiving flask, hold in place with ring and wire gauze.

9. Readjust all clamps to align.

10. Press joints together tightly.

11. Tighten clamps.

12. Add stopper.

13. Add thermometer adapter, position thermometer.

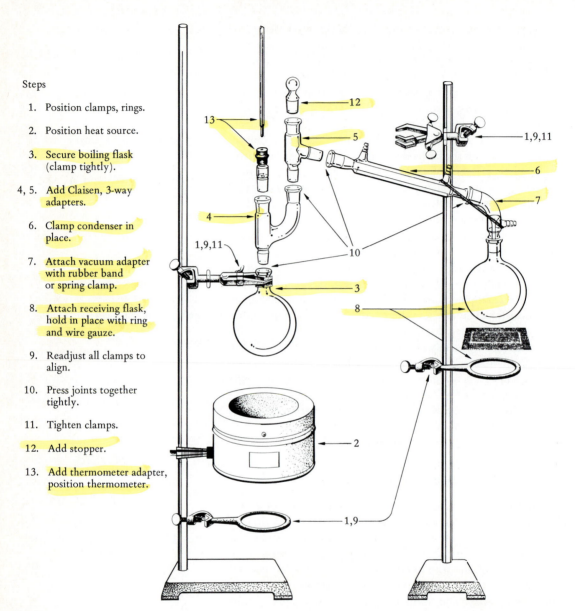

Figure 2.   Steps in the assembly of a typical ground-glass apparatus

When it is clean and dry, each component should be returned to its proper location in the lab kit or to the stockroom, as directed by the instructor.

Using Corks and Rubber Stoppers

Corks and rubber stoppers have been largely replaced by ground-glass joints for many operations such as refluxing and distillation, but they still have some important applications in organic chemistry laboratories. Rubber stoppers sometimes swell in contact with organic liquids and vapors and may contaminate organic solvents, but they provide a tighter seal than corks do and should be used for vacuum filtration ⟨OP-12⟩ and other reduced-pressure operations. Corks should generally be used to stopper test tubes or flasks and in other applications that do not require airtight seals. Ground-glass stoppers should be used to stopper standard-taper glassware; materials such as Parafilm or aluminum foil are often preferred over corks for sealing flasks and other containers when contamination may be a problem.

Boring Corks

The cork borer should be sharp (if it is not, ask the instructor to show you how to use a cork-borer sharpener) and slightly smaller in diameter than the object to be inserted in the cork. The borer should *not* be pushed forcefully through the cork; allow it to *cut* its way through by using a rotating motion and only a small amount of pressure. The borer and cork should be rotated in opposite directions, and the alignment should be checked frequently to make sure that the borer is going in straight. To avoid breaking off large chunks of cork, the borer should be removed when it is about halfway through and applied to the other end until the two holes meet.

Boring Rubber Stoppers

Rubber stoppers are bored in about the same manner as corks, but the borer must be *very* sharp and a lubricant such as glycerol should be applied to the cutting edge to reduce friction. It is particularly important to avoid using excessive force in boring rubber stoppers; trying to "punch" out a hole rather than cutting it out will result in a hole that is much smaller than the diameter of the borer. It is difficult to enlarge a hole already cut in a stopper, so if a large hole is required it should be made in a

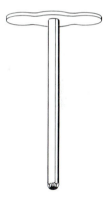

Figure 1. Cork borer

$$\begin{array}{ccc} OH & OH & OH \\ | & | & | \\ CH_2CH & - & CH_2 \end{array}$$

glycerol
(glycerine)

solid stopper and not in a stopper that already has one or more holes.

Inserting Glass Tubing and Thermometers

Glass tubing and thermometers can generally be inserted through corks and rubber stoppers without difficulty if a lubricant such as glycerol is first applied *generously* to the glass. Then grasp the tubing *close to the stopper* and *twist* through the hole with firm, steady pressure. Holding the tube too far from the stopper is dangerous, since the glass may break and lacerate or puncture the hand. Holding the glass tubing in a towel or other piece of cloth offers partial protection to the hands. It is important that force be applied with gentle but firm pressure *directly along the axis* of the tubing; any sideways force may cause the tube to break. Forcing a tube through a hole too small for it can also cause it to break. After the tube or thermometer is correctly positioned in the stopper, the glycerol should be rinsed off thoroughly with water.

SAFETY PRECAUTION

Improper insertion of glass tubing is one of the most frequent causes of laboratory accidents. The resulting cuts and puncture wounds can be very severe, requiring medical treatment and sometimes causing the victim to go into shock. Thermometers are particularly easy to break (especially at the 76-mm immersion line) and are expensive to replace.

Removing Tubing

In most cases, a tube or thermometer can be removed by twisting it out of the stopper with a firm, continuous motion. Hold the tubing close to the stopper, and avoid applying any sideways force that could cause it to break. A little glycerol applied to the part of the glass that will pass through the stopper can facilitate the process. If the glass tubing or thermometer cannot be removed easily by this method, the stopper can be cut off with a single-edged razor blade. Alternatively, a cork borer with a diameter slightly larger than the tube can be placed around it and twisted gently through the stopper until the glass can be dislodged.

Basic Glass Working

Glass connecting tubes and other simple glass items are required for many operations in organic chemistry. Soft-glass tubing, which is suitable for most applications of this type, can be worked easily with a Bunsen burner. Pyrex tubing may require the hotter flame provided by an oxygen torch or by some large burners.

Equipment and supplies:
  burner
  flame spreader
  glass tubing or glass rod

Cutting Glass Rod and Tubing

Most glass rods and tubes can be "cut" by scoring them at the desired location and snapping them in two. The rod or tube is scored by drawing a sharp triangular file across the surface at a right angle to the axis of the tubing. Often only a single stroke is needed to make a deep scratch in the surface. The file should not be used like a saw, and the scratch should be on only one side of the glass. Moisten the scratch with water or saliva; then break the rod or tube by placing the thumbs about 1 cm apart on the side opposite the scratch and, while holding the glass firmly in both hands, pushing out quickly with the thumbs and pulling the hands apart simultaneously. A towel placed over the glass helps to protect the hands.

Special glass-scoring knives are available.

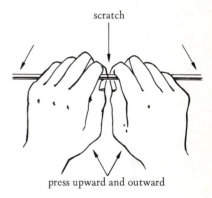

scratch

press upward and outward

Figure 1.   Breaking glass tubing

Working Glass Rod

The ends of a glass rod can be rounded by rotating the rod in a burner flame, holding it at a 45° angle with its tip near the inner blue cone of the flame. An end can be flattened by rotating it in the flame until it is incandescent and very soft, then pressing the rod straight down onto a hard surface such as the base of a ringstand.

See OP–7 for directions on using a burner.

Be careful not to burn yourself on the hot end of the glass rod or tube you are working with. Do not lay hot glass directly on the desk top or onto combustible materials.

**SAFETY
PRECAUTION**

Fire Polishing and Sealing Glass Tubing

Cut glass tubing (and glass rod) should always be *fire polished* to remove sharp edges and prevent accidental cuts. The cut end of

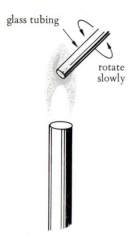

glass tubing

rotate
slowly

Figure 2.   Fire polishing

the tube is rotated slowly in a burner flame at about a 45° angle to the flame until the edge becomes rounded and smooth.

The end of a soft-glass tube can be closed by rotating it in a flame until the soft edges come together and eventually merge. To obtain a sealed end of uniform thickness, remove the tube from the flame as soon as it is closed and blow gently into the open end. Occasionally, a tiny hole will remain in the closed end. This can be detected by allowing the tube to cool to room temperature, attaching the open end to an aspirator, and placing the closed end in a test tube containing a small amount of acetone **[caution: flammable!]** or methylene chloride. If the tube is not properly sealed, the liquid will leak into it when you apply suction.

A capillary tube can be sealed at one end by holding it at a 45° angle to a small flame and rotating the tip of the tube in the outer edge of the flame. When a tube is to be used for a sealed-tube reaction, as in Experiment 47, it should be heated gently in the flame just before sealing the open end; this results in a partial vacuum inside and ensures a good seal. Otherwise, expanding air may cause the newly sealed end to bulge outward, weakening the glass so that it may burst during the reaction.

Bending Glass Tubing

Unless a large-barreled burner (Meker burner, etc.) is available, a flame spreader should be used for all glass-bending operations. The tubing is held over the burner flame parallel to the long axis of the flame spreader and rotated constantly at a slow, even rate until it is soft enough to bend easily. (At this point, it

The flame will turn yellow as the glass begins to soften.

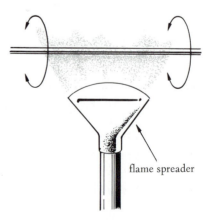

flame spreader

Figure 3.   Bending tubing

should be nearly soft enough to bend under its own weight.) The hot tubing is then removed from the flame and immediately bent to the desired shape with a firm, even motion. If the glass is soft enough, it should not be necessary to use much force to bend it, and the bend should follow a smooth curve with no constrictions.

Bend the glass in a vertical plane, with the ends up and the bend at the bottom.

Weighing

Many different kinds of balances are used in chemistry laboratories, including triple-beam balances, which require manual adjustment of weights on each of three balance beams; top-loading automatic balances; and analytical balances, which are accurate to closer than 1 mg. Strictly speaking, such balances measure mass rather than weight; however, the word "weight" is commonly used for "mass" when this will not cause error or confusion. Your instructor will explain or demonstrate the operation of the balances used in your laboratory.

Mass is a measure of the quantity of matter in an object; weight is a measure of the earth's gravitational attraction for the object.

Weighing Solids

Most solids can be weighed in glass containers (such as vials or beakers), in aluminum weighing dishes, or on special glazed weighing papers. Filter paper and other absorbent papers should not be used for accurate weighings, since a few particles will always remain in the fibers of the paper. If you are weighing an indeterminate amount of a solid (such as the product of a reaction), its container or the weighing paper should first be weighed separately and the mass recorded (unless the balance has a taring system). Then the solid is added, the total mass is recorded, and the mass of the container or weighing paper is subtracted. If the balance has a taring system, the mass readout scale is set to zero with the taring control while the container or weighing paper is on the balance pan. Then the solid is added, and its mass is read directly from the scale.

If you are measuring out a specified quantity of a solid (such as a solid reactant) on a balance that lacks a taring system, the expected total mass of the solid plus its container should be calculated, and the balance should be set to approximately that value. Then the solid is added in small portions, using a clean

Hygroscopic solids (those that absorb water from the atmosphere) should be weighed in closed containers.

See J. Chem. Educ. 55, 455 (1978) for a description of a convenient solids dispenser.

scoop or spatula, until the desired reading is attained. If the balance has a taring system, its readout scale can be zeroed with the container on the pan and then set to the desired mass of the solid. Modern electronic balances have no dials for presetting the expected mass. The weighing container is placed on the pan, and a button or bar is pressed to zero the digital readout; then the material being weighed is added until the desired mass appears on the readout.

Most products obtained from a preparation are transferred to vials or other small containers, which should be weighed empty and then reweighed after the product has been added. As a general practice, the container should be weighed with its cap and label on and this **tare** weight recorded. Then the mass of the contents at any given time can be obtained by subtracting the tare weight from the total mass of container and contents.

Weighing Liquids

The mass of an indeterminate quantity of a liquid is measured as described above for a solid. A tared container should be used, and it should be kept stoppered during the weighing to avoid loss by evaporation. To weigh out a specified quantity of a liquid from a reagent bottle, you should first measure out the approximate quantity of the liquid *by volume* and then weigh that quantity accurately in a closed container. For example, if 9.0 g of 1-butanol ($d = 0.81$ g/mL) is required for a synthesis, about 11.1 mL (9.0 g $\div$ 0.81 g/mL) of the liquid is measured into a graduated cylinder and transferred to a tared vial. If the measured weight is not close enough to that required, liquid can be added or removed with a clean dropper. Be careful not to get any liquid on the balance; a modern balance is a precision instrument that can easily be damaged by contaminants.

Before you leave the balance area be sure to replace the caps on all reagent bottles, return them to their proper locations, and clean up any spills on or near the balance.

Allow a moistened label time to dry before weighing a labeled container.

tare: allowance for the weight of a container

Never withdraw liquid directly from a reagent bottle with your dropper or pipet — you may contaminate the liquid.

OPERATION 6

Measuring Volume

A given volume of a liquid can be measured using a graduated cylinder, a pipet, or a syringe, depending on the quantity and accuracy required. Burets and volumetric flasks are also used to measure liquid volumes accurately. Their use is discussed in most general and analytical chemistry laboratory manuals.

Graduated Cylinders

Graduated cylinders are not highly accurate, but they are adequate for measuring specified quantities of solvents and wash liquids as well as liquid reactants that are present in excess. The liquid volume should always be read from the bottom of the liquid meniscus as shown in Figure 1.

Pipets

Graduated or volumetric pipets can be used to measure relatively small quantities of a liquid accurately. Suction is required to draw the liquid into a pipet; using mouth suction is unwise because of the danger of drawing toxic or corrosive liquids into the mouth. An ordinary ear syringe works quite well as a pipetting bulb. The convenient pipetting-bulb assembly illustrated in Figure 2 is operated as follows:

1. The top end of the pipet is inserted into the pinchcock valve tubing.

2. The pinchcock is opened by pinching it at the glass bead, and the bulb is squeezed to eject the air.

3. The pipet tip is placed in the liquid, and the pinchcock is squeezed open to fill the pipet to just above the calibration mark.

4. The bulb is removed from the narrow end of the dropper, and the pinchcock is carefully squeezed open until the liquid falls to the calibration mark.

5. The liquid is then delivered into another container by opening the pinchcock valve.

Most volumetric pipets are calibrated "to deliver" (TD) a given volume, meaning that the measured liquid is allowed to drain out by gravity, leaving a small amount of liquid in the tip of the pipet. This liquid is not removed, since it is accounted for in the calibration. Graduated pipets are generally filled to the top (zero) calibration mark and then drained into a separate container until the calibration mark for the desired volume is reached (see Figure 3 on page 610). The maximum indicated capacity of some graduated pipets is delivered by draining to a given calibration mark; others are to be drained completely. It is important not to confuse the two, since draining the first

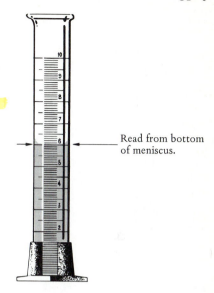

Read from bottom of meniscus.

Figure 1. Reading the volume contained in a graduated cylinder—in this case, 6.0 mL

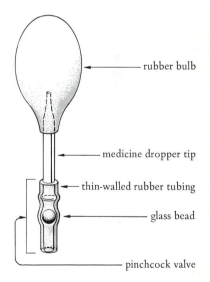

rubber bulb

medicine dropper tip

thin-walled rubber tubing

glass bead

pinchcock valve

Figure 2. Pipetting bulb [*J. Chem. Educ. 51,* 467 (1974)]

Figure 3.   Reading the volume delivered from a graduated pipet—in this case, 0.30 mL

type completely will deliver a greater volume than the indicated capacity of the pipet.

Syringes

Syringes are most often used for the precise measurement and delivery of very small volumes of liquid, as in gas-chromatographic analysis ⟨OP-32⟩. A syringe is filled by placing the needle in the liquid and slowly pulling out the plunger until the barrel contains a little more than the required volume of liquid. Then the syringe is held with the needle pointed up and the plunger is pushed in to eject the excess sample. Excess liquid is wiped off the needle with a tissue.

Syringes should be cleaned immediately after use by rinsing them several times with a volatile solvent, then removing the plunger and letting the barrel dry. Microsyringes can be dried rapidly by aspiration using the apparatus illustrated in Figure 5. The needle is inserted carefully through the dropper bulb and the aspirator is turned on for a minute or so. The pinchclamp is then opened to release the vacuum, the aspirator is turned off, and the syringe is removed.

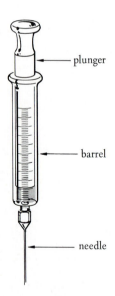

Figure 4.   Syringe

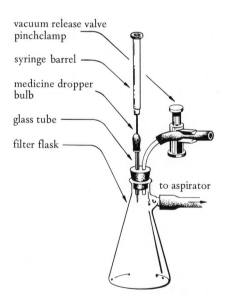

Figure 5.   Apparatus for drying microsyringes

Operations for Conducting
Chemical Reactions

Heating

Heat Sources

Many kinds of heat sources are available for such applications as heating reaction mixtures, evaporating volatile solvents, and carrying out distillations. The choice of a heat source for a particular application depends on such factors as the temperature required, the flammability of a liquid being heated, the need for simultaneous stirring ⟨OP–9⟩, and the cost and convenience of the heating device.

Burners. Bunsen burners and similar heat sources are simple and convenient to operate and can bring a liquid rapidly to its boiling point; however, they have some serious disadvantages that limit their use. It is difficult to control the temperature of a burner flame precisely, although rough adjustments can be made by varying its distance from the vessel being heated and its air/gas mix. The temperature at different parts of the vessel can vary widely; generally it is hottest at the bottom. The resultant local superheating can cause reaction mixtures to decompose or undergo unwanted side reactions. Finally, burners create a serious risk of fire in laboratories where volatile solvents are in use.

Burners should *never* be used to heat flammable liquids in open containers. Some flammable liquids with relatively high boiling points can be safely heated under reflux or in distillation setups; however, all joints must be tight, water condensers must be working properly, and the flame must be extinguished when the liquid is being introduced into, or removed from, the apparatus. Before using a burner, you must make sure that other students will not be handling flammable solvents nearby.

For refluxing or distilling liquid mixtures, a burner with an asbestos-centered wire gauze should be used to spread the flame and promote even heating. The gauze should be placed

Never leave a burner flame unattended; it may change in intensity, or even go out and possibly cause an explosion.

Heating a flask directly can strain the glass and cause it to break.

on a ring about 2–3 cm from the barrel of the burner, and the reaction flask should be clamped to the ringstand with its bottom *just* above the center of the gauze. If necessary, open the gas valve (usually on the bottom of the burner). Then ignite the burner and adjust the flame by rotating the barrel counterclockwise for a hotter, more oxygen-rich flame or clockwise for a cooler flame. The size of the flame can be controlled by readjusting the gas valve and oxygen control in sequence.

Local superheating will be reduced if the flask is not actually touching the gauze.

On some burners, the air flow is regulated by a sliding valve on the barrel.

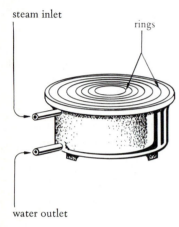

steam inlet

rings

water outlet

Figure 1. Steam bath

Steam Baths. A steam bath is a safe, convenient heat source that is somewhat limited by the fact that it has only one operating temperature, 100°C. It is particularly useful for heating recrystallization mixtures, evaporating volatile solvents, and refluxing low-boiling liquids. A steam bath *cannot* be used to boil water or aqueous solutions. The condensation of steam in the vicinity of a steam bath may be a nuisance, but this can be reduced by using enough rings to bridge any gaps between a flask and the steam bath and by maintaining a slow rate of steam flow. Beyond a certain point there is no advantage to increasing the steam-flow rate, since the temperature is constant. If the flask is placed correctly, heating is comparatively even and efficient, and the low operating temperature helps prevent the decomposition of heat-sensitive substances.

Before a steam bath is used, the steam valve should be opened fully **[avoid contact with the steam!]** to purge the steam line of water, and then closed and connected to the *steam inlet* with a length of rubber tubing. Another length of tubing should extend from the *water outlet* to a sink for drainage. The steam valve can then be partially opened and adjusted to provide the desired rate of heating. Metal water baths that have an inlet tube can serve as steam baths if they are emptied of water periodically.

A steam bath can be used to heat a reaction mixture at reflux ⟨OP–7a⟩ when the boiling point of the solvent is below 100°C. Just enough rings should be removed to accommodate the boiling flask, which is clamped in position so that its liquid level is just below the level of the rings and then fitted with a reflux condenser. At the end of the reaction period, the steam is turned off completely, and the reaction apparatus is allowed to cool. Raising the apparatus out of the steam bath (or lowering the steam bath, if it is on a support) will shorten the cooling time.

Flat-bottomed containers such as beakers and Erlenmeyer flasks can be heated on top of the steam bath by leaving on just

The steam bath should be drained of water after use.

enough rings to support them safely. If a low rate of heating is desired, more rings can be left on the steam bath to reduce the area of direct contact between the steam and the container. If a higher heating rate is needed, enough rings can be removed so that the flask or beaker can be held or clamped inside the steam bath, with its bottom below the level of the rings.

Heating Mantles. A heating mantle is a good general-purpose heating device that can be used in a variety of applications. Unlike a Bunsen burner, a heating mantle can be used with magnetic stirrers, and its heat output can be controlled precisely. Unfortunately, most heating mantles are costly and slow to warm up and accommodate only a narrow range of flask sizes. It is also difficult to monitor the operating temperature of a heating mantle. Despite these disadvantages, heating mantles are preferred for many heating operations.

A heating mantle must be used in conjunction with a variable transformer or a time-cycling heat control to regulate the heat output. Since the temperature of the mantle itself can be measured only by a thermocouple, it is difficult to set a heating mantle for operation at a specific temperature. A mantle will generally not be at thermal equilibrium with the contents of a flask, so if the flask is not filled to a level near the top of the mantle, the part of the flask above the liquid level will be hotter than its contents, and this can cause decomposition of materials splashed onto it. When possible, the mantle should have a well of nearly the same diameter as the flask being heated. Some kinds of all-purpose mantles are intended for operation with a range of flask sizes; however, heating efficiency is reduced and the chance of superheating is increased when a small flask is heated in a large mantle.

The mantle should be mounted on a lab jack, ring, set of wood blocks, or some other support so that it can be lowered and removed quickly if the rate of heating becomes too rapid. The flask is clamped in place so that it is in direct contact with the heating well, and the heating control dial is adjusted until the desired rate of heating is attained. Because a heating mantle responds slowly to changes in the control setting, it is easy to overshoot the desired temperature by turning the control too high at the start. If this occurs, the mantle should be lowered so that it is no longer in contact with the flask. The voltage input should then be reduced, and sufficient time should be allowed for the temperature to drop before the mantle is raised again. Further adjustment may be required to maintain heating at the desired rate.

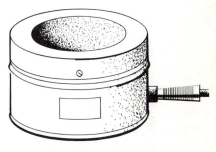

Figure 2. Heating mantle

Figure 3. Heat control for mantle

A 250-mL Briskeat beaker heater with integral heat control works quite well with small flasks. Its heating rate can be increased by lowering the flask in the well and by using glass wool for insulation.

Never heat an empty flask; this may burn out the heating element on the heating mantle.

Oil Baths. A good oil bath can supply uniform heating and precise temperature control. Unlike the preceding heat sources, an oil bath can operate at thermal equilibrium with the contents of a reaction flask. Accordingly, decomposition and side reactions caused by local overheating are less likely, and the reaction temperature can be determined by suspending a thermometer in the bath liquid.

The main disadvantages of oil baths are that they are messy to work with and difficult to clean and can cause dangerous fires or severe burns. You should not heat an oil bath above the *flash point* of the heating oil, since above this temperature the vapors of the oil can be ignited by a spark or burner flame. (Most oil fires can be extinguished by dry-chemical fire extinguishers or powdered sodium bicarbonate.) Old, dark oil is more likely to flash than new oil, and most bath oils will start to smoke and decompose at elevated temperatures. Hot bath oil can cause severe injury if accidentally spilled on the skin— the oil, being difficult to remove and slow to cool, remains in contact with the skin long enough to produce deep burns. Water should be kept away from hot oil baths since it causes dangerous splattering; oil that contains water should not be used until the water is removed. A bath liquid that is dark and contains gummy residues should be replaced.

Useful heating-bath liquids:

Mineral oil
   b.p. 360°C
   flash pt. 193°C
   Potential fire hazard

Glycerol
   b.p. 290°C
   flash pt. 160°C
   Water-soluble, nontoxic, viscous

Dibutyl phthalate
   b.p. 340°C
   flash pt. 157°C
   Viscous at low temperatures

Triethylene glycol
   b.p. 276°C
   flash pt. 177°C
   Water-soluble

Polyethylene glycols (Carbowaxes)
   b.p. and flash points vary, depending on M. W. range. Some are solid at room temperature. Water-soluble

Dow Corning silicone oils
   DC 330: flash pt. 290°C
   DC 550: flash pt. 310°C
   Expensive. Decomposition products are very toxic.

Some commonly used bath liquids are listed in the margin. Mineral oil is useful for most applications, although it presents a potential fire hazard and is hard to clean up. Polyethylene glycol oils such as Carbowax 600 are water-soluble (which makes cleanup much easier) and can be used at relatively high temperatures without appreciable decomposition. Silicone oils can be used at even higher temperatures.

An oil bath can be heated by a coil of resistance wire, a power resistor, a Calrod heating element, or some other device that can be safely immersed in the bath liquid. External heat sources such as hot plates are sometimes used, but they may cause a fire if the oil spills onto the hot surface. The output of a heating element is controlled by a variable transformer and the temperature of the bath is measured ⟨OP–7c⟩ with a thermometer suspended in the liquid. A large porcelain casserole makes a convenient bath container, since it is less easily broken than a glass container and has a handle for convenient placement and removal.

The oil bath should be placed on a lab jack, a set of wood blocks, or some other support that will allow it to be lowered quickly when necessary. It should not be set on a ring support because of the danger of spilling hot oil when the ring is raised or lowered. The flask to be heated is placed in the bath so that

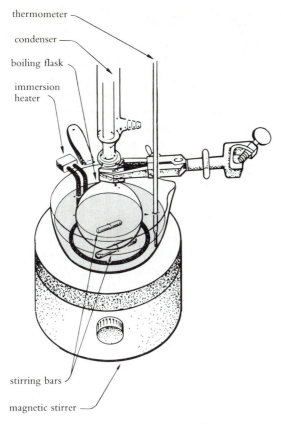

thermometer

condenser

boiling flask

immersion
heater

stirring bars

magnetic stirrer

Figure 4.   Oil bath assembly

the liquid level inside it is just below the oil level. The heating control unit is turned up until the desired temperature is obtained, then adjusted to maintain that temperature. Very uniform heating is possible if the oil bath is stirred magnetically; a reaction mixture can be stirred simultaneously if desired, as illustrated in Figure 4.

Water Baths.   Water baths are useful for evaporating volatile solvents ⟨OP–14⟩ and in other applications requiring gentle, even heating. When precise temperature control is not essential, a water bath can simply be a large beaker, porcelain casserole, or other container filled with preheated water (from a hot-water tap or another source); its temperature can be adjusted by adding hot or cold water. As the bath cools, some of the bath water can be drained or siphoned off and replaced by fresh hot water. When there is no significant fire hazard, a water bath may be heated with a burner or hot plate. Like an oil bath, a water bath can be stirred for more uniform heating, and

Figure 5.   Metal water bath

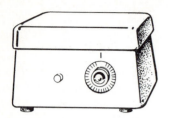

Figure 6.   Hot plate

Asbestos gloves or beaker tongs should be kept handy so that a vessel can be quickly removed from a hot plate if necessary.

Figure 7.   Heat lamp

J. Chem. Educ. 53, 50 (1976) and *J. Chem. Educ. 54,* 611 (1977) describe the construction and use of a "boiling tube" to prevent bumping.

its temperature can be monitored by suspending a thermometer in the water. Commercial water baths with drain tubes and removable metal rings are available.

Hot Plates.   A hot plate can be used to heat Erlenmeyer flasks or beakers, unless they contain low-boiling flammable liquids that could splatter on the hot surface and ignite. Hot plates can also be used to heat oil baths or water baths for reactions or distillations. A hot plate should never be used to heat a round-bottomed flask directly, since local overheating and breakage may result. Hot plates with built-in magnetic stirrers are available.

Other Heat Sources.   Other heating devices such as infrared heat lamps, electric forced-air heaters (heat guns), and electrically heated "air baths" can be used in some heating applications. A heat lamp plugged into a variable transformer provides a safe and convenient way to heat comparatively low-boiling liquids. The boiling flask is usually fitted with an aluminum-foil heat shield to concentrate the heat on the reaction mixture.

Smooth Boiling Devices

When liquids are heated to their boiling points or above, they may erupt violently as large bubbles of superheated vapor are discharged from the solution. A porous object such as a boiling chip or boiling stick can prevent this "bumping" by emitting a steady stream of small bubbles that breaks up the large vapor bubbles. Boiling chips (or boiling stones) are generally made from pieces of pumice, carborundum, marble, or glass. Acid- or base-resistant boiling chips (carborundum, for example) should be used when heating strongly acidic or alkaline mixtures, since ordinary boiling chips may break down in such mixtures.

Microporous boiling chips, made from a special grade of anthracite, can be used when liquids are distilled under reduced pressure ⟨OP–26⟩. Wooden applicator sticks can be broken in two and the broken ends used to promote smooth boiling in nonreactive solvents; they should not be used in reaction mixtures because of the possibility of contamination.

When boiling stops, liquid is drawn into the pores of a boiling chip, making it less efficient.

It is important to add boiling chips *before* heating begins, since a liquid may froth violently and boil over if they are added when the liquid is hot. When a liquid is cooled below its boiling point and then reheated, one or two new boiling chips should be added before heating is resumed.

Stirring causes turbulence that breaks up large bubbles, so boiling chips are not needed when a liquid is stirred constantly at the boiling point. Unless you are instructed differently, you should *always* add boiling chips to an unstirred reaction mixture or a liquid being distilled.

Refluxing

Heating accelerates the rate of a chemical reaction by increasing the average kinetic energy of the molecules, so that a larger fraction of molecules will have sufficient energy to react at a given instant. The temperature of a reaction mixture can be controlled in several ways, the simplest and most convenient being to use a reaction solvent that has a boiling point within the desired temperature range for the reaction. The reaction is then conducted at the boiling point of the solvent, using a water-cooled condenser to return the solvent vapors to the reaction vessel and prevent their escape. This process of boiling a reaction mixture and condensing the solvent vapors back into the reaction vessel is known as *refluxing* or heating under reflux. It is the most widely used technique for carrying out organic reactions at elevated temperatures.

Operational Procedure

Position the heating device ⟨OP–7⟩ at the proper location on or near a ringstand so that it can be quickly removed if the flask should break or the reaction become too vigorous. Select a round-bottomed flask of the right size (the reactants should fill it about half full, or a little less) and clamp it onto the ringstand at the proper location in relation to the heat source (see OP–7). Solids should be added to the flask through a powder funnel or a makeshift "funnel" constructed from a piece of glazed paper; liquids should be added through a stemmed funnel of some kind. Mix the reactants by swirling or stirring, and add a few boiling chips; then insert a reflux condenser into the flask and clamp it to the ringstand. (Do not mistake a jacketed distilling column for a reflux condenser; the condenser has a smaller diameter.)

Connect the water inlet (the lower connector) on the condenser jacket to a cold-water tap with a length of rubber tubing, and run another length of tubing from the water outlet (the upper connector) to a sink, making sure that it is long enough to prevent splashing when the water is turned on.

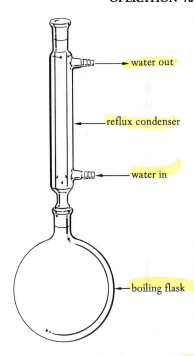

Figure 1. Simple reflux assembly

A clamp on the condenser keeps the apparatus from toppling over if it is jarred. Be sure the ground-glass joints don't separate when the clamp is tightened.

If the rubber tubing slips off when pulled with moderate force, replace it by tubing of smaller diameter or secure it with wire.

SAFETY PRECAUTION

===

Never begin heating before the condenser water is turned on; solvent vapors may escape and cause a fire or health hazard.

===

High water pressure may force the tubing off the condenser or cause a "flood."

Equipment and supplies:
  round-bottomed flask
  reflux condenser
  boiling chips
  two lengths of rubber tubing (about 1 m each)
  heat source

Turn on the water carefully so that the condenser jacket slowly fills with water from the bottom up, and adjust the water pressure so that a narrow stream flows from the outlet. The flow rate should be just great enough to (1) maintain a continuous flow of water in spite of pressure changes in the water line and (2) keep the condenser at the temperature of the tap water during the reaction.

Turn on the heat source, and adjust it to maintain gentle boiling of the solvent. A continuous stream of bubbles should emerge from the liquid, but bumping and excessive foaming should be avoided. Refluxing has started when liquid begins to drip into the flask from the condenser. The vapors passing into the condenser will then form a *reflux ring* of condensate that should be clearly visible. Below this point, solvent will be seen flowing back into the flask; above it, the condenser should be dry. If the reflux ring rises more than halfway up the condenser, the heating rate should be reduced to prevent the escape of solvent vapors.

If the reflux ring is too high, it may be necessary to change to a more efficient condenser or use two condensers in tandem.

At the end of the reaction period, turn off the heat source and remove it from contact with the flask. Let the apparatus cool; then turn off the condenser water, and pour the reactants into a container suitable for the next operation. Clean the reaction flask ⟨OP–1⟩ as soon as possible so that residues do not dry on the glass.

Cooling can be accelerated by passing an air stream over the flask.

Summary

1. Position heat source.

2. Clamp flask at proper location in relation to heat source.

3. Add reactants and solvents to flask through funnel, and mix.

4. Add boiling chips.

5. Insert reflux condenser, clamp in place, and attach rubber tubing.

6. Turn on and adjust water flow.

See OP–9 if the reaction mixture must be stirred while refluxing and OP–22a or OP–22b if a dry atmosphere or a gas trap is required.

7. Commence heating; adjust heat until reaction mixture boils gently.

8. Check position of reflux ring; readjust water flow or heating rate, if necessary.

9. At end of reflux period, remove heat source, allow flask to cool, transfer reactants, and disassemble and clean apparatus.

Semimicro Refluxing

OPERATION 7*b*

Small amounts of reactants can be heated under reflux using a *cold-finger condenser* inserted into a test tube or small flask. When water is passed through the condenser, it cools the surrounding area enough to condense the rising vapors. To prevent pressure buildup in the container, a stopper with a groove in one side (or a two-hole stopper) must be used. The reflux ring that appears on the sides of the container should be well below the top of the test tube or flask so that vapors will not escape. Some microcondensers are available that resemble an ordinary reflux condenser (except in size) and are used in the same way.

The operational procedure for small-scale refluxing is essentially the same as that for ordinary refluxing ⟨OP–7*a*⟩, except that the cooling water goes into the *upper* connector of a cold-finger condenser and comes out the lower one.

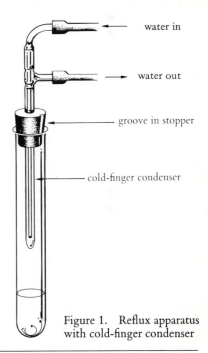

water in

water out

groove in stopper

cold-finger condenser

Figure 1. Reflux apparatus with cold-finger condenser

Temperature Monitoring

OPERATION 7*c*

The temperature of a liquid in a reaction flask, oil bath, or other container can be monitored with a thermometer positioned so that its bulb is *entirely immersed* in the liquid. If the thermometer is used in an open container, it can be held in place by a three-fingered clamp or inserted into a stopper that is held by an ordinary utility clamp. If it is used in a container that must be closed off (as in a distillation assembly), the thermom-

If possible, the thermometer should be immersed down to an inscribed immersion mark (usually 76 mm).

eter is usually inserted into a rubber stopper or (preferably) a thermometer adapter.

Some reactions are conducted at temperatures below the boiling point of the solvent, either because decomposition or unwanted side reactions occur at the boiling point or because the reaction proceeds at a more convenient rate at a lower temperature. When water or other high-boiling solvents are used, such a reaction might be run in an Erlenmeyer flask into which a thermometer has been inserted to monitor the reaction temperature. (The thermometer should not be used as a stirring rod, since the bulb is fragile and breaks easily.) If necessary, the reactants can be mixed ⟨OP–9⟩ by either magnetic stirring or manual mixing. If the temperature must be monitored continuously as you swirl the reaction flask, it is important to hold the thermometer securely so that it doesn't break. This can be done by holding the neck of the flask between your thumb and fingers and nesting the thermometer stem in the "vee" between the thumb and index finger so that the bulb of the thermometer is held securely inside the flask. (The technique is described in more detail in Experiment 19.) With a little practice, you should be able to mix the contents of the flask quite vigorously without damage to the thermometer.

If the reaction solvent is volatile at the reaction temperature, and particularly if it poses a health or fire hazard, a reflux condenser should be used to keep its vapors out of the atmosphere. Stirring is often necessary in this case, as the absence of boiling action prevents good mixing. An example of an apparatus suitable for simultaneous heating, magnetic stirring, and temperature monitoring is shown in Figure 1. If mechanical stirring ⟨OP–9⟩ or addition of reagents ⟨OP–10⟩ is also required during the reaction period, a three-necked flask can be used. One neck supports the reflux condenser, a second holds a thermometer, and a third receives an addition funnel or stirrer sleeve. Some flasks are provided with a thermometer well, which frees one neck of the flask for other purposes.

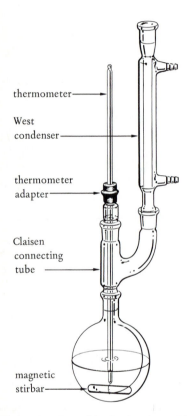

thermometer

West condenser

thermometer adapter

Claisen connecting tube

magnetic stirbar

Figure 1. Assembly for heating with temperature monitoring

OPERATION 8

Cooling

Occasionally a reaction proceeds too violently at room temperature or involves reactants or products that decompose at room temperature. In such cases, it may be necessary to cool the reaction mixture using some kind of cold bath, which can be anything from a beaker filled with cold water to an electrically refrigerated bath.

A mixture of ice and water in a suitable container can be used for cooling down to 0°C. The ice should be finely divided; snow is ideal. Enough water should be present to form a thick slurry, since ice alone is not an efficient heat-transfer medium. A freezing mixture consisting of three parts of finely crushed ice or snow with one part sodium chloride is good for temperatures down to about −20°C; and mixtures of $CaCl_2 \cdot 6H_2O$ containing up to 1.4 g of the calcium salt per 1 g of ice or snow can provide temperatures down to −55°C. In practice, the minimum values may be difficult to attain; the actual temperature of an ice bath depends on the fineness of the ice and salt, the rate of stirring, and the insulating ability of the container.

Equipment and supplies:
bath container (beaker, crystallization dish, Dewar flask etc.)
ice water or other cooling mixture
thermometer

Ice-salt combinations must be mixed thoroughly and stirred frequently.

Do not handle dry ice with bare hands.

SAFETY PRECAUTION

Lower temperatures can be reached by mixing small chunks of dry ice (solid carbon dioxide) with a suitable solvent (for example, acetone, chloroform, or ethanol) in a Dewar flask or other insulated container. Such a bath can reach minimum temperatures around −75°C. Specific temperatures between about −26° and −72°C can be attained by using dry ice in mixtures of *ortho* and *meta*-xylene of varying composition [*J. Chem. Educ.* **45**, 664 (1968)]. Temperatures below those attainable with dry ice are possible using liquid nitrogen (b.p. −196°C), either alone or with an appropriate solvent.

A beaker, crystallization dish, or metal water bath is usually satisfactory for a cold bath; the latter two have a more convenient shape. The useful life of a cold bath can be extended by wrapping the bath container with glass wool and placing it inside a larger container. If a cold bath is needed for a long period of time or at a particularly low temperature, an insulated container such as a Dewar flask or Thermos jar should be used. The temperature of the bath (and of the reaction mixture, if desired) can be monitored with a suitable thermometer. For the most efficient cooling, the flask (or other container) should be immersed deeply in the cooling bath, and its contents should be swirled or stirred frequently.

Temperatures below −40°C cannot be measured using a mercury thermometer; mercury freezes at that temperature.

OPERATION 9

Mixing

Reaction mixtures are frequently stirred, shaken, or agitated in some other way to promote efficient heat transfer, improve contact between the components of a heterogeneous mixture, or mix in a reactant that is being added during the course of a reaction. If the reaction is being carried out in an Erlenmeyer flask, mixing can be accomplished by manual shaking and swirling or by using a stirring rod. If the apparatus is not too unwieldy and the reaction time is comparatively short, ground-glass assemblies can sometimes be manually shaken for adequate mixing. This is most easily done by clamping the assembly *securely* to the ringstand and carefully sliding the base of the ringstand back and forth. But when more efficient and convenient mixing is required, particularly over a long period of time, it is necessary to use some kind of magnetic or mechanical stirring device.

Magnetic Stirring. A magnetic stirrer consists of an enclosed unit containing a motor attached to a magnet, underneath a platform. As the magnet inside the unit rotates, it can in turn rotate a Teflon- or glass-covered stirring bar (called a *stirbar* or *spinbar*) inside a container placed on (or above) the platform. The rate of stirring is controlled by a dial on the stirring unit. Since no moving parts extend outside of this unit, a reaction assembly that is to be stirred magnetically can be completely enclosed if necessary.

Magnetic stirrers can be used with heating mantles or heating baths that are constructed of nonferrous metals. They work particularly well with oil baths, since they can be used to stir the oil and a reaction mixture simultaneously. The reaction flask must be positioned close enough to the bottom of the oil bath to allow sufficient transfer of magnetic torque from motor to stirring bar. When a copper or aluminum steam bath is used for heating, the flask should be clamped inside the rings, close to the bottom of the steam bath. Hot plates that have an integral magnetic stirrer can be used (with a heating bath) to simultaneously heat and stir a reaction mixture.

When magnetic stirring is used during a reaction, the heat source is set directly on the stirring unit, and a stirbar is placed in the reaction flask in place of boiling chips. The stirring motor should be started and cooling water for the reflux condenser turned on (if applicable) before heating is begun.

A motion combining shaking with swirling is more effective than swirling alone.

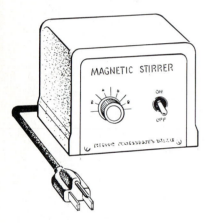

Figure 1.   Magnetic stirring unit

See *J. Chem. Educ.* **54**, 229 (1977) for a description of an easily constructed magnetic stirrer.

The stirring bar in the bath should be larger than the one in the reaction mixture.

Boiling chips are not used with magnetic or mechanical stirring, since the stirring action prevents bumping.

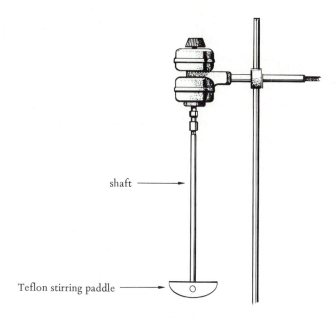

shaft

Teflon stirring paddle

Figure 2.   Mechanical stirrer

Mechanical Stirring.   A mechanical stirrer consists of a stirring motor connected to a paddle or agitator by means of a shaft that extends through the neck of the reaction vessel. A glass sleeve or bearing is used to align the shaft, which is ordinarily made of glass to reduce the likelihood of contamination. Mechanical stirrers can exert more torque than magnetic stirrers and are preferred when viscous liquids or large quantities of suspended solids must be stirred. A variety of stirring paddles made of Teflon, glass, and chemically resistant wire are available.

Addition

In many organic syntheses, the reactants are not all mixed together at the start of the reaction; instead, one or more of them is added over a period of time. This is necessary when the reaction is strongly exothermic or when one of the reactants must be kept in excess to prevent side reactions. Cylindrical addition funnels incorporating a pressure-release tube are specially constructed for this purpose. However, an ordinary separatory funnel can be used to perform the same function.

Equipment and supplies:
  boiling flask
  Claisen connecting tube
  separatory/addition funnel
  reflux condenser
  stopper

If a reaction is run in an open container such as an Erlenmeyer flask, the funnel can simply be clamped to a ringstand above the flask, which is shaken and swirled during the addition. More frequently, a reaction will be conducted with *addition under reflux,* using an apparatus such as that illustrated in Figure 1. The addition funnel should be placed on the straight arm of the Claisen connecting tube, directly over the reaction flask, and the reflux condenser should be on the bent arm. The reactant can be added in portions or drop by drop. In portionwise addition, small portions of the reactant are added at regular intervals, with shaking or stirring after each addition. To allow for pressure equalization between the addition funnel and the reaction flask, a strip of filter paper can be placed between the stopper and the top of the addition funnel, or the stopper can be loosened during the addition. In dropwise (continuous) addition, the stopcock is kept open just far enough to bring about the desired rate of addition, until all of the reactant has been added. The reaction mixture can be stirred magnetically ⟨OP–9⟩ or shaken and swirled manually to provide adequate mixing during the addition.

The stopper can be omitted if the liquid being added is quite involatile (for example, an aqueous solution).

Figure 1.   Apparatus for addition and reflux

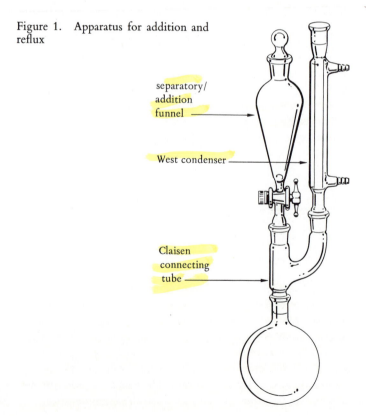

separatory/
addition
funnel

West condenser

Claisen
connecting
tube

The apparatus can be modified to provide for temperature monitoring ⟨OP–7*c*⟩ or mechanical stirring if a three-necked flask is available. In the latter case, one neck of the flask is used for the addition funnel, one for the reflux condenser, and the third for another function.

Semimicro Addition

OPERATION 10*a*

An apparatus suitable for the addition of small quantities of liquid is illustrated in Figure 1. The pinchcock valve (constructed by inserting a solid glass bead into a length of thin-walled rubber tubing) is attached to the volumetric pipet, which is filled with the specified volume of reactant using the squeeze-bulb assembly pictured in Figure 2 of OP–6. After the rubber bulb has been removed, the reactant can be added to the reaction mixture by squeezing the pinchcock valve. As in OP–10, the reactants should be mixed during the addition by shaking and swirling.

An ordinary test tube can be used as a reaction vessel if the stopper is notched for pressure release. However, the sidearm test tube allows a gas trap ⟨OP–22*b*⟩ or drying tube ⟨OP–22*a*⟩ to be attached when necessary. The assembly can be modified for semimicro addition under reflux by using a two-holed rubber stopper with a cold-finger condenser and an addition pipet inserted in the holes.

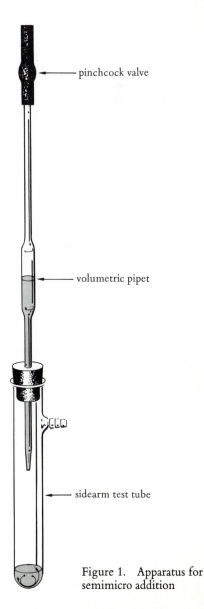

Figure 1. Apparatus for semimicro addition

Gravity Filtration

Filtration is used for two main purposes in organic chemistry: (1) to remove solid impurities from a liquid or solution, and (2) to isolate an organic solid from a reaction mixture or a crystallization solvent. In most instances, gravity filtration is preferred for the first operation and vacuum filtration for the second. Gravity filtration is often used to remove drying agents such as magnesium sulfate from dried organic liquids or solutions and to remove solid impurities from hot recrystallization solutions.

Gravity filtration of most organic liquids is best performed using a funnel with a short, wide stem (such as a powder or filling funnel) and a relatively fast, fluted filter paper. Ordinary filter paper can be folded as shown in Figure 1, but good commercial fluted filter papers are available at a reasonable price from chemical supply houses. Aqueous solutions can be filtered using a long-stemmed funnel and an ordinary folded filter cone.

Glass wool is sometimes used for very fast filtration of coarse solids. A thin layer of glass wool is placed over the outlet of a short-stemmed funnel, and the mixture to be filtered is poured directly onto the glass wool. Because fine particles will pass on through, this method is most often used for prefiltration of mixtures that will be refiltered later.

Operational Procedure

If you are filtering the liquid into a narrow-necked container such as an Erlenmeyer flask, support the funnel on the neck of the flask, with a bent wire or a paper clip between them to provide space for pressure equalization (see Figure 2). Otherwise, support the funnel on a ring or a funnel support directly over the collecting container. Place the fluted filter paper inside the funnel, and add the liquid fast enough to keep the paper well filled throughout the filtration. If the mixture to be

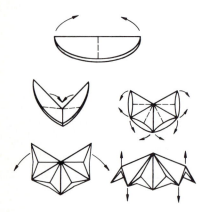

Figure 1.   Making a fluted filter paper

Figure 2.   Apparatus for gravity filtration

Equipment and supplies:
  powder funnel
  fluted filter paper
  collecting beaker or flask

fluted filter paper
powder funnel
bent wire
collecting flask

filtered contains an appreciable amount of finely divided solid, allow the solid matter to settle and decant the liquid carefully so that most of the solid will remain behind until the end of the filtration. (Otherwise, the pores of the filter paper may become clogged and retard the filtration.) If you are filtering a solid from a solution, it is a good practice to stir the solid on the filter paper with a small amount of fresh solvent and drain this wash solvent into the **filtrate.** This will reduce losses due to adsorption of dissolved organic materials by the solid.

Allowing a hot recrystallization solution to stand for too long before filtering can cause premature crystallization.

Be careful not to tear the filter paper while stirring.

filtrate: the liquid that has been filtered

Vacuum Filtration

OPERATION 12

Vacuum filtration (also called suction filtration) offers a fast, convenient method for isolating a solid from a solid-liquid mixture or for removing impurities from large quantities of a liquid. A circle of filter paper is laid flat on a perforated plate inside a *Buchner funnel,* which is supported on a filter flask (see Figure 1). When a partial vacuum is created in the flask by a water aspirator, liquid is rapidly forced through the filter paper by the unbalanced external pressure. [See *J. Chem. Educ. 53,* 45 (1976) for the construction of a unitized filtration stand.] A filter trap interposed between the trap and the aspirator helps to prevent water from backing up into the flask when the water pressure changes. The filter trap can be a thick-walled Pyrex

Figure 1.   Apparatus for vacuum filtration

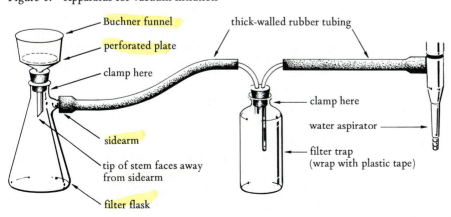

thick-walled rubber tubing

Buchner funnel

perforated plate

clamp here

sidearm

tip of stem faces away from sidearm

filter flask

clamp here

water aspirator

filter trap
(wrap with plastic tape)

jar, wrapped with transparent plastic tape to reduce the chance of injury in case of implosion. (A better trap that uses a pressure-release valve is illustrated in Figure 2 of OP–26, but it requires a large filter flask for construction.) The shorter bent tube on the trap should be connected to the filter flask, so that water cannot back up into the filter flask without filling the trap first.

With vacuum filtration, because of the pressure on the mixture being filtered, solid particles are more likely to pass through the filter paper than with gravity filtration, so a finer-grained ("slower") filter paper should be used. An all-purpose paper such as Whatman #1 is adequate for filtration of most solids. When filtering finely divided solids from a liquid, it is sometimes necessary to use a *filtering aid* (such as Celite) to keep the solid from plugging the pores in the filter paper. The filtering aid is mixed with a solvent to form a thin paste or slurry, which is poured into the filter under suction until a bed about 2–3 mm thick has been deposited. The solvent is then removed from the filter flask before continuing with the filtration. This technique should *not* be used when the solid is to be saved, since it would be contaminated with the filtering aid.

Unless otherwise instructed, always *wash* the solid on the filter paper with an appropriate solvent, usually the same as the one from which it was filtered. For example, if the solid was filtered from an aqueous solution, use distilled water as the wash solvent. If it was filtered from a solvent mixture (as in a mixed-solvent recrystallization), use the component of that mixture in which the solid is least soluble. Sometimes a solid is washed a second time with a lower-boiling solvent to facilitate drying; if so, the second wash solvent must be miscible with the first one. Thus, a solid filtered from an aqueous solution might be washed with water and then with methanol, and one filtered from toluene could be washed with toluene and then with petroleum ether, but *only* if the solids were not appreciably soluble in those second solvents. To reduce losses, it is best in all cases to cool the wash solvent in ice before using it.

Operational Procedure

Clamp the filter flask and filter trap to a ringstand, connect them as illustrated in Figure 1, and connect the trap to a vacuum line or water aspirator. Use thick-walled rubber tubing for all connections. Insert a Buchner funnel with a snug-fitting rubber stopper into the filter flask, and place a circle of filter paper inside the funnel. The diameter of the filter paper should

Filtering aids are made of diatomaceous earth, which consists of the microscopic shells of deceased diatoms.

Equipment and supplies:
  Buchner funnel
  filter flask
  water trap
  filter paper
  thick-walled rubber tubing
  flat-bottomed stirring rod
  flat-bladed spatula
  washing solvent

be slightly less than that of the perforated plate, so that the paper covers all the holes but does not fit too snugly. Moisten the paper with a few drops of solvent (preferably that present in the filtration mixture); then apply suction by turning on the aspirator. The tap on the water aspirator should be as far open as it will go throughout the filtration, because operating the aspirator at a lower flow rate greatly increases the chance of water backup. The water stream can be directed into a large beaker or another container to prevent splashing.

The filter paper should not extend up the sides of the funnel.

Add the filtration mixture in portions, keeping the funnel nearly full throughout. Stir and swirl the filtration mixture near the end of the filtration to get most of the solid onto the filter. Transfer any remaining solid to the filter paper with a flat-bladed spatula, using a little of the filtrate or some additional pure solvent to facilitate the transfer. If the trap has a pressure-release valve, open it before turning off the aspirator; otherwise, disconnect the rubber tubing at the aspirator. This will keep water from backing up into the system.

If the solid is finely divided, it may be advisable to decant the liquid carefully after the solid has settled and wait until the end of the filtration to transfer the bulk of the solid.

With the aspirator *off*, add just enough of the cold wash solvent to cover the solid. Being careful not to disturb the filter paper, stir the mixture gently with a spatula or a flat-bottomed stirring rod until the solid is suspended in the liquid. Apply suction to drain off the wash liquid; then turn off the aspirator and repeat the process with at least one more portion of fresh solvent. After the last washing, leave the aspirator on a few minutes to *air-dry* the solid on the filter and make it easier to handle. Run the tip of a small flat-bladed spatula around the circumference of the filter paper to dislodge the *filter cake,* then invert the funnel carefully over a large filter paper or a watch glass (or another container) to remove the filter cake and paper. The filter paper can be scraped clean using the spatula. The solid should be dried by one of the methods described in OP–21.

Any fresh solvent used in the transfer can be used for washing as well.

Usually, 1–2 mL of wash solvent per gram of solid is sufficient.

Sometimes the solid is compressed with the top of a clean cork or another flat object to remove excess water.

The filter cake is the compressed solid on the filter paper.

Any particles remaining in the funnel should be scraped out.

Summary

1. Assemble apparatus for vacuum filtration.

2. Position and moisten filter paper; turn on aspirator.

3. Add filtration mixture (in portions).

4. Transfer remaining solid to funnel, using additional solvent if necessary.

5. Disconnect and turn off aspirator.

6. Wash solid on filter with cold solvent; turn on aspirator to drain solvent. Repeat as necessary.

7. Air-dry solid on filter; remove filter cake.

8. Remove solvent from filter flask; disassemble and clean apparatus.

Semimicro Vacuum Filtration

Small quantities of solids can be filtered by the general method described in OP–12, but using a Hirsch funnel and a small filter flask or a sidearm test tube. Since the perforated plate of a Hirsch funnel is small in diameter, filter paper of the right size is not generally available. Small circles can be cut from an ordinary filter paper using a sharp cork borer on a flat cutting surface, such as the bottom of a large cork.

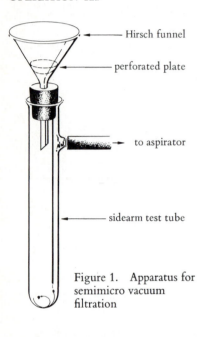

Hirsch funnel

perforated plate

to aspirator

sidearm test tube

Figure 1. Apparatus for semimicro vacuum filtration

Extraction

Principles and Applications

Extraction is a convenient method for separating an organic substance from a mixture, such as an aqueous reaction mixture or a steam distillate. The *extraction solvent* is usually a volatile organic liquid that can be removed by evaporation ⟨OP–14⟩ after the desired component has been extracted. (The related process of *washing* liquids, in which impurities are extracted from an organic liquid, is described in OP–19.)

The extraction technique is based on the fact that if a substance is soluble to some extent in two immiscible liquids, it can be transferred from one liquid to the other by shaking it together with the two liquids. For example, acetanilide is partly soluble in both water and ethyl ether. If a solution of acetanilide in water is shaken with a portion of ethyl ether (which is immiscible with water), some of the acetanilide will be transferred to the ether layer. The ether layer, being less dense than water, separates above the water layer and can be removed and replaced with another portion of ether. When this is in turn shaken with the aqueous solution, more acetanilide passes into the new ether layer. This new layer can then be removed and combined with the first. By repeating this process enough times, virtually all of the acetanilide can be transferred from the water to the ether.

The ability of an extraction solvent, S_2, to remove a solute A from another solvent, S_1, depends on the *partition coefficient* (K) of solute A in the two solvents, as defined in Equation 1:

$$K = \frac{\text{concentration of A in } S_2}{\text{concentration of A in } S_1} \quad (1)$$

In the example of acetanilide in water and ethyl ether, the partition coefficient is given by

$$K = \frac{[\text{acetanilide}]_{\text{ether}}}{[\text{acetanilide}]_{\text{water}}}$$

The larger the value of K, the more solute will be transferred to the ether with each extraction, and the fewer portions of ether will be required for essentially complete removal of the solute.

Extraction Solvents

If a solvent is to be used to extract an organic compound from an aqueous mixture or solution, it must be virtually insoluble in water, and it should have a low boiling point so that the solvent can be evaporated after the extraction ⟨OP–14⟩. The solute should also be more soluble in the extraction solvent than in water, since otherwise too many extraction steps will be required to remove all of the solute.

Ethyl ether is the most commonly used extraction solvent. It has a very low boiling point (34.5 °C) and can dissolve a large number of organic compounds, both polar and nonpolar. However, ethyl ether must be used with great care, since it is

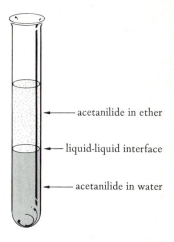

Figure 1. Distribution of a solute between two liquids

A very rough estimate of K can be obtained by using the ratio of the solubilities of the solute in the two solvents, that is:

$$K \approx \frac{\text{solubility of A in } S_2}{\text{solubility of A in } S_1}$$

This approximate relationship can be helpful in choosing a suitable extraction solvent.

Table 1. Properties of commonly used extraction solvents

Solvent	b.p.	d.	Comments
ethyl acetate	77°	0.90	Absorbs much water
ethyl ether	34.5°	0.71	Good general solvent; absorbs some water; easy to remove; very flammable; vapors should not be inhaled
methylene chloride	40°	1.34	Good general solvent; easy to dry and remove; suspected carcinogen
chloroform	62°	1.48	Can form emulsions; easy to dry and remove; health hazard; suspected carcinogen
1,1,2-trichlorotrifluoroethane (Freon TF)	48°	1.58	May be substituted for carbon tetrachloride
pentane	36°	0.63	Easy to dry and remove; very flammable
hexane	69°	0.66	Easily dried

If the extraction solvent is less dense than water, the aqueous layer must be drained out after each extraction and returned to the separatory funnel for the next extraction. This is not necessary if the solvent is more dense than water.

extremely flammable and tends to form explosive peroxides on standing. Methylene chloride (dichloromethane) has most of the advantages of ethyl ether; in addition, it is nonflammable and more dense than water. However, it has a tendency to form emulsions, which can make it difficult to separate the layers cleanly. Other useful solvents and their properties are listed in Table 1. Various grades of petroleum ether (a mixture of low-boiling hydrocarbons) can be used in place of pentane and hexane.

Potential hazards must always be considered in selecting and using an extraction solvent. Flames must not be allowed in the laboratory when highly flammable solvents, such as ethyl ether, pentane, and petroleum ether, are in use. Because of the toxicity and apparent carcinogenic potential of chloroform, you should wear protective gloves and work under a hood when handling this solvent (or preferably, use a less toxic substitute). Precautions must be taken with all solvents to minimize skin and eye contact, inhalation of vapors, and exposure to possible ignition sources.

See pages 12–14 for explanation of hazard symbols.

Hazards

ethyl
acetate

ethyl
ether

methylene
chloride

chloroform

pentane

hexane

Experimental Considerations

Separatory funnels are very expensive and break easily. Never prop such a funnel on its stem; support it on a ring, a funnel support, or some other stable support. Glass stopcocks should be lubricated by applying thin bands of stopcock grease on both sides, leaving the center (where the drain hole is located) free of grease to prevent contamination (see Figure 2C). A Teflon stopcock should not be treated with stopcock grease.

The volume of extraction solvent and the number of extraction steps are frequently specified in an experimental procedure. If they are not, it is usually sufficient to use a volume of extraction solvent about equal to the volume of liquid being extracted, divided into at least two portions. For example, 60 mL of an aqueous solution can be extracted with two 30-mL (or three 20-mL) portions of ethyl ether. It is more efficient to use several small portions of extraction solvent than one large portion of the same total volume.

Rule of thumb: total volume of extraction solvent ≈ volume of liquid being extracted. (There are many exceptions to this "rule.")

Figure 2. Extraction techniques

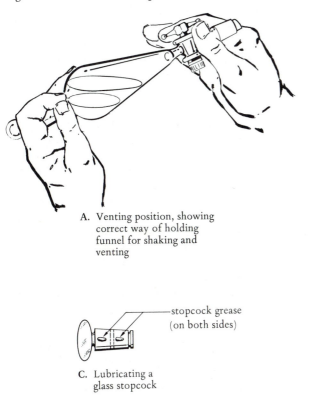

A. Venting position, showing correct way of holding funnel for shaking and venting

C. Lubricating a glass stopcock

— stopcock grease (on both sides)

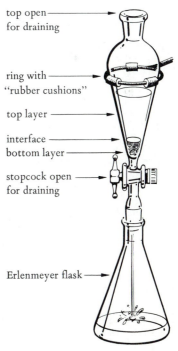

top open for draining

ring with "rubber cushions"

top layer

interface
bottom layer

stopcock open for draining

Erlenmeyer flask

B. Draining lower layer from separatory funnel

A liquid emulsion contains microscopic droplets of one liquid suspended in another.

Avoid using too much salt; excess solid can clog the stopcock drain hole.

When certain kinds of solutes are present, an emulsion may form at the interface between the two liquids, making it impossible to separate them sharply. Emulsions can sometimes be broken up by gently swirling or stirring the liquids and allowing the funnel to stand open and undisturbed for a time, or by dissolving a little sodium chloride in the aqueous layer with stirring. Small amounts of insoluble material that sometimes form at the interface can be removed by filtering the mixture through a loose pad of glass wool. After each extraction, the extraction solvent should be transferred to a collecting flask, which is kept stoppered to prevent evaporation and reduce any possible fire hazard.

Operational Procedure

Equipment and supplies:
  separatory funnel
  two flasks
  extraction solvent
  graduated cylinder
  ring support
  ringstand
  closure for solvent flask

The liquid in the funnel must be at room temperature or below if the extraction solvent is very volatile.

Support a separatory funnel on a ring of suitable diameter, equipped with some lengths of split rubber tubing (see Figure 2B) to cushion the funnel and prevent damage. (A special funnel support can also be used.) Close the stopcock by turning the handle to a horizontal position *(important!)*, and add the liquid to be extracted through the top of the funnel. Measure the required volume of extraction solvent using a graduated cylinder or some other graduated container (the exact volume is not crucial), and pour it into the funnel. The total volume of both liquids should not exceed three-quarters of the funnel's capacity; if it does, obtain a larger funnel or carry out the extraction in two or more steps, using a fraction of the liquid in each step. Moisten the stopper with water (or other solvent) and insert it firmly. Then pick up the funnel in both hands and partly invert it, with the right hand holding the stopcock (or the left hand if you are a southpaw) and the first two fingers of the left hand holding the stopper in place (see Figure 2A). *Vent the funnel by slowly opening the stopcock to release any pressure buildup.* Close the stopcock, shake or swirl the funnel gently for a few seconds, then vent it again (be sure the funnel is inverted). Shake the funnel more vigorously, with occasional venting, for 2–3 minutes. Venting should not be necessary after there is no longer an audible hiss of escaping vapors when the stopcock is opened.

Be sure that the stem of the funnel is pointed away from you and your neighbors when you are venting it.

A combined shaking and swirling motion is more efficient than swirling alone. Avoid vigorous shaking if the system is prone to emulsions.

Gentle stirring with a stirring rod or wooden applicator stick can sometimes accelerate the consolidation of layers.

Replace the funnel on its support, remove the stopper, and allow the funnel to stand until there is a sharp demarcation line between the two layers. Drain the bottom layer into an Erlenmeyer flask (flask A) by opening the stopcock fully; turn it to slow the drainage rate as the interface approaches the bottom of the funnel. When the interface *just* reaches the outlet, quickly close the stopcock to separate the layers cleanly.

If the extraction solvent is *more* dense than the aqueous solution being extracted, the aqueous layer will remain in the separatory funnel after each extraction and can be extracted with a fresh portion of solvent, which is then combined with the first extract in flask A. Additional extractions with fresh solvent can be performed as needed. After the last extraction is finished, pour the aqueous layer out of the *top* of the funnel into a separate container (flask B).

If the extraction solvent is *less* dense than the aqueous layer, it will remain in the separatory funnel after the bottom (aqueous) layer has been drained into flask A. Pour it out of the *top* of the funnel into flask B after each extraction. Return the aqueous solution to the funnel from flask A, and extract it with another portion of solvent; then drain it into flask A and again pour the solvent into flask B. Repeat the process, as necessary, with fresh extraction solvent.

Most aqueous solutions have densities close to 1 g/mL or a little higher.

If there is any doubt about which layer should be discarded, save both *until the correct one is identified. Mixing a drop or two of water with a little of each layer will establish which is the aqueous layer.*

Summary

1. Place separatory funnel on support; add liquid to be extracted.

2. Add extraction solvent to separatory funnel, stopper funnel, and invert and vent funnel.

3. Shake and swirl funnel, with occasional venting, to extract solute into extraction solvent.

4. Remove stopper, drain lower layer into flask A.
IF extraction solvent is the *lower* layer, GO TO 5.
IF extraction solvent is the *upper* layer, GO TO 6.

5. Stopper flask A.
IF another extraction step is needed, GO TO 2.
IF extraction is complete, empty and clean funnel; STOP.

Usually two or three extraction steps should be performed.

6. Pour upper layer into flask B and stopper it.
IF another extraction step is needed, return contents of flask A to separatory funnel, and then GO TO 2.
IF extraction is complete, clean funnel; STOP.

Semimicro Extraction

Small-scale extractions can be carried out by shaking the aqueous solution with the extraction solvent in a stoppered test tube (with occasional venting) and separating the layers using a capillary pipet or a volumetric pipet. The basic extraction pro-

See OP–6 on the use of volumetric pipets.

cedures resemble those described in OP–13, except that *either* the top or the bottom layer—depending on the density of the extraction solvent—can be withdrawn in order to leave the liquid being extracted in the test tube after each extraction step.

Extractions involving *very* small quantities of liquid (as in Experiment 47) can be carried out as follows. Put the solution being extracted into a small test tube or centrifuge tube (the extraction tube), and add the extraction solvent. Draw both liquid layers into a capillary pipet fitted with a large rubber bulb, and eject the mixture forcefully into the extraction tube so that the layers are well mixed. Repeat the operation a half-dozen times or more, and let the mixture in the extraction tube settle until the layers are sharply separated. Remove the extraction solvent by drawing all of it into the capillary pipet, along with a little of the other layer. When the layers are well separated, transfer the extraction solvent to a different container, and return the other layer to the extraction tube, taking care to separate the layers sharply. Fresh extraction solvent can then be added to the extraction tube, and the process can be repeated.

OPERATION 13*b*

Separation of Liquids

Sometimes a distillate or the product of a reaction consists of two immiscible liquids that must be separated for subsequent operations. This can be accomplished by transferring the liquid mixture to a separatory funnel and allowing it to stand until a sharp demarcation line forms between the liquids. The lower layer is then drained into one container and the upper layer poured out the top into another. For small amounts of liquids, the separation method described in OP–13*a* can be used. See OP–13 for information on dealing with emulsions, identifying the aqueous layer, etc.

OPERATION 13*c*

Salting Out

Adding an inorganic salt (such as sodium chloride or potassium carbonate) to an aqueous solution containing an organic solute often reduces the solubility of the organic compound in the water, and thus promotes its separation. This *salting out* technique is often used in extractions and liquid-liquid separations to maximize the transfer of an organic solute from the aqueous to the organic layer or to separate an organic liquid from its aqueous solution. Usually enough of the salt is added to saturate the aqueous solution, which is stirred or shaken to dissolve the salt. The mixture is filtered by gravity ⟨OP–11⟩ if necessary and transferred to a separatory funnel for separation ⟨OP–13b⟩ or extraction ⟨OP–13⟩.

Filtering the solution through a loose plug of glass wool may be satisfactory.

Solid Extraction

Some substances, particularly natural products, contain components that can be separated from a solid residue by extraction. The simplest technique for solid extraction is to mix the solid intimately with an appropriate solvent using a flat-bottomed stirring rod in a beaker and then filter off the solid by gravity ⟨OP–11⟩ or vacuum filtration ⟨OP–12⟩, repeating the process as many times as necessary. The mixing should be as thorough as possible; use the flat end of the stirring rod to press down or crush the solid in contact with the solvent, in order to extract more of the desired component. After each filtration, the solid is returned to the beaker for the next extraction. The liquid extracts are combined in a single collecting flask.

A mortar and pestle can also be used.

This crushing process is called triturating.

If the material being extracted is not heat-sensitive, it may be extracted with hot solvent. A portion of the solvent is mixed with the solid in a boiling flask. The mixture is heated under reflux ⟨OP–7a⟩ for 15 minutes or more, then filtered by gravity after cooling. Repeated extractions can be carried out using fresh solvent, if necessary. Alternatively, the extraction of a solid can be carried out using an automatic extraction apparatus called a Soxhlet extractor, in which the solid is repeatedly extracted with small portions of fresh hot solvent and the extracts are combined in the solvent flask.

Evaporation

Evaporation refers to the vaporization of a liquid at or below its boiling point. It can be used to remove a volatile solvent from a comparatively involatile residue. Partial solvent removal *(concentration)* is frequently used to bring a recrystallization solution to its saturation point (see OP–23). Complete

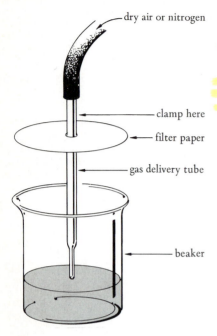

dry air or nitrogen

clamp here

filter paper

gas delivery tube

beaker

Figure 1. Apparatus for small-scale evaporation

Hazard signs for many of the solvents commonly encountered are given in OP–13, page 632, and OP–23c, page 686.

Equipment and supplies:
  evaporation vessel (Figure 2)
  rubber tubing
  solvent trap
  aspirator
  water bath or steam bath

solvent removal is used to isolate an organic solute after such operations as extraction ⟨OP–13⟩ or column chromatography ⟨OP–16⟩.

Small quantities of solvent can be evaporated simply by leaving the solution in an evaporating dish under a fume hood or by passing a slow, dry stream (see OP–22) of air or nitrogen over it (Figure 1). The solution can be protected from airborne particles by supporting a filter paper some distance above it. Relatively large quantities of solvent can be removed by simple distillation ⟨OP–25⟩ or vacuum distillation ⟨OP–26⟩ when convenient. This procedure is most useful when the residue is to be distilled immediately after solvent removal. Commercial flash evaporators are available for fast removal of solvents under reduced pressure, but these are expensive and not generally available for use by undergraduates.

Experimental Considerations

For most purposes, moderate quantities of solvent can be evaporated efficiently with one of the setups pictured in Figure 2. Evaporation is accelerated by using reduced pressure and gentle heating from a water bath or steam bath ⟨OP–7⟩. Swirling the solution during evaporation will speed up the process and reduce foaming and bumping. A solvent trap similar to that pictured in Figure 1 of OP–12 or Figure 2 of OP–26 should always be interposed between the evaporation vessel and the aspirator to collect the solvent, which can then be returned to a used-solvent container for recycling. To recover a low-boiling solvent such as ethyl ether, you should use a *cold trap.* The simplest cold trap is an ordinary solvent trap clamped securely inside an ice bath ⟨OP–8⟩. More elaborate cold traps, which are immersed in an insulated container filled with dry ice and acetone or another cooling mixture, are also available.

Because of possible health and fire hazards, it is not advisable to evaporate solvents by heating open containers in the laboratory. Even when using the method described below, you should know and allow for the hazards associated with each solvent.

Operational Procedure

Assemble one of the three setups pictured in Figure 2. Add the liquid to be evaporated and connect the apparatus to a solvent trap and aspirator. Turn on the aspirator and swirl the evaporation vessel over a steam bath or in a hot-water bath; it should be swirled throughout the evaporation to minimize foaming and bumping. (Be ready to remove it from the heat source immedi-

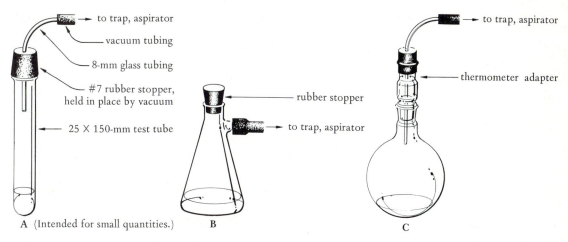

to trap, aspirator
vacuum tubing
8-mm glass tubing
#7 rubber stopper, held in place by vacuum
25 × 150-mm test tube

rubber stopper
to trap, aspirator

to trap, aspirator
thermometer adapter

A (Intended for small quantities.)   **B**   **C**

Figure 2. Apparatus for solvent evaporation

ately if it starts to foam up.) If necessary, use several layers of paper towels or other insulating material to protect your hands from the steam. Adjust the steam-flow rate (or the temperature of a water bath) to attain a satisfactory rate of evaporation; the liquid should boil but must not foam up and out of the flask. If you wish merely to concentrate the solution, stop the evaporation when sufficient solvent has been removed; otherwise, continue until all the solvent has been removed. At this point, the liquid should stop boiling, and no more bubbles of solvent will appear at its surface.

It should be possible to tell from the odor and volume of the residue whether all of the solvent has been removed.

Break the vacuum by opening the pressure-release valve on the trap (if it has one) or by removing the hose from the aspirator (or, in the case of apparatus A, by sliding the stopper off the mouth of the test tube), and turn off the aspirator. If the product is a solid, it can be removed with a flat-bladed spatula. The last traces of residue are sometimes removed by rinsing the container with a little volatile solvent and allowing the solvent to evaporate under a hood or in a dry-air stream. Place the solvent from the trap in a waste-solvent container or dispose of it as recommended by the instructor.

Some residues may solidify only after cooling.

Ethyl ether is often used for this rinsing.

Summary

1. Assemble apparatus for solvent evaporation.

2. Add liquid, stopper evaporation vessel, and turn on aspirator.

3. Apply heat and swirl to maintain rapid evaporation.

4. Discontinue heating, disconnect aspirator vacuum, and recover residue and solvent.

OPERATION 15

Codistillation

Principles and Applications

When a *homogeneous solution* of two liquids is distilled, the vapor pressure of each liquid is lowered by an amount proportional to the mole fraction of the other liquid present. This usually results in a solution boiling point somewhere between the boiling points of the components. When a *heterogeneous mixture* of two immiscible liquids is distilled, however, each liquid exerts its vapor pressure independently of the other, so that the total vapor pressure over the mixture is approximately equal to the sum of the vapor pressures that would be exerted by the separate pure liquids at the same temperature. This has several important consequences: (1) the vapor pressure of a mixture of immiscible components must be *higher* than that of the most volatile component; therefore, (2) the boiling point of the mixture must be *lower* than that of the lowest-boiling component, and (3) the boiling point will remain constant during distillation of such a mixture, as long as each component is present in significant quantity.

For example, suppose you are distilling a mixture of the immiscible liquids toluene and water at standard atmospheric pressure. The mixture will start to boil when the sum of the vapor pressures of the two liquids is equal to the external pressure, 760 torr. This occurs at 85°C, where the vapor pressure of water is 434 torr and that of toluene is 326 torr. Because the vapor pressures of the two components are additive, the mixture distills considerably below the normal boiling point of either toluene (b.p. 111°C) or water. According to Avogadro's law, the number of moles of a component in a mixture of gases is proportional to its partial pressure in the mixture, so the mole fraction of toluene in the vapor should be about $326/760 = 0.43$ and that of water should be $434/760 = 0.57$. (Because the vapors are not ideal gases, these calculations are approximate.) In other words, about 43% of the molecules in the vapor are toluene molecules. Because toluene molecules (M.W. = 92) are heavier than water molecules (M.W. = 18), they make a greater contribution to the total mass of the vapor. In 1 mol of vapor, there will be 0.43 mol of toluene and 0.57 mol of water; the mass of toluene in the vapor will be 0.43 mol × 92 g/mol = 40 g, and the mass of water will be 0.57 mol × 18.0 g/mol = 10 g. So the mass of a mole of the vapor

Distillation of a mixture of two or more immiscible liquids is called *codistillation*.

Vapor pressure of a mixture of two immiscible components:

$$P \approx P_A^\circ + P_B^\circ$$

(P° stands for the vapor pressure of the pure liquid A or B)

is about 50 g, of which toluene makes up 40 g, or 80%. Distilling a mixture of toluene and water will therefore yield a distillate (the condensed vapor) that contains about 80% toluene, by mass.

Because of its relatively low molecular weight and its immiscibility with many organic compounds, water is nearly always one of the components used in a codistillation. In general, the higher the boiling point of the organic component in a mixture containing water, the lower will be its proportion in the vapor (and therefore in the distillate) and the closer the mixture boiling point will be to 100°C (see Table 1). If the vapor pressure of the organic component at 100°C is much below 5 torr, codistillation becomes impractical, since it would require the distillation of a large amount of water to recover a small quantity of the organic component. Besides having an appreciable vapor pressure at that temperature, the organic component in a codistillation must be insoluble enough in water to form a separate phase, and must not react with (or be decomposed by) hot water or steam.

The term *steam distillation* is generally used to refer to a codistillation in which externally generated steam is passed through the mixture being distilled. The term can also apply to any codistillation in which water is one of the components. Because the external-steam process requires a different kind of apparatus than that used for other codistillations, it is described in a separate section ⟨OP–15a⟩. Both processes are useful for isolating organic components from reaction mixtures or natural products, by separating them from high-boiling residues such as tars, inorganic salts, or other relatively involatile components. Because distillation occurs well below the normal boiling point of the organic component, decomposition caused by excessive heat is minimized.

Codistillation is of little use for the final purification of a liquid, since it cannot effectively separate components with similar boiling points.

Table 1. Boiling points and compositions of heterogeneous mixtures with water

Component A	b.p. of A	b.p. of mixture	mass % of A
toluene	111°	85°	80%
chlorobenzene	132°	90°	71%
bromobenzene	156°	95°	62%
iodobenzene	188°	98°	43%
quinoline	237°	99.6°	10%

Note: Component B is water.

Experimental Considerations

Refer to OP–25 for experimental details.

A mixture of water with an organic liquid can be distilled by essentially the same method used for simple distillation ⟨OP–25⟩, except that additional water can be added during the distillation, if necessary. If water is to be added, a setup like the one illustrated in Figure 1 should be used; otherwise, an ordinary simple distillation apparatus is adequate. The thermometer is not essential, but it can sometimes indicate the end of the distillation, since the temperature should rise to 100°C at that point.

Operational Procedure

Equipment and supplies:
  Boiling flask
  Claisen connecting tube (optional)
  three-way connecting tube
  West condenser
  distillation adapter
  receiving flask
  separatory/addition funnel (optional)
  thermometer
  thermometer adapter
  heat source
  condenser tubing

If it will be necessary to add water during the codistillation, assemble the apparatus in Figure 1; otherwise, assemble the apparatus pictured in Figure 2 of OP–25. Add the organic material, boiling chips, and enough water to fill the boiling flask about one-third to one-half full (unless enough water is already present). Heat the flask ⟨OP–7⟩ to maintain a rapid rate of distillation. Do not allow vapors to escape at the re-

Figure 1.   Apparatus for codistillation of immiscible liquids

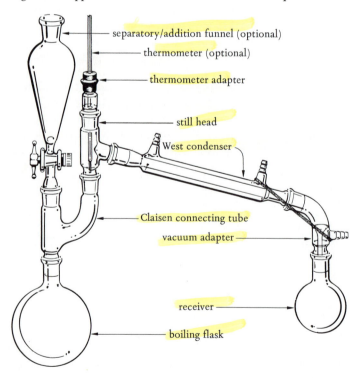

separatory/addition funnel (optional)
thermometer (optional)
thermometer adapter
still head
West condenser
Claisen connecting tube
vacuum adapter
receiver
boiling flask

ceiver (it may be necessary to cool the receiver in an ice bath to prevent this). Add water if necessary to replace that lost during the distillation. Discontinue heating when the distillate contains no more of the organic component *and* the distillation temperature is about 100°C. The distillate should no longer be cloudy or contain droplets of organic liquid at this point.

Steam Distillation

OPERATION 15a

Experimental Considerations

The use of live steam is preferred for many distillations, especially those involving solids or comparatively involatile liquids. The steam can be obtained from a steam line or by boiling water in a steam generator like the one pictured in Figure 1. The safety tube, which should be about 50 cm long, allows pressure release in case there is an obstruction in the apparatus, such as a solid plugging up the steam-delivery tube. It also sounds a warning when the water in the generator is getting low.

When steam is obtained from a steam line, a steam trap should be used to remove condensed water and foreign matter such as grease or rust. An ordinary Pyrex filter trap can be used in some cases, but the trap illustrated in Figure 2 is preferable, since it includes a valve for draining off excess water. A steam trap is also desirable (but not essential) when a steam generator is used.

The external steam helps prevent bumping caused by solids or tars.

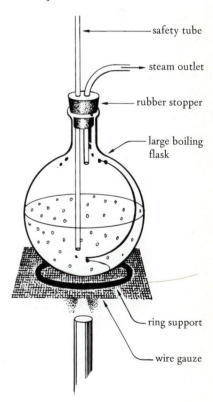

- safety tube
- steam outlet
- rubber stopper
- large boiling flask
- ring support
- wire gauze

Figure 1.   Steam generator

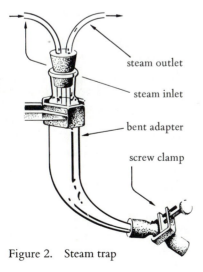

- steam outlet
- steam inlet
- bent adapter
- screw clamp

Figure 2.   Steam trap

The boiling flask should be large enough so that the liquid will not fill it much more than half full throughout the distillation. Some steam will condense in the flask during the distillation and can be removed by external heating, if necessary, during extended distillations. The Claisen connecting tube is used to prevent mechanical transfer of materials in the spray expelled from the liquid surface. A thermometer can be used to indicate the end of the distillation if the organic distillate is quite volatile. For example, in a toluene-water mixture, the temperature will rise rather rapidly from 85°C to about 100°C when the toluene is nearly gone. With less volatile materials, the temperature may be close to 100°C throughout the distillation, so a thermometer will be of little use. The distillation should be carried out rapidly to reduce condensation in the distilling flask and to compensate for the large volume of water-laden distillate that may have to be collected to yield much of the organic component. Due to the rapid distillation rate and the high heat content of steam, efficient condensing is essential, so the cooling water should run faster than for ordinary distillations. Other ways of increasing the cooling rate are using two condensers in tandem and placing the receiving flask in an ice bath.

The distillation adapter should be cool to the touch throughout the distillation; no vapor should escape from the outlet.

Operational Procedure

Assemble the apparatus pictured in Figure 3, using a large boiling flask and, as the steam inlet tube, a 6-mm O.D. glass tube extending to within about 0.5 cm of the bottom of the flask. Connect this tube through a trap to the steam line or to a steam generator. Add the organic mixture and a small amount of water (unless the mixture already contains water) to the distilling flask. Make sure that the condenser water is running rapidly enough to condense the distillate efficiently. Turn on the steam, and after distillation begins, adjust the steam flow to maintain a rapid rate of distillation. Do not distill so rapidly that vapor escapes at the receiver; use an ice bath, if necessary, to prevent this. Drain the trap periodically to remove condensed water. If the flask begins to fill up excessively (it should not be much more than half full) heat it with a steam bath or other heat source to reduce the condensation of steam. If you must interrupt the distillation for any reason, open the steam-trap valve (or raise the steam inlet tube out of the liquid) before turning off the steam. If a solid is being distilled, check the receiver outlet frequently to make sure that it does not become

Position the boiling flask high enough that heat can be applied if necessary. In some cases, it may be advisable to apply heat from the beginning of the distillation.

The flask should be no more than one-third full at the start.

The steam should be turned on for a while with the steam-trap valve open to remove condensed water from the line.

Check the connection between the condenser and three-way connecting tube frequently to make sure no vapor is escaping; these joints sometimes separate because of the violent action of the steam.

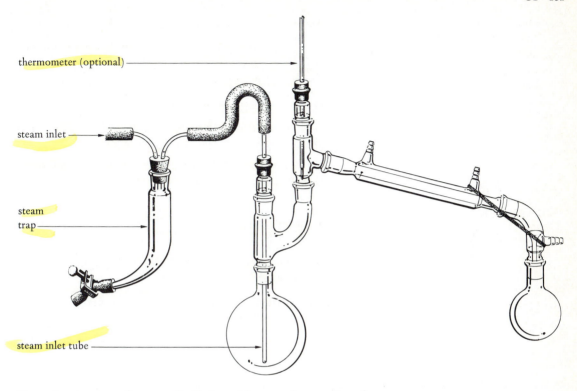

thermometer (optional)

steam inlet

steam trap

steam inlet tube

Figure 3.   Apparatus for steam distillation with external generation of steam

obstructed. If it begins to plug up, turn the condenser water off *momentarily,* and drain the condenser jacket so that the hot vapors will melt the solid and carry it into the receiver; then immediately turn the water on again.

When the distillate appears water-clear and the temperature is about 100°C, collect and examine a few drops of the distillate on a watch glass to see whether it is cloudy or contains oily droplets. If so, continue distilling, and collect and examine the distillate at 5- or 10-minute intervals. When you are certain the distillation is complete, open the steam-trap valve fully (or raise the steam inlet tube out of the liquid); *then* turn off the steam. This will keep the liquid in the boiling flask from backing up into the steam line.

The organic liquid can be obtained by separation in a separatory funnel ⟨OP–13*b*⟩ or by extraction with ether or another suitable solvent ⟨OP–13⟩. Extraction is necessary if the volume of the organic liquid is small compared to that of the water.

When all of the organic component has distilled, the distillate should be pure water.

If the aqueous layer appears cloudy, you can dissolve some NaCl or other salt in it ⟨OP–13c⟩ to improve the separation.

Equipment and supplies:
   boiling flask
   Claisen connecting tube
   three-way connecting tube
   condenser
   distillation adapter
   receiving flask
   thermometer adapter
   thermometer (optional)
   steam-delivery tube
   condenser tubing
   steam source
   trap made of bent adapter, screw clamp, two-hole rubber stopper, bent glass tubes, and rubber tubing

Summary

1. Assemble apparatus for steam distillation (Figure 3). IF steam line is available, GO TO 3. IF no steam line is available, GO TO 2.

2. Assemble steam generator (Figure 1).

3. Assemble steam trap (Figure 2); connect trap to steam line (or steam generator) and to steam-delivery tube.

4. Add water to boiling flask (if necessary), turn on condenser water, and turn on steam line (or boil water in steam generator).

5. Distill rapidly, using external heat if necessary, until distillate is clear. Drain trap periodically.

6. Open steam-trap valve (or raise steam-inlet tube out of liquid); turn off steam.

7. Separate or extract organic liquid from distillate; disassemble and clean apparatus.

OPERATION 16

Column Chromatography

Principles and Applications

Column chromatography is a simple, efficient method for separating the components of a mixture. We will be concerned primarily with the form of column chromatography known as *liquid-solid adsorption chromatography,* which operates on the principle that different substances are adsorbed on the surface of a solid adsorbent (such as alumina) to differing extents that depend on their polarity and other structural features. Since some compounds are more strongly adsorbed than others, they will be washed down a column of adsorbent at a slower rate and thus become separated from those less strongly adsorbed.

Liquid-solid adsorption chromatography involves the use of a solid **stationary phase** (such as alumina or silica) and a liquid **mobile phase** (such as hexane or methylene chloride). The stationary phase — called the **adsorbent** in adsorption chromatography — is packed firmly into a glass tube called the *column.* The sample, consisting of two or more components in

stationary phase: in column chromatography, a solid or liquid that remains immobile in the column and separates components by adsorption, partition, etc.

mobile phase: a fluid that passes through the column and carries the components with it at varying rates

adsorbent: a substance that retains molecules of other substances by surface attraction

a neat liquid or solution, is placed on top of the adsorbent in a narrow band and washed down the column (eluted) by a suitable mobile phase, called the *eluent*. As the eluent passes down the column, the components of the sample spread out to form separate bands of solute, some passing down the column rapidly with the solvent, others lagging behind.

For example, consider a separation of limonene and carvone on a silica gel adsorbent. At any given time, a molecule of one component will either be adsorbed on the silica (stationary phase) or dissolved in the mobile phase. While it is adsorbed, the molecule will stay put; while dissolved, it will move down the column with the eluent. A relatively polar molecule of carvone is strongly attracted to the polar adsorbent and spends more time adsorbed on the silica than dissolved in a nonpolar eluent such as petroleum ether. It will therefore pass down the column very slowly with this solvent. On the other hand, a nonpolar molecule of limonene is very soluble in petroleum ether and only weakly attracted to the adsorbent; with this solvent, it will therefore spend less time sitting still and more time moving than a carvone molecule. As a result, the limonene passes down the column rapidly and will soon separate from the slow-moving carvone. If a more polar solvent such as methylene chloride is then added, the carvone will spend a greater fraction of its time in solution and be washed down the column in turn. The kind of separation attained by column chromatography thus depends on a number of factors, including the quantity and kind of adsorbent used, the polarity of the mobile phase, and the nature of the components in the mixture.

Experimental Considerations

Adsorbents. A number of different adsorbents are used for column chromatography, but alumina and silica gel are the most popular. Adsorbents are available in a wide variety of activity grades and particle-size ranges; alumina can be obtained in acidic, basic, or neutral forms as well. The *activity* of an adsorbent is a measure of its attraction for solute molecules, the most active grade of a given adsorbent being one from which all water has been removed. The most active grade is not always the best for a given application, since too active an adsorbent may catalyze a reaction or cause solute bands to move too slowly. Alumina is *deactivated* by mixing in 3–15% water (see Table 1 on page 648), whereas silica gel is generally deactivated with 10–20% water.

Approximate strength of adsorption of different functional groups in order of increasing adsorption power:

—Cl, —Br, —I	weakest
$-\overset{\textstyle\vert}{C}{=}\overset{\textstyle\vert}{C}-$	
—OR	
—COOR	
$-\overset{\textstyle\vert}{C}{=}O$	
—CHO	
—SH	
—NH$_2$	
—OH	
—COOH	strongest

limonene   carvone

Common chromatography adsorbents (in approximate order of adsorbent strength):

Alumina (Al$_2$O$_3$)	strong
Charcoal (C)	
Florisil (MgO/SiO$_2$)	
Silica (SiO$_2$)	
Magnesia (MgO)	weak

Table 1.   Alumina
activity grades

Grade	wt% water
I	0
II	3
III	6
IV	10
V	15

Since all adsorbents are deactivated by water, it is important to keep their containers tightly closed and to minimize their exposure to atmospheric moisture.

Elution solvents should be as pure and water-free as possible.

Table 2.   Eluotropic series for alumina and silica

alumina	silica
Pentane	Cyclohexane
Petroleum ether (light)	Petroleum ether
	Pentane
Hexane	Chloroform
Cyclohexane	
	Ethyl ether
Ethyl ether	Ethyl acetate
Chloroform	
	Ethanol
Methylene chloride	Water
	Acetone
Ethyl acetate	Acetic acid
2-Propanol	Methanol
Ethanol	
Methanol	
Acetic acid	

Note: For additional solvents, see p. 375 of *The Chemist's Companion*, 〈Bib–A23〉.

Some mixtures should not be separated on certain kinds of adsorbents. For example, basic alumina would be a poor choice to separate a mixture containing aldehydes or ketones, which might undergo aldol condensation reactions on the column; it is also unsuitable for carboxylic acids, which bond so strongly to alumina that they cannot be easily desorbed. Deactivated silica gel, although less active than alumina, is a good all-purpose adsorbent that can be used with most kinds of functional groups.

The amount of adsorbent required for a given application depends on the sample size and the difficulty of the separation. If the components of a mixture differ greatly in polarity, a long column of adsorbent should not be necessary, since the separation will be easy. The more difficult the separation, the more adsorbent will be needed. About 20–50 g of adsorbent per 1 g of sample is sufficient for most separations, but ratios of 200 : 1 or higher are occasionally required.

Eluents.   In a typical elution process, the eluent acts primarily as a solvent to differentially remove molecules of solute from the surface of the adsorbent. In some cases, polar solvent molecules will also *displace* solute molecules from the adsorbent by becoming adsorbed themselves. If the solvent is too strongly adsorbed, the components of a mixture will spend most of their time in the mobile phase and will not separate efficiently. It is generally best to start with a solvent of low polarity and then (if necessary) increase the polarity gradually to elute the more strongly adsorbed components. Table 2 lists a series of common chromatographic solvents in order of increasing eluting power from alumina and silica. Such a listing is called an *eluotropic series.* The eluotropic series for a nonpolar adsorbent such as charcoal is nearly the reverse of the one for alumina; the less polar solvents are more effective eluents with such adsorbents.

Elution Techniques.   Many chromatographic separations cannot be performed efficiently with a single solvent, so several solvents or solvent mixtures are used in sequence, starting with the weaker eluents, those near the top of the eluotropic series for the adsorbent being used. These eluents will wash down only the most weakly adsorbed components while strongly adsorbed solutes remain near the top of the column. By adding more powerful eluents, the remaining solute bands can then be washed off the column one by one.

In practice, it is best to change eluents gradually by using solvent mixtures rather than to change directly from one solvent to another. In *stepwise elution,* the strength of the eluting solvent is changed in stages by adding small portions of a stronger eluent to the weaker one. Because subsequent portions of the stronger eluent have less effect on elution power than the first one, the proportion of that eluent is increased more or less exponentially. For example, 5% methylene chloride in hexane may be followed by 15% and 50% mixtures of these solvents. One rule of thumb suggests that the eluent composition should be changed after three column volumes of the previous eluent have passed through; for example, if the packed volume of the adsorbent is 15 mL, then the eluent composition should be changed with every 45 mL or so of eluent.

Columns. In choosing a column for a particular chromatographic separation, an experimenter must first consider the amount of adsorbent needed for a given amount of sample and then choose a column that will completely contain the adsorbent with about 10–15 cm to spare. Ordinarily, the height of the column packing should be at least 10 times its diameter.

There are many different kinds of chromatography columns, from a simple glass tube with a constriction at one end to an elaborate column with a porous plate to support the packing and a detachable base. A buret will suffice for some purposes, particularly if it has a Teflon stopcock, but the lack of a detachable base makes it difficult to remove the adsorbent afterward. If the column does not have a stopcock, the tip can be closed with a piece of flexible tubing equipped with a screw clamp. Unless the tubing is resistant to the eluents (polyethylene and Teflon will not contaminate most solvents) it should be removed before elution begins. If the column contains a porous plate to support the packing, no additional support will be necessary; otherwise, the column packing should be supported on a layer of glass wool and clean sand.

Flow Rate. The rate of eluent flow through the column should be slow enough so that the solute can attain equilibrium, but not so slow that the solute bands will broaden appreciably by diffusion. For most purposes, a flow rate of between 5 and 50 drops per minute should be suitable — difficult separations require the slower rates. The flow rate can be reduced, if necessary, by partly closing the stopcock on the column or by reducing the *solvent head* (the depth of the eluent layer in the column above the adsorbent). The flow rate can be increased by increasing the solvent head or by applying uniform air pressure

See *J. Chem. Educ. 50,* 401 (1973) for a description of an alternative elution technique called dry-column chromatography.

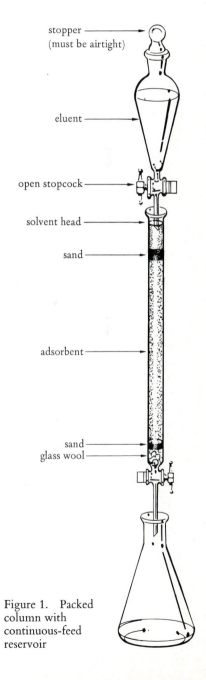

Figure 1. Packed column with continuous-feed reservoir

It is not advisable to try increasing the flow rate by applying suction to the bottom of the column; this procedure can cause the column adsorbent to separate.

to the top of the column through a large ballast volume (for example, with a rubber bulb connected to a carboy).

High-Performance Liquid Chromatography

Most column chromatography packings have comparatively large particles with diameters in the range of $75-175$ μm. Reducing the particle size by a factor of ten or more can increase the efficiency of a chromatographic separation dramatically. This is impractical for ordinary (gravity) column chromatography, however, because the flow rate of the eluent through the finer particles is extremely slow. High-performance liquid chromatography (HPLC) is a modern development in column chromatography in which a powerful pump is used to force eluents through a column filled with extremely fine particles ($3-10$ μm) of a packing material. A typical HPLC unit has several large solvent reservoirs, each of which can be filled with a different solvent so that the eluent composition can be varied continually throughout an elution, according to a programmed sequence. This greatly improves the separation of components in a complex mixture. Samples are introduced onto the column by means of a syringe or sampling valve and are carried through it by the pressurized eluent mixture, as illustrated by the schematic diagram in Figure 2.

A typical stainless steel HPLC column is about 25 cm long, has an inside diameter of 4.6 mm, and costs around $300; very efficient microcolumns with lengths as small as 3 cm and diameters of $1-4$ mm are also available. Although adsorption chromatography is used in some HPLC applications, most HPLC columns separate mixtures by *partition chromatography*, in which a liquid stationary phase is adsorbed on or chemically bonded to the solid particles of the column packing. One popular HPLC packing has octadecyl ($C_{18}H_{37}$) groups chemically bonded to silica gel; the silica particles are, in effect, coated with a very thin layer of liquid hydrocarbon. Partition chromatography is based on the same principle as extraction $\langle OP-13 \rangle$, in which a solute is distributed (partitioned) between two liquids. As the molecules of a component pass down the column, they are partitioned between the eluent and the liquid stationary phase according to a ratio—the partition coefficient—that depends on the solubilities of that component in the two liquids. The various components in a mixture generally have different partition coefficients, so they pass down the column at different rates.

In adsorption chromatography, the stationary phase is usually a polar substance such as alumina or silica that is much

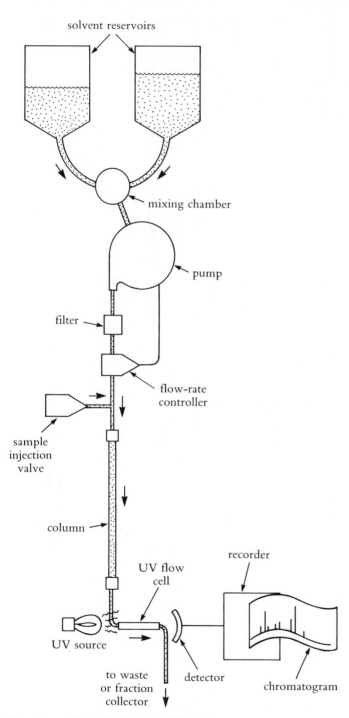

Figure 2.   Schematic diagram of HPLC system

more polar than the eluent; nonpolar components are therefore eluted faster than polar ones. HPLC separations frequently use column packings whose liquid stationary phase is *less* polar than the eluent, which may be a mixture of water with another solvent such as methanol or tetrahydrofuran. In that case, polar components will spend more time in the eluent than in the stationary phase and will be eluted faster than nonpolar ones, reversing the "normal" order of elution. This mode of separation is called *reverse-phase chromatography*.

As a component of a mixture leaves the column, its presence is revealed by a *detector* that responds to some property of the component, such as its refractive index or its ability to absorb ultraviolet radiation. The detector sends an electrical signal to a recorder, which traces a peak on a chart; the record of all such peaks for a given sample is called its chromatogram. Preparative HPLC instruments, which use special wide-bore columns, are equipped with fraction collectors to collect the eluent as it comes off the column; the fractions are then evaporated to yield the pure components. Analytical HPLC instruments, which are used to determine the compositions of mixtures, require much smaller samples, but the components are not generally recovered.

High-performance liquid chromatography is much faster than gravity column chromatography and more sensitive and efficient than either gravity column or thin-layer chromatography ⟨OP – 17⟩. It can be used to separate mixtures containing proteins, nucleic acids, steroids, antibiotics, pesticides, inorganic compounds, and many other substances whose volatilities are too low for gas chromatography ⟨OP – 32⟩. Because HPLC separations can be run at room temperature, there is little danger of decomposition reactions or other chemical changes that sometimes occur in the heated column of a gas chromatograph. These advantages have made HPLC the fastest-growing separation technique in chemistry, but its use in undergraduate laboratories is still very limited because of the high cost of the instruments, columns, and high-purity solvents required (HPLC-grade water costs about $30 a gallon!). The following operational procedures therefore apply only to gravity column chromatography. If you are fortunate enough to have access to an HPLC instrument, your instructor will describe its operation.

Packing the Column

To achieve good separation with a chromatographic column, it must be packed properly. The packing must be uniform, without air bubbles or channels, and its surface must be even and horizontal. Columns using alumina are generally packed with the dry adsorbent; those using silica gel are packed with a *slurry* containing the adsorbent in a solvent. The following directions are for packing a column that lacks a porous plate (if there is one, omit the glass wool and sand support).

Packing a Column with Silica Gel. First fill the column about half full (close the outlet!) with the least powerful eluting solvent to be used in the separation (often petroleum ether or hexane). Push a plug of glass wool to the bottom of the column with a glass rod, and tamp it down enough to form a level surface and press out any air bubbles. At this time, be sure the column is absolutely vertical and clamp it tightly so that it remains that way. Carefully pour clean sand (enough to form a 1-cm layer) through a funnel into the column. As it filters down through the solvent, tap the column gently and continuously with a "tapper" made of (for example) a rubber stopper on the end of a pencil, so that the sand forms a level, uniform deposit at the bottom of the column. Mix the measured amount of silica gel (or another slurry-packed adsorbent) thoroughly with enough solvent to make a fairly thick, but pourable, slurry. Pour a little of this slurry into the column, with tapping, through a dry funnel, so that the adsorbent gently filters down to form a layer about 2 cm thick at the bottom. Then open the column outlet to let the solvent drain into a flask while you slowly add the rest of the slurry, tapping constantly to help settle and pack the adsorbent. There should be enough solvent in the column so that the solvent level is always well above the adsorbent level; if necessary, add more solvent. When all of the adsorbent has been added, close the outlet. The surface of the adsorbent should be as level as possible, so continue tapping gently until it has completely settled. (Stirring the top of the solvent layer while the adsorbent is settling can help form a level surface.) Rinse down any adsorbent adhering to the sides of the column using an eyedropper or capillary pipet filled with solvent. If the column will not be used immediately, fill it with solvent, taking care not to disturb the adsorbent surface, and stopper it tightly.

Packing a Column with Alumina. Fill the column about two-thirds full with the appropriate solvent, and add glass wool and a layer of sand as described above. With the stopcock closed, pour enough *dry* alumina through a dry funnel, with tapping, to form a 2-cm layer at the bottom of the column. Then open

Equipment and supplies:
chromatography column
stopper
powder funnel
Pasteur pipet
tapper
adsorbent
eluent
clean sand
glass wool
filter paper

The column should be tapped gently near the middle (where it is clamped) to avoid displacing it from the vertical.

Keep the adsorbent container tightly capped!

If the slurry becomes too thick to pour, add more solvent to it.

Important: *No part of the column packing should ever be allowed to dry out—the entire column of adsorbent must be wet with solvent.*

the outlet and add the rest of the dry alumina while the solvent drains, tapping constantly so that the alumina settles uniformly. Finish preparing the column as described above for a silica column.

Operational Procedure

Equipment and supplies:
  packed chromatography column
  eluent(s)
  sample
  collecting tubes or flasks
  capillary pipet
  separatory/addition funnel

Be careful not to disturb the surface of the adsorbent while applying the sample.

The sand protects the adsorbent surface while eluents are being added.

The stopper of a continuous-feed reservoir should be moistened with solvent to create an airtight seal.

Some components can be observed under ultraviolet light, particularly if special inorganic phosphors have been added to the adsorbent.

J. Chem. Educ. 50, 401 (1973) describes a specially constructed solvent evaporator for column chromatography.

If the sample is a solid, dissolve it in a minimum amount of a suitable nonpolar solvent; use liquid samples without dilution. Open the column outlet until the solvent level comes down to the top of the adsorbent (no lower). Apply the sample carefully around the circumference of the adsorbent using a capillary pipet, so that it spreads evenly over the surface of the adsorbent. Then open the outlet until the liquid level again falls to the top of the adsorbent. Pipet a little of the initial eluent around the inside of the column to rinse down any adherent sample, and again open the outlet to bring the liquid level to the top of the adsorbent. Carefully add 3–5 mL of eluent, and sprinkle enough clean sand through the liquid to provide a uniform protective layer about 0.5 cm thick.

Clamp a separatory/addition funnel over the column, and measure the initial eluent into it; then add enough eluent to cover the adsorbent with 10–15 cm or more of liquid. If you have no suitable addition funnel to use as a reservoir, add the eluents through an ordinary funnel. Place a collecting vessel at the column outlet, open the outlet, and continue adding eluent (or use a continuous-feed reservoir as shown in Figure 1) to keep the liquid level nearly constant throughout the elution. When the eluent is being changed, allow the previous eluent to drain to the level of the sand before adding the next eluent from the reservoir.

If the components are colored or can be observed on the column by some visualization method, change collectors each time a new band of solute begins to come off the column *and* when it has about disappeared from the column. If two or more bands overlap, collect the overlapping regions in separate containers to avoid contaminating the pure fractions. When all of the desired components have been eluted from the column, evaporate the solvent ⟨OP–14⟩ from the pure fractions unless the experimental procedure indicates otherwise.

If the components are not visible and the procedure does not specify the volume of eluent to be collected, collect equal fractions (usually 25–100 mL each) in tared collecting vessels. Evaporate the solvent from each fraction, weigh the collectors, and plot the mass of each residue versus the fraction number to obtain an elution curve such as that illustrated in Figure 3. From the elution curve, it should be possible to identify separate components and decide which fractions can be safely combined.

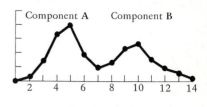

Figure 3. Elution curve

Summary

1. Pack column using designated adsorbent and solvent.

2. Drain column until the liquid level reaches top of the adsorbent, add sample, drain, rinse down sample, and drain again.

3. Add a little solvent and 0.5 cm of sand; drain to top of sand.

4. Add eluent from reservoir, put collector in place, and open column outlet; keep eluent level in column nearly constant.

5. To change elution solvents, drain previous solvent to top of sand layer, and add next eluent from reservoir.
IF components are visible, GO TO 6.
IF components are not visible, GO TO 7.

6. Change collectors when new band starts or ends and also for overlapping areas between bands.
GO TO 8.

7. Change collectors after a predetermined volume has been collected.

8. When appropriate, evaporate and weigh desired fractions; combine fractions containing the same component.

Thin-Layer Chromatography

OPERATION 17

Principles and Applications

As its name suggests, thin-layer chromatography (TLC) is carried out on a very thin layer of chromatographically active material dispersed on the surface of an inert support. In TLC, as in column chromatography ⟨OP–16⟩, a liquid mobile phase moves along a solid stationary phase, carrying with it the

The principles of liquid-solid adsorption chromatography are discussed in OP-16.

components of a mixture. In the process, the components are separated by differential partitioning between the solid and liquid phases. Usually, the stationary phase is an adsorbent such as alumina or silica gel, so separation is effected by the same basic mechanism as in adsorption chromatography on a column. Although TLC is not applicable to large-scale preparative separations, it is much faster than column chromatography and can be carried out with very small amounts of material, so that little is wasted. It also offers greater speed and resolution than the related technique of paper chromatography ⟨OP – 18⟩ and can be applied to a broader range of organic compounds.

TLC is often used by organic chemists to help identify unknown compounds, analyze reaction mixtures, determine the purity of products, and monitor various processes. For example, by running thin-layer chromatograms of a reaction mixture at regular intervals, an investigator can determine the optimum reaction time by comparing the relative amounts of product, reactants and by-products appearing on each chromatogram. If a product mixture is to be separated by column chromatography, TLC can help determine the best solvent for the separation; minute amounts of the mixture are spotted on TLC plates having the same adsorbent as the column, and different solvents are applied to each spot to see which gives the best separation. The fractions eluted from a column can also be analyzed by TLC to determine which component is in each fraction; then the fractions containing the same component can be combined and evaporated. Finally, if a product is further purified by recrystallization, TLC analysis can quickly show whether the "pure" product contains appreciable amounts of impurities.

The procedure is described in the section titled "Choosing a Developing Solvent" (pages 658 and 659).

Experimental Considerations

Adsorbents. In contrast to column chromatography, TLC is generally carried out with the mobile phase moving *up* the thin layer of adsorbent by capillary action, rather than *down* through the adsorbent by gravity. The adsorbent for TLC is more finely divided than that used in a column and is usually provided with a *binder* to make it stick to the support. The most commonly used binder for silica gel and alumina is plaster of Paris ($CaSO_4 \cdot \frac{1}{2}H_2O$), which hardens when combined with water or another hydroxylic solvent by forming gypsum ($CaSO_4 \cdot 2H_2O$) or other solvated forms of calcium sulfate. The most popular backing for "do-it-yourself" TLC plates is

TLC-grade adsorbents containing $CaSO_4 \cdot \frac{1}{2}H_2O$ are sold under trade names such as Silica Gel G (for "gypsum") and Alumina G. It is much simpler to use the premixed adsorbent than to prepare it yourself.

glass, in form of either microscope slides or glass plates measuring (typically) 20 × 20 cm. Commercial plates on glass, plastic, and aluminum backings are available in a variety of sizes and compositions.

TLC Plates. Microscope-slide TLC plates are usually made by dipping two slides at a time into a slurry of the adsorbent in water or an organic solvent. Although water gives a more durable layer, organic solvents such as chloroform or a chloroform-methanol mixture are satisfactory and considerably easier to work with (see Table 1 for the compositions of several of these slurries). With some practice, the dipping technique can produce a reasonably uniform layer of adsorbent on microscope slides; however, it cannot be used conveniently for large plates, which are coated by spreading, pouring, or spraying techniques. After the layer has been applied, a plate should be inspected by holding it up to a strong light to see that there are no streaks, thin spots, lumps, or loose particles. If the coating is not uniform or is otherwise imperfect, the plate should be scraped clean, washed, and coated again.

If a plate has been prepared with water or another hydroxylic solvent, the adsorbent must be *activated* by heating it for a period of time, usually at 110°C, to remove the solvent and increase the adsorptive power. The activated slides should be used immediately or stored in a desiccator over an efficient drying agent (blue silica gel is recommended), since they lose their activity rapidly in a moist atmosphere.

Spotting. The activated plate is prepared for development by *spotting* it with a solution of the mixture to be analyzed. It is important to position the spots accurately; incorrectly placed spots may run into each other or onto the edge of the adsorbent layer. This is most conveniently done with a template, although a ruler (preferably a transparent plastic one) will work if it can be supported just above the surface of the plate so that it does not touch the adsorbent. On a microscope-slide TLC plate, the starting point is marked on both edges about 1 cm from the bottom (see Figure 1). A finish line, if desired, can be inscribed across the width of the slide about 0.5 cm from the end of the adsorbent layer. Two spots can be placed about 1 cm apart between the starting marks and equidistant from the edges of the plate. If three spots are to be developed, one is placed in the center and the other two about 8 mm from the edges.

Table 1. Composition of nonaqueous TLC slurries for dipping

Adsorbent	Grams of adsorbent	mL of CHCl$_3$	mL of CH$_3$OH
Silica Gel G	35	67	33
Alumina G	60	70	30
cellulose	50	50	50

Note: No binder is required for cellulose.

Aqueous slurries of adsorbents containing a binder will harden in a few minutes; chloroform slurries can be kept for several days.

When preparing and handling TLC plates, it is very important to avoid touching the surface of the glass or adsorbent and to protect the plate from foreign materials.

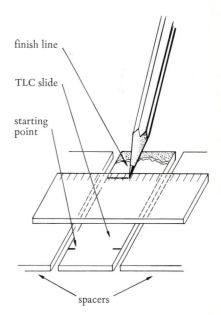

finish line

TLC slide

starting point

spacers

Figure 1. Scoring a TLC slide

The finish line is not essential, but it can prevent overdevelopment by stopping further migration of the solvent front.

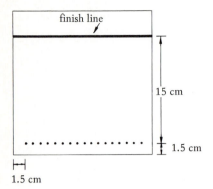

Figure 2.   Spotted TLC plate (20 × 20 cm)

The solvent should, if possible, be quite nonpolar and have a boiling point of 50–100°C. Chloroform is a good choice if the solute is sufficiently soluble in it.

Commercial TLC sheets usually have thinner coatings and require less sample than homemade plates do.

Capillary micropipets can be reused a few times if they are rinsed with solvent between applications; however, it is best to use a new pipet for each different mixture.

It is important not to dig a hole in the adsorbent surface while spotting; this will obstruct solvent flow and distort the chromatogram.

You can practice your spotting technique on a used plate or on the portion of a plate above the finish line.

Solvents should be chromatography-grade or recently distilled for good results.

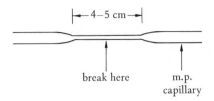

Figure 3.   Drawing a capillary micropipet

A 20 × 20 cm TLC plate can accommodate up to 18 spots spaced 1 cm apart along the starting line, with a distance of 1.5 cm between the end spots and the edges of the plate. The starting line is 1–1.5 cm from the bottom of the plate; the finish line is usually 15 cm or more from the starting line.

The substance to be analyzed is dissolved in a suitable solvent to make an approximately 1% solution. Column chromatography fractions and other solutions can often be used as is, if the solute is present at a concentration of about 0.1–2%. The spots are best applied with a syringe or micropipet, delivering about 1–5 μL of liquid to form a spot 1–3 mm in diameter. To keep the diameter of the spot small, the solution can be delivered by successive applications on the same spot, letting the solvent dry after each application.

A pair of capillary micropipets can be prepared by heating an open-ended melting-point capillary in the middle over a small flame and drawing it out to form a very fine capillary about 4–5 cm long. The tube is allowed to cool, and then is scored and snapped apart in the middle (see Figure 3). To spot a plate, the narrow capillary tip is dipped in the solution, then very gently applied to the surface of the TLC plate, at the proper location, for only an instant. With a little practice, it should be possible to deliver just the right volume of solution. Too much solution can result in "tailing" (a zone of diffuse solute following the spot), "bearding" (a zone of diffuse solute preceding the spot), overlapping of components, or unreliable R_f values. Too little solution will make it difficult to detect some of the components. It is often worthwhile to try one, two, and three applications at three separate points on the TLC plate to determine which quantity gives the best results.

Choosing a Developing Solvent. Solvents that are suitable eluents for column chromatography are equally suitable as TLC developers; the eluotropic series of Table 2 in OP–16 may be of help in choosing a solvent for a particular application. It is desirable to use the least polar solvent that will give a satisfactory separation. A quick way to find a suitable solvent is to spot a TLC plate with as many spots as you have solvents to test (you can use a grid pattern with the spots 1.5 – 2 cm apart), then to apply enough solvent directly to each spot to form a circle of solvent 1 – 2 cm in diameter. A solvent whose chromatogram (after visualization) shows well-separated rings, with the outermost ring about 50 – 75% of the distance from the center to the solvent front, should be satisfactory. If no single solvent is suitable, you can choose two mutually miscible solvents whose outermost rings bracket this range (one with too little solute migration, the other with too much) and test them in varying proportions.

Mark the circumference of each circle before the solvent dries.

Hexane, toluene, chloroform, and methanol or ethanol (alone or in binary combinations) can effect most of the commonly encountered separations.

Development. TLC plates are *developed* by placing them in a *developing chamber* containing the developing solvent and a paper liner, which helps saturate the atmosphere with the solvent vapors. A suitable chamber can be a jar with either a screw-cap or ground-glass lid (Figure 4), a beaker covered by aluminum foil or plastic (polyethylene) wrap, or any of a variety of cylindrical or rectangular containers with tight-fitting closures. The container should be as small as possible so that it will quickly fill with solvent vapors. Development should be carried out in a place away from direct sunlight or drafts, to prevent temperature gradients. A solvent advance of 10 – 15 cm may take 15 – 45 minutes; a microscope-slide plate can often be developed in 5 – 10 minutes. When the solvent front reaches the finish line or approaches the top of the absorbent layer, the TLC plate should be removed and allowed to dry. The plate should not be left in the solvent after development is complete, since the spots will spread by diffusion.

A 250-mL Erlenmeyer flask works well for 2 × 10 cm strips cut from commercial TLC plates.

4-oz. (125-mL) jar

liner

TLC slide

developing solvent

Figure 4. Developing chamber for TLC slides

Visualization. If the spots are colored, they can be observed immediately; otherwise, they must be *visualized* by some method that will allow them to be distinguished from the background. A good general visualization procedure is to place the dry plate in a closed chamber such as a wide-mouthed jar with a screw-cap lid, add a few crystals of iodine, and gently heat the chamber on a steam bath so that the iodine vapors sublime onto the adsorbent. Most organic compounds will form brown spots; unsaturated compounds will often show up as white spots against the dark background. Another simple method is to hold the plate under an ultraviolet light which will reveal fluorescent compounds on an ordinary adsorbent or compounds that quench fluorescence if the adsorbent contains an added phosphor.

The iodine color fades in time, so the spots should be marked promptly. Saturated hydrocarbons and alkyl halides may not form spots with iodine.

Spots can often be located by spraying a special visualizing reagent onto the TLC plate. Visualizing reagents for specific classes of compounds include ninhydrin (for amino acids) and 2,4-dinitrophenylhydrazine (for aldehydes and ketones). All spraying should be done under a hood or in a spraying chamber. An aerosol can or special spraying bottle is used, and a thin spray is applied from about 2 feet away. Large plates are sprayed by criss-crossing them with horizontal and vertical passes.

Analysis. To identify the components on a developed chromatogram, the R_f values of the spots must be determined. This is done by measuring the perpendicular distance from the starting line to the center of each spot (or its "center of gravity" if it is irregular) and dividing this by the distance traveled by the solvent. Although R_f values are reported in the literature for certain compounds, the absolute values can rarely be used to prove the identity of a substance, since they depend on a number of factors that are difficult to standardize. They are, however, a useful guide to the relative migration distances of various compounds, since with the same solvent and adsorbent, components should migrate in the same sequence. The only way to be reasonably sure that an unknown is identical to a known compound is to run the known beside it on the same TLC plate and compare the distances traveled by the two. Even then, its identity must often be confirmed by an independent method.

$$R_f = \frac{\text{distance traveled by spot}}{\text{distance traveled by solvent}}$$

The R_f value of a given component depends on the nature, thickness, and activity of the adsorbent; the identity and purity of the solvent; the size of the sample; and the temperature of the developing chamber.

Preparation of TLC Plates

TLC plates can be prepared by either one of the following methods.

A. *Dipping.* *Under the hood,* combine 33 mL of methanol and 67 mL of chloroform **[vapor hazard! suspected carcinogen!]** in a 4-oz. (125-mL) screw-cap jar, stir in 35 g of Silica Gel G, and shake the capped jar vigorously for about a minute. Stack two *clean* microscope slides back-to-back, holding them together at the top. Without delay, dip them into the slurry for about 2 seconds, using a smooth, unhurried, paddle-like motion (see Figure 5A) to coat them uniformly with the adsorbent. (Dip the slides shortly after shaking the slurry, so that the adsorbent does not have time to settle, and immerse them deeply enough so that only the top 1 cm or so remains uncoated.) Touch the bottom of the stacked slides to the jar to

Several students should work together so they can use the same slurry. Add more slurry (or dilute it to replace evaporated solvent) as necessary.

Handle the clean slides by the edges or at the very top, or use forceps. The slides can be cleaned with detergent and water, then rinsed with distilled water and 50% aqueous methanol before dipping.

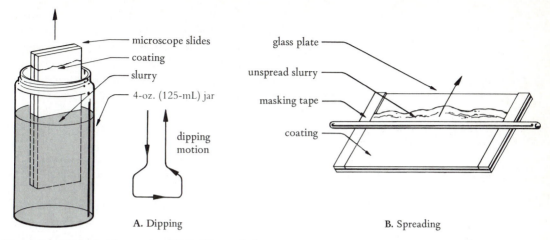

Figure 5.   Methods of preparing TLC slides and plates

drain off the excess slurry, let them air-dry a minute or so to evaporate the solvent, then separate them and wipe the excess adsorbent off the edges with a tissue paper. Activate the slides by heating them in a 110°C oven for 15 minutes.

Coated slides can be stored in a microscope-slide box inside a dessicator.

B. *Spreading.*   Put down some newspapers or other paper on your bench top. Place two strips of masking tape along two parallel edges of a *clean* glass plate (20 × 20 cm is standard) so that each strip of tape covers a 0.5-cm strip along the edge of the plate. Mix about 10 g of Silica Gel G with 20 mL of water (stirring and shaking well to get out any lumps), and pour some of it along one of the untaped ends. Spread it out by drawing a heavy glass rod across the plate (see Figure 5B) using one firm, continuous motion. Let the plate air-dry for 10 minutes or more, remove the tape, and set the plate vertically in a well-ventilated 110°C oven for at least 30 minutes to activate the adsorbent.

Additional layers of tape can be used for thicker coatings.

The exact amount of slurry has to be determined by trial and error.

Since the binder sets rapidly, the mixing and spreading must be completed in 2–3 minutes. Making more than one pass with the rod is not desirable.

Operational Procedure

Prepare a covered developing chamber containing a liner made of filter paper (or chromatography paper) that extends at least halfway around the circumference. Pour in enough developing solvent to form a layer 5 mm deep on the bottom. Tip the chamber and slosh the solvent around to soak the liner; let it stand for a half hour or so to saturate the atmosphere with solvent vapors. Keeping the apparatus saturated with solvent vapors reduces the development time and results in better shaped spots.

Equipment and supplies:
   TLC plate(s)
   developing chamber with lid
   filter-paper liner
   developing solvent
   micropipet or syringe
   pencil and ruler
   unknown solution
   standard solution(s)
   visualizing agent (spraying bottle, iodine chamber, UV lamp, etc.)

Do not actually draw a starting line *across the plate—just mark it at the edges.*

Mark the starting point (and the finish line, if desired) on a TLC plate, and spot it with (1) the solution to be chromatographed and (2) any solutions of known standards (if used). When the spots are dry, place the slide or plate in the chamber, spotted end down, propping the back (uncoated) side against the liner. Cover the chamber immediately.

When the solvent front is within 0.5–1 cm of the top of the plate (or when it reaches the finish line), carefully remove the plate and mark the centers of any visible spots with a pencil. Let the plate dry thoroughly, and visualize the spots (if necessary) by one of the methods described above. Measure the distance from the center of each spot to the starting line, and calculate the R_f values. If requested, make a permanent record of the chromatogram by (1) tracing it on semitransparent paper, (2) photographing or photocopying it, or (3) pressing strips of transparent tape onto the surface and stripping off the adsorbent layer.

Summary

1. Obtain or prepare TLC plate(s).

2. Spot TLC plate with solutions of the unknown and standard(s).

3. Place TLC plate in developing chamber containing solvent and liner; cover chamber.

4. Remove plate when development is complete; allow it to dry.

5. Visualize spots, if necessary, and mark their centers.

6. Measure R_f values; copy or otherwise preserve the chromatogram.

OPERATION 18

Paper Chromatography

Principles and Applications

Cellulose chromatography paper contains up to 22% water, and the developing solvents also contain water.

Paper chromatography is similar to thin-layer chromatography ⟨OP–17⟩ in practice, but quite different in principle. Although paper consists mainly of cellulose, the stationary phase is not cellulose, but the water that is adsorbed or chemically bound to it. Development is carried out by passing a comparatively nonpolar mobile phase through the cellulose fibers, partitioning the solutes between the bound water and the mobile phase. Paper chromatography thus operates by a liquid-liquid partitioning process rather than by adsorption on the surface of a solid.

Since only polar compounds are appreciably soluble in water, paper chromatography is most frequently used to separate polar substances such as amino acids and sugars. Manufactured chromatography paper is quite uniform and doesn't have the activation requirements of TLC adsorbents, so R_f values obtained by paper chromatography are generally more reproducible than those from TLC. However, the resolution of spots is often poorer, and the development times are usually much longer. Nevertheless, paper chromatography is a valuable analytical technique that can be applied to a large variety of separations.

The principle of paper chromatography is the same as that of liquid-liquid extraction ⟨OP-13⟩.

Paper can be specially treated to separate less polar compounds by impregnating it with other stationary phases, such as formamide or paraffin oil.

Experimental Considerations

Paper. Ordinary filter paper can be used for paper chromatography; however, the same paper is manufactured in convenient rectangular sheets, strips, and many other shapes for chromatographic use. For good results, the paper surface must be kept clean. Paper or plastic liners should be laid on the bench top to prevent contamination, and the chromatography paper should be handled only by the edges or along the top, beyond the anticipated finish line. Forceps are helpful; it is sometimes a good idea to leave a "handle" (which can be cut off just before development) on one (or both) ends of the paper. For much qualitative work, the paper can be 10–15 cm high (in the direction of development) and wide enough to accommodate the desired number of spots, spaced 1.5–2.0 cm apart. Narrow strips accommodating two or more spots can be developed in test tubes, bottles, cylinders, or Erlenmeyer flasks. A wider sheet can be rolled into a cylinder and developed in a beaker. The paper should be cut so that its grain is parallel to the direction of development, and the starting line should be marked in pencil about 2 cm above the bottom edge. The positions of the spots can also be marked lightly with pencil, but the finish line is not drawn until development is complete.

Whatman #1 and S&S 4043b are standard chromatography papers.

The spots tend to spread out more with paper chromatography than with TLC, so they should be applied farther apart.

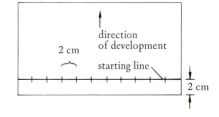

Figure 1. 11 × 22-cm paper marked for spotting

Spotting. As in TLC, the substance to be analyzed should be dissolved in a suitable solvent, usually at a concentration of about 1%. The spotting can be performed by the same techniques as for TLC (see OP–17) using a micropipet or syringe. However, for many purposes, a platinum wire loop 2–3 mm

Acetone, ethyl ether, and chloroform are frequently used as solvents.

During the spotting, nothing should touch the underside of the paper at the point of application.

Practice your spotting technique on a piece of filter paper before attempting to spot the chromatogram. Keep in mind that too little is usually better than too much.

The water is needed to maintain the composition of the aqueous stationary phase on the cellulose.

in diameter—or even a round toothpick saturated with the solution—is quite adequate. Each spot can be made with up to 10 μL of solution and should be 2–5 mm (no more) in diameter. Large volumes of solution should be applied in steps, drying the spot after each application.

Developing Solvents. The developing solvents used for paper chromatography are nearly always mixtures containing an organic solvent, water, and often a third component to increase the solubility of the water in the organic solvent or to provide an acidic or basic medium. The solvent mixtures are frequently prepared by saturating the organic solvent(s) with water in a separatory funnel, separating the two phases, and using the organic phase for development. Alternatively, monophase mixtures having about the same composition as the organic phase of the saturated mixtures can be used; these yield about the same results. Some typical monophase solvent mixtures are listed in Table 1. Most solvent mixtures should be made up fresh each time they are used and not kept more than a day or two.

Development. Paper chromatograms can be developed by ascending, descending, horizontal, and radial techniques. An ascending development is carried out in much the same way as for TLC. However, more time is required to saturate the developing chamber with solvent vapors, and the paper liner can be omitted.

Visualization. Visualization of spots by ultraviolet light is quite useful in paper chromatography, since paper fluoresces dimly in a dark room and many organic compounds will quench that fluorescence. Paper chromatograms can also be visualized by spraying them or by dipping them into a solution of a suitable visualizing reagent.

J. Chem. Educ. 49, 20 (1972) describes "an easily constructed aerosol sprayer."

Table 1.   Monophase solvent mixtures for paper chromatography

Solvents	Composition	Equivalent two-phase system
2-propanol, ammonia, water	9:1:2 (by volume)	none
1-butanol, acetic acid, water	12:3:5 (by volume)	4:1:5
phenol and water	500 g of PhOH, 125 mL of H_2O	saturated solution
ethyl acetate, 1-propanol, water	14:2:4 (by volume)	6:1:3

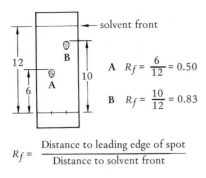

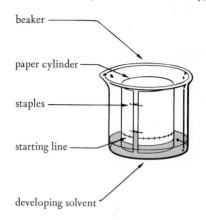

(cover omitted for clarity)

Figure 2. Measuring R_f values in paper chromatography

$$R_f = \frac{\text{Distance to leading edge of spot}}{\text{Distance to solvent front}}$$

Figure 3. Apparatus for paper chromatography

Analysis. Although R_f values in paper chromatography are somewhat more reproducible than those obtained from TLC, it is usually necessary to run a standard along with the unknown for qualitative analysis. If possible, the solvent and concentration should be the same for the standard as for the unknown. Spot migration distances are customarily measured from the starting line to the *front* of each spot, rather than to the center as for TLC.

Operational Procedure

The following procedure can be used when a number of samples or standards are to be chromatographed. For fewer samples, narrow strips of chromatography paper in jars, test tubes, or other appropriate containers can be used. Narrow strips can be folded in the middle or hung from a wire embedded in a cork for support.

Obtain chromatography paper, and cut it to form an 11 × 22 cm rectangle (or another appropriate size). Without touching the surface of the paper, use a pencil to draw a starting line 2 cm from one long edge (the "bottom" edge) and lightly mark the positions for spots, spacing them about 2 cm from each side and 1.5–2.0 cm apart. Spot the paper with the solutions and standards to be chromatographed; then roll it into a cylinder with the starting line at the bottom and staple the ends together, leaving a small gap between them. Add enough developing solvent to a 600-mL beaker (or another developing chamber) to give a liquid depth of about 1 cm on the bottom. Cover the chamber tightly with aluminum foil or polyethylene film, and slosh the solvent around in it for a short time. After allowing sufficient time (usually an hour or so) for the

The distance from the starting line to each spot should be measured along a line passing through the spot perpendicular to the starting line.

Equipment and supplies:
    chromatography paper
    pencil and ruler
    capillary micropipet(s) (or tooth-picks, etc.)
    solutions to be spotted
    developing chamber
    developing solvent
    visualizing reagent and equipment

The paper must not touch the sides of the beaker.

Development may take an hour or more, depending on the solvent and the distance traveled.

Smaller chromatograms can be prepunched and hung on a wooden applicator stick (clamped in a horizontal position) to air-dry.

solvent to saturate the developing chamber, quickly uncover the chamber, place the paper cylinder inside (spotted end down), and cover tightly again. Leave the apparatus in an area away from sunlight and strong drafts.

When the solvent front is a centimeter or less from the top of the paper, carefully remove the cylinder and separate the edges, then accurately draw a line along the *entire* solvent front with a pencil. If any spots are visible at this time, outline them with a pencil (carefully, to avoid tearing the wet paper), as they may fade in time. Again roll the paper into a cylinder, and stand it on edge to air-dry. When it is completely dry, visualize the spots by an applicable method. Measure the appropriate spot and solvent migration distances, and calculate the R_f values.

Summary

1. Obtain chromatography paper and cut it to size.

2. Mark starting line; apply spots.

3. Develop chromatogram in appropriate solvent and chamber.

4. Remove chromatogram, mark solvent front and visible spots, and dry chromatogram.

5. Visualize spots, if necessary.

6. Measure R_f values.

Washing and Drying Operations

Washing Liquids

Organic liquids or solutions often contain impurities that can be removed by extracting them into a suitable wash liquid. The organic substance being washed may be either a *neat* liquid (without solvent) or a compound dissolved in a solvent. In either case, it must not dissolve appreciably in the wash liquid or it will be extracted along with the impurities.

Water and saturated aqueous sodium chloride are used to remove water-soluble impurities such as salts and low-boiling polar organic compounds. Saturated sodium chloride is frequently used for the last washing before a liquid is dried ⟨OP–20⟩ because it removes excess water from the organic liquid by the salting-out effect ⟨OP–13c⟩. It is preferred to water in some other cases because it helps prevent emulsions at the interface between the liquids. (Emulsions can also be broken up or prevented by the methods described in OP–13.)

Some wash liquids contain chemically reactive solutes that are used to convert water-insoluble impurities into soluble salts. Aqueous solutions of sodium hydroxide, sodium carbonate, and sodium bicarbonate remove acidic impurities by converting them into soluble sodium salts. Dilute aqueous solutions of hydrochloric acid or sodium hydrogen sulfate are useful for removing alkaline impurities such as amines. Aqueous sodium hydrogen sulfite can remove certain aldehyde and ketone impurities by forming soluble bisulfite addition compounds. When a chemically reactive wash liquid is used, it is usually advisable to perform a preliminary washing with water or aqueous sodium chloride to remove most of the water-soluble impurities. This may prevent a potentially violent reaction between the reactive wash liquid and the impurities.

The effectiveness of an extraction using a given volume of solvent increases if the extraction is carried out in several steps —the more steps the better. Unless otherwise indicated, a liquid should be washed in two or three stages using equal

Reactions of washing solutions

$NaHCO_3$ + HA (acidic impurity)

Na^+A^- + H_2O + CO_2

Na_2CO_3 + 2HA

$2Na^+A^-$ + H_2O + CO_2

HCl + RNH_2 (amine)

$RNH_3^+Cl^-$ (amine salt)

$NaHSO_3$ + RCHO (aldehyde)

$$RCHSO_3^-Na^+ \quad \overset{OH}{\underset{|}{}}$$

(bisulfite addition compound)

volumes of wash liquid for each stage, and the total volume of wash liquid should be roughly equal to the volume of the liquid being washed. For example, if you are washing 30 mL of an organic liquid, you can use two 15-mL or three 10-mL portions of the wash liquid. When several different wash liquids are used in succession, one or two washings with each liquid may be sufficient, but the total volume of all the wash liquids should usually equal or exceed the volume of the liquid being washed.

Operational Procedure

See OP–13 for experimental details and a summary.

The procedure for washing liquids is essentially the same as that for extracting solvents ⟨OP–13⟩, except that the extraction solvent (wash liquid) is discarded and the liquid being washed is retained. If the liquid being washed is less dense than the wash liquid (for example, if it is an ether solution), the wash liquid is drained out after each washing and discarded, and the next portion of wash liquid is added to the organic layer, which remains in the separatory funnel. If the organic layer is more dense than the wash liquid (for example, if it is a methylene chloride solution), the organic layer is drained into a flask after each washing, the wash liquid is poured out the top of the separatory funnel and discarded, and the organic layer is returned to the funnel for the next washing, if necessary. If the wash liquid contains sodium carbonate, sodium bicarbonate, or another reactive solute that generates a gas, it should first be stirred vigorously with the organic layer until gas evolution subsides. Only then should the separatory funnel be stoppered. Also, the funnel should be shaken very gently and vented frequently at first — otherwise, a pressure buildup might cause the stopper to pop out and your product to spray all over the lab.

OPERATION 20

Drying Liquids

Organic liquids and solutions that have been separated from a reaction mixture or isolated from a natural product often retain traces of water. This water must often be removed before subsequent operations are performed. Liquids can be dried conveniently by allowing them to stand in contact with a *drying agent,* which is then removed by decanting or filtration.

Drying Agents

Most drying agents are inorganic salts that form hydrates by combining chemically with water. A drying agent is evaluated according to the amount of water it can absorb per unit of mass (capacity), the degree of dryness it can bring about (intensity), and how fast the drying takes place (speed). An ideal drying agent should have a high capacity, high intensity, and short drying time; it also must not react with or dissolve in the liquid being dried. Anhydrous magnesium sulfate is a good general-purpose drying agent, although other drying agents have advantages for specific applications. For example, sodium sulfate can be used to pre-dry very wet solutions because it has a high capacity. The final drying can then be accomplished with a high-intensity drying agent such as calcium sulfate. Table 1 summarizes the properties of the most common drying agents.

Hydration of magnesium sulfate:

$$MgSO_4 + 3H_2O \leftrightarrows MgSO_4 \cdot 3H_2O$$

(other hydrates are also formed)

Experimental Considerations

Ethyl ether, ethyl acetate, and other comparatively polar liquids absorb appreciable quantities of water and are often washed with saturated sodium chloride ⟨OP–19⟩ to remove excess water just before drying. They should then be dried with a high-capacity drying agent such as magnesium sulfate. Less polar liquids, such as petroleum ether and chlorinated hydrocarbons, absorb little water and may be dried by calcium sulfate or calcium chloride.

Table 1.   Properties of commonly used drying agents

Drying Agent	Speed	Capacity	Intensity	Comments
magnesium sulfate	fast	15–75%	medium	good general drying agent, suitable for nearly all organic liquids
calcium sulfate (Drierite)	very fast	7%	high	very fast and efficient, but low capacity
sodium sulfate	slow	126%	low	inefficient and slow; often used for pre-drying; loses water above 32.4°C
calcium chloride	slow to fast	15–30%	medium	removes traces of water quickly, larger amounts slowly; reacts with many oxygen and nitrogen-containing compounds
potassium carbonate	fast	16%	medium	cannot be used to dry acidic compounds
potassium hydroxide	fast	very high	high	used mainly to dry amines; reacts with many other compounds; causes severe burns

Note: Capacity is given as the percentage of its weight that a drying agent can absorb.

Small amounts of liquid can be dissolved in ether or another volatile solvent before drying to reduce adsorption losses. The solvent is subsequently removed by evaporation (OP–14).

The *quantity* of drying agent needed depends on the capacity and particle size of the drying agent and on the amount of water present. As a rough rule of thumb, about 1 g of drying agent should be used for each 25 mL of liquid; more should be added if the drying agent has a low capacity or a large particle size and if the liquid has a high water content. It is best to start with a small amount of drying agent and then add more if necessary, since starting with too much results in excessive losses by adsorption. The appearance of the drying agent after the drying time is up often suggests whether more is required. Magnesium sulfate clumps together in large crystals as it absorbs water, calcium chloride takes on a glassy surface appearance, and indicating Drierite changes color from blue to pink. If it appears that most of the drying agent is exhausted, it is preferable to remove it by decanting or filtering before adding fresh drying agent. (Spent drying agent contains hydrates that reduce drying efficiency.)

The *time* required for drying depends on the speed of the drying agent and the amount of water present. With magnesium sulfate or Drierite, 5 minutes is usually sufficient; calcium chloride and sodium sulfate take somewhat longer. Most drying agents attain at least 80% of their ultimate drying capacity within 15 minutes, so longer drying times are seldom necessary. When more complete drying is required, it is better to replace the spent drying agent with fresh agent or to use a more efficient drying agent than to extend the drying time.

The drying agent should always be removed when the drying period is over. A coarse-grained drying agent such as granular calcium chloride or Drierite can be removed by carefully decanting the liquid. Other drying agents should be removed by gravity filtration through a coarse fluted filter paper. When a solution (an ether extract, for example) is being dried, the drying agent should be washed with a little of the pure solvent to recover adsorbed solute that would otherwise be discarded with the spent drying agent.

Operational Procedure

Equipment and supplies:
  Erlenmeyer flask
  stopper
  drying agent
  funnel
  fluted filter paper

Select an Erlenmeyer flask that will hold the liquid with plenty of room to spare, and add the liquid and the estimated quantity of a suitable drying agent (keep its container tightly closed!). If a second (aqueous) phase forms, remove it with a capillary pipet and add more drying agent. Stopper and shake the flask; then set it aside for 5–15 minutes, shaking it intermittently throughout the drying period. Examine the drying agent carefully; if most of it is exhausted, add more drying agent or replace it with fresh drying agent. When drying appears complete, remove the drying agent by decanting or gravity filtration ⟨OP–11⟩. If you are drying a solution, wash the drying agent with a small volume of the pure solvent, and combine the wash liquid with the filtrate.

Summary

1. Choose drying agent; estimate quantity required (if necessary).

2. Mix drying agent with liquid in flask, stopper and shake, and set aside.

3. Swirl occasionally until drying time is up.
IF drying agent is exhausted *or* aqueous phase separates, GO TO 4.
IF not, GO TO 5.

4. Remove aqueous phase or used-up drying agent; add fresh drying agent.
GO TO 3.

5. IF liquid is not sufficiently dry, GO TO 6.
IF liquid is sufficiently dry, GO TO 7.

6. Add more drying agent, *or* filter and add a more efficient drying agent.
GO TO 3.

7. Filter or decant to remove drying agent.

8. Wash drying agent with fresh solvent; combine wash solvent with filtrate.

Step 8 applies to solutions, not to neat liquids.

Drying Solids

OPERATION 21

Solids that have been separated from a reaction mixture or isolated from other sources usually retain traces of water or other solvents used in the separation. Solvents can be removed by a number of drying methods, depending on the kind of solvent being removed, the amount of material to be dried, and the melting point and thermal stability of the solid compound.

Experimental Considerations

Solids that have been collected by vacuum filtration ⟨OP – 12⟩ are usually air-dried on the filter by leaving the aspirator turned on for a few minutes after filtration is complete. Unless the solvent is very volatile, further drying is required. Comparatively volatile solvents can be removed by simply spreading the solid out on a watch glass (covered to keep out airborne particles) and placing the glass in a location with good air circulation (such as a hood) for a sufficient period of time. If faster drying is necessary and the compound is not heat-sensitive, it can be placed in an evaporating dish and warmed over a steam bath, using a low rate of steam flow to prevent condensation inside the container. Clamping an inverted funnel over the evaporating dish and passing a gentle, dry ⟨OP – 22⟩ stream of air through it *(be careful: don't blow the crystals away!)* will accelerate the drying rate. Solids can be dried rapidly in a laboratory oven set at 110°C or another suitable temperature. The container used for oven-drying should be wide and shallow so that the solid can be spread out in a thin, uniform layer; the container should be labeled or otherwise marked to prevent mixups. Aluminum weighing dishes and other commercially available containers made of heavy aluminum foil are ideal because the aluminum conducts heat well, cools quickly, and is not likely to burn your fingers. Porcelain or Pyrex containers such as evaporating dishes, watch glasses, and Petri dishes can also be used. You should always make sure that the oven temperature is 20 – 30°C below the melting point of your product before you put it inside. An oven should *not* be used to dry heat-sensitive or low-melting solids or solids that sublime readily at oven temperatures. It is not unusual for a student to open an oven door and discover that the product he or she worked many hours to prepare has just turned into a charred or

The compound must not melt, decompose, or sublime near 100°C if a steam bath is used.

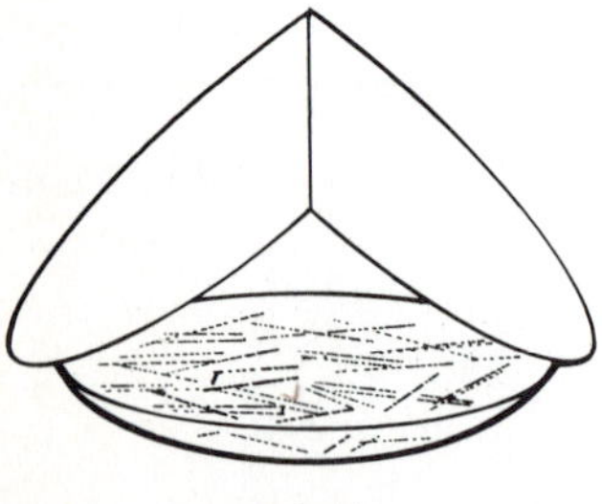

folded filter paper

watch glass

Figure 1.   Covered watch glass for drying crystals

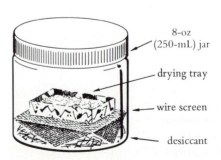

8-oz (250-mL) jar

drying tray

wire screen

desiccant

Figure 2.   Desiccator jar

molten mass, or disappeared entirely. If you are not sure whether your product can be oven-dried safely, consult your instructor.

When time permits, the safest way of drying a solid is to leave it in a *desiccator* overnight or longer. A desiccator is simply a tightly sealed container partly filled with a *desiccant* (drying agent) that absorbs water vapor, creating a moisture-free environment in which the solid will dry quite rapidly and thoroughly. Anhydrous calcium chloride is a good (if rather slow-working) desiccant because it is inexpensive and has a high water capacity. Drierite (anhydrous calcium sulfate) is faster and more efficient than calcium chloride but has a much lower capacity. A combination of calcium chloride with a little indicating Drierite works better than either desiccant separately; the Drierite removes traces of moisture that calcium chloride cannot, and it turns from blue to pink to indicate when the desiccant needs to be replaced or reactivated. Although commercial desiccators are available, a simple desiccator jar for efficient drying of small amounts of solid can be constructed from an 8-oz. (about 250-mL) wide-mouthed jar with a screw cap (Figure 2). The solid desiccant is added to a depth of 1 cm or so, and a piece of wire screen is cut and bent to fit on top of it, providing a stable, horizontal surface for the sample. The wire screen can be omitted if you use a fairly rigid flat-bottomed container, such as an aluminum weighing dish, that is not likely to tip over and spill the product. The jar must be kept tightly closed until ready for use, and the desiccant should be replaced when it is exhausted; unless otherwise indicated, put used desiccants into a designated container for reactivation. Drierite can be reactivated by heating it in a 225°C oven overnight; calcium chloride should be heated at 250–350°C.

Sulfuric acid and phosphorus pentoxide are also widely used as drying agents but are somewhat hazardous for use in undergraduate laboratories.

Vacuum desiccators combine the use of a desiccant with low air pressure for very fast, efficient drying.

Operational Procedure

The following procedure can be used to remove water from solids when quick drying is not essential. If faster drying is necessary, use one of the methods described above. If the solid is quite wet, transfer it to a large circle of filter paper on a clean surface and blot it with another filter paper to remove excess water. Break up the solid with a large flat-bladed spatula and rub it against the filter paper with the blade of the spatula until it is finely divided and friable. You may have to blot it with fresh filter papers or pulverize it with a flat-bottomed stirring rod.

Spread out the solid in a shallow container, and place it in a desiccator charged with fresh desiccant. Leave the solid overnight or longer in the tightly closed desiccator; then remove and weigh it. Put it back in the desiccator for another hour or so (add fresh desiccant if necessary), and then reweigh it to see if drying is complete. Drying should be sufficient for most purposes when two successive weighings differ by 0.5% or less.

OPERATION 22

Drying Gases

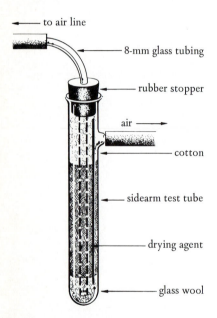

— to air line

— 8-mm glass tubing

— rubber stopper

air —

— cotton

— sidearm test tube

— drying agent

— glass wool

Figure 1.   Trap for drying air and other gases

Indicating silica gel and granular alumina are very efficient drying agents; indicating Drierite and calcium chloride are satisfactory for many purposes.

For very efficient drying, a drying train containing one or more drying towers and gas scrubbing bottles is used. Sulfuric acid is the preferred desiccant for gas scrubbing bottles, which must always be preceded by a trap.

The gas most often dried in organic chemistry experiments is air. For instance, it may be necessary to exclude moisture ⟨OP–22a⟩ from a reaction mixture or to evaporate a solvent in a stream of dry air. Other gases are available in cylinders and can be purchased in a form that is dry enough for most applications. In some cases, however, a cylinder gas or a gas generated in the laboratory may also require drying before use.

Air from an air line is sometimes dried by passing it slowly through an ordinary calcium chloride drying tube or a U-tube filled with a drying agent. The gas drying trap in Figure 1 is more efficient for this purpose; it removes oil, rust, and other impurities from the air line as well. About 2–3 cm of glass wool is loosely placed in the bottom of the sidearm test tube, then the glass tube is inserted far enough to contact it. The tube is held there to keep the open end clear while the drying agent is added. Another layer of cotton or glass wool is added to absorb residual impurities and keep the sidearm free of desiccant particles. The stopper is then inserted tightly and wired in place. Since most drying agents require a few seconds of contact time for adequate drying, air should be passed through the trap in a *slow* stream.

Gases other than air can be dried by the same general method. Some gases react with certain drying agents, so it is important to make the right choice. Obviously, an acidic gas such as HCl should not be dried with a basic substance such as silica or alumina, and gaseous ammonia should not be dried with sulfuric acid. Care must be taken when drying flammable gases such as methane or hydrogen to avoid fires and explosions.

Excluding Moisture

If it is necessary to exclude moisture from an experimental setup during an operation such as refluxing or distillation, a *drying tube* containing a suitable drying agent should be attached to any part of the apparatus that is open to the atmosphere. Granular alumina, silica gel, calcium sulfate (Drierite), and calcium chloride are commonly used for this purpose. Calcium chloride is the least efficient of the four, but it is still adequate in many cases. The drying tube is filled by tamping a small plug of dry cotton or glass wool into the bottom opening, adding the desiccant, and inserting another plug of cotton or glass wool to prevent spills. The tube must never be stoppered, as this would result in a closed system that might explode or fly apart when heated. The drying tube can be inserted in the top of a reflux condenser through a rubber stopper or thermometer adapter, or it can be attached to the sidearm of a vacuum adapter and similar devices with a short length of rubber tubing. When a very dry atmosphere is required, the apparatus should be swept out before the operation is begun with dry nitrogen or another suitable gas introduced through a gas delivery tube, preferably one with fused, porous glass (fritted glass) at the outlet.

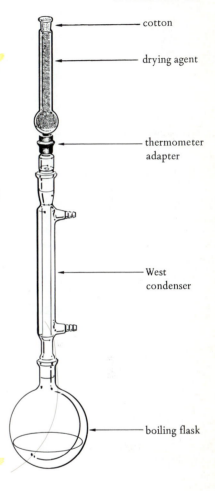

cotton

drying agent

thermometer adapter

West condenser

boiling flask

Figure 1.   Assembly for refluxing in a dry atmosphere

Trapping Gases

Some reactions release toxic or corrosive gases that must be *trapped* to keep them out of the laboratory atmosphere. An efficient *gas trap* can be constructed by inverting a narrow-stemmed funnel into a beaker containing a suitable gas-absorbing liquid, so that the rim of the funnel just touches the liquid surface. Water alone is suitable for absorbing many gases, but dilute aqueous sodium hydroxide (1 – 2 M) is generally used to absorb acidic gases such as hydrogen bromide or hydrogen chloride. For small amounts of gases, a straight glass tube inserted into a large test tube is often adequate. In this case, however, the outlet of the tube must be about a millimeter *above* the surface of the liquid to keep the liquid from backing up into the system when gas evolution ceases or heating is discontinued. A gas trap is connected to the reaction apparatus at any point that is open to the atmosphere (usually the top of the reflux condenser) by means of a length of tubing and a stopper or thermometer adapter.

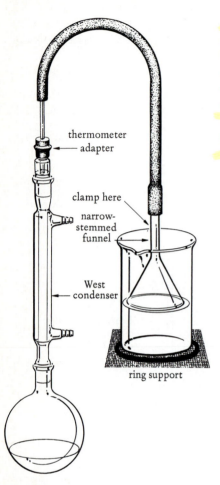

thermometer adapter

clamp here

narrow-stemmed funnel

West condenser

ring support

Figure 1.   Apparatus for trapping gases during reflux

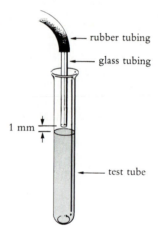

rubber tubing

glass tubing

1 mm

test tube

Figure 2.   Semimicro gas trap

Purification Operations

Recrystallization

Principles and Applications

Recrystallization is the most frequently used operation for purifying organic solids. This technique is based on the fact that the solubility of an organic compound in a given solvent will often increase greatly as the solvent is heated to its boiling point. When an impure organic solid is first heated in such a solvent until it dissolves, then cooled to room temperature or below, it will usually **recrystallize** from solution in a much purer form. Most of the impurities will either fail to dissolve in the hot solution (from which they can be filtered) or remain dissolved in the cooled solution (from which the pure crystals are filtered).

Salicylic acid dissolves in water to the extent of 0.20 g per 100 mL at 20°C and 7.5 g per 100 mL at 100°C, whereas acetanilide has a solubility in water of 0.50 g per 100 mL at 20°C. Suppose a 5.00-g sample of salicylic acid contaminated by 0.25 g of acetanilide is dissolved in and recrystallized from boiling (100°C) water. The amount of water required to just dissolve all of the salicylic acid at the boiling point is 67 mL (see calculations); all of the acetanilide impurity will also dissolve in the hot solution. When the solution is cooled down to 20°C, about 0.13 g of salicylic acid will remain dissolved; the other 4.87 g will crystallize out if sufficient time is allowed. *All* of the acetanilide should remain in solution, since up to 0.35 g of acetanilide can dissolve in 67 mL of water at 20°C. Therefore one should obtain, under ideal conditions, about a 97% recovery of essentially pure salicylic acid (see calculations).

This is a simplified description of a complex process; a number of factors, including the following, may bring about results different from those predicted.

1. Very slow crystallization yields large crystals that may trap or *occlude* impurities within the crystal lattice; fast

crystallize: to obtain crystals of a substance

recrystallize: to *again* crystallize substances that were initially in crystalline form and then dissolved

Calculations:

$$\frac{5.0\ \text{g}}{7.5\ \text{g}} \times 100\ \text{mL} = 67\ \text{mL of water}$$

$$0.20\ \text{g} \times \frac{67\ \text{mL}}{100\ \text{mL}}$$
$$= 0.13\ \text{g of salicylic acid}$$

$$0.50\ \text{g} \times \frac{67\ \text{mL}}{100\ \text{mL}}$$
$$= 0.35\ \text{g of acetanilide}$$

$$\frac{4.87\ \text{g}}{5.00\ \text{g}} \times 100 = 97\%\ \text{recovery}$$

Medium-sized crystals are usually the purest.

crystallization results in small crystals that readily adsorb impurities on their surfaces and are difficult to wash.

2. The solubility of one solute in a saturated solution of another solute is not usually the same as its solubility in pure solvent.

3. Adding only enough solvent to dissolve a solid may result in premature crystallization from the saturated solution and consequent losses. In practice, an excess of solvent is often used to prevent this.

Experimental Considerations

In its simplest form, the recrystallization of a solid is carried out by dissolving the solid in the hot (usually boiling) recrystallization solvent, cooling the resulting solution to room temperature or below, and collecting the crystals by vacuum filtration ⟨OP–12⟩. In practice, experimental factors will often require additional steps such as filtering or decolorizing the hot solution.

If the solid is unusually heat-sensitive or prone to oiling out, heating can be carried out below the boiling point of the solvent.

J. Chem. Educ. 54, 639 (1977) describes a novel "gauze bandage method" for recrystallization.

The size of the crystals formed depends on the rate of cooling; rapid cooling yields small crystals and slow cooling on a thermally-nonconducting surface yields large ones. Large crystals are not necessarily the purest, but slow cooling is recommended when the solid is initially quite impure. If further purification is required, it should be carried out with more rapid cooling to form smaller crystals. In some instances, it may be desirable to collect a second or third crop of crystals by concentrating ⟨OP–14⟩ the **mother liquor** from the first crop. These crystals will contain more impurities than the first crop and may require recrystallization from fresh solvent. A melting-point determination ⟨OP–28⟩ or TLC analysis ⟨OP–17⟩ can be used to determine whether a solid is sufficiently pure after recrystallization.

mother liquor: the liquid from which the crystals are filtered after crystallization

Dust, laboratory debris, filter-paper fibers, and other organic or inorganic impurities may be among the insoluble contaminants.

If too much solvent is used, the product yield may be greatly reduced, or the compound may not crystallize at all. Excess solvent can be removed by evaporation ⟨OP–14⟩.

Filtering the Hot Solution. Some impurities in a substance being crystallized may be insoluble in the boiling solvent and should be removed by filtration after the organic compound has dissolved. It is important not to mistake such impurities for the compound being purified, and thus add too much solvent in an attempt to dissolve them. If, after most of the crude solid has dissolved, the addition of another portion of hot solvent does not appreciably reduce the amount of solid in the flask, that solid is probably an impurity — particularly if it is different in appearance from the remainder of the solid.

The hot solution is filtered by the usual procedure for gravity filtration ⟨OP – 11⟩, which should be carried out as quickly as possible so that the solution does not have time to cool appreciably. Vacuum filtration ⟨OP – 12⟩ can be used when large quantities of solvent are involved. However, reduced pressure cools the filtrate by evaporation, which can cause premature crystallization and make it necessary to reheat the filtrate or add more solvent to redissolve the product after filtration. The funnel and collecting flask should be preheated to keep crystals from forming in them during filtration. A convenient way of doing this is to set the funnel on the hot solvent flask to be heated by boiling solvent vapors; this flask is then used as the collecting flask after unused solvent has been poured out. The funnel can instead be heated in an oven or inverted inside a large beaker set on a steam bath. Crystals that form on the filter paper or in the funnel stem can sometimes be dissolved and combined with the filtrate by pouring a *small* amount of hot recrystallization solvent through them. If a large quantity of solid crystallizes in the filter, it should be scraped out and combined with the filtrate, then redissolved by heating the mixture with a little additional recrystallization solvent; the extra solvent will help prevent premature crystallization when the hot solution is again filtered. Any cloudiness or precipitate that forms in the collecting flask during the filtering operation should be redissolved by heating before the flask is set aside to cool.

The hot solution should be returned to the heat source after each addition to the funnel.

Removing Colored Impurities. If a crude sample of a compound known to be white yields a colored recrystallization solution, *activated carbon* (Norit) can often be used to remove the colored impurity. The hot solution is removed from the heat to cool it down a few degrees below the boiling point, and a small quantity of activated carbon (about 0.2 g per 100 mL of solution) is stirred in, along with an equal weight of a filtering aid such as Celite. The solution is then heated back to the boiling point with stirring and filtered by gravity. If the solution is still colored, the treatment can be repeated with fresh carbon.

Using too much carbon results in adsorption of the product as well as the impurities.

Never add the carbon to a solution near the boiling point — it may boil over.

**SAFETY
PRECAUTION**

If any decolorizing carbon passes through the filter paper, it can sometimes be removed by vacuum filtration through a bed

The recrystallization solvent cannot be water, alcohol, or any other polar solvent because it would solvate the magnesium sulfate.

of anhydrous magnesium sulfate in a small Buchner funnel. The magnesium sulfate is mixed with enough low-boiling solvent (methylene chloride, for example) to form a thin slurry, and the slurry is poured into the funnel, while the aspirator is running, so as to form a uniform layer several millimeters thick on the filter paper. When the magnesium sulfate layer is completely dry and all of the solvent has been removed from the filter flask, the reheated solution is poured through the layer under vacuum. The magnesium sulfate should be washed with a little hot recrystallization solvent, and any crystals that form in the filtrate should be dissolved by heating before setting the filtrate aside to cool. It may be necessary to add more recrystallization solvent to replace any that evaporates during this filtration.

Some compounds have been known to remain in supersaturated solutions for years before crystallizing!

Inducing Crystallization. If no crystals form after the hot solution is cooled to room temperature, it is likely that the solution is supersaturated. If so, crystallization can often be induced by the following procedure. The tip of a glass rod is rubbed against the side of the flask with an up-and-down motion just above the liquid surface, so that it touches the liquid on the downstroke. If several minutes of scratching does not effect crystallization, then a few seed crystals of the compound (if available) are dropped into the solution with cooling ⟨OP–8⟩ and stirring. If crystals do not form after these measures have been taken and after the solution has been allowed to stand overnight or longer, it is likely that the recrystallization solvent is unsuitable or that too much was used. In the first case, the solvent must be removed by evaporation ⟨OP–14⟩ and a new solvent tried; in the second, the solution should be concentrated by evaporation ⟨OP–14⟩ until it reaches the saturation point, and then cooled to effect crystallization.

Crystallization may also be effected by adding a compatible solvent in which the compound is less soluble, then proceeding as described in OP–23b.

Some compounds, such as acetanilide in water, separate as a second liquid phase when a certain saturation level is exceeded. This can be prevented by using additional solvent.

Dealing with Oils and Colloidal Suspensions. When a compound being recrystallized is quite impure or has a low melting point, it may separate as a second liquid phase (an "oil") on cooling. Oils are undesirable, since even if they solidify on cooling the solid retains most of the original impurities. The following general approach is recommended unless the cause of the oiling is known and it can be prevented by another method. Heat the solution until the oil dissolves completely; then cool it slowly with constant stirring, adding a seed crystal or two (if available) at the approximate temperature where oiling occurred previously. If this is not successful, add about 25% more solvent and repeat the process. If oiling still occurs, cool the solution ⟨OP–8⟩ until the oil crystallizes (seeding and rub-

Add more solvent, if necessary, to dissolve the oil.

bing the oil with a stirring rod may help); then collect it by vacuum filtration and recrystallize it from the same solvent or a more suitable one, using the above techniques as necessary to prevent further oiling.

If a compound separates from solution as a fine colloidal suspension, the colloid can often be coagulated to form normal crystals by extended heating on a water bath or by adding an electrolyte such as sodium sulfate. Colloids can sometimes be prevented by treating the solution with activated carbon as described above, or by cooling it very slowly.

Crystallization can sometimes be induced by removing all the solvent, adding a seed crystal to the oil, and letting it stand in a refrigerator or freezer.

Colloids cannot be filtered — they will pass through the filter paper.

Electrolytes are useful only in solvents polar enough to dissolve them.

See OP–23c for hazards associated with common recrystallization solvents. Do not heat flammable solvents with a burner; extinguish all nearby flames before filtering the hot solution.

SAFETY PRECAUTION

Operational Procedure

Measure the estimated volume of solvent needed for recrystallization (plus a little more to allow for error and evaporation) into an Erlenmeyer flask (the *solvent flask*). Add a boiling stick or a few boiling chips, and heat the solvent to boiling on an appropriate heat source ⟨OP–7⟩. Place the solid to be purified in a second Erlenmeyer flask (the *boiling flask*), and add about half to three-quarters of the estimated volume of hot solvent, or 2–3 mL per gram of crude solid if the amount of solvent is not known. Heat the mixture at the boiling point with constant swirling or stirring, breaking up any large particles with a flat-bottomed stirring rod, until it appears that no more solid will go into solution with further heating. If undissolved solid remains, add more hot solvent in small portions (5–10% of the total volume), heating the solution at the boiling point with swirling or stirring after each addition, until the solid is completely dissolved — or until no more solid dissolves when a fresh portion of solvent is added and it appears that only solid impurities remain. If it is necessary to remove colored impurities or coagulate a colloidal suspension, treat the solution with activated carbon as described above. Have a preheated wide-necked funnel and collecting flask ready, and filter the hot solution through a coarse fluted filter paper while it is still near the boiling point. Keep the remaining unfiltered solution hot between additions. If some residue remains in the filter paper, dissolve it into the collecting flask with a little hot recrystallization solvent; if a considerable amount of the product has

If the amount of solvent is not known, start with 5–10 mL per gram of solid and use more if needed.

A steam bath is preferred for volatile organic solvents; water can be boiled using a burner or hot plate. Some hazardous solvents should be heated only under reflux or in a fume hood.

If the solvent is quite volatile or extended heating is required, place a small watch glass on the flask or insert a cold-finger condenser (see OP–7b) to reduce evaporation.

Equipment and supplies:
  two Erlenmeyer flasks
  recrystallization solvent
  graduated cylinder
  heat source
  boiling sticks or boiling chips
  flat-bottomed stirring rod
  funnel
  fluted filter paper
  small watch glass
  vacuum filtration apparatus ⟨OP–12⟩
  cold washing solvent
  optional: decolorizing carbon and filtering aid

crystallized in the filter, redissolve it as described above, in the section titled "Filtering the Hot Solution."

If small crystals are desired, swirl the flask under cold running water until crystals begin to form; then let it stand until crystallization is complete.

Set the flask on the bench top, cover it to keep out airborne particles, and let it stand undisturbed until crystallization is complete. If no crystals form by the time the solution reaches room temperature, use one of the methods described above to induce crystallization. Once crystals have begun to form, 15–30 minutes more at room temperature or below is usually sufficient time for crystallization to go to completion, but some compounds may require more time. The yield of product can be increased by cooling the mixture in ice once it reaches room temperature, but the purity of the product may decrease slightly as a result.

Collect the crystals by vacuum filtration ⟨OP–12⟩, using a little ice-cold recrystallization solvent (or another appropriate solvent) to transfer the last remaining crystals and to wash the solid on the filter. It is a good idea to turn off the aspirator and add the first portion of wash solvent while the crystals are still moist; otherwise impurities in the mother liquor may adhere to them. Air-dry the crystals by keeping the aspirator running a few minutes after the last washing; then dry them further ⟨OP–21⟩ as necessary.

If crystals form in the filtrate during the vacuum filtration, crystallization was incomplete. The filtrate should be allowed to cool longer and then refiltered. If a second crop of crystals is to be collected, retain the mother liquor in the filter flask and concentrate it by evaporation ⟨OP–14⟩ to the saturation point. Heat the solution (or add a little solvent) to discharge any turbidity; then cool it to induce crystallization, and proceed as before.

Summary

1. Measure estimated volume of solvent (plus a small excess) into solvent flask, and heat to boiling.

2. Add part of solvent to solid in boiling flask; boil with stirring.

3. Add more hot solvent in portions (if necessary) until organic solid dissolves.
IF solution contains colored impurities, GO TO 4.
IF not, GO TO 5.

4. Cool below boiling point, stir in decolorizing carbon and filtering aid (optional), and heat to boiling.

5. IF solution contains undissolved impurities, GO TO 6.

IF not, GO TO 7.

6. Filter hot solution by gravity.

7. Cover flask and set aside to cool until crystallization is complete.

Consult the Experimental Considerations section if crystals do not form, or if an oil or colloid forms.

8. Collect crystals by vacuum filtration; wash and air-dry on filter.

Semimicro Recrystallization

Recrystallization of quantities from a few tenths of a gram up to a gram or two should be carried out using small-scale apparatus to avoid excessive losses. Test tubes of the appropriate size can be used in place of Erlenmeyer flasks for dissolving and recrystallizing the solute, and small-scale filtration equipment can be used to filter the hot solution and the crystals. A very simple filter tube for this purpose can be constructed from soft glass tubing of a diameter small enough to nest inside a test tube. A plug of cotton is tamped into the filter tube (which has a small opening at the bottom) with a wooden applicator stick and washed by forcing some hot recrystallization solvent through it with an ear syringe (or another suitable rubber bulb). The preheated filter tube is then supported on the lip of the collecting tube and filled with the hot recrystallization solution (added in portions, if necessary), which is forced through the cotton plug with an ear syringe. An excess of the recrystallization solvent is generally used to prevent premature crystallization in the filter; the excess solvent is easily removed by evaporation after filtration.

The following procedure is suggested for semimicro recrystallization, although other kinds of apparatus and techniques can also be used. (See OP–23 for general information about recrystallization.)

Operational Procedure

Dissolve the solid in a test tube large enough to accommodate the recrystallization solvent by adding the solvent in small portions, boiling the mixture *gently* after each addition, and stirring constantly to prevent bumping or boilover. When the solid has dissolved, add a moderate excess of the recrystalliza-

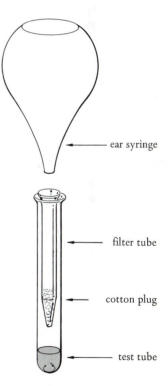

ear syringe

filter tube

cotton plug

test tube

Figure 1. Using a filter tube

tion solvent (10–20%), heat the mixture to boiling, and filter it through a preheated filter tube or another small-scale filtration apparatus. Rinse the filter with a small portion of hot recrystallization solvent, and combine the rinse liquid with the filtrate. Concentrate the solution to the saturation point by evaporating the excess solvent under a hood (or under aspirator vacuum), stirring or shaking to prevent boilover. When the solution becomes cloudy or crystals start to form, add more recrystallization solvent drop by drop, with heating and stirring, until the solution clears up or the crystals dissolve. Set the solution in an insulated container, such as a covered Styrofoam cup *or* a small flask set inside a beaker and covered with a larger beaker; let the solution cool slowly to room temperature. Cool it further in an ice bath to improve the yield, if desired.

When crystallization is complete, collect the product by semimicro vacuum filtration ⟨OP–12a⟩, washing it on the filter with a small quantity of cold recrystallization solvent. Air-dry the crystals on the filter; then dry them further ⟨OP–21⟩, if necessary.

OPERATION 23b

Recrystallization from Mixed Solvents

Table 1.   Some compatible solvent pairs

ethanol–water
methanol–water
acetic acid–water
acetone–water
ethyl ether–methanol
ethanol–acetone
ethanol–petroleum ether
ethyl acetate–cyclohexane
chloroform–petroleum ether

Solids that cannot be recrystallized readily from any of the common solvents can usually be purified by recrystallization from a mixture of two compatible solvents. The solvents must be miscible in one another, and the compound should be quite soluble in one and relatively insoluble in the other. If the composition of a suitable solvent mixture is known beforehand (such as 40% ethanol in water, for example), the recrystallization may be carried out with the premixed solvent in the same manner as for a single-solvent recrystallization ⟨OP–23⟩. But a mixed-solvent recrystallization is usually performed by heating the compound in the solvent in which it is most soluble (which we will call solvent A) until it dissolves, then adding enough of the second solvent (solvent B) to bring the solution to the saturation point.

If the compound is *very* soluble in solvent A, the total volume of solvent may be quite small compared to that of the crystals, which may then separate as a dense slurry. In such a case, it is desirable to use more of solvent A than is needed to just dissolve the compound, and to add correspondingly more of solvent B to bring about saturation. (Recrystallization from a very small volume of solvent is not advisable.) One must be careful to avoid adding so much of solvent A that *no* amount of solvent B will result in saturation; if that occurs, it will be necessary to concentrate ⟨OP – 14⟩ the solution or to remove all of the solvent and start over.

See *J. Chem. Educ.* 51, 602 (1974) for an alternative mixed-solvent recrystallization method.

Operational Procedure

Place the crude solid in a flask, and add solvent A (previously heated to boiling) in portions, with heating and stirring, until the solid is dissolved. Add a small excess of solvent A (about 10%) to prevent premature crystallization, and heat the mixture to boiling. If the solution contains colored impurities, decolorize it with activated carbon ⟨OP – 23⟩. Filter the hot solution by gravity, and add solvent B in small portions with stirring, keeping the mixture boiling after each addition. When a persistent cloudiness appears or crystals start to form, add just enough hot solvent A drop by drop to clear it up or dissolve the precipitate. Set the mixture aside to cool to room temperature; then cool it further in ice if desired. When crystallization is complete, collect the product by vacuum filtration ⟨OP – 12⟩, washing it on the filter with cold solvent B or another suitable solvent.

See OP – 23 for general recrystallization procedures and OP – 23c for hazards associated with recrystallization solvents.

Choosing a
Recrystallization Solvent

OPERATION 23*c*

Several factors should be considered when choosing a recrystallization solvent for an organic compound.

1.   The compound being purified must dissolve to a substantially greater extent in the boiling solvent than in the cold solvent.

2.   The boiling point of the solvent should be high enough to take advantage of the compound's temperature coefficient of solubility, but not so high that it cannot be easily removed from the crystals or concentrated by evaporation. Most good recrystallization solvents boil in the 50 – 100°C range.

3.   Ideally, the impurities should either be insoluble in the recrystallization solvent or more soluble than the compound being purified. If this is not the case, recrystallization can still be effective unless the amount of impurity is quite large.

4. The solvent must not react with the compound being purified and should not be excessively hazardous to work with. When possible, solvents such as benzene and carbon tetrachloride should be avoided in favor of other, less toxic solvents.

5. The high freezing points of solvents such as acetic acid (17°C) can be disadvantageous, since they limit the minimum temperature attainable on cooling.

Some common recrystallization solvents and their properties are listed in Table 1.

Table 1.   Properties of common recrystallization solvents

Solvent	b.p.	f.p.	Comments
water	100°	0°	good solvent for polar compounds; crystals dry slowly
acetic acid	118°	17°	good solvent for polar compounds but hard to remove; unpleasant to work with
ethanol	78°	−116°	good general solvent; 95% ethanol is commonly used
2-butanone	80°	−86°	acetone has similar solvent properties but is limited by lower boiling point
ethyl acetate	77°	−84°	
chloroform	62°	−64°	good general solvent that is easily removed; suspected carcinogen
toluene	111°	−95°	good solvent for hydrocarbons, aromatic compounds; more difficult than most to remove
petroleum ether	~50–85°	low	high-boiling petroleum ether is preferred; ligroin is similar; very flammable
hexane	69°	−94°	good solvent for less polar compounds

Note: Solvents are listed in approximate order of decreasing polarity.

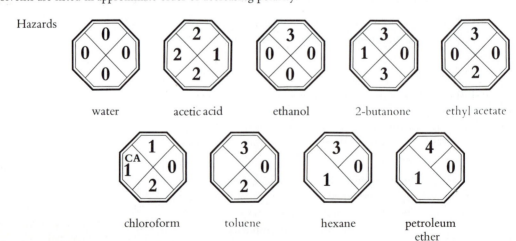

In choosing a suitable solvent, information available in reference books may be helpful. For example, the entry under "crystalline form" for naphthalene in the *CRC Handbook* reads "mcl pl (al)" meaning that naphthalene crystallizes from ethyl alcohol in the form of monoclinic plates. If no recrystallization solvent is listed, solubility data should be useful. A solvent in which the compound is designated as *sparingly soluble* or *insoluble* when cold and *very soluble* or *soluble* when hot may be suitable for recrystallization. Alternatively, a mixture of two compatible solvents may work if the compound is insoluble or sparingly soluble in one and soluble or very soluble in the other. If solubility data are not available, a solvent may have to be chosen by trial and error from a selection such as that in Table 1. Ordinarily the recrystallization solvent should be either somewhat *more* or somewhat *less* polar than the solute, since a solvent of similar polarity may dissolve too much of it. Once a possible solvent or solvent mixture is chosen, it should be tested by the following procedure.

CRC Handbook of Chemistry and Physics, Lange's Handbook of Chemistry, Merck Index, Dictionary of Organic Compounds, Beilstein, and others list recrystallization solvents or solubility data for many organic compounds. See Bibliography, category A.

If possible, the compound should have a solubility of about 5–25 g/100 mL in the hot solvent and less than 2 g/100 mL in the cold solvent, with at least a 5 : 1 ratio between the two values

A table on pp. 442–3 of *The Chemist's Companion* ⟨Bib–A23⟩ gives the classes of compounds that can be re-crystallized from different solvents.

Testing Potential Recrystallization Solvents. Weigh about 0.3 g of the finely divided solid into a test tube, add 3 mL of the solvent, and shake and stir the mixture to see if the solid will dissolve. If it does not, warm the mixture very gently, with stirring and shaking. If the solid dissolves in the cold solvent or with gentle warming, the solvent is unsuitable. Carefully heat the mixture to boiling, with stirring. If the solid does not dissolve readily, add more solvent in 1-mL portions, gently boiling and stirring the mixture after each addition, until it dissolves *or* until the total volume of added solvent is about 10 mL. If the solid does not dissolve (or nearly so) by this time, the solvent is probably not suitable; if it does dissolve, cool the solution with scratching (see the section titled "Inducing Crystallization," page 680) to see if crystallization occurs. If crystals separate, examine them visually for apparent yield and evidence of purity (absence of extraneous color, good crystal structure). If necessary, repeat the process with other solvents until a suitable one is identified. If no single solvent is satisfactory, choose one solvent in which the compound is quite soluble and another in which it is comparatively insoluble (be sure that the two are miscible!). Dissolve about 0.1 g of the solid in the first solvent with heating and stirring; then add the second hot solvent drop by drop until saturation occurs. Cool the resulting mixture to induce crystallization. If crystals form, examine them as before; if saturation cannot be attained or no crystals form, try another solvent pair.

The quantities can be scaled down if you can spare only a little solid for testing.

A flat-bottomed stirring rod and watch glass can be used to grind the solid.

Use a water or steam bath if the solvent is flammable. Heat gently to reduce solvent evaporation.

Use an ice or ice-salt bath if necessary ⟨OP–8⟩.

It is more efficient to test several solvents at one time and choose the best among them.

See Table 1 of OP–23b for a list of compatible solvents.

Sublimation

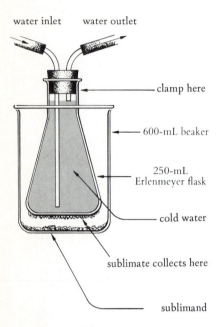

water inlet   water outlet

clamp here

600-mL beaker

250-mL
Erlenmeyer flask

cold water

sublimate collects here

sublimand

Figure 1.   Sublimation apparatus

sublimand: the solid before it has sublimed

sublimate: the solid after it has sublimed and condensed

Good beaker combinations are 250 mL/400 mL and 400 mL/600 mL.

Equipment and supplies:
  sublimation apparatus
  heat source
  cooling fluid
  flat-bladed spatula

Principles and Applications

Sublimation is a phase change in which a solid passes directly into the vapor phase without going through an intermediate liquid phase. Many solids with appreciable vapor pressures below their melting points can be purified by sublimation, either at atmospheric pressure or under vacuum. Such purification is most effective if the impurities have either low or very high vapor pressures at the sublimation temperature, so that they either do not sublime appreciably or fail to condense after vaporization. Sublimation is not as selective as recrystallization or chromatography, but it offers the advantages that no solvent is required and losses in transfer can be kept low. Sublimation can be accelerated if it is performed under reduced pressure (vacuum sublimation, OP–24a) or in an air stream (entrainer sublimation).

Experimental Considerations

Sublimation is usually carried out by heating the **sublimand** with an oil bath or other uniform, easily controllable heat source and collecting the **sublimate** on a cool surface. For best results, the sublimand should be finely divided and free of solvent. A very simple sublimator consists of two nested beakers with a space of about 1–2 cm between them at the bottom. The sublimand is spread out inside the larger beaker, and the condensing (inner) beaker is partially filled with ice water; ice can be added periodically to maintain the temperature. As the outer beaker is heated, crystals of sublimate collect on the bottom of the condensing beaker. Figure 1 illustrates a sublimator operating on the same principle, except that an Erlenmeyer flask is used as a condenser and the temperature is controlled by flowing water.

Operational Procedure

Construct one of the sublimation setups described above (or one suggested by your instructor). Powder the dry sublimand finely (if necessary), and spread it in a thin, even layer over the bottom of the beaker or other container. Assemble the apparatus and turn on the cooling water (or partly fill a condensing beaker with ice water). Heat the sublimand with an appropriate heat source until sublimate begins to collect on the condenser; then adjust the temperature to attain a suitable rate of sublimation without melting or charring the sublimand.

If the condenser becomes overloaded with sublimate, cool and disassemble the apparatus to remove the sublimate, then reassemble it and resume heating. When all of the compound has sublimed or only a nonvolatile residue remains, remove the apparatus from the heat source and let it cool. *Carefully* remove the condenser (avoid dislodging any sublimate), and scrape the crystals into a suitable container using a flat-bladed spatula.

Melting the sublimand will decrease the sublimation rate by reducing surface area; also, molten sublimand may splatter onto the condenser.

The last traces of sublimate can be removed by rinsing the collector with a volatile solvent and evaporating the solvent ⟨OP–14⟩.

Summary

1. Construct sublimation apparatus; spread solid out in sublimand container.

2. Add cooling fluid to condenser *or* turn on cooling water.

3. Heat until sublimation begins, and control heating to maintain sublimation rate.

4. When sublimation is complete (or collector reaches its capacity), discontinue heating, cool and disassemble apparatus, and remove sublimate.

Vacuum Sublimation

OPERATION 24a

Many solids that do not sublime rapidly at atmospheric pressure will do so under reduced pressure. The sublimation assemblies pictured in Figure 1 (on page 690) can be evacuated for that purpose. Apparatus A requires a sleeve, which can be a rubber stopper bored out with a large cork borer or a length of rubber tubing of a suitable diameter. For larger quantities of solid, a filter flask can be used in place of the sidearm test tube. Sublimators similar to apparatus B are available commercially. A vacuum sublimation is carried out as in OP–24, except that the vacuum is turned on before heating is begun and turned off after sublimation is complete *and* the heat source has been removed. If aspirator vacuum is used, a trap must be placed between the sublimation apparatus and the aspirator. The trap should have a pressure-release valve (see Figure 2 of OP–26) so that the vacuum can be broken slowly; otherwise, air entering the sublimator may disturb the crystals of sublimate.

J. Chem. Educ. 52, 720 (1975) describes the construction of a vacuum sublimator for student use.

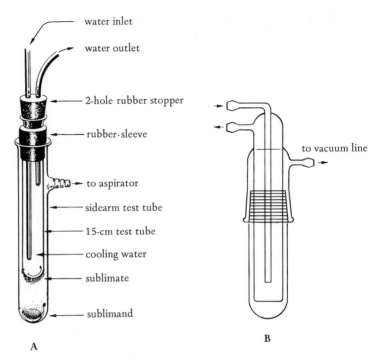

Figure 1.   Apparatus for vacuum sublimation

OPERATION 25

Simple Distillation

Principles and Applications

Boiling occurs when the vapor
pressure over a liquid equals the external pressure (usually about 1 atmosphere in simple distillations).

When a mixture of several liquids (or of a liquid with solid
impurities) is heated in a boiling flask to its boiling point, the
resulting vapor ordinarily has a composition different from
that of the liquid itself. If this vapor is condensed into a separate
vessel, the resulting liquid will contain a high proportion of
the most volatile component present in the boiling flask. This
process of vaporizing a liquid in one vessel and condensing the
vapor into another is called *distillation;* it is an important
method for purifying organic liquids.

  Simple distillation involves only a single vaporization-condensation cycle. It is useful for purifying a liquid that contains
either involatile impurities or small amounts of higher- or
lower-boiling liquids. *Fractional distillation* ⟨OP – 27⟩ is a more

efficient way of separating miscible liquids with comparable volatilities and boiling points, since it allows several cycles of vaporization and condensation to occur in a single operation. *Vacuum distillation* ⟨OP-26⟩ is used to purify high-boiling liquids and liquids that decompose when distilled at atmospheric pressure.

If a pure liquid (or a liquid containing involatile impurities) is heated at standard pressure (1 atm) in a distillation apparatus such as that pictured in Figure 2 (page 694), it will begin to boil when its vapor pressure equals 1 atmosphere, at the temperature called its *normal boiling point*. Its vapors will then remain at that temperature throughout its distillation. When a mixture of liquids, each exerting an appreciable vapor pressure at the boiling temperature, is distilled, the process is more complicated. To simplify the explanation somewhat we shall consider a mixture of ideal liquids that have ideal vapors, which obey both Raoult's Law (Equation 1) and Dalton's Law (Equation 2).

$$P_A = X_A P_A^\circ \qquad (1)$$

$$P_A = Y_A P \qquad (2)$$

P_A = partial pressure of component A in a mixture
P_A° = vapor pressure of pure A at the same temperature
P = total pressure over the mixture
X_A = mole fraction of A in the liquid
Y_A = mole fraction of A in the vapor
(The same relationships hold for any other component in the mixture, such as B.)

Unfortunately, there are no *real* liquids that obey these laws perfectly, so we shall invent some imaginary hydrocarbons. *Entane* ($C_{5.5}H_{13}$) is a pseudoalkane having a normal boiling point of 50°C, whereas *orctane* ($C_{7.5}H_{17}$) boils at 100°C at 1 atmosphere. The equilibrium vapor pressures of entane and orctane at several temperatures between their boiling points are listed in Table 1. As expected, the vapor pressure of entane

Entane and orctane presumably exist only in J. R. R. Tolkien's Middle-Earth, where they are used for fuel by Ents and Orcs, respectively.

Table 1. Equilibrium vapor pressures and mole fractions of entane and orctane at different temperatures

T, °C	Entane			Orctane		
	P°, torr	X	Y	P°, torr	X	Y
50°	760	1.00	1.00	160	0.00	0.00
60°	1030	0.67	0.90	227	0.33	0.10
70°	1370	0.42	0.76	315	0.58	0.24
80°	1790	0.24	0.57	430	0.76	0.43
90°	2300	0.11	0.32	576	0.89	0.68
100°	2930	0.00	0.00	760	1.00	1.00

Note: P° = equilibrium vapor pressure of the pure liquid; *X =* mole fraction in liquid mixture; *Y =* mole fraction in vapor

at its normal boiling point of 50°C is 760 torr (1 atm); the less volatile orctane has a much lower vapor pressure at that temperature. As the temperature increases, the vapor pressure of orctane rises, attaining a value of 760 torr at its normal boiling point of 100°C.

If a mixture of entane and orctane is heated at normal atmospheric pressure, it will begin to boil at a temperature determined by the composition of the liquid mixture, producing vapor of a given composition. For example, a mixture containing an equal number of moles of entane and orctane will start to boil at $66\frac{1}{2}$°C and the vapor will contain more than 4 moles of entane for every mole of orctane (81.4 mole percent entane). The liquid and vapor composition at any temperature can be calculated using Equations 3 and 4, which are derived from Dalton's Law and Raoult's Law; compositions at several temperatures are given in Table 1.

$$X_A = \frac{P - P_B^{\circ}}{P_A^{\circ} - P_B^{\circ}} \tag{3}$$

$$Y_A = \frac{P_A^{\circ}}{P} \cdot X_A \tag{4}$$

It is apparent that the *vapor* over such a mixture of liquids is richer in the more volatile component (entane) than is the liquid mixture itself; this is the key to the purification effected by distillation. For a better understanding of the process, refer to Figure 1, in which the composition of liquid and vapor are plotted against the boiling temperature of the mixture. Suppose we wish to purify a mixture containing 2 moles of entane for every mole of orctane (67 mole percent entane). From the graph (and Table 1) it can be seen that such a mixture will boil at 60°C (point A) and that its vapor will contain 90 mole percent entane (point A'). Thus the liquid that is condensed from this vapor (point A''), which is called the *distillate,* will be much richer in entane than was the original mixture in the boiling flask. As the distillation continues, however, the more volatile component will boil away faster, and the boiling flask will contain progressively less entane. The vapor will also contain less entane, and the boiling temperature will rise. When the mole fraction of entane has fallen to 0.42 (point B), the boiling temperature will have risen to 70°C and the distillate will contain only 76 mole percent entane (point B'). Only when nearly all of the entane has been distilled will the distillate be richer in the less volatile component; at 90°C, for example, over two-thirds of the distillate will be orctane.

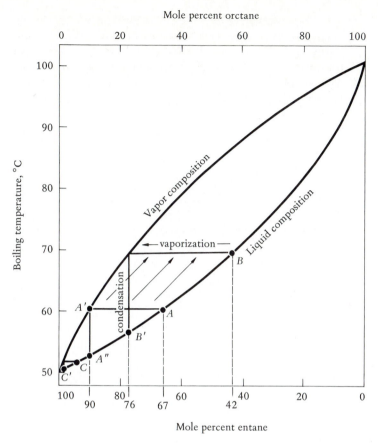

Figure 1. Temperature-composition diagram for entane-orctane mixtures

It can be seen that the purification effected by the simple distillation of such a mixture of volatile liquids is very imperfect. In the example given, the distillate never contains more than 90 mole percent entane, and it may be considerably less pure than that, depending on the temperature range over which it is collected. However, if we had started with a mixture containing only 5 mole percent orctane (point C), considerably greater purity could have been achieved. The initial distillate would be 99 mole percent entane (point C') at 51°C, and if the distillation were continued until the temperature rose to 55°C, the final distillate would be 95 mole percent entane. The average composition of the distillate would lie somewhere between those values; so, if the distillation were stopped at 55°C, most of the orctane would remain in the boiling flask along with some unrecovered entane, and the distillate would be relatively pure entane.

Thus, simple distillation can be used to purify a liquid containing *small* amounts of volatile impurities if (1) the impurities have boiling points appreciably higher or lower than that of the liquid, and (2) the distillate is collected over a narrow range (usually 4 – 6°C) starting at a temperature that is within a few degrees of the liquid's normal boiling point.

Experimental Considerations

Apparatus.　A typical setup for simple distillation is pictured in Figure 2. The size of the components should be consistent with the volume of the **distilland** — otherwise, excessive losses will occur. For example, suppose you recovered 20 mL of crude isoamyl acetate from Experiment 5 and decided to distill it using a 250-mL boiling flask. At the end of the distillation, the flask would be filled with 250 mL of undistilled vapor at the distillation temperature of 142°C; an ideal gas-law calculation shows that this is about 7.3 mmol of vapor. The vapor would condense to about 1.1 mL of the liquid ester, which, along with some additional liquid adhering to the inner walls of the flask, would not be recovered. By comparison, a 50-mL boiling flask will retain only one-fifth as much vapor and adsorb less than one-tenth as much liquid on its inner surface. Therefore, using the smaller flask will cut losses considerably.

distilland: the liquid in the boiling flask, which is to be purified by distillation

Figure 2.　Apparatus for simple distillation

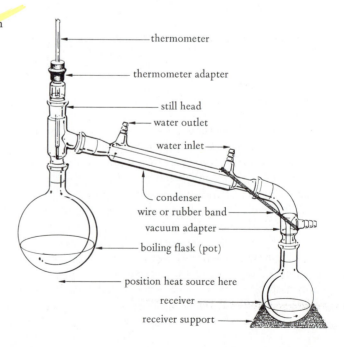

- thermometer
- thermometer adapter
- still head
- water outlet
- water inlet
- condenser
- wire or rubber band
- vacuum adapter
- boiling flask (pot)
- position heat source here
- receiver
- receiver support

The boiling flask should be about one-third to one-half full of the distilland and should also contain several boiling chips or other smooth boiling devices to prevent bumping (see OP–7a). The thermometer is inserted into the *still head* (a three-way connecting tube) through a stopper or thermometer adapter. It should be well centered in the still head (not closer to one wall than another) with the entire bulb below a line extending from the bottom of the sidearm. In other words, the *top* of the bulb should be aligned with the *bottom* of the sidearm, as illustrated in Figure 3. In this way, the entire bulb will become moistened by condensing vapors of the distillate. The placement of the thermometer is extremely important, since otherwise the observed distillation temperature will be inaccurate (usually too low) and the distillate will be collected over the wrong temperature range.

The condenser should have a straight inner section of comparatively small diameter. The Liebig-West condenser (also used for refluxing) found in a typical organic lab kit may be used. (Do not mistakenly use a jacketed distillation column as a condenser; its larger inner diameter results in less efficient condensation.) The vacuum adapter should be secured to the condenser by a rubber band or a spring clamp. (See OP–22a if moisture must be excluded from the system during distillation.) If the distillate is quite volatile, or if it is flammable and a burner is used for heating, the receiver should be a ground-joint flask that fits snugly on the vacuum adapter. If not, another container such as an Erlenmeyer flask or a tared vial can be used. If a ground-joint receiver is used, the vacuum-adapter sidearm must be open to the atmosphere — otherwise, heating the system will build up pressure that could result in an explosion and severe injury from flying glass.

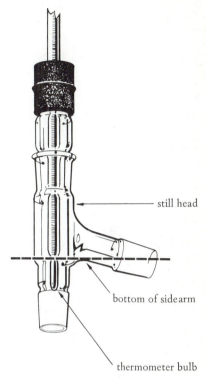

still head

bottom of sidearm

thermometer bulb

Figure 3.   Thermometer placement

Alternatively, a special collecting adapter can be used. See *J. Chem. Educ.* *53,* 39 (1976).

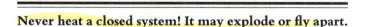

Never heat a closed system! It may explode or fly apart.

SAFETY PRECAUTION

Heat Sources.   Almost any of the heat sources described in OP–7 can be used for simple distillation. Heating mantles and oil baths are preferred because they are less likely to start a fire and because they can provide constant, even heating over a wide temperature range. Steam baths can be used to distill some low-boiling liquids; a burner used with a wire gauze is adequate for higher boiling, less flammable liquids. It is important to keep the heating rate constant — temperature drops can cause the vapor level to fall so that the observed boiling temperature will fluctuate and distillation may cease. An excessive heating rate can reduce the efficiency of the distillation and cause decomposition or mechanical carryover of liquid to the receiver. If a burner is used, it should be positioned close enough to the wire gauze so that drafts will not cause appreciable temperature fluctuations. In most simple distillations, a distilling rate of 1–3 drops per second (about 3–10 mL per minute) will provide adequate distilling efficiency; higher rates may reduce the purity of the distillate.

Boiling Range.   An appropriate boiling range for a distillation may be specified in the experimental procedure, as in Experiment 5, where it is designated as 136–143°C for isoamyl acetate. Because of the presence in that case of the more volatile alcohol as an impurity, much of the crude ester distills below its normal boiling point of 142°C. If the boiling range is *not* given, you should collect the distillate over a narrow range (usually 4–6°C) that brackets its boiling point. The initial temperature of the range can be lowered if the product is likely to be contaminated by a lower-boiling impurity, or the final temperature can be raised if the product is contaminated by a higher-boiling one.

Operational Procedure

Assemble the apparatus pictured in Figure 2, using a boiling flask of appropriate size and taking great care to position the thermometer correctly *(important! see Figure 3)*. Add the distilland through a stemmed funnel, drop in a few boiling chips, replace the thermometer, and turn on the condenser water to provide a slow but steady stream of cooling water. Apply heat at such a rate that the liquid begins to boil gently and the reflux ring of condensing vapors rises slowly into the still head. Shortly after the reflux ring reaches the thermometer, the temperature reading should rise sharply and vapors will begin passing through the sidearm into the condenser, coalescing into droplets that run into the receiving flask. As the first few droplets come over, the thermometer reading should rise to an equilibrium value. (At this time, the entire thermometer bulb should be bathed in condensing liquid that drips off the end of the bulb into the pot.) Record the temperature where the reading stabilizes, and observe it frequently throughout the distillation. Distill the liquid at a rate of about 1–3 drops per second.

If the initial thermometer reading is below the expected boiling range, carry out the distillation until the *lower* end of

Some burners are provided with a "chimney" to keep out drafts. It is not a good idea to hold the burner in your hand and try to control the heating rate by moving it around.

In order to maintain a good distilling rate as the more volatile components distill over, it may be necessary to increase the heating rate.

Review OP–2 for directions on the proper assembly of ground-glass apparatus.

Condenser water should flow in the lower end of the condenser and out the upper end.

The temperature will usually stabilize after 5–10 drops of distillate have been collected. Check the thermometer placement if the initial boiling temperature is lower than expected.

See OP–14 if a low-boiling solvent must be removed before the main component is distilled.

the range is attained; then quickly replace the receiver by another one. If necessary, heating can be discontinued during the switch; but with a normal distillation rate it should be possible to make the change "between drops." The **forerun** in the first receiver contains volatile impurities and is ordinarily discarded. If the initial thermometer reading is within the expected boiling range, there is no need to change receivers.

forerun: a volatile liquid fraction that distills before the main fraction

Collect the distillate until the *upper* end of the expected boiling range is reached or until only a small volume of liquid remains in the boiling flask. Remove the heat source, and (if necessary) transfer the distillate to a tared vial or another suitable container. If another fraction is to be collected, change receivers at the specified temperature without interrupting the distillation, and collect the second fraction over its expected boiling range. Disassemble and clean the apparatus as soon as possible after the distillation is completed.

It is important to stop heating when the bottom of the pot is still moist. Heating a dry flask might decompose the residue and cause tar formation or even an explosion; it might also break the flask.

Summary

1. Assemble distillation apparatus.

2. Add liquid to pot through funnel, and add boiling chips.

3. Turn on and adjust condenser water flow.

4. Commence heating; adjust heat so that vapors rise slowly into still head.

5. Record temperature after distillation begins and thermometer reading stabilizes.
IF initial temperature is below expected range, GO TO 6.
IF initial temperature is within expected range, GO TO 8.

6. Check thermometer placement, adjust if necessary.

7. Collect distillate until temperature rises to lower end of expected range; change receivers.

8. Distill liquid until temperature rises to upper end of expected range *or* until nearly all the liquid has distilled.
IF another fraction is to be collected, GO TO 9.
IF not, GO TO 10.

9. Change receivers; increase heating rate if necessary. GO TO 5.

10. Remove heat source, retain distillate(s) (except forerun), and disassemble and clean apparatus.

Equipment and supplies:
   boiling flask
   three-way connecting tube (still head)
   thermometer adapter
   thermometer
   condenser
   vacuum adapter
   receiving vessel
   boiling chips
   condenser tubing
   heat source
   utility clamp
   condenser clamp
   two ringstands
   rubber band or spring clamp(s)
   supports for receiver and heat source

OPERATION 25*a*

Semimicro Distillation

The distillation of small quantities of liquids is best carried out in a special semimicro apparatus using a microcondenser or a cold-finger condenser. However, the apparatus pictured in Figure 1 can be constructed from equipment found in an organic lab kit, and it works reasonably well. The water-cooled condenser of an ordinary simple distillation apparatus is replaced by a cooling bath surrounding the receiver. If a bulky heat source is used, it may be necessary to slant the vacuum adapter and receiver to allow room for the heat source. If the distilland boils at over 150°C and the receiver is well insulated from the heat source, it may be possible to omit the cooling bath. Cold tap water is a suitable coolant for liquids boiling near 100°C or above; ice water should be used for liquids distilling between 50° and 100°C. Below 50°C an ice-salt bath (see OP–8) should be used. When an ice bath is used, the vacuum-adapter sidearm should be connected to a drying tube ⟨OP–22*a*⟩ so that moisture will not condense inside the receiver. If a burner must be used for distillation and the distillate is flammable, a length of rubber tubing should be run from the vacuum-adapter sidearm to a sink to conduct flammable vapors away from the burner.

The vacuum adapter should be secured by a wire or spring clamp—a rubber band exposed to the heat of the vapors may break.

The receiver can be attached to the vacuum adapter by a length of wire or clamped in place.

Use of a burner should be avoided unless no other suitable heat source is available.

Both flasks must be clamped or otherwise supported!

Figure 1.   Apparatus for semimicro distillation

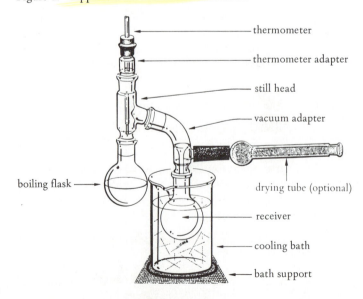

- thermometer
- thermometer adapter
- still head
- vacuum adapter
- boiling flask
- drying tube (optional)
- receiver
- cooling bath
- bath support

Distillation is carried out as described in the Operational Procedure section of OP–25, except that the distillation rate can be reduced to 1 drop per second or slower since less liquid is being distilled. With a cooling bath, it is necessary to remove the heat source to stop the distillation while changing receivers.

Distillation of Solids

OPERATION 25b

Low-melting solids can be distilled using the apparatus pictured in Figure 1 or with a special apparatus designed for that purpose. No condenser is necessary, although the condensation of low-*boiling* solids must sometimes be induced by passing an air or water stream over the receiver. It may be necessary to heat the vacuum adapter with a burner or heat lamp to keep solids from plugging up the vacuum-adapter outlet. It is *very* important to keep this outlet free from solid—if it becomes plugged, the resulting pressure buildup could result in an explosion. When distillation is complete, the distillate is usually melted (if necessary) by heating the receiver in a water or steam bath so that the liquefied solid can be transferred to a suitable container. The last traces of solid can be transferred with a small amount of ethyl ether or another volatile solvent, which is then evaporated under a hood.

The apparatus pictured in Figure 1 can be used to distill viscous or high-boiling liquids as well as solids; it can also be adapted for the vacuum distillation of low-melting solids, if a suitable smooth-boiling device is used. Because the narrow vacuum-adapter drip tube is particularly easy to plug up, this apparatus should *not* be used to distill high-melting solids.

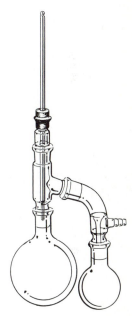

Figure 1. Apparatus for distillation of solids

Water Separation

OPERATION 25c

Water may have to be removed during the course of a reaction to increase the yield of a reversible reaction or to prevent the decomposition of water-sensitive compounds. This can be done by simple distillation, but it is often more convenient to use a device called a *water separator*. Commercial water separators such as the Dean-Stark trap (Figure 1 on page 700) can be used, but they are generally not available in undergraduate laboratories. A "homemade" Dean-Stark trap can

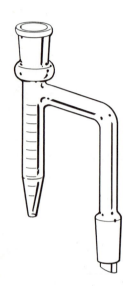

Figure 1.   Dean-Stark trap

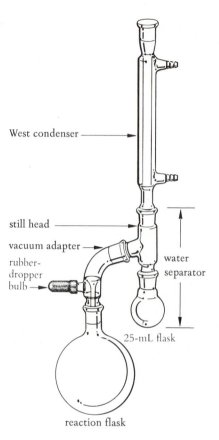

West condenser

still head

vacuum adapter

rubber-
dropper
bulb

water
separator

25-mL flask

reaction flask

Figure 2.   Apparatus for water separation

easily be constructed from a large test tube and a length of glass tubing, as described in *J. Chem. Educ. 40,* 349 (1963). If neither of these alternatives is available, the parts from an organic lab kit can be assembled as shown in Figure 2 so that the 25-mL flask with adapter functions as a water separator.

The reaction is generally run in a solvent (such as toluene) that is less dense than water, and the trap is filled with that solvent beforehand. As water forms during the reaction, its vapors and those of the solvent pass into the reflux condenser, where they condense and drip down into the trap. The organic condensate overflows through the sidearm and returns to the reaction flask, while the water collects at the bottom of the trap. The volume of water expected from the reaction can be estimated beforehand, so that by measuring the amount of water in the trap at any given time, one can tell whether the reaction has reached completion.

When using the apparatus in Figure 2, both flasks should be clamped securely and the water separator should be filled through a funnel to a point just below the sidearm (be sure to close off the vacuum adapter outlet!). The reaction mixture must be refluxed gently enough that the adapter drip tip does not become flooded with liquid. When disassembling the apparatus after the reaction, leave the water separator clamped while the other components are removed; then cautiously tilt the water separator to pour liquid from the sidearm into a beaker.

Vacuum Distillation

OPERATION 26

Principles and Applications

Distillation under reduced pressure, or *vacuum distillation* as it is frequently called, is based on the fact that lowering the external pressure about a liquid also lowers its boiling point. For example, a liquid that boils at 200°C at a pressure of 1 atmosphere (760 torr) will boil near 100°C when the pressure is lowered to 25 torr and around 40°C at 1 torr. This is important when high-boiling liquids are to be distilled, since many of them decompose or undergo other chemical transformations at high temperatures (especially in the presence of impurities that may be present in a reaction mixture). As a general rule, most liquids that boil at over 200°C at atmospheric pressure should be purified by distillation under reduced pressure to prevent losses due to such transformations.

Vacuum distillation has certain inherent features that can cause experimental difficulties or hazards. The volume of the vapor generated when a given amount of liquid vaporizes is much greater at low pressures; this can cause excessive bumping in the boiling flask and mechanical carryover of liquid to the receiver. The vapor velocity is also greater because there are fewer molecules around to bump into; this can cause superheating of the vapor at the still head and a pressure differential throughout the system, so that the observed distillation temperature may be too high and the pressure reading too low. These difficulties can be overcome (or at least lived with) by using the right kind of apparatus and by carrying out the distillation slowly. Even then, the separation attainable under vac-

Table 1. Approximate boiling points of liquids at 25 torr

normal b.p. at 760 torr	b.p. at 25 torr
150°C	60°C
200°C	100°C
250°C	140°C
300°C	180°C

These chemical transformations may include condensation, dehydration, isomerization, oxidation, pyrolysis, polymerization, or rearrangement reactions.

The vapor bubble that occupies a volume of 0.25 mL at 760 torr will expand to nearly 10 mL at 20 torr.

A Claisen head helps prevent mechanical carryover; many devices can be used to reduce bumping.

Table 2. Vapor pressure of water below 30°C

T, °C	P, torr	T, °C	P, torr
30	31.8	18	15.5
28	28.3	16	13.6
26	25.2	14	12.0
24	22.4	12	10.5
22	19.8	10	9.2
20	17.5	8	8.0

If for any reason a pressure above the value attainable by an aspirator is desired, the screw clamp can be replaced by a bleed valve consisting of the bottom part of a Bunsen burner. The pressure is then varied by adjusting the needle valve of the burner.

A cold trap consists of a chamber cooled by dry ice in isopropyl alcohol (or some other solvent) that condenses the vapors and keeps them out of the pump.

When reporting a boiling range from a vacuum distillation, report the pressure also.

Vapors from the distillation may also migrate to the closed end of the manometer, invalidating the pressure readings. Therefore, it is advisable to open the manometer to the system only when a pressure reading is being made.

uum distillation does not equal that possible at atmospheric pressure, and there is always the danger that the apparatus may shatter (implode) because of the unbalanced external pressure on the system. Accordingly, a vacuum distillation must be performed with great care and attention to detail to obtain satisfactory results and prevent accidents.

Experimental Considerations

Vacuum Pumps. The most convenient vacuum source for most undergraduate laboratories is the water aspirator, which is capable (under optimum conditions) of pulling a vacuum of 10–25 torr, depending on the water temperature. In theory, an aspirator should be able to attain a vacuum equal to the vapor pressure of the water flowing through it, which is a function of the water temperature (Table 2). In practice, the pressure is often 5–10 torr higher because of insufficient water pressure, leaks in the system, or deficiencies in the aspirator itself.

An aspirator pump must be provided with a water trap to prevent backup of water into the receiving flask (due to changes of water pressure) and to reduce pressure fluctuations throughout the system. A thick-walled filter flask of the largest convenient size makes a suitable trap, if it is provided with a pressure release valve and hooked up to the aspirator and distillation apparatus (as illustrated in Figure 2 on page 706).

For lower pressures, an oil-diffusion pump can be used. Most oil pumps will routinely pull a 1-torr vacuum if they are in good condition, and 2–10 torr otherwise. Since the oil is subject to dilution or decomposition from vapors passing out of the distillation apparatus, an oil pump must always be used with one or more cold traps; it should also be provided with a manostat for adjusting the pressure.

Pressure Measurement. In order to know when the desired product is distilling and how pure it is, you must know the pressure inside the system and the approximate boiling point at that pressure. The pressure can be measured by a *manometer* (such as that illustrated in Figure 2). The pressure (in torr) is equal to the vertical distance between the two columns of mercury, in millimeters, as measured by means of a movable scale or a sheet of metric graph paper. The mercury in the manometer can present a hazard when the vacuum is broken suddenly — air rushing in can push the mercury column forcefully to the closed end of the tube, breaking it and spraying poisonous mercury all over the laboratory. For this and other reasons, the vacuum must always be released *slowly*.

If a manometer is not available, it may still be possible to carry out a vacuum distillation successfully when the distilland does not contain impurities with boiling points near that of the desired product. From the temperature of the aspirator water and Table 2, you can determine the lowest attainable pressure, from which the *minimum* boiling temperature of the product can be estimated, as described below. The product should distill somewhat above that temperature; its actual boiling range will depend on the efficiency of the aspirator and the airtightness of your apparatus.

Boiling Points at Reduced Pressure. If the boiling point of a substance under reduced pressure is not known, it can be estimated from Equation 1 (with fair accuracy) for liquids that are not associated by hydrogen bonding. Associated liquids, such as alcohols and carboxylic acids, often have boiling-point reductions about 10% less than the calculated decrease.

$$T^P \approx \frac{5.46 \cdot T^{760}}{8.34 - \log p} \qquad \textbf{(1)} \qquad (T \text{ is in kelvins, } p \text{ in torr})$$

For example, the boiling point of *p*-nitrotoluene is 211°C at 760 torr. Using Equation 1, its boiling point at 10 torr is calculated to be 87°C, compared to the literature value of 85°C.

$$T^{10} = \frac{5.46 \times (211 + 273)\ \text{K}}{8.34 - \log 10} = 360\ \text{K}\ (87°\text{C})$$

However, the 10-torr boiling point of 1-octanol (normal b.p. = 195°C) is calculated to be 75°C, compared to an experimental value of 88°C; thus, the actual reduction is 13°C less than the calculated reduction. Reduced-pressure boiling points can also be estimated by using the vapor pressure–temperature nomograph in Figure 1 (on page 704). More precise estimates can be made by referring to tables such as those in R. R. Dreisbach, *Pressure-Volume-Temperature Relationships of Organic Compounds* (New York: McGraw-Hill, 1952), by applying Equation 2 using A and B values given in older editions of the *CRC Handbook of Chemistry and Physics* ⟨Bib–A1⟩ (for example, 50th ed., 1969, p. 167), or by using various empirical relationships.

The reduced-pressure boiling range can be estimated as described on page 696.

$$\log p = -\frac{0.05223\text{A}}{T} + \text{B} \qquad \textbf{(2)} \qquad (T \text{ is in kelvins, } p \text{ in torr})$$

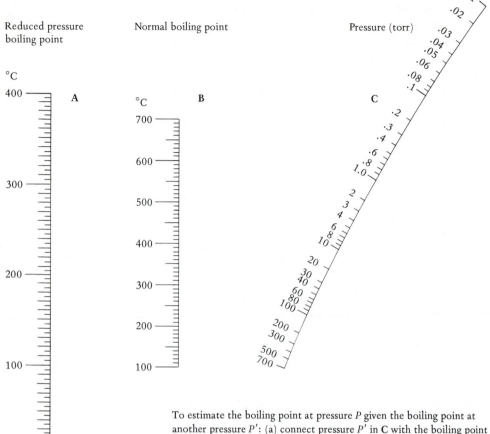

Reduced pressure
boiling point

Normal boiling point

Pressure (torr)

To estimate the boiling point at pressure P given the boiling point at
another pressure P': (a) connect pressure P' in **C** with the boiling point
at P' in **A** using a transparent ruler and insert a needle or sharp pin at
the point where the ruler intersects line **B**; (b) pivot the needle around
the ruler until it reaches the desired pressure (P) in **C**, then read the boiling point at P from **A**.

Example: the boiling point of dibutyl phthalate is reported to be 236°
at 40 torr. Connecting these values in **A** and **C** causes the ruler to inter-
sect line **B** at about 345° (which roughly equals the normal boiling point),
and pivoting about that point to 10 torr on **C** yields a value (from **A**) of
about 197° for the boiling point of dibutyl phthalate at 10 torr.

Figure 1.   Reduced-pressure boiling point nomograph

Heat Sources.   The heat source should be capable of providing
constant, uniform heating to prevent bumping and superheating and to maintain a constant distillation rate. An oil bath
⟨OP–7⟩ is best, especially if it can be electrically heated and
stirred. Heating mantles and heat lamps may also be satisfactory.

Assembling the Apparatus. The glassware used in assembling the apparatus *must* be inspected very closely to make certain that it is free of cracks and star fractures. Joints should be well lubricated with silicone vacuum grease to prevent leaks, and all rubber tubing should be stretched or bent to make sure it is pliable and free of cracks. All connecting tubing in the vacuum line should be heavy-walled to prevent collapse *and* as short as possible to reduce the pressure differential between vacuum pump and distilling flask. Connections between rubber and glass, as well as ground-joint connections, must be secure and airtight.

Never use thin-walled glassware, such as Erlenmeyer flasks, in an evacuated system.

Receiving flasks can be tared to facilitate weighing the distillate.

Smooth-Boiling Devices. Numerous techniques are used to prevent bumping in vacuum distillations. One simple method is to use a magnetic stirbar in the distilling flask (possibly coupled with a larger one in the heating bath). Ordinary boiling chips are of little use since they are soon exhausted under vacuum, but special microporous boiling chips are suitable for most vacuum distillations.

The classical method of vacuum distillation uses a flexible capillary bubbler about as fine as a cat's whisker, drawn from a length of thick-walled capillary tubing. This kind of bubbler is inserted into the boiling flask through a thermometer adapter (or a rubber stopper if necessary) so that its tip extends to within a millimeter or two of the bottom. Under vacuum, the capillary delivers a very fine stream of air bubbles that prevents the development of the large bubbles that cause bumping. Aside from the difficulty of constructing a capillary bubbler, using one has the following disadvantages: the air entering the system raises the pressure slightly and decreases the accuracy of the manometer reading; air may oxidize the product at high temperatures (this can be prevented by using nitrogen or another inert gas); and the capillary may plug up when the vacuum is broken. Nevertheless, a well-constructed capillary bubbler often works better than any of the other devices mentioned.

Foaming can be dealt with by using a larger boiling flask, distilling slowly, or using an antifoaming agent.

The microporous chips that are available from the Todd Scientific Company are made of specially purified anthracite coal.

A screw-clamp valve can be used to regulate the rate of boiling.

Capillaries drawn from ordinary glass tubing tend to be inflexible and can break easily.

Because of the danger of an implosion, safety glasses *MUST* be worn during a vacuum distillation.

SAFETY PRECAUTION

Operational Procedure

Inspect your glassware and rubber tubing for imperfections, then assemble the apparatus illustrated in Figure 2. If a ma-

Equipment and supplies are listed on page 708.

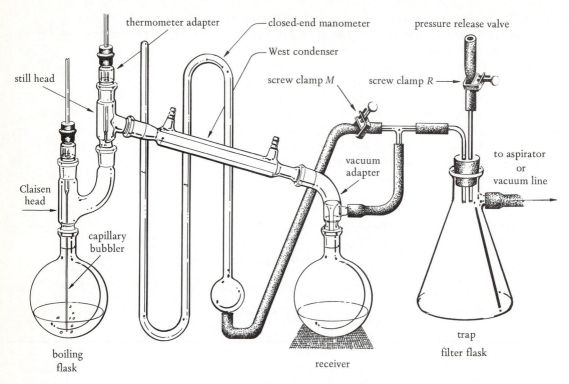

thermometer adapter — closed-end manometer   pressure release valve

West condenser

still head   screw clamp M   screw clamp R →

vacuum adapter   to aspirator or vacuum line

Claisen head

capillary bubbler   trap filter flask

boiling flask   receiver

Figure 2.   Apparatus for distillation at reduced pressure

With an oil pump, add one or more cold traps, a manostat, and a pressure-release valve such as a three-way stopcock. Omit the filter flask trap assembly.

It may be necessary to insulate the still head with glass wool or aluminum foil for high-boiling liquids.

If the liquid may contain a volatile solvent, remove it first by distilling at atmospheric pressure, then let the apparatus cool before proceeding.

If the pressure is within 10 torr of that estimated from Table 2, the apparatus should be satisfactory.

nometer is not available, omit the glass tee and connect the vacuum adapter directly to the trap (also, disregard any instructions below that refer to a manometer). If you are using microporous boiling chips or a stirbar to prevent bumping, replace the capillary bubbler assembly with a stopper.

Make sure that all joints and connections are tight; then add the material to be distilled through a funnel. Drop in a few microporous boiling chips, or use one of the other methods described to prevent bumping. Raise the heat source into position, but do not begin heating. Turn on the cooling water for the condenser, open the screw clamp(s), and turn on the aspirator to its maximum flow rate. *Slowly* close clamp R on the pressure-release valve. (If bumping and foaming occur, there is probably some residual solvent in the distilland. Open the pressure-release valve, and then close it down to a point where the solvent can be evaporated without excessive bumping.) With clamp R completely closed, wait a minute or two until the pressure equilibrates, then read the manometer. If the pressure is higher than expected, check the system for leaks; if necessary, release the vacuum by opening clamp *R* and repair them. If the pressure is satisfactory, use the nomograph in Figure 1 (or one of the other methods described above) to estimate the boiling range at that pressure. Commence heating (and magnetic stirring, if it is to be used) until distillation begins. Adjust the heat so that a distillation rate of about 1 drop per second is attained; then record the temperature and pressure readings. Clamp *M* should then be closed to protect the manometer from vapors and opened slowly when another pressure reading is made. The pressure may fluctuate as a result of changes in the water flow rate; this will make heating adjustments necessary to reestablish the distillation rate.

Any leaks should also be revealed by the appearance of bubbles around the joints during the distillation.

If possible, only a few students should use the aspirators at the same time, to maintain pressure and minimize fluctuations.

A rapid pressure increase, accompanied by a thick fog in the distilling flask, indicates decomposition of the distilland. If this occurs, remove the heat source *immediately* and get away (warning others to do so also) until the flask cools. Then try to determine the cause of the problem.

SAFETY PRECAUTION

If the initial vapor temperature is lower than the estimated boiling point of the product, distill the volatile forerun until the temperature reaches the expected value. Then change receivers by the following procedure:

1. Lower the heat source, and wait about 5 minutes to let the system cool down.

2. Open clamp *R* *slowly* until the system is at atmospheric pressure.

3. Replace the receiver with a clean one (preferably tared so the product can be weighed directly). If you are using microporous boiling chips, add another chip or two to the boiling flask.

Do not turn off the aspirator or oil pump while changing collectors.

4. Open clamp *M*, and then slowly close clamp *R* and read the manometer to be sure the vacuum is adequate.

A capillary bubbler will sometimes become plugged when the vacuum is broken; if that happens, you may have to replace it with a new one.

5. Commence heating until distillation begins; then record the temperature and pressure.

6. Close clamp *M*.

Continue distilling until the upper end of the anticipated temperature range is reached or until a significant drop in

A temperature drop may also occur because the pressure has changed, the boiling chips (or bubbler) are not working, or the heating source is not hot enough. It may be necessary to raise the temperature of the heating bath at the end of the distillation to drive over the last milliliter or so of distilland.

Equipment and supplies:

boiling flask
Claisen connecting tube
still head (three-way connecting tube)
stopper (optional)
thermometer
thermometer adapter
condenser
vacuum adapter
receiving flask(s)
microporous boiling chips (or capillary bubbler, stirbar, etc.)
rubber band or spring clamp(s)
utility clamp
condenser clamp
heat source
supports for receiver and heat source
condenser tubing
heavy-walled rubber tubing
glass tee
screw clamps
manometer
aspirator or oil pump
trap: large filter flask, utility clamp, two-hole rubber stopper, two short glass tubes (8 mm O.D.), screw clamp, and short rubber tube

temperature indicates that the product is completely distilled. If a higher-boiling fraction is to be distilled, change collectors by the same procedure as before. Stop the distillation before the boiling flask is completely dry.

When the distillation is completed, follow steps 1 and 2 above for bringing the system back to atmospheric pressure; then turn off the aspirator or oil pump. Disassemble the apparatus, and clean the glassware promptly.

Summary

1. Assemble apparatus, and check connections and joints.

2. Add distilland and microporous boiling chips (or use another smooth-boiling device).

3. Position heat source, start cooling water, open pressure-release valve and manometer valve, and turn on aspirator or oil pump.

4. Close pressure-release valve slowly; let pressure equilibrate.
IF bumping and foaming occur, adjust pressure-release valve to evaporate solvent.

5. Read manometer, and estimate boiling range.
IF pressure is too high, check system for leaks and repair as necessary.

6. Adjust heat to attain desired distillation rate.

7. Record temperature and pressure; close manometer valve.
IF temperature is below expected boiling range, GO TO 8.
IF temperature is within expected boiling range, GO TO 12.

8. Distill until temperature reaches lower end of expected boiling range.

9. Lower heat source, let cool 5 minutes, open pressure-release valve slowly, and change receivers.

10. Open manometer valve, close pressure-release valve slowly, and read manometer.

11. Heat to resume distillation, record temperature and pressure, and close manometer valve.

12. Distill until upper end of temperature range is attained *or* only a little distilland remains.

13. Lower heat source, let cool 5 minutes, and open pressure release valve slowly.

IF last fraction has been collected, GO TO 14.

IF additional fractions are to be collected, change receivers, and GO TO 4.

14. Turn off vacuum source, and disassemble and clean apparatus.

Semimicro Vacuum Distillation

The apparatus for semimicro distillation pictured in Figure 1 of OP-25a can be used under reduced pressure if the vacuum adapter sidearm is connected to a vacuum line (rather than to a drying tube) through a trap, as illustrated in Figure 2 of OP-26. Except that a cooling bath is used instead of condenser water, the procedure is essentially the same as that used for ordinary vacuum distillation ⟨OP-26⟩. It may be necessary to insert a Claisen head between the boiling flask and the still head to prevent mechanical carryover of the distilland or to allow for insertion of a capillary bubbler.

Fractional Distillation

Principles and Applications

Simple distillation is not a very efficient way of separating liquids whose boiling points are less than 80–100°C apart. With closer-boiling liquids, the distillate will be a mixture of constantly changing composition and boiling point, containing an excess of the more volatile component at the beginning of the distillation and an excess of the less volatile component at the end (see the volume-temperature graph in Figure 1). The separation could be improved, of course, by redistilling the distillate fractions and combining fractions of similar composition, and then repeating the process until a reasonably

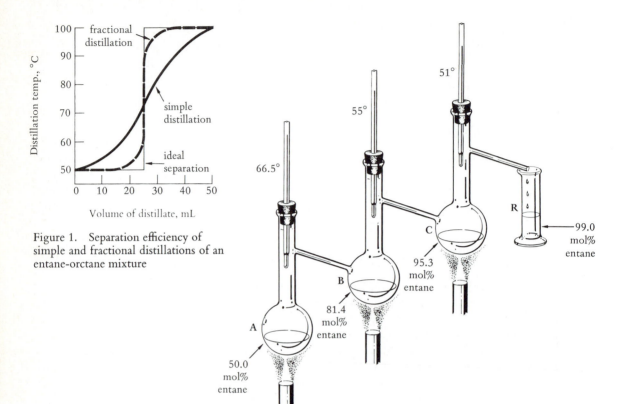

Figure 1.   Separation efficiency of simple and fractional distillations of an entane-orctane mixture

Figure 2.   Hypothetical multi-stage distilling apparatus

The boiling point in flask B is lower than in flask A because the liquid in B is richer in the more volatile component.

an apparatus such as that in Figure 2, in which each distillate is delivered directly to another distilling flask that redistills the condensed vapors and delivers them to another flask, and so on. This process can be understood by referring to the temperature-composition diagram in Figure 3 for the imaginary entane-orctane mixture (see OP–25).

If we boil a 50:50 mole percent mixture of entane and orctane in flask A (point *A* on the diagram in Figure 3), the *vapor* over that mixture will have a composition of 81.4 mol% entane and only 18.6 mol% orctane (point *A'*). This is because the entane has a considerably higher vapor pressure at the boiling point of the mixture (66.5°C) than does orctane. This vapor is then condensed into flask B (line *A'-B*), where it is boiled at 55°C to yield a vapor containing 95.3 mol% entane (line *B-B'*), which is condensed (*B'-C*) into flask C. The condensed liquid in C boils at 51°C to yield a vapor that is 99.0

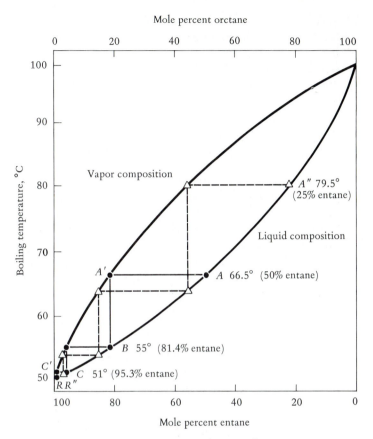

Figure 3. Temperature-composition diagram for entane-orctane mixture

mol% entane (*C-C'*), which is delivered to receiver R as a liquid of the same composition (line *C'-R*).

Unfortunately, only the first few drops of distillate will attain this degree of purity; because as the more volatile entane is removed, the less volatile orctane will accumulate in the distilling flasks, reducing the proportion of entane in the vapor. The efficiency of this apparatus would be improved considerably if some of the orctane-enriched liquid in each flask were drained back into the preceding flask through an overflow tube, so that orctane would not accumulate as rapidly in the upper stages. Even then, the purity of the distillate would decrease with time. For example, when the amount of entane in flask A had decreased to 25 mol% (point *A''*), the vapor condensing into the receiver (point *R''*) would be only 96.6 mol% entane, as shown by the broken lines in Figure 3. In other words, during a distillation the distilland always climbs

Fractional distillation is, in effect, a multistage distillation process performed in a single operation.

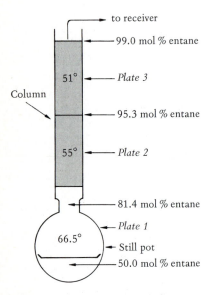

to receiver

99.0 mol % entane

51° — *Plate 3*

Column

95.3 mol % entane

55° — *Plate 2*

81.4 mol % entane

Plate 1

66.5° — Still pot

50.0 mol % entane

Figure 4.   Operation of a fractionating column

In certain multistage columns, the "theoretical" plates are real. A Bruun column consists of a series of horizontal plates stacked at intervals inside a vertical tube, and one vaporization-condensation cycle occurs on each plate.

inexorably up the temperature-composition graph, and the composition of the distillate changes accordingly. Nevertheless, the separation effected by several distillation stages is considerably better than by only one, as can be seen by comparing the curves for simple distillation and fractional distillation in Figure 1.

The apparatus described above would be very cumbersome and difficult to operate efficiently. A much simpler alternative is to interpose a vertical *distilling column* between the boiling flask and still head, and to allow the successive distillation operations to take place on the column. Such a column is usually filled with some kind of *column packing*, which provides a large surface area from which repeated vaporization and condensation operations can take place. Suppose our 50:50 mol% mixture of entane and orctane is distilled through such a column. The vapor entering the column from the still pot should have the same composition (81.4 mol% entane) as before, but as it passes onto the column it will cool, condense on the packing surface, and begin to trickle down the column on its way back to the pot. Since the temperature is higher near the pot, part of the condensate will vaporize on the way down, yielding a vapor still richer in entane. This vapor passes up the column until, at a higher level than before (since its boiling point is lower), it cools enough to recondense. This process of vaporization and condensation may be repeated a number of times on the way to the top of the column, so that when the vapor finally arrives at that point it is nearly pure entane. Although this is a continuous process involving the simultaneous upward flow of vapor and downward flow of liquid, the net result is the same as that produced by a process of successive discrete distillations.

Each section of the distillation apparatus that provides a separation equivalent to one cycle of vaporization and condensation (one "step" on the temperature-composition graph) is called a *theoretical plate*. Since the first cycle occurs in the pot, it provides one theoretical plate; the column illustrated in Figure 4 contributes two more. Note that there is a continuous variation in both temperature and vapor composition as one proceeds up the column, and that the temperature is fixed by the liquid-vapor composition. For instance, the entane-rich liquid near the top of the column boils at a lower temperature than the original mixture, so the average temperature of plate 3 is lower than that of the plates below it.

The efficiency of a particular column can be measured in terms of its theoretical plate rating. If the *volatility* (*v*) of liquid

C is defined as the ratio of its vapor composition to its liquid composition at a given temperature, then the *relative volatility* of two liquids C and D is just the ratio of their respective volatilities at that temperature (Equation 1).

$$v_C = \frac{Y_C}{X_C}$$

$$\alpha = \frac{v_C}{v_D} = \frac{Y_C X_D}{Y_D X_C} \tag{1}$$

v_C = volatility of component C
Y_C = mole fraction of component in vapor
X_C = mole fraction of component in liquid

By convention, the volatility of the more volatile component is put in the numerator; thus, the relative volatility (α) is a measure of the enrichment of the vapor in the more volatile component after each vaporization-condensation cycle. The enrichment after two cycles will be α^2, and after n cycles, α^n, where n is equal to the number of theoretical plates provided by the distillation apparatus. If the mole fractions of C and D in the vapor as it emerges from the top of the column are given by Z_C and Z_D, then the number of plates, n, can be expressed in terms of α, as shown in Equation 2, the Fenske equation.

A more convenient form of the Fenske equation can be derived by taking the logarithm of both sides (see Experiment 7).

$$\alpha^n = \frac{Z_C X_D}{Z_D X_C} \tag{2}$$

Therefore, the theoretical plate rating of a given column can be determined by distilling a liquid mixture for which α is known, measuring the composition of the distillate and of the liquid in the pot, and plugging these values into the Fenske equation.

Another useful measure of column efficiency is the *height equivalent to a theoretical plate (HETP)*, which is equal to the length of the column divided by the number of theoretical plates. For example, a 48-cm Vigreux column with 6 theoretical plates has a HETP value of 8 cm.

The lower the HETP, the more efficient the column.

In practice, a number of factors limit the efficiency of a given column. Its efficiency is highest under the equilibrium condition of *total reflux,* in which all the vapors are returned to the pot. In practice, however, part of the vapor is always being drawn off and condensed into the receiver, which disturbs the equilibrium and reduces the column's efficiency. To maintain a reasonably high efficiency, the ratio of liquid returned to liquid distilled measured over the same time interval (the reflux ratio, R) must be carefully controlled.

$$R = \frac{\text{liquid vol. returning to pot}}{\text{liquid vol. distilled}}$$

Reflux ratios of 5–10 are common for routine separations.

holdup: the amount of liquid that adheres to column's surface and packing
throughput: the distillation rate, usually expressed in mL of distillate per minute

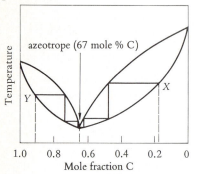

Figure 5. Temperature-composition graph for a mixture that forms a minimum-boiling azeotrope

Figure 6. Vigreux column

According to one rule of thumb, the reflux ratio should at least equal the number of theoretical plates for efficient operation. To a lesser extent, the **holdup** of a column and the **throughput** rate can also limit the efficiency of a distillation.

To this point, the discussion has dealt only with ideal liquids. No real liquid systems display ideal behavior, although some may come very close. The greatest deviations from ideal behavior are displayed by liquid mixtures that form maximum-boiling or minimum-boiling azeotropes. Mixtures that yield minimum-boiling azeotropes have boiling temperature–composition graphs similar to the one in Figure 5. From that graph, it can readily be shown that distilling such a mixture cannot possibly yield *both* components in pure form. For example, distillation of a mixture having the composition X will yield a distillate of the azeotropic composition (indicated by the minimum in the graph). Since the azeotrope contains more of component C than does the mixture in the pot, the pot mixture will become richer in the second component, D, until it is eventually pure D and the distillate is pure azeotrope. Starting at point Y gives the same kind of result, except that the pot liquid then becomes pure C instead of D. It can be shown by similar reasoning that distilling a maximum-boiling azeotrope will yield C or D in the distillate and azeotrope in the pot.

Azeotrope formation can cause difficulties in purifying liquids. For example, it is impossible to obtain pure ethyl alcohol by distilling 50% aqueous alcohol, since no distillate containing more alcohol than the azeotrope (95% ethanol) can be obtained. In this case, the problem is solved by "fighting fire with fire." Benzene and water form an even lower-boiling azeotrope with ethanol than does water alone (b.p. 64.9°C, compared to 78.1°C for 95% ethanol); so by boiling 95% ethanol with some benzene, the benzene-water-ethanol azeotrope can be distilled off until all the water is removed. Other azeotropes can sometimes be purified by chemical means or by such separation methods as chromatography.

Experimental Considerations

Columns. The column is the most important part of any fractional distillation apparatus; many kinds are available. One of the simplest is the Vigreux column, which has a series of indentations or "drip points" in a more-or-less spiral arrangement down the length of the column. The drip points provide some surface for condensation-vaporization cycles, but the total surface area is comparatively small. This kind of column is not very efficient, having HETP values in the range of

8–12 cm. The most commonly used column for laboratory fractional distillation is simply a straight tube (usually jacketed) with indentations at the bottom, filled with a suitable packing material.

Column Packing. Most columns are provided with some kind of packing, which can come in the form of rings, saddles, helices, beads, or metal turnings. Some packing materials such as Heli-Pak can provide HETP ratings of 1 cm or lower, but are quite expensive. Glass beads (or rings) and stainless steel "sponge" are more practical for use in undergraduate laboratories. The metal packing can be corroded by halogen compounds, or it may catalyze decomposition reactions of sensitive liquids. Packed columns are more efficient than Vigreux columns, but they suffer from higher holdup and lower optimum throughput rates. As a result, distillations must be conducted more slowly and material losses due to adsorption on the column packing must be allowed for.

Insulation. If either of the components to be separated boils at a temperature much above 100°C, the column and still head should be insulated to prevent heat losses that may reduce efficiency or even prevent distillation entirely. The column and still head can be covered by one or more layers of crumpled aluminum foil, or glass wool can be wrapped around them and held in place by aluminum foil. The insulation should be secured in such a way that the packing at the bottom of the column can be watched for evidence of flooding and the vapors in the still head can be observed.

Heat Sources. For good results, the heat source must provide constant, uniform heating. An oil bath is preferred because of its ability to provide uniform heating at a preset temperature. A heating mantle can be used successfully if its heat output is carefully controlled to bring about the desired distillation rate. Because of the potential fire hazard and the difficulty of maintaining a constant heating rate, a burner should be used only if no other suitable heat source is available, and then appropriate precautions should be taken to prevent fires.

Flooding. One problem often encountered in a fractional distillation is *flooding,* in which the column becomes partly or entirely filled with liquid. Flooding can be prevented by using the proper insulation and heating rate. If flooding occurs, the heat source must be removed until all of the excess liquid has returned to the pot before the distillation is resumed. Flooding can greatly decrease the efficiency of a separation, since it

Stainless-steel sponge pads, obtained from household goods stores, can be stretched and cut into 6–8 inch lengths for use in a typical column. The packing should be pulled into the column using a rod with a hook at one end.

Losses of 3–5 mL can be expected from fractional distillation with the apparatus illustrated in Figure 7 (on page 717). Therefore, it is not practical to distill less than about 25 mL from such a setup.

Flooding may also be caused by unsuitable packing or packing supports. For example, a glass-wool plug used as a packing support can hold up enough liquid to cause flooding.

reduces the surface area of packing available for the separation.

Operational Procedure

Fill the column with packing to within a centimeter or less of the upper ground joint. To introduce loose packing such as glass beads, hold the column nearly horizontal with your hand over the bottom opening and place a few large beads in the top of the column. Then quickly turn the column upright so that (with a little luck) the beads will jam together at the constriction and support the remainder of the packing. (A small plug of stainless steel sponge can also serve as a support.) Then slowly pour in the rest of the packing from a beaker, with continuous shaking, so that it is as uniform as possible.

Assemble the apparatus illustrated in Figure 7, using a vial, a graduated cylinder, a flask, or another appropriate container as the receiver. Make sure that all joints are tight, the column is perpendicular to the bench top, the boiling flask and receiver are supported properly, and none of the joints are under excessive strain. If the distilland will boil at much above 100°C, insulate the apparatus from the bottom of the column to the top of the distilling head. Fill the distilling flask one-third to one-half full, and add a few boiling chips.

Raise the heat source into place, and begin heating to bring the mixture to the boiling point. When it boils, adjust the heating rate so that the reflux ring of condensing vapor passes up the column at a *slow,* even rate (it should take 5–10 minutes to reach the top of the column). Watch the packing at the bottom of the column closely for evidence of flooding. If flooding occurs, remove the heat source immediately and let the liquid drain into the boiling flask; then resume heating at a lower temperature. When the vapors rise above the column packing, lower the heat enough to keep the reflux ring between the packing and the sidearm for a few minutes, so that the column has time to equilibrate and all of the packing becomes moistened with liquid. Adjust the heat to initiate distillation, and read the thermometer when the temperature stabilizes. Distill at a rate of about 1 drop every 1–3 seconds, or at a rate that gives the desired reflux ratio. (Estimate the ratio by counting the drops returning to the flask and those distilling over a short period of time.) If the initial distillate is cloudy (due to dissolved water), change receivers when it becomes clear. Continue the distillation until the temperature at the still head drops (indicating that all of the more volatile component has distilled) or until a predetermined temperature is reached. If another fraction is to be collected, change receivers and

Equipment and supplies are listed on page 718.

Glass helices should be dropped into the column one at a time.

The receivers should be numbered and (if the fractions are to be weighed) tared. The vacuum-adapter drip tube should extend into the receiver to prevent losses by evaporation. For very volatile distillates, a ground-joint flask can be used as a receiver.

With stainless-steel packing, the column is sometimes deliberately flooded by strong heating to wet the packing. Then the heat is reduced to drain the column before distillation is begun.

For most packed columns, 1–3 mL/min is a good rate. A Vigreux column may operate efficiently at 5–10 mL/min.

The still-head temperature may also drop when the pot temperature is not high enough to bring the vapors of the less volatile component up to the still head.

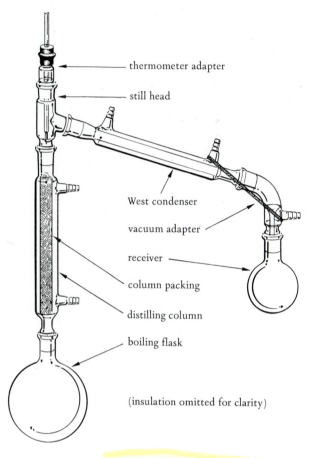

thermometer adapter

still head

West condenser

vacuum adapter

receiver

column packing

distilling column

boiling flask

(insulation omitted for clarity)

Figure 7. Apparatus for fractional distillation

increase the pot temperature (if necessary) until that next fraction starts to distill.

If the separation is not sharp, the temperature may rise only gradually throughout the distillation; in that case, it is best to collect fractions continuously at regular temperature intervals and redistill them. Otherwise, continue heating until the still-head temperature stabilizes at a higher value; then record the temperature, change receivers, and collect the next component. Repeat this process as necessary until the pot is nearly dry (or until all the desired components have been collected); then lower the heat source and let the column drain. The intermediate fractions collected while the temperature was rising rapidly are impure; unless you wish to redistill them, they can be discarded. Disassemble the apparatus, and clean it promptly.

Equipment and supplies:
 boiling flask
 distilling column
 column packing
 insulating material (optional)
 still head (three-way connecting tube)
 thermometer
 thermometer adapter
 condenser
 vacuum adapter
 receivers
 rubber band or spring clamp(s)
 boiling chips
 condenser tubing
 heat source
 two utility clamps
 condenser clamp
 two ringstands
 supports for receiver and heat source

Summary

1. Pack column; assemble and (if necessary) insulate apparatus.

2. Add liquid and boiling chips; position heat source.

3. Adjust heat so that reflux ring passes slowly up the column; hold it at the top for a few minutes.

4. Distill slowly until head temperature drops *or* until predetermined temperature (or fraction volume) is reached *or* until pot is nearly empty.
IF another fraction is to be collected, GO TO 5.
IF distillation is complete, GO TO 7.

5. Change receivers.
IF fractions are to be collected continuously at predetermined intervals, GO TO 4.
IF not, GO TO 6.

6. Increase heating rate until distillation resumes; change receivers when temperature stabilizes.
GO TO 4.

7. Remove heat, drain column, and disassemble and clean apparatus.

Measurement
of Physical Constants

Melting-Point Determination

Principles and Applications

The melting point of a pure substance is defined as the temperature at which the solid and liquid phases of the substance are in equilibrium at a pressure of 1 atmosphere. At a temperature slightly lower than the melting point, a mixture of the two phases solidifies; at a temperature slightly above the melting point, the mixture liquefies. Melting points can be used to identify organic compounds and to assess their purity.

    The melting point of a pure compound is a unique property of that compound, which is essentially independent of its source and method of purification. This is not to say that no two compounds will have the same melting point. If two pure samples have *different* melting points, however, they are almost certainly different compounds. If a substance is thought to be a certain compound, its identity can be confirmed beyond reasonable doubt by mixing it with a sample of the known compound and obtaining a melting point of the mixture. If the two substances are not identical, the melting point will generally be lowered and the melting point range broadened; otherwise, the melting point will be essentially the same as that for the two samples measured separately.

    A pure substance usually melts within a range of no more than $1-2°C$; that is, the transition from a crystalline solid to a clear, mobile liquid occurs within a degree or two if the rate of heating is sufficiently slow and the sample is properly prepared. The presence of impurities in a substance lowers its melting temperature and broadens its melting range. Since the melting point decreases in a nearly linear fashion as the degree of impurity increases (up to a point, at least), the difference between the observed and expected values can give an approximate indication of the compound's purity.

    Melting phenomena can be better understood by referring to a *phase diagram* such as the one for phenol *(P)* and diphenyl-

Allotropic forms of an element or different crystalline forms of a compound *may* differ in melting point.

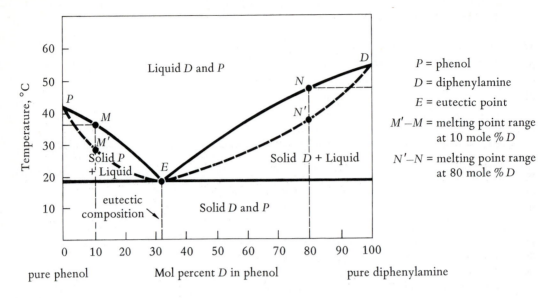

Figure 1. Phase diagram for the phenol-diphenylamine system

For example, the melting point of a mixture containing 10 mol% diphenylamine in phenol is given by the point *M*; that of 20 mol% phenol in diphenylamine is given by point *N*.

The situation is more complicated when formation of a complex occurs between two compounds. In some cases, the melting point of a complex may be higher than the melting points of one or both components. For this and other reasons, melting points should be interpreted with caution.

amine *(D)* in Figure 1. Pure phenol melts at 41°C and pure diphenylamine at 53°C. If a sample of phenol contains a small amount of diphenylamine as an impurity, its melting point will decrease approximately in proportion to the mole percent of diphenylamine present; likewise, the melting point of diphenylamine will decrease on addition of phenol. If more than 32 mol% of diphenylamine is added to phenol, the melting point will begin to rise from its minimum value as the *eutectic point E* is passed. Since the organic compounds you are likely to encounter will contain much smaller amounts of impurities than this, you can expect that the impurities will lower the melting point and not raise it.

Pure phenol and diphenylamine should both have sharp melting points; on the other hand, mixtures of the two (except the eutectic mixture) will ordinarily exhibit broad melting-point ranges, depending on their composition. The observable melting-point range for a mixture is given roughly by the distance between the broken line connecting points *P* and *E* or points *E* and *D* and the solid lines connecting the same points. For instance, the melting-point range of a 10 mol% diphenylamine mixture is given in the figure by the distance between *M'* and *M*, and for an 80 mol% mixture by the distance between *N'* and *N*.

Experimental Considerations

Melting Behavior. The *melting-point range* of a substance is reported as the range between the temperature at which the crystals first begin to liquefy and that at which they are completely liquid. In many cases, the crystals will soften and shrink before they begin to liquefy. The melting-point range does not begin until the first free liquid is visible.

The softening process begins at the eutectic temperature.

Occasionally, traces of solvent will be present in the crystals, due to either insufficient drying or chemical interactions with a solvent. This may be indicated by "sweating" of solvent by the crystals or by the presence of bubbles in the melt, which may resolidify when all of the solvent is driven off. If the sample is obviously wet, it should be dried and the melting point remeasured. In any case, sweating or the dissolution of crystals in the solvent should not be mistaken for melting behavior.

Bubbling and resolidification may also result from decomposition of the sample.

Some compounds can be converted directly from the solid to the vapor state by heating them in an open container. Such *sublimation* can be detected during a melting-point determination by a shrinking of the sample accompanied by the appearance of crystals higher up inside the melting-point tube (see Figure 2 on page 722). The melting point of a sample that sublimes at or below its melting temperature can be determined using a sealed capillary tube. An ordinary melting-point tube should be cut ⟨OP–4⟩ short enough so that it does not project above the block of a melting-point apparatus such as the one in Figure 4 (on page 722), or so that it can be entirely immersed in a melting bath. The sample is introduced and the open end is sealed with a burner ⟨OP–4⟩. Then the melting point is measured by one of the methods described below.

If a melting bath is used, the sealed tube should be secured to the thermometer by a fine wire (not by a rubber band!).

Compounds that *decompose* before a true melting point can be reached do not readily give reproducible "melting" points. They usually liquefy over a broad temperature range, at temperatures that vary with the rate of heating. The "melting" point of such a compound, which is better called its *decomposition point,* is measured by heating the melting-point bath (or block) to within a few degrees of the expected decomposition temperature before inserting the sample, then raising the temperature at a rate of about 3–6°C per minute.

Decomposition should be suspected when a compound becomes discolored and liquefies over an unusually broad range.

Apparatus for Measuring Melting Points. Melting points can be determined with good accuracy using a special melting-point tube (Thiele or Thiele-Dennis tube) filled with a heating-bath liquid such as mineral oil, as illustrated in Figure 5. (See OP–7

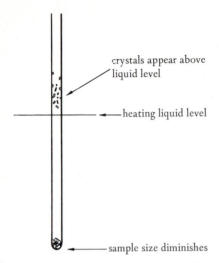

crysrals appear above liquid level

heating liquid level

sample size diminishes

Figure 2.   Sublimation of a sample in a capillary tube

clamp here

heat

stir with a gentle, up-and-down motion

Figure 3.   Stirred heating bath

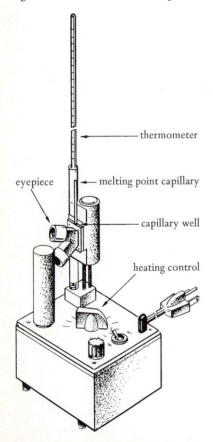

thermometer

eyepiece

melting point capillary

capillary well

heating control

Figure 4.   Capillary melting-point apparatus

for information about bath liquids.) The design of such tubes promotes good circulation of the heating liquid without stirring. A capillary tube containing the solid is secured to a thermometer, which is then immersed in the bath liquid. The solid is observed carefully for evidence of melting as the apparatus is heated. Melting points can also be measured in an ordinary beaker if the bath liquid is stirred constantly, either manually or with a magnetic stirrer ⟨OP–9⟩. Heating coils or tapes which eliminate the need for an external heat source, can be used.

Commercial melting-point instruments of the "hot-stage" type are available, as well as ones that use capillary melting-point tubes. Hot-stage instruments employ an electrically heated metal block containing a thermometer. The sample is placed between two thin glass discs, positioned on top of the block, and observed through a magnifying lens or microscope while the rate of heating is controlled with a dial. Because the sample is out in the open, it may be at a lower temperature than that measured by the thermometer bulb, which is centered in the block. Therefore, melting points obtained by this method may be too high.

Melting-point devices using capillary tubes are quite accurate if operated properly, and they can be used to make several measurements at once. The latter feature is particularly useful when a *mixture melting point* is being determined, since the melting points of the unknown compound, the known, and the mixture can be measured at the same time and compared. The heating rate is adjusted by a knob controlling the voltage input to the heating coil. The voltage required to obtain a given heating rate at the melting point of the sample can be estimated from the heating-rate chart furnished with the instrument. The heating-rate control is initially set higher than this estimated voltage to bring the temperature within 15–20°C of the expected melting point, and then reduced to maintain the desired heating rate at the melting point.

Melting-Point Corrections. The observed melting point of a compound may be inaccurate because of defects in the thermometer used or because of the "emergent stem error" that results when a thermometer is not immersed in the bath to its intended depth. Many thermometers are designed to be immersed to the depth indicated by an engraved line on the stem, usually 76 mm from the bottom of the bulb; slight deviations from this depth will not result in serious error. Other thermometers are designed for total immersion; if they are used under other circumstances, the temperature readings will be in error. Such an error can be compensated for by adding an *emergent stem correction* calculated with the following equation. This requires that a second thermometer be held opposite the middle of the exposed part of the mercury column.

emergent stem correction (to be added to t_1) = $0.00017 \cdot N(t_1 - t_2)$

N = length in degrees of exposed mercury column
t_1 = observed temperature
t_2 = temperature at middle of exposed column

   Errors due to either thermometer defects or an emergent stem can be corrected by calibrating the thermometer under the conditions in which it is to be used. A thermometer used for melting-point determinations, for example, is calibrated by measuring the melting points of a series of known compounds, subtracting the observed melting point from the true melting point of each compound, and plotting this correction as a function of temperature. The correction at the melting point of any other compound is then read from the graph and added to its observed value. A list of pure compounds that can be used for melting-point calibrations is given in Table 1. Sets of pure calibration substances are available from chemical supply houses.

Melting points that have been corrected by adding a correction factor should be reported as, for example, "m.p. 123–124.5° (corr.)."

Table 1. Substances used for melting-point calibrations

Substance	m.p., °C
ice	0
diphenylamine	54
m-dinitrobenzene	90
benzoic acid	122.5
salicylic acid	159
3,5-dinitrobenzoic acid	205

Mixture Melting Points. A mixture melting point is obtained by grinding together approximately equal quantities of two solids (a few milligrams of each) until they are thoroughly intermixed, then measuring the melting point of the mixture by the usual method. Usually one of the compounds (X) is an "unknown" that is believed to be identical to a known compound, Y. A sample of pure Y is mixed with X and the melting points of this mixture and of pure Y (and sometimes of X as well) are measured. If the melting-point ranges of pure Y and of the mixture are identical (within a degree or so), then X is probably identical to Y. If the mixture melts at a lower temperature than pure Y does and has a much broader melting range, then X and Y are different compounds.

Operational Procedure (Thiele Tube Method)

If a commercial instrument is to be used for melting point measurements, your instructor will demonstrate its operation.

Clamp the Thiele tube or Thiele-Dennis tube to a ringstand, and add enough mineral oil (or other bath liquid) to just cover the top of the sidearm outlet. Put a few milligrams of the dry solid on a watch glass (or a glass plate), and grind it to a fine powder with a flat-bottomed stirring rod (or a flat-bladed spatula). Press the open end of a capillary melting-point tube into a pile of the powder until enough has entered the tube to form a column 1–2 mm high. Tap the closed end of the tube gently on the bench top (or rub its sides with a file); then drop it (open end up) through a 1-meter length of small-diameter glass tubing onto a hard surface, such as the bench top. Repeat the process several times to pack the solid firmly into the bottom of the tube. Secure the capillary tube at the top to a broad-range thermometer (for example, $-10°C$ to $360°C$) by means of a 3-mm-thick rubber band (cut from $\frac{1}{4}$-inch I.D. thin-walled rubber tubing) so that the sample is adjacent to the middle of the thermometer bulb (see Figure 5c for placement). Snap the thermometer into a cut-away cork (Figure 5b) at a point above the rubber band, with the capillary tube and the degree markings on the same side as the opening in the cork. Insert this assembly into the bath liquid, and move the thermometer, if necessary, until its bulb is centered in the Thiele tube about 3 cm below the sidearm junction and the temperature can be read through the opening in the cork (Figure 5a). The rubber band should be 2–3 cm above the liquid level so that the bath liquid, as it expands on heating, will not contact it.

If the approximate melting point of the compound is known, heat the bottom of the Thiele tube with a burner until the temperature is about $15°C$ below that temperature; then turn down the flame and apply it at the sidearm to reduce the heating rate. The temperature should rise about $1–2°C$ per minute while the compound is melting, so that the sample will be in thermal equilibrium with the bath and thermometer. Record the temperatures (1) when the first free liquid appears in the melting-point tube and (2) when the sample is completely liquid, as the limits of melting point range (see the section titled "Melting Behavior" above).

The capillary melting-point tube can be constructed by sealing one end of a 10-cm length of 1-mm (I.D.) capillary tubing.

Using too much sample can result in too broad a melting-point range and a high value for the melting point.

Pinch the rubber band between your fingers when inserting the melting-point tube.

The cork is bored out to accommodate the thermometer, then cut with a single-edged razor blade (or sharp knife) so that the thermometer can be snapped into place from the side rather than inserted from the top.

If the hot oil contacts the rubber band, it may soften and allow the capillary tube to drop out.

A micro burner (or a Bunsen burner with the barrel removed) allows more precise heat control.

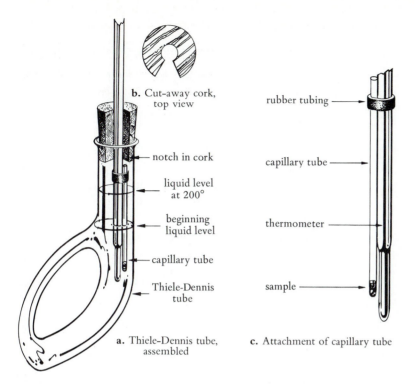

b. Cut-away cork, top view

notch in cork

liquid level at 200°

beginning liquid level

capillary tube

Thiele-Dennis tube

a. Thiele-Dennis tube, assembled

rubber tubing

capillary tube

thermometer

sample

c. Attachment of capillary tube

Figure 5.   Apparatus for determining melting points

Mineral oil begins to smoke and discolor below 200°C and can burst into flames at higher temperatures. Oil fires can be extinguished with solid-chemical fire extinguishers or powdered sodium bicarbonate.

SAFETY
PRECAUTION

If the approximate melting point of the compound is not known, it is advisable to make a rough initial determination with fast heating (6°C per minute or more), and then to measure an accurate melting point with a second sample. If two or more determinations are to be carried out in succession on the same compound, the bath should be cooled to 10–15°C below the melting point and a new sample used for each additional measurement. Cooling can be accelerated by passing an air stream over the Thiele tube. Unless the oil is too dark for further use, the Thiele tube can be stoppered and stored with the oil inside.

It is advisable to do at least two measurements on each compound.

Mineral oil can be removed from glassware by rinsing the glassware with ligroin (or petroleum ether) followed by acetone, then washing it with a detergent and water.

Summary

1. Assemble apparatus for melting-point determination.

2. Grind solid to powder, and fill capillary tube(s) to depth of 1–2 mm.

3. Secure capillary tube to thermometer; insert assembly in heating bath.
IF approximate m.p. is known, GO TO 5.
IF not, GO TO 4.

4. Heat rapidly until sample melts, and record approximate melting temperature; select another sample.
GO TO 3.

5. Heat rapidly to about 15°C below m.p.; then reduce heating rate to 1–2°C per minute.

6. Record temperatures at beginning and end of melting range.
IF another determination is needed, GO TO 7.
IF not, disassemble apparatus, and STOP.

7. Cool bath to 10–15°C below m.p.; select another sample.
GO TO 3.

Equipment and supplies (Thiele Tube Method):
  Thiele or Thiele-Dennis tube
  heating oil
  thermometer
  capillary melting-point tube
  burner
  cut-away cork
  3-mm slice of rubber tubing
  watch glass
  flat-bottomed stirring rod
  1-m length of glass tubing
  clamp
  ringstand

OPERATION 29

Boiling-Point Determination

The boiling point of a liquid is defined as the temperature at which the vapor pressure of the liquid is equal to the external pressure exerted at the surface of the liquid. The *normal boiling point* of a liquid is its boiling point at an external pressure of 1 atmosphere (760 torr), which is the temperature at which the vapor pressure of the liquid equals 1 atmosphere. Since the atmospheric pressure at the time of a boiling-point determination is seldom exactly 760 torr, observed boiling points may differ somewhat from values reported in the literature and should be corrected. Boiling-point values can be used to identify liquid organic compounds and assess their purity.

Experimental Considerations

Boiling points can be measured with good accuracy by distilling a small quantity of liquid or by using a micro method such as that described in OP–29a. During a carefully performed distillation, it can be assumed that the vapors surrounding the thermometer bulb are in equilibrium (or nearly so) with the liquid condensing on the bulb. Since the boiling point of a liquid is (by one definition) the temperature at which its liquid and gaseous forms are in equilibrium, the temperature recorded during the distillation of a pure liquid should equal its boiling point. If the liquid is contaminated by impurities, the boiling point obtained by distillation may be either too high or too low, depending on the nature of the impurity. Therefore, if there is any doubt about the purity of a liquid, it should be distilled or otherwise purified prior to a boiling-point determination.

Volatile impurities in a liquid lower its boiling point, whereas involatile impurities raise it. In either case, the distillation boiling-point range will be broadened.

If a laboratory boiling-point determination is carried out at a location reasonably close to sea level, atmospheric pressure will rarely vary by more than 30 torr from 760. For deviations of this magnitude, a *boiling-point correction, Δt, can be estimated* using Equation 1.

$$\Delta t \approx y(760 - P)(273.1 + t) \tag{1}$$

Δt = temperature correction, to be added to observed boiling point
P = barometric pressure (torr)
t = observed boiling point (°C)

The value 1.0×10^{-4} is used for the constant y if the liquid is water, an alcohol, a carboxylic acid, or another associated liquid; otherwise, y is assigned the value 1.2×10^{-4}. For example, the boiling point of water at 730 torr is 98.9°C. Use of Equation 1 leads to a correction factor of $(1.0 \times 10^{-4})(760 - 730)(273.1 + 98.9) = 1.1$°C, which yields the correct normal boiling point of 100.0°C.

At high altitudes, the atmospheric pressure may be considerably lower than 1 atmosphere, resulting in observed boiling points substantially lower than the normal values. For example, water boils at 93°C on the campus of the University of Wyoming (elevation 7520 feet) and at 81°C at the top of Mount Evans in Colorado (elevation 14,264 feet). For major deviations from atmospheric pressure, Equation 2 can be used in conjunction with approximate entropy of vaporization values listed in the *CRC Handbook of Chemistry and Physics* (on p. D-229 in the 59th edition) or obtained from published data.

Equation 2 is a simplified form of the Hass-Newton equation found in the CRC Handbook, *59th ed., p. D-228.*

$$\Delta t \approx \frac{(273 + t)}{\phi} \log \frac{760}{P} \tag{2}$$

ϕ = entropy of vaporization at normal boiling point (760 torr)

For example, hexane boils at 49.6 °C at 400 torr; the value of ϕ for alkanes is found (from the *CRC Handbook* tables) to be about 4.65 at that temperature. Substituting into Equation 2, we obtain a correction factor of:

$$\Delta t = \frac{(273 + 49.6)}{4.65} \log \frac{760}{400} = 19.3\,°C$$

For a second approximation, the value of ϕ at 68.9 °C would be estimated from the tables and resubstituted into Equation 2. This results in a value of 68.8 °C.

which gives a normal boiling point of 68.9 °C as a first approximation. This compares very favorably to the reported value of 68.7 °C.

As in melting-point determinations, it may be necessary to correct the boiling point for thermometer error, especially when working with high-boiling liquids. See OP – 28 for details.

Operational Procedure

If a special semimicro distillation flask is available, use it for a boiling-point determination as directed by your instructor. Otherwise assemble the apparatus pictured in Figure 1 of OP – 25a, using a 25-mL distilling flask and following the instructions given in that operation. Be sure the still head is well insulated from the heat source to prevent superheating of the vapors. Introduce 5 – 10 mL of the pure liquid and a boiling chip or two into the pot, and distill the liquid *slowly* (1 drop per second or less), recording the temperature after the first 1 – 2 mL has distilled and again when only 1 – 2 mL remains in the pot. It may also be desirable to record a median boiling point, the temperature at which half of the liquid has distilled. Record the barometric pressure so you can make a pressure correction, if necessary. If the boiling range is more than 2 °C, it may be advisable to purify the liquid by distillation ⟨OP – 27 or OP – 25⟩ and repeat the determination.

Correct thermometer placement is essential; see Figure 3 of OP – 25.

Low- or high-boiling impurities may distort the boiling range at the beginning or end of a distillation.

OPERATION 29a

Micro Boiling-Point Determination

When a liquid is heated to its boiling point, the pressure exerted by its vapor becomes just equal to the external pressure at the liquid's surface. If a tube that is closed at one end is filled with a liquid and immersed open end down in a reservoir containing the same liquid, it will begin to fill with vapor as the liquid is heated to its boiling point. At the boiling point,

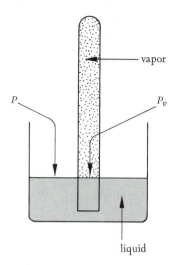

Figure 1. Micro boiling-point principle: P_v = pressure exerted by vapor on liquid surface; P = pressure exerted by atmosphere on liquid surface; at boiling point, $P = P_v$

the vapor pressure inside the tube will be balanced by the pressure exerted on the liquid surface by the surrounding atmosphere, so that the tube will be just filled with vapor. If the temperature is raised above the boiling point, vapor will begin to escape in the form of bubbles; if the tube is cooled below the boiling point, it will begin to again fill with liquid. This behavior is the basis of a micro method for determining boiling points.

Operational Procedure

Add 2–4 drops of the liquid to a boiling tube constructed of a 10-cm length of 4–5 mm O.D. glass tubing sealed at one end ⟨OP–4⟩. Insert a capillary melting-point tube (sealed at one end) into the boiling tube with its open end down, and secure the assembly to a thermometer by means of a rubber band cut from thin-walled rubber tubing, as illustrated in Figure 2. Insert this assembly into a Thiele tube as for a melting-point determination (see Figure 5 of OP–28) with the rubber band 2 cm or more above the liquid level.

Heat the Thiele tube until a *rapid*, continuous stream of bubbles (of the vaporized liquid) emerges from the capillary tube. Remove the heat source, let the bath cool slowly until the bubbling stops, and record the temperature when liquid just begins to enter the capillary tube. Let the temperature drop a few degrees so that the liquid partly fills the capillary; then

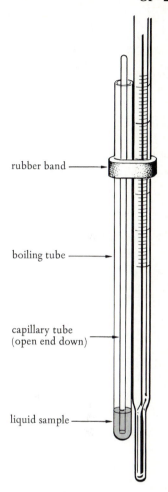

Figure 2. Micro boiling-point assembly

Avoid overheating or the liquid will boil away. It may be necessary to add more liquid if the sample is very volatile.

Sometimes the capillary tube will stick to the bottom of the boiling tube. This can be prevented by cutting a small nick in the open end of the capillary tube with a triangular file.

heat very slowly until the first bubble of vapor emerges from the mouth of the capillary tube. Record the temperature at that point also. The two temperatures represent the boiling-point range; they should be within a degree or two of each other. Cool the bath until liquid again enters the capillary tube; then repeat the determination. If repeated determinations on the same sample give appreciably different (usually higher) values for the boiling point, the sample is probably impure and should be distilled. Record the barometric pressure so that a pressure correction can be made. Clean ⟨OP–1⟩ and dry the boiling tube, saving it for future determinations.

Summary

1. Assemble apparatus for micro boiling-point measurement; add liquid to boiling tube.

2. Heat until continuous stream of bubbles emerges from capillary, and then remove heat.

3. Record temperature when liquid enters capillary.

4. Heat slowly until first vapor bubble emerges from capillary; record temperature.

5. Let cool below boiling point.
IF another measurement is required, GO TO 2.
IF not, GO TO 6.

6. Record barometric pressure; correct observed boiling point.

Equipment and supplies:
  Thiele tube
  mineral oil
  thermometer
  capillary tube
  boiling tube
  rubber band
  burner

OPERATION 30

Measuring the Refractive Index

Note that the index of refraction is a dimensionless ratio.

The *index of refraction* (refractive index) of a substance is defined as the ratio of the speed of light in a vacuum to its speed in the substance in question. When a beam of light passes into a liquid, its velocity is reduced, causing it to bend downward. The index of refraction is related to the angles that the incident and refracted beams make with a line perpendicular to the liquid surface (see Figure 1) by Equation 1.

n_λ^t = index of refraction at temperature
   t using light of wavelength λ
c = velocity of light

$$n_\lambda^t = \frac{c_{vac}}{c_{liq}} = \frac{\sin \theta_{vac}}{\sin \theta_{liq}} \tag{1}$$

The refractive index of a liquid is a unique physical property that can be measured with great accuracy (up to eight decimal places), and it is thus very useful in identifying pure organic compounds. Refractive-index measurements are also used to assess the purity of known liquids and to determine the composition of solutions.

Experimental Considerations

Because of experimental difficulties associated with measurements in a vacuum, refractive-index measurements are generally made in air ($n_D = 1.0003$), and compensation is made for the difference between the speed of light in air and its speed in a vacuum. Refractive-index values are dependent both on the wavelength of the light used for their measurement and on the density (and therefore temperature) of the liquid. Most values are reported with reference to light from the yellow D line of the sodium emission spectrum, which has a wavelength of 589.3 nm. A refractive index in the literature may be reported as $n_D^{20} = 1.3330$, for example, where the superscript is the temperature in °C and D refers to the sodium D line. (Since most refractive index readings are made at this wavelength, the D is often omitted.)

The most commonly used instrument for measuring refractive indexes is probably the Abbe refractometer (Figure 3), which measures the critical angle of reflection at the boundary between the liquid and a glass prism and converts it to refractive index. The instrument employs a set of compensating prisms so that white light can be used to give refractive-index values corresponding to the sodium D line.

If a refractive index is measured at a temperature other than 20°C, the temperature should be read from the thermometer on the instrument and the refractive index corrected by one of the following methods:

1. A correction factor Δn is estimated from the equation $\Delta n \approx 0.00045 \times (t - 20.0)$, where t is the temperature of the measurement in °C. The correction factor (including its sign) is then added to the observed refractive index.

2. The refractive index of a *reference liquid* is measured and a correction factor is calculated as follows: $\Delta n = n^{20} - n^t$, where n^t is the measured value at temperature t for the pure reference liquid and n^{20} is its literature value at 20°C. The correction factor is then added to the observed refractive index of the sample. This method can correct for experimental errors (improper calibration of instrument, etc.) as well as temperature differences, but it is accurate only when the reference liquid is similar in structure and properties to the liquid being measured.

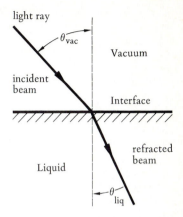

Figure 1. Refraction of light in a liquid

If another light source is used, its wavelength is usually specified in nm.

The correction factor Δn is only an approximation; the actual change in refractive index with temperature may be larger or smaller for a given liquid.

A properly calibrated B&L Abbe 3L refractometer should be reliable to ±0.0002. The calibration of the instrument can be checked by measuring the refractive index of distilled water, which should be 1.3330 at 20°C and 1.3325 at 25°C. The refractive index of distilled water decreases by about 1×10^{-4} for each degree increase in temperature between 20 and 30°C.

For example, the refractive index of an unknown aliphatic hydrocarbon is measured at 25.1°C and found to be 1.3874. By the first method, the corrected value would be $n^{20} = 1.3874 + 0.00045 \times (25.1 - 20.0) = 1.3897$. If the refractive index of hexane, a reference liquid, ($n^{20} = 1.3751$) is measured at 25°C and found to be 1.3726, a correction factor can be calculated as follows: $\Delta n = 1.3751 - 1.3726 = +0.0025$. Adding this to the observed value for the unknown gives a corrected refractive index of 1.3899 by the second method.

The effect of a contaminant on the refractive index of a liquid varies with the difference in the refractive indexes of the two substances.

Small amounts of impurities can cause substantial errors in the refractive index. For example, the presence of just 1% (by weight) of acetone in chloroform reduces the refractive index of the latter by 0.0015. Such errors can be critical when refractive-index values are being used for qualitative analysis, so it is essential that an unknown liquid be pure when its refractive index is measured.

Operational Procedure (for B&L Abbe Refractometer)

Never allow the tip of a dropper or other hard object to touch the prisms—they are easily damaged.

Using an eyedropper, place about two drops of the sample between the prisms (Figure 3), either by raising the *hinged prism* and dropping the liquid on the middle of the *fixed prism* below it or (for a volatile and free-flowing liquid) by introducing it into the channel alongside the closed prisms. With the prism assembly closed, switch on the *lamp* and move it toward the prisms to illuminate the visual field viewed through the eyepiece. Rotate the *handwheel* until two distinct fields (light and dark) are visible in the eyepiece, and reposition the lamp for the best contrast and definition at the borderline between the fields. Rotate the *compensating drum* on the front of the instrument until the borderline is sharp and achromatic (black and white) where it intersects the vertical reticle mark. Rotate the handwheel or the fine-adjustment knob (if there is one) to center the borderline exactly on the crosshairs (Figure 2). Depress and hold down the *display switch* (not shown, but usually on the left side of the instrument) to display an optical scale in the eyepiece. Read the refractive index from this scale (estimate the fourth decimal place), and record the temperature (if it is different from 20°C). Open the prism assembly, and remove the sample by gently blotting it with a soft tissue (do not rub!). Wash the prisms by moistening a tissue or cotton ball with a suitable solvent (acetone, methanol, etc.) or with a non-ionic detergent solution and blotting them gently. When any residual solvent has evaporated, close the prism assembly and turn off the instrument.

Equipment and supplies:
  refractometer
  dropper
  washing liquid
  soft tissues

If the borderline cannot be made sharp and achromatic, the sample may have evaporated.

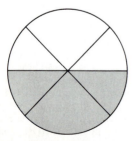

Figure 2.   Visual field for properly adjusted refractometer

If an instrument other than the B&L Abbe Refractometer is to be used, your instructor will demonstrate its operation.

Summary

1. Start water circulating in refractometer jacket at 20°C (optional).

2. Insert sample between prisms.

3. Switch on and position lamp.

4. Rotate handwheel until two fields are visible; reposition lamp for best contrast.

5. Rotate compensating drum until borderline is sharp and achromatic.

6. Rotate handwheel or fine-adjustment knob until line is centered on crosshairs.

7. Depress display switch to display optical scale, estimate refractive index to four decimal places, and record temperature.

8. Clean and dry prisms (a small piece of tissue paper may be placed between prisms to prevent abrasion), close prism assembly, and turn off refractometer.

9. Make temperature correction to observed refractive index, if necessary.

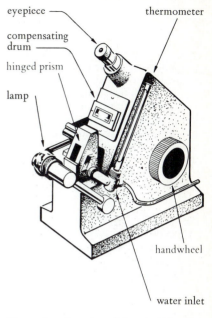

Figure 3.   Bausch and Lomb Abbe 3L Refractometer

Measuring Optical Rotation

OPERATION 31

Ordinary light can be considered a wave phenomenon with vibrations occurring in an infinite number of planes perpendicular to the direction of propagation. Plane-polarized light, which is generated when ordinary light passes through a polarizer (such as a Nicol prism), is light with vibrations restricted to a single plane (see Figure 1). When plane-polarized light passes through an optically active substance, molecules of the substance are capable of rotating its plane of polarization. The net rotational angle depends on the number of molecules in the light path, and therefore on the sample size and concentration. The angle of rotation is measured with an instrument called a *polarimeter* (see Figure 2, on page 734).

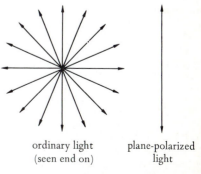

ordinary light   plane-polarized
(seen end on)   light

Figure 1.   Schematic representations of ordinary and plane-polarized light

Experimental Considerations

A polarimeter consists of a polarizer, a cell to contain the sample, and an analyzer (a second polarizing prism) that can be rotated so as to compensate for the rotation induced by the sample. Usually the analyzer is rotated until the extinction point (the angle of minimum illumination) is reached. At this point, the axis of the analyzer is perpendicular to the axis of the polarized light, and therefore all the light is blocked out. By measuring the angle at which extinction occurs, the optical rotation of the sample is determined. Since the amount of rotation depends on factors that are not inherent properties of the sample itself (such as the length of the polarimeter tube and the concentration of the sample), this observed rotation is converted to specific rotation by means of Equation 1.

The analyzer in precision instruments contains two or more prisms set at a small angle to each other. The reading is made when the prisms bracket the extinction point, transmitting equal light intensities. (It is easier for the human eye to match intensities than to estimate the point of minimum intensity.)

$$[\alpha] = \frac{\alpha}{l \cdot c} \qquad (1)$$

$[\alpha]$ = specific rotation
α = observed rotation
l = length of polarimeter cell, in decimeters
c = concentration of solution, in grams of solute per milliliter solution (for a neat liquid, substitute the density of the liquid in g/mL)

The *specific rotation* of a pure substance under a given set of conditions is an invariable property of the substance and can be used to characterize it. Specific rotation can also be used to measure the optical purity of an enantiomer or the composition of a mixture of optically active substances.

The *optical purity* of an enantiomer can be defined as the ratio of its measured specific rotation to the specific rotation of the pure enantiomer, multiplied by 100. Thus, an alcoholic solution of (+)-camphor ($[\alpha] = 44.3°$ in ethanol) with a specific rotation of 36.4° is $(36.4/44.3) \times 100 = 82.2\%$ optically pure. This does *not* mean that it contains 82.2% (+)-camphor

Calculation of optical purity:

$$\text{optical purity} = \frac{[\alpha] \text{ (observed)} \times 100}{[\alpha] \text{ (pure substance)}} \%$$

Figure 2.   Schematic diagram of a polarimeter

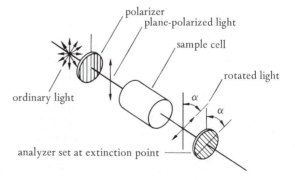

polarizer
plane-polarized light
sample cell
rotated light
ordinary light
α
α
analyzer set at extinction point

and 17.8% (−)-camphor; the "impurity" in the camphor is considered to be the racemate, (±)-camphor, of which half is (+) and half (−). Of course, optically active or inactive impurities other than the racemate can also alter the specific rotation of an optically active compound.

An 82.2% optically pure sample of (+)-camphor actually contains about 91.1% (82.2 + 8.9) of the (+)-camphor and 8.9% (−)-camphor, if the racemate is the only impurity present.

The composition of a mixture containing two optically active components can be determined by solving Equation 2 for X_a (the mole fraction of component a); $[\alpha]$ is the specific rotation of the mixture, and $[\alpha_a]$ and $[\alpha_b]$ are the specific rotations of the pure components.

$$[\alpha] = [\alpha_a]X_a + [\alpha_b](1 - X_a) \qquad (2)$$

For example, an equilibrium mixture of α-D-glucose ($[\alpha] = 112°$) and β-D-glucose ($[\alpha] = 18.7°$) has a specific rotation of 52.7°. The mole fraction of α-D-glucose in the mixture can be calculated as follows:

$$52.7° = (112°)X_\alpha + 18.7°(1 - X_\alpha)$$
$$X_\alpha = 0.364$$

Since both forms of glucose have the same molecular weight, the equilibrium mixture contains 36.4% α-D-glucose and 63.6% β-D-glucose by mass.

The optical rotation of a substance is usually measured in solution. Water and alcohol are common solvents for polar compounds; chloroform is used for less polar ones. Often a suitable solvent will be listed in the literature, along with the light source and temperature used for the measurement. The volume of solution required is dependent on the size of the polarimeter cell, but 10–25 mL is usually sufficient. A solution for polarimetry ordinarily contains about 1–10 g of solute per 100 mL of solution and should be prepared using an accurate balance and a volumetric flask. If the solution contains particles of dust or other solid impurities, it should be filtered ⟨OP–11⟩. When possible, the concentration should be comparable to that reported in the literature. For example, the *CRC Handbook* reports the optical rotation of (+)-menthol as "+49.2 (al, c = 5)," where the concentration is given in grams per 100-mL of solution (c is defined differently here than in Equation 1). Thus, a menthol solution for polarimetry could be prepared by accurately weighing about 1.25 g of (+)-menthol, dissolving it in ethyl alcohol (al) in a 25-mL volumetric flask, and adding more alcohol up to the calibration mark.

This procedure applies to a Zeiss-type polarimeter with a split-field image. Experimental details may vary for different instruments.

Operational Procedure

Remove the screw cap and glass endplate from one end of a clean 1- or 2-decimeter polarimeter cell *(do not get fingerprints on the endplate),* and rinse the cell with a small amount of the solution to be analyzed. Stand the polarimeter cell vertically on the bench top and overfill it with the solution (rocking the tube to shake loose any air bubbles), adding the last milliliter or so with a dropper so that the liquid surface is convex. Carefully slide the glass endplate on so that there are no air bubbles trapped inside. If the tube has a bulge at one end, a small bubble can be tolerated; in that case the tube should be tilted so that the bubble migrates to the bulge and stays out of the light path. Screw the cap on just tightly enough to provide a leakproof seal—overtightening it may strain the glass and cause erroneous readings. If the light source is a sodium lamp, make sure it has ample time to warm up (some require 30 minutes or more). Place the sample cell in the polarimeter trough, close the cover, and see that the light source is oriented so as to provide the maximum illumination in the eyepiece. Set the analyzer scale to zero, and if necessary rotate it a few degrees in either direction until a dark and a light field are clearly visible. (You may see a vertical bar down the middle and a background field on both sides.) Focus the eyepiece so that the lines separating the fields are as sharp as possible. Starting from zero, rotate the analyzer scale clockwise or counterclockwise (usually not more than 10° in either direction) until both fields are of nearly equal intensity. Then back off a degree or so toward zero and use the fine-adjustment knob (if there is one) to rotate the scale *away* from zero until the entire visual field is as uniform as possible. If you overshoot the final reading, move the scale back a few degrees so that you again approach it going away from zero. Read the optical rotation from the analyzer scale, using the Vernier scale (if there is one) to read fractions of a degree, and record the direction of rotation (+ or −). Obtain another reading of the optical rotation, this time approaching the final reading from the *opposite* direction (going toward zero). For accurate work, you should take a half-dozen readings or more, reversing direction each time to compensate for mechanical play in the instrument, and average them.

Rinse the cell with the solvent used in preparing the solution; then fill the cell with that solvent, and record its optical rotation by the same procedure as before. (This value is the solvent blank.) Remove the solvent and let the cell drain dry, or clean it as directed by your instructor. Subtract the

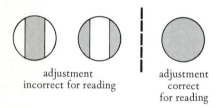

adjustment
incorrect for reading

adjustment
correct
for reading

Figure 3.   Split-field image of polarimeter

If the field is very bright and a slight turn of the micrometer knob has little effect on it, you are probably 90° off. Rotate the scale 90° back toward zero.

Alternatively, your instructor may provide a solvent blank in another polarimeter cell.

optical rotation of the solvent blank (remember + and − signs) from that of the sample to obtain the optical rotation of the sample, and calculate its specific rotation using Equation 1.

In reporting the specific rotation, specify the temperature, light source, solvent, and concentration.

Summary

1. Prepare solution of compound to be analyzed (filter if necessary).

2. Rinse and fill sample cell with solution, place cell in polarimeter, and adjust light source.

3. Focus polarimeter.

4. Rotate analyzer scale until field is uniform.

5. Read optical rotation.

6. Repeat steps 4 and 5 several times, reversing direction each time, and average readings.

7. Rinse and fill sample cell with solvent; place cell in polarimeter.
GO TO 3.

8. Clean cell, and drain dry.

9. Calculate optical rotation and specific rotation of sample.

Equipment and supplies:
  polarimeter
  polarimeter cell(s)
  balance
  volumetric flask (10 or 25 mL)
  dropper

Instrumental Analysis

OPERATION 32

Gas Chromatography

Gas chromatography (GC) is used to analyze mixtures by separating and detecting their components. The samples analyzed by gas chromatography must be volatile enough that they vaporize without decomposing when heated, so they are usually liquids; some volatile solids can be analyzed if they are first dissolved in a low-boiling solvent. A typical analytical gas chromatograph requires only a minute amount of sample (usually a microliter or so), and records a graph called a *gas chromatogram* from which the identities and quantities of the components of the sample can often be determined. A preparative gas chromatograph can separate larger samples and deliver appreciable quantities of their pure components.

The term gas-liquid chromatography (GLC) is often used interchangeably with other terms such as vapor-phase chromatography (VPC) and gas-liquid partition chromatography (GLPC).

In liquid column chromatography ⟨OP–16⟩, the components of a mixture are partitioned between a solid or liquid stationary phase and a liquid mobile phase. In gas-liquid chromatography, which is the most common form of GC, the components are partitioned between a liquid stationary phase and a gaseous mobile phase called the *carrier gas*. The time it takes for a component to pass through the column, called its *retention time*, depends on its relative concentrations in the stationary and mobile phases (its partition coefficient). While the molecules of a component are in the gas phase, they pass through the column at the speed of the carrier gas; while they are in the liquid phase, they remain stationary. Any component that spends most of its time in the gas phase will pass through the column rapidly and thus have a short retention time; a component that spends most of its time in the liquid phase will have a longer retention time. The time a component spends in the gas phase depends on its volatility (and thus on its boiling point) and on the temperature of the column. The time it spends in the liquid phase depends on its solubility in that phase, which is a function of the intermolecular forces between its molecules and the molecules of the liquid phase.

In order for any two components of a mixture to be separated sharply, they must have significantly different retention times and their bands (the regions of the column they occupy) must be narrow enough that they do not overlap appreciably. The degree of separation depends on a number of factors, including the length and efficiency of the column, the carrier-gas flow rate, and the temperature at which the separation is carried out. Modern gas chromatographs provide the means to vary these factors precisely, and they offer an almost endless variety of applications, from analyzing automobile emissions for noxious gases to detecting PCBs in lake trout.

Instrumentation

In addition to the column and carrier gas, a gas chromatograph must have some means of vaporizing the sample, controlling the temperature of the separation, detecting each constituent of the sample as it leaves the column, and recording data from which the composition of the sample can be determined. A typical gas chromatograph includes the following components, as diagrammed in Figure 1.

Injector. The injector is a port where the sample is introduced, generally by means of a microsyringe. The injector port must be heated to a temperature sufficient to vaporize the sample.

Column. The column is a long tube containing an inert solid *support,* which is coated with a high-boiling *liquid phase.* After a

Figure 1.   Schematic diagram of gas chromatograph

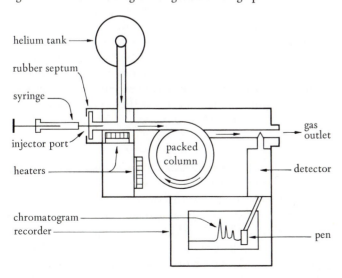

The mesh number is the number of meshes per inch in the finest sieve that will pass the particles; 80/100 mesh particles have effective diameters in the 150–175 μm range.

liquid mixture is vaporized in the injector port, it passes onto the column, where its components are separated. A typical glass or stainless steel column may be 2 m long and 0.25 or 0.5 cm ($\frac{1}{4}$ or $\frac{1}{8}$ inch) in diameter, and is usually bent into a coil to fit inside a column oven, where its temperature is controlled. The column is packed with a finely ground support medium such as Chromosorb W (white diatomaceous earth), whose particles are precoated with the liquid phase. The composition of the packing is described on a tag attached to each column; a column packing described as "10% DEGS/Chromosorb W 80/100" contains 10% by weight of a diethylene glycol succinate liquid phase on a Chromosorb W support with a mesh range of 80 to 100.

In recent years, *open tubular columns* have begun to replace packed columns in many analytical gas chromatographs. A very thin film of the liquid phase (about 0.1–1.0 μm) is adsorbed on or chemically bonded to the inner wall of the column, which is a long capillary tube made of glass or fused silica. Fused silica open tubular (FSOT) columns up to 100 m long are flexible enough to be bent into coils 15 cm in diameter. A typical FSOT capillary column with an inner diameter of 0.25 mm and a length of 50 m may be hundreds of times more efficient than a packed column containing the same liquid phase—and may cost about ten times as much.

Column Oven. The column is mounted inside a heated chamber that controls its temperature. The temperature of the column oven may either be kept constant during a separation (isothermal operation) or varied in accordance with a programmed sequence (program operation). Increasing the temperature causes components to spend a greater fraction of their time in the gas phase, thus reducing their retention times. With program operation, it is possible to elute the more volatile (or less polar) components at a lower temperature, and then increase the temperature gradually so that less volatile (or more polar) components are eluted within a reasonable period of time.

Carrier Gas. An inert gas is used to sweep the sample through the column. Helium is the most widely used carrier gas, although nitrogen and argon are also used with some detectors. Carrier gases come in large gas cylinders, and various devices are used to measure and control their flow rates.

Detector. The detector is a device that detects the presence of each component as it leaves the column and sends an electrical signal to a recorder or other output device. A separate oven is used to heat the detector enclosure so that the sample molecules remain in the vapor phase as they pass the detector. The most widely used detectors for routine gas chromatography are called *flame ionization (FI) detectors* and *thermal conductivity (TC) detectors.* With an FI detector, the component molecules are pyrolyzed in a hydrogen-oxygen flame, producing short-lived ions that can be captured and used to generate an electrical current. A TC conductivity detector responds to changes in the thermal conductivity of the carrier gas stream as molecules of the sample pass through. TC detectors are comparatively simple and inexpensive, and they respond to a wide variety of organic and inorganic species. FI detectors are more sensitive than TC detectors, but they require pure hydrogen and air to produce the flame. Also, since an FI detector destroys the sample, it cannot be used for preparative gas chromatography.

Recorder. The recorder is an instrument that records a peak on moving chart paper as each component passes the detector. The resulting gas chromatogram consists of a series of peaks of different sizes, each produced by a different component of the mixture (assuming complete separation). The recorder may be provided with an *integrator* that measures the area under each component peak. Some modern GC instruments are interfaced with computers that process their output signals and provide digital readouts of peak areas, retention times, and various instrumental parameters.

Liquid Phases

The factor that most frequently determines the success or failure of a gas chromatographic separation is the choice of the liquid phase. In general, polar liquid phases are best for separating polar compounds, and nonpolar liquid phases are best for separating nonpolar compounds. However, there are hundreds of liquid phases available, and choosing the right one for a particular separation may not be an easy task. A few widely used liquid phases and their maximum operating temperatures are listed in Table 1 (on page 742). General-purpose liquid phases such as OV-1 can separate a variety of compounds successfully, usually in approximate order of their boiling points. Other liquid phases have more specialized applications; diethylene glycol succinate is a high-boiling ester used to separate the esters of long-chain fatty acids (as in Experiment 45), and OV-17 can be used to separate drugs, steroids, and pesticides.

Heating a column above its maximum operating temperature will cause the liquid phase to vaporize.

Instrumental Analysis

Table 1. Selected stationary phases for gas chromatography

Stationary phase	Maximum temperature, °C	McReynolds constants		
		X'	Y'	Z'
squalane (hydrocarbon)	150	0	0	0
OV–1 (dimethylsilicone)	350	16	55	44
OV–17 (phenylmethylsilicone)	350	119	158	162
Carbowax 20M (polyethylene glycol)	250	322	536	368
DEGS (diethylene glycol succinate)	225	496	746	590
OV–275 (dicyanoallylsilicone)	275	629	872	763

Note: The chemical name or class of the stationary phase is given in parentheses.

The selection of a liquid phase may be facilitated by the use of McReynolds constants, which indicate the affinity of a liquid phase for different kinds of solutes. X' is a McReynolds constant measuring the relative affinity of a liquid phase for aromatic compounds and alkenes; Y' measures the liquid phase's affinity for alcohols, phenols, and carboxylic acids; and Z' measures its affinity for aldehydes, ketones, ethers, esters, and related compounds. Thus, the very polar liquid phase OV-275 ($Z' = 763$), for example, will retain an ester much longer than will OV-1 ($Z' = 44$). McReynolds constants can help in selecting a column to separate compounds of different chemical classes, such as alcohols and esters. Thus, Carbowax 20M should separate isoamyl acetate from isoamyl alcohol satisfactorily because its Y' and Z' values are very different, but OV-17 would obviously not be the best choice for that particular separation.

Qualitative Analysis

The retention time of a component corresponds to the distance on the chromatogram (parallel to the baseline) from the injection line to the top of its peak, as shown in Figure 2; this distance can be converted to units of time if the chart speed is known. If all instrumental parameters (temperature, column, flow rate, etc.) are kept constant, the retention time of a com-

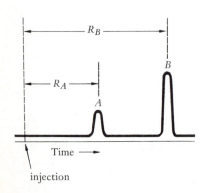

Figure 2. Retention times of two components

ponent will be characteristic of that compound and may be used to identify it. It is seldom (if ever) possible to identify a compound with certainty from its retention time alone, but it is usually possible to confirm the presence of a compound you already suspect is there, if an authentic sample of that compound is available. For example, a crude product mixture from the esterification of acetic acid by ethanol might be expected to contain some unreacted starting materials along with the product, ethyl acetate. To detect the presence of unreacted ethanol in the sample, it is "spiked" with a little pure ethanol, and another gas chromatogram is recorded; the peak that increases in relative area is the ethanol peak, as shown in Figure 3. The process can be repeated to identify the other components, if necessary.

Quantitative Analysis

The use of gas chromatography for quantitative analysis is based on the fact that over a wide range of concentrations a detector's response to a given component is proportional to the amount of that component in the sample. Thus, the area under a component's peak can be used to determine its mass percentage in the sample. The same mass of two different components will not necessarily result in the same detector response, however. FI detectors respond mainly to ions produced by certain "reduced" carbon atoms, such as those in methylene and methyl groups. Because an ethanol molecule contains about three-fifths as much carbon by weight as a heptane molecule does, the response of an FI detector to a gram of ethanol is only about three-fifths as great as its response to a gram of heptane. TC detectors also respond differently to different substances, but the variations are usually not as large as those of FI detectors.

In analyzing a mixture of similar compounds, such as a series of long-chain fatty acid esters, using uncorrected peak areas to calculate mass percentages may be acceptable if small errors can be tolerated. Otherwise, *correction factors,* obtained from the literature or by direct measurement, should be applied. If carefully weighed amounts of component A and reference compound R are combined and chromatographed, the correction factor for A can be determined from the peak areas using the following relationship:

$$\text{correction factor for A} = \frac{\text{area}_R}{\text{area}_A} \times \frac{\text{mass}_A}{\text{mass}_R}$$

Comparing the retention values of a compound with retention values from the literature may help to identify it, but the usefulness of such data is limited by the number of variables that must be controlled to obtain reproducible results.

(1)

(2)

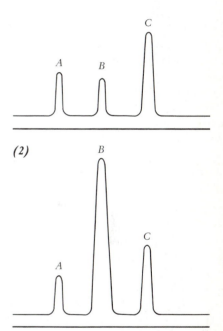

Figure 3. Use of gas chromatography for qualitative analysis: *(1)* chromatogram before adding ethanol; *(2)* chromatogram after adding ethanol

The reference compound is arbitrarily assigned a correction factor of 1.

As long as their peaks are well resolved, weighed amounts of a number of components can be mixed with the reference compound, and the correction factors of all the components can be determined from a single chromatogram. Any correction factor derived in this way is *multiplied* by the area of a component peak to give its corrected area. For example, TC correction factors for cyclohexane and toluene are reported to be 0.942 and 1.02, respectively (relative to benzene). If the chromatogram of a distillation fraction from Experiment 7 has peak areas of 78.0 mm^2 for cyclohexane and 14.6 mm^2 for toluene, the corrected peak areas are 0.942 $\times$ 78.0 mm^2 = 73.5 mm^2 and 1.02 $\times$ 14.6 mm^2 = 14.9 mm^2, respectively. The percentage composition of the sample is then determined by dividing each corrected area by the sum of the corrected areas and multiplying by 100. Thus, the mass percentage of cyclohexane in the sample is 100 $\times$ 73.5/88.4 = 83.1%.

If a peak on a gas chromatogram is symmetrical, its area can be calculated reasonably accurately by multiplying its height in millimeters (measured from the baseline) by its width at a point exactly halfway between the top of the peak and the baseline. Peak areas can also be measured by making a photocopy of the chromatogram, accurately cutting out each peak with a sharp knife or razor blade (using a straight-edge to cut across the baseline), and weighing the peaks to the nearest milligram or (preferably) tenth of a milligram. Some recorders are equipped with an integrator that draws a line as each component is eluted from the column; the vertical distance traveled by the integrator pen is proportional to the area of the component peak. Figure 4 shows the zigzag trace produced by one type of integrator; the number of chart units the pen crosses in all of its up-and-down traverses gives the relative area under the peak. More sophisticated integrators provide a digital readout of relative peak areas.

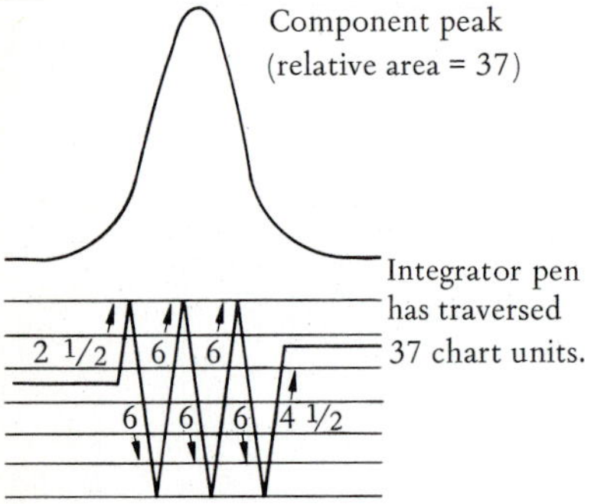

Figure 4.   Use of the integrator

Equipment and supplies:
  gas chromatograph with recorder
  microsyringe

Operational Procedure

Do not touch any of the controls on the instrument except at your instructor's request and under his or her supervision. Consult the instructor if the instrument does not seem to be working properly or if you have questions about its operation. It will be assumed that all instrumental parameters have been preset, that an appropriate column has been installed, and that the column oven will be operated isothermally. If not, the instructor will show you what to do.

If the sample is a volatile solid, dissolve it in the minimum volume of a suitable low-boiling solvent; otherwise, use the undiluted liquid. Rinse a microsyringe with the sample a few times; then partially fill it with the sample. If there are air bubbles inside the syringe, eject the sample and refill the syringe more slowly (also, tapping the barrel with the needle pointing up may help to expel bubbles). Hold the syringe with the needle up, and expel excess liquid until the desired volume of sample (usually $1-2$ μL) is left inside. Wipe the needle dry with a tissue, and pull the plunger back a centimeter or so to prevent prevaporization of the sample.

Set the chart speed, if necessary, and switch on the recorder-chart drive (and the integrator, if there is one). Carefully insert the syringe needle into the injector port [**hot!**] by holding the needle with its tip at the center of the rubber septum and pushing the barrel slowly (but firmly) with the other hand until the needle is all the way in. *Be careful not to bend the needle!* Inject the sample by *gently* pushing the plunger all the way in just as the recorder pen crosses a chart line (the starting line); use as little force as possible, to avoid bending the plunger. Withdraw the needle and mark the starting line, from which all retention times will be measured. Let the recorder run until all the anticipated component peaks have appeared; then turn off the chart drive, and tear off the chart paper using a straight-edge. Inspect the chromatogram carefully; if it is unsuitable because the significant peaks are too small, poorly resolved, or off-scale, repeat the analysis after taking measures to remedy the problem. If there is evidence of prevaporization (as indicated by an extraneous small peak preceding each major peak at a fixed interval), be sure the plunger is pulled back before you inject the sample for the next run.

Before you analyze a different sample, clean the syringe and rinse it thoroughly with that sample. When you are finished, rinse the syringe with an appropriate low-boiling solvent such as methanol or methylene chloride, remove the plunger, and set it on a clean surface to dry, or dry it using an apparatus such as the one shown in Figure 5 of OP–6.

A microsyringe is a very delicate instrument, so it is essential that you handle it carefully and avoid using excessive force that might bend the needle or plunger. See OP–6 for further information about the use and cleaning of syringes.

Injecting the sample immediately after you insert the needle in the injector port will also prevent prevaporization, but this technique requires good timing.

Summary

1. Rinse syringe with sample.

2. Fill syringe with desired volume of sample.

3. Start chart drive and integrator (if applicable).

4. Insert needle into injector port, and inject sample when pen crosses a chart line.

5. Withdraw needle; mark starting line.

6. Stop chart drive when chromatogram is complete.

7. Remove chromatogram, and clean and dry syringe.

8. Measure peak areas and retention times.

Infrared Spectrometry

Figure 1. "Ball-and-spring" model of a chemical bond

The frequency of infrared radiation is usually reported as a wavenumber ($\bar{v}$), the number of peak-to-peak waves per centimeter. The traditional unit of measurement is reciprocal centimeters, cm^{-1}. The relationship between wavenumber and frequency in hertz (v) is given by:

$$v = c \cdot \bar{v}$$

where c is the speed of light (about 3.0×10^{10} cm/sec).

The correlation chart on the back endpaper gives the approximate positions of commonly encountered infrared absorption bands. See Part III for information on the use of IR for qualitative analysis.

Principles and Applications

The atoms of a molecule behave as if they were connected by flexible springs, rather than by rigid bonds resembling the connectors of a ball-and-stick model. A molecule's component parts can oscillate in different *vibrational modes,* designated by such terms as rocking, scissoring, twisting, wagging, and symmetrical and asymmetrical stretching. When infrared radiation is passed through a sample of a given compound, its molecules can absorb radiation of the energy (and frequency) needed to bring about transitions between vibrational ground states and vibrational excited states. For example, a C—H bond that vibrates 90 trillion times a second must absorb infrared radiation of *just* that frequency (9.0×10^{13} Hz, or 3000 cm^{-1}) to jump to its first vibrational excited state. This absorption of energy at various frequencies can be detected by an *infrared spectrophotometer,* which plots the amount of infrared radiation transmitted through the sample as a function of the frequency (or wavelength) of the radiation. Because vibrational transitions are usually accompanied by rotational transitions in the region scanned by a typical infrared spectrophotometer (about 4000–600 cm^{-1} or 2.5–17 μm), an infrared spectrum consists of comparatively broad *absorption bands* rather than sharp peaks such as those seen in NMR spectra. The bands are also usually "inverted"; that is, strong absorption is represented by a deep valley rather than a peak.

Infrared spectrometry is extremely useful in qualitative analysis. It can be used both to detect the presence of specific functional groups and other structural features from band positions and intensities and to establish the identity of an unknown compound with a known standard. The *fingerprint region* of an infrared spectrum (1250–670 cm^{-1}, 8–15 μm) is best for showing that two substances are identical, since the distinctive patterns found in this region are usually characteristic of the whole molecule and not of isolated groups. Infrared spectra can also be used in establishing the purity of compounds, monitoring reaction rates, measuring the concentrations of solutions, determining the structures of complex molecules, and carrying out theoretical studies of hydrogen bonding and other phenomena.

Infrared spectra cannot be used to distinguish between enantiomers, since vibrational frequency is not a function of left- or right-handedness.

Instrumentation

Most infrared spectrometers used for routine applications are optical-null double-beam instruments, whose operation can be understood by referring to the schematic diagram in Figure 2. The sample cell contains the substance being analyzed. If this substance is in solution, the reference cell is filled with the solvent; otherwise, it is empty. Radiation from the infrared source (usually a heated rod or filament) is directed through both cells by mirrors, and the *sample beam* and *reference beam* are

A spectrometer is an instrument that measures and records the components of a spectrum in order of wavelength, mass, or some other property.

Figure 2. Schematic diagram of double-beam infrared spectrometer

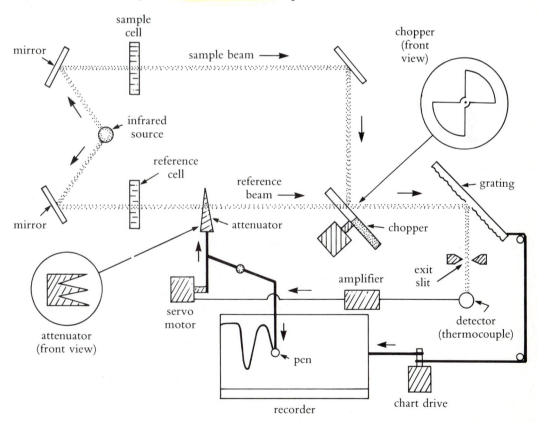

Note: Dots represent the light path, and heavy black lines represent mechanical linkages.

both sent to a rotating *chopper* with a mirrored surface, which alternately blocks one beam while passing the other. The resulting radiation is a rapidly alternating composite beam, consisting of the sample beam one instant and the reference beam the next. This combination beam is then reflected from a *diffraction grating* that breaks it up into its component wavelengths, allowing only monochromatic radiation (actually a narrow band of wavelengths) to pass through the *exit slit*. When this monochromatic beam impinges on the *detector,* an amplified and processed electrical signal is sent to an alternating-current *servo motor,* whose function is to maintain an "optical-null" status by equalizing the intensities of the sample and reference components of the beam. When the sample and reference beam intensities are equal, the resulting electrical signal is a direct current that will not drive the motor. When the sample beam is less intense than the reference beam, an alternating current is sent to the servo motor, which pushes an *attenuator* into the reference beam and simultaneously moves the recorder pen down the chart. Attenuating the reference beam reduces its intensity and brings it back into balance with the sample beam, and the pen movement records the absorption of radiation. Meanwhile the chart motor is rotating the diffraction grating as it moves the chart along, so that the wavelength of the combination sample and reference beam is constantly changing. When at some point the sample beam becomes more intense than the attenuated reference beam, a reversed-phase alternating current is generated that drives the servo motor in the opposite direction, moving the attenuator out of the reference beam and the pen back up the chart. The distance of the pen from the top of the chart at any wavelength is proportional to the amount of light absorbed by the sample molecules at that wavelength. Thus, the resulting spectrum is a plot of the intensity of the transmitted radiation (recorded as percent transmittance) versus the wavelength (or wavenumber) of that radiation.

Some modern infrared instruments analyze the sample beam at all wavelengths simultaneously rather than one wavelength at a time. With a sophisticated computer and a mathematical technique called Fourier transform analysis, the signal from such instruments can be "decoded" to generate a complete infrared spectrum. A Fourier transform infrared (FTIR) spectrometer can record a complete spectrum in a fraction of a second, so a large number of spectra of the same sample can be run in a few seconds. The data from many spectra are then averaged to yield a composite spectrum with a much improved signal-to-noise ratio. FTIR instruments also offer better resolution and more reproducible frequency measurements than scanning infrared instruments, but at a considerably higher cost.

Experimental Considerations

Frequency Alignment. A typical infrared spectrometer may have the chart paper wrapped around a drum or laid flat on a recorder bed (see Figure 3). In either case, the paper must be carefully aligned so that the absorption bands appear at the right place; otherwise, the spectrum may be interpreted incorrectly. The chart paper is usually aligned by matching a given chart frequency (for instance, 4000 cm⁻¹) with an alignment mark on the recorder bed or drum. If precise values of the band frequencies are required (or if the instrument requires wavelength calibration), a calibration spectrum can be run using a standard thin film of polystyrene. The tips of a few polystyrene peaks (usually including those at 2850, 1603, and 906 cm⁻¹) are recorded directly over the sample spectrum and used to calculate correction factors for the absorption bands of interest.

Setting the Baseline. For maximum resolution, the "baseline" of an infrared spectrum should stay close to the top of the chart paper, but should not go off-scale (above 100%*T*). This can

An infrared spectrum can be linear in either wavenumber (cm⁻¹) or wavelength (μm), depending on the instrument used. These units can be interconverted by the equation

$$\lambda \ (\text{in } \mu\text{m}) = \frac{10,000}{\bar{v}}$$

$\%T$ = percent transmittance
  = $100 \times T$ (transmittance)

Figure 3. Perkin-Elmer Model 710b infrared spectrometer

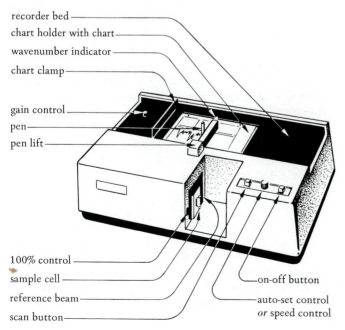

recorder bed
chart holder with chart
wavenumber indicator
chart clamp

gain control
pen
pen lift

100% control
sample cell
reference beam
scan button

on-off button

auto-set control
or speed control

usually be accomplished by setting the manual attenuator (100% control) so that the pen is at 85–90%T with the sample cell in place and the recorder at its starting position (about 4000 cm^{-1}). It may occasionally be necessary to run a preliminary scan over the spectrum (manually, if the instrument allows this) and set the 100% control so that the highest part of the baseline is just below 100%T.

It is important not to run over 100%—this can damage the instrument.

Spectrum Intensity. An infrared spectrum should not, obviously, have absorption bands so weak that they are hard to identify or so strong that they crowd the 0% T line and are not well resolved. Ideally, the spectrum should be recorded so that the strongest absorption band has a maximum transmittance of 5–10%. This may be difficult to accomplish with precision, but with the right kind of cell and careful sample preparation, you should be able to obtain a good spectrum. If your first spectrum is not acceptable, try varying the following parameters, depending on the sample preparation method:

Sometimes a satisfactory infrared spectrum is obtained only by trial and error.

1. *Neat liquid:* vary cell path length or film thickness.

2. *Solution:* vary concentration or cell path length.

3. *KBr disc:* vary amount of sample or thickness of disc.

4. *Mull:* vary amount of sample or film thickness.

Precautions. Infrared spectrophotometers are comparatively sensitive instruments that do not respond well to rough handling, so certain precautions should be observed in their use. *Never* move the chart holder, drum, or pen carriage while the instrument is in operation or before it has been properly reset at the end of a run—this can damage mechanical components. *Never* try to move the pen manually. Do not leave objects lying on the bed of a recorder—they might jam the chart drive. Do not unplug the instrument—some infrared spectrophotometers contain heaters to keep certain components at thermal equilibrium even when the instrument is not in operation. Do not use water or aqueous solutions near a spectrophotometer, since some of its components may be constructed of water-soluble alkali metal salts—a spill could be disastrous! Do not insert objects into the sample or reference beam openings where they can damage the attenuators and other fragile components. An infrared spectrophotometer is a precision instrument that can easily cost as much as a BMW (or at least a Subaru). You must treat it with respect and follow instructions carefully to prevent damage and avoid the need for costly repairs.

Sample Preparation

Infrared spectra of liquids can be run with the liquid either undiluted or in solution; spectra of solids can be run using a solution of the solid, a mull, or a pressed disc. The following outline should serve as a general guide to sample preparation.

Care of Infrared Cells. Most sample cells contain metal halide windows (sodium chloride, silver chloride, etc.), which are either very fragile or water-soluble, or both. A window should be touched only on the edges with clean, dry hands or gloves and handled with great care to avoid damage; sodium chloride windows must not be exposed to moisture. Therefore, samples and solvents run in sodium chloride cells must be dry, and the cells should be kept in an oven or desiccator when not in use. When assembling demountable cells, the cinch nuts must not be overtightened, as this can fracture the windows.

Thin Films. To prepare a thin film of an undiluted ("neat") liquid using a *demountable cell,* disassemble the cell carefully according to the instructor's directions, and place 1 – 2 drops of the liquid on the lower window. Position the upper window by touching an edge to the corresponding edge of the lower window and carefully lowering it into place. Press the plates

Since sample cells and other sampling accessories vary widely in construction and application, most sampling techniques should be demonstrated by the instructor.

Even breathing on a NaCl window can cause some etching because of moisture in your breath.

Thin films cannot be used for some low-boiling liquids since they may evaporate before the spectrum is complete.

Do not use a spacer for thin films.

Figure 4.   Demountable cell

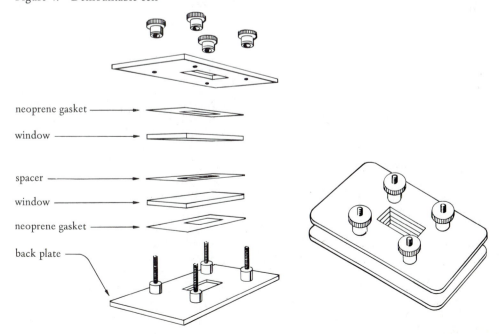

neoprene gasket

window

spacer

window

neoprene gasket

back plate

together so that the liquid fills the space between them, taking care to exclude air bubbles. Assemble the cell as directed, and run the spectrum. When you have finished, disassemble the cell and rinse the windows with a *dry* volatile solvent (chloroform, methylene chloride, and NaCl-saturated 100% ethanol are often used). Let the solvent evaporate, and return the cells to a desiccator or oven.

Volatile Neat Liquids.   If a liquid is comparatively volatile, its spectrum can be run in a cell with a spacer approximately 0.015 – 0.030 mm thick. A demountable cell (Figure 4) is filled by disassembling the cell, placing the spacer on the lower window, adding sufficient liquid to fill the cavity in the spacer, positioning the upper window, and reassembling the cell. After the spectrum has been run, the cell should be disassembled and cleaned as described above.

A *sealed cell* or *sealed demountable cell* (Figure 5) is filled by injecting the sample into an injection (filling) port with a Luer-lock syringe body. **[*Be careful:* The syringe tips are fragile and break easily.]** Draw about 0.5 mL of the liquid into the syringe, and carefully insert the tip into one of the ports with a half twist (remove the plugs first!). Holding the cell upright with the syringe port at the bottom, depress the plunger until the space between the windows is filled and a

> A sealed demountable cell with a thin spacer works well for most neat liquids, whether volatile or not.

Figure 5.   Sealed demountable cell

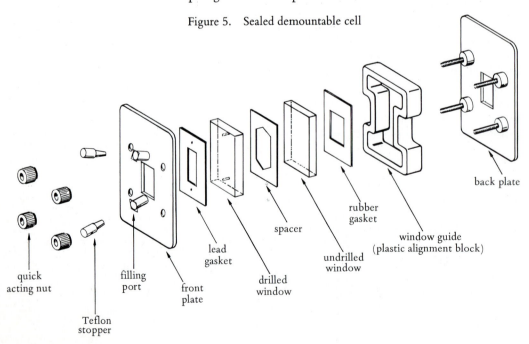

quick acting nut

Teflon stopper

filling port

front plate

lead gasket

drilled window

spacer

undrilled window

rubber gasket

window guide (plastic alignment block)

back plate

little liquid appears at the upper port. Put the cell on a flat surface, remove the syringe, and insert a plug in the upper port with a slight twist. Remove any excess solvent in the lower port with a piece of tissue paper or cotton, and close that port with another plug.

After the spectrum has been run, clean the cell by removing most of the liquid with a syringe and flushing the cell several times with a volatile solvent. A convenient way of flushing a cell is to lay it on a flat surface, insert a syringe filled with solvent in one port and an empty syringe in the other port, and slowly push on the one plunger while pulling on the other (see Figure 6). This will draw washing liquid through the cell from one syringe to the other. Do this several times; then remove the excess liquid with an empty syringe, and dry the cell by passing dry, clean air or nitrogen between the windows. An ear syringe or a special cell-drying syringe can be used to force air (gently!) through the cell, or it can be dried by attaching a trap and aspirator to one port and a drying tube filled with desiccant ⟨OP–22a⟩ to the other.

Solutions. Solutions of liquids or solids in a suitable solvent can be analyzed in a sealed or sealed demountable cell that has a spacer 0.1 mm or more in thickness. It is best to use specially purified spectral-grade solvents, so that impurities will not give rise to extraneous peaks. Solution spectra must always be run with a reference cell in the reference beam of the spectrophotometer so as to subtract out absorption due to the solvent. Even then, strong solvent peaks will obscure certain portions of the spectrum. For example, the spectrum of a solute run in chloroform will yield no useful information in the 1250–1200 cm^{-1} and 800–650 cm^{-1} regions because the chloroform absorbs nearly all of the infrared radiation there. Sometimes a sample can be run separately in two solvents to yield a complete spectrum. Chloroform and carbon disulfide constitute such a pair; the first yields a good spectrum above 1300 cm^{-1}, and the second is virtually transparent below that frequency. Prepare the sample by making up a 5–10% solution of the substance to be analyzed and filling the sample cell using a Luer-lock syringe body, as described above for volatile neat liquids. Fill an identical cell (with a spacer of the same thickness) with pure solvent, and place it in the reference beam when the spectrum is run. Clean the sample cell by removing the excess solution and flushing it (as described above) with the pure solvent used in preparing the solution. If necessary, rinse the sample cell with a more volatile solvent before drying.

If there is much resistance to filling, use the push-pull technique described below. If there are air bubbles between the windows, they can sometimes be removed by tapping gently on the metal frame of the cell.

Twist the syringes slightly as you insert them so they will not pull out too easily.

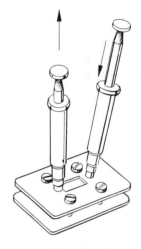

Figure 6. Push-pull technique for flushing a sealed or sealed demountable cell

The solvent should be relatively nonpolar and must not react with the solute. It should ordinarily dissolve enough of the solute to yield a 5–10% solution; however, more dilute solutions can be used with cells having long path lengths.

Usually 0.1–0.5 mL of solution will be required. A 0.1-mm spacer can be used unless the solution is quite dilute.

It is a good idea to run a preliminary spectrum using pure solvent in both cells, to see that the cells are clean and matched.

The particles must be ground to an average diameter of about 1 μm or less to avoid excessive radiation loss by scattering.

The effect of scattering can be partly compensated for by using an attenuator accessory in the reference beam.

Nujol is essentially transparent at frequencies lower than 1300 cm⁻¹; Fluorlube is transparent above 1300 cm⁻¹.

Clean the cell windows with petroleum ether or another suitable solvent after use.

Mulls.   A solid sample can be prepared as a *mull* in a suitable oil for infrared analysis. Grind about 10 – 20 mg of the solid in an agate or mullite mortar until it coats the entire inner surface of the mortar and has a glassy appearance (about 5 – 10 minutes of grinding will be required). Add a drop or two of mulling oil (usually a mineral-oil preparation called Nujol); then grind the mixture until it has about the consistency of Vaseline. Transfer most of the mixture to the lower window of a demountable cell using a "rubber policeman" or a plastic-coated spatula. Spread the mull evenly with the top window, taking care to exclude air bubbles. Assemble the cell, run the spectrum, and clean the windows as for a neat liquid.

Some unavoidable *scattering* of the infrared radiation may reduce the transmittance at the high-frequency end of the spectrum. For this reason, you should rapidly scan the spectrum with the pen raised to locate the frequency at which the transmittance is highest, and set the baseline just below 100%T at that point. If after this adjustment the transmittance at 4000 cm⁻¹ is 50% or lower, the sample is too coarsely ground; a new mull should be prepared. The sample spectrum should be compared with a spectrum of the mulling agent so that peaks due to the oil can be identified and disregarded during interpretation. When Nujol is used, the aliphatic C—H stretching and bending regions (3000 – 2850, 1470, 1380 cm⁻¹) cannot be interpreted, but most functional groups and other structural features can be identified. If it is necessary to examine the entire spectrum, another sample can be prepared using a complementary mulling agent such as Fluorlube.

KBr Discs.   A potassium bromide *disc* is prepared by mixing a solid with *dry* spectral-grade KBr and using a die to press the mixture into a more or less transparent wafer. If a KBr Mini-Press (Figure 7) is used, grind 0.5 – 2 mg of the solid very finely in a dry agate or mullite mortar as if you were preparing a mull; then add about 100 mg of the special potassium bromide, and mix it thoroughly with the sample. Screw in the bottom bolt of the Mini-Press five full turns and introduce about half of the sample mixture into the barrel. Keeping the open end of the barrel pointed up, tap it gently against the bench top to level the mixture, brush down any material on the threads with a camel's-hair brush, and screw in the top bolt. Alternately tap the bottom bolt on the bench top and screw in the top bolt with your fingers to level the sample further. When the bolt is "finger tight," clamp the bottom bolt in a vise (or other

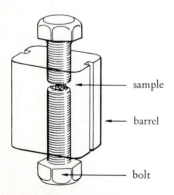

Figure 7.   KBr Mini-Press

holder) and gradually tighten the top bolt as far as you can with a heavy wrench (or to 20 ft-lbs with a torque wrench). Leave the die under pressure for a minute or two. Remove both bolts, leaving the KBr disc in the center of the barrel, and check to see that it is reasonably transparent; if not, make up a new disc. Place the block containing the disc on a holder (provided with the Mini-Press) in the spectrophotometer sample beam, orient it for maximum transmittance, set the baseline by the method described above for a mull, and run the spectrum. Punch out the pellet with a pencil or other suitable instrument. Wash the barrel and bolts with water; then rinse them with acetone or methanol, and store the clean, dry press in a desiccator.

    If another kind of press is to be used, follow the manufacturer's or your instructor's directions for preparing the disc.

Melts. Infrared spectra of low-melting solids can sometimes be obtained by spreading a thin, uniform layer of the finely powdered solid on a silver chloride window, covering it with another silver chloride window, and heating the assembly *slowly* on a hot plate or the hot stage of a Fisher-Johns melting-point apparatus. As soon as the solid melts to form a uniform film between the plates, the windows should be removed and pressed together with forceps until solidification occurs. The windows are then installed in a cell, and the spectrum is run as for a thin film. This method works well only when the solid crystallizes as a glassy film or very small crystals; larger crystals produce excessive light scattering.

Operational Procedure

The construction and operation of commercial infrared spectrophotometers vary widely, so the following is meant only as a general guide to assist you in recording an infrared spectrum. Specific operating techniques must be learned from the instructor, the operating manual, or both.

    Do not touch any controls other than those specified in the following procedure except at the instructor's request and under his or her supervision. It will be assumed that the gain, slit width, and balance controls have been adjusted beforehand; if not, your instructor will show you how to make the necessary adjustments.

    Prepare a sample of the substance to be analyzed by one of the methods described above and place the cell or KBr disc assembly in the sample beam of the instrument. If the sample is being run in a solvent, place a reference cell containing the

---

KBr adsorbs water readily and should be dried before use by heating it at 105–110°C for several hours. It should be kept in a tightly capped container and stored in an oven or desiccator.

The disc may be cloudy or inhomogeneous for any of the following reasons:

  (1) the KBr or sample may be wet;

  (2) the two components may not be completely mixed;

  (3) the pellet may be too thick;

  (4) the sample may be low melting;

  (5) the sample size may be too large; or

  (6) the die may not have been tightened enough.

See *J. Chem. Educ. 54,* 287 (1977) for additional suggestions regarding the preparation of KBr discs.

Equipment and supplies:

  General: infrared spectrophotometer, desiccator or oven, sample, cells and other accessories, chart paper

  Neat liquids: Demountable or sealed-demountable cell (with or without thin spacer), washing solvent

  Solutions: Two sealed or sealed demountable cells with spacers, two Luer-lock syringes, washing solvent(s), ear syringe (or other cell-drying apparatus)

  Mulls: Agate or mullite mortar and pestle, mulling oil, demountable or sealed demountable cell with no spacer, washing solvent

  KBR discs: Dried spectral-grade potassium bromide, agate or mullite mortar and pestle, KBr press assembly, washing and drying solvents

  Melts: AgCl windows and cell assembly, hot plate or Fisher-Johns melting-point apparatus

solvent in the reference beam. An attenuator accessory can be placed in the reference beam if a mull or KBr disc is being run; with neat liquids and melts, that compartment remains empty. If necessary, place chart paper on the recorder bed (or wrap it around a drum); then align the paper properly and move the drum or carriage to the starting position. *(Important: Make sure the instrument has been reset correctly before attempting to move the drum or carriage.)* Set the 100% control and any other necessary controls, lower the pen onto the chart paper, and scan the spectrum at "normal" or "fast" speed. Examine the spectrum to see that the absorption bands show satisfactory intensity and resolution. If the spectrum is not satisfactory, prepare another sample (varying concentration, path length, solvent, etc.) or consult the instructor to see if the instrument requires adjustment. Then, if desired, run another spectrum at a slower speed.

If the spectrum is to be calibrated, replace the sample with a polystyrene film, reset the instrument, move the bed or drum to the appropriate frequencies, and superimpose the polystyrene peaks over the spectrum.

Remove the sample, reset the instrument (if necessary), and tear off the chart paper. Write all appropriate data on the chart paper, including the following (as applicable): identity and source of the sample, sampling method used, spacer thickness, concentration and solvent (for a solution spectrum), mulling agent (if one was used), scanning speed, current date, and name of the operator.

For information that will help you to interpret the infrared spectrum, read about infrared spectra in the Spectral Analysis section in Part III. You may wish to consult your lecture textbook and appropriate references in category F of the Bibliography for additional help.

OPERATION 34

Nuclear Magnetic Resonance Spectrometry

Information about the theoretical principles underlying nuclear magnetic resonance (NMR) spectrometry can be found in most textbooks of organic chemistry and in appropriate sources listed in category F of the Bibliography in this book. A description of NMR spectral parameters and their use in structure analysis is provided in the Spectral Analysis section in Part III. The following discussion stresses NMR instrumentation and the practical aspects of recording a proton NMR (PMR) spectrum and includes brief sections on carbon-13 (CMR) spectrometry and shift reagents.

Instrumentation

The schematic diagram in Figure 1 illustrates the major components of a typical NMR spectrometer, which is equipped with a powerful *magnet* to produce a continuous magnetic field around the sample, a *transmitter coil* to irradiate the sample with radio-frequency (RF) radiation, and a *receiver coil* to pick up signals from the sample. Such an instrument will have several other coils to stabilize the magnetic field and keep it homogeneous throughout the sample, as well as output devices to process the signals and record a spectrum. Most NMR spectrometers used for routine work are "field-sweep" instruments, in which the intensity of the magnetic field is varied by *sweep coils* and the RF frequency is kept constant; instruments that vary the RF frequency and keep the magnetic field constant are also available.

Before an NMR spectrum can be recorded, the substance being analyzed must first be dissolved in a suitable solvent and transferred to a special thin-walled sample tube; a little tetramethylsilane (TMS) is usually added to produce a reference signal. The sample tube is then inserted into a *sample probe* between the poles of a magnet, which produces a magnetic field whose strength is represented by $\mathbf{H_0}$. A rapidly alternating current is generated by an oscillator in the RF transmitter and sent through the *transmitter coil,* producing an oscillating electromagnetic field of intensity $\mathbf{H_1}$ and frequency v (usually 30 – 360 MHz) at the sample probe. At certain values of $\mathbf{H_0}$ (or of v, if the frequency is varied), magnetic nuclei in the sample

Figure 1. Schematic diagram of a cross-coil NMR spectrometer

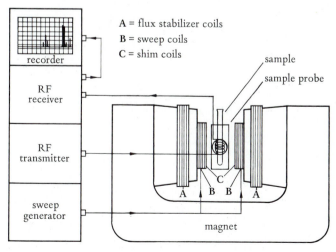

can interact with this electromagnetic radiation and "flip over" — going from a spin state in which their magnetic moments are aligned with the external magnetic field to a higher spin state in which their magnetic moments are aligned against the external magnetic field. As they reverse their spins, these nuclei generate an electromagnetic signal that is picked up by the receiver coil and sent to an RF receiver, where it is amplified and then used to produce a peak or set of peaks (called a *signal*) at the recorder. Because of the different kinds of environments existing in a molecule, not all nuclei of the same element will undergo transitions at the same value of H_0 (or of v), so a typical NMR spectrum will consist of a series of signals generated by atoms from different molecular environments.

The oscillating electromagnetic field H_1 can be left on continuously while a spectrum is being recorded, as it is in a continuous-wave (CW) NMR spectrometer, or applied in a series of short intense pulses, as it is in a Fourier transform (FT) NMR spectrometer. Fourier transform instruments are used for recording virtually all carbon-13 spectra (see below) and for proton spectra when high sensitivity is required. Most of the NMR spectrometers available for use by undergraduate students are continuous-wave instruments that operate at an RF frequency of 30–60 MHz and a magnetic field strength (H_0) of 7000–14,000 gauss. In a typical run using a 60-MHz spectrometer, the magnetic field is swept over a very narrow range of about 0.14 gauss, which is only 10 millionths of the external field strength (10 parts per million). This sweep range can be extended to 0.21 gauss (15 ppm) or so, if necessary, to detect protons with chemical shifts higher than 10 ppm. On an NMR spectrum, the sweep range is usually calibrated in equivalent frequency units (Hz) and in parts per million rather than gauss. A range of 0.14 gauss is equivalent to 600 Hz on a 60-MHz instrument.

See the Spectral Analysis section in Part III for a description of the chemical shift and other parameters that characterize an NMR spectrum.

Carbon-13 NMR Spectrometry

Isotopes whose nuclei are not magnetic do not produce NMR spectra.

The nuclei of carbon-12 atoms, which make up about 99% of the carbon in organic molecules, are not magnetic because they have a net spin of zero. A carbon-13 nucleus, like a hydrogen nucleus, has a spin of $\frac{1}{2}$ and thus a magnetic moment. Because of the scarcity of carbon-13 in nature (about 1.1% of carbon atoms are carbon-13), the ^{13}C signals recorded by a continuous-wave spectrometer are so weak that they can hardly be distinguished from the electronic noise associated with the spectrum. For that reason, carbon-13 NMR (CMR) spectra are now obtained using Fourier transform spectrometers to improve the signal-to-noise ratio. In such instruments, the sample is subjected to very short pulses ($1-10\,\mu s$) of RF radiation at the resonant frequency of carbon-13 nuclei. The sample responds to each pulse by generating a *free-induction decay (FID) signal* that contains all the information needed to reconstruct an entire CMR spectrum. By applying a series of short pulses (one every second or so) to the same sample, hundreds or even thousands of FID signals can be accumulated and averaged in a relatively short time. Because electronic noise is random and the ^{13}C signals are coherent, averaging the FID signals causes the noise to level out and the signals to be enhanced. The averaged signal is then decoded by a computer to produce a low-noise CMR spectrum.

As ^{13}C nuclei in the higher spin state return to the ground state, they generate a fluctuating electromagnetic signal (the FID signal) whose magnitude quickly decays to zero.

A CMR spectrum is usually much simpler and easier to interpret than a PMR spectrum of the same compound. Carbon-carbon splitting is unimportant because carbon-13 nuclei cannot couple with nonmagnetic carbon-12 nuclei, and there is only a slight chance that two carbon-13 atoms will be next to one another in the same molecule. Carbon-13 nuclei *can* couple strongly with hydrogen nuclei, and since such coupling results in very complex spectra, it is often prevented by a *spin-decoupling technique.* If the sample is subjected to continuous RF radiation tuned to the resonant frequencies of its protons, all the protons attached to a given carbon atom will flip over (change spin states) so rapidly that their coupling effect on the carbon will average out, and each ^{13}C signal will be a single peak rather than being split into a multiplet. Thus, a spin-decoupled CMR spectrum consists of a series of sharp singlets, one for each set of magnetically equivalent carbon atoms. Unlike a proton signal, the size of a ^{13}C signal is not directly proportional to the number of carbon atoms responsible for the signal, so CMR spectra are not integrated. The chemical shift of a carbon-13 atom depends on its hybridization and molecular environment; as in PMR spectrometry, electron-withdrawing groups cause downfield chemical shifts. Chemical shifts for carbon-13 atoms vary over a range of about 250 ppm, so the peaks on a CMR spectrum are usually well separated. CMR spectra can be very useful because they give information about the carbon "backbone" of a molecule that may not be available from a PMR spectrum.

The improvement in the signal-to-noise ratio is proportional to $\sqrt{n}$, where n is the number of accumulated FID signals.

In the absence of electronegative atoms, the range of chemical shifts for sp^3, sp^2, and sp carbons is about $0-60$ ppm, $100-160$ ppm, and $65-105$ ppm, respectively.

Chemical-Shift Reagents

The amount of structural information that can be obtained from a conventional NMR spectrum is often limited by the presence of overlapping signals. The distance (measured in

Hz) separating any two NMR signals increases in direct proportion to the magnetic field strength; therefore, powerful superconducting magnets can be used to spread out the signals in an NMR spectrum enough to reduce or eliminate overlapping. A much less expensive way to increase signal separation is to use a *chemical-shift reagent.* Organometallic complexes of certain paramagnetic rare earth metals can coordinate with the functional groups of alcohols, amines, carbonyl compounds, and other polar organic compounds. The local magnetic field produced by the metal atom shifts the signals of nearby protons to an extent that varies with distance; the closer a proton is to the metal atom, the more its chemical shift will change. The chemical-shift reagent called *tris*(dipivaloylmethanato)europium(III), whose formula is abbreviated as $Eu(dpm)_3$, causes downfield chemical shifts; the corresponding complex of praseodymium, $Pr(dpm)_3$, induces upfield shifts. Adding one of these chemical-shift reagents to a sample often separates its overlapping signals enough that they can be readily interpreted.

Experimental Considerations

Sample Preparation. The sample is usually analyzed in solution in a specially fabricated NMR tube, which is placed in a sample probe containing the transmitter and receiver coils. A typical NMR tube is 4–5 mm in outer diameter and 6–8 in. long. For routine NMR analysis, samples can be made up by *(1)* dissolving 20–50 mg of a compound in 0.4–0.5 mL of a suitable solvent, *(2)* filtering the solution directly into the NMR tube through a Pasteur (capillary) pipet containing a small plug of glass wool, *(3)* adding 5–15 μL of a reference standard such as TMS, *(4)* stoppering the tube carefully (NMR tubes are fragile and expensive), and *(5)* inverting the tube several times to mix the components thoroughly.

NMR Solvents. An ideal NMR solvent should have no interfering protons, a low viscosity, a high solvent strength, and no appreciable interactions with the solute. Chloroform (in deuterated form) is the most widely used NMR solvent because its polarity is low enough to prevent significant solute-solvent interactions, and most organic compounds are sufficiently soluble in it for NMR analysis. When a more polar solvent is required, dimethyl sulfoxide is often used, sometimes in mixtures with chloroform. Protic solvents can be used for NMR spectrometry when their proton signals do not interfere with

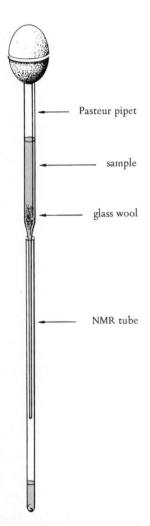

Pasteur pipet

sample

glass wool

NMR tube

Figure 2.   Filtering an NMR solution

those of the solute, but more often the corresponding deuterated solvents are used. Thus, deuterochloroform ($CDCl_3$) is preferred to protic chloroform and DMSO-d_6 (CD_3SOCD_3) to protic dimethyl sulfoxide. Since deuterium (^2H) resonates at about $6\frac{1}{2}$ times the field strength required for ^1H, deuterated solvents do not interfere significantly with proton NMR spectra; unfortunately, they are quite expensive.

Table 1 compares the properties of some commonly used solvents and gives the approximate positions of their proton NMR signals. It is often convenient to add 1–3% TMS to the bulk solvent so that it does not have to be added during sample preparation. Otherwise, the TMS should be kept in a refrigerator and added with a *cold* syringe or fine-tipped dropper.

Deuterated forms of the solvents listed in Table 1 usually have ^1H peaks about 0.5–2% as strong as their nondeuterated counterparts, due to isotopic impurity.

Commercial deuterated solvents often have 1–3% TMS added.

TMS is highly volatile, so both the liquid and the syringe should be refrigerated.

Sample Spinning. To average out the effect of magnetic-field variations in the plane perpendicular to the axis of the NMR tube, the sample is rotated at a rate of 30–60 revolutions per second while the NMR spectrum is being recorded. It is important to set the spinning rate correctly—excessively high rates can cause a vortex extending into the region of the receiver coil, and low rates can cause large *spinning sidebands* or signal distortion. Spinning sidebands are small peaks on either side of a main peak at a distance equal to the spinning rate from

Table 1. Properties of some NMR solvents

Solvent	^1H Chemical Shift(s) (δ, ppm)	Solvent Strength	Freedom from Solute Interactions	Viscosity
carbon disulfide	none	good	good	low
cyclohexane	1.4	poor	good	medium
acetonitrile	2.0	good	fair	low
acetone	2.1	good	poor	low
dimethyl sulfoxide	2.5	very good	poor	high
dioxane	3.5	good	fair	medium
water	~5.2 (v)	good	poor	medium
chloroform	7.3	very good	fair	low
pyridine	7.0–8.7	good	poor	medium
trifluoroacetic acid	~12.5 (v)	good	poor	medium

Note: Most solvents are used in their deuterated forms. Solvent strength refers to the ability to dissolve a broad spectrum of organic compounds; v = variable.

the main peak. An NMR tube spun at 30 cycles per second gives rise to sidebands 30 Hz from each main peak. To find out whether small signals are spinning sidebands or impurity peaks, change the spinning rate and scan again to see if their positions change.

Field Homogeneity.   The most important homogeneity control is the one used to produce a uniform field along the axis of the sample tube. This is usually called the *Y control. Never touch any homogeneity control but the Y or Fine Y control without permission; any other alterations may make a lengthy retuning operation necessary.* For routine work on a previously tuned instrument, the Y control can often be set by placing a blank sheet of paper on the instrument's platen and repeatedly scanning a strong peak in the spectrum of the sample, each time making small adjustments in the Y control until the peak is as tall and narrow as possible and shows a good "ringing" (beat) pattern. Figure 4 shows an excellent ringing pattern, characterized by the high amplitude, long duration, and exponential decay of the "wiggles" following the main peaks.

Signal Amplitude.   The amplitude of NMR signals can be altered by two controls. One of them, usually called the *spectrum amplitude control,* changes the amplitude of the signals *and* the baseline "noise." It is usually set so that the strongest peak in the spectrum extends nearly to the top of the chart paper. The *RF power control* can increase the signal height without increasing baseline noise, up to the point where *saturation* sets in. Increasing RF power beyond that point eventually reduces the height and distorts the shape of a signal. Unless high sensitivity is required, the RF level is usually set at about mid-range or at some other value where there is little likelihood of saturation. Since saturation is a function of sweep time and sweep width as well as of RF power, saturation at high power can sometimes be avoided by using a short sweep time and a wide sweep range.

Sweep Controls.   There are four important sweep controls on a typical continuous-wave NMR spectrometer; these control the reference point of the spectrum, the sweep rate, and the portion of the spectrum to be scanned. The *sweep zero control* is used to set the signal for the reference compound to the proper value (0.000 for TMS). The *sweep width (sweep range) dial* sets the total chemical-shift range to be scanned, usually 600–1000 Hz when the entire spectrum is being scanned on a

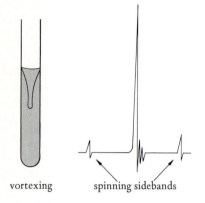

vortexing   spinning sidebands

Figure 3.   Effects of spinning rate

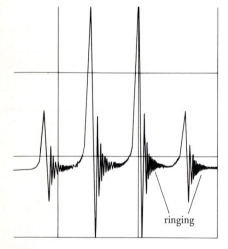

ringing

Figure 4.   Ringing pattern for the quartet of acetaldehyde

Saturation occurs when the excess proton population in the lower nuclear spin state is so depleted that increasing the RF field intensity no longer increases the number of transitions.

60-MHz instrument. Since many continuous-wave spectra are scanned at a rate of about 1 Hz per second, the *sweep time* is often set so that it is numerically equal to the sweep width (for example, 600 seconds for a 600-Hz sweep width). The *sweep offset dial* is used when only a specific portion of the spectrum is to be scanned; it sets the upfield limit of the scan. For example, if a scan between 350 and 500 Hz (about 5.8–8.3 δ on a 60-MHz instrument) is desired, the sweep offset should be 350 Hz and the sweep range 150 Hz.

Other Adjustments. A few additional controls will require adjustment in some cases. The *filter response time* (sometimes called the time constant) can be lengthened to reduce noise; however, too long a filter response time will distort the line shapes. According to a general rule of thumb, the magnitude of the filter response time (in sec) should be no longer than the scan rate (in Hz/sec). For reasonably concentrated samples, the filter response time is often set lower than this. The *phasing control* affects the shape of the baseline before and after a signal. It should be adjusted to obtain a straight baseline. Correct phasing is much more important when an NMR spectrum is being integrated than when it is being recorded.

Integration. If you will be expected to integrate your NMR spectrum, you should consult your instructor for specific instructions; only a brief outline of the steps in the process will be given here.

1. The RF power is optimized to provide an acceptable signal-to-noise ratio.

2. The instrument is switched to the integral mode.

3. The integral amplitude control is set, while scanning the spectrum rapidly, so that the integrator trace spans the vertical axis of the chart.

4. With the sweep offset and sweep width controls set to scan a region free from NMR signals, the balance control is adjusted during a slow scan of that region to give a horizontal line.

5. The phasing control is adjusted while scanning over a signal to make the integrator traces before and after the signal as nearly horizontal as possible.

6. The integral over the entire spectrum is recorded several times (preferably in both directions) using a sweep time

On a 60-MHz instrument, 1000 Hz is about 17 ppm, which encompasses virtually all protons likely to be encountered in routine work. A 600-Hz (10-ppm) range can be used if the sample is known to contain no protons absorbing downfield of 10δ.

The scan rate is obtained by dividing sweep width by sweep time.

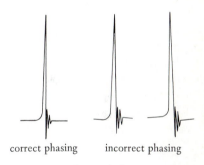

correct phasing   incorrect phasing

Figure 5. Effects of phasing on baseline

Phasing should be adjusted as accurately as possible in the normal mode before starting the integration operation. It should then be readjusted in the integral mode.

When the signals on a spectrum are comparatively far apart, better integrals may be obtained with a nonspinning sample. The field homogeneity controls must be optimized on the stationary sample before the integral is recorded.

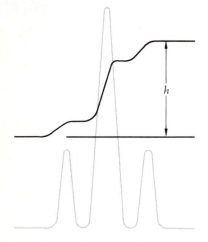

(*h* is proportional to the total area under
the triplet)

Figure 6.   Measuring signal area

Typical settings for a 60-MHz
instrument:
    sweep offset: 0
    sweep width: 600 Hz
    sweep time: 600 sec

*If you obtain no spectrum at this point,
recheck the settings and make sure the
sample is spinning. Then, if necessary, run
a standard sample such as TMS/chloro-
form to make sure the instrument is
working properly.*

Equipment and supplies:
    sample
    NMR solvent
    TMS
    NMR tube
    Pasteur pipet
    glass wool
    tissues
    washing solvent

about one-fifth to one-tenth that for the normal spectrum. The
pen should be returned to the baseline after each scan.

7.   The relative peak areas are determined by measuring
the vertical distances between the integrator traces before and
after each signal, and the results for successive scans are averaged. (See Figure 6.)

Operational Procedure

Do not touch any controls other than the ones specified below, except
at the instructor's request and under his or her supervision. The
following procedure applies to a typical 60-MHz continuous-
wave NMR spectrometer; operating procedures for other instruments may vary considerably.

Fill the NMR tube to a depth of about 4 cm with a 5–20%
(w/v) solution of the compound in a suitable solvent containing 1–3% TMS. Wipe the outside of the tube carefully with a
tissue paper, insert it in the sample *spinneret* using a depth gauge
to adjust its position, wipe it again, and gently place the assembly into the sample probe between the magnet pole faces.
Adjust the air flow to spin the sample at 30–60 Hz. Set the
sweep width, sweep offset, and sweep time controls to scan the
desired range at a rate of about 1 Hz/sec. Set the RF power to
about mid-range and the filter response time to a low value.
Place a blank paper on the recorder bed, set the spectrum
amplitude control to about mid-range, and scan the spectrum.
Readjust the spectrum amplitude to obtain good peak heights.
Optimize peak shape and ringing by repeatedly scanning a
strong, sharp peak (the TMS peak may be suitable) and adjusting the Y control each time. Adjust the phasing, if necessary,
and set the TMS peak to 0 δ with the sweep zero control (the
reference peak should be very sharp and close to its expected
position when the sweep offset is at zero). Record the spectrum
and (if desired) the integral on a single sheet of chart paper.

If the sample is dilute, it may be necessary to optimize the
RF power and filter settings to obtain a suitable signal-to-noise
ratio.

Clean the sample tube immediately with an appropriate
solvent. Usually, the same solvent is used for cleaning as was
used in preparing the sample. (If the NMR solvent was deuterated, use the *protic* form for cleaning!) A Pasteur pipet can be
used to rinse out the tube, or a device such as the one described
in *J. Chem. Educ. 53,* 127 (1976) can be constructed for the
purpose. Invert the tube in a suitable rack, and let it drain dry. If
necessary, it can be dried more rapidly with a Pasteur pipet
attached to an aspirator or to a source of clean, dry air.

Ultraviolet-Visible Spectrometry

Principles and Applications

Spectrophotometers that measure the absorption of radiation in the visible (~400–800 nm) and near ultraviolet (~200–400 nm) regions of the electromagnetic spectrum are useful in both the qualitative and quantitative analysis of certain organic compounds. Radiation in these regions induces molecular transitions in which electrons are promoted from an electronic ground state (a bonding or nonbonding orbital) to one or more excited states (antibonding orbitals). The energy required for electronic transitions is much greater than that needed to induce vibrational or nuclear magnetic transitions, so the wavelength of the radiation used is much shorter: 0.2–0.8 μm compared to about 2.5–50 μm for IR and several meters for NMR. The most important of these molecular transitions involve pi electrons in conjugated systems, so ultraviolet-visible spectrometry is used mainly to analyze aromatic compounds and aliphatic compounds that contain two or more conjugated double bonds.

> 1 nanometer (nm) = 10^{-9} m = 10 Å. The older unit "millimicrons" (mμ) is sometimes used for nm.

An electronic spectrum in either the ultraviolet or visible region is often quite featureless compared to an infrared or NMR spectrum; it may consist of only one or two broad absorption bands. The band structure is caused by rotational and vibrational transitions that accompany each electronic transition. Each different combination of vibrational and rotational transitions has a different energy; collectively, they span a broad range of wavelengths centered about the wavelength of the "pure" electronic transition. An electronic absorption band is characterized by a maximum intensity measured along the y-axis of the spectrum (usually in terms of *absorbance, A*) and a wavelength of maximum intensity, λ_{max}, measured along the x-axis. For example, the absorption band illustrated in Figure 1 has an absorbance of 0.80 and a λ_{max} of 350 nm. Absorbance is related to transmittance by this equation:

> Electronic transitions of single bonds and isolated double bonds generally occur at wavelengths below 200 nm, which is outside the range of most ultraviolet-visible spectrophotometers.

$$A = \log(1/T) = -\log T \qquad (1)$$

Experimental Considerations

The construction of ultraviolet-visible spectrophotometers varies widely; both single- and double-beam instruments are available, with or without recording capability. A typical re-

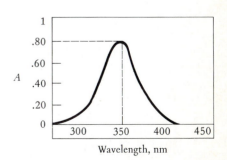

Figure 1.   Ultraviolet absorption band

cording instrument contains a closed *sample compartment* with places for sample and reference cells, a *zero adjustment control* for setting the absorbance to zero before the spectrum is recorded, a *wavelength control* to fix the starting wavelength, a *scan switch* to start and stop the scanning of the desired wavelength range; a lever or switch to change the *radiation source;* and a *recorder,* either as an integral part of the instrument or as a separate unit. Some chart papers, premarked with wavelength and absorbance values, are wrapped around a recording drum; others may come in the form of a roll that feeds onto a flat recorder bed during a scan, requiring that wavelength and absorbance values be written in by the user. Most instruments have a control that can be used to vary the absorbance range (along the y-axis of the chart paper). Although the slit, gain, and balance controls do not usually require readjustment for routine work, they can affect the quality of a spectrum and should be adjusted periodically. Setting the slit width too low, for example, will decrease the signal-to-noise ratio; setting it too high will reduce the height and resolution of absorption bands.

> A typical absorbance range is from 0 to 1 or 2 absorbance units.

Once a spectrum has been recorded, the data it contains can be presented as a tabulation of λ_{max} values giving either the absorbance, molar absorptivity (ϵ), or log ϵ at each wavelength specified. For example, the *CRC Handbook of Chemistry and Physics* reports the ultraviolet spectrum of cinnamic acid as "λ^{al} 210 (4.24), 215 (4.28), 221 (4.18), 268 (4.31)." The numbers in parentheses are log ϵ values for peaks having the λ_{max} values (in nanometers) given, and "al" indicates that the spectrum was run in ethyl alcohol.

Sample Preparation

Although ultraviolet-visible spectra can be obtained using gaseous samples and KBr pellets, routine spectra are nearly always obtained in solution. The solvent must be transparent (or nearly so) in the regions to be scanned. Water, 95% ethanol, methanol, dioxane, acetonitrile, and cyclohexane are suitable down to about 210–220 nm; many other solvents can be used at higher wavelengths. 95% ethanol is the preferred solvent, in part because it does not require additional purification; most of the others must be purified or purchased as spectral-grade solvents. The solvent must not, of course, react with the solute — for example, alcohols should not be used as solvents for aldehydes.

> *In making up a solution, great care must be taken to avoid contamination by impurities.*

> Absolute ethanol *does* require purification, because it contains some benzene.

The optimum *concentration* of a solution depends on the molar absorptivity of the bands to be recorded, the cell path length (*b*), and the absorbance ranges available on the instrument. From Beer's law (Equation 2) it can be seen that, in order to record a spectrum with a maximum absorbance of 1 using a 1-cm cell, the molar concentration (*c*) of the solution must be less than or equal to $1/\epsilon_{max}$, where ϵ_{max} is the molar absorptivity of the strongest peak in the spectrum.

$$c = \frac{A}{\epsilon \cdot b}$$ (2)

(ϵ is in L mol^{-1} cm^{-1} and *b* is in cm)

For example, if the maximum molar absorptivity of a sample is 10,000 (a fairly typical value), the solution concentration should be 1×10^{-4} M or less. To make up such dilute solutions accurately, it is usually necessary to prepare a solution that is too concentrated by several powers of 10, and then measure out an aliquot of this solution and dilute it. For example, to prepare a $\sim 1 \times 10^{-4}$ M solution of cinnamic acid (M.W. = 148), one could measure 0.15 g of the solid into a 100-mL volumetric flask and make it up to volume with solvent; then pipet a 1-mL aliquot of this 0.01 M solution into another 100-mL volumetric flask and fill it to the mark with solvent. If the molar absorptivity of the sample is not known, it may be necessary to find the optimum concentration by trial and error, starting with a more concentrated solution and diluting it as needed.

If there is great variation in the intensities of different peaks, it may be necessary to analyze several solutions of differing concentrations. For example, 3-buten-2-one has a band at 210 nm with $\epsilon = 11,500$ and one at 315 nm with $\epsilon = 26$; the latter peak would be virtually undetectable at a concentration suitable for observing the former.

The most commonly used *sample cell* for ultraviolet-visible spectrometry is a transparent rectangular container with a square cross-section, having a path length of 1.00 cm and a capacity of about 3 mL. Two sides of such a cell are ground (nontransparent) and the other two are transparent. The cell must be held by its ground sides and placed in the sample compartment so that the transparent sides are perpendicular to the light beam. Cells should *never* be touched on the transparent sides, since even a fingerprint can yield a spectrum. Cells must be cleaned thoroughly after use by rinsing them several times with the solvent. In some cases, it may be necessary to use detergent or a special cleaning solution to remove all of a sample.

Fused quartz cells are used for the ultraviolet region; glass or plastic cells are suitable in the visible region.

Operating Procedure

Do not touch any controls other than the ones specified below without your instructor's permission and under his or her supervision. Ultraviolet-visible spectrophotometers vary widely in construction and operation, so the following is intended only as a general guide and may not be applicable to all instruments. Specific operating instructions should be learned from in-class demonstrations or the operator's manual.

Be sure that the instrument's *power* and *lamp switches* have been turned on and that adequate time has been allowed for warmup. Select the appropriate radiation source for the desired wavelength range. A tungsten lamp can generally be used between 300 and 800 nm and a hydrogen lamp between 190 and 350 nm. If applicable, set the absorbance range to $0-1$ (or another appropriate value) and the wavelength dial to the starting wavelength. If the instrument scans from high to low wavelength, the starting wavelength will be the highest wavelength of the range to be scanned. Place the reference cell and sample cell in the appropriate locations in the sample compartment; then close the compartment, and set the absorbance to 0 with the zero adjustment control. Fill the sample cell with the solution to be analyzed and the reference cell with the pure solvent, return them to the sample compartment, and close the compartment. Position the chart paper (if necessary) so that the scan starts on an ordinate line; label this line with the starting wavelength, lower the pen to the paper, and begin to scan the spectrum. Stop the scan when the other end of the wavelength range has been reached. If both ultraviolet and visible regions are to be scanned, change the radiation source if necessary (some instruments do this automatically), and scan the spectrum in the other region. Raise the pen from the chart and tear off the chart paper using a straight-edge (there is usually one on the recorder). Write down the wavelength and absorbance ranges immediately, along with the wavelength interval between chart units; if you forget to record this information, it may be difficult or impossible to interpret the spectrum. (It is a good idea to mark down the wavelengths at intervals along the chart.) Rinse the sample cell several times with the solvent; then drain and dry both sample and reference cells. Measure and write down the λ_{max} and absorbance values of any peaks and shoulders of interest. If necessary, calculate the absorptivity or log ϵ at each λ_{max} value as well.

To find out if the sample concentration and absorbance range are satisfactory, scan the wavelength range manually (if the instrument allows this) before recording a spectrum.

Some instruments allow the operator to apply a baseline correction, which subtracts out any absorbance differences between the sample and reference cells over the entire wavelength range. A baseline is usually run with both cells filled with the solvent.

If one of the peaks goes off scale, change the absorbance range or dilute the sample.

Cells can be cleaned and partly dried using cotton swabs, then dried with an ear syringe or a stream of dry air ⟨OP–22⟩.

OPERATION 35a

Colorimetry

Instruments operating in the near-ultraviolet and visible regions of the spectrum can be very useful for the quantitative analysis of compounds in solution, provided that the solute either absorbs sufficient radiation in the appropriate range or can be made to absorb by being treated with a suitable reagent. Although a recording ultraviolet-visible spectrophotometer can be used for this purpose, a simple colorimeter such as the Bausch and Lomb Spectronic 20 will suffice.

Concentrations can be determined from absorbance values using either Beer's law (see Equation 1) or a calibration curve of absorbance versus molar concentration. In kinetic studies (as in Experiment 42), absorbance values can be used to calculate rate constants directly, without first being converted to concentrations.

$$c = \frac{A}{\epsilon \cdot b} \qquad (1)$$

c = molar concentration
A = absorbance
b = cell path length, in cm
ϵ = molar absorptivity

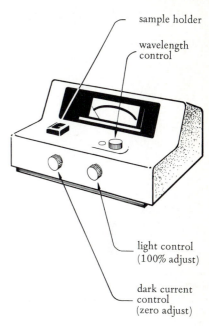

Figure 1. Bausch and Lomb Spectronic 20 spectrophotometer

Operational Procedure

Make sure that the instrument has been switched on and that adequate time has been allowed for warmup. Set the *wavelength control* to the desired value and set the transmittance (read from the scale) to zero with the *dark current control*. Insert a clean *cuvette* containing the pure solvent into the sample holder, and adjust the *100% control* until the transmittance reading is 100%. Make sure that the alignment mark on the cuvette is opposite the mark on the cell holder. Rinse the cuvette with a little of the sample to be analyzed. Fill it with fresh sample; then replace it in the cell holder, being very careful to position it in exactly the same alignment as before. Different cuvettes are sometimes used for the solvent and sample, but errors will result if the cuvettes are not well matched; for precise work, it is best to use the same cuvette for all measurements. Read the percentage transmittance as accurately as possible (absorbance values may also be shown, but it is better to read %*T* if the absorbance scale is nonlinear). Convert the percent transmittance values to absorbance, if necessary, using the equation $A = \log(100/\%T)$.

If another solution containing the same solvent and solute is to be analyzed, drain the cuvette and rinse it with that solution before you make the next measurement. If another solution containing a different solute or solvent is to be analyzed, rinse the cuvette with the solvent used for preparing the first solution (clean the cuvette if necessary) and dry it before running the next solution. After the last measurement, rinse the cuvette with pure solvent, clean it, and let it air-dry.

Use the same solvent as you used to prepare the solution being analyzed.

These directions are for operation of the B & L Spectronic 20 or a similar instrument.

OPERATION 36

Mass Spectrometry

Principles and Applications

When a compound is bombarded with a beam of high-energy electrons (or other particles) in a mass spectrometer, each of its molecules (M) can lose an electron and form a *molecular ion* ($M \cdot ^+$). If the energy of the electron beam is high enough, the molecular ions will have enough excess vibrational and electronic energy to break apart into fragments, each pair of fragments consisting of another positive ion and a neutral molecule. Each positive ion *(daughter ion)* may in turn break down, losing a neutral fragment to form yet another positive ion, and so on. The positive ions are accelerated into an evacuated chamber (the *mass analyzer*), where they are separated according to their mass:charge ratios (m/e), usually by means of strong electric and magnetic fields. As a beam of ions with a given mass:charge ratio impinges on an *ion collector,* it gives rise to an *ion current,* which is amplified and recorded on a *mass spectrum.* A mass spectrum recorded on a strip chart (or a computer-generated bar graph) consists of a series of peaks of different heights arranged in order of their m/e values. Each peak corresponds to a particular ion, and its height is proportional to the relative abundance of that ion. Because most of the ions will have a charge of $+1$, the m/e value associated with a peak is nearly always equal to the mass of the ion that gave rise to that peak — or in rare cases to one-half of its mass. A typical large molecule may be fragmented into several hundred different ions, each with a different m/e value and relative abundance, so modern mass spectrometers are commonly interfaced with computers that record, store, and process the data.

Mass spectrometry is an extremely valuable tool for structural analysis; it can be used to identify or characterize a host of organic (and inorganic) compounds, including biologically active substances with very complex molecular structures. The mass spectrum of a compound gives the masses of its structural components, which can often be identified with the help of published tables of molecular fragments. Usually a mass spectrum also reveals the molecular weight of the compound and data that can be used to determine its molecular formula. Such information may make it possible to piece together the compound's molecular structure, or at least to learn more about its structural features.

Fragmentation of molecular ion and daughter ion

$$M \cdot ^+ \longrightarrow A^+ + X$$
$$A^+ \longrightarrow B^+ + Y$$
etc.

(The location of the unpaired electron varies — it may be on either A^+ or X, depending on the kind of fragmentation.)

Different compounds yield distinctively different patterns of ion fragments, so an unknown compound can sometimes be identified by a comparison of its mass spectrum with mass spectral data from the literature. Unfortunately, peak intensities on mass spectra are quite sensitive to instrumental parameters such as the energy of the electron beam. Thus, there may be significant differences in the mass spectra recorded on different instruments — or even by different operators using the same instrument. Nevertheless, it is often possible to make a tentative identification from a literature comparison, and then to confirm it by recording mass spectra of the unknown and the most likely known compounds under identical operating conditions. This process can be facilitated by using a computer to compare the spectrum of the unknown with a "library" of mass spectral data in its memory. A mass spectrometer can also be used as a kind of "superdetector" for a gas or liquid chromatograph, characterizing each component of a mixture by its mass spectrum as it comes off the column.

Experimental Considerations

Mass spectrometers are costly, highly complex instruments ordinarily operated by skilled technicians; most undergraduate students will not have an opportunity to obtain hands-on experience with one. For that reason, and because the instruments vary so widely in construction and operation, no attempt will be made here to describe specific operating techniques; these must be learned by special instruction and by studying the manufacturer's operating manual.

Samples prepared for mass spectrometry should be very pure, since traces of impurities can make interpretation difficult. No special sample preparation is required. Liquids are inserted directly into the sample inlet with a hypodermic syringe, micropipet, or break-off device, and solids can be introduced by means of a melting-point capillary. The samples vaporize in the sample inlet system, after which their molecules flow through a controlled "molecular leak" (usually a minute hole in a piece of gold foil) into the evacuated ionization chamber.

Sample sizes for mass spectrometry range from less than a microgram to several milligrams.

Interpretation of Mass Spectra

Each ion recorded on a mass spectrum is characterized by its mass : charge ratio and relative abundance, the latter being proportional to the height of its peak. This information can be

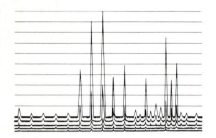

Figure 1. Part of a mass spectrum traced by a five-galvanometer recorder

The molecular-ion peak is sometimes, but not always, the base peak.

The molecular-ion peak is often called the *parent peak*. Its mass is essentially equal to the molecular mass of the sample in its most abundant isotopic variation.

Since benzene contains six carbon atoms, the chance that any one of the six will be carbon-13 is 6 × 1.08, or 6.48%.

These intensities are given as percentages of the molecular-ion (parent) peak intensity, not of the base-peak intensity.

displayed by a variety of output devices including oscilloscopes, digital readouts, strip-chart recorders, and computer printers. One kind of recorder consists of five galvanometers of varying sensitivity that produce a spectrum consisting of five separate traces, increasing in amplitude from bottom to top. Since the uppermost trace is recorded by the most sensitive galvanometer, it is assigned a sensitivity factor of 1; the others are, in order, usually assigned sensitivity factors of 3, 10, 30, and 100, respectively. The height of each peak is measured from the baseline of the trace in which it has the highest on-scale amplitude, and then multiplied by the appropriate sensitivity factor to give its relative intensity. The most intense peak in a mass spectrum, called the *base peak,* is assigned a relative intensity of 100 (or sometimes 1000), and the intensities of all other peaks are reported as percentages of the base-peak intensity. The mass number (actually m/e) of each peak can be determined by counting from the low-mass end of the spectrum along the upper trace, since there will usually be at least a small "blip" at each mass unit. Computerized instruments generally provide a direct readout of mass number, based on calibration with a known substance such as perfluorotributylamine (PFTBA).

To determine the molecular weight and molecular formula of the sample, the *molecular-ion peak* must be found. This is usually the last strong peak on the spectrum, since no fragment can have a higher mass than the original molecular ion. There will usually, however, be at least two discernible low-intensity peaks following the molecular ion peak. These are generated by isotopic variations of the molecular ion. Most elements occurring in organic compounds have at least one isotope of higher mass number than the common form. For example, just over one carbon atom (1.08) in a hundred is carbon-13 rather than the more abundant carbon-12, so the molecular ion peak of benzene (C_6H_6, $m/e = 78$) should be followed by a peak for "heavy" benzene ($^{13}C^{12}C_5H_6$, $m/e = 79$) with an intensity that is 6.48% of the parent peak intensity. Actually, the $m/e = 79$ peak, called the $M + 1$ peak, has a slightly higher intensity than this because ordinary benzene also contains minute quantities of benzene-d_1 (C_6H_5D). Likewise, oxygen-18 occurs naturally to the extent of about 0.20 atom for every 100 atoms of oxygen-16, so formaldehyde shows an $M + 2$ peak (corresponding to $CH_2{}^{18}O$) with an intensity that is 0.20% of the parent-peak intensity.

Relative intensities of the $M + 1$ and $M + 2$ peaks for a compound $C_wH_xN_yO_z$ can be calculated using Equations 1 and 2:

$$\%(M + 1) = 1.08w + 0.015x + 0.37y + 0.037z \qquad (1)$$

$$\%(M + 2) = 0.006w(w - 1) + 0.0002wx + 0.004wy + 0.20z \qquad (2)$$

For example, quinine (as the hydrate) has the molecular formula $C_{20}H_{30}N_2O_3$. From the equations, the intensity of its $M + 1$ peak should be 22.9% of the parent-peak intensity and that of its $M + 2$ peak should be 3.16% of the parent-peak intensity. Tabulations of these values are available (see reference F7 or F28 in the Bibliography).

No two combinations of atoms with a given molecular mass are likely to yield $M + 1$ and $M + 2$ peaks of exactly the same intensity. Therefore, if the intensities of these peaks are measured from the mass spectrum of an unknown compound, its molecular formula can usually be determined from published tables. A general procedure for determining molecular formulas is as follows:

1. Locate the molecular-ion peak, and determine its mass, M.

2. Measure the intensities of the M, $M + 1$, and $M + 2$ peaks and express the latter two as a percentage of the intensity of the M peak.

3. Find the formula (or formulas) listed under its M value that give $M + 1$ and $M + 2$ intensities close to the experimental values and that make sense from a chemical standpoint.

Some formulas may be eliminated immediately because they do not correspond to stable molecules or have the expected degree of hydrogen deficiency. For compounds with the general formula $C_wH_xN_yO_z$, an *index of hydrogen deficiency* (equal to the number of rings plus the number of pi bonds in each molecule) can be calculated from Equation 3.

$$\text{I.H.D.} = \tfrac{1}{2}(2w - x + y + 2) \qquad (3)$$

For example, the calculated I.H.D. of compound *1* is 7, which is consistent with its molecular structure (one ring and six pi bonds). Another useful generalization is the *nitrogen rule,* which says that a stable compound with an even-numbered value of M can have only zero or an even number of nitrogen atoms, whereas one with an odd value of M can have only an odd number of nitrogen atoms. Finally, some possibilities can be eliminated because compounds with those formulas would be impossible to obtain from a given reaction or source.

Calculation:

$$\%(M + 1) = 1.08(20) + 0.015(30) + 0.37(2) + 0.037(3)$$
$$= 22.9$$

$$\%(M + 2) = 0.006(20)(19) + 0.0002(20)(30) + 0.004(20)(2) + 0.20(3)$$
$$= 3.16$$

Example:

$$\text{I.H.D.} = \tfrac{1}{2}(18 - 7 + 1 + 2) = 7$$

The nitrogen rule applies only to organic compounds containing the more commonly encountered elements.

Table 1.   Formulas of some neutral species lost from molecular ions

Mass of resulting ion	Possible formulas
$M-1$	H·
$M-15$	·CH$_3$
$M-16$	·NH$_2$
$M-17$	·OH, NH$_3$
$M-18$	H$_2$O
$M-26$	C$_2$H$_2$, ·CN
$M-27$	·C$_2$H$_3$, HCN
$M-28$	CO, C$_2$H$_4$
$M-29$	·CHO, ·C$_2$H$_5$
$M-30$	H$_2$CO·, NO
$M-31$	CH$_3$O·, ·CH$_2$OH
$M-35$	Cl·
$M-36$	HCl
$M-42$	CH$_2$CO
$M-43$	CH$_3$CO·, ·C$_3$H$_7$
$M-44$	CO$_2$, ·CONH$_2$
$M-45$	·CO$_2$H, C$_2$H$_5$O·
$M-46$	NO$_2$
$M-49$	·CH$_2$Cl
$M-57$	CH$_3$COCH$_2$·
$M-59$	·CO$_2$CH$_3$
$M-77$	C$_6$H$_5$·
$M-79$	Br·

The use of *fragmentation patterns* to determine molecular structures is a very broad subject covered in detail elsewhere; only a few generalizations will be given here. A molecular ion (or its daughter ions) often fragments by eliminating small neutral molecules or free radicals such as CO, H$_2$O, HCN, C$_2$H$_2$, H·, or CH$_3$·, yielding ions with masses equal to $M - X$, where X is the mass of the neutral molecule (see Table 1). For example, the mass spectrum of methyl benzoate in Figure 2 has its base peak at $m/e = 105$ and other strong peaks at $m/e = 77$ and $m/e = 51$. As is often the case, the base peak belongs to a fragment ion rather than the molecular ion itself, whose peak is at $m/e = 136$. The molecular ion is believed to be a radical cation with the structure shown below. Since the mass numbers of the base peak and molecular-ion peak differ by 31 mass units, a neutral species with a mass number of 31 must have been lost from the molecular ion to produce the base-peak ion. The OCH$_3$ group of methyl benzoate has just that mass number; so the base-peak ion must be a benzoyl cation, formed by loss of a methoxyl radical from the molecular ion.

$$
\begin{array}{c}
\overset{O^{+}}{\underset{\parallel}{}} \\
\text{C}_6\text{H}_5\overset{}{\text{C}}\text{OCH}_3 \xrightarrow{-\text{CH}_3\text{O·}} \text{C}_6\text{H}_5\text{C}{\equiv}\text{O}^+ \xrightarrow{-\text{CO}} \text{C}_6\text{H}_5{}^+ \xrightarrow{-\text{C}_2\text{H}_2} \text{C}_4\text{H}_3{}^+
\end{array}
$$

molecular ion $\quad$ benzoyl ion $\quad$ phenyl ion $\quad$ $(m/e = 51)$
$(m/e = 136)$ $\quad$ $(m/e = 105)$ $\quad$ $(m/e = 77)$

Figure 2.   Mass spectrum of methyl benzoate

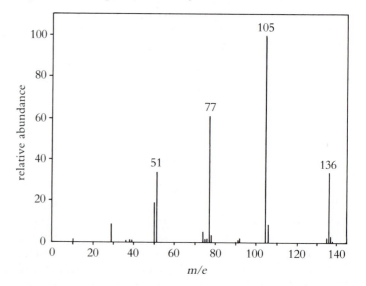

The difference between the mass number of the base peak and that of the next major peak is $105 - 77 = 28$, which is just the mass number of the carbonyl (CO) group; loss of a carbonyl group (as carbon monoxide) from the benzoyl cation should yield a phenyl cation with the expected mass number of 77. Formation of the next major species ($m/e = 51$) requires a loss of 26 mass units, which can be accounted for if acetylene (C_2H_2) is lost from the phenyl cation to form an ion with the formula $C_4H_3^+$; this species is often encountered in the mass spectra of aromatic compounds. There are a number of smaller peaks that may provide additional structural information, but usually it is not possible (or necessary) to characterize all the peaks in a mass spectrum.

In the above example, the interpretation was simplified because each major ion was produced from the preceding one along a single reaction path. Often there are fragmentation paths leading directly from the molecular ion to a number of different daughter ions. It is therefore a common practice to compare the mass number of the molecular ion with those of the significant daughter ions before attempting to compare the mass numbers of individual daughter ions.

Table 1 gives the formulas of some neutral species that are frequently lost from molecular ions, and Table 2 lists some common fragment ions. Structures have been determined for some (but not all) of these fragment ions. For example, the $C_7H_7^+$ ion with mass number 91, which results from the cleavage of alkylbenzenes, has been formulated as either a benzyl cation or a tropylium ion; in most cases, it appears to have the latter structure.

Before you can interpret a mass spectrum using such tables, you must learn something about the characteristic fragmentation patterns and mechanisms for different classes of organic compounds. This kind of information and some general rules for interpretation of mass spectra are given in reference F8 and other references on mass spectrometry listed in category F of the Bibliography.

Table 2.   Some common fragment ions

m/e	Possible formulas
15	CH_3^+
17	OH^+
18	H_2O^+, NH_4^+
26	$C_2H_2^+$
27	$C_2H_3^+$
28	CO^+, $C_2H_4^+$
29	CHO^+, $C_2H_5^+$
30	$CH_2NH_2^+$, NO^+
31	CH_2OH^+, CH_3O^+
35	Cl^+ (also 37)
39	$C_3H_3^+$
41	$C_3H_5^+$
43	CH_3CO^+, $C_3H_7^+$
44	CO_2^+, $C_3H_8^+$
45	$CH_3OCH_2^+$, CO_2H^+
46	NO_2^+
49	CH_2Cl^+
51	$C_4H_3^+$
57	$C_4H_9^+$, $C_2H_5CO^+$
59	$COOCH_3^+$
65	$C_5H_5^+$
66	$C_5H_6^+$
71	$C_5H_{11}^+$, $C_3H_7CO^+$
76	$C_6H_4^+$
77	$C_6H_5^+$
78	$C_6H_6^+$
79	Br^+ (also 81)
91	$C_7H_7^+$, $C_6H_5N^+$
93	$C_6H_5O^+$
94	$C_6H_6O^+$
105	$C_6H_5CO^+$

APPENDIXES AND BIBLIOGRAPHY

CHEMICAL GLASSWARE

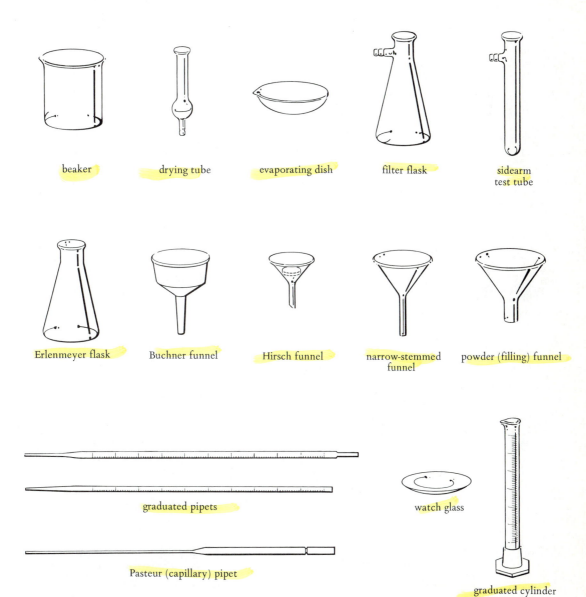

beaker   drying tube   evaporating dish   filter flask   sidearm
test tube

Erlenmeyer flask   Buchner funnel   Hirsch funnel   narrow-stemmed
funnel   powder (filling) funnel

graduated pipets   watch glass

Pasteur (capillary) pipet   graduated cylinder

STANDARD TAPER GLASSWARE

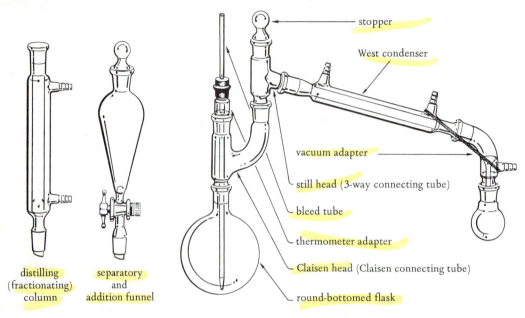

stopper

West condenser

vacuum adapter

still head (3-way connecting tube)

bleed tube

thermometer adapter

Claisen head (Claisen connecting tube)

round-bottomed flask

distilling
(fractionating)
column

separatory
and
addition funnel

HARDWARE

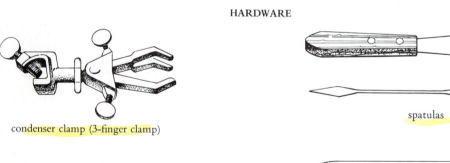

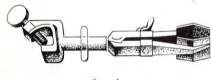

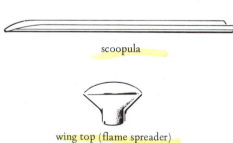

condenser clamp (3-finger clamp)

utility clamp

ring support

spatulas

scoopula

wing top (flame spreader)

wire gauze

A laboratory notebook is essentially a factual account of work performed in the laboratory; it may also include the writer's interpretation of the results. Your laboratory work may not require the degree of documentation expected of a research chemist, but you should at least be aware of the characteristics of a good laboratory notebook. Most laboratory notebooks are bound, and their pages are quadrilled (marked with a square grid of lines). A notebook with duplicate pages will permit you to turn in a carbon copy of your notes to the instructor after each laboratory period. The first page or two should be reserved for a table of contents, and the pages should be numbered sequentially so that you can find information quickly. Insofar as possible, each entry should be written *immediately* after the work is performed, and it should be properly dated and signed by the experimenter (notebooks of research chemists are usually signed by a witness as well). Each section of the notebook should have a clear, descriptive heading, and the writing must be grammatically correct and sufficiently legible to be read and understood by any knowledgeable individual.

Before each experiment, you should write your experimental plan (see Appendix V) and other relevant information in the notebook. Some or all of the following items should be included:

1. The experiment number and title

2. A clear, concise statement of what the experiment is intended to accomplish

3. Balanced equations for all significant reactions, including possible side reactions that might reduce the yield

4. A table of the reactants, products, solvents, and other chemicals to be used, listing any physical properties or other information (M.W., m.p., solubility, potential hazards, source, purity, etc.) that may be relevant to the experiment

5. Calculations of the quantities of reactants required and the theoretical yield of product

6. A list of the chemicals, supplies, and equipment needed for the experiment

7. Sketches of any nonstandard apparatus to be used

8. A checklist, flowchart, or other kind of outline summarizing the procedure

Some of the above information will be given in the experiment itself, and certain items may not apply to some kinds of experiments, such as kinetic studies or qualitative analyses.

During the experiment, you should keep a detailed account of your work, reporting everything of importance that you actually *did* and *saw.* Your notes should not simply restate the textbook procedure but should describe in your own words how you carried out the experiment. You should include all relevant data such as the quantities of materials that you actually used (not the theoretical quantities that you calculated, unless they are exactly the same) and the results of any analyses you performed. Raw data should be recorded with particular care; if you forget to record data at the time you measure it, or if you record it incorrectly or illegibly, the results of an entire experiment may be invalidated.

You may find it helpful to think of your lab notebook as telling a story about your accomplishments (or misadventures) in the laboratory. Although professional journal articles are generally written in a dry, impersonal style (with some notable exceptions), this need not be true of a lab notebook. The American Chemical Society's publication *Writing the Laboratory Notebook* ⟨Bib – L42⟩ suggests a more personal approach, stating that "the use of the active voice in the first person tells the story and clearly indicates who did the work." (*Example:* "I added 10.3 mL of benzaldehyde to the solution . . . .") Other chemists think personal pronouns should be avoided and recommend using the passive voice instead. (*Example:* "10.3 mL of benzaldehyde were added to the solution . . . .") If your instructor has a strong preference either way, you should follow his or her recommendation.

APPENDIX III

Writing a Laboratory Report

You will be expected to submit a report for each experiment (1) on a report form provided by your instructor, (2) in your laboratory notebook, or (3) on separate sheets of paper, stapled together or bound in a folder. Your instructor may recommend a "professional" format such as the one described in Experiment 50 or may suggest a different format. In any case, each report should include information under some or all of the following headings.

1. *Prelab:* Your experimental plan, calculations, and any other written material specified in the Prelab Assignments.

2. *Observations:* Any significant observations made during the course of the experiment. You should record all observations that might be of help to you (or another experimenter) if you were to repeat the experiment at a later time. These should include the quantities of solvents or drying agents used, the reaction times, the distillation ranges, and a description of any experimental difficulties you encounter. You should also make note of phenomena that might provide clues about the nature of chemical or physical transformations taking place during the experiment, such as color changes, phase separations, tar formation, and gas evolution. If you keep a detailed record of your observations in a laboratory notebook, it should not be necessary to include them in a separate report.

3. *Raw Data:* All numerical data obtained directly from an experiment, before it is graphed, used in calculations, or otherwise processed. This can include quantities of reactants and products, titration volumes, kinetic data, gas chromatographic retention times and integrated peak areas, and spectrometric absorbance values.

4. *Calculations:* Yield and stoichiometry calculations and any other calculations based on the raw data. If a number of repetitive calculations are required, one or two sample calculations of each type may suffice.

5. *Results:* A list, graph, tabulation, or verbal description of the significant results of the experiment. For a preparation, this should include a physical description of the product (color, physical state, evidence of purity, etc.), the percentage yield, and all physical constants, spectra, and analytical data obtained for the product. For the qualitative analysis of an unknown compound, the results of all tests and derivative preparations should be included, along with spectra and physical constants. The results of calculations based on the raw data should be reported in this section as well.

6. *Discussion:* A discussion of the significant experimental results that attempts to explain and interpret those results. The discussion should describe possible sources of error or material losses, compare numerical results obtained with those expected from theory (if applicable), or discuss the theoretical significance of the results. This section should also include your interpretation of spectra or chromatograms, which should be attached to the report.

7. *Conclusions:* Any conclusions to be drawn as a result of the performance of an experiment, as suggested by the objectives for the experiment. In a synthetic experiment, for example, you should be able to conclude (based on physical evidence and spectra) whether or not you have prepared the expected product, or which of several possible structures is the correct one.

8. *Exercises:* Your answers to all exercises assigned by your instructor, showing your calculations and describing your reasoning when applicable

The Report section following most experiments in this book lists specific items that should be addressed in an appropriate section of your report, but it is by no means complete. For example, you should calculate the percentage yield from a preparation whether or not you are specifically requested to. Each report should also include (on the first page or cover) the name and number of the experiment, your name, and the date on which the experiment was turned in to the instructor. The date(s) on which the experiment was performed may also be required. The sample laboratory report on the following pages illustrates the kind of writeup that might be prepared by a conscientious student.

Name: Cynthia Sizer
Date: Sept. 21, 1988

EXPERIMENT NO. 5: PREPARATION OF TETRAHEDRANOL

Prelab

Tetrahedryl benzoate needed:

$$\text{mass} = 0.075 \text{ mol} \times 172.2 \text{ g/mol} = 12.9 \text{ g}$$
$$\text{volume} = \frac{12.9 \text{ g}}{0.951 \text{ g/mL}} = 13.6 \text{ mL}$$

Aqueous sodium hydroxide (5 M) required:

$$\text{volume} = \frac{0.50 \text{ mol}}{5.0 \text{ mol/L}} \times 1000 \text{ mL/L} = 100 \text{ mL}$$

Observations

The reaction was carried out under reflux for 1 hour and 10 minutes, during which time the organic layer dissolved in the aqueous sodium hydroxide and the odor of the ester disappeared. The acidified reaction mixture was extracted with two 50-mL portions of ethyl ether, and the extracts were dried over 3.0 g of anhydrous magnesium sulfate. Evaporation of the ether under aspirator vacuum (steam bath) required about 25 minutes. The residue was distilled over a 4°C boiling range from 78–82°C, and solidified on cooling. None of the product was seen to solidify in the vacuum adapter during the distillation, although the procedure warned against this possibility. The infrared spectrum of the product was recorded using a thin film between heated silver bromide plates.

The original reaction mixture, when acidified with 88 mL of 3 M sulfuric acid (~0.26 mol), yielded a white precipitate presumed to be benzoic acid, which was dried in a desiccator for 48 hours before weighing. During the precipitation of benzoic acid, a white solid appeared each time a portion of sulfuric acid was added and then disappeared when the mixture was stirred, right up to the last portion — then it all precipitated at once! It formed long white needles on cooling.

Data

Mass of tetrahedryl benzoate: 12.87 g
Mass of tetrahedranol recovered: 3.27 g
Mass of benzoic acid recovered: 8.70 g

Calculations

(TetOBz = tetrahedryl benzoate; TetOH = tetrahedranol;
BzOH = benzoic acid)

Theoretical yield of tetrahedranol:

$$\frac{12.87 \text{ g TetOBz}}{172.1 \text{ g/mol TetOBz}} \times 68.1 \text{ g/mol TetOH} = 5.09 \text{ g TetOH}$$

Percentage yield of tetrahedranol:

$$\frac{3.27 \text{ g}}{5.09 \text{ g}} \times 100 = 64.2\%$$

Theoretical yield of benzoic acid:

$$\frac{12.87 \text{ g TetOBz}}{172.1 \text{ g/mol TetOBz}} \times 122.1 \text{ g/mol BzOH} = 9.13 \text{ g BzOH}$$

Percentage yield of benzoic acid:

$$\frac{8.70 \text{ g}}{9.13 \text{ g}} \times 100 = 95.3\%$$

Results

A product assumed to be tetrahedranol was obtained in 64.2%
yield from the alkaline hydrolysis of tetrahedryl benzoate.
Tetrahedranol at room temperature was a colorless, almost
transparent solid with a mild "spiritous" odor. It could easily
be melted to form a clear, colorless liquid. It had a distillation
boiling range of 78–82°C and a micro boiling point of 81°C.
Its infrared spectrum displayed absorption bands at 3370,
2975, 2885, 1465, 1375, and 1194 cm^{-1}. Benzoic acid, obtained in 95.3% yield from the reaction, was a white solid
crystallizing in the form of small needles that melted at 120–
122°C. The crystals were slightly discolored, but appeared dry
and otherwise free from impurities.

Discussion

Probable infrared band assignments for the product boiling at 81°C are as follows: 3370 cm^{-1}, O—H stretch; 2975 and 2885 cm^{-1}, C—H stretch; 1465 and 1375 cm^{-1}, C—H bend; and 1194 cm^{-1}, C—O stretch. The broad, strong O—H stretch band suggests an alcohol or phenol, and the position of the C—O band is consistent with a tertiary alcohol. The product's infrared spectrum and micro boiling point of 81°C (literature value = 82°C) provide strong evidence that this product is indeed tetrahedranol. The facts that it melted just above room temperature (reported m.p. = 28°C) and that it was synthesized from tetrahedryl benzoate by a reaction designed to hydrolyze such an ester lend support to this assumption.

The product melting at 120–122°C would appear to be benzoic acid from its melting point (literature value = 122.4°C) and its origin as a product of the hydrolysis of tetrahedryl benzoate. From the melting point, it seems to be reasonably pure. The slight discoloration could probably be removed by recrystallizing it from water, which should also narrow the melting-point range.

The yield of tetrahedranol was 1.82 g less than the theoretical value. Losses could have occurred as a result of (1) incomplete reaction; (2) incomplete extraction of the alcohol from the aqueous layer; (3) transfers of the reaction mixture to the separatory funnel, the ether extract from the separatory funnel to the evaporating flask, and the residue from the evaporating flask to the distilling flask; or (4) distillation of the product. Since the yield of benzoic acid was 95.2%, the reaction must have been at least 95% complete, so incomplete reaction could account for *at most* 0.25 g (5% of 5.09) of the loss. More likely the reaction was nearly 100% complete, since some benzoic acid must have been lost during the workup (see below). I tried to keep losses in transfers to a minimum by using additional solvent (ether or water) in each transfer, so losses during transfers should not be a major reason for the low yield. About 0.5 mL of residue remained in the distilling flask, which can account for about 0.5 g of the loss, assuming the residue was mostly tetrahedranol. It seems likely that most of the loss (up to 1.3 g) occurred during the extraction operation, particularly since tetrahedranol is reported to be partially soluble in water. This loss might have been reduced by saturating the aqueous layer with potassium carbonate to salt out the alcohol or by carrying out several more ether extractions.

The yield of benzoic acid was very high; it should also be reliable since the product dried to a constant mass (three weighings were performed at approximately 1-hour intervals). The loss of 0.43 g can be accounted for by (1) losses in transfers and (2) incomplete precipitation from solution. Traces of benzoic acid remained on the Buchner funnel and filter paper used for the vacuum filtration. About 0.34 g might have stayed in the 188 mL of ice-cold solution from which the acid was precipitated (its solubility in water at 4°C is 0.18 g/100 mL), but the high salt (Na_2SO_4) concentration in this solution probably decreased its solubility somewhat.

Conclusion

It can be concluded from the experimental results that the reaction of tetrahedryl benzoate with 5 M aqueous sodium hydroxide does in fact yield tetrahedranol and benzoic acid in acceptable yield, and therefore the experimental objective "to prepare tetrahedranol by the alkaline hydrolysis of tetrahedryl benzoate and isolate benzoic acid as a by-product of the reaction" has been fulfilled.

(Attached: Prelab checklist, infrared spectrum of tetrahedranol, and worked exercises)

Calculations for Organic Synthesis

APPENDIX IV

In this book, the quantities of many reactants are given in units of "chemical amount," that is, in moles or millimoles. You will have to convert such quantities to units of mass or volume before you can begin a synthetic experiment. In any experiment, it is the relationship between these *chemical* quantities that is significant, not the relationship between such *physical* quantities as mass and volume. Because there are no "molemeters" that can measure molar amounts directly, we are forced to use balances and volumetric glassware for that purpose. But that should not obscure the fact that the chemical units are fundamental; only by knowing the chemical amounts of reactants involved in a preparation, for example, can one recognize the stoichiometric relationships between them or predict the yield of the expected product.

It is often helpful to regard a chemical calculation as a process by which a given quantity is "converted" to the required quantity by being multiplied by a series of ratios that can be regarded as unit or dimensional *conversion factors.* Unit conversions are carried out by using conversion factors, such as 454 g/lb, that are written as ratios between two quantities whose quotient is unity (for example, 454 g = 1 lb, so 454 g/1 lb = 1). Dimensional conversions are carried out by using conversion factors that are ratios of quantities in different *dimensions,* such as mass, volume, and chemical amount (amount of substance). For example, the density of a substance can be regarded as a conversion factor linking the two dimensions of mass and volume, and it is used to convert the mass of a given quantity of the substance to units of volume, or vice versa. Any conversion factor can be used *as written* or *in inverted form;* thus, the molar mass of butyl acetate, written as 116 g/1 mol, will convert moles of butyl acetate to grams; the inverse ratio, 1 mol/116 g, will convert grams to moles.

The following examples illustrate some fundamental types of calculations that you will encounter:

The chemical amount (in moles or millimoles) of a substance is converted to its mass by multiplying by its molar mass. Remember that the molar mass of a substance is obtained by simply appending the units g/mol to its molecular weight, which is a dimensionless quantity. For example, the mass of 15.0 mmol of butyl acetate (M.W. = 116) is

$$15.0 \text{ mmol} \times \frac{1 \text{ mol}}{1000 \text{ mmol}} \times \frac{116 \text{ g}}{1 \text{ mol}} = 1.74 \text{ g}$$

Note that the chemical amount in millimoles must be converted to moles before the conversion factor is applied; otherwise, the units will not cancel out.

The chemical amount (moles or millimoles) of a liquid is converted to volume by multiplying by the liquid substance's molar mass and by the *inverse* of its density. For example, the volume of 25.0 mmol of acetic acid (M.W. = 60.1; d = 1.049 g/mL) is

$$25.0 \text{ mmol} \times \frac{1 \text{ mol}}{1000 \text{ mmol}} \times \frac{60.1 \text{ g}}{1 \text{ mol}} \times \frac{1 \text{ mL}}{1.049 \text{ g}} = 14.4 \text{ mL}$$

The volume of a solution needed to provide a specified chemical amount of solute is calculated by multiplying the number of moles required by the inverse of the solution's molar concentration. For example, the volume of 6.0 M HCl (which contains 6.0 mol of HCl per liter of solution) needed to provide 18 mmol of HCl is

$$18 \text{ mmol} \times \frac{1 \text{ mol}}{1000 \text{ mmol}} \times \frac{1 \text{ liter}}{6.0 \text{ mol}} \times \frac{1000 \text{ mL}}{1 \text{ liter}} = 3.0 \text{ mL}$$

Remember that concentrations expressed in mol/L and mmol/mL have the same numerical value. Thus, a 6.0 M solution also has a concentration of 6.0 mmol/mL; using these units simplifies the above calculation considerably:

$$18 \text{ mmol} \times \frac{1 \text{ mL}}{6.0 \text{ mmol}} = 3.0 \text{ mL}$$

All calculations should be checked by making sure that the units involved cancel out to yield the correct units in the answer. This does not ensure that your answer is correct; but if the units do *not* cancel, it is almost certainly wrong.

The maximum amount of product that could be attained from a reaction is called the *theoretical yield* of the reaction. Theoretical yields can be calculated using *stoichiometric factors*—ratios derived from the coefficients (expressed in moles) of the products and reactants in a balanced equation for the reaction. For example, the stoichiometric factors relating the chemical amount of the product to the chemical amounts of the two reactants in the following reaction are 1 mol picric acid/1 mol phenol and 1 mol picric acid/3 mol HNO_3.

phenol   picric acid

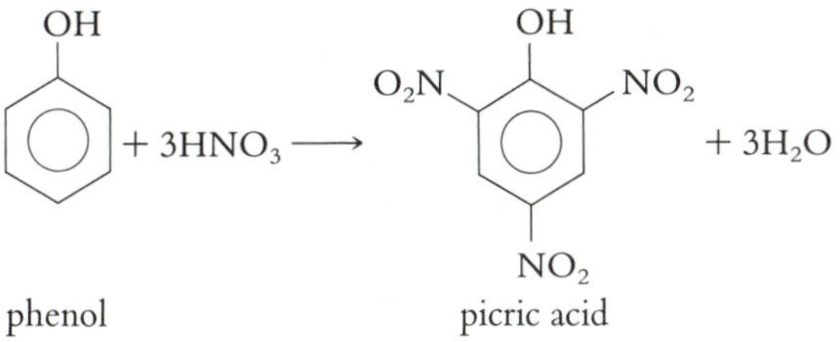

Suppose you were preparing picric acid (M.W. = 229) starting with 5.00 g of phenol (M.W. = 94.1) and 15.0 mL of concentrated (15.9 M) nitric acid. You can calculate the maximum amount of product that could be formed from each reactant by converting the given quantity to moles, then applying the appropriate stoichiometric factor:

$$5.00 \text{ g phenol} \times \frac{1 \text{ mol}}{94.1 \text{ g}} \times \frac{1 \text{ mol picric acid}}{1 \text{ mol phenol}} = 0.0531 \text{ mol picric acid}$$

$$15.0 \text{ mL } HNO_3 \times \frac{15.9 \text{ mmol}}{1 \text{ mL}} \times \frac{1 \text{ mol}}{1000 \text{ mmol}} \times \frac{1 \text{ mol picric acid}}{3 \text{ mol } HNO_3} = 0.0795 \text{ mol picric acid}$$

Since there is only enough phenol to produce 0.0531 mol of picric acid, it is impossible to obtain more than that from the specified quantities of reactants. Once that much picric acid has been formed, the reaction mixture will have run out of phenol, and the *excess* (leftover) nitric acid will have nothing to react with. Therefore, phenol is the *limiting reactant* on which the yield calculations must be based. (The limiting reactant is always the one that would produce the *least* amount of product.) The theoretical yield of picric acid, in grams, is then $0.0531 \text{ mol} \times 229 \text{ g/mol} = 12.2 \text{ g}$.

It is seldom, if ever, possible to attain the theoretical yield from an organic preparation; the reaction may not go to completion, there may be side reactions that reduce the yield of product, and there are invariably material losses when the product is separated from the reaction mixture and purified. The *percentage yield* of a preparation compares the actual yield to the theoretical yield as follows:

$$\text{Percentage yield} = \frac{\text{actual yield}}{\text{theoretical yield}} \times 100\%$$

For example, if you prepare 10.4 g of picric acid from 5.00 g of phenol and 15.0 mL of nitric acid, the percentage yield of your synthesis would be about $(10.4 \text{ g}/12.2 \text{ g}) \times 100\% = 85\%$.

In many experiments, you will estimate the theoretical yield of a preparation based on reactant quantities given in the Prelab Assignment (usually in millimoles). This will allow you to quickly assess your performance by comparing your actual yield with an estimate of the "ideal" yield. But the percentage yield that you report should be based on the quantities of reactants that you actually used in the synthesis, not on the values calculated for the Prelab Assignment (unless they are exactly the same).

APPENDIX V   # Planning an Experiment

Before starting any project, whether you are making a bookshelf, duck à l'orange, or isoamyl acetate, you must have a plan. An *experimental plan* should tell clearly what you expect to do in the laboratory and how you intend to go about it. You should state how the work is to be done in simple sentences, without excessive detail. You can always refer to the textbook procedure and operation descriptions for the details, but as you become more proficient in the laboratory, you should find yourself relying less on the textbook and more on your experimental plan. A good plan should give you quick access to the essential information you will need while performing the experiment: quantities of chemicals (including wash solvents, drying agents, etc.), reflux times, physical properties, hazard warnings, and other useful data should be included. Much information of this kind, such as the quantities and properties of chemicals, can be summarized in tables. You can list the supplies and equipment you will need for each operation (these are usually specified in the Operations) so that you can have them cleaned and ready when you need them. You may also wish to sketch the apparatus you will be using, so that you can assemble it quickly in the laboratory.

An experimental plan should help you organize your time efficiently by listing tasks in the approximate order in which you expect to accomplish them. For example, whenever a reflux period is specified in a procedure, you will have some free time to set up the apparatus for the next step, reorganize your work area, review an operation, start a Minilab, take a melting point, record the spectrum of a previous product, or tie up other loose ends. Your plan should be flexible enough that you can alter it or deviate from it during the experiment, if there is good reason to do so. One simple and effective way of organizing your time is to use a laboratory checklist, such as the one shown in Experiment 4. You should refer to the Methodology and Procedure sections of the experiment as you prepare your checklist, as well as to the relevant Operations. It is important to read the Methodology section first, because it will help you understand the experimental procedure and may tell you how best to carry out certain operations. Leave enough space between the items on your checklist so that you can add new ones, as necessary, during the experiment. As you complete each task in the laboratory, simply check it off the list and go on to the next one.

Properties of Organic Compounds   APPENDIX VI

The following tables are to be used in conjunction with the procedures described in Part III, Systematic Organic Qualitative Analysis. Compounds that melt below ordinary ambient temperatures (about 25°C) are listed in order of increasing boiling point; those that (when pure) are generally solid at room temperature are listed in order of increasing melting point. Both melting and boiling points are specified for some borderline cases. Derivative preparations are described on pages 565–580 and are referred to by number at the head of each column of these tables. Melting points in parentheses are for derivatives that exist in more than one crystalline form or for which significantly different melting points have been reported in the literature. Sometimes the recrystallization solvent will determine the form in which a derivative crystallizes, so significant deviations from the listed melting points should not be considered conclusive proof that a sample is not the expected derivative. In cases where a given compound can form more than one product (for instance, nitro and bromo derivatives), the reaction conditions for the preparation may determine the derivative isolated (a mixture of derivatives may also result). A blank (—) in a derivative column indicates either that the derivative has not been reported in the literature or that it is not suitable for identification. See references A1–A4, A8–A11, B4, G1, or G4 in the Bibliography for physical constants and derivative melting points not listed in the tables.

Most compounds are listed by their systematic (IUPAC) names except when those names would be too lengthy.

Abbreviations Used in Tables:

d = decomposes on melting

s = sublimes at or below melting point

m = monosubstituted derivative (such as a mononitrated aromatic hydrocarbon)

di = disubstituted derivative

t = trisubstituted derivative

tet = tetrasubstituted derivative

List of tables

Table 1. Alcohols

			Deriv. m.p., °C			
Compound	b.p., °C	m.p., °C	D–1 3,5-Dinitrobenzoate	D–2 1-Naphthylurethane	D–2 Phenylurethane	D–1 4-Nitrobenzoate
Methanol	65		108	124	47	96
Ethanol	78		93	79	52	57
2-Propanol	82		123	106	88	110
2-Methyl-2-propanol*	83	*26*	142	—	136	—
2-Propen-1-ol	97		49	108	70	28
1-Propanol	97		74	105	57	35
2-Butanol	99		76	97	65	26
2-Methyl-2-butanol	102		116	72	42	85
2-Methyl-1-propanol	108		87	104	86	69
3-Pentanol	116		101	95	48	17
1-Butanol	118		64	71	61	36
2-Pentanol	120		62	74	—	24
3-Methyl-3-pentanol	123		94 (62)	104	43	69
3-Methyl-1-butanol	132		61	68	57	21
4-Methyl-2-pentanol	132		65	88	143	26
1-Pentanol	138		46	68	46	11
Cyclopentanol	141		115	118	132	62

Table 1.   Alcohols (*continued*)

Compound	b.p., °C	m.p., °C	Deriv. m.p., °C			
			D-1 3,5-Dinitro-benzoate	D-2 1-Naphthyl-urethane	D-2 Phenyl-urethane	D-1 4-Nitro-benzoate
2-Ethyl-1-butanol	148		51	60	—	—
1-Hexanol	157		58	59	42	5
Cyclohexanol★	161	*25*	113	129	82	50
Furfuryl alcohol	172		80	130	45	76
1-Heptanol	177		47	62	60	10
2-Octanol	179		32	63	114	28
1-Octanol	195		61	67	74	12
1-Phenylethanol	202		95	106	92	43
Benzyl alcohol	205		113	134	77	85
2-Phenylethanol	219		108	119	78	62
1-Decanol	231		57	73	60	30
3-Phenylpropanol	236		45	—	92	47
1-Dodecanol★	259	*24*	60	80	74	45 (42)
1-Tetradecanol		39	67	82	74	51
(−)-Menthol		44	153	119	111	62
1-Hexadecanol		49	66	82	73	58
1-Octadecanol		59	77	89	79	64
Diphenylmethanol		68	141	136	139	132
Cholesterol		148	—	176	168	185
(+)-Borneol		208	154	132 (127)	138	137 (153)

★ May be solid at or just below room temperature.

Table 2.   Aldehydes

Compound	b.p., °C	m.p., °C	Deriv. m.p., °C		
			D-4 Semi-carbazone	D-3 2,4-Dinitrophenyl-hydrazone	D-5 Oxime
Ethanal	21		162	168 (157)	47
Propanal	48		154	148 (155)	40
Propenal	52		171	165	—
2-Methylpropanal	64		125 (119)	187 (183)	—
Butanal	75		106	123	—

Table 2.　Aldehydes (*continued*)

| Compound | b.p., °C | m.p., °C | Deriv. m.p., °C | | |
			D–4 Semi-carbazone	D–3 2,4-Dinitrophenyl-hydrazone	D–5 Oxime
3-Methylbutanal	92		107	123	48
Pentanal	103		—	106 (98)	52
2-Butenal	104		199	190	119
2-Ethylbutanal	117		99	95 (130)	—
Hexanal	130		106	104	51
Heptanal	153		109	106	57
2-Furaldehyde	162		202	212 (230)	91
2-Ethylhexanal	163		254d	114 (120)	—
Octanal	171		101	106	60
Benzaldehyde	179		222	239	35
4-Methylbenzaldehyde	204		234 (215)	234	80
3,7-Dimethyl-6-Octenal	207		84 (91)	77	—
2-Chlorobenzaldehyde	213		229 (146)	213 (209)	76 (101)
4-Methoxybenzaldehyde	248		210	253d	133
Phenylethanal	*195*	33	153 (156)	121 (110)	99
2-Methoxybenzaldehyde		38	215	254	92
4-Chlorobenzaldehyde		48	230	265	110 (146)
3-Nitrobenzaldehyde		58	246	290	120
4-Nitrobenzaldehyde		106	221 (211)	320	133 (182)

Table 3.　Ketones

| Compound | b.p., °C | m.p., °C | Deriv. m.p., °C | | |
			D–4 Semi-carbazone	D–3 2,4-Dinitrophenyl-hydrazone	D–5 Oxime
Acetone	56		187	126	59
2-Butanone	80		146	118	—
3-Methyl-2-butanone	94		113	124	—
2-Pentanone	102		112 (106)	143	58
3-Pentanone	102		138	156	69
3,3-Dimethyl-2-butanone	106		157	125	75 (79)
4-Methyl-2-pentanone	117		132	95 (81)	58
2,4-Dimethyl-3-pentanone	124		160	95 (88)	34

Table 3.   Ketones (*continued*)

Compound	b.p., °C	m.p., °C	Deriv. m.p., °C		
			D–4 Semi-carbazone	D–3 2,4-Dinitrophenyl-hydrazone	D–5 Oxime
2-Hexanone	128		125	110	49
4-Methyl-3-penten-2-one	130		164 (133)	205	48
Cyclopentanone	131		210 (203)	146	56
4-Heptanone	144		132	75	—
2-Heptanone	151		123	89	—
Cyclohexanone	156		166	162	91
2,6-Dimethyl-4-heptanone	168		122	92	210
2-Octanone	173		124	58	—
Cycloheptanone	181		163	148	23
2,5-Hexanedione	194		185 (m); 224 (di)	257 (di)	137 (di)
Acetophenone	202	20	198 (203)	238	60
2-Methylacetophenone	214		205	159	61
Propiophenone	218	21	182 (174)	191	54
3-Methylacetophenone	220		203	207	57
2-Undecanone	228		122	63	44
4-Phenyl-2-butanone	235		142	127	87
3-Methoxyacetophenone	240		196	—	—
2-Methoxyacetophenone	245		183	—	83 (96)
4-Methylacetophenone	226	28	205	258	88
4-Methoxyacetophenone		38	198	228	87
4-Phenyl-3-buten-2-one		42	187	227 (223)	117
Benzophenone		48	167	238	144
2-Acetonaphthone		54	235	262d	145
3-Nitroacetophenone		80	257	228	132
9-Fluorenone		83	234	283	195
(±)-Camphor		179	237	177	118

Table 4.   Amides

Compound	b.p., °C	m.p., °C	Deriv. m.p., °C	
			D–6 Carboxylic acid	D–7 N–Xanthylamide
Formamide	195d		—	184
Propanamide		81	—	211

Table 4.   Amides (*continued*)

Compound	b.p., °C	m.p., °C	Deriv. m.p., °C D–6 Carboxylic acid	D–7 N–Xanthylamide
Ethanamide		82	17	240
Heptanamide		96	—	155
Nonanamide		99	12	148
Hexanamide		100	—	160
Hexadecanamide		106	63	142
Pentanamide		106	—	167
Octadecanamide		109	69	141
Butanamide		115	—	187
Chloroacetamide		120	61 (53)	209
4-Methylpentanamide		121	—	160
Succinimide		126	185	247
2-Methylpropanamide		129	—	211
Benzamide		130	122	223
3-Methylbutanamide		136	—	183
o-Toluamide		143	104 (108)	200
Furamide		143	133	210
Phenylacetamide		156	77s	195
p-Toluamide		159	180s	225
4-Nitrobenzamide		201	240	233
Phthalimide		238	210d	177

Table 5.   Primary and secondary amines

Compound	b.p., °C	m.p., °C	D–8 Benzamide	D–9 *p*-Toluene-sulfonamide	D–10 Phenylthio-urea	D–12 Picrate
t-Butylamine	44		134	—	120	198
Propylamine	48		84	52	63	135
Diethylamine	56		42	60	34	155
sec-Butylamine	63		76	55	101	140
2-Methylpropylamine	69		57	78	82	150
Butylamine	77		42	—	65	151
Diisopropylamine	84	—	—	—	—	140

Table 5.   Primary and secondary amines (*continued*)

Compound	b.p., °C	m.p., °C	D–8 Benzamide	D–9 p-Toluene-sulfonamide	D–10 Phenylthio-urea	D–12 Picrate
Pyrrolidine	89		—	123	—	112 (164)
3-Methylbutylamine	95		—	65	102	138
Pentylamine	104		—	—	69	139
Piperidine	106		48	96	101	152
Dipropylamine	109		—	—	69	75
Morpholine	128		75	147	136	146
Pyrrole	131		—	—	143	69d
Hexylamine	132		40	—	77	126
Cyclohexylamine	134		149	—	148	—
Diisobutylamine	139		—	—	113	121
N-Methylcyclohexylamine	147		86	—	—	170
Dibutylamine	159		—	—	86	59
N-Ethylbenzylamine	181		—	95	—	118
Aniline	184		163	103	154	198 (180)
Benzylamine	185		105	116 (185)	156	199 (194)
N-Methylaniline	196		63	94	87	145
o-Toluidine	200		144	108	136	213
m-Toluidine	203		125	114	104 (92)	200
N-Ethylaniline	205		60	87	89	138
2-Chloroaniline	209		99	105 (193)	156	134
2,6-Dimethylaniline	215		168	212	204	180
2-Methoxyaniline	225		60	127	136	200
2-Ethoxyaniline	229		104	164	137	—
3-Chloroaniline	230		120	138 (210)	124 (116)	177
4-Ethoxyaniline	250		173	106	136	69
Dicyclohexylamine	255d		153 (57)	—	—	173
N-Benzylaniline		37	107	149	103	48
p-Toluidine		44	158	118	141	182d
Diphenylamine		54	180 (109)	141	152	182
4-Methoxyaniline		58	154	114	157 (171)	170
4-Bromoaniline		66	204	101	148	180
2-Nitroaniline		71	110 (98)	142	—	73

Table 5.   Primary and secondary amines (*continued*)

Compound	b.p., °C	m.p., °C	D–8 Benzamide	Deriv. m.p., °C D–9 p-Toluene-sulfonamide	D–10 Phenylthio-urea	D–12 Picrate
4-Chloroaniline		72	192	95 (119)	152	178
1,2-Diaminobenzene		102	301 (di)	260 (di)	—	208
3-Nitroaniline		114	155 (di)	138	160	143
1,4-Diaminobenzene		142	300 (di)	266 (di)	—	—
4-Nitroaniline		147	199 (di)	191	—	100

Note: See other Qualitative Analysis references listed in the Bibliography for the melting points of benzene-sulfonamides and 1-naphthylthioureas.

Table 6.   Tertiary amines

Compound	b.p., °C	m.p., °C	Deriv. m.p., °C D–12 Picrate	D–11 Methiodide
Triethylamine	89		173	280
Pyridine	115		167	117
2-Methylpyridine	129		169	230
3-Methylpyridine	143		150	92
4-Methylpyridine	143		167	152
Tripropylamine	157		116	207
2,4-Dimethylpyridine	159		183 (169)	113
N,N-Dimethylbenzylamine	183		93	179
N,N-Dimethylaniline	193		163	228d
Tributylamine	216 (211)		105	186
N,N-Diethylaniline	217		142	102
Quinoline	237		203	133 (72)
Isoquinoline	243	26	222	159
Tribenzylamine	—	91	190	184
Acridine	—	111	208	224

Table 7. Carboxylic acids

Compound	b.p., °C	m.p., °C	Deriv. m.p., °C			
			D–14 p-Toluidide	D–14 Anilide	D–13 Amide	D–15 p-Nitrobenzyl Ester
Formic acid	101		53	50	43	31
Acetic acid	118		148	114	82	78
Propenoic acid	139		141	104	85	—
Propanoic acid	141		124	103	81	31
2-Methylpropanoic acid	154		107	105	128	—
Butanoic acid	164		75	95	115	35
3-Methylbutanoic acid	176		107	109	135	—
Pentanoic acid	186		74	63	106	—
2-Chloropropanoic acid	186		124	92	80	—
Dichloroacetic acid	194		153	118	98s	—
2-Methylpentanoic acid	196		80	95	79	—
Hexanoic acid	205		75	95	101	—
2-Bromopropanoic acid	205d	*24*	125	99	123	—
Octanoic acid	239		70	57	107	—
Nonanoic acid	254		84	57	99	—
Decanoic acid		32	78	70	108	—
2,2-Dimethylpropanoic acid	*164*	35	—	129 (133)	178 (154)	—
Dodecanoic acid		44	87	78	110 (99)	—
3-Phenylpropanoic acid		48	135	98	105	36
Tetradecanoic acid		54	93	84	103	—
Hexadecanoic acid		62	98	90	106	42
Chloroacetic acid		63	162	137	120	—
Octadecanoic acid		70	102	95	109	—
trans-2-Butenoic acid		72	132	118	160	67
Phenylacetic acid		77	136	118	156	65
2-Methoxybenzoic acid		101	—	131	129	113
Oxalic acid (dihydrate)		101	268 (di)	254 (246) (di)	419d (di)	204 (di)
2-Methylbenzoic acid		104	144	125	142	91
Nonanedioic acid		106	201 (di)	186 (di)	175 (di)	44
3-Methylbenzoic acid		112	118	126	94	87
Benzoic acid		122	158	163	130	89

Table 7.   Carboxylic acids (*continued*)

Compound	b.p., °C	m.p., °C	Deriv. m.p., °C			
			D–14 p-Toluidide	D–14 Anilide	D–13 Amide	D–15 p-Nitrobenzyl Ester
Maleic acid		130	142 (di)	187 (di)	181 (m) 266 (di)	91
Decanedioic acid		133	201 (di)	122 (m) 200 (di)	170 (m) 210 (di)	73 (di)
Cinnamic acid		133	168	153	147	117
Propanedioic acid		135	86 (m) 253 (di)	132 (m) 230 (di)	50 (m) 170 (di)	86
2-Chlorobenzoic acid		140	131	118	140	106
3-Nitrobenzoic acid		140	162	155	143	141
Diphenylacetic acid		148	172	180	167	—
2-Bromobenzoic acid		150	—	141	155	110
Hexanedioic acid		152	238	151 (m) 241 (di)	125 (m) 224 (di)	106
4-Methylbenzoic acid		180s	160 (165)	145	160	104
4-Methoxybenzoic acid		184	186	169	167 (163)	132
Butanedioic acid		188	180 (m) 255 (di)	143 (m) 230 (di)	157 (m) 260 (di)	—
3,5-Dinitrobenzoic acid		205	—	234	183	157
Phthalic acid		210d	201 (di)	253 (di)	220 (di)	155
4-Nitrobenzoic acid		240	204	211	201	168
4-Chlorobenzoic acid		242	—	194	179	129
Terephthalic acid		>300s	—	337	—	263 (di)

Table 8.   Esters

Compound	b.p., °C	m.p., °C	Deriv. m.p., °C			
			D–16 Carboxylic acid	D–16 Alcohol or phenol	D–18 3,5-Dinitro-benzoate	D–17 N-Benzyl-amide
Ethyl formate	54		8	—	93	60
Methyl acetate	57		17	—	108	61
Ethyl acetate	77		17	—	93	61
Methyl propanoate	80		—	—	108	43
Methyl acrylate	80		13	—	108	237
Isopropyl acetate	91		17	—	123	61

Table 8. Esters (*continued*)

Compound	b.p., °C	m.p., °C	D-16 Carboxylic acid	D-16 Alcohol or phenol	D-18 3,5-Dinitro-benzoate	D-17 N-Benzyl-amide
t-Butyl acetate	98		17	26	142	61
Ethyl propanoate	99		—	—	93	43
Methyl 2,2-dimethylpropanoate	101		35	—	108	—
Propyl acetate	102		17	—	74	61
Methyl butanoate	102		—	—	108	38
Ethyl 2-methylpropanoate	111		—	—	93	87
sec-Butyl acetate	112		17	—	76	61
Methyl 3-methylbutanoate	117		—	—	108	54
Isobutyl acetate	117		17	—	87	61
Ethyl butanoate	122		—	—	93	38
Butyl acetate	126		17	—	64	61
Methyl pentanoate	128		—	—	108	43
Ethyl 3-methylbutanoate	135		—	—	93	54
3-Methylbutyl acetate	142		17	—	61	61
Ethyl chloroacetate	145		63	—	93	—
Pentyl acetate	149		17	—	46	61
Ethyl hexanoate	168		—	—	93	53
Hexyl acetate	172		17	—	58	61
Cyclohexyl acetate	175		17	25	113	61
Dimethyl malonate	182		135	—	108	142
Diethyl oxalate	185		101★	—	93	223
Heptyl acetate	192		17	—	47	61
Phenyl acetate	197		17	42	146	61
Methyl benzoate	199		122	—	108	105
Diethyl malonate	199		135	—	93	142
o-Tolyl acetate	208		17	31	135	61
m-Tolyl acetate	212		17	12	165	61
Ethyl benzoate	213		122	—	93	105
p-Tolyl acetate	213		17	36	189	61
Methyl *o*-toluate	215		104	—	108	—
Benzyl acetate	217		17	—	113	61
Diethyl succinate	218		188	—	93	206

Table 8.   Esters (*continued*)

Compound	b.p., °C	m.p., °C	Deriv. m.p., °C			
			D–16 Carboxylic acid	D–16 Alcohol or phenol	D–18 3,5-Dinitro-benzoate	D–17 N-Benzyl-amide
Isopropyl benzoate	218		122	—	123	105
Methyl phenylacetate	220		77s	—	108	122
Diethyl maleate	223		137	—	93	150
Ethyl phenylacetate	228		77s	—	93	122
Propyl benzoate	230		122	—	74	105
Diethyl adipate	245		152	—	93	189
Butyl benzoate	250		122	—	64	105
Ethyl cinnamate	271		133	—	93	226
Dimethyl phthalate	284		210d	—	108	179
(+)-Bornyl acetate	*226*	27	17	208	154	61
Methyl p-toluate		33	180s	—	108	133
Methyl cinnamate		36	133	—	108	226
Benzyl cinnamate		39	133	—	113	226
1-Naphthyl acetate		49	17	94	217	61
Ethyl p-nitrobenzoate		56	240	—	93	—
Phenyl benzoate		69	122	42	146	105
2-Naphthyl acetate		71	17	123	210	61
p-Tolyl benzoate		71	122	36	189	105
Methyl m-nitrobenzoate		78	140	—	108	101
Methyl p-nitrobenzoate		96	240	—	108	142

Note: Additional derivatives of the acid and alcohol portions of most esters can be found in Tables 1 and 7.
★ dihydrate; the anhydrous acid melts at 190°C.

Table 9.   Alkyl halides

Compound	b.p., °C	Density C–10 d_4^{20}, g/ml	Deriv. m.p., °C D–19 S-Alkylthiuronium-picrate
Bromoethane	38	1.461	188
2-Bromopropane	60	1.314	196
1-Chloro-2-methylpropane	69	0.879	167 (174)
3-Bromopropene	71	1.398	155
1-Bromopropane	71	1.354	177
Iodoethane	72	1.936	188

Table 9.   Alkyl halides (*continued*)

Compound	b.p., °C	Density C–10 d_4^{20}, g/ml	Deriv. m.p., °C D–19 S-Alkylthiuronium-picrate
1-Chlorobutane	78	0.884	177
2-Iodopropane	89	1.703	196
1-Bromo-2-methylpropane	93	1.264	167 (174)
1-Chloro-3-methylbutane	100	0.875	173
1-Bromobutane	101	1.274	177
1-Iodopropane	102	1.749	177
3-Iodopropene	102	1.848	155
1-Chloropentane	108	0.882	154
1-Bromo-3-methylbutane	119	1.207	173 (179)
2-Iodobutane	119	1.595	166
1-Iodo-2-methylpropane	120	1.606	167 (174)
1-Bromopentane	129	1.218	154
1-Iodobutane	131	1.617	177
1-Chlorohexane	134	0.876	157
1-Iodo-3-methylbutane	148	1.503	173
1-Iodopentane	155	1.516	154
1-Bromohexane	155	1.173	157
1-Iodohexane	181	1.439	157
1-Bromooctane	201	1.112	134
1-Iodooctane	225	1.330	134

Table 10.   Aryl halides

Compound	b.p., °C	m.p., °C	Density C–10 d_4^{20}, g/ml	Deriv. m.p., °C	
				D–20 Nitro derivative	D–21 Carboxylic acid
Chlorobenzene	132		1.106	52	—
Bromobenzene	156		1.495	75 (70)	—
2-Chlorotoluene	159		1.083	63	140
3-Chlorotoluene	162		1.072	91	158
4-Chlorotoluene	162		1.071	38 (m)	242
1,3-Dichlorobenzene	173		1.288	103	—
1,2-Dichlorobenzene	181		1.306	110	—
2-Bromotoluene	182		1.423	82	150

Table 10.   Aryl halides (*continued*)

Compound	b.p., °C	m.p., °C	Density C–10 d_4^{20}, g/ml	Deriv. m.p., °C D–20 Nitro derivative	D–21 Carboxylic acid
3-Bromotoluene	184		1.410	103	155
Iodobenzene	188		1.831	171 (m)	—
2,6-Dichlorotoluene	199		1.269	50 (m)	139
2,4-Dichlorotoluene	200		1.249	104	164
3-Iodotoluene	204		1.698	108	187
2-Iodotoluene	211		1.698	103 (m)	162
1-Chloronaphthalene	259		1.191	180	—
4-Bromotoluene	*184*	28	—	—	251
4-Iodotoluene		35	—	—	270
1,4-Dichlorobenzene		53	—	106 54 (m)	—
2-Chloronaphthalene		56	—	175	—
1,4-Dibromobenzene		89	—	84	—

Note: Nitro derivatives signified by (m) are mononitro compounds; all others are dinitro derivatives.

Table 11.   Aromatic hydrocarbons

Compound	b.p., °C	m.p., °C	Deriv. m.p., °C D–20 Nitro derivative	D–21 Carboxylic acid	D–12 Picrate
Benzene	80		89 (di)	—	84u
Toluene	111		70 (di)	122	88u
Ethylbenzene	136		37 (t)	122	96u
1,4-Xylene	138		139 (t)	300s	90u
1,3-Xylene	139		183 (t)	330s	91u
1,2-Xylene	144		118 (di)	210d	88u
Isopropylbenzene	152		109 (t)	122	—
Propylbenzene	159		—	122	103u
1,3,5-Trimethylbenzene	165		86 (di) 235 (t)	350 (t)	97u
t-Butylbenzene	169		62 (di)	122	—
4-Isopropyltoluene	177		54 (di)	300s	—
1,3-Diethylbenzene	181		62 (t)	330s	—

Table 11.   Aromatic hydrocarbons (*continued*)

| Compound | b.p., °C | m.p., °C | Deriv. m.p., °C | | |
			D–21 Nitro derivative	D–21 Carboxylic acid	D–12 Picrate
1,2,3,4–Tetrahydronaphthalene	206		96 (di)	210d	—
Diphenylmethane	*262*	26	172 (tet)	—	—
1,2-Diphenylethane		53	180 (di) 169 (tet)	—	—
Naphthalene		80	61 (m)	—	149
Triphenylmethane		92	206 (t)	—	—
Acenaphthene		96	101 (m)	—	161
Phenanthrene			—	—	144 (133)
Fluorene		114	199 (di) 156 (m)	—	87 (77)
Anthracene		216	—	—	138u

Note: Picrates designated u are unstable and cannot easily be purified by recrystallization.

Table 12.   Phenols

| Compound | b.p., °C | m.p., °C | Deriv. m.p., °C | | |
			D–24 1-Naphthyl-urethane	D–23 Bromo derivative	D–22 Aryloxy-acetic acid
2-Chlorophenol	176		120	49 (m); 76 (di)	145
3-Methylphenol	202		128	84 (t)	—
2-Methylphenol	*192*	31	142	56 (di)	152
4-Methylphenol	*232*	36	146	49 (di); 108 (tet)	135
Phenol	*182*	42	133	95 (t)	99
4-Chlorophenol		43	166	33 (m); 90 (di)	156
2,4-Dichlorophenol		43	—	68	135 (141)
2-Nitrophenol		45	113	117 (di)	158
4-Ethylphenol		47	128	—	97
5-Methyl-2-isopropylphenol		50	160	55 (m)	149
3,4-Dimethylphenol		63	142	171 (t)	163
4-Bromophenol		64	169	95 (t)	157
3,5-Dichlorophenol		68	—	189 (t)	—
2,5-Dimethylphenol		75	173	178 (t)	118
1-Naphthol		94	152	105 (di)	194

Table 12. Phenols (*continued*)

| Compound | b.p., °C | m.p., °C | Deriv. m.p., °C | | |
			D–24 1–Naphthylurethane	D–23 Bromo derivative	D–22 Aryloxyacetic acid
3-Nitrophenol		97	167	91 (di)	156
4-*t*-Butylphenol		100	110	50 (m)	86
1,2-Dihydroxybenzene		105	175	193 (tet)	—
1,3-Dihydroxybenzene		110	206	112 (di)	195
4-Nitrophenol		114	150	142 (di)	187
2-Naphthol		123	157	84 (m)	154
1,2,3-Trihydroxybenzene		133	—	158 (di)	198
1,4-Dihydroxybenzene		172	—	186 (di)	250

APPENDIX VII   The Chemical Literature

==

Some important literature sources for organic chemistry are listed in the following Bibliography under twelve general categories. Each source is referred to here by the category letter and its number within the category. The Background section of Experiment 50 tells how to carry out a literature search and how to use *Beilstein* ⟨A9⟩, *Chemical Abstracts* ⟨K1⟩, and *Fieser* ⟨B20⟩.

Category A: General Reference Works

References A1–A9 provide information of varying depth about organic compounds. The *CRC Handbook of Chemistry and Physics* ⟨A1⟩ and *Lange's Handbook of Chemistry* ⟨A2⟩ tabulate physical properties and other data for many common organic compounds and contain a large amount of useful information about chemistry. The *Merck Index* ⟨A3⟩ is an excellent source of information on approximately 10,000 organic and inorganic compounds; it describes their uses and hazardous properties and provides detailed physical and structural data. The *CRC Atlas of Spectral Data* ⟨A4⟩ gives physical constants and detailed spectral parameters (IR, NMR, UV, and MS) for about 21,000 organic compounds. The more recent *CRC Handbook of Data on Organic Compounds* ⟨A5⟩ provides simlar information in a more compact form, but only gives references to published spectra. The *Aldrich Catalog* ⟨A6⟩ lists the physical properties of reagent-grade chemicals manufactured by the Aldrich Chemical Company and, for many other chemicals, also gives hazard warnings, procedures for safe disposal, references to published Aldrich spectra, and references to listings in the *Merck Index, Beilstein* ⟨A9⟩, and Fieser ⟨B20⟩. The *Encyclopedia of the Terpenoids* ⟨A7⟩ provides information that may not be readily available elsewhere about some natural compounds; the *Encyclopedia of the Alkaloids* by the same author covers nitrogen-containing natural products. The *Dictionary of Organic Compounds* ⟨A8⟩ is an important six-volume set, updated by annual supplements, that gives structures, physical constants, hazard descriptions, sources, uses, derivatives, and bibliographic references for more than 150,000 organic compounds. Figure 1 (see page A-30) shows the kind of information provided by some of these reference works.

Beilstein ⟨A9⟩ reports on the structure, occurrence, preparation, and properties of virtually every organic compound that appeared in the chemical literature during the periods covered. The fifth supplemental series of *Beilstein* is being published in an English-language edition; when it is complete, *Beilstein* will cover the chemical literature through 1979. The basic series and four previous supplemental series are published in German and cover the literature only through 1959.

References A10 and A11 list properties that can be used to identify organic compounds. Reference A12 is a multivolume compilation of solubility data that can be used to locate suitable reaction or recrystallization solvents. The *Kirk-Othmer Encyclopedia of Chemical Technology* ⟨A13⟩ contains comprehensive and up-to-date articles on a variety of chemical topics; it offers particularly good coverage of industrial chemistry and commercial products. Reference A14 is a condensed version of *Kirk-Othmer,* and references A15–A17 are useful sources of general information about chemistry. References A18 and A19 define and illustrate many of the terms used in organic chemistry. The *Ring Systems Handbook* ⟨A20⟩ is an up-to-date catalog of nearly 60,000 ring and cage systems that have appeared in *Chemical Abstracts* ⟨K1⟩, with their names, structures, preferred numbering systems, and other information. German-English dictionaries for chemists, such as references A21 and A22, can help you cope with *Beilstein* or *Houben-Weyl* ⟨B8⟩.

You should always read the introduction or explanatory material in a general reference work or abstracting journal before you try to use it; this will invariably save you time in the long run. The introductory section of a reference work may (1) describe the content and organization of the material, (2) list symbols and abbreviations, and (3) describe the system of nomenclature used. Before you can find information about a compound in any reference work, you must know how it is listed, and different sources often use very different naming systems. For example, the *Merck Index* emphasizes therapeutic uses of compounds, so it lists aspirin under that name; but in *Lange's Handbook* you will find aspirin listed as acetylsalicylic acid, and in the *CRC Handbook* ⟨A1⟩ as salicylic acid acetate. To locate most compounds in the *CRC Handbook,* you must also know that they are entered under the name of the parent compound; thus, 2,4-dinitrobenzene is listed as "benzene, 2,4-dinitro."

Indexes will often provide the quickest and most reliable means of access to a given entry in a reference work. When using the *Merck Index* or the

Figure 1.   Entries for benzoic acid from some reference works

A. *Dictionary of Organic Compounds,* 5th ed. (Reprinted with permission from *Dictionary of Organic Compounds,* 5th ed., edited by J. R. A. Pollock and R. Stevens. Copyright Chapman & Hall, London, 1982. The 5th edition is updated by annual supplements that contain both revised versions of existing entries and completely new entries.)

Benzoic acid, ✕   **B-00378**
[65-85-0]

PhCOOH

$C_7H_6O_2$ M 122
Preservative in the food industry. Used in manuf. of preservatives, plasticisers, alkyd resin coatings and caprolactam. Leaflets or needles (H_2O). V. spar. sol. H_2O. Mp 122° (ca. 100° subl.). Bp 249°, Bp_{10} 133°. pK_a 4.2 (25°). Steam-volatile.

▷Toxic by skin absorption. DG0875000.

Me ester: [93-58-3]. Used in perfumery and flavourings. Liq. d_{23}^{25} 1.09. Fp −12.3°. Bp 199.6°, Bp_{24} 96-8°.

▷DH3850000.

Et ester: [93-89-0]. Polymerisation catalyst. Used in perfumery and flavourings. Liq. d_4^{25} 1.04. Fp −34°. Bp 212.9°, Bp_{10} 87.2°.

▷Mod. toxic. DH0200000.

Propyl ester: [2315-68-6]. Flavour ingredient. d_{15}^{15} 1.03. Bp 230°.

Isopropyl ester: [939-48-0]. Polymerisation catalyst, flavour ingredient. d_{15}^{15} 1.02. Bp 218-9°.

▷DH3150000.

Butyl ester: [136-60-7]. Dye carrier; used in perfumery. d_{15}^{15} 1.01. Bp 248-9°.

▷DG4925000.

tert-Butyl ester: [774-65-2]. Bp_2 96°.

Benzyl ester: [120-51-4]. *Benzyl benzoate.* Contained in Peru balsam. Tobacco flavouring agent, insect repellant component. Leaflets. d^{18} 1.11. Mp 21°. Bp 323-4° (316-7°).

▷Mod. toxic. DG4200000.

Fluoride: [455-32-3]. *Benzoyl fluoride.* Fuming liq. Bp 159-61°. Hydrolysed by hot H_2O.

Chloride: [98-88-4]. *Benzoyl chloride.* Polymerisation catalyst, benzoylating agent. Can be used for synth. of aliphatic acid chlorides. Fuming liq. d_{15}^{15} 1.22. Fp −1°. Bp 197°.

▷Highly irritant, causes burns, violent reaction with dimethyl sulphoxide. DM6600000.

Bromide: [618-32-6]. *Benzoyl bromide.* Fuming liq. d^{15} 1.57. Fp −24°. Bp 218-9°, $Bp_{0.05}$ 48-50°.

Iodide: [618-38-2]. *Benzoyl iodide.* Needles. Mp 3°. Bp_{20} 128°.

Anhydride: [93-97-0]. *Benzoic anhydride.* Cross-linking agent for polymers. Acylation and decarboxylating agent, can be used in polymer-linked form. Rhombic prisms. d_4^{15} 1.99. Mp 42°. Bp 360°.

▷Mild irritant and allergen

Amide: see Benzamide, B-00140 [55-21-0].

▷CU8700000.

Ethylamide: [614-17-5]. Needles (H_2O). Mp 70-1°. Bp 298-300°.

▷CV4920000.

Anilide: Benzanilide. Benzoylaniline. Leaflets. Mp 163°. Bp_{10} 117-9°.

Toluidide: see 4-Methylaniline, M-00814
Xylidide: see 2,5-Dimethylaniline, D-05527
N-Chloroanilide: [5014-47-1]. N-Chlorobenzanilide. Needles (ligroin). Mp 81.5-82°.

Nitrile: [100-47-0]. *Phenyl cyanide. Cyanobenzene.* d_{15}^{15} 1.01. Fp −13°. Bp 190.7°, Bp_{10} 69°.

▷D12450000.

Hydrazide: Plates. Mp 112.5°.

Hydroxamate: see N-Hydroxybenzamide, H-01254
Azide: Benzazide. Benzoylazimide. Plates. Mp 32°.

▷Explodes on heating

Org. Synth., Coll.Vol., **1**, 75, 361 (*synth, deriv*)
Jesson, J.P. *et al, Proc. R. Soc. London, Ser. A,* 1962, **268**, 68 (*raman*)
Beynon, J.H. *et al, Z. Naturforsch., A,* 1965, **20**, 883 (*ms*)
Evans, H.B. *et al, J. Phys. Chem.,* 1968, **72**, 2552 (*pmr*)
Fauvet, G. *et al, Acta Crystallogr., Sect. B,* 1978, **34**, 1376 (*cryst struct, nitrile*)
Fieser, M. *et al, Reagents for Organic Synthesis,* Wiley, 1967-78, **1**, 49, 1004; **5**, 23, 24
Bretherick, L., *Handbook of Reactive Chemical Hazards,* 2nd Ed., Butterworths, London and Boston, 1979, 394, 622
Sax, N.I., *Dangerous Properties of Industrial Materials,* 5th Ed., Van Nostrand-Reinhold, 1979, 407, 408, 410, 651
Hazards in the Chemical Laboratory, (Bretherick, L., Ed.), 3rd Ed., Royal Society of Chemistry, London, 1981, 193

Figure 1.   *(continued)*

B. *The Merck Index,* 10th ed. (Reprinted with permission from *The Merck Index,* 10th ed., edited by M. Windholtz. Copyright Merck and Co., Inc., Rahway, N.J., 1983.)

1093. Benzoic Acid. Benzenecarboxylic acid; phenylformic acid; dracylic acid. $C_7H_6O_2$; mol wt 122.12. C 68.84%, H 4.95%, O 26.20%. Occurs in nature in free and combined forms. Gum benzoin may contain as much as 20%. Most berries contain appreciable amounts (around 0.05%). Excreted mainly as hippuric acid by almost all vertebrates, except fowl. Mfg processes include the air oxidation of toluene, the hydrolysis of benzotrichloride, and the decarboxylation of phthalic anhydride: Faith, Keyes & Clark's *Industrial Chemicals,* F. A. Lowenheim, M. K. Moran, Eds. (Wiley-Interscience, New York, 4th ed., 1975) pp 138–144. Lab prepn from benzyl chloride: A. I. Vogel, *Practical Organic Chemistry* (Longmans, Londonα, 3rd ed, 1959) p 755; from benzaldehyde: Gattermann-Wieland, *Praxis des organischen Chemikers* (de Gruyter, Berlin, 40th ed, 1961) p 193. Prepn of ultra-pure benzoic acid for use as titrimetric and calorimetric standard: Schwab, Wicher, *J. Res. Nat. Bur. Standards* **25**, 747 (1940). *Review:* A. E. Williams in Kirk-Othmer *Encyclopedia of Chemical Technology* **vol. 3** (Wiley-Interscience, New York, 3rd ed., 1978) pp 778–792.

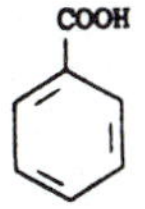

Monoclinic tablets, plates, leaflets. d 1.321 (also reported as 1.266). mp 122.4°. Begins to sublime at around 100°. bp$_{760}$ 249.2°; bp$_{400}$ 227°; bp$_{200}$ 205.8°; bp$_{100}$ 186.2°; bp$_{60}$ 172.8°; bp$_{40}$ 162.6°; bp$_{20}$ 146.7°; bp$_{10}$ 132.1°. Volatile with steam. Flash pt 121–131°. K at 25°: 6.40 × 10^{-5}; pH of satd soln at 25°: 2.8. Soly in water (g/l) at 0° = 1.7; at 10° = 2.1; at 20° = 2.9; at 25° = 3.4; at 30° = 4.2; at 40° = 6.0; at 50° = 9.5; at 60° = 12.0; at 70° = 17.7; at 80° = 27.5; at 90° = 45.5; at 95° = 68.0. Mixtures of excess benzoic acid and water form two liquid phases beginning at 89.7°. The two liquid phases unite at the critical soln temp of 117.2°. Composition of critical mixture: 32.34% benzoic acid, 67.66% water: see Ward, Cooper, *J. Phys. Chem.* **34**, 1484

(1930). One gram dissolves in 2.3 ml cold alc, 1.5 ml boiling alc, 4.5 ml chloroform, 3 ml ether, 3 ml acetone, 30 ml carbon tetrachloride, 10 ml benzene, 30 ml carbon disulfide, 23 ml oil of turpentine; also sol in volatile and fixed oils, slightly sol in petr ether. The soly in water is increased by alkaline substances, such as borax or trisodium phosphate, *see also* Sodium Benzoate.

Barium salt dihydrate, $C_{14}H_{10}BaO_4.2H_2O$, *barium benzoate.* Nacreous leaflets. *Poisonous!* Soluble in about 20 parts water; slightly sol in alc.

Calcium salt trihydrate, $C_{14}H_{10}CaO_4.3H_2O$, *calcium benzoate.* Orthorhombic crystals or powder. d 1.44. Soluble in 25 parts water; very sol in boiling water.

Cerium salt trihydrate, $C_{21}H_{15}CeO_6.3H_2O$, *cerous benzoate.* White to reddish-white powder. Sol in hot water or hot alc.

Copper salt dihydrate, $C_{14}H_{10}CuO_4.2H_2O$, *cupric benzoate.* Light blue, cryst powder. Slightly soluble in cold water, more in hot water; sol in alc or in dil acids with separation of benzoic acid.

Lead salt dihydrate, $C_{14}H_{10}O_4Pb.2H_2O$, *lead benzoate.* Cryst powder. *Poisonous!* Slightly sol in water.

Manganese salt tetrahydrate, $C_{14}H_{10}MnO_4.4H_2O$, *manganese benzoate.* Pale-red powder. Sol in water, alc. Also occurs with $3H_2O$.

Nickel salt trihydrate, $C_{14}H_{10}NiO_4.3H_2O$, *nickel benzoate.* Light-green odorless powder. Slightly sol in water; sol in ammonia; dec by acids.

Potassium salt trihydrate, $C_7H_5KO_2.3H_2O$, *potassium benzoate.* Cryst powder. Sol in water, alc.

Silver salt, $C_7H_5AgO_2$, *silver benzoate.* Light-sensitive powder. Sol in 385 parts cold water, more sol in hot water; very slightly sol in alc.

Uranium salt, $C_{14}H_{10}O_6U$, *uranium benzoate, uranyl benzoate.* Yellow powder. Slightly sol in water, alc.

Toxicity: Mild irritant to skin, eyes, mucous membranes.

USE: Preserving foods, fats, fruit juices, alkaloidal solns, etc; manuf benzoates and benzoyl compds, dyes; as a mordant in calico printing; for curing tobacco. As standard in volumetric and calorimetric analysis.

THERAP CAT: Pharmaceutic aid (antifungal agent).

THERAP CAT (VET): Has been used with salicylic acid as a topical antifungal.

C. *CRC Handbook of Data on Organic Compounds.* (Reprinted with permission from *CRC Handbook of Data on Organic Compounds.* Copyright CRC Press, Inc., Boca Raton, Fla.)

HODOC No. Beilstein No. CAS No.	Name Synonym Line formula	Refractive index Mol. wt. Mol. formula	Melting point Boiling point Density	Crystalline form Color Specific rotation	Solubility, > 10%
B01467 9^4, 273 65-85-0	Benzoic acid $C_6H_5CO_2H$	1.504^{132} 122.12 $C_7H_6O_2$	122.4 249, 133^{10} 1.2659$^{15/4}$, 1.0749^{130}	mcl lf or nd	AC, Bz, Et$_2$O, EtOH

Figure 1. (*continued*)

D. *Lange's Handbook of Chemistry*, 13th ed. (Reprinted with permission from *Lange's Handbook of Chemistry*, 13th ed., by N. A. Lange, edited by J. A. Dean. Copyright McGraw-Hill, Inc., New York, 1985.)

No.	Name	Formula	Formula weight	Beilstein reference	Density	Refractive index	Melting point	Boiling point	Flash point	Solubility in 100 parts solvent
b55	Benzoic acid	C_6H_5COOH	122.12	9, 92	1.080		122.4	133^{10mm}	121	0.29 aq; 43 alc; 10 bz; 22 chl; 33 eth, acet

E. *CRC Atlas of Spectral Data and Physical Constants for Organic Compounds*. (Reprinted with permission from *CRC Atlas of Spectral Data and Physical Constants*. Copyright CRC Press, Inc., Boca Raton, Fla.)

Atlas No. CAS No. References	Name Formula Line Formula	Wiswesser Line Notation	m.p. b.p. °C	Density Ref. Ind., n_D Sp. Rot., [α]	Solubility†	Infrared Raman	Ultraviolet	¹H NMR ¹³C NMR	Mass Spec.	M Wt. m/e
b1664 65-85-0 **B9**², 72	Benzoic acid $C_7H_6O_2$ $C_6H_5CO_2H$	QVR	122.4 249⁷⁶⁰ 133¹⁰	1.26591^5 1.0749^{130} 1.504^{132}	H_2O 2; al 4 eth 4; ace 3 bz 3; ctc 3 chl 3; lig 2 tol 3 MeOH 3	μCOB 6994 3130 1750 1590 1450 1350 1280 1180 1080 1030 720 SAD 6 3070 3000 1630 1600 1450 1330 1290 1180 1160 1130 1030 1000 850 810 790 710 660 620 420 250 190 110	SAD 252 279(729 272(893) 228(11900) MeOH	SAD 57 7.5 8.1 12.8 CCl₄ JJ 230 FT 128.4 129.4 130.2 133.7 172.6 CDCl₃	WILEY 500 105(100) 77(85) 122(74) 51(60) 50(39) 39(15) 74(12) 78(10)	122.12 122.04

Dictionary of Organic Compounds, you should ordinarily consult the name index first to find the name (or number) under which a compound is listed. If you can't find a compound listed under a particular name, you may be able to locate it in a formula index. In most formula indexes, carbon and hydrogen are listed first, followed by the other elements in alphabetical order; the formulas themselves are listed in order of increasing subscript numbers, going from left to right.

Category B: Organic Synthesis

Organic Syntheses ⟨B1⟩ is a continuing series that provides a good selection of carefully tested synthetic procedures. *Organic Reactions* ⟨B2⟩ contains very comprehensive monographs dealing with many of the reactions used for organic syntheses; some representative procedures and numerous references to literature procedures are included in each monograph. Both references B3 and B4 give detailed procedures for a number of selected organic syntheses, and the latter is also an excellent source of information about laboratory techniques. References B5–B7 describe synthetic procedures for photochemical and polymer syntheses. *Houben-Weyl* ⟨B8⟩ is a multivolume set (in German) that includes descriptions of laboratory techniques as well as synthetic methods. References B9–B12 describe a number of synthetic methods and give some general or representative procedures; *Weygand/Hilgetag* and *Organicum* provide good coverage of "classical" syntheses, whereas the works by Sandler and Karo and by Buehler and Pearson include modern synthetic reactions. Reference B13 is a very comprehensive guide (published annually) to synthetic procedures from the primary literature; each entry gives a brief outline of the procedure and cites its source. References B14 and B15 describe a number of modern synthetic methods in detail. References B16–B19 serve primarily as guides to synthetic procedures in the literature; some advanced textbooks, such as references J8 and J9, also give up-to-date procedural references. The work by Fieser and Fieser ⟨B20⟩ is an important guide to virtually every significant reagent used by organic chemists and includes numerous applications from the literature. *Synthetic Reagents* ⟨B21⟩ covers a much smaller number of reagents (perhaps half a dozen per volume) in more detail. Reference B22 tells how to purify many solvents used in organic synthesis and gives detailed information about their properties. References B23–B29 are more specialized works covering different aspects of organic synthesis. For example, B24 tells how to use protective groups in synthetic reactions, and B26 or B29 can help you work out a synthetic pathway on paper before you carry it out in the laboratory.

Category C: Laboratory Safety and Chemical Hazards

The titles of these works are generally descriptive of their contents. The listings include guides to safe laboratory practices ⟨C1, C3, C11⟩, a laboratory first-aid manual ⟨C4⟩, a guide to the safe disposal of chemicals ⟨C7⟩, a comprehensive listing of carcinogenic chemicals ⟨C10⟩, and a general discussion of chemical carcinogens ⟨C12⟩. The *Sigma/Aldrich Library of Chemical Safety Data* ⟨C9⟩ provides detailed health and safety data for over 7000 chemicals and describes safe-handling, storage, and disposal procedures. Chemical hazard information can also be found in references C2, C5, C6, C8, and C11, and in A3 and A8. The 9th edition of the *Merck Index* contains a section on first aid for poisoning and chemical burns and a list of poison control centers in the United States.

Category D: General Laboratory Techniques

Weissberger's *Technique of Organic Chemistry* ⟨D1⟩ and the more recent *Techniques of Chemistry* ⟨D2⟩ are comprehensive, multivolume sets containing information about a wide variety of experimental methods. Information on classical laboratory techniques such as distillation and recrystallization can be found in Volume I of reference D1, which is subtitled *Physical Methods of Organic Chemistry*. References D3–D7 describe laboratory apparatus and techniques at different levels; D5 and D7 are intended primarily for beginning students, whereas the others deal with more advanced techniques. References D8 and D9 cover the theory and practice of separation methods; D10 gives specific procedures for purifying a large number of chemicals. The textbook by Vogel ⟨B4⟩ and the first four volumes of *Houben-Weyl* ⟨B8⟩ also deal with laboratory methods for organic chemistry.

Category E: Chromatography

References E1–E3 are good general textbooks on the theory and practice of chromatography. Reference E4 is a multivolume set (updated periodically) that provides detailed discussions of the principles, techniques, and applications of chromatographic methods, along with tables of chromatographic data. Thin-layer chromatography is covered in depth by references E5–E7, gas chromatography by references E8–E11, and liquid chromatography by references E12–E16. Chapters on the theory and practice of gas chromatography and high pressure liquid chromatography can also be found in references F1–F4.

Category F: Spectrometry

The instrumental analysis textbooks ⟨F1–F4⟩ are good general sources of information about the principles and applications of spectrometric and other instrumental methods. Reference F5 should be consulted by anyone who wants to find out how scientific instruments work. References F6 and F7 are general guides to a variety of spectrometric methods; F7 is probably the best one-volume introduction available that covers all four major spectrometric methods (IR, NMR, UV-VIS, MS). Reference F8 is a brief article that may help beginning students interpret IR and NMR spectra. Reference F9 is a multivolume set containing a wide variety of data for use by practising spectrometrists. Infrared spectrometry is covered in depth by references

F10–F13, NMR spectrometry by F14–F22, ultraviolet-visible spectrometry by F23–F24, and mass spectrometry by F25–F27. The remaining sources in category F include compilations of information used in the interpretation of mass spectra ⟨F28–F29⟩ and collections of published spectra or spectral data ⟨F30–F35⟩. The *Sadtler Standard Spectra* ⟨F30⟩ consist of a large number of IR, NMR, and UV-VIS spectra in ring binders; although they are not arranged systematically, individual spectra can be located by using the index volumes. Spectra in the Aldrich collections ⟨F32–F34⟩ are arranged by functional classes and in order of increasing molecular complexity within a functional class, making it possible to observe the effect of various structural features on the spectra. Reference F35 gives mass numbers and intensities for the eight strongest peaks in the mass spectra of over 30,000 compounds. The *CRC Atlas* ⟨A4⟩ also reports useful spectral data for a number of compounds and gives references to published spectra.

Category G: Qualitative Analysis and Structure Determination

References G1–G4 are general textbooks of organic qualitative analysis; the last three of these cover some spectrometric methods as well as the traditional chemical methods. Reference G5 describes a number of chemical tests that can be used to characterize organic compounds. References G6–G8 describe methods used to determine the structures of organic compounds. Collections of spectra or spectral data (such as references A4 and F30–F35) and textbooks covering spectral interpretation (such as F7) can also facilitate the identification of unknown organic compounds.

Category H: Reaction Mechanisms and Theoretical Topics

Organic Reaction Mechanisms ⟨H1⟩ is an annual survey of recent developments in the field. References H2–H6 are general textbooks that cover a large number of reaction mechanisms. The monographs in the series *Reaction Mechanisms in Organic Chemistry* ⟨H7⟩ are published under different titles, such as *Nucleophilic Substitution at a Saturated Carbon Atom;* each book covers one general type of reaction in depth, as does reference H8. References H9–H12 contain considerable information about reaction kinetics, equilibria, mechanisms, and other theoretical aspects of organic chemistry. Reference H13 is a classic textbook of reaction kinetics, and H14 deals with the mechanisms of concerted reactions.

Category J: Comprehensive and Advanced Textbooks

Rodd's Chemistry of Carbon Compounds ⟨J1⟩ and *Comprehensive Organic Chemistry* ⟨J2⟩ are multivolume treatises that attempt to cover the whole field of organic chemistry; both sets provide in-depth coverage of organic reactions and syntheses, but the latter is more up-to-date. References J3–J6 are older textbooks that include enough information about organic compounds, reactions, and other topics to be useful reference works. References J7–J9 are modern textbooks with good coverage of current topics in organic chemistry and numerous literature references. The book by March ⟨J8⟩ is a particularly valuable one-volume source of information about the mechanisms and synthetic applications of a large number of organic reactions. For advanced textbooks covering the theoretical aspects of organic chemistry, see references H9–H11.

Category K: Reports of Chemical Research

Chemical Abstracts ⟨K1⟩ and *Current Abstracts of Chemistry* ⟨K2⟩ are used mainly to locate articles about particular chemical compounds or topics, although they also provide useful information in the form of brief summaries of the articles. The remaining references in this category ⟨K3–K9⟩ summarize or review current research in various fields of chemistry. Review articles on many chemical topics can also be found in such serial publications as *Chemical Reviews, Quarterly Reviews, Angewandte Chemie,* and *Synthesis.*

   Chemical Abstracts ⟨K1⟩ is the most important key to the chemical literature; it cites and summarizes virtually every original paper about chemistry that appears anywhere in the world, shortly after publication. *Current Abstracts of Chemistry* ⟨K2⟩ reports weekly on new compounds and chemical reactions that appear in some of the more important chemical journals. It is essentially a "current awareness" publication to inform researchers quickly of recent advances in their fields, whereas *Chemical Abstracts* serves as a repository of information about past work as well as a medium for reporting current work. *Annual Reports on the Progress of Chemistry* ⟨K5⟩ summarizes important advances in chemistry each year; section B is devoted to organic chemistry. References K8 and K9 are examples of the *Specialist Periodical Reports* published yearly by the Royal Society of Chemistry; each series is devoted to a specialized field of chemistry and summarizes the important work done in that field during the year.

Category L: Sources on Selected Topics

In this category are listed a number of books and articles that may be of interest to students of organic chemistry. Most are about topics related to the experiments, such as reference L15 on enamines, L25 on adamantane, and L65 on sugar chemistry. Other references are listed here because they don't fit into any of the other categories; for example, L18 and L42 are excellent guides for writing a scientific paper or a laboratory notebook.

Category M: Guides to the Chemical Literature

The books listed in this category in the Bibliography tell how to use such tools as *Chemical Abstracts* and *Beilstein* to gain access to the primary literature of chemistry. They may also list and critically evaluate books and publications that are useful to chemists and chemistry students. Also, two brief but excellent guides to the chemical literature can be found in references J8 (pp. 1121–1145) and D4 (pp. 1–27).

Bibliography

A. General Reference Works

1. Weast, R. C., ed., *CRC Handbook of Chemistry and Physics,* 67th ed. (and other editions). Boca Raton, Fla.: CRC Press, 1986.

2. Lange, N. A., *Lange's Handbook of Chemistry,* 13th ed., ed. by J. A. Dean. New York: McGraw-Hill, 1985.

3. Windholz, M., ed., *The Merck Index,* 10th ed. Rahway, N.J.: Merck and Co., 1983.

4. Grasselli, J. G., and Ritchey, W. R., eds., *CRC Atlas of Spectral Data and Physical Constants for Organic Compounds,* 2nd ed. Cleveland: CRC Press, 1975.

5. Weast, R. C., and Astle, M. J., eds., *CRC Handbook of Data on Organic Compounds.* Boca Raton, Fla.: CRC Press, 1985.

6. *Aldrich Catalog Handbook of Fine Chemicals.* Milwaukee, Wisc.: Aldrich Chemical Co., 1986–87 (and other years).

7. Glasby, J. S., *Encyclopedia of the Terpenoids.* New York: Wiley, 1982.

8. Pollock, J. R. A., and Stevens, R., eds., *Dictionary of Organic Compounds,* 5th ed. New York: Chapman and Hall, 1982.

9. *Beilstein's Handbook of Organic Chemistry.* New York: Springer-Verlag, 1918 to date.

10. Rappoport, Z., ed., *CRC Handbook of Tables for Organic Compound Identification,* 3rd ed. Cleveland: Chemical Rubber Co., 1967.

11. Utermark, W., and Schicke, W., *Melting Point Tables of Organic Compounds,* 2nd ed. New York: Interscience Publishers, 1963.

12. *Solubilities of Inorganic and Organic Compounds.* New York: Macmillan, 1963–64; Pergamon Press, 1979.

13. *Kirk-Othmer Encyclopedia of Chemical Technology,* 3rd ed., ed. by H. F. Marks et al. New York: Wiley, 1978–84.

14. *Kirk-Othmer Concise Encyclopedia of Chemical Technology.* New York: Wiley, 1985.

15. Considine, D. M., and Considine, G. D., eds., *Van Nostrand Reinhold Encyclopedia of Chemistry,* 4th ed. New York: Van Nostrand Reinhold, 1984.

16. Parker, S. P., ed., *McGraw-Hill Encyclopedia of Chemistry.* New York: McGraw-Hill, 1983.

17. Thorpe, J. F., and Whitely, M. A., *Thorpe's Dictionary of Applied Chemistry,* 4th ed. London: Longmans, Green and Co., 1937–56.

18. Orchin, M., et al., *The Vocabulary of Organic Chemistry.* New York: Wiley, 1980.

19. Parker, S. P., *McGraw-Hill Dictionary of Chemical Terms*. New York: McGraw-Hill, 1985.

20. Chemical Abstracts Service, *Ring Systems Handbook*. Columbus, Ohio: American Chemical Society, 1984.

21. *Beilstein Dictionary: German-English: For the Users of the Beilstein Handbook of Organic Chemistry*. New York: Springer-Verlag, 1979.

22. Gross, H., ed., *Dictionary of Chemistry and Chemical Technology: German-English*. Amsterdam, New York: Elsevier, 1984.

23. Gordon, A. J., and Ford, R. A., *The Chemist's Companion*. New York: Wiley-Interscience, 1972.

B. Organic Synthesis

1. *Organic Synthesis,* Collective Volumes 1–6 and annual volumes. New York: Wiley, 1932 to date.

2. *Organic Reactions.* New York: Wiley, 1942 to date.

3. Shirley, D. A., *Preparation of Organic Intermediates*. New York: Wiley, 1951.

4. Vogel, A. I., *Vogel's Textbook of Practical Organic Chemistry,* 4th ed., rev. by B. Furniss et al. London, New York: Longman, 1978.

5. Srinivasan, R., ed., *Organic Photochemical Synthesis*. New York: Wiley, 1971 to date.

6. Sandler, S. R., and Karo, W., *Polymer Syntheses*. New York: Academic Press, 1974, 1977, 1980.

7. Sorensen, W. R., and Campbell, T. W., *Preparative Methods of Polymer Chemistry,* 2nd ed. New York: Interscience Publishers, 1968.

8. *Methoden der Organischen Chemie (Houben-Weyl),* 4th ed. Stuttgart: Georg Thieme, 1952 to date.

9. Weygand, C., *Weygand/Hilgetag Preparative Organic Chemistry,* ed. by G. Hilgetag and A. Martini. New York: Wiley-Interscience, 1972.

10. Becker, H. et al., *Organicum: Practical Handbook of Organic Chemistry,* trans. by B. J. Hazzard. Reading, Mass.: Addison-Wesley Publishing Co., 1973.

11. Sandler, S. R., and Karo, W., *Organic Functional Group Preparations,* 2nd ed. New York: Academic Press, 1983 to date.

12. Buehler, C. A., and Pearson, D. E., *Survey of Organic Syntheses*. New York: Wiley-Interscience, 1970–77.

13. Theilheimer, W., ed., *Synthetic Methods of Organic Chemistry*. Basel: S. Karger, 1946 to date.

14. House, H. O., *Modern Synthetic Reactions,* 2nd ed. Menlo Park, Calif.: W. A. Benjamin, 1972.

15. Carruthers, W., *Some Modern Methods of Organic Synthesis,* 2nd ed. Cambridge: Cambridge University Press, 1978.

16. Wagner, R. B., and Zook, H. D., *Synthetic Organic Chemistry.* New York: Wiley, 1953.

17. Migridichian, V., *Organic Synthesis.* New York: Reinhold, 1957.

18. Sugasawa, S., and Nakai, S., *Reaction Index of Organic Syntheses,* rev. ed. New York: Wiley, 1967.

19. *Compendium of Organic Synthetic Methods.* New York: Wiley-Interscience, 1971–84.

20. Fieser, L. F., and Fieser, M., *Reagents for Organic Synthesis.* New York: Wiley-Interscience, 1967 to date.

21. Pizey, J. S., *Synthetic Reagents.* New York: Halsted Press, 1974 to date.

22. Riddick, J. A., and Bunger, W. M., *Organic Solvents: Physical Properties and Methods of Purification (Techniques of Chemistry,* Vol. II). New York: Wiley-Interscience, 1970.

23. Brown, H. C., *Organic Synthesis via Boranes.* New York: Wiley, 1975.

24. Greene, T. W., *Protective Groups in Organic Synthesis.* New York: Wiley, 1981.

25. Horspool, W. M., ed., *Synthetic Organic Photochemistry.* New York: Plenum Press, 1984.

26. Mackie, R. K., and Smith, D. M., *Guidebook to Organic Synthesis.* New York: Longman, 1982.

27. Negishi, E-I., *Organometallics in Organic Synthesis.* New York: Wiley, 1980 to date.

28. Stowell, J. C., *Carbanions in Organic Synthesis.* New York: Wiley, 1979.

29. Stuart, W., *Organic Synthesis: The Disconnection Approach.* New York: Wiley, 1982.

C. Laboratory Safety and Chemical Hazards

1. ACS Committee on Chemical Safety, *Safety in Academic Chemistry Laboratories.* Washington: American Chemical Society, 1974.

2. Bretherick, L., *Handbook of Reactive Chemical Hazards,* 3rd ed. London, Boston: Butterworths, 1985.

3. Green, M. E., and Turk, A., *Safety in Working with Chemicals.* New York: Macmillan, 1978.

4. LeFevre, M. J., *First-Aid Manual for Chemical Accidents.* Stroudsburg, Pa.: Dowden, Hutchinson & Ross, 1980.

5. Meyer, E., *Chemistry of Hazardous Materials.* Englewood Cliffs, N.J.: Prentice-Hall, 1977.

6. Muir, G. D., ed., *Hazards in the Chemical Laboratory,* 3rd ed. London: The Chemical Society, 1981.

7. National Research Council, *Prudent Practices for Disposal of Chemicals from Laboratories.* Washington: National Academic Press, 1983.

8. Sax, N. I., *Dangerous Properties of Industrial Materials,* 6th ed. New York: Van Nostrand Reinhold, 1984.

9. Lenga, R. E., ed., *The Sigma-Aldrich Library of Chemical Safety Data.* Milwaukee, Wisc.: Aldrich Chemical Co., 1986.

10. Soderman, J. V., *Handbook of Identified Carcinogens and Noncarcinogens.* Boca Raton, Fla.: CRC Press, 1982.

11. Steere, N. V., ed., *CRC Handbook of Laboratory Safety,* 2nd ed. Cleveland: Chemical Rubber Company, 1971.

12. Searle, C. E., ed., *Chemical Carcinogens,* 2nd ed. Washington: American Chemical Society, 1984.

D. General Laboratory Techniques

1. Weissberger, A., ed., *Technique of Organic Chemistry,* 3rd ed. New York: Wiley-Interscience, 1959 to date.

2. Weissberger, A., ed., *Techniques of Chemistry.* New York: Wiley-Interscience, 1971 to date.

3. Bates, R. B., and Schaefer, J. P., *Research Techniques in Organic Chemistry.* Englewood Cliffs, N.J.: Prentice-Hall, 1971.

4. Loewenthal, H. J. E., *Guide for the Perplexed Organic Experimentalist.* Philadelphia: Heyden, 1978.

5. Marmor, S., *Laboratory Methods in Organic Chemistry.* Minneapolis: Burgess, 1981.

6. Wiberg, K. B., *Laboratory Technique in Organic Chemistry.* New York: McGraw-Hill, 1960.

7. Zubrick, J. W., *The Organic Chem Lab Survival Manual: A Student's Guide to Techniques.* New York: Wiley, 1984.

8. Dean, J. A., *Chemical Separation Methods.* New York: Van Nostrand Reinhold Co., 1969.

9. Karger, B. L., Snyder, L. R., and Horvath, C., *An Introduction to Separation Science.* New York: Wiley, 1973.

10. Perrin, D. D., Armarego, W. L. F., and Perrin, D. R., *Purification of Laboratory Chemicals,* 2nd ed. New York: Pergamon Press, 1980.

E. Chromatography

1. Bobbitt, J. M., and Schwarting, A. E., *Introduction to Chromatography.* New York: Van Nostrand Reinhold, 1968.

2. Stock, R., and Rice, C. B. F., *Chromatographic Methods,* 3rd ed. London: Chapman and Hall, 1974.

3. Heftmann, E., ed., *Chromatography: A Laboratory Handbook of Chromatographic and Electrophoretic Methods,* 3rd ed. New York: Van Nostrand Reinhold, 1975.

4. *CRC Handbook of Chromatography.* Cleveland: CRC Press, 1972 to date.

5. Kirchner, J. G., *Thin-Layer Chromatography,* 2nd ed. (*Techniques of Chemistry,* Vol. XIV). New York: Wiley, 1978.

6. Stahl, E., ed., *Thin Layer Chromatography: A Laboratory Handbook,* 2nd ed. New York: Springer-Verlag, 1969.

7. Touchstone, J. F., and Dobbins, M. F., *Practice of Thin Layer Chromatography,* 2nd ed. Somerset, N.J.: Wiley, 1983.

8. Grob, R. L., ed., *Modern Practice of Gas Chromatography.* New York: Wiley, 1977.

9. Lee, M. L., Yang, F. J., and Bartle, K. D., *Open Tubular Gas Chromatography: Theory and Practice.* New York: Wiley, 1984.

10. McNair, H. M., *Basic Gas Chromatography,* 5th ed. Walnut Creek, Calif.: Varian Aerograph, 1969.

11. Perry, J. A., *Introduction to Analytical Gas Chromatography.* New York: Marcel Dekker, 1981.

12. Englehardt, H., ed., *Practice of High Performance Liquid Chromatography: Applications, Equipment, and Quantitative Analysis.* New York: Springer-Verlag, 1986.

13. Hadden, N., et al., *Basic Liquid Chromatography.* Walnut Creek, Calif.: Varian Aerograph, 1971.

14. Hamilton, R. J., and Sewell, P. A., *Introduction to High Performance Liquid Chromatography,* 2nd ed. New York: Chapman and Hall, 1982.

15. Simpson, C. F., ed., *Techniques in Liquid Chromatography.* New York: Wiley, 1982.

16. Snyder, L. R., and Kirkland, J. J., *Introduction to Modern Liquid Chromatography,* 2nd ed. New York: Wiley, 1979.

F. Spectrometry

1. Bauer, H. H., Christian, G. D., and O'Reilly, J. E., *Instrumental Analysis.* Boston: Allyn and Bacon, 1978.

2. Ewing, G. W., *Instrumental Methods of Chemical Analysis,* 5th ed. New York: McGraw-Hill, 1985.

3. Skoog, D. A., *Principles of Instrumental Analysis,* 3rd ed. New York: Saunders, 1985.

4. Willard, H. H., Merritt, L. L., and Dean, J. A., *Instrumental Methods of Analysis,* 5th ed. New York: Van Nostrand Reinhold Co., 1974.

5. Malmstadt, H. V., Enke, C. G., and Crouch, S. R., *Electronics and Instrumentation for Scientists.* Reading, Mass.: Benjamin/Cummings, 1981.

6. Cooper, J. W., *Spectroscopic Techniques for Organic Chemists.* New York: Wiley, 1980.

7. Silverstein, R. M., Bassler, G. C., and Morrill, T. C., *Spectrometric Identification of Organic Compounds,* 4th ed. New York: Wiley, 1981.

8. Ingham, A. M., and Henson, R. C., "Interpreting Infrared and Nuclear Magnetic Resonance Spectra of Simple Organic Compounds for the Beginner." *Journal of Chemical Education,* 61 (1984), p. 704.

9. *CRC Handbook of Spectroscopy.* Cleveland: CRC Press, 1974 to date.

10. Bellamy, L. J., *The Infra-red Spectra of Complex Molecules,* 3rd ed. New York: Wiley, 1975.

11. Colthup, N. B., Daly, L. H., and Wiberley, S. E. *Introduction to Infrared and Raman Spectroscopy,* 2nd ed. New York: Academic Press, 1975.

12. Griffiths, P. R., and de Haseth, J. A., *Fourier Transform Infrared Spectrometry.* New York: Wiley, 1986.

13. Smith, A. L., *Applied Infrared Spectrometry.* New York: Wiley, 1979.

14. Akitt, J. W., *NMR and Chemistry: An Introduction to the Fourier Transform Multinuclear Era,* 2nd ed. New York: Chapman and Hall, 1983.

15. Ault, A., and Dudek, G. O., *NMR: An Introduction to Proton Nuclear Magnetic Resonance Spectroscopy.* San Francisco: Holden-Day, 1976.

16. Becker, E. D., *High Resolution NMR: Theory and Chemical Applications,* 2nd ed. Orlando, Fla.: Academic Press, 1980.

17. Harris, R., *Nuclear Magnetic Resonance Spectroscopy.* Marshfield, Mass.: Pitman Publishing, 1983.

18. Haws, E. J., Hill, R. R., and Mowthorpe, D. J., *The Interpretation of Proton Magnetic Resonance Spectra: A Programmed Introduction.* New York: Heyden, 1973.

19. Jackman, L. M., and Sternhell, S., *Applications of Nuclear Magnetic Resonance Spectrometry in Organic Chemistry,* 2nd ed. Oxford: Pergamon Press, 1969.

20. Levy, G. C., Lichter, R. L., and Nelson, G. L., *Carbon-13 Nuclear Magnetic Resonance Spectroscopy,* 2nd ed. New York: Wiley, 1980.

21. Martin, M. L., and Martin, G. J., *Practical NMR Spectroscopy.* Philadelphia: Heyden, 1980.

22. Wehrli, F. W., and Wirthlin, T., *Interpretation of Carbon-13 NMR Spectra.* Philadelphia: Heyden, 1980.

23. Jaffe, H. H. and Orchin, M., *Theory and Applications of Ultraviolet Spectroscopy.* New York: Wiley, 1962.

24. Rao, C. N. R., *Ultra-Violet Spectroscopy: Chemical Applications,* 3rd ed. New York: Plenum Press, 1975.

25. Budzikiewicz, H., Djerassi, C., and Williams, D. H., *Mass Spectrometry of Organic Compounds.* San Francisco: Holden-Day, 1967.

26. McLafferty, F. W., *Interpretation of Mass Spectra,* 3rd ed. Mill Valley, Cal.: University Science Books, 1980.

27. Middleditch, B. S., ed., *Practical Mass Spectroscopy: A Contemporary Introduction.* New York: Plenum Press, 1979.

28. Beynon, J. H. and Williams, A. E., *Mass and Abundance Tables for Use in Mass Spectrometry.* Amsterdam: Elsevier Publishing Co., 1963.

29. McLafferty, F. W. and Venkataraghavan, R., *Mass Spectral Correlations,* 2nd ed. Washington: American Chemical Society, 1982.

30. *Sadtler Standard Spectra.* (Collections of infrared, ultraviolet, and NMR spectra.) Philadelphia: Sadtler Research Laboratories.

31. *Infrared Spectra Handbook of Common Organic Solvents.* Philadelphia: Sadtler Research Laboratories, 1983.

32. Pouchert, C. J., *The Aldrich Library of Infrared Spectra,* 3rd ed. Milwaukee, Wisc.: Aldrich Chemical Co., 1981.

33. Pouchert, C. J., *The Aldrich Library of FT-IR Spectra.* Milwaukee, Wisc.: Aldrich Chemical Co., 1985.

34. Pouchert, C. J. and Campbell, J. R. *The Aldrich Library of NMR Spectra,* 2nd ed. Milwaukee, Wisc.: Aldrich Chemical Co., 1983.

35. *Eight Peak Index of Mass Spectra,* 3rd ed. Nottingham: Royal Society of Chemistry, 1983.

G. Qualitative Analysis and Structure Determination

1. Cheronis, N. D., Entrikin, J. B., and Hodnett, E. M., *Semimicro Qualitative Organic Analysis,* 3rd ed. New York: Interscience Publishers, 1965.

2. Criddle, W. J. *Spectral and Chemical Characterization of Organic Compounds: A Laboratory Handbook.* New York: Wiley, 1980.

3. Pasto, D. J. and Johnson, C. R., *Organic Structure Determination.* Englewood Cliffs, N.J.: Prentice-Hall, 1969.

4. Shriner, R. L., Fuson, R. C., Curtin, D. Y., and Morrill, T. C. *The Systematic Identification of Organic Compounds: A Laboratory Manual,* 6th ed. New York: Wiley, 1980.

5. Feigl, F. and Anger, V., *Spot Tests in Organic Analysis,* 7th ed. Amsterdam, New York: Elsevier Publishing Co., 1966.

6. Bentley, K. W., and Kirby, G. W., *Elucidation of Organic Structures by Physical and Chemical Methods,* 2nd ed. (*Techniques of Chemistry,* Vol. IV). New York: Wiley-Interscience, 1972–73.

7. Lambert, J. B., et al., *Organic Structural Analysis.* New York: Macmillan, 1976.

8. Nachod, F. C., et al., eds., *Determination of Organic Structures by Physical Methods.* New York: Academic Press, 1955–76..

H. Reaction Mechanisms and Theoretical Topics

1. *Organic Reaction Mechanisms.* New York: Wiley, 1965 to date.

2. Benfey, O. T., *Introduction to Organic Reaction Mechanisms.* Melbourne, Fla.: Krieger, 1982.

3. Breslow, R., *Organic Reaction Mechanisms: An Introduction,* 2nd ed. New York: W. A. Benjamin, 1969.

4. Harris, J. M., and Wamser, C. C., *Fundamentals of Organic Reaction Mechanisms.* New York: Wiley, 1976.

5. Jones, R. A. Y., *Physical and Mechanistic Organic Chemistry,* 2nd ed. Cambridge, New York: Cambridge University Press, 1984.

6. Sykes, P., *A Guidebook to Mechanism in Organic Chemistry,* 5th ed. London: Longman, 1981.

7. *Reaction Mechanisms in Organic Chemistry.* (A series of monographs.) Amsterdam, New York: Elsevier, 1963 to date.

8. Streitwieser, A., *Solvolytic Displacement Reactions.* New York: McGraw-Hill, 1962.

9. Hammet, L. P., *Physical Organic Chemistry: Reaction Rates, Equilibria, and Mechanisms,* 2nd ed. New York: McGraw-Hill, 1970.

10. Ingold, C. K., *Structure and Mechanism in Organic Chemistry,* 2nd ed. Ithaca, N.Y.: Cornell University Press, 1969.

11. Lowry, T. H., and Richardson, K. S., *Mechanism and Theory in Organic Chemistry,* 2nd ed. New York: Harper & Row, 1981.

12. Hine, J. S., *Structural Effects on Equilibria in Organic Chemistry.* New York: Wiley, 1975.

13. Laidler, K. J., *Chemical Kinetics,* 2nd ed. New York: McGraw-Hill, 1965.

14. Woodward, R. B., and Hoffman, R., *The Conservation of Orbital Symmetry.* New York: Academic Press, 1970.

J. Comprehensive and Advanced Textbooks

1. Rodd, E. H., *Rodd's Chemistry Carbon Compounds,* 2nd ed., ed. by S. Coffey. Amsterdam: Elsevier Publishing Co., 1964–80.

2. Burton, D., and Ollis, W. D., *Comprehensive Organic Chemistry.* Oxford, New York: Pergamon Press, 1979.

3. Fieser, L. F., and Fieser, M., *Advanced Organic Chemistry.* New York: Reinhold Publishing Corp., 1961.

4. Fieser, L. F., and Fieser, M., *Topics in Organic Chemistry.* New York: Reinhold Publishing Corp., 1963.

5. Karrer, P., *Organic Chemistry,* 4th English ed. New York: Elsevier Publishing Co., 1950.

6. Noller, C. R., *Chemistry of Organic Compounds,* 3rd ed. Philadelphia: Saunders, 1965.

7. Carey, F. A., and Sundberg, R. J., *Advanced Organic Chemistry,* 2nd ed. New York: Plenum Press, 1983, 1984.

8. March, J., *Advanced Organic Chemistry: Reactions, Mechanisms and Structure,* 3rd ed. New York: McGraw-Hill, 1985.

9. Le Noble, W. J., *Highlights of Organic Chemistry: An Advanced Textbook.* New York: Marcel Dekker, 1974.

K. Reports of Chemical Research

1. *Chemical Abstracts.* Columbus, Ohio: CA Service, American Chemical Society, 1907 to date.

2. *Current Abstracts of Chemistry and Index Chemicus.* Philadelphia: ISI Press, 1960 to date.

3. *Advances in Organic Chemistry: Methods and Results.* New York: Interscience Publishers, 1960 to date.

4. *Annual Reports in Organic Synthesis.* New York: Academic Press, 1970 to date.

5. *Annual Reports on the Progress of Chemistry, Section B: Organic Chemistry.* London: Royal Society of Chemistry, 1904 to date.

6. *International Review of Science: Organic Chemistry,* Series 1 and 2. Boston: Butterworths, 1973–75.

7. *Progress in Organic Chemistry.* New York: Academic Press, 1952 to date.

8. *Specialist Periodical Reports: The Alkaloids.* London: Royal Society of Chemistry, 1969 to date.

9. *Specialist Periodical Reports: The Terpenoids.* London: Royal Society of Chemistry, 1969 to date.

L. Sources on Selected Topics

1. Allen, R. L. M., *Colour Chemistry.* New York: Appleton-Century-Crofts, 1971.

2. AMA Council on Drugs, *AMA Drug Evaluations.* Acton, Mass.: Publishing Sciences Group, 1971 to date.

3. Amundsen, L. H., "Sulfanilamide and Related Chemotherapeutic Agents." *Journal of Chemical Education, 19* (1942), p. 167.

4. Baker, A. A. *Unsaturation in Organic Chemistry.* Boston: Houghton Mifflin, 1968.

5. Batra, S. W. T., "Polyester-Making Bees and Other Innovative Insect Chemists." *Journal of Chemical Education, 62* (1985), p. 121.

6. Bethell, D., and Gold, V., *Carbonium Ions: An Introduction.* New York: Academic Press, 1967.

7. Billmeyer, F. W., *Textbook of Polymer Sciences,* 3rd ed. New York: Wiley-Interscience, 1984.

8. Blackburn, S., *Protein Sequence Determination: Methods and Techniques.* New York: Marcel Dekker, 1970.

9. Bragg, R. W., et al., "Sweet Organic Chemistry." *Journal of Chemical Education, 55* (1978), p. 281.

10. Brewster, J. H., "Stereochemistry and the Origins of Life." *Journal of Chemical Education, 63* (1986), p. 667.

11. *Burger's Medicinal Chemistry,* 4th ed., ed. by M. E. Wolff. New York: Wiley, 1980.

12. Cahn, R. S., *Introduction to Chemical Nomenclature,* 5th ed. Woburn, Mass.: Butterworths, 1979.

13. Cleary, J., "Diapers and Polymers." *Journal of Chemical Education, 63* (1986), p. 422.

14. Coates, G. E., et al., *Organometallic Compounds,* 4th ed. New York: Wiley, 1979.

15. Cook, A. G., ed., *Enamines: Synthesis, Structure and Reactions.* New York: Marcel Dekker, 1969.

16. Cremlyn, R. J. W., and Still, R. H., *Named and Miscellaneous Reactions in Practical Organic Chemistry.* London: Heinemann, 1967.

17. Dehmlow, E. W., and Dehmlow, S. S., *Phase Transfer Catalysis.* Weinheim: Verlag Chemie, 1980.

18. Dodd, J. S., ed., *The ACS Style Guide: A Manual for Authors and Editors.* Washington: American Chemical Society, 1986.

19. Eliel, E. L., *Stereochemistry of Carbon Compounds.* New York: McGraw-Hill, 1962.

20. Eliel, E. L., "Recent Advances in Stereochemical Nomenclature." *Journal of Chemical Education, 48* (1971), p. 163.

21. Farber, E., ed., *Great Chemists.* New York: Interscience Publishers, 1961.

22. Fernelius, W. C., and Renfrew, E. E., "Indigo." *Journal of Chemical Education, 60* (1983), p. 633.

23. Fletcher, J. H., Dermer, O. C., and Fox, R. B., eds., *Nomenclature of Organic Compounds: Principles and Practice (Advances in Chemistry Series,* Vol. 126). Washington: American Chemical Society, 1974.

24. Forrester, A. R., Hay, J. M., and Thomson, R. H., *Organic Chemistry of Stable Free Radicals.* New York: Academic Press, 1968.

25. Fort, R. G., *Adamantane: The Chemistry of Diamond Molecules.* New York: Marcel Dekker, 1976.

26. Foye, W. O., ed., *Principles of Medicinal Chemistry,* 2nd ed. Philadelphia: Lea and Febiger, 1981.

27. Garratt, P. J., *Aromaticity.* New York: McGraw-Hill, 1971.

28. Geissman, T. A., and Grout, D. H. G., *Organic Chemistry of Secondary Plant Metabolism.* San Francisco: Freeman, Cooper, 1969.

29. Glidewell, C., and Lloyd, D., "The Arithmetic of Aromaticity." *Journal of Chemical Education, 63* (1986), p. 306.

30. Gokel, G. W., and Weber, W. P., "Phase Transfer Catalysis." *Journal of Chemical Education, 55* (1978), pp. 350, 429.

31. Goodwin, T. W. ed., *Chemistry and Biochemistry of Plant Pigments,* 2nd ed. New York: Academic Press, 1976.

32. Grant, N., and Naves, R. G., "Perfumes and the Art of Perfumery." *Journal of Chemical Education, 49* (1972), p. 526.

33. Greenberg, A., and Liebman, J. F., *Strained Organic Molecules.* New York: Academic Press, 1978.

34. Grieve, M., *A Modern Herbal.* New York: Dover Publications, 1971.

35. Guenther, E., *The Essential Oils.* New York: D. Van Nostrand, 1948–52.

36. Guild, W., "Theory of Sweet Taste." *Journal of Chemical Education, 49* (1972), p. 171.

37. Hansch, C., "Drug Research and the Luck of the Draw." *Journal of Chemical Education, 51* (1974), p. 360.

38. Hill, J. W., and Jones, S. W., "Consumer Applications of Chemical Principles: Drugs." *Journal of Chemical Education, 62* (1985), p. 328.

39. Hubbard, R., and Kropf, A., "Molecular Isomers in Vision." *Scientific American, 216* (June 1967), p. 64.

40. Jones, M., Jr., "Carbenes." *Scientific American, 234* (Feb. 1976), p. 101.

41. Juster, N. J., "Color and Chemical Constitution." *Journal of Chemical Education, 39* (1962), p. 596.

42. Kanare, H. M., *Writing the Laboratory Notebook.* Washington: American Chemical Society, 1985.

43. Kauffman, G. B., "Pittacal — The First Synthetic Dyestuff." *Journal of Chemical Education, 54* (1977), p. 753.

44. Kharasch, M. S., and Reinmuth, O., *Grignard Reactions of Nonmetallic Substances.* New York: Prentice-Hall, 1954.

45. Kirmse, W., *Carbene Chemistry,* 2nd ed. New York: Academic Press, 1971.

46. Krauch, H., and Kunz, W., *Organic Name Reactions.* New York: John Wiley & Sons, 1964.

47. Lednicer, D., and Mitscher, L. A., *Organic Chemistry of Drug Synthesis.* New York: Wiley-Interscience, 1977, 1980.

48. McCullough, T., "Furfural — Ubiquitous Natural Product." *Journal of Chemical Education, 49* (1972), p. 836.

49. Mead, J. F., and Fulco, A. J., *The Unsaturated and Polyunsaturated Fatty Acids in Health and Disease.* Springfield, Ill.: Thomas, 1976.

50. *Modern Drug Encyclopedia and Therapeutic Index: A Compendium,* 16th ed. New York: Yorke Medical Books, 1981.

51. Moncrieff, R. W., *The Chemical Senses,* 3rd ed. London: L. Hill, 1967.

52. Mosher, M. W., and Ansell, J. M., "Preparation and Color of Azo-Dyes." *Journal of Chemical Education, 52* (1975), p. 195.

53. Nakanishi, K., ed., *Natural Products Chemistry*. New York: Academic Press, 1974–75.

54. Olah, G. A., *Carbocations and Electrophilic Reactions*. New York: Wiley, 1974.

55. Olah, G. A., *Friedel-Crafts Chemistry*. New York: Wiley-Interscience, 1973.

56. Olah, G. A., *Halonium Ions*. New York: Wiley, 1975.

57. Pauli, G. H., "Chemistry of Food Additives." *Journal of Chemical Education, 61* (1984), p. 332.

58. Pryor, W. A., *Free Radicals*. New York: McGraw-Hill, 1966.

59. *Riegel's Handbook of Industrial Chemistry,* 8th ed., ed. by J. A. Kent. New York: Van Nostrand Reinhold, 1983.

60. Robinson, B., *The Fischer Indole Synthesis*. New York: Wiley, 1982.

61. Sarkanen, K. V., and Ludwig, C. H., eds. *Lignins: Occurrence, Formation, Structure and Reactions*. New York: Wiley-Interscience, 1971.

62. Scientific American, *Bio-Organic Chemistry*. San Francisco: W. H. Freeman, 1968.

63. Scientific American, *The Molecular Basis of Life: An Introduction to Molecular Biology*. San Francisco: W. H. Freeman, 1968.

64. Senozan, N. M., and Hunt, R. L., "Hemoglobin: Its Occurrence, Structure, and Adaptation." *Journal of Chemical Education, 59* (1982), p. 173.

65. Shallenberger, R. S., and Birch, G. G., *Sugar Chemistry*. Westport, Conn.: Avi Pub. Co., 1975.

66. Shreve, R. N., and Brink, J. A., Jr., *Chemical Process Industries,* 4th ed. New York: McGraw-Hill, 1977.

67. Simonsen, J. L., and Owen, L. N., *The Terpenes,* 2nd ed. Cambridge: University Press, 1947–57.

68. Starks, C. M., and Liotta, C., *Phase Transfer Catalysis: Principles and Techniques*. New York: Academic Press, 1978.

69. Sundberg, R. J., *The Chemistry of Indoles*. New York: Academic Press, 1970.

70. Swan, G. A., *An Introduction to the Alkaloids*. New York: Wiley, 1967.

71. Treptow, R. S., "Determination of Alcohol in Breath for Law Enforcement." *Journal of Chemical Education, 51* (1974), p. 651.

72. Turro, N. J., *Modern Molecular Photochemistry*. Menlo Park, Calif: Benjamin/Cummings, 1978.

73. Venkataraman, K., ed., *The Chemistry of Synthetic Dyes*. New York: Academic Press, 1952–74.

74. Vogler, A., and Kunkley, H., "Photochemistry and Beer." *Journal of Chemical Education, 59* (1982), p. 25.

75. Waddell, T. G., Jones, H., and Keith, A. L., "Legendary Chemical Aphrodisiacs." *Journal of Chemical Education, 57* (1980), p. 341.

76. Walling, C., "The Development of Free Radical Chemistry." *Journal of Chemical Education, 63* (1986), p. 99.

77. Wallis, H. J., *Forensic Science,* 2nd ed. New York: Praeger, 1974.

78. Wassermann, A., *Diels-Alder Reactions: Organic Background and Physico-chemical Aspects.* Amsterdam, New York: Elsevier Publishing Co., 1965.

79. Weber, W. P., and Gokel, G. W., *Phase Transfer Catalysis in Organic Synthesis.* Berlin: Springer-Verlag, 1977.

80. West, J. A., "The Chemistry of Enamines." *Journal of Chemical Education, 40* (1963), p. 194.

81. Wilson, E. O., "Pheromones." *Scientific American, 208* (May 1963), p. 100.

82. Wood, W. F., "Chemical Ecology: Chemical Communication in Nature." *Journal of Chemical Education, 60* (1983), p. 531.

83. Wood, W. F., "The Sex Pheromones of the Gypsy Moth and the American Cockroach." *Journal of Chemical Education, 59* (1982), p. 35.

M. Guides to the Chemical Literature

1. *CA Search for Beginners: An Introduction to On-Line Access to CA SEARCH.* (Editions for DIALOG and ORBIT search systems.) Columbus, Ohio: CA Service, American Chemical Society, 1980.

2. Hancock, J. E. H., "An Introduction to the Literature of Organic Chemistry." *Journal of Chemical Education, 45* (1968), pp. 193, 260, 336.

3. Maizell, R. W., *How to Find Chemical Information.* New York: Wiley-Interscience, 1979.

4. Mellon, M. G., *Chemical Publications: Their Nature and Use,* 5th ed. New York: McGraw-Hill, 1982.

5. Mellon, M. G., *Searching the Chemical Literature.* Washington: American Chemical Society, 1964.

6. Skolnik, H., *The Literature Matrix of Chemistry.* New York: Wiley, 1982.

7. Weissbach, O., *The Beilstein Guide: A Manual for the Use of Beilstein's Handbuch der Organischen Chemie.* Berlin: Springer-Verlag, 1976.

8. Wilen, S. H. *Use of Chemical Literature: An Introduction to Chemical Information Retrieval.* Washington: American Chemical Society, 1978.

9. Wolman, Y., *Chemical Information: A Practical Guide to Utilization.* New York: Wiley, 1983.

Index

INFRARED SPECTRUM-STRUCTURE CORRELATION CHART

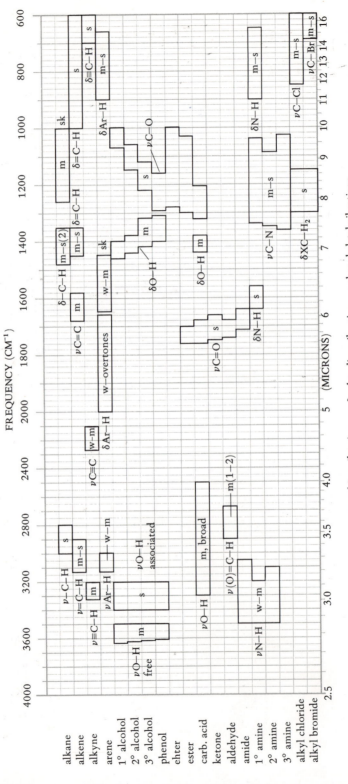

ν = stretching vibration δ = bending vibration   sk = skeletal vibrations

w = weak   m = medium intensity   s = strong